SPACE RESEARCH XIII

Vol. 1

COSPAR

SPACE RESEARCH XIII

Proceedings of Open Meetings of Working Groups on Physical Sciences of the Fifteenth Plenary Meeting of COSPAR

MADRID, SPAIN, 10—24 May, 1972

Organized by

THE COMMITTEE ON SPACE RESEARCH—COSPAR

and

THE "COMISION NACIONAL DE INVESTIGACION DEL ESPACIO" OF SPAIN

Edited by

M. J. RYCROFT
S. K. RUNCORN

Volume 1

AKADEMIE-VERLAG · BERLIN
1973

Executive Editor:
Dr. A. C. Stickland

Library of Congress Catalog Card Number 60-50878

202 · 100/474/73
Gesamtherstellung: VEB Druckhaus „Maxim Gorki", Altenburg
Bestellnummer: 761 751 8 (3059/XIII) · ES 18 D 1, 18 E 2
Printed in German Democratic Republic

FOREWORD

This impressive volume results from the energetic cooperation of many individuals, all over the world. Space Research XIII includes the papers presented at the open meetings of the COSPAR Working Groups, during the XVth Plenary Meeting of COSPAR, held in Madrid from May 16th through 24th, 1972. The volume, like the two preceding ones, was edited by Dr. M. J. Rycroft, U. K., except for the lunar papers, which were edited by Professor S. K. Runcorn, U. K. COSPAR wishes to extend to them its most sincere gratitude for their devoted work, as well as to the Programme Committees of the open sessions, and to the technical editor Dr. A. C. Stickland. Likewise, it is our pleasure to extend our thanks to the publishers, Akademie-Verlag, Berlin, G. D. R., for the high quality of this book — the third volume of Space Research prepared by them. Finally our thanks go to the many authors of the papers presented in this volume, for their cordial cooperation.

This volume does not contain the proceedings of the joint specialized symposia, held concurrently with COSPAR's Plenary Meeting, which are to be published as follows:

a) the URSI/IAU/COSPAR Symposium on 'Planetary Atmospheres and Surfaces', in special issues of *Icarus* (17/2, 17/3 and 18/1);

b) the COSPAR/IAGA(IUGG)/URSI Symposium on 'Critical Problems of Magnetospheric Physics', to appear as a special publication of the IUCSTP Secretariat;

c) the IAU/COSPAR Symposium on 'X and Gamma-Ray Astronomy', to appear as IAU Symposium Proceedings No. 55, editor Prof. H. Bradt, publisher Reidel, Dordrecht.

The Programme Committees of these Symposia, chaired respectively by Prof. G. H. Pettengill, Dr. N. F. Ness and Dr. R. Giacconi are to be congratulated for the excellent scientific content of the contributions.

The Meeting in Madrid was the tenth in succession held under the Presidency of Professor M. Roy. Here it is also my sincere pleasure to thank him for the invaluable services rendered by him to COSPAR during so long a period. That the Meeting in Madrid was highly successful owes a great deal to his efforts, and to the continual excellent services of the COSPAR Secretariat, headed so diligently and efficiently by Mr. Z. Niemirowicz. It is our particular pleasure to mention here the support of so many Spanish authorities, under the sponsorship of the highest Spanish state officials, grouped in the Honorary Committee, chaired by His Excellency the Chief of State, and by His Royal Highness, the Prince of

Spain. The Spanish Organizing Committee was chaired by Excmo. Sr. D. L. de Azcárraga, and the Spanish Executive Committee by Excmo. Sr. D. A. Pérez-Marín y Castro, who were assisted by the secretaries of the respective committees, Drs. L. Pueyo Panduro and J. M. Carballal Prado.

The present volume appears at a time when scientific research in outer space is moving rapidly from being a science that could be largely carried out on a national basis to one in which the magnitude of the enterprises undertaken demands international cooperation; this is instanced in the preparations now being made for very large partly manned space vehicles, in the proposed space shuttle service, and in the agreement between the U. S. and Soviet authorities on rendezvous in space. This all points to the necessity for a very actively guided international, inter-continental, and indeed global, cooperation in space. COSPAR should find ways to strengthen this cooperation between *all* large space countries and should at the same time help to enable scientists from the smaller countries of this planet to participate most fruitfully in really *advanced* space research. It will be a matter of continual concern to COSPAR to search for means to reach these goals.

Cornelis de Jager
President of COSPAR

PREFACE

"The desire of knowledge increases ever with acquisition of it"

Sterne, 1759.

"In the field of outer space, COSPAR has been, from the inception of the Space Age, the catalyst and the rallying point for international scientific cooperation in the peaceful exploration and use of outer space. The work of COSPAR therefore has a bearing on the international relationships, especially more appropriate in terms of scientific cooperation and collaboration. Based on more open and optimistic objectives overshadowing national and political differences and motivations, COSPAR presents the culmination of the possibilities of mutual cooperation between men of all nations." This is an excerpt from the 1972 message to COSPAR of the Secretary-General of the United Nations, who continues "... The development of space technology presents not only resourcefulness of the innovative scientists to achieve elaborate accomplishments but also manifold practical applications with a wide spectrum of uses that will have profound consequences on the entire fabric of our economic, social and educational patterns."

The dual aspects of the work of COSPAR, in the complementary fields of fundamental scientific research and technological uses of space, were discussed at the Fifteenth Plenary Meeting of COSPAR. This was held, from the tenth to the twenty fourth of May 1972, in the Palacio de Congresos y Exposiciones, Madrid, Spain. 839 scientists from 32 countries, including 130 from Spain, met as seven Working Groups of COSPAR to discuss their latest significant results.

Of the papers presented, those concerned with Physical Sciences and Space Research are published in this work, which is comprised of ten main sections:

Tracking of artificial earth satellites
Remote sensing of the earth
The neutral atmosphere
Dynamics of the thermosphere and ionosphere
The ionosphere
The magnetosphere
The sun
Astronomical measurements
The moon
Cosmic dust.

Included herein are four of the eight lectures in the Annual Reviews of Space Research series. The papers on thermosphere and ionosphere dynamics were written after the meeting, for the presentation took the form of a Workshop

Meeting; each paper is a summary of the points made by an invited speaker and a number of scientists who participated in the subsequent discussions.

Of the remaining papers, those to be published elsewhere are listed at the end of the second volume, with appropriate references where these are available. Papers on Life Sciences and Space Research delivered at the Meeting are published in a companion volume.

In organizing the papers for publication we, as Scientific Editors, are most grateful to the Chairmen and Reporters of the different Sessions of the Working Groups for their invaluable comments and recommendations. We also express our thanks to colleagues at our own Universities and to other referees for helping us with this task.

Three specialised Symposia were also held on the occasion of the Assembly in Madrid. These were on:

Critical problems of magnetospheric physics
X-ray and gamma ray astronomy
Planetary atmospheres and surfaces.

The papers presented at these Symposia are to be published separately, by IUCSTP, by IAU, and as a special issue of the journal *Icarus* respectively.

To close the preface to this work, we wish to acknowledge with gratitude the assistance rendered by Mr. Z. Niemirowicz, Executive Secretary of COSPAR, and by other members of the Secretariat. To Dr. A. C. Stickland who, as Executive Editor, willingly takes on and unfailingly completes a mammoth task, we cannot adequately express our thanks. Our thanks also go to all members of Akademie-Verlag GmbH for their excellent culmination of all our efforts in the resulting publication.

M. J. Rycroft
S. K. Runcorn

CONTENTS LIST

Volume 1

Page

Foreword V
Preface VII

Tracking of Artificial Earth Satellites

Earth's Gravitational Field

CH. REIGBER
Generalized Fourier Analysis of Resonant Orbits 3
D. E. SMITH, F. J. LERCH and C. A. WAGNER
A Gravitational Field Model for the Earth 11
D. G. KING-HELE
15th-Order Harmonics in the Geopotential, from analysis of Decaying Satellite Orbits 21
J. L. PIEPLU et M. LEFEBVRE
Détermination d'équations de conditions entre des harmoniques de resonnance d'ordre 14 à partir des observations du satellite EOLE 31
R. H. MERSON
The Longitude Acceleration of the Skynet Satellite March 1970—April 1971 35
C. SIMÓ
Secular Perturbations due to Potential: Second Approximation 45

Space Metrology

L. SEHNAL
Satellite Laser Ranging Instrument at the Ondřejov Observatory 51
M. LEFEBVRE and J. GAIGNEBET
Utilisation scientifique possible de longues bases laser 53
T. J. KUKKAMÄKI
Stellar Triangulation measured with Balloon-borne Beacons, its Scale and Position . 59
B. B. HOLLAND
Uses of Geoceiver as a Geodetic Instrument 65
J. M. MORAN
Some Characteristics of an Operational System for measuring UT 1 using Very Long Baseline Interferometry 73
J. F. JORDAN, W. G. MELBOURNE and J. D. ANDERSON
Testing Relativistic Gravity Theories using Radio Tracking Data from Planetary Orbiting Spacecraft 83

Radio Beacons

R. LEITINGER
The Use of Radio Beacons in Geophysics and their Applications 93

Page

Remote Sensing of the Earth

Earth's Surface

A. A. Grigoryev
A Study of Lineaments from a Zond 5 Photograph of Northern Africa 101
V. V. Melentyev and Yu. I. Rabinovich
Remote Sounding of Water Surface Conditions from aboard Artificial Satellites . . 105
B. S. Yaplee, A. Shapiro, E. A. Uliana, D. L. Hammond and K. J. Craig
Experimental Verification of Proposed Skylab Altimeter Measurements 107

Earth Resources

K. Kuusela and S. Poso
Multi-stage Acquisition of Forest Information from Space and Aircraft Imagery and Ground Sampling. 113
J. H. Joseph
Use in Israel of Remote Sensing from Satellites 119
D. D. Clark
UK Activity in Earth Surveys using Remote Sensing Techniques. 123

Space Meteorology

P. Morel
Satellite Techniques for Automatic Platforms Location and Data Relay 129
F. G. Finger and M. E. Gelman
Use of Meteorological Rocketsonde and Satellite Radiation Data for Constant-Pressure Analyses at levels between 5 and 0.4 mb 147
A. I. Herman, V. M. Zakharov, O. K. Kostko, V. E. Rokotyan and E. A. Chayanova
Some Results of Laser Radar Studies of Clouds and Underlying Surfaces 153

The Neutral Atmosphere

Stratospheric Balloons

M. Desbois
Preliminary Analysis of EOLE Meteorological Data 157
G. Necco
Estimation du champ de divergence à partir de traceurs Lagrangiens 163
M. Larcheveque
Turbulent Dispersion — EOLE Experiment 169

Stratosphere

S. S. Gaigerov and Yu. P. Koshelkov
Temperature and Wind in the Upper Atmosphere over the Antarctic 173
A. Azcárraga and J. M. Cisneros
The Establishment of the Winter Polar Vortex in Middle Latitudes in 1971 177
J. M. Cisneros
The Stratospheric Circulation over Middle Atlantic Latitudes in the 1966—1971 Period 181
S. S. Gaigerov, V. G. Kidiyarova, L. M. Kolomiitseva and D. A. Tarasenko
Deviations of Atmospheric Parameters from Model Values in the Stratosphere and Mesosphere at High Latitudes 185

Mesopheric and Lower Thermospheric Structure

L. A. Andreeva, B. O. Vugmeister, Yu. D. Ilyichev, E. S. Kazimirovsky, L. A. Katasev, V. D. Kokourov, N. S. Lifshitz, S. V. Pahomov and D. B. Uvarov
Results of Simultaneous Wind Measurements in the Stratosphere, Mesosphere and Lower Thermosphere 191

Page

A. C. Faire, K. S. W. Champion and E. A. Murphy
Variability in Density and Temperature Measurements at White Sands during the 1971 Winter . . . .
S. P. Zimmerman and C. A. Trowbridge 197
The Measurement of Turbulent Spectra and Diffusion Coefficients in the Altitude Region 95 to ~110 km 203
S. P. Zimmerman, G. P. Pereira, E. A. Murphy and J. Theon
Internal Gravity Waves and Turbulence in Simultaneous Upper Atmosphere Temperature and Wind Measurements 209
H. Teitelbaum
La polarisation des marées et des ondes de gravité 217

Lower Thermospheric Density and Composition

K. S. W. Champion, F. A. Marcos and R. A. Schweinfurth
Atmospheric Density Values from Orbital Drag on the Cannon Ball 2 and Musket Ball Satellites 223
K. S. W. Champion and F. A. Marcos
Lower Thermosphere Density Variations determined from Accelerometers on the Cannon Ball 2 Satellite 229
A. O. Nier and D. R. Hickman
Study of Neutral Composition of Lower Thermosphere at Fort Churchill. 235
B. R. May
Molecular Oxygen Density in the Lower Thermosphere from May to November 1967 243
R. E. Good and D. Golomb
Atomic Oxygen Profiles in the Lower Thermosphere 249
C. R. Philbrick, G. A. Faucher and E. Trzcinski
Rocket Measurements of Mesospheric and Lower Thermospheric Composition . . . 255
G. M. Martynkevich and E. D. Byuro
Minor Constituents of the Arctic Thermosphere during Disturbances 261

Thermospheric Models

J. E. Salah and J. V. Evans
Measurements of Thermospheric Temperatures by Incoherent Scatter Radar 267
G. P. Newton, D. T. Pelz and W. T. Kasprzak
Equatorial Thermospheric Composition and its Variations 287
M. N. Izakov, S. K. Morozov and I. A. Yashchenko
On the Diurnal Variations in the Thermosphere 291
K. K. Mahajan and S. Kumari
Neutral Composition and its Variations in the Lower Thermosphere. 299
A. P. Mitra and P. Chakrabarty
Estimates of Thermospheric Neutral Constituents from Ion Composition Measurements 307

Upper Thermosphere, including Seasonal Variations

A. E. Hedin, H. G. Mayr, C. A. Reber, G. R. Carignan and N. W. Spencer
A Global Empirical Model of Thermospheric Composition based on OGO 6 Mass Spectrometer Measurements 315
C. R. Philbrick, R. S. Narcisi, D. W. Baker, E. Trzcinski and M. E. Gardner
Satellite Measurements of Neutral Composition with a Velocity Mass Spectrometer . 321
G. M. Keating, D. S. McDougal, E. J. Prior and J. S. Levine
North—South Asymmetry of the Neutral Exosphere 327
M. Ya. Marov and A. M. Alpherov
Diurnal Variations in the Thermosphere. 337

Page

L. G. Jacchia and J. W. Slowey
A Study of the Variations in the Thermosphere related to Solar Activity 343

F. Barlier, J. L. Falin, M. Ill and C. Jaeck
Structure of the Neutral Atmosphere between 150 and 500 km 349

Upper Thermospheric Disturbance

R. R. Allan
Further Studies of Upper Atmosphere Densities at High Latitudes 357

I. Almár and E. Illés-Almár
An Analysis of the Altitude Dependence of the Geomagnetic Effect, by means of "Equivalent Durations" 363

P. W. Blum and I. Harris
The Global Wind System in the Thermosphere 369

Dynamics of the Thermosphere and Ionosphere

F. S. Johnson and P. Bauer
Variations in Density and Chemical Composition at 120 km from Chemical and Dynamical Processes 379

J. Testud
Waves and Tides and their Observation from Ground and Space 393

S. Matsushita and F. S. Mozer
Origin of Currents and Electric Fields in the Dynamo Region 397

R. E. Dickinson and H. Rishbeth
Planetary-scale Motions at F Layer Heights 413

The Ionosphere

S. A. Bowhill
Review of Recent Progress in Ionospheric Research 423

Neutral and Ionic Constituents

N. W. Rosenberg, D. Golomb, S. P. Zimmerman, W. K. Vickery and J. S. Theon
The ALADDIN Experiment — Part I, Dynamics 435

C. R. Philbrick, R. S. Narcisi, R. E. Good, H. S. Hoffman, T. J. Keneshea, M. A. MacLeod, S. P. Zimmerman and B. W. Reinisch
The ALADDIN Experiment — Part II, Composition 441

D. Rees, T. L. Aggson, K. Burrows, G. Haerendel, E. Rieger, K. H. Lloyd, J. W. G. Wilson, E. B. Dorling and G. L. Wrenn
Investigation of Mid-Latitude Ionospheric Currents by Combined Rocket Techniques 449

A. D. Danilov and M. N. Vlasov
N and NO Distribution based on E-Region Ion Composition 455

T. R. Tyagi, K. K. Mittal, A. P. Mitra and S. Prakash
Seasonal Variation of Atmospheric Composition in the F Region as a Function of Solar Activity 461

Solar X-radiation

R. W. Kreplin, D. M. Horan and K. P. Dere
Reduction of Solar X-ray Flux Measurements for Use in Ionospheric Research . . . 469

Ionospheric Studies using Rockets

J. D. Mitchell and L. C. Hale
Observations of the Lowest Ionosphere 471

Page

Y. V. Somayajulu, M. B. Avadhanulu and K. S. Zalpuri
A Study of the Equatorial D Region 477

E. Neske and R. Kist
Rocket Observations in the Equatorial Ionosphere 485

K. Hirao and K. Oyama
Comparison of Cleaned and Uncleaned Probes on board a Rocket 489

A. D. Danilov and V. K. Semenov
Atmospheric Ion Composition Measurements up to a height of 170 km 493

J. R. Sanmartin
The Determination of Ionospheric Charged Particle Temperatures from *in situ* Measurements 497

B. W. Reinisch
Burnt-out Rocket Punches Hole into Ionosphere 503

Disturbance Phenomena at High Latitudes

V. V. Mikhnevich and R. F. Usmanov
Tropospheric and Stratospheric Response to Solar Influence during Geomagnetic Disturbances. 507

D. Rees, T. L. Aggson, K. Burrows, G. Haerendel and J. W. G. Wilson
Diurnal and Seasonal Variations of Neutral Winds and Electric Fields above 90 km in the vicinity of the Auroral Electrojet 511

J. P. Treilhou, G. Gasset and I. Zhulin
Subauroral X-rays and Conjugate Electrojets 519

K. Wilhelm
The Distribution Function of Energetic Electrons measured during a Slowly Varying Ionospheric Absorption Event 527

N. M. Klyueva, V. G. Khryukin, Yu. K. Chasovitin, G. P. Komrakov and I. V. Popkov
Rocket Measurements of Electron Concentration and Electron Temperature in the Polar Ionosphere 535

M. Ahmed and R. C. Sagalyn
Thermal Positive Ions in the Dayside Polar Cusp measured on the ISIS 1 Satellite 541

K. I. Gringauz, G. L. Gdalevich, M. Z. Khokhlov, A. P. Remizov, Yu. I. Logachev, V. G. Stolpovsky, B. M. Gorozhankin, V. V. Afonin and S. M. Sheronova
Observations of Electron Fluxes and Related Variations of Ionospheric Plasma Parameters in the South Polar Cusp 549

L.-Å. Holmgren and B. Aparicio
Field-aligned Electron Anisotropies observed by the ESRO 1A (Aurorae) Satellite 555

Effect of Trapped Electrons from Nuclear Explosion

J. Oksman
A3 Absorption associated with Artificial Radiation Belts 563

Observations during PCA Events

M. B. Baker, P. R. Satterblom, A. J. Masley and A. D. Goedeke
Simultaneous Satellite and Riometer Measurements of Particles during Solar Cosmic Ray Events 569

W. F. Grieder and D. A. Burt
Rocket Measurements of Production and Ionization during a PCA Event 575

J. C. Ulwick
Steady State Coefficients in the D Region during Solar Particle Events 581

Page

New Observational Techniques

G. S. Ivanov-Kholodny, A. A. Korchak, I. P. Stakhanov and B. M. Chikhachev †
Phase Method of Investigation of Short-time-scale Disturbances and Motions in Space Plasma 587
R. Grabowski
Plasma Parameter Determination by Measurement of Electric Noise Fields due to Thermodynamic Plasma Fluctuations 593

Volume 2

The Magnetosphere

G. Haerendel
Results from Barium Cloud Releases in the Ionosphere and Magnetosphere 601

Geomagnetic Field

R. D. Regan, W. M. Davis and J. C. Cain
The Detection of "Intermediate Size" Magnetic Anomalies in Cosmos 49 and OGO 2, 4, 6 Data 619
H. H. Hilton
Secular Magnetic Variation and Inner Zone Protons 625

Magnetospheric Configuration

D. E. Page, V. Domingo, D. Köhn, B. G. Taylor, K.-P. Wenzel and P. C. Hedgecock
High-energy Electrons at the Magnetopause above the North Pole: Preliminary Results from the HEOS 2 Satellite 631
A. E. Antonova
Topology of the Magnetosphere and some Geophysical Phenomena 639
I. M. Podgorny, I. V. Andrianov and E. M. Dubinin
Plasma Characteristics near the Magnetosphere Boundary 643
M. K. Bird
Observational Comparisons with a Self-Consistent Model of the Geomagnetic Tail 655
A. Berthelier and C. Guerin
Influence of the Polarity of the Interplanetary Magnetic Field on Magnetic Activity at High Latitudes 661

Trapped Protons

J. Moritz
Energetic Protons at Low L-values of the Equatorial Magnetosphere 669
P. Valot and J. Engelmann
Pitch Angle Distribution of Geomagnetically Trapped Protons for $1.2 < L < 2.1$ 675
R. S. White, S. Moon, A. M. Preszler and G. M. Simnett
Earth Albedo and Solar Neutrons 683

Very Low Frequency Radiophysics

Ya. I. Likhter, O. A. Molchanov, V. M. Chmyrev, V. O. Rapoport, V. Yu. Trakhtengerts and V. A. Cherepovitsky
Modulation of Spectrum and Amplitude of VLF Signal in the Magnetosphere . . . 689
Ya. I. Likhter, Ya. P. Sobolev, F. Jiříček and P. Tříska
Early Results from the Intercosmos 5 Satellite VLF Experiment 695

Page

F. LEFEUVRE and K. BULLOUGH
Ariel 3 Evidence of Zones of VLF Emission at Medium and Low Invariant Latitudes which co-rotate with the Earth 699
M. J. RYCROFT
Enhanced Energetic Electron Intensity associated with a Whistler propagating through the Plasmasphere 707

The Sun

R. TOUSEY
The Solar Corona 713

Solar Electrons

A. A. KORCHAK
Sources of Accelerated Electrons on the Sun 731
G. M. SIMNETT
Relativistic Electrons in Space 745

Solar Protons

Y. HAKURA, Y. TAKENOSHITA and M. OHSHIO
Solar—Interplanetary Disturbances during 5—18 June 1969 (the PFP Interval/IASY) 763
M. A. SHEA and D. F. SMART
Prediction of the End of Solar Proton Events 771

X- and UV-radiation

H. F. VAN BEEK and L. D. DE FEITER
First Results of the Solar Hard X-ray Spectrometer on board the ESRO TD-1A Satellite 777
L. W. ACTON, R. C. CATURA and J. L. CULHANE
X-ray Study of Solar Plage Regions and a Small Flare 781
Z. KORDYLEWSKI, J. MERGENTALER, J. JAKIMIEC, B. SYLWESTER and J. SYLWESTER
Observations of the X-Ray Emission of Solar Active Regions on 28 November 1970 and 20 August 1971 787
G. E. KOCHAROV and YU. E. CHARIKOV
The Impulsive Increase in the Intensity of Solar X-Rays 793
C. B. OPAL
Enhancements of the Photoelectron-excited Dayglow during Solar Flares 797

Solar Flare Models and Forecasting

V. BUMBA and J. SÝKORA
Particle Emitting Flares and the Large-Scale Distribution of Solar Magnetic Fields and Green Corona 803
A. B. SEVERNY and N. V. STESHENKO
Development of Methods of Solar Flare Forecasting 809
R. T. PODSIADLO
An Objective Baseline for Flare Prediction 817

Solar Wind

J. W. FREEMAN, JR.
Suprathermal Ion Detector Results from Apollo Missions 821
A. A. FAUS
The Effect of Electron Pressure on the Wake of the Moon 831

Page

H. J. Fahr
The Annual Variation of the Solar Wind He^+ Content 837
H. J. Fahr and G. Lay
Interplanetary He I 584 Å Background Radiation 843

Astronomical Measurements

X-ray Observations

S. Miyamoto, Y. Ogawara, M. Matsuoka and M. Oda
Wide Band Energy Spectrum of Sco X-1 851
A. Bui-Van and G. Vedrenne
Hard X-Rays from Coma Cluster and Virgo Cluster 855
M. Landini and B. C. Monsignori Fossi
X-ray and Radio Emission from Stellar Coronae 859
U. R. Rao, D. P. Sharma, A. K. Jain, S. C. Chakravarty, K. Kasturirangan and K. R. Ramanathan
On the Detection of X-rays from Celestial Sources through their Ionization of the Terrestrial Atmosphere 861

B Type Stars

A. B. Underhill
The Absolute Energy Flux Envelopes of B Type Stars 867
R. Hoekstra, K. A. van der Hucht, C. de Jager, Th. Kamperman and H. J. Lamers
Ultraviolet Stellar Spectra obtained with the Utrecht Orbiting Stellar Spectrophotometer S59 aboard the ESRO TD-1A Satellite 871

Interstellar Hydrogen

P. W. Blum
Neutral Interstellar Hydrogen and Extraterrestrial Lyman Alpha Radiation 879

The Moon

Global Investigations

R. A. Petrone and W. T. O'Bryant
An Overview of Apollo Program of Lunar Exploration 887
G. F. Esenwein and F. I. Roberson
Apollo 15 Orbital Science Summary. 903
J. R. Arnold, J. I. Trombka, L. E. Peterson, R. C. Reedy and A. E. Metzger
Lunar Orbital Gamma Ray Measurements from Apollo 15 and Apollo 16 927

Physical Properties of Lunar Surface

D. A. Beattie and J. B. Hanley
Apollo 15 Surface Science Summary 935
C. T. Russell, P. J. Coleman, Jr., B. R. Lichtenstein, G. Schubert and L. R. Sharp
Surface and Orbital Magnetic Results from Apollo 15 951
E. S. Gorshkov, E. G. Gus'kova and V. I. Pochtarev
Magnetic Investigations of Lunar Soil delivered by AIS Luna 16 961
M. V. Akhmanova, B. V. Dementyev, M. N. Markov and M. M. Sushchinsky
Infrared Reflection Characteristics of Lunar Mare Regolith 965
N. N. Kroupenio
Results of Radar Experiments performed on Automatic Stations Luna 16 and Luna 17 969

Page

V. A. KOCHNEV, M. D. NUSSINOV, E. M. RUDNITZKI, YU. A. SURKOV, V. V. SHVAREV and I. L. ETTINGER
Sorption Processes on the Lunar Surface 975

Geochemical Properties of Lunar Soil

G. E. KOCHAROV and S. V. VICTOROV
Experimental Data on the Investigation of Lunar Surface Chemical Composition . . . 979
YU. A. SURKOV, G. A. FEDOSEYEV, O. P. SOBORNOV, G. B. NAZARKINA and L. P. BACHINA
Gamma-Spectrometric Analysis of Lunar Samples from Luna 16 985
A. K. LAVRUKHINA and V. D. GORIN
^{26}Al and ^{22}Na Measurements on Luna 16 Samples by non-destructive Gamma—Gamma Coincidence Spectrometry 991
I. I. ANTIPOVA-KARATAEVA, YU. I. STACHEEV and L. S. TARASOV
Absorption Spectra of Luna Sections from Different Lunar Areas 997

Lunar Laser Ranging

A. ORSZAG
Aspects récents de la télémétrie laser lune 1001
J. D. MULHOLLAND, H. H. PLOTKIN, E. C. SILVERBERG, D. T. WILKINSON, C. O. ALLEY, P. L. BENDER, D. G. CURRIE, R. H. DICKE, J. E. FALLER, W. M. KAULA and J. G. WILLIAMS
A Self-Consistent Set of Surface Coordinates for the Apollo Lunar Laser Retroreflectors deduced from Laser Range Measures 1009
I. I. MUELLER, F. A. FAJEMIROKUN and H. B. PAPO
The Influence of Laser Ranging on Selenodetic Control 1015

Cosmic Dust

Cosmic Dust Data from Satellites and Space Probes

R. D. MERCER, L. DUNKELMAN, C. L. ROSS and A. WORDEN
Lunar Orbital Photography of Astronomical Phenomena 1025
T. N. NAZAROVA, A. K. RYBAKOV, S. I. BAZAZYANTS and A. I. KUZMICH
Investigations of Meteoritic Matter in the Vicinity of the Earth and the Moon from the Orbiting Station Salyut and the Moon Satellite Luna 19 1033
W. M. ALEXANDER, C. W. ARTHUR, J. L. BOHN and J. C. SMITH
Four Years of Dust Particle Measurements in Cislunar Selenocentric Space from Lunar Explorer 35 and OGO 3 1037
O. E. BERG and E. GRÜN
Evidence of Hyperbolic Cosmic Dust Particles 1047
E. GRÜN, O. E. BERG and J. S. DOHNANYI
Reliability of Cosmic Dust Data from Pioneers 8 and 9 1057
D. K. BEDFORD and J. SAYERS
The Near Earth Micrometeoroid Flux from the Satellite Prospero. 1063

Erosion Processes of Lunar Material including Micrometeorites

D. G. ASHWORTH and J. A. M. MCDONNELL
Lunar Surface Micro-erosion related to Interplanetary Dust Particle Distributions . . 1071
D. E. GAULT, F. HÖRZ and J. B. HARTUNG
Abrasion and Catastrophic Rupture of Lunar Rocks: Some Implications to the Micrometeoroid Flux at 1 AU. 1085

Page

F. Di Benedetto, A. Carusi and A. Corradini
Preliminary Measurements of Spherules of the Pontina Plain and of Micrometeorites of Apollo 12 and related Impact Studies 1095

Cosmic Dust from Rockets and near the Earth

D. S. Hallgren, C. L. Hemenway, V. A. Mohnen and C. D. Tackett
Preliminary Results from the Noctilucent Sampling from Kiruna in 1970 1099
D. S. Hallgren, D. C. Schmalberger and C. L. Hemenway
Noctilucent Cloud Sampling by a Multi-experiment Payload 1105
B. A. Lindblad, G. Arinder and T. Wiesel
Continued Rocket Observations of Micrometeorites 1113
C. L. Hemenway, J. W. Erkes, J. M. Greenberg, D. S. Hallgren and D. C. Schmalberger
Do some of the Sub-Micrometer Cosmic Dust Particles come from the Sun? 1121
P. Rauser and H. Fechtig
Dust Measurements in the Upper Atmosphere during and in the absence of Noctilucent Cloud Display 1127
F. Link
Cosmic Dust in the Upper Atmosphere 1135
D. E. Brownlee and P. W. Hodge
Ablation Debris and Primary Micrometeoroids in the Stratosphere 1139
N. H. Farlow, G. V. Ferry and H. Y. Lem
Analysis of Individual Particles collected from the Stratosphere 1153

Distribution and Comparative Studies of Interplanetary, Cometary and Interstellar Dust

G. S. Hawkins
Micrometeorite and Cosmic Dust Data near the Earth's Orbit 1159
R. H. Giese
Optical and Mechanical Models of Interplanetary Dust 1165
V. Vanýsek
The Composition of Cosmic Dust in Comets 1173
W. Kokott
Improved Model of Asteroidal Dust Population 1181
J. L. Weinberg, M. S. Hanner, H. M. Mann, P. B. Hutchison and R. Fimmel
Observations of Zodiacal Light from the Pioneer 10 Asteroid—Jupiter Probe: Preliminary Results 1187

List of Papers presented at the XVth COSPAR Plenary Meeting and published, but not in this Volume 1193

Index of Authors 1195

Tracking of Artificial Earth Satellites

Space Research XIII — Akademie-Verlag, Berlin 1973

GENERALIZED FOURIER ANALYSIS OF RESONANT ORBITS*

Ch. Reigber

Institut für Astronomische und Physikalische Geodäsie, Technische Universität München, FRG

Using optical data for two arcs of GEOS 2 (1968 2A) and one arc of BE-C (1965 32A) and analysing the effect of the 13th-order terms of the geopotential on the mean longitude of these satellites three condition equations were derived. It is shown that the 13th-order sets of the SAO 1969 and SAO 1970 solutions satisfactorily model the resonance effect for BE-C and one arc of GEOS 2 respectively, whereas the odd degree coefficients (13,13)—(17,13) of Yionoulis most accurately model the effect for both GEOS 2 arcs.

Using the values $\bar{C}_{17,13} = 6.0 \times 10^{-8}$, $\bar{S}_{17,13} = -4.5 \times 10^{-9}$, obtained from the three condition equations, together with the other 13th-order values of the SAO 1970 solution an improved fit to all three amplitudes can be achieved.

1. Method

If one treats the motion of a dynamical system in the sense of a boundary value problem the solution can be given as [1]

$$\boldsymbol{a}(t) = \boldsymbol{a}_{\mathrm{A}} + (\boldsymbol{a}_{\mathrm{B}} - \boldsymbol{a}_{\mathrm{A}})t + \sum_{\nu=1}^{\infty} \boldsymbol{a}_\nu \sin \nu\pi t \tag{1}$$

where the motion is represented by the n-dimensional vector $\boldsymbol{a}(t)$, t being the time projected on the interval [0, 1], $\boldsymbol{a}_{\mathrm{A}}$, $\boldsymbol{a}_{\mathrm{B}}$ being the boundary values and the coefficients $\boldsymbol{a}_\nu$ fulfilling the infinite system of condition equations

$$\boldsymbol{a}_\nu = -\frac{2T^2}{\nu^2\pi^2} \int_0^1 \sin \nu\pi t' \, \boldsymbol{R}(t; \boldsymbol{a}; \dot{\boldsymbol{a}}) \, dt' \tag{2}$$

with $T = t_{\mathrm{B}} - t_{\mathrm{A}}$ the period of the problem.

If the n-dimensional vector $\boldsymbol{R}$ can be represented by a linear expression of the form

$$\boldsymbol{R} = \sum_{\mu=1}^{N} \boldsymbol{P}_\mu \boldsymbol{C}_\mu + \boldsymbol{Z} \tag{3}$$

where the $\boldsymbol{P}_\mu$ are any parameters, $\boldsymbol{C}_\mu$ are functions depending only on t; $\boldsymbol{a}$, $\dot{\boldsymbol{a}}$ and $\boldsymbol{Z}$ is the completely known remainder of $\boldsymbol{R}$ then substituting Eq. (3) into Eq. (2)

---

* Contribution No. 89 of the Institute of Astronomical and Physical Geodesy, Technical University Munich.

parameters P_μ can be determined from a linear system, where the coefficients $\boldsymbol{a}_\nu$ on the left-hand side are given by the aid of Eq. (1) from observations.

This method has been applied for setting up some condition equations for the 13th order resonant harmonics of the geopotential, where the coefficients $\boldsymbol{a}_\nu$ come from an analysis of the orbits of GEOS 2 (1968-2A) and BE-C (1965-32A) using precisely reduced Baker-Nunn observations.

2. Application of the Method

2.1. Analysis of Simulated Observations

In order to answer the question which elements are most appropriate for the analysis a great number of trial computations have been performed using numerically integrated orbits of GEOS 2, BE-C and D1D over periods which nearly correspond to the beat periods for these satellites (Table 1). As earth potential model the SAO 1969 solution [2] has been used with (orbit $\boldsymbol{r}(t)$ in the "actual" force field $\boldsymbol{J}$) and without (orbit $\boldsymbol{r}_g(t)$ in the reference field $\boldsymbol{J}_g$) the 13th order resonant harmonics.

Table 1

Satellite	Epoch (MJD)	a (km)	e	$i(°)$	$\omega(°)$	$\Omega(°)$	B. P. (days)
GEOS 2	40028.979	7713.037	0.031	105.8	265.8	270.5	– 6.3
BE-C	41015.0	7506.158	0.024	41.2	299.8	158.0	5.6
D1D	41001.0	7606.241	0.085	39.5	127.3	108.0	–20.5

Though the effect of these resonant harmonics can be observed in different form and size in a variety of elements which are appropriate for this analysis, e.g. cartesian coordinates $\boldsymbol{a} = (x_1, x_2, x_3)$, the absolute value of the deviation $|\boldsymbol{d}| = |\boldsymbol{r} - \boldsymbol{r}_g|$ or the Keplerian elements $\boldsymbol{a} = (a, e, i, \omega, \Omega, M)$ the simulations have shown that it is not a question of taste which elements will be used.

Using the spherical harmonics expansion of the potential and neglecting all nongravitational forces the determining equations for the $\overline{m}$th order potential coefficients $\overline{C}_{l,\overline{m}}$, $\overline{S}_{l,\overline{m}}$ in the case of cartesian coordinates read [1]

$$\boldsymbol{d}^\nu + \frac{2T^2}{\nu^2\pi^2} GM \int_0^1 \sin \nu\pi t' \left(\frac{\boldsymbol{r}_g}{r_g^3} - \frac{\boldsymbol{r}}{r^3}\right) dt' + \sum_{l=2}^{N} \sum_{\substack{m=0 \\ m \neq \overline{m}}}^{n} \Big(\overline{C}_{l,m} (\tilde{\boldsymbol{A}}^\nu_{l,m} - \boldsymbol{A}^\nu_{l,m})$$
$$+ \overline{S}_{l,m} (\tilde{\boldsymbol{B}}^\nu_{l,m} - \boldsymbol{B}^\nu_{l,m})\Big) = \sum_{l=\overline{m}}^{N_{\max}} (\overline{C}_{l,\overline{m}} \boldsymbol{A}^\nu_{l,\overline{m}} + \overline{S}_{l,\overline{m}} \boldsymbol{B}^\nu_{l,\overline{m}}) \tag{4}$$

where GM is the gravitational constant times mass of the earth and the expressions for the coefficients $\boldsymbol{A}_{l,m}$, $\boldsymbol{B}_{l,m}$ are given in [1].

The first difficulty in this case arises due to the fact that one has to determine a great number of amplitudes $\boldsymbol{d}^\nu$ in order to get the deviations $\boldsymbol{d}(t)$ caused by the $\overline{m}$th order harmonics to a good accuracy. This is because, besides some long-period oscillations, the components of $\boldsymbol{d}$ show high-amplitude oscillations with near orbit period for the orbits in question. But even if one assumes that the deviations $\boldsymbol{d}$ and thereby the geocentric positions $\boldsymbol{r} = \boldsymbol{r}_g + \boldsymbol{d}$ are given to an accuracy of 1 m a simple error analysis of the integral on the left-hand side of Eq. (4) shows that such an error in $\boldsymbol{r}$ can falsify the numerical value of the integral by some orders of magnitude depending on the length of the period T. This has been verified by all trial computations where an accurate determination of the theoretical amplitudes $\boldsymbol{d}^\nu$ was only possible by using the deviations $\boldsymbol{d}$ as given directly by numerical integration. The same is true if one analyses $d = \boldsymbol{d} \cdot \boldsymbol{d}_0$, the absolute value of the deviation, because the integral in question is given then as [3]

$$\frac{2T^2}{\nu^2\pi^2} GM \int_0^1 \sin \nu\pi t' \left\{\left(\frac{\boldsymbol{r}_g}{r_g^3} - \frac{\boldsymbol{r}}{r^3}\right) \cdot \boldsymbol{d}_0 - d\dot{\boldsymbol{d}}_0^2\right\} dt'$$

with the additional difficulty that one needs the time derivative of the unit vector $\boldsymbol{d}_0$.

These problems can be avoided by applying the method to Keplerian orbit elements. For this the disturbing function $R_{l,m}$ is used in terms of Keplerian elements $\alpha_\sigma = (a, e, i, \omega, \Omega, M)$ as [4]

$$R_{lm} = N_{lm} \sum_{pq} \frac{GM}{a} \left(\frac{a_e}{a}\right)^l F_{lmp}(i)\, G_{lpq}(e)\, \bar{J}_{lm} \begin{Bmatrix}\cos\\ \sin\end{Bmatrix} \psi_{lmpq} \begin{matrix} l - m \text{ even} \\ l - m \text{ odd}\end{matrix}$$

where $\psi_{lmpq} = [(l - 2p)\,\omega + (l - 2p + q) M + m(\Omega - \Theta - \lambda_{lm})]$, Θ is the Greenwich hour angle, $\bar{J}_{lm}$ and λ_{lm} are the harmonic amplitudes and phase angles which are related to the potential coefficients by $\bar{C}_{lm} = \bar{J}_{lm} \cos m\lambda_{lm}$, $\bar{S}_{lm} = \bar{J}_{lm} \sin m\lambda_{lm}$, N_{lm} is a normalization factor and $F(i)$ and $G(e)$ are the inclination and eccentricity function respectively as defined in Kaula's text [4]. With the disturbing function in this form the six components of the vector $\boldsymbol{R}$ in Eq. (2) are given by differentiating the Lagrange planetary equations for the elements $\alpha_\sigma = (a, e, i, \omega, \Omega, M)$ or in the case of small eccentricities for the modified elements $\alpha_\sigma = (a, h_1, i, h_2, \Omega, \lambda)$ [5] with respect to time for the particular resonant $lmpq$-set. Thus $K_{\sigma lmpq}$ will have the form

$$K_{\sigma lmpq} = \frac{\partial \dot{\alpha}_{\sigma lmpq}}{\partial t} + \sum_{j=1}^{6} \frac{\partial \dot{\alpha}_{\sigma lmpq}}{\partial \alpha_j} \dot{\alpha}_{jlmpq}$$

where with the exception of α_6 the products in the summation are of the order $\bar{J}^2_{lm}$.

According to Eq. (2) the condition equations for the mth order resonant harmonics then read [3]

$$\delta\alpha_\sigma^\nu = \sum_{l=m}^{N_{\max}} ({}^{c}A^\nu_{\sigma lm} \bar{C}_{lm} + {}^{s}A^\nu_{\sigma lm} \bar{S}_{lm}) \qquad \begin{matrix} \sigma = 1(1)6 \\ \nu = 1(1)\mu \end{matrix} \tag{5}$$

The coefficients ${}^{c,s}A^\nu_{\sigma lm}$ are computed from integrals of the form

$${}^{c,s}A^\nu_{\sigma lm} = -\frac{2T^2}{\nu^2\pi^2} \int_0^1 \sin \nu\pi t' \sum_{pq \text{ critical}} ({}^{c,s}K_{\sigma lmpq})\, dt' \tag{6}$$

where the pq summation in the integrand runs over all values for which $(r - 2p + q) = m/s$ (s = integer number of satellite revolutions per day). Explicit expressions for these integrals are given in [3].

In order to obtain the amplitudes $\delta\alpha_\sigma^\nu$ the residuals $d\alpha$, $d\delta$, $d\varrho$ in the direction and range observations from the corresponding values in the reference orbit are expressed according to Eq. (1) by the corrections

$$\delta\alpha_\sigma = \delta\alpha_\sigma^A + (\delta\alpha_\sigma^B - \delta\alpha_\sigma^A)t + \sum_{\nu=1}^{K_{\max}} \delta\alpha_\sigma^\nu \sin \nu\pi t$$

in the orbital elements α_σ as

$$(d\alpha, d\delta, d\varrho)^T = B\,\delta\alpha_\sigma \qquad \sigma = 1(1)6$$

where the elements of the matrix B are functions of the orbital elements and time. After assigning appropriate weights to the equations one gets the boundary values $\delta\alpha_\sigma^A$, $\delta\alpha_\sigma^B$ and the amplitudes $\delta\alpha_\sigma^\nu$ by a standard least squares adjustment. If the reference orbit is close enough to the real orbit the solution vector is obtained in one step. Otherwise an iterative scheme may be used.

This analysis has been applied to the simulated orbits. The measurements, station locations and the gravity model were assumed to be free of errors and the measurements $(\alpha, \delta, \varrho)$ assumed to exist every twenty minutes and nearly according to the distribution of the real observations (α, δ) as used in § 2.2. Solving for corrections in $\alpha_1 = a$ (accomplished by a correction in the mean motion n), $\alpha_2 = e$ (or $h_1 = e \sin \omega$), $\alpha_3 = i$, $\alpha_4 = \omega$ (or $h_2 = e \cos \omega$), $\alpha_5 = \Omega$ and for a correction $\delta\lambda$ in the mean longitude $\lambda = M + \omega + \Omega$ one gets the "observed" amplitude spectrum ${}^o\delta\alpha_\sigma^\nu$ of the long period perturbations (ν much smaller than the orbital frequency) of each element due to the SAO 1969 13th-order terms.

For GEOS 2 these amplitudes ($\nu \leq 3$) and additionally the theoretical amplitudes ${}^c\delta\alpha_\sigma^\nu$ as computed according to Eq. (5) using the SAO 1969 13th order harmonics are given in Table 2. The coefficients ${}^{c,s}A_{\sigma lm}^\nu$ in this equation are

Table 2

GEOS 2 (simulation), amplitudes of sin $\nu\pi t$ terms

ν	Case	$\delta a^\nu (m)$	$\delta e^\nu \times 10^7$	$\delta i^\nu \times 10^3$	$\delta\omega^\nu \times 10^3$	$\delta\Omega^\nu \times 10^3$	$\delta\lambda^\nu \times 10^3$
	a	5.20	3.59	−0.24	3.96	−0.04	0.10
1	b	6.78	2.74	−0.25	4.29	−0.03	0.06
	c	5.43	3.63	−0.24	3.69	−0.03	0.09
	a	−0.76	−3.59	0.00	1.79	−0.09	2.71
2	b	−1.05	−3.51	0.00	1.61	−0.09	2.64
	c	−0.73	−3.60	0.00	1.77	−0.10	2.70
	a	−0.98	−0.65	0.05	−0.73	0.01	−0.02
3	b	−0.32	−0.84	0.05	−0.55	0.01	−0.04
	c	−0.91	−0.72	0.05	−0.75	0.01	−0.02

period T = 6.278 days; units are degrees except for e and a

Case a, from 452 equidistant spaced $(\alpha, \delta, \varrho)$ observations, 44 stations.

Case b, from 65 (α, δ) observations, 10 stations.

Case c, theoretical amplitudes as computed by Eq. (5).

determined according to Eq. (6) by the aid of a 10-point Gauss—Legendre quadrature formula. Table 2 shows the well-known fact that the effect of resonance can be observed on all orbital elements though the dominant amplitudes for most of the elements are small. A pronounced effect shows up on ω and M or the mean longitude λ except for very small eccentricities, where the superposition of the amplitudes δM^{ν} and $\delta\omega^{\nu}$ may nearly lead to a cancellation. The dominant amplitude $\delta\lambda^{\nu}$ of this non-singular angle quantity λ can be estimated with satisfactory accuracy from observations even if there are great gaps in their distribution.

2.2 Analysis of Real Observations

Up to now two arcs of GEOS 2 and one arc of BE-C have been analysed using Baker-Nunn observations. Satellite characteristics, number of stations, number of observations, periods of analysis and osculating elements as computed with the CNES-orbit improvement programme GIN [6] are given in Table 3. Going out from these elements reference orbits were generated by numerical integration of the motion equation where the force field was given by the SAO 1970 earth gravity

Table 3
Summary of data

Satellite	Mass/ Area (kg m^{-2})	Number of stations	Number of observations	Epoch (MJD)	a (km)	e
BE-C (1965 32A)	94	5	82	39206.0	7509.7433	0.02382
GEOS 2 (1968 2A) 1st arc	175	11	272	40028.9792425	7713.0346	0.03102
GEOS 2 (1968 2A) 2nd arc	175	15	510	40115.0296161	7702.3690	0.03298

Satellite	$i(°)$	$\omega(°)$	$\Omega(°)$	$M(°)$	Period $T = t_B - t_A$ (days)
BE-C (1965 32A)	41.2002	311.4353	291.0146	211.7663	5.892318024
GEOS 2 (1968 2A) 1st arc	105.7750	265.8204	270.4536	93.6886	5.771111272
GEOS 2 (1968 2A) 2nd arc	105.7982	123.1115	30.8526	119.3055	5.753611337

field model [7] (neglecting 13th-order coefficients), lunisolar attraction, solar radiation pressure acceleration (including earth shadowing effect) and the atmospheric drag acceleration (CIRA 71 model [8]). Using these reference orbits and taking the SAO 1970 values for the station coordinates the boundary values $\delta\alpha_\sigma{}^A$, $\delta\alpha_\sigma{}^B$ and a limited number of amplitudes ${}^{\circ}\delta\alpha_\sigma{}^{\nu}$ were computed by a differential orbit correction programme after careful reduction of the observations to the SAO system [9]. After the differential correction the standard errors of unit weight are 2.0″ for BE-C, 1.8″ for GEOS 2 first arc and 1.9″ for GEOS 2 second arc.

From the amplitudes obtained only the dominant $\delta\lambda^{\nu}$ values as given in Table 4 were used as observed quantities. Though it seems to be possible to use some amplitudes of other elements if the satellites are tracked by a sufficient number of stations such values were rejected in this study because of the small amount of

Table 4

Equations of condition, observed amplitudes and residuals according to the 13th-order terms of different authors

Satellite	Element	ν	13/13		14/13		15/13		16/13		17/13		18/13		19/13	
			$\bar{C}$	$\bar{S}$	$\bar{C}$	$\bar{S}$	$\bar{C}$	$\bar{S}$	$\bar{C}$	$\bar{S}$	$\bar{C}$	$\bar{S}$	$\bar{C}$	$\bar{S}$	$\bar{C}$	$\bar{S}$
1965 32A	λ	1	− 6	+ 20	−17	− 1	+ 38	−126	+ 64	+ 12	− 94	+304	−121	−40	+136	−421
1968 2A(1)	λ	2	−106	+428	−64	+ 64	+136	−518	− 19	+ 96	+ 54	−226	+ 11	+22	− 11	+ 25
1968 2A(2)	λ	1	−671	+605	−85	−328	+778	−769	+149	−109	+357	−311	+ 90	+51	− 19	+ 60

Satellite	Element	ν	20/13		21/13		22/13		23/13		24/13		25/13	
			$\bar{C}$	$\bar{S}$	$\bar{C}$	$\bar{S}$	$\bar{C}$	$\bar{S}$	$\bar{C}$	$\bar{S}$	$\bar{C}$	$\bar{S}$	$\bar{C}$	$\bar{S}$
1965 32A	λ	1	+141	+ 75	−117	+332	−101	−88	+ 36	− 71	+53	+73	+71	−223
1968 2A(1)	λ	2	+ 20	− 36	− 30	+107	+ 16	−49	− 22	+ 84	+ 5	−33	−14	+ 61
1968 2A(2)	λ	1	− 10	+104	−150	+168	− 55	+83	−124	+121	−56	+30	−99	+ 80

Satellite	Element	ν	$\delta\lambda^{\nu} \times 10^7$	$(O-C) \times 10^7$						
				SAO 70	SAO 69	SAO 66	Kaula	Anderle	Yiounoulis	SAO 70 + corr.*
1965 32A	λ	1	302 ± 31	− 52	− 50	+247	+226	+234	+250	−12
1968 2A(1)	λ	2	300 ± 16	− 3	−116	−252	− 19	−380	− 36	−27
1968 2A(2)	λ	1	604 ± 38	+111	+ 30	−160	+302	−252	− 15	− 1

units are radians

* correction (17.13).

information for the 13th-order coefficient determination. Computing the coefficients ${}^{c,s}A^{\nu}_{\sigma lm}$ of the system (5) with the aid of the integrals (6) and assuming that the amplitudes computed reflect only the effect of the 13th-order terms of the geopotential three condition equations were derived as given in Table 4. Using the 13th-order sets of the SAO 1970 [7], SAO 1969 [2] and SAO 1966 [9] solutions and the 13th-order coefficients of some other investigators [10] $(O - C)$ values are computed as shown in Table 4. It can be seen from these figures that the odd degree coefficients (13, 13), (15, 13), (17, 13) of Yionoulis most accurately model the observed values for the two GEOS 2 arcs. This is not true for BE-C because of the missing (19, 13), (21, 13) coefficients in his set. For all three arcs the SAO 1970 and SAO 1969 13th-order sets most satisfactorily model the resonance effect, however, with a remarkable discrepancy between the two sets for the two GEOS 2 arcs. The reason for this discrepancy seems to be the difference in the (15, 13), (17, 13) coefficients for these two sets. A correction for these terms seems to be necessary. Due to the lack of more equations only the (17, 13) coefficients were corrected fixing the other coefficients to the SAO 1970 values. Assigning a weight (inversely proportional to the square of the standard deviation of the observed quantities) to each equation preliminary values $\bar{C}_{17,13} = (6.0 \pm 1.1)\, 10^{-8}$, $\bar{S}_{17,13} = (-4.5 \pm 9.4) \times 10^{-9}$ can be derived which are not very accurate and surely are composite coefficients because of the omitted correction to the (15,13) coefficients. Using, however, these values with the other SAO 1970 13th-order coefficients an improved fit to all three observed amplitudes can be achieved as shown in the last column of Table 4.

3. Conclusion

The results have shown that it is possible to get some information about the resonant harmonics by the method employed. In order to be able to separate the different coefficients from each other and to obtain more accurate results, observations for a greater number of satisfactorily tracked arcs and more satellites which are in resonance with the 13th-order terms have to be used. An analysis of a 20-day D1D arc has recently been started.

Acknowledgment

This study was carried out in the Sonderforschungsbereich 78, Satellitengeodäsie, Technische Universität München and supported by the Deutsche Forschungsgemeinschaft. All computations were performed on the IBM 360/91 computer of the Institut für Plasmaphysik, Max-Planck-Gesellschaft, Garching.

The author wishes to thank Dr. Gaposchkin (SAO) and Dr. Barlier (GRGS Meudon) for the observations, Prof. Kovalevsky and Mr. Lefebvre (GRGS Meudon/Brétigny) for the orbit improvement programme GIN, Mr. Brachet (GRGS Brétigny) for some data reduction sub-routines and Mr. Nagel (IAPG) for the data preparation.

References

[1] M. Schneider and Ch. Reigber, in: Dynamics of Satellites 1969, Springer-Verlag, Berlin 1970 (p. 181).

[2] E. M. Gaposchkin, in: Dynamics of Satellites 1969, Springer-Verlag, Berlin 1970 (p. 109).

[3] CH. REIGBER, Bestimmung von Resonanztermen der Ordnung 13 aus Satellitenbahnstörungen, Deutsche Geodätische Kommission, Reihe C (in preparation).
[4] W. M. KAULA, Theory of Satellite Geodesy, Blaisdell Publ. Co., Waltham, Mass. 1966.
[5] G. A. CHEBOTAREV, AIAA J. 2, No. 1 (1964).
[6] J. P. CHAISSAING and A. ROUMIGIER, CNES Rep. AR/JPCH/0.339/CB/MT.
[7] E. M. GAPOSCHKIN and K. LAMBECK, Smithson. Astrophys. Obs. Spec. Rep. No. 315 (1970).
[8] L. G. JACCHIA, Smithson. Astrophys. Obs. Spec. Rep. No. 332 (1971).
[9] C. A. LUNDQUIST and G. VEIS, Smithson. Astrophys. Obs. Spec. Rep. No. 200 (1966).
[10] B. C. DOUGLAS, J. G. MARSH and R. G. WILLIAMSON, Goddard Space Flight Center Rep. No. TMX-63637 (1969).

Space Research XIII — Akademie-Verlag, Berlin 1973

A GRAVITATIONAL FIELD MODEL FOR THE EARTH

D. E. Smith, F. J. Lerch and C. A. Wagner

Goddard Space Flight Center, Greenbelt, Md, USA

Two models of the earth's gravitational field have been computed at Goddard Space Flight Center. The first, Goddard Earth Model 1 (GEM 1), has been derived from satellite tracking data. The second, Goddard Earth Model 2 (GEM 2), has been derived from a combination of satellite tracking and surface gravimetric data. The geopotential models are represented in spherical harmonics complete to degree and order 16 for the combined solution and complete to degree and order 12 for the satellite solution. Both solutions include zonal terms to degree 21 and related satellite resonant coefficients to degree 22. The satellite data consisted primarily of optical data processed on 300 weekly orbital arcs for 25 close earth satellites. Surface gravity data were employed in the form of $5° \times 5°$ mean free-air gravity anomalies providing about 70% world coverage. Station locations were obtained for 46 tracking sites by combining electronic, laser, and additional optical tracking data with the above satellite data. Analysis of the radial positions of these stations and a value of mean gravity on the geoid indicated a mean equatorial radius for the earth of about 6378 145 meters. Results of geopotential tests on satellite data not used in the solution show that better agreement was obtained with the GEM 1 and GEM 2 models than with the 1969 Smithsonian Standard Earth II model.

1. Introduction

The establishment of an accurate earth model (geometric and gravimetric) is an essential requirement of any earth physics program. Progress in this area has been made at the Smithsonian Astrophysical Observatory using optical tracking data (mainly) and analytic techniques in the computation of orbits. Work at Goddard Space Flight Center has proceeded with the object of extending the use of these optical data and also of including highly accurate and dense electronic satellite tracking data available since 1965 (i.e. Tranet Doppler, Goddard Range and Range-Rate, and laser). Numerical integration, providing a precise solution for modeled forces, has been used extensively at GSFC for the computation of orbits and with this approach two geopotential models have been obtained. The coordinates of 46 stations have also been determined and consistent results have been obtained for an adjusted scale of the reference ellipsoid from analysis of both station positions and surface gravimetric data.

2. Description of Solutions

A geopotential solution (GEM 1) has been derived from tracking data (primarily optical) on 25 close earth satellites. Spherical harmonic terms in the geopotential are complete to degree and order 12 with zonals extending to degree

21 and selected satellite resonant coefficients to degree 22. Surface gravity data, consisting of 21000 $1° \times 1°$ mean free-air gravity anomalies and used in the form of $5° \times 5°$ mean anomalies, have been combined with the satellite data to derive a geopotential (GEM 2) complete to degree and order 16 with similar higher degree terms as in the satellite solution.

Station locations were obtained for 46 tracking sites by combining satellite electronic, laser, and additional optical data with the data used in the satellite geopotential solution described above. Tracking systems consisted of 13 Baker-Nunn cameras, 23 Minitrack Optical Tracking System cameras (MOTS), 2 lasers, 3 Goddard Range and Range-Rate (GRARR) systems, and 5 NWL Tranet Doppler systems. A value of mean gravity for the reference system was determined through a simultaneous adjustment of this quantity with the geopotential coefficients using both the satellite and surface gravity data.

Data used in the Solutions

a. *Satellite Data.* Data used in the satellite geopotential solution (GEM 1) consisted mainly of approximately 300 weekly orbital arcs of optical data on 23 satellites. Also included in the solution were 21 weekly arcs of Minitrack interferometer data on TIROS 9 and Alouette 2 satellites, which were principally used to support the determination of satellite resonant coefficients. In the solution for the 46 station locations, the data in the above 300 weekly orbital arcs (120000 optical observations) were combined with 15 weekly GEOS 1 and 2 orbital arcs of densely covered electronic and laser data (150000 observations) including some additional optical data. Also, some 66 one- and two-day arcs of GEOS 1 and 2 flashing light optical data were combined with the above arcs, mainly to strengthen the solution of the MOTS stations.

b. *Surface Gravity Data.* The source of most of the gravimetry data was the US Aeronautical Chart and Information Center which provided 19000 $1° \times 1°$ mean free-air gravity anomalies. A further set of 2000 mean gravity anomalies was obtained from a number of other sources. These data were used to form 1707 $5° \times 5°$ mean gravity anomalies by a straight averaging of the $1° \times 1°$ mean anomalies, and provided a total coverage of about 70% of the earth's surface.

The reference ellipsoid that was adopted for the solution was the same as that used by SAO in the Standard Earth II (1969) [1], viz.

Mean equatorial radius, $a_e = 6378155$ meters,

Flattening, $f = 1/298.255$.

Product of mass of the earth and gravitational constant,

$$GM = 3.986013 \times 10^{14} \text{ m}^3 \text{ s}^{-2}.$$

3. Analysis Techniques

The motion of the satellite was obtained by numerical integration of the equations of motion with the following forces being modeled: the earth's gravity field, atmospheric drag, solar radiation pressure, and lunar and solar gravity. The integration was performed in a cartesian, geocentric, inertial coordinate system referenced to the true equator and equinox of the epoch. The time frame was A1 time and modeling in the reference system included luni-solar precession and nutation of the earth and polar motion, the latter being obtained from the US Naval Observatory who also provided the adjustment from UT1 to A1. The origin of the coordinate system for the pole positions was the mean pole of 1900 to 1905.

The surface gravity data was given in terms of $1° \times 1°$ mean gravity anomalies (mgal) referenced to the International System. These data were used as described below in terms of the reference system defined by GM, a_e, f, and ω (rotation rate of the earth). Let W be the gravity potential of the earth so that

$$W = V + \Phi \tag{1}$$

where V is the gravitational potential, and Φ the centrifugal potential, and where

$$V = \frac{GM}{r}\left[1 + \sum_{n=0}^{\infty}\sum_{m=0}^{n}\left(\frac{a_e}{r}\right)^n \mathrm{P}_n^m(\sin\varphi)\,(\mathrm{C}_{nm}\cos m\lambda + \mathrm{S}_{nm}\sin m\lambda)\right]$$

$$\Phi = {}^1/_2\,(r\omega\cos\varphi)^2$$

r, φ, λ spherical coordinates of radial distance, latitude, and longitude, $\mathrm{P}_n^m(\sin\varphi)$ associated Legendre polynomial of degree n, order m and argument $\sin\varphi$, C_{nm}, S_{nm} spherical harmonic coefficients.

On an equipotential surface (i.e. mean sea level), we may write $W = W_0$ so that gravity g_0, on this surface is given by

$$g_0 = |\operatorname{grad} W_0|.$$

The quantity g_0 is related to the International System for gravity measurement as follows

$$g_0' = \Delta g_{\mathrm{I}} + \gamma_{\mathrm{I}}$$

$$g_0 = g_0' - 13.7 \tag{2}$$

where γ_{I} is normal gravity on the International Ellipsoid, Δg_{I} is gravity anomaly for the International System, -13.7 mgal is the Potsdam correction.

The $5° \times 5°$ mean gravity anomalies were formed from a straight average of the given $1° \times 1°$ mean anomalies for each $5° \times 5°$ section and were represented at the mid-point of the section in latitude and longitude. A value of g_0 was then computed from (2) using each of the $5° \times 5°$ mean anomalies, and the resultant g_0 was taken as the observed quantity. A residual, $g_0 - g_c$, was then formed for the equation of condition and expressed in terms of linear adjustments to the starting values of the spherical harmonic coefficients. The quantity g_c was computed as

$$g_c = \left\{\left(\frac{\partial W}{\partial r}\right)^2 + \frac{1}{r^2}\left(\frac{\partial W}{\partial \varphi}\right)^2\right\}^{1/2} \tag{3}$$

where the partial differentials are obtained from (1) with the use of our starting coefficients and reference system parameters. A constant term $\Delta\bar{g}_0$ [2], variation of mean gravity from the reference system, was added to the equation of condition for simultaneous adjustment with the spherical harmonic coefficients. Weighted least squares normal equations were then formed for simultaneous adjustment with the satellite normal equations.

In forming the weighted least squares normal equations the variance σ^2 of g_0 (observed gravity value corresponding to the $5° \times 5°$ mean anomalies) was obtained from

$$\sigma^2 = \frac{\sigma_0^2}{N'+1}. \tag{4}$$

N' is the number of $1° \times 1°$ mean anomalies within a $5° \times 5°$ section and σ_0^2, the computed variance of the 21000 $1° \times 1°$ observed anomalies, is 33^2 ($mgal^2$). For an unobserved $5° \times 5°$ area, a value of zero was adopted with a variance of σ_0^2. The weighting of the gravity data was according to the formula

$$w = \frac{\cos \varphi}{\sigma^2} \tag{5}$$

where the cosine of the latitude was introduced to account for the unequal area covered by the $5° \times 5°$ section. For a $5° \times 5°$ mean anomaly derived from 25 $1° \times 1°$ means, $\sigma = 6$ mgal.

4. Results

4.1. Solutions

The geopotential and station coordinate solutions were described in a previous section. The values of the geopotential coefficients, for the satellite solution (GEM 1) and the combined satellite/surface gravity (GEM 2) solution, are listed in Tables 1 a—c and the station coordinates for the 46 tracking stations in Table 2. The ellipsoid heights for the 13 Baker-Nunn stations have been lowered by 15 meters to account for an error in the computer program for the parallactic refraction correction. The solution value for $\Delta \bar{g}_0$, the variation of mean gravity from the reference system, was 3.28 mgal.

4.2. Zonal Coefficients

Zonal coefficients complete to degree 21 in the GEM 2 solution are obtained from the combined effects of both satellite and surface gravity data. These zonal values have been compared with the SAO 1966 (M 1) and 1969 Standard Earth II values derived by Kozai from secular and long period orbit perturbations and show remarkable agreement. Generally high correlation exists between consecutive odd and consecutive even zonal coefficients in satellite solutions from secular and long term orbital effects. Correlations, although very large (generally greater than 0.9) for consecutive odd or even zonals, fall off as odd or even pairs separate in degree. However, in our combination solution the coefficients do not exhibit these high correlations and have essentially become decoupled through the surface gravity data.

4.3. Geoid Height Maps (Undulations)

Geoid height contour maps at ten meter intervals for the GEM 2 and the Standard Earth II solutions compare very favorably but differences of 8 meters can be seen between certain highs and lows. An rms of 5 meters has been obtained from a more detailed analysis of GEM 2 and Standard Earth II solutions.

Table 1a

Geopotential solutions (normalized coefficients × 10^6)

l m	GEM 2	GEM 1	*l m*	GEM 2	GEM 1	*l m*	GEM 2	GEM 1
C 2 0	484.1671	–484.1766	C 2 1	–0.0016	–0.0010	C 8 2	0.0473	0.0444
S 2 0	0.0	0.0	S 2 1	–0.0030	–0.0106	S 8 2	0.0843	0.0688
C 3 0	0.9548	0.9619	C 3 1	1.9929	2.0015	C 9 2	0.0512	0.0021
S 3 0	0.0	0.0	S 3 1	0.2366	0.2381	S 9 2	–0.0002	0.0092
C 4 0	0.5368	0.5572	C 4 1	–0.5343	–0.5326	C 10 2	–0.0500	–0.0537
S 4 0	0.0	0.0	S 4 1	–0.4640	–0.4426	S 10 2	–0.0599	–0.0335
C 5 0	0.0726	0.0623	C 5 1	–0.0688	–0.0600	C 11 2	0.0098	0.0297
S 5 0	0.0	0.0	S 5 1	–0.0926	–0.0992	S 11 2	–0.1132	–0.1093
C 6 0	–0.1452	–0.1775	C 6 1	–0.0776	–0.0798	C 12 2	–0.0370	–0.0122
S 6 0	0.0	0.0	S 6 1	0.0103	–0.0306	S 12 2	0.0443	0.0104
C 7 0	0.0869	0.1051	C 7 1	0.2510	0.2533	C 13 2	0.0211	0.0
S 7 0	0.0	0.0	S 7 1	0.1204	0.1134	S 13 2	–0.1134	0.0
C 8 0	0.0397	0.0799	C 8 1	0.0253	0.0245	C 14 2	–0.0313	0.0
S 8 0	0.0	0.0	S 8 1	0.0533	0.0981	S 14 2	0.0969	0.0
C 9 0	0.0329	0.0076	C 9 1	0.1521	0.1637	C 15 2	–0.0039	0.0
S 9 0	0.0	0.0	S 9 1	0.0100	–0.0018	S 15 2	–0.0672	0.0
C 10 0	0.0648	0.0211	C 10 1	0.0873	0.0817	C 16 2	0.0091	0.0
S 10 0	0.0	0.0	S 10 1	–0.1401	–0.1919	S 16 2	0.0220	0.0
C 11 0	–0.0551	–0.0198	C 11 1	–0.0260	–0.0191	C 3 3	0.6876	0.7351
S 11 0	0.0	0.0	S 11 1	0.0255	0.0245	S 3 3	1.4571	1.3977
C 12 0	0.0211	0.0594	C 12 1	–0.0695	–0.0766	C 4 3	0.9830	0.9875
S 12 0	0.0	0.0	S 12 1	–0.0538	–0.0108	S 4 3	–0.2080	–0.2134
C 13 0	0.0432	0.0017	C 13 1	0.0408	0.0	C 5 3	–0.4316	–0.4022
S 13 0	0.0	0.0	S 13 1	–0.0606	0.0	S 5 3	–0.1785	–0.2487
C 14 0	–0.0089	–0.0369	C 14 1	–0.0296	–0.0150	C 6 3	0.0313	0.0365
S 14 0	0.0	0.0	S 14 1	0.0423	0.0053	S 6 3	0.0191	–0.0060
C 15 0	0.0039	0.0470	C 15 1	0.0739	0.0	C 7 3	0.2458	0.2909
S 15 0	0.0	0.0	S 15 1	0.0202	0.0	S 7 3	–0.1728	–0.2299
C 16 0	–0.0260	–0.0131	C 16 1	–0.0291	0.0	C 8 3	–0.0198	–0.0232
S 16 0	0.0	0.0	S 16 1	0.0147	0.0	S 8 3	–0.0460	–0.0687
C 17 0	0.0072	–0.0350	C 2 2	2.4230	2.4268	C 9 3	–0.1381	–0.1040
S 17 0	0.0	0.0	S 2 2	–1.3825	–1.3741	S 9 3	–0.0365	–0.1377
C 18 0	0.0228	0.0180	C 3 2	0.9244	0.9287	C 10 3	–0.0380	–0.0465
S 18 0	0.0	0.0	S 3 2	–0.6213	–0.6084	S 10 3	–0.0851	–0.1164
C 19 0	0.0153	0.0450	C 4 2	0.3428	0.3461	C 11 3	–0.0173	0.0006
S 19 0	0.0	0.0	S 4 2	0.6755	0.6777	S 11 3	–0.0577	–0.1283
C 20 0	–0.0006	–0.0016	C 5 2	0.6649	0.6403	C 12 3	0.1264	0.0872
S 20 0	0.0	0.0	S 5 2	–0.3065	–0.3038	S 12 3	0.0273	0.0810
C 21 0	–0.0117	–0.0305	C 6 2	0.0592	0.0474	C 13 3	–0.0215	0.0
S 21 0	0.0	0.0	S 6 2	–0.3613	–0.3478	S 13 3	0.0214	0.0
C 22 0	0.0	0.0	C 7 2	0.3430	0.3283	C 14 3	0.0408	0.0
S 22 0	0.0	0.0	S 7 2	0.0890	0.0922	S 14 3	–0.0065	0.0

Table 1b
Geopotential solutions (normalized coefficients × 10^6)

l m	GEM 2	GEM 1	l m	GEM 2	GEM 1	l m	GEM 2	GEM 1
C 15 3	0.0044	0.0	C 11 5	0.0873	0.0228	C 11 7	0.0071	0.0060
S 15 3	0.0402	0.0	S 11 5	0.0571	0.0338	S 11 7	−0.1018	−0.1115
C 16 3	0.0252	0.0	C 12 5	0.0203	0.0176	C 12 7	−0.0298	−0.0102
S 16 3	−0.0203	0.0	S 12 5	−0.0126	−0.0027	S 12 7	−0.0049	−0.0080
C 4 4	−0.1703	−0.1708	C 13 5	0.0343	0.0	C 13 7	−0.0416	0.0
S 4 4	0.3116	0.3023	S 13 5	0.0528	0.0	S 13 7	0.1067	0.0
C 5 4	−0.2511	−0.2946	C 14 5	0.0411	0.0	C 14 7	0.1021	0.0
S 5 4	0.0410	0.0359	S 14 5	−0.0309	0.0	S 14 7	−0.0687	0.0
C 6 4	−0.0778	−0.0974	C 15 5	0.0212	0.0	C 15 7	−0.0116	0.0
S 6 4	−0.4524	−0.4491	S 15 5	−0.0101	0.0	S 15 7	0.0696	0.0
C 7 4	−0.2591	−0.2795	C 16 5	0.0126	0.0	C 16 7	0.0296	0.0
S 7 4	−0.1424	−0.0968	S 16 5	0.0260	0.0	S 16 7	−0.0317	0.0
C 8 4	−0.2135	−0.2330	C 6 6	0.0186	0.0276	C 8 8	−0.1237	−0.0897
S 8 4	0.0308	0.0527	S 6 6	−0.2498	−0.2430	S 8 8	0.0935	0.0848
C 9 4	0.0313	0.0213	C 7 6	−0.3030	−0.3205	C 9 8	0.2420	0.1765
S 9 4	0.0186	0.0463	S 7 6	0.1784	0.1675	S 9 8	0.0467	−0.0214
C 10 4	−0.0748	−0.0713	C 8 6	−0.0632	−0.0599	C 10 8	0.0157	0.0286
S 10 4	−0.1073	−0.1127	S 8 6	0.2871	0.3092	S 10 8	−0.1318	−0.1346
C 11 4	0.0053	0.0516	C 9 6	0.0339	0.0615	C 11 8	0.0210	−0.0167
S 11 4	−0.0854	−0.0746	S 9 6	0.2071	0.2565	S 11 8	0.0855	0.0579
C 12 4	−0.0392	−0.0052	C 10 6	−0.0346	−0.0125	C 12 8	−0.0304	−0.0269
S 12 4	−0.0224	−0.0421	S 10 6	−0.0948	−0.0658	S 12 8	0.0164	−0.0240
C 13 4	−0.0326	0.0	C 11 6	0.0045	−0.0058	C 13 8	0.0396	0.0
S 13 4	−0.0616	0.0	S 11 6	0.0157	0.0809	S 13 8	−0.0074	0.0
C 14 4	0.0292	0.0	C 12 6	0.0522	0.0594	C 14 8	−0.0041	0.0
S 14 4	0.0142	0.0	S 12 6	0.0103	0.0546	S 14 8	−0.0518	0.0
C 15 4	−0.0006	0.0	C 13 6	−0.1120	0.0	C 15 8	−0.1339	0.0
S 15 4	−0.0146	0.0	S 13 6	0.0042	0.0	S 15 8	0.0228	0.0
C 16 4	0.0333	0.0	C 14 6	0.0234	0.0	C 16 8	−0.0318	0.0
S 16 4	0.0732	0.0	S 14 6	−0.0094	0.0	S 16 8	−0.0141	0.0
C 5 5	0.2023	0.1338	C 15 6	−0.0018	0.0	C 9 9	−0.0233	−0.0362
S 5 5	−0.6721	−0.6662	S 15 6	−0.0569	0.0	S 9 9	0.0935	0.0886
C 6 5	−0.2600	−0.2613	C 16 6	−0.0354	0.0	C 10 9	0.0909	0.1172
S 6 5	−0.5639	−0.5310	S 16 6	−0.0219	0.0	S 10 9	−0.0755	−0.0321
C 7 5	0.0395	−0.0017	C 7 7	0.0406	0.0153	C 11 9	−0.0433	0.0329
S 7 5	0.0526	0.0580	S 7 7	0.0194	0.0980	S 11 9	0.0911	0.0484
C 8 5	−0.0631	−0.0740	C 8 7	0.0640	0.0788	C 12 9	−0.0105	0.0330
S 8 5	0.0649	0.0588	S 8 7	0.0805	0.0691	S 12 9	0.0151	0.0631
C 9 5	−0.0125	−0.0278	C 9 7	−0.0810	−0.0760	C 13 9	0.0239	0.1087
S 9 5	−0.0249	−0.0933	S 9 7	−0.0456	−0.0469	S 13 9	0.1181	0.0679
C 10 5	−0.0800	−0.1134	C 10 7	0.0394	0.0009	C 14 9	−0.0026	0.0413
S 10 5	−0.0535	−0.0745	S 10 7	−0.0153	−0.0330	S 14 9	0.0447	0.0758

Table 1c

Geopotential solutions (normalized coefficients $\times 10^6$)

	l	*m*	GEM 2	GEM 1
C	15	9	0.0128	0.0713
S	15	9	0.0681	0.0483
C	16	9	0.0204	0.0
S	16	9	-0.0620	0.0
C	17	9	0.0	0.0
S	17	9	0.0	0.0
C	18	9	0.0	0.0
S	18	9	0.0	0.0
C	19	9	0.0	0.0
S	19	9	0.0	0.0
C	20	9	0.0	0.0
S	20	9	0.0	0.0
C	21	9	0.0	0.0
S	21	9	0.0	0.0
C	22	9	0.0	0.0
S	22	9	0.0	0.0
C	10	10	0.1044	0.0574
S	10	10	-0.0370	-0.0398
C	11	10	-0.0567	-0.1206
S	11	10	-0.0219	-0.0067
C	12	10	0.0261	-0.0014
S	12	10	-0.0084	0.0309
C	13	10	-0.0120	0.0
S	13	10	0.0128	0.0
C	14	10	0.0208	0.0
S	14	10	-0.1030	0.0
C	15	10	0.0383	0.0
S	15	10	0.0124	0.0
C	16	10	-0.0456	0.0
S	16	10	-0.0013	0.0
C	11	11	0.0683	0.0539
S	11	11	-0.0319	-0.0360
C	12	11	-0.0058	-0.0028
S	12	11	0.0276	0.0402
C	13	11	-0.0680	0.0
S	13	11	-0.0084	0.0
C	14	11	0.1045	0.0
S	14	11	-0.0438	0.0
C	15	11	-0.0618	0.0
S	15	11	0.0574	0.0
C	16	11	0.0089	0.0
S	16	11	-0.0061	0.0
C	17	11	0.0	0.0
S	17	11	0.0	0.0
C	18	11	0.0	0.0
S	18	11	0.0	0.0
C	19	11	0.0	0.0
S	19	11	0.0	0.0
C	20	11	0.0	0.0
S	20	11	0.0	0.0
C	21	11	0.0	0.0
S	21	11	0.0	0.0
C	22	11	0.0	0.0
S	22	11	0.0	0.0
C	12	12	-0.0183	-0.0180
S	12	12	-0.0128	-0.0114
C	13	12	-0.0290	-0.0288
S	13	12	0.1005	0.1004
C	14	12	0.0031	0.0026
S	14	12	-0.0431	-0.0401
C	15	12	-0.0389	-0.0371
S	15	12	0.0142	0.0145
C	16	12	0.0232	0.0202
S	16	12	-0.0153	-0.0119
C	17	12	0.0058	0.0108
S	17	12	-0.0053	-0.0043
C	18	12	-0.0611	-0.0606
S	18	12	-0.0264	-0.0230
C	19	12	-0.0347	-0.0310
S	19	12	-0.0380	-0.0364
C	20	12	0.0024	-0.0003
S	20	12	-0.0115	-0.0054
C	21	12	-0.0464	-0.0355
S	21	12	-0.0204	-0.0204
C	22	12	-0.0414	-0.0433
S	22	12	0.0111	0.0053
C	13	13	-0.0620	-0.0633
S	13	13	0.0725	0.0704
C	14	13	0.0225	0.0174
S	14	13	0.0524	0.0464
C	15	13	-0.0257	-0.0264
S	15	13	0.0020	0.0005
C	16	13	0.0203	0.0163
S	16	13	-0.0016	-0.0034
C	17	13	0.0072	0.0056
S	17	13	0.0116	0.0111
C	18	13	-0.0024	-0.0087
S	18	13	-0.0293	-0.0380
C	19	13	-0.0268	-0.0285
S	19	13	-0.0459	-0.0454
C	20	13	0.0124	0.0021
S	20	13	0.0140	0.0015
C	21	13	-0.0275	-0.0290
S	21	13	0.0007	0.0010
C	22	13	-0.0334	-0.0413
S	22	13	0.0121	0.0050
C	14	14	-0.0459	-0.0457
S	14	14	-0.0034	-0.0045
C	15	14	0.0047	0.0053
S	15	14	-0.0203	-0.0185
C	16	14	-0.0160	-0.0208
S	16	14	-0.0385	-0.0387
C	17	14	-0.0128	-0.0137
S	17	14	-0.0007	-0.0029
C	18	14	-0.0118	-0.0084
S	18	14	-0.0049	-0.0055
C	19	14	0.0012	0.0023
S	19	14	-0.0001	0.0032
C	20	14	0.0065	0.0069
S	20	14	-0.0124	-0.0096
C	21	14	0.0112	0.0122
S	21	14	0.0042	0.0069
C	22	14	-0.0095	-0.0124
S	22	14	0.0006	0.0028
C	15	15	-0.0575	0.0
S	15	15	0.0249	0.0
C	16	15	-0.0522	0.0
S	16	15	0.0179	0.0
C	17	15	0.0	0.0
S	17	15	0.0	0.0
C	18	15	0.0	0.0
S	18	15	0.0	0.0
C	16	16	-0.0169	0.0
S	16	16	-0.0115	0.0

Table 2
Station coordinates

	Stn	Latitude DD	MM	SS.SS	Longitude DDD	MM	SS.SS	Height (m)
Anchor	14	61	17	0.51	210	10	29.69	68.7
Lacres	103	32	16	44.68	253	14	45.42	1149.0
Lashm 2	106	51	11	9.30	358	58	26.32	221.5
Aplmnd	111	39	9	48.67	283	6	12.78	97.3
1 Bpoin	1021	38	25	49.77	282	54	48.71	−40.1
1 Ftmyr	1022	26	32	53.45	278	8	4.33	−31.1
1 Oomfr	1024	−31	23	25.19	136	52	15.89	129.9
1 Satag	1028	−33	8	58.49	289	19	53.60	707.2
1 Mojav	1030	35	19	47.90	243	5	59.30	888.8
1 Jobur	1031	−25	53	0.66	27	42	26.59	1539.0
1 Newfl	1032	47	44	29.84	307	16	46.34	69.0
1 Gfork	1034	48	1	21.27	262	59	19.87	219.4
1 Wnkfl	1035	51	26	46.48	359	18	8.38	104.0
1 Rosmn	1037	35	12	7.13	277	7	41.75	865.4
1 Ororl	1038	−35	37	32.15	148	57	14.80	939.5
1 Rosma	1042	35	12	7.33	277	7	41.44	865.5
1 Tanan	1043	−19	0	31.72	47	17	59.73	1364.4
Madgas	1123	−19	1	14.19	47	18	11.75	1390.8
Rosran	1126	35	11	45.51	277	7	26.20	828.9
Ulaskr	1128	64	58	19.05	212	29	12.44	341.6
Pretor	2115	−25	56	48.21	28	20	51.88	1579.8
1 Undak	7034	48	1	21.41	262	59	19.60	217.3
1 Edinb	7036	26	22	46.76	261	40	7.73	21.4
1 Colba	7037	38	53	36.18	267	47	41.11	227.3
1 Bermd	7039	32	21	49.86	295	20	35.37	−15.2
1 Purio	7040	18	15	28.94	294	0	23.77	−5.7
1 Gsfcp	7043	39	1	15.32	283	10	20.33	0.6
1 Denvr	7045	39	38	48.02	255	23	38.87	1759.3
Godlas	7050	39	1	14.53	283	10	18.33	12.3
Wallas	7052	37	51	36.19	284	29	23.76	−42.1
1 Jum 40	7072	27	1	14.54	279	53	13.07	−30.6
1 Jubc 4	7074	27	1	14.79	279	53	13.11	−34.2
1 Sudbr	7075	46	27	20.40	279	3	10.99	233.3
1 Jamac	7076	18	4	34.78	283	11	27.12	422.4
1 Organ	9001	32	25	25.04	253	26	49.06	1616.4
1 Olfan	9002	−25	57	35.95	28	14	52.83	1553.5
Woomer	9003	−31	6	1.98	136	47	3.63	151.0
1 Spain	9004	36	27	46.74	353	47	37.13	56.9
1 Tokyo	9005	35	40	23.00	139	32	16.65	81.0
1 Natal	9006	29	21	34.71	79	27	27.60	1869.6
1 Quipa	9007	−16	27	56.74	288	30	24.82	2476.5
1 Shraz	9008	29	38	13.87	52	31	11.52	1578.2
1 Curac	9009	12	5	25.19	291	9	44.71	−28.8
1 Juptr	9010	27	1	14.15	279	53	13.55	−26.8
1 Vildo	9011	−31	56	34.67	294	53	36.93	618.9
1 Mauio	9012	20	42	26.15	203	44	33.98	3041.1
Ausbak	9023	−31	23	25.82	136	52	43.96	128.4

4.4. Comparison of Station Heights with Local Survey

The station heights above the geoid have been obtained for 46 stations using our combined satellite—surface gravity geopotential and station coordinate solutions. These heights were then compared with the mean sea level heights given by local survey. The mean sea level height, MSLH, for each of the stations was obtained from the survey sheets in the NASA Directory of Observation Station Locations published at Goddard Space Flight Center.

The quantity calculated for each of the 46 stations was

$$\Delta H = h - N - \mathrm{MSLH} \tag{6}$$

where h is the station height and N the geoid height from the GEM 2 solution. Since the geoid heights N and the station heights h are referenced to our ellipsoid of $a_e = 6378\,155$ m, the average ΔH, which is close to -10 meters, implies that $a_e = 6378\,145$ m is a better reference radius for this set of stations. This value of $\Delta H = -10$ meters is based upon the MOTS, laser, GRARR, and Doppler coordinates; the Baker-Nunn coordinates were excluded from this calculation because of the adjustment for parallactic refraction made to the station heights, as noted in § 4.1.

4.5. Adjusted Ellipsoid Scale (a_e) based upon the Mean Value of Gravity Variation (Δg_0)

The mean value of gravity variation $\Delta \bar{g}_0$ from the reference system was obtained as part of a general solution as previously described. The value of $\Delta \bar{g}_0 = 3.28$ mgal was obtained. Using the simple relation

$$g_e \simeq \frac{GM}{a_e^2} \tag{7}$$

the variational relationship becomes

$$\frac{\Delta g_e}{g_e} = \frac{\Delta GM}{GM} - \frac{2\Delta a_e}{a_e}. \tag{8}$$

Since the GM of the reference ellipsoid includes the atmosphere [3], and the relative mass of the atmosphere to the earth is 0.85×10^{-6} and further, since

$$\frac{\Delta g_e}{g_e} = \frac{\Delta g_0}{g_e} = 3.35 \times 10^{-6}$$

where $g_e = 9.780291$ m s^{-2}, we can obtain

$$\Delta a_e = -\frac{a_e}{10^6}\frac{(3.35 + 0.85)}{2}$$

or -13 meters. This implies that $a_e = 6378142$ meters is a better scale for our reference ellipsoid.

5. Tests of Solutions

Two tests have been applied to the solutions that have been obtained so far. Both concern the ability of the gravity fields to represent the motion of close earth satellites. The first test was on 22 six-hour orbital arcs of the Beacon Explorer C (BE-C) spacecraft composed of laser range data from the Goddard experimental laser. The rms of fit to the data on the 22 arcs using the Standard Earth II gravity field varied between 0.9 and 4.3 m with a mean of about 3 m. With the GEM 1 solution these rms values dropped to between 0.6 and 2.0 m with a mean of about 1.3 m. This considerable improvement was obtained even though no laser data on BE-C was used in the solution. The second test involved the long-term behavior of Intelsat 2-F1, Cosmos 41 rocket and Cosmos 382 rocket. The variation of mean elements over periods of hundreds of days have been compared using the Standard Earth II and GEM 2 gravity fields. The results show that a three-fold improvement has been obtained with the GEM 2 solution. Further, it should be remembered that *none* of these satellites was used in the solutions for the gravity field.

6. Summary and Conclusions

The first comprehensive GSFC Earth Model has been obtained. This consists of a combined solution for the geopotential field using satellite and surface gravity data (GEM 2) as well as the associated satellite-only solution (GEM 1). It also includes a set of 46 geocentric station positions. Tests and comparisons indicate encouraging results for these solutions, particularly in their capability for improved orbit determination. Very reasonable agreement has been obtained between analyses of surface gravity data and station height data to provide for an adjusted scale of the equatorial radius a_e of a mean earth reference ellipsoid. A value of $a_e = 6378\,145$ meters is suggested.

References

[1] E. M. Gaposchkin and K. Lambeck, Smithson. Astrophys. Obs. Spec. Rep. No. 315 (1970).
[2] R. H. Rapp, paper presented at Amer. Geophys. Un. Ann. Meeting, Washington, D.C. 1972.
[3] W. L. Sjøgren et al., JPL Tech. Rep. No. 32-1057 (1966).

Space Research XIII — Akademie-Verlag, Berlin 1973

15TH-ORDER HARMONICS IN THE GEOPOTENTIAL, FROM ANALYSIS OF DECAYING SATELLITE ORBITS

D. G. King-Hele

Royal Aircraft Establishment, Farnborough, Hants., England

A satellite with an initial orbital period greater than about 96 minutes will, as it decays under the action of air drag, pass through 15th-order resonance when successive equator crossings are exactly 24° apart, so that each day the satellite follows a fixed set of 15 ground tracks. If the decay is slow enough, the inclination i of the orbit to the equator changes appreciably as the satellite passes through resonance, because of the build-up of perturbations caused by the tesseral harmonics in the earth's gravitational potential of order 15 and degree 15, 17, 19, 21 ... Equations between the coefficients of these harmonics can be obtained by analysing the changes in orbits at a variety of inclinations, offering a method of evaluating 15th-order harmonics almost free of contamination from other harmonics.

This paper describes the results so far obtained in these studies, from the satellites Ariel 3 ($i = 80°$), Transit 1B ($i = 51°$), Cosmos 373 ($i = 63°$) and Cosmos 387 ($i = 74°$). A set of values is derived for the coefficients of order 15 and degree 15, 17, 19 and 21.

1. Introduction

In the past few years there have been great advances in the general mapping of the geoid on a world scale from observations of earth satellites, and the determination of the geopotential complete up to harmonics of order (16,16), as in the Smithsonian Standard Earth 1969, is a fine achievement [1]. With the accumulation of more accurate observations, and particularly laser observations, improved solutions will no doubt be obtained, perhaps up to even higher orders. However, these comprehensive solutions including 300 or more harmonic coefficients suffer from the difficulties caused by correlations between the coefficients, which throw doubt on the reliability of some of the terms of high order. It is therefore important to develop alternative methods for evaluating coefficients of a particular order, which are not subject to the same uncertainties, so that these coefficients can be taken as fixed constants in the comprehensive solution. This procedure is of course already used for the zonal harmonics, those of order zero. Also, the determination of harmonics of order 9, 12, 13 and 14 has been greatly strengthened by the use of stable near-resonant orbits, i.e. satellites almost unaffected by air drag which are close to 9th, 12th, 13th or 14th-order resonance and are sensitive to harmonics of these orders.

The possibility of utilizing orbits at 15th-order resonance, with ground tracks exactly 24° apart, may not seem so promising at first sight, because the appropriate orbital period is between 94 and 96 minutes if the inclination is less than 90°, corresponding to orbits with average heights between 470 and 550 km, where air drag is substantial, especially if the orbit is non-circular. However, all satellites with initial periods greater than 98 minutes eventually pass through exact 15th-order resonance as their orbital periods decrease under the action of air drag, and it has been shown, in theory by Allan [2,3] and in practice by Gooding [4], that appreciable changes in some of the orbital parameters, particularly the inclination of the orbit to the equator, occur if the orbit passes through 15th-order resonance slowly. By analysing the variation in inclination at times near resonance, an equation between the harmonic coefficients of order 15 and degree 15, 17, 19 ... can be obtained. By deriving such equations for an number of satellites at different inclinations, it should be possible to determine a set of individual coefficients of order 15. If the passage through resonance is slow enough, quite accurate results can be obtained using observations of only moderate accuracy: Gooding [4] used Minitrack observations, accurate to about 2 minutes of arc.

With the aim of evaluating 15th-order coefficients, several satellites that have recently passed through 15th-order resonance, or are about to do so [5], are being observed by the British tracking network, which includes the accurate Hewitt cameras at Malvern and Edinburgh and the kinetheodolite at the Cape Observatory, as well as radar sensors and about 50 skilled visual observers.

This paper reviews the progress so far in these studies. Similar analyses may be possible for 16th-order resonance or for 29:2 and 31:2 resonance, but these possibilities are not further discussed here.

2. Theory

The gravitational potential U of the earth at an exterior point (r, θ, λ) may be written in the form

$$U = \frac{\mu}{r}\left\{1 - \sum_{l=2}^{\infty} J_l \left(\frac{R}{r}\right)^l \mathrm{P}_l(\cos\theta) + \sum_{l=2}^{\infty}\sum_{m=1}^{l} \mathrm{J}_{lm}\left(\frac{R}{r}\right)^l \mathrm{P}_l^m(\cos\theta)\cos m(\lambda - \lambda_{lm})\right\} \tag{1}$$

where r is distance from the earth's centre, θ is co-latitude, λ is longitude (positive to the east), μ is the gravitational constant for the earth (398 602 km^3 s^{-2}), R is the earth's equatorial radius (6378.2 km), the J_l, J_{lm} and λ_{lm} are constants to be determined, $\mathrm{P}_l(\cos\theta)$ is the Legendre polynomial of degree l, and $\mathrm{P}_l^m(\cos\theta)$ is the associated Legendre function of order m.

The rate of change of inclination i due to the effect of 15th-order resonant terms in U may be written [6]

$$\frac{di}{dt} = 0.5877\, n(15 - \cos i)(1 + \cos i)\sin^{13} i \left(\frac{R}{a}\right)^{15} \{\bar{C}_{15}\sin\Phi - \bar{S}_{15}\cos\Phi\}. \tag{2}$$

In Eq. (2) the 'resonance angle' Φ is given by

$$\Phi = \omega + M + 15(\Omega - \nu) \tag{3}$$

where ω is the argument of perigee, M the mean anomaly, Ω the right ascension of the node, and ν is the sidereal angle. The quantity $\bar{C}_{15}$ in Eq. (2) is defined by [3]:

$$\bar{C}_{15} = \frac{1}{\bar{F}_{15,15,7}} \sum_{l=15,17\ldots} (-1)^{(l+1)/2} \left(\frac{R}{a}\right)^{l-15} \bar{F}_{l,15,\frac{1}{2}(l-1)} \bar{C}_{l,15} \tag{4}$$

where

$$N_{l,15}\,\bar{C}_{l,15} = \mathrm{J}_{l,15} \cos 15\lambda_{l,15} \tag{5}$$

$$N_{l,15} = \left\{\frac{2(l-15)!\,(2l+1)}{(l+15)!}\right\}^{1/2} \tag{6}$$

and $\bar{F}_{l,15,\frac{1}{2}(l-1)}$ is a function dependent on inclination, values being given by Allan [3]. The quantity $\bar{S}_{15}$ is given by Eq. (4) with C replaced by S, and

$$N_{l,15}\,\bar{S}_{l,15} = \mathrm{J}_{l,15} \sin 15\lambda_{l,15}\,. \tag{7}$$

If we rewrite Eq. (2) in the form

$$\frac{\Delta i}{\Delta t} = K(\bar{C}_{15} \sin \Phi - \bar{S}_{15} \cos \Phi) \tag{8}$$

the quantity K may be taken as constant during the time when the main change in inclination is occurring. Since Φ is known accurately, the observed change Δi in i, between two times Δt apart, yields one equation between $\bar{C}_{15}$ and $\bar{S}_{15}$. (In calculating Δi the values of i must of course be corrected for odd-harmonic, luni-solar and any other relevant perturbations.) Individual values of $\bar{C}_{15}$ and $\bar{S}_{15}$ can therefore be obtained from three reliable values of i, provided they are not symmetrically disposed about the resonance point. If there are more than three values of i available, the best values of $\bar{C}_{15}$ and $\bar{S}_{15}$ can be found by least-squares fitting.

Thus each resonant satellite should yield a value of $\bar{C}_{15}$ and of $\bar{S}_{15}$, and these are expressible as linear functions of the individual coefficients $\bar{C}_{l,15}$ and $\bar{S}_{l,15}$ through Eq. (4). The numerical values of the quantities $\bar{F}$ in these linear functions are strongly dependent on inclination [3] and values of $\bar{C}_{15}$ and $\bar{S}_{15}$ at several different inclinations should therefore give several non-singular equations between the $\bar{C}_{l,15}$ and $\bar{S}_{l,15}$.

3. Results from Individual Satellites

3.1. Ariel 3

The satellite Ariel 3 (1927-42 A), with inclination 80.2°, passed through 15th-order resonance between December 1967 and April 1968, and the inclination decreased by about 0.02° during this time. Fig. 1 shows the observed values of inclination, as obtained by Gooding [4] from Minitrack observations, after correction for luni-solar perturbations, and the curve given by Eq. (2) with the best-fitting values of the coefficients. The fit is good and the equations obtained are as follows:

$$\left.\begin{aligned} \bar{C}_{15,15} + 0.35\bar{C}_{17,15} + 0.06\bar{C}_{19,15} - 0.10\bar{C}_{21,15} &= -(19.9 \pm 1.2) \times 10^{-9} \\ \bar{S}_{15,15} + 0.35\bar{S}_{17,15} + 0.06\bar{S}_{19,15} - 0.10\bar{S}_{21,15} &= -(\ 7.7 \pm 0.8) \times 10^{-9} \end{aligned}\right\}. \tag{9}$$

In these, and in the equations for subsequent satellites, the (23,15) and higher-degree terms are omitted: their effect will be considered later.

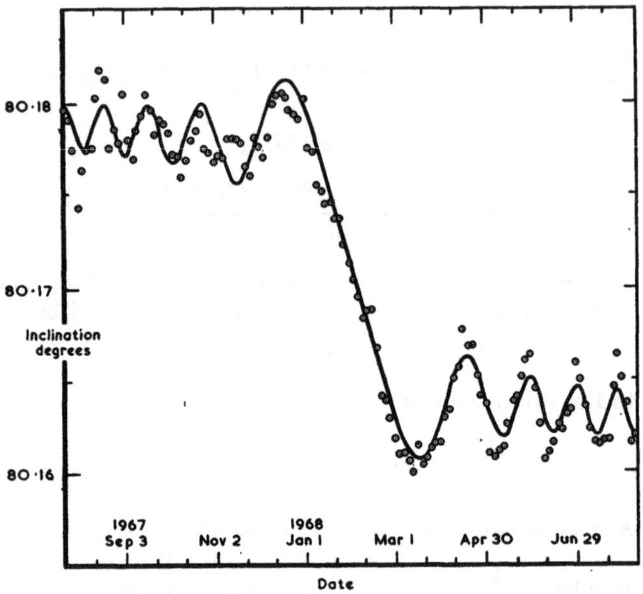

Fig. 1. Values of inclination for Ariel 3 at 15th-order resonance, with fitted theoretical curve [4].

3.2. Transit 1B

The satellite Transit 1B (1960 γ2), with inclination near 51°, passed through 15th-order resonance very slowly during 1962. Unfortunately, the observations at this time were few in number and poor in accuracy. Nevertheless it has been possible [6] to determine some reasonably accurate orbits which show that the inclination decreased by nearly 0.04° between February 1962 and February 1963. The observational values of i, after correction for perturbations, and the curve fitted are shown in Fig. 2. The equations obtained for the 15th-order coefficients are:

$$\left.\begin{aligned} 0.56\bar{C}_{17,15} - \bar{C}_{19,15} + 0.74\bar{C}_{21,15} &= (36 \pm 4) \times 10^{-9} \\ 0.56\bar{S}_{17,15} - \bar{S}_{19,15} + 0.74\bar{S}_{21,15} &= (35 \pm 6) \times 10^{-9} \end{aligned}\right\}. \qquad (10)$$

3.3. Cosmos 316

Cosmos 316 (1969-108A), with inclination near 49.5°, was a most unsuitable satellite for studies of 15th-order harmonics, because its perigee height was near 140 km and decay was very rapid. Nevertheless, analysis of the orbit [7] has shown a distinct increase in inclination of about 0.004° at the time of the 15th-order resonance, from which one equation has been obtained, namely

$$0.34\bar{S}_{17,15} - 0.88\bar{S}_{19,15} + \bar{S}_{21,15} = (36 \pm 12) \times 10^{-9}. \qquad (11)$$

Although less accurate, this equation is consistent with the second of Eq. (10), in which the numerical coefficients are not very different because the inclinations of Transit 1B and Cosmos 316 differ by less than 2°. Eq. (11) therefore provides useful confirmation of Eqs. (10).

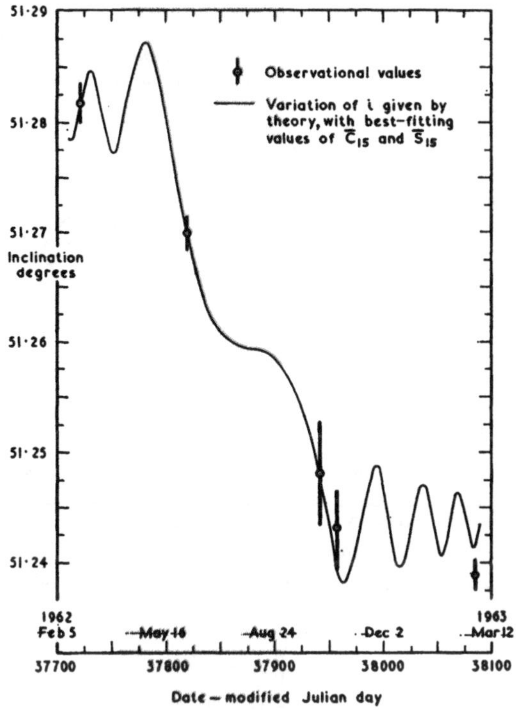

Fig. 2. Transit 1 B: change in inclination at resonance [6].

3.4. Cosmos 373

Cosmos 373 (1970-87 A), with inclination 62.9°, was in exact 15th-order resonance on 10 November 1970, 20 days after launch. Fig. 3 shows the values of inclination from US Navy orbits during the subsequent 7 months, after correction for luni-solar perturbations, together with the best-fitting theoretical curve, determined with the aid of Gooding's computer program THROE. The perturbations taken into account were those due to luni-solar attraction, zonal harmonics in the geopotential and an atmospheric rotation rate of 0.8 rev/day. This last perturbation was very small ($<0.0003°$).

Although the orbital data do not begin until 20 days after resonance, the change in inclination is well-defined, and the following equations for the coefficients are obtained:

$$\left.\begin{aligned} 0.40\bar{C}_{15,15} - \bar{C}_{17,15} + 0.20\bar{C}_{19,15} + 0.54\bar{C}_{21,15} &= -(1.0 \pm 1.4)\times 10^{-9} \\ 0.40\bar{S}_{15,15} - \bar{S}_{17,15} + 0.20\bar{S}_{19,15} + 0.54\bar{S}_{21,15} &= -(12.5 \pm 1.1)\times 10^{-9} \end{aligned}\right\}. \quad (12)$$

3.5. Cosmos 387

Cosmos 387 (1970-111 A), with inclination 74.0° is the first satellite to be observed specially for the purpose of evaluating 15th-order harmonics. It experienced exact 15th-order resonance on 3 November 1971 and the change in inclination

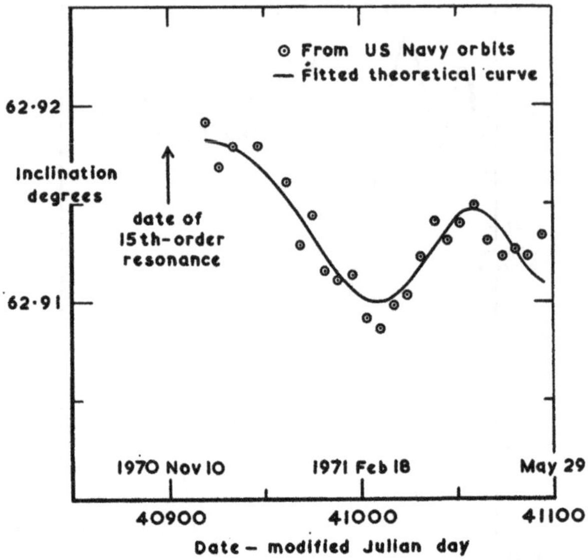

Fig. 3. Cosmos 373: values of inclination after resonance, with fitted curve.

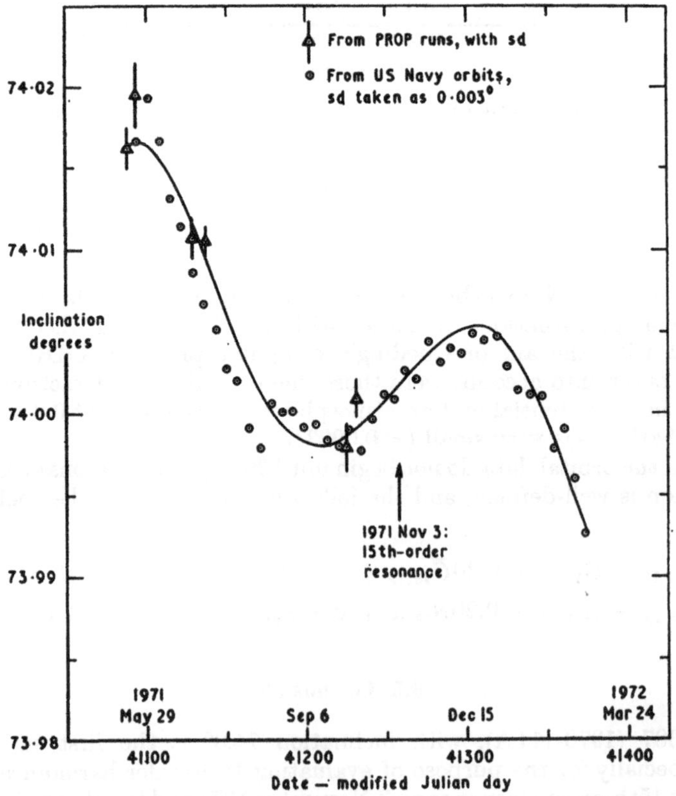

Fig. 4. Cosmos 387: values of inclination at resonance, with fitted curve.

resulting from the resonance was still continuing strongly in April 1972. It should be possible to obtain better results when the change in inclination is complete.

Fig. 4 shows the values of inclination given by US Navy orbits, which are usually accurate to about 0.003°, and also six more accurate values obtained using the RAE orbit determination program PROP 5 with Hewitt camera and visual observations in addition to US Navy observations. The values are corrected for luni-solar and atmospheric perturbations, and the PROP values are also corrected for $J_{2,2}$ perturbations. Fig. 4 also shows the best-fitting theoretical curve, again determined with the aid of the THROE computer program. The fit is excellent, the rms scatter of the points about the curve being 0.66 sd, and the equations obtained are:

$$\left.\begin{aligned}\bar{C}_{15,15} - 0.33\bar{C}_{17,15} - 0.59\bar{C}_{19,15} - 0.45\bar{C}_{21,15} &= -(24.0 \pm 1.0)\times 10^{-9}\\ \bar{S}_{15,15} - 0.33\bar{S}_{17,15} - 0.59\bar{S}_{19,15} - 0.45\bar{S}_{21,15} &= -\ (3.9 \pm 0.5)\times 10^{-9}\end{aligned}\right\}. \quad (13)$$

For Cosmos 387 the resonance angle Φ had the value $-15.7°$ at resonance, and remained between $-16°$ and $+16°$ between MJD 41200 and 41310, so that the average value of $\sin \Phi$ during this time was nearly zero. Hence, from Eq. (8), the variation of inclination for 55 days on each side of resonance is controlled largely by $\bar{S}_{15}$, and the rather small and positive change in i during this time interval implies that $\bar{S}_{15}$ is small and negative, as confirmed by the second of Eqs. (13), in which the right-hand side is $\bar{S}_{15}$.

4. Solutions for the Individual Coefficients

If coefficients of degree 23 and higher are ignored, the formal solution of Eqs. (9), (10), (12) and (13) gives the following values for the $\bar{C}_{l,15}$ and $\bar{S}_{l,15}$:

l	$10^9\bar{C}_{l,15}$	$10^9\bar{S}_{l,15}$
15	-18.3 ± 1.0	$-\ 8.9 \pm 0.7$
17	5.3 ± 1.4	10.1 ± 1.4
19	-13.5 ± 2.3	-21.8 ± 3.0
21	26.5 ± 2.5	10.2 ± 3.3

To obtain realistic values for the likely errors, it is necessary to estimate the effect of neglecting the terms of degree 23 and higher. The errors from this source are assessed by making the reasonable but unprovable assumptions that the main error comes from neglecting the coefficients of degree 23, and that their numerical values are 2/3 of the average of the eight numerical values above, i.e. 10×10^{-9}, since the magnitude of the coefficients tends [1] to fall off as $1/l^2$. The neglect of these terms introduces additional errors of $\pm$ 1.7, 1.3, 1.3 and 2.0 on the right-hand sides of Eqs. (9), (10), (12) and (13) respectively, giving total errors on the right-hand sides of these equations of 2.1 and 1.9; 4 and 6; 1.9 and 1.7; and 2.2 and 2.1 respectively. This leads to the following set of values for the coefficients, all multiplied by 10^{-9}:

$$\left.\begin{aligned} \bar{C}_{15,15} &= -18.3 \pm 1.7 & \bar{S}_{15,15} &= -\ \ 8.9 \pm 1.6 \\ \bar{C}_{17,15} &= \quad 5.3 \pm 2.1 & \bar{S}_{17,15} &= \quad 10.1 \pm 2.1 \\ \bar{C}_{19,15} &= -13.5 \pm 2.8 & \bar{S}_{19,15} &= -21.8 \pm 3.5 \\ \bar{C}_{21,15} &= \quad 26.5 \pm 3.0 & \bar{S}_{21,15} &= \quad 10.2 \pm 3.7 \end{aligned}\right\}. \tag{14}$$

The 'errors' given here, since they allow for the effect of ignoring the (23,15) terms, should provide realistic estimates of the actual errors in the values, unless there are important contributions from terms of still higher degree, e.g. (25,15), or higher order (30 or 45). Both these possibilities seem rather unlikely. The neglect of terms involving the eccentricity e should not be important, except possibly for Transit 1B ($e = 0.025$), for which however the assigned errors are already between 11 and 17%. The value of e at resonance was 0.007 for Ariel 3, 0.007 for Cosmos 373, and 0.001 for Cosmos 387.

5. Discussion

The values for the coefficients obtained in Section 4 show the feasibility of the method, and provide a first determination of the coefficients of degree 17, 19 and 21. But it should be emphasized that the orbital information used was rather unsatisfactory for two of the four satellites: the results from Transit 1B are, in effect, based on only four values of inclination, each determined from a rather small number of observations of limited accuracy made in 1962; and the data for Cosmos 373 do not begin until three weeks after resonance. Consequently it should soon be possible to obtain much better sets of values, using more accurate orbits obtained from the observations of satellites now being tracked for this purpose by numerous observing stations including the Hewitt cameras at Malvern and Edinburgh [5].

However, the values obtained here ought to be reliable. The alternation of signs as the degree increases, although always slightly disquieting, is a familiar phenomenon in such evaluations: for example in the Smithsonian Standard Earth (SSE) solution [1] the (7,7), (9,7), (11,7), (13,7) and (15,7) C-coefficients alternate in sign, and so do the (9,9), (11,9), (13,9) and (15,9) C-coefficients. Only two of the eight coefficients evaluated here are included in the SSE solution — the (15,15) terms. These are likely to be among the least reliable of the SSE coefficients, because, being the 'last of the line', they are most exposed to contamination from unevaluated coefficients. The SSE values for $\bar{C}_{15,15}(1.5 \times 10^{-9})$ and $\bar{S}_{15,15}(-14.8 \times 10^{-9})$ are not consistent with those in Eqs. (14). In deriving the values (14), Eq. (11) from Cosmos 316 was ignored, because it is less accurate than Eqs. (10). However, the coefficients given in (14) do satisfy Eq. (11): when they are substituted into the left-hand side of Eq. (11), the value obtained is 33×10^{-9}.

The value of $(\bar{C}^2_{l,15} + \bar{S}^2_{l,15})^{1/2}$ is more significant than $\bar{C}_{l,15}$ or $\bar{S}_{l,15}$ alone, and it will be seen that the coefficients of degree 17, when combined in this way, are considerably smaller than those of degree 15, 19 or 21. One consequence is that the lumped 15th-order coefficients quoted by Gooding [4] provide quite good approximations to the values for the (15,15) coefficients.

6. Conclusions

A set of values for the geopotential coefficients of order 15 and degree 15, 17, 19 and 21 is derived for the first time, and estimates of the accuracy are given: see Eqs. (14). It should soon be possible to improve the values, utilizing satellites now being observed for the purpose.

Acknowledgments

The author thanks R. H. Gooding, R. H. Merson and R. Clarke for assistance in the computational work. He is also most grateful to R. L. Easton and R. R. Zirm of the US Naval Research Laboratory for kindly supplying US Navy data.

References

[1] E. M. Gaposhkin and K. Lambeck, **1969** Smithsonian Standard Earth, Smithsonian Astrophys. Obs. Spec. Rep. No. 315 (1970).
[2] R. R. Allan, Planet. Space Sci. **15**, 53 (1967).
[3] R. R. Allan, RAE Tech. Rep. No. 71245 (1971).
[4] R. H. Gooding, Nature Physical Science **231**, 168 (1971).
[5] R. D. Eberst, Nature Physical Science **235**, 130 (1972).
[6] H. Hiller and D. G. King-Hele, Planet. Space Sci. **20**, 1213 (1972).
[7] D. G. King-Hele, Proc. Roy. Soc. A **330**, 467 (1972).

Space Research XIII — Akademie-Verlag, Berlin 1973

DETERMINATION D'EQUATIONS DE CONDITIONS ENTRE HARMONIQUES DE RESONNANCE D'ORDRE 14 A PARTIR DES OBSERVATIONS DU SATELLITE EOLE

J. L. PIEPLU et M. LEFEBVRE

Groupe de Recherche de Géodésie Spatiale, Meudon, France

Le satellite EOLE d'inclinaison 50° présente une résonnance avec les termes d'ordre 14. L'amplitude sur la composante le long de l'orbite est de l'ordre de 500 mètres et la période de 10j 3.

Le satellite EOLE lancé en août 1971 dans le cadre d'une coopération CNES—NASA a pour éléments d'orbite $a = 7170$ km, $e = 0{,}016$, $i = 50°$.

Les valeurs des variations séculaires de l'argument du périgée et de l'ascension droite du noeud ascendant et du moyen mouvement sont telles qu'il y a une quasi-commensurabilité avec la rotation de la terre et que l'effet des termes d'ordre 14 est amplifié par un phénomène de résonnance. La période de la raie centrale est de 9j 12 et l'amplitude du terme périodique sur l'anomalie moyenne de l'ordre de 400 mètres.

Le satellite a été observé à partir des stations interférométriques Diane situées à Prétoria et à Kourou. La précision des mesures angulaires restituée est de 2 à $4 \cdot 10^{-4}$. Elle peut toutefois être un peu moins bonne à Kourou (anomalie ionosphérique). La position des stations est connue à 50 mètres près dans un repère géocentrique. La couverture de l'orbite est de 3 à 4 passages par jour. Le traitement se fait tous les jours sur trois jours de mesures avec recouvrement par un programme de correction différentielle utilisant pour le modèle de potentiel, la théorie de Brouwer limitée à J_2, J_2^2, J_4 pour les termes séculaires et de J_2 à J_5 pour les termes à longue période. On obtient donc tous les jours un jeu de paramètres moyens au sens de cette théorie.

On voit clairement sur la courbe du demi-grand axe une dérive moyenne due au frottement atmosphérique de l'ordre de 2 m/jour, une variation longue période de ce même et des variations à plus courte période qui apparaissent également sur l'excentricité. Les variations des éléments orbitaux au premier ordre dues à une harmonique C_{lm}, S_{lm} sont [1]:

$$\Delta a = \mu \frac{a_e^l \, 2F_{lmp} G_{lpq} (l-2p+q)\, S_{lmpq}}{n a^{l+2}[(l-2p)\dot{\omega} + (l-2p+q)\dot{M} + m(\dot{r}-\dot{\theta})]}$$

et

$$\Delta e = \mu \frac{a_e^l F_{lmp} G_{lpq} (1-e^2)^{1/2} [(1-e^2)^{1/2}(l-2p+q) - (l-2p)]\, S_{lmpq}}{n a^{l+3} e[(l-2p)\dot{\omega} + (l-2p+q)\dot{M} + m(\dot{r}-\dot{\theta})]}.$$

Les valeurs des dérives des éléments angulaires sont: $\dot{M} = 14{,}302224$ tours/jour, $\dot{\omega} = +\,3{,}4815$ °/jour, $\dot{\Omega} = -\,4{,}2448$ °/jour. On voit qu'on a résonnance

Equations de condition

sur *e*

1) 8j 46

$$-0{,}58 \cdot 10^{-15} \begin{cases} C_{14,14} \\ S_{14,14} \end{cases} + 0{,}54 \cdot 10^{17} \begin{cases} C_{16,14} \\ S_{16,14} \end{cases} - 0{,}67 \cdot 10^{18} \begin{cases} C_{18,14} \\ S_{18,14} \end{cases} + 0{,}13 \cdot 10^{19} \begin{cases} C_{20,14} \\ S_{20,14} \end{cases} = \begin{matrix} -\,0{,}82 \cdot 10^{-5} \\ +\,1{,}058 \cdot 10^{-5} \end{matrix}$$

2) 10j 12

$$0{,}23 \cdot 10^{15} \begin{cases} S_{14,14} \\ C_{14,14} \end{cases} - 0{,}34 \cdot 10^{17} \begin{cases} S_{16,14} \\ C_{16,14} \end{cases} + 0{,}72 \cdot 10^{18} \begin{cases} S_{18,14} \\ C_{18,14} \end{cases} + 0{,}48 \cdot 10^{19} \begin{cases} S_{20,24} \\ C_{20,14} \end{cases} = \begin{matrix} -\,1{,}029 \cdot 10^{-5} \\ +\,0{,}892 \cdot 10^{-5} \end{matrix}$$

sur *a* (en kilométres)

9j 214

$$-0{,}12 \cdot 10^{20} \begin{cases} C_{15,14} \\ S_{15,14} \end{cases} + 0{,}39 \cdot 10^{21} \begin{cases} C_{17,14} \\ S_{17,14} \end{cases} - 0{,}30 \cdot 10^{22} \begin{cases} C_{19,14} \\ S_{19,14} \end{cases} + 0{,}45 \cdot 10^{22} \begin{cases} C_{21,14} \\ S_{21,14} \end{cases} = \begin{matrix} -\,1{,}789 \cdot 10^{-3} \\ +\,1{,}660 \cdot 10^{-3} \end{matrix}$$

lorsque

$$1 - 2p + q = 1$$

$$\text{car} \quad M \sim m(\dot{\Omega} - \dot{\theta}).$$

Les termes avec l impair vont donner des résonnances sur a, avec une période de 9,213 jours. Les termes avec l pair vont donner deux raies symétriques sur l'excentricité de périodes 8,46 j et 10,12 j, et également deux petites raies sur a, mais e fois plus faibles que la raie centrale avec l impair.

Notre méthode a consisté à prendre les éléments moyens sur 180 jours à retirer des termes polynomiaux ou à longue période. Une analyse harmonique a montré l'appartition des raies prévues et un ajustement en considérant la fréquence connue a permis de déterminer les amplitudes des raies et leur erreur quadratique moyenne.

On a décomposé chaque raie en:

$$A \sin \psi + B \sin \psi$$

et les valeurs de A et de B nous donnent des équations liant les C_{nm} et S_{nm} respectivement. Pour le demi-grand axe, on obtient (période 9,213 j)

$$A_a = -1{,}789 \text{ mètres} \pm 0{,}340$$

$$B_a = +1{,}660 \text{ mètres} \pm 0{,}327$$

(amplitude de la raie 2,50 m $\pm$ 0,4 m).

Pour l'excentricité, les deux raies ne sont pas exactement symétriques:
Periode de 8j 46:

$$10^5 \times \Delta A_{1e} = -0{,}8201 \pm 0{,}27$$

$$10^5 \times \Delta B_{1e} = +1{,}058 \quad \pm 0{,}27.$$

Période de 10j 11:

$$10^5 \times \Delta A_{2e} = -1{,}029 \quad \pm 0{,}27$$

$$10^5 \times \Delta B_{2e} = +0{,}892 \quad \pm 0{,}26.$$

On trouvera in fine les équations de conditions correspondantes.

Plus que les résultats eux-même, il faut surtout considérer la méthode utilisée, c'est-à-dire qui a permis la détermination assez précise des termes périodiques, malgré la qualité médiocre des mesures. Une telle méthode appliquée à des orbites de bonne qualité pourrait permettre une amélioration sensible de ces harmoniques et pourrait même être utilisée de manière plus générale même si la couverture est moins bonne pourvu que le satellite soit drag-free.

Il va de soi que la méthode doit être améliorée et que des itérations successives sur les différents termes périodiques doivent être effectuées pour se débarasser totalement des effets de couplage.

Bibliographie

[1] W. M. Kaula, Theory of Satellite Geodesy, Blaisdell Publ. Co., Waltham, Mass. 1966.

Space Research XIII — Akademie-Verlag, Berlin 1973

THE LONGITUDE ACCELERATION OF THE SKYNET SATELLITE MARCH 1970—APRIL 1971

R. H. Merson

Royal Aircraft Establishment, Farnborough, Hants, UK

Accurate orbit determinations of the synchronous Skynet 1 satellite have been made at regular intervals using range data from four tracking stations (rms range residual ~ 10 metres). From these orbital data the longitude acceleration due to tesseral harmonics of earth gravity alone has been calculated at a number of longitudes in the range 40° E to 50° E. The results have been incorporated in a new NASA determination of the low-order tesseral harmonics.

1. Introduction

In November 1966 Space Department, RAE, undertook the task of providing, and subsequently operating, the computer software for orbit and attitude control of the satellites of the Skynet communication system. Skynet 1 was launched in November 1969 and placed in a near-equatorial, geostationary orbit at 42° E longitude. Shortly afterwards it was decided to keep the launching characteristics for the second satellite identical with those for the first and consequently Skynet 1 was allowed to drift to a new position centred on 48 °E longitude.

This paper summarizes a study of the orbit of Skynet 1 during three drift periods, culminating in very good estimates of the longitude acceleration over the useful longitude range 40° E—50° E. When combined with similar existing data from other satellites the results should lead to an improved accuracy of the low-order tesseral harmonics of the earth's gravity potential field.

2. Tracking

The operational network for tracking the satellite consisted of the 40ft dishes of the three earth terminals situated in Cyprus, at Singapore and at the control station at Oakhanger (UK). In addition, tracking data were provided by a similar dish at the Signals Research and Development Establishment, Bournemouth (UK). The raw data consisted of signal transit times to the satellite and back, quantized to the nearest 0.1 μs. This is equivalent to the range being quantized in 15 m steps. The associated observation time is accurate to about 0.1 s, equivalent to 3 m in maximum range rate conditions.

Normally each station obtained a burst of data once every 8 hours, a burst consisting of 20—30 range observations at intervals of 15 or 30 seconds, and the data arrived at RAE, via Telex, on 5-hole paper tape.

3. Initial Data Processing

Each burst of range data was smoothed on an ICL 1907 computer by a least-squares quadratic fitting procedure incorporating the rejection of rogues. In the absence of noise, the expected rms departure from the quadratic is 4.3 m. In practice, once some initial teething troubles had been eliminated, the observed rms errors have been mostly about 5 m, indicating that there is little, if any, noise. Three points were picked off each smoothed curve, the corresponding range, time and station identifier being output on punched cards.

4. Orbit Generator

Given initial values of position and velocity in cartesian coordinates, the orbit of the satellite is generated by integrating forwards and/or backwards the second-order differential equations

$$\ddot{x}_i = -\mu x_i/r^3 + F_i, \quad i = 1, 2, 3, \quad r = (\Sigma x_i^2)^{1/2}.$$

The perturbing acceleration F_i includes the J_2, J_3 and J_4 zonal harmonic earth gravity terms, lunar and solar gravity terms, the effect of solar radiation pressure, the first-order effect of earth spin-axis precession and a $J_{2,2}$ term representing the lumped effect of tesseral harmonic earth gravity terms. A full list of the acceleration formulae is given in the Appendix.

5. Time and Coordinate Systems

The time and coordinate systems used in the Skynet computer programs are the same as those used in earlier RAE programmes [1] and in the DOI program of the Smithsonian Institution Astrophysical Observatory [2]. Calendar dates are reckoned in Modified Julian Days (MJD), related to ordinary Julian Days by the formula MJD = JD − 2400000.5, e.g. 1972 May 1.0 = MJD 41438.

6. Orbit Determination

The orbit determination program starts with initial estimates of the position and velocity at a given Epoch, together with estimates of other quantities to be determined (solar radiation coefficient $\varkappa$ and/or lumped tesseral harmonic coefficient $J^*_{2,2}$, if required). Using selected range data over 10—20 days the estimates are improved by a differential correction procedure. The partial derivatives required are generated by integrating their second-order time derivatives based on the equations of Goodyear [3].

7. Development of the Orbital Model

The first Skynet orbits obtained gave rms range residuals of 80—100 m. At that time a constant pre-launch estimate of $\varkappa$ was being used and $J^*_{2,2}$, $\lambda^*_{2,2}$ were based on the Guier—Newton 1965 model [4]. Over a period of time various improvements to the program and the model were introduced, as follows.

Station bias. The facility to add a constant value to the range data from each given station was introduced into the program. Eventually it was determined that best results could be obtained by adding approximately 60 m to the Singapore range data, and giving small biases to the UK data.

Station coordinate correction. Levallois and Dufour [5] give three estimates of the position of the earth's centre of mass relative to the origin of the European datum, due to Veis, IGN and Fischer. Taking the average of these estimates, $(\Delta x, \Delta y, \Delta z) = (-70, -94, -137)$ metres have been added to survey values of station coordinates.

Earth spin-axis precession. A small but consistent discrepancy between predicted and determined orbital inclination was traced to the precession of the earth's spin axis. After insertion of first-order correction terms in the acceleration (see Appendix) the discrepancy virtually disappeared.

Variation of the solar radiation coefficient $\varkappa$. The variation of $\varkappa$ over the year May 1970 to May 1971 was obtained using the orbit determination programme [6]. The variation was seen to be approximately in the form of a constant plus sinusoid, the latter having a half-year period with minima at the autumn and spring equinoxes. This form, with appropriate parameters, was inserted in the orbit determination programme in place of the previously used constant value.

The lumped tesseral harmonic coefficients $J^*_{2,2}$, $\lambda^*_{2,2}$. Keeping $\lambda^*_{2,2}$ fixed, $J^*_{2,2}$ was determined at several points in the free-drift periods. The results of these runs form the main part of this paper. Using the orbit determination programme in its final form, the rms range residuals have been reduced to about 8 m, which is as close as can ever be expected to the basic accuracy of 4 m.

8. Determination of $J^*_{2,2}$

Having established the effects of solar radiation pressure on the orbit the only parameters remaining to be determined are $J^*_{2,2}$ and $\lambda^*_{2,2}$. Unfortunately, given data for one satellite at virtually one longitude, it is not possible accurately to separate $J^*_{2,2}$ and $\lambda^*_{2,2}$ as they are very highly correlated. It seemed best, therefore, to select a fixed value for $\lambda^*_{2,2}$ and to determine $J^*_{2,2}$, together with the standard orbital parameters (position and velocity), by a least-squares fitting procedure.

8.1. Choice of Time Spans

The requirements governing the choice of the time spans for each run are: (a) avoidance of orbit and attitude manoeuvres; (b) a long enough span to 'establish' the acceleration; (c) a span sufficiently short to keep the longitude essentially constant; (d) an even spread of observations from each station over the whole span.

The first step in making the choice is to look at the longitude variation. This is shown in Fig. 1, together with the dates of orbit and attitude manoeuvres. From the point of view of (b) and (c) the best periods are those near longitude turning

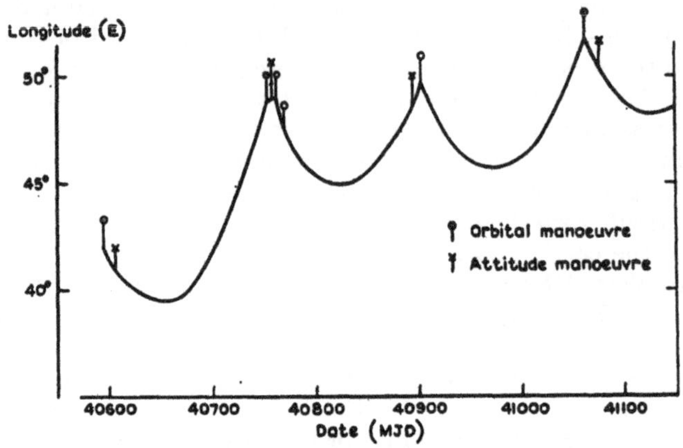

Fig. 1. The longitude of Skynet 1, Jan. 1970—May 1971.

points, of which there are four, and experiments have shown that a span of 20 days can be used in these cases. Where the longitude is changing rapidly it is found that a 10-day span is near the optimum. The selected spans, with their epochs (time average of data used) and other pertinent data, are given in Table 1.

Table 1

Skynet 1 orbit computation results ($\lambda^*_{2,2} = -12.5804°$)

Epoch MJD	Data span MJD	λ (°)	$10^6 J^*_{2,2}$	i (°)	$\frac{a}{R}$	$10^5 \ddot{\lambda}$ (rad. (sid. day)$^{-2}$)
40652.98	40641—40663	39.58	1.9407 ± 0.0016	2.174	6.6108	3.0553 ± 0.0025
40700.71	40695—40708	42.10	1.9482 ± 0.0010	2.089	6.6098	2.9875 ± 0.0015
40816.01	40808—40827	45.04	1.9528 ± 0.0011	1.759	6.6111	2.8705 ± 0.0016
40924.42	40918—40930	47.67	1.9509 ± 0.0006	1.529	6.6120	2.7312 ± 0.0008
40976.90	40970—40983	45.64	1.9519 ± 0.0006	1.364	6.6109	2.8409 ± 0.0008
41043.57	41036—41050	49.42	1.9512 ± 0.0005	1.222	6.6096	2.6306 ± 0.0007
41049.36	41042—41055	50.04	1.9473 ± 0.0007	1.211	6.6095	2.5865 ± 0.0009
41054.86	41050—41060	50.84	1.9424 ± 0.0005	1.209	6.6094	2.5282 ± 0.0007
41139.91	41130—41149	47.88	1.9503 ± 0.0005	0.994	6.6109	2.7200 ± 0.0007
		sensitivity runs ($\lambda^*_{2,2} = -10°.0268$)				
40976.90	40970—40983	45.64	1.8763 ± 0.0004	1.364	6.6109	2.8421 ± 0.0005
41054.86	41050—41060	50.84	1.8295 ± 0.0015	1.209	6.6094	2.5306 ± 0.0021

8.2. Choice of $\lambda^*_{2,2}$

For all the computer runs, except two sensitivity runs, the value used was $\lambda^*_{2,2} = -0.21957$ rad. This was based on the Guier—Newton 1965 model [4].

8.3. Results

The determined values of $J^*_{2,2}$, with standard deviations, together with time-averages of longitude λ, inclination i and the ratio of semi-major axis a to earth's equatorial radius R are given in Table 1. The eccentricity e of the orbit never exceeded 2×10^{-4}, and is negligible in the present context.

9. The Longitude Acceleration

Using the formulae of Allan [7], the mean longitude acceleration due to the earth's field alone is given by

$$\ddot{\lambda} = 18n^2 J^*_{2,2} \left(\frac{R}{a}\right)^2 c^4 \sin 2(\lambda - \lambda^*_{2,2}) - 12n\dot{\lambda} J^*_{2,2} \left(\frac{R}{a}\right)^2 c^2(4c^2 - 1) \sin 2(\lambda - \lambda^*_{2,2}),$$

where n is the mean motion and $c = \cos {}^1/_2\, i$.

For i close to zero and $\dot{\lambda}$ small, this reduces to approximately

$$\ddot{\lambda} = 18n_s^2\, J^*_{2,2} \left(\frac{R}{a}\right)^2 c^4 \sin 2(\lambda - \lambda^*_{2,2}),$$

where n_s is the synchronous mean motion. For $i < 2°$ and $\dot{\lambda} < 0.1°$ per day it has been checked that the error in this formula is less than 1 part in 10^5.

Finally, if the units are taken as radians and sidereal days,

$$\ddot{\lambda} = 72\pi^2 J^*_{2,2} \left(\frac{R}{a}\right)^2 c^4 \sin 2(\lambda - \lambda^*_{2,2}) \qquad \text{rad. (sid. day)}^{-2}. \tag{1}$$

The values of $\ddot{\lambda}$ obtained with this formula are given in the final column of Table 1.

10. Sensitivity of $\ddot{\lambda}$ to $\lambda^*_{2,2}$

Two additional runs were done to test the sensitivity of the determined acceleration to the choice of the fixed value of $\lambda^*_{2,2}$. The results are appended to Table 1. Apart from $J^*_{2,2}$, the values obtained for the orbit parameters were the same as before, and there was no observable difference in the range residuals. It will be seen that the $1\,\sigma$ levels in $\ddot{\lambda}$ overlap those from the previous runs with $\lambda^*_{2,2} = -12.58°$ indicating that there is no appreciable distortion.

11. $\ddot{\lambda}$ Comparisons

The acceleration has been calculated for Wagner's constrained set of coefficients [8] and for the 1969 SAO set [9]. The results are shown in Fig. 2 together with the Skynet 1 accelerations of Table 1 adjusted to $i = 0$ and $a/R = 6.6108$ (though this adjustment is too small to be visible with the scale used). All the Skynet 1 values lie neatly between the other two, and indicate that Wagner's constrained set is better than the SAO set, which is not surprising since the determination of the latter did not include any data from synchronous satellites.

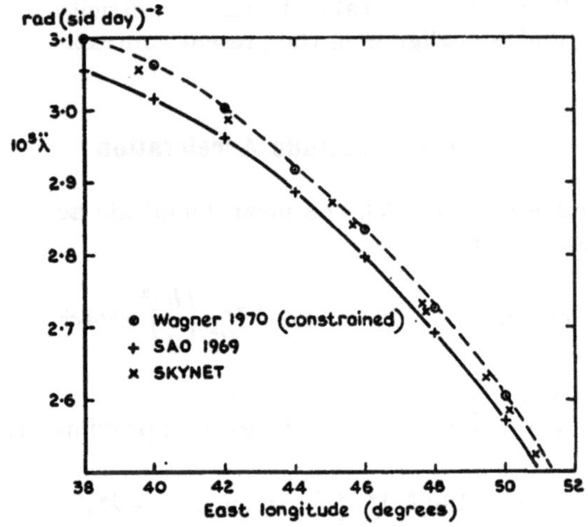

Fig. 2. Geosynchronous mean longitude acceleration.

12. Revised Geopotential Coefficients

A preliminary table of Skynet accelerations was sent to C. A. Wagner, of NASA, who very kindly agreed to include them, together with other available data, in a special computer run of his program for determining geopotential coefficients. In this run, the 1969 Smithsonian Standard Earth [9] coefficients for (5,1), (5,3) and (5,5) were included as fixed quantities, and the coefficients for (2,2), (3,1), (3,3), (4,2) and (4,4) were determined. The resulting coefficients, in normalized form, are given in Table 2. Using these new values the longitude acceleration of Skynet 1 has been calculated for the nine epochs of Table 1. These calculated accelerations are compared with the observed accelerations in Table 3

Table 2

Wagner 1971 geopotential coefficients (normalized 10^{-6} units)

l	m	$\bar{C}_{lm}$	$\bar{S}_{lm}$
2	2	2.4369	−1.4005
3	1	2.0192	0.2278
3	3	0.7387	1.4343
4	2	0.3444	0.7021
4	4	−0.1846	0.2508

plus SAO (standard earth) 2 coefficients for (5,1), (5,3), (5,5)

Table 3
Skynet acceleration

λ (°)	$10^5 \ddot{\lambda}$ (rad. (sid. day)$^{-2}$) Wagner 71	meas.	diff.
39.58	3.0569	3.0553	16
42.10	2.9864	2.9875	−11
45.04	2.8684	2.8705	−21
45.64	2.8398	2.8409	11
47.67	2.7319	2.7312	7
47.88	2.7213	2.7200	13
49.42	2.6282	2.6306	−24
50.04	2.5877	2.5865	12
50.84	2.5309	2.5282	25

rms residual = 0.0017

and it will be seen that the fit is extremely good. It should be emphasized, however, that the Skynet 1 data, being so much more accurate than those for other satellites, were relatively heavily weighted in the fitting procedure. This means, in effect, that the resulting geopotential coefficients represent the geopotential very accurately in the region 40°—50° E, but not necessarily at other longitudes.

13. Conclusions

The longitude acceleration of Skynet 1 due to the earth's gravitational field has been determined at a number of longitudes between 39.5° E and 50.8° E to an accuracy better than 1 part in 1000. These results not only fill a gap in the longitude coverage of earlier synchronous satellites but are also the most accurate yet obtained. The observed Skynet accelerations lie between those estimated from Wagner's 1970 (constrained) and the 1969 SAO sets of geopotential coefficients. A revised set of geopotential coefficients obtained by Wagner and based on the Skynet observations, give a very accurate representation of the geopotential in the longitude region 40° E—50° E.

Appendix
Equations of Motion of a Geostationary Satellite

The cartesian equations of motion of a satellite, as used in the Skynet programs are as follows:

$$\ddot{x} = Ax + Bc - \mu_s(\Delta_s^{-3}(x - x_s) + r_s^{-3}x_s) - \mu_m(\Delta_m^{-3}(x - x_m) + r_m^{-3}x_m)$$
$$+ \varkappa\Delta_s^{-3}(x - x_s) - p\dot{x},$$

$$\ddot{y} = Ay + Bs - \mu_s(\Delta_s^{-3}(y - y_s) + r_s^{-3}y_s) - \mu_m(\Delta_m^{-3}(y - y_m) + r_m^{-3}y_m)$$
$$+ \varkappa\Delta_s^{-3}(y - y_s),$$

$$\ddot{z} = (A + C)z + D - \mu_s(\Delta_s^{-3}(z - z_s) + r_s^{-3}z_s) - \mu_m(\Delta_m^{-3}(z - z_m) + r_m^{-3}z_m) + \varkappa\Delta_s^{-3}(z - z_s) + p\dot{z},$$

$$\text{where } A = \mu_e r^{-3}\left\{-1 + \frac{1}{2}J_2R_1^2(15z_1^2 - 3) + \frac{1}{2}J_3R_1^3(35z_1^3 - 15z_1) + \frac{1}{8}J_4R_1^4(315z_1^4 - 210z_1^2 + 15) + J_{2,2}R_1^2(9 - 15z_1^2 - 30(cx + sy)^2)\right\},$$

$$C = 6\mu_e r^{-3}J_{2,2}R_1^2,$$

$$B = 2C(cx + sy),$$

$$D = -\mu_e r^{-2}\left\{3J_2R_1^2z_1 + \frac{1}{2}J_3R_1^3(15z_1^2 - 3) + \frac{1}{2}J_4R_1^4(35z_1^3 - 15z_1)\right\};$$

$$r = (x^2 + y^2 + z^2)^{1/2},\ R_1 = R/r,\ z_1 = z/r,$$

$$c = \cos(\alpha + \lambda_{2,2}),\ s = \sin(\alpha + \lambda_{2,2});$$

α is the Modified Sidereal Angle, given by

$$\alpha = (0.277987616 + 0.00273781191(\mathrm{MJD} - 33282) + 1.0027378119(\mathrm{TIME}))\ \text{revs},$$

where MJD + TIME is the date/time, MJD is the modified Julian day number (33282 ≡ 1 JAN 1950); x_s, y_s, z_s are the cartesian coordinates of the sun, x_m, y_m, z_m are the cartesian coordinates of the moon, obtained from the J. P. L. Ephemeris;

$$r_s = (x_s^2 + y_s^2 + z_s^2)^{1/2},\ \Delta_s^2 = (x - x_s)^2 + (y - y_s)^2 + (z - z_s)^2,$$
$$r_m = (x_m^2 + y_m^2 + z_m^2)^{1/2},\ \Delta_m^2 = (x - x_m)^2 + (y - y_m)^2 + (z - z_m)^2;$$

μ_e, μ_s, μ_m are the gravitational constants of earth, sun and moon, $\mu_e = 398602$ km^3 s^{-2}, $\mu_s = 1.327127 \times 10^{11}$ km^3 s^{-2}, $\mu_m = 4902.756$ km^3 s^{-2}; R is the mean equatorial radius of earth, $R = 6378.163$ km; J_2, J_3, J_4 are zonal harmonic coefficients, $10^6J_2 = 1082.68$, $10^6J_3 = -2.53$, $10^6J_4 = -1.61$; $J_{2,2}, \lambda_{2,2}$ are lumped tesseral harmonic parameters dependent on longitude; $\varkappa$ is the solar radiation parameter, currently represented in the form

$$\varkappa = \{c_1 + c_2\cos(c_3(t - t_s))\} \times (1 - \pi^{-1}\psi),$$

where t is the date, c_1, c_2, c_3, t_s are constants; $\psi = 0$ except during the eclipse seasons, where ψ/π is the fraction of time in eclipse during the day. Specifically, let

$$r_s r v \sin\delta = x_s(y\dot{z} - \dot{y}z) + y_s(z\dot{x} - \dot{z}x) + z_s(x\dot{y} - \dot{x}y),$$

where $v = (\dot{x}^2 + \dot{y}^2 + \dot{z}^2)^{1/2}$, so that δ is the angle the earth—sun line makes with the satellite orbit plane, and let a be the osculating semi-major axis of the orbit,

$$a^{-1} = 2r^{-1} - \mu_e^{-1}(\dot{x}^2 + \dot{y}^2 + \dot{z}^2),$$

then, if $\sin^2 \delta \geqq (R/a)^2$, set $\psi = 0$
and if $\sin^2 \delta \leqq (R/a)^2$, set

$$\psi = \arctan [\{(R/a)^2 - \sin^2 \delta\}^{1/2}/\{1 - (R/a)^2\}^{1/2}].$$

$$c_3 = 0.03440485 \text{ rad (day)}^{-1}.$$

For Skynet 1, fitted values of the other three parameters are

$$c_1 = 1.2886 \times 10^6 \text{ km}^3 \text{ s}^{-2}, \qquad c_2 = 8.4887 \times 10^4 \text{ km}^3 \text{ s}^{-1}, \qquad t_{sol} = 40946.5 \text{ MJD}.$$

$\varkappa$ is kept constant during each day, and changed when the time passes through midnight. For Skynet this introduces a negligible discontinuity. The final p-terms in the equations of motion express the precession of the earth's spin-axis relative to the ecliptic: $p = 6.079 \times 10^{-12}$ rad s^{-1}.

References

[1] R. H. Gooding and R. J. Tayler, RAE Tech. Rep. No. 68299 (1968).
[2] E. M. Gaposchkin, Smithson. Astrophys. Obs. Spec. Rep. No. 161 (1964).
[3] W. H. Goodyear, Astron. J. 70, 189 (1965).
[4] W. H. Guier and R. R. Newton, J. Geophys. Res. 70, 4613 (1965).
[5] J. J. Levallois and H. M. Dufour, paper presented at XIVème Assemblée Générale, Assoc. Int. de Geodesie, Lucerne 1967.
[6] G. J. Davison and R. H. Merson, RAE Tech. Rep. No. 72044 (1972).
[7] R. R. Allan, Proc. Roy. Soc. A 288, 60 (1965).
[8] C. A. Wagner, Goddard Space Flight Center Rep. X-552-70-402 (1970).
[9] E. M. Gaposchkin and K. Lambeck, Smithson. Astrophys. Obs. Spec. Rep. No. 315 (1970); also J. Geophys. Res. 76, 4855 (1971).

Space Research XIII — Akademie-Verlag, Berlin 1973

SECULAR PERTURBATIONS DUE TO POTENTIAL: SECOND APPROXIMATION

C. Simó

Cátedra de Astronomía, Universidad de Barcelona, Barcelona, Spain

The expression of the earth's potential allows us to obtain formally the secular and periodic perturbations of any order. To achieve an accurate analytic theory we must introduce the second approximation. The secular terms are specially interesting due to their cumulative effect.

Using Lagrange's equations and the potential as given by Challe and Laclaverie, we obtain the perturbations arising from the second approximation. The use of the complex variable makes the calculations less tedious and allows an easy analysis showing the terms responsible for secular perturbations. Due to the different order of magnitude of the harmonics ($J_{2,0}$ first order, $J_{n,q}$ second order if $n,q \neq 2, 0$) we obtain perturbations ranging from second to fourth order.

The contribution of this work is the compact form in which the results are obtained, including all the harmonics. As we know no small divisors in e, $\sin I$ are found in the terms in J_2^2.

Comparison is made with formulae given by Batrakov and by Kovalevsky, and with the results obtained from a different approach, the von Zeipel procedure, as given by Brouwer.

1. Introduction

The perturbing part of the terrestrial potential can be expressed by means of the osculating orbital elements as given by Challe and Laclaverie [1]:

$$R = \sum_{n=2}^{\infty} \sum_{q=0}^{n} \sum_{k=-n}^{n} \sum_{p=-\infty}^{\infty} (-1)^{\delta_q^0} \mu J_{n,q} (a_e^n/a^{n+1}) A_{n,q}^k(I) B_{n+1,p}^k(e) \times \mathrm{Re}\left[i^{n-q} \exp i\left(pM + k\omega + q(\Omega - T - \lambda_{n,q})\right)\right]$$

where $\delta_q^0 = 1$ if $q = 0$, $= 0$ if $q \neq 0$

$\mu = GM_\oplus =$ gravitational coefficient $= 398601.3\ \mathrm{km^3\,s^{-2}}$

$J_{n,q} =$ coefficient of the (n, q) spherical harmonic

$a_e =$ equatorial radius $= 6378.145$ km

$a =$ semimajor axis

$A_{n,q}^k(I) =$ a function of the inclination

$B_{n+1,p}^k(e) =$ a function of the eccentricity $=$ Hansen's coefficient

$M =$ mean anomaly

$\Omega =$ argument of the node

$\omega =$ argument of the perigee

$T =$ Greenwich sidereal time

$\lambda_{n,q} =$ angle associated with $J_{n,q}$ (also $C_{n,q} = J_{n,q} \cos q\lambda_{n,q}$, $S_{n,q} = J_{n,q} \sin q\lambda_{n,q}$).

We put

$$J = (-1)^{\delta_q^0} \mu \mathrm{J}_{n,q}, \quad C = a_e^n / a^{n+1}, \quad A = A_{n,q}^k(I), \quad B = B_{n+1,p}^k(e)$$

$$\mathrm{Arg} = pM + k\omega + q(\Omega - T - \lambda_{n,q}), \quad \sum \text{ for } \sum_{n=2}^{\infty} \sum_{q=0}^{n} \sum_{k=-n}^{n} \sum_{p=-\infty}^{\infty}$$

$$AP = \frac{d}{dI} A(I), \quad AS = \frac{d^2}{dI^2} A(I), \quad BP = \frac{d}{de} B(e), \quad BS = \frac{d^2}{de^2} B(e)$$

$$CP = \frac{d}{da} C(a), \quad CS = \frac{d^2}{da^2} C(a), \quad \eta = (1 - e^2)^{1/2}, \quad s = \sin I, \quad c = \cos I.$$

We recall that

$$\frac{d}{dt}\left[\mathrm{Re}\left(i^m \exp(i\,\mathrm{Arg})\right)\right] = \mathrm{Re}\left(i^{m+1} \exp(i\,\mathrm{Arg})\right) \cdot \dot{\mathrm{Arg}}$$

$$\int dt \left[\mathrm{Re}\left(i^m \exp(i\,\mathrm{Arg})\right)\right] = \mathrm{Re}\left(i^{m-1} \exp(i\,\mathrm{Arg})\right) \cdot \dot{\mathrm{Arg}}^{-1}$$

$$\mathrm{Re}\, F \cdot \mathrm{Re}\, G = {}^1/_2\, \mathrm{Re}\,(F \cdot G + F \cdot G^{-1}) \text{ provided that } |F| = |G| = 1.$$

With the conventions made we write the Lagrange equations:

$$\begin{aligned}
\dot{a} &= 2\mu^{-1/2} \sum pJCa^{1/2}AB\,\mathrm{Re}\,(i^{n-q+1} \exp i\,\mathrm{Arg}) \\
\dot{e} &= \mu^{-1/2} \sum (-k\eta + p\eta^2) JCa^{-1/2}ABe^{-1}\,\mathrm{Re}\,(i^{n-q+1} \exp i\,\mathrm{Arg}) \\
\dot{I} &= \mu^{-1/2} \sum (-q + kc) JCa^{-1/2}As^{-1}B\eta^{-1}\,\mathrm{Re}\,(i^{n-q+1} \exp i\,\mathrm{Arg}) \\
\dot{\Omega} &= \mu^{-1/2} \sum JCa^{-1/2}APs^{-1}B\eta^{-1}\,\mathrm{Re}\,(i^{n-q} \exp i\,\mathrm{Arg}) \qquad (1) \\
\dot{\omega} &= \mu^{-1/2} \sum JCa^{-1/2}(ABP\eta e^{-1} - cs^{-1}APB\eta^{-1})\,\mathrm{Re}\,(i^{n-q} \exp i\,\mathrm{Arg}) \\
\dot{M} &= n_0 - \mu^{-1/2} \sum JA\,(2a^{1/2}CPB + Ca^{-1/2}\eta^2BPe^{-1} + 3Ca^{-1/2}B) \\
&\quad \times \mathrm{Re}\,(i^{n-q} \exp i\,\mathrm{Arg}).
\end{aligned}$$

In $\dot{M}$ we include the effect due to the variation of a:

$$n^2a^3 = \mu \quad \text{implies} \quad 2n\dot{n} + 3a^2\dot{a} = 0 \quad \text{and thus}$$

$$n = n_0 + \int \dot{n}\, dt = n_0 - 1.5a^2n^{-1} \int \dot{a}\, dt.$$

For the integration of (1) we remember that

$$\dot{\mathrm{Arg}} = k\dot{\omega} + q(\dot{\Omega} - \dot{T}) + p\dot{M}$$

where $\dot{\omega}$, $\dot{\Omega}$, $\dot{M}$ are obtained at the so-called zero-th approximation:

$$\dot{\Omega} = -\left(\frac{\mu}{a^3}\right)^{1/2} \eta^{-1}s^{-1} \sum_{m=1}^{\infty} \mathrm{J}_{2m,0}(-1)^m \left(\frac{a_e}{a}\right)^{2m} AP_{2m,0}^0 B_{2m+1,0}^0$$

$$\dot{\omega} = -\left(\frac{\mu}{a^3}\right)^{1/2} \eta^{-1}s^{-1}e^{-1} \sum_{m=1}^{\infty} \mathrm{J}_{2m,0}(-1)^m \left(\frac{a_e}{a}\right)^{2m} [A_{2m,0}^0 BP_{2m+1,0}^0 s\eta^2 - AP_{2m,0}^0 B_{2m+1,0}^0 ce]$$

$$\dot{M} = n_0 + \left(\frac{\mu}{a^3}\right)^{1/2} \sum_{m=1}^{\infty} \mathrm{J}_{2m,0}(-1)^m \left(\frac{a_e}{a}\right)^{2m} [-6B_{2m+1,0}^0 + \eta^2 BP_{2m+1,0}^0 e^{-1}] A_{2m,0}^0.$$

System (1) can be written: $\dot{\sigma} = \varphi(\sigma),\ \sigma \in R^6$. The first approximation is given by: $\sigma_1 = \sigma_0 + \Delta\sigma_1 = \sigma_0 + \int_{t_0}^{t_1} \varphi(\sigma_0)\, dt$. The second one by $\sigma_2 = \sigma_1 + \Delta\sigma_2 = \sigma_1 + \int_{t_0}^{t} \left(D\varphi(\sigma_0) \cdot \int_{t_0}^{\tau} \varphi(\sigma_0)\, d\tau\right) dt$. For instance

$$\Delta\dot{\Omega}_2 = \frac{\partial\dot{\Omega}}{\partial a}\,\Delta a_1 + \cdots + \frac{\partial\dot{\Omega}}{\partial M}\,\Delta M_1. \tag{2}$$

Let us to obtain a compact form for (2).

2. The Case of the Argument of the Node

To avoid confusion we denote by $'$ the terms in (2) arising from $\Delta\sigma_1$. Then:

$$\begin{aligned}
\Delta\dot{\Omega}_2 = \frac{1}{\mu}\sum\sum{}' JJ' \Bigg\{ \Bigg[& \left(\frac{CP}{a^{1/2}} - \frac{C}{2a\,a^{1/2}}\right)\frac{AP}{s}\,\frac{B}{\eta}\,2p'C'a^{1/2}A'B' \\
& + \frac{C}{a^{1/2}}\,\frac{AP}{s}\left(\frac{BP}{\eta} + \frac{Be}{\eta^3}\right)(-k'\eta + p'\eta^2)\,\frac{C'}{a^{1/2}}\,A'\,\frac{B'}{e} \\
& + \frac{C}{a^{1/2}}\left(\frac{AS}{s} - \frac{APc}{s^2}\right)\frac{B}{\eta}\,(-q' + k'c)\,\frac{C'}{a^{1/2}}\,\frac{A'}{s}\,\frac{B'}{\eta}\Bigg]\frac{1}{2\,\dot{\mathrm{Arg}}'} \\
& \times \mathrm{Re}\left(i^{n-q+n'-q'}\exp i\,(\mathrm{Arg} + \mathrm{Arg}') + i^{n-q-n'+q'}\exp i\,(\mathrm{Arg} - \mathrm{Arg}')\right) \qquad (3) \\
& + \frac{C}{a^{1/2}}\,\frac{AP}{s}\,\frac{B}{\eta}\Bigg[q\,\frac{C'}{a^{1/2}}\,\frac{AP'}{s}\,\frac{B'}{\eta} + k\,\frac{C'}{a^{1/2}}\left(A'\,\frac{BP'\eta}{e} - \frac{c}{s}\,AP'\,\frac{B'}{\eta}\right) \\
& + p\left(-3p'\,\frac{C'}{a^2}\,A'B'\,\frac{\mu^{1/2}}{\dot{\mathrm{Arg}}'} - A'\left(2a^{1/2}CP'B' + \frac{C'}{a^{1/2}}\,\frac{\eta^2BP'}{e}\right)\right)\Bigg]\frac{1}{2\,\dot{\mathrm{Arg}}'} \\
& \times \mathrm{Re}\left(i^{n-q+1+n'-q'-1}\exp i\,(\mathrm{Arg} + \mathrm{Arg}') + i^{n-q+1-n'+q'+1}\exp i\,(\mathrm{Arg} - \mathrm{Arg}')\right)\Bigg\}.
\end{aligned}$$

The secular terms are present if and only if Arg + Arg′ or Arg − Arg′ are zero. Thus

$$(q \pm q')\,(\Omega - T - \lambda_{n,q}) + (k \pm k')\,(\omega) + (p \pm p')\,(M) = 0,$$

and provided that resonance does not occur it follows that

$$q \pm q' = k \pm k' = p \pm p' = 0.$$

Furthermore for the real part not to be zero we must have

$$n + n' - q - q' = \dot{2}.$$

Let us examine several cases:

a) If $p \neq 0$, $\dot{\mathrm{Arg}}'$ is not small. The term in $J_{2,0}^2$ is of second order. Terms in $J_{2,0}J_{n',0}$, $n' = 2m$, $m > 1$, are of third order, and the ones in $J_{n,q}J_{n',q}$, $n + n' = \dot{2}$, are of fourth order.

b) If $p = 0$, $q = 0$, $\dot{\mathrm{Arg}}' = k'\dot{\omega}$, there is no term in $J_{2,0}^2$ because $B_{3,0}^{\pm 2} = 0$ (i.e. the long-period terms in $\Delta a_1, \Delta e_1, \ldots$ do not introduce secular terms in $J_{2,0}^2$). The remaining harmonics give terms of third order in $J_{n,0} J_{n',0} J_{2,0}^{-1}$, $n, n' \neq 2$, $n + n' = \dot{2}$, the denominator being originated by the integration.

c) If $p = 0$, $q \neq 0$ there are terms in $J_{n,q} J_{n',q} \dot{T}^{-1}$, $n + n' = \dot{2}$, of order 3.5, say.

3. The Remaining Elements

We proceed first with the metric elements a, e, I. Let us denote σ one of them. In the expression for $\Delta\dot{\sigma}_2$ analogous to (3) we must take the real part of

$$i^{n-q+1+n'-q'} \exp i\,(\mathrm{Arg} + \mathrm{Arg}'), \qquad i^{n-q+1-n'-q'} \exp i\,(\mathrm{Arg} - \mathrm{Arg}'),$$

$$i^{n-q+n'-q'-1} \exp i\,(\mathrm{Arg} + \mathrm{Arg}'), \qquad i^{n-q-n'+q'+1} \exp i\,(\mathrm{Arg} - \mathrm{Arg}').$$

With the assumptions made in § 2, secular terms occur if $n + n' - q - q' \neq \dot{2}$, $q = q'$, $k = \pm k'$, $p = \pm p'$. This implies $n + n' \neq \dot{2}$. Nevertheless $A_{n,q}^k = 0$ if $n - k \neq \dot{2}$ implies $n - k = \dot{2}$, $n' - k' = \dot{2}$, $k + k' = \dot{2}$, and it follows that $n + n' = \dot{2}$. We conclude that no secular terms in a, e, I, are found at the second approximation.

The formulae and analysis for $\Delta\dot{\omega}_2$, $\Delta\dot{M}_2$ are similar to these found for $\Delta\dot{\Omega}_2$ and we will not display them because they do not bring additional interest.

4. Comparisons

W. Kaula [2, p. 42] says that the evaluation of the secular terms arising from $J_{2,0}^2$ in $\Delta\dot{\Omega}_2$, $\Delta\dot{\omega}_2$, $\Delta\dot{M}_2$ to the order e^{2k} involves $180k$ differentiations and an equal number of multiplications. However, the treatment given above allows us to avoid a great part of this task, being reduced after some elementary manipulations to:

$$\Delta\dot{\Omega}_2 \text{ secular due to } J_{2,0}^2 = \frac{3}{16} n J_{2,0}^2 \left(\frac{a_e}{a}\right)^4 c \sum_{\substack{p\in Z\\ p\neq 0}} \left\{ \frac{36}{\eta} (B_{3p}^0)^2 - \frac{8\eta}{e} B_{3p}^0 BP_{3p}^0 \right.$$
$$+ \frac{12}{p\eta^2} (B_{3p}^2)^2 + s^2 \left[-\frac{54}{\eta} (B_{3p}^0)^2 + \frac{12\eta}{e} B_{3p}^0 BP_{3p}^0 \right.$$
$$\left.\left. + \left(-\frac{27}{\eta} - \frac{24}{p\eta^2}\right) (B_{3p}^2)^2 + \left(\frac{6\eta}{e} - \frac{12}{pe}\right) B_{3p}^2 BP_{3p}^2 \right]\right\}. \qquad (4)$$

(We have made use of $A_{2,0}^0 = (2 - 3s^2)/4$, $A_{2,0}^2 = 3s^2/8$, $A_{n,0}^k = A_{n,0}^{-k}$, $B_{n,p}^k = B_{n,-p}^{-k}$).

The small divisor e in (4) is fictitious because $B_{n,p}^k = e^{|k-p|} \sum_{i>0} \alpha_i e^{2i}$ (D'Alembert characteristic) and $B \cdot BP$ is of degree at least 1.

Expressions similar to (4) are found for ω and M. Only terms in s^0, s^2, s^4 are present. However they include terms in e^{-2} that cancel out in substituting $B_{n,p}^k$ for their values.

We note that only sums

$$\sum_{p\neq 0} (B^0_{3p})^2, \quad \sum_{p\neq 0} (B^2_{3p})^2, \quad \sum_{p\neq 0} \frac{1}{p} (B^2_{3p})^2$$

and their first and second derivatives are needed in the evaluations.

Hansen's coefficients can be found in some lists of elliptic expansions (Cayley tables) or computed by the formula [3]:

$$B^k_{n,p}(e) = (1+\beta^2)^{n-1} \sum_{h\geqq 0} \sum_{j\geqq 0} \binom{k-n+1}{h} \binom{-k-n+1}{j} (-\beta)^{h+j} \mathrm{J}_{h-k+p-j}(pe)$$

where $\beta = e/(1+\eta)$ and J_m denotes the Bessel function of first kind and order m.

The final result from (4)

$$\frac{3}{16}\, n \mathrm{J}^2_{2,0} \left(\frac{a_e}{a}\right)^4 c \left\{-30 - 116e^2 + s^2 \left(38 + \frac{281}{2} e^2\right) + \mathrm{O}(e^4)\right\}$$

agrees with Kovalevsky and Levallois [3], Batrakov [4] and Brouwer and Clemence [5].

The equivalent formula for $\Delta \dot{w}_2$ agrees with [3] and [5], but the term in $e^2 s^2$ is reduced to one half of that present in [4].

For the $\Delta \dot{M}_2$ expression we agree with [4], and the discrepancy with [5] is explained in the last paragraph of [4]. [3] do not offers the corresponding formula due to their lack of interest in geodetic purposes.

References

[1] A. Challe and J. J. Laclaverie, Astron. Astrophys. 3, 5 (1969).
[2] W. Kaula, Theory of Satellite Geodesy, Blaisdell Publ. Co., Waltham, Mass. 1966.
[3] J. Kovalevsky and J. J. Levallois, Géodésie générale: IV Géodésie Spatiale, Eyrolles, Paris 1971.
[4] Y. V. Batrakov, in: Dynamics of Satellites, Springer, Berlin 1963 (p. 74).
[5] D. Brouwer and G. M. Clemence, Methods of Celestial Mechanics, Academic Press, New York 1961.

Space Research XIII — Akademie-Verlag, Berlin 1973

SATELLITE LASER RANGING INSTRUMENT AT THE ONDŘEJOV OBSERVATORY

L. Sehnal

Astronomical Institute of the Czechoslovak Academy of Sciences, Ondřejov, Czechoslovakia

The laser ranging instrument which was put into experimental operation at the Ondřejov Observatory is described. The system enables distances to a satellite to be measured with a precision of 2—3 m. The device is supported on a four-axial mount and the observer tracks the satellite with a guiding telescope. The main technical data are tabulated.

A satellite laser ranging system has been put into experimental operation at the Observatory of the Astronomical Institute of the Czechoslovak Academy of

Fig. 1. The satellite laser ranging instrument of the Ondřejov Observatory.

Sciences in Ondřejov. The instrument, which measures the distances of the artificial satellites of the earth, is the joint work of all the East-European countries. The partners participated in the construction of individual parts of the device and they share also the results of the observations.

The whole instrument is placed on a four-axial mounting and the observer follows the satellite using a guiding telescope. The mount thus supports the laser transmitter, the receiving optics with photomultiplier and filter and the guiding telescope. Guiding can be done with a precision better than 1 mrad. The first two axes put the instrument into the satellite's orbital plane and the device moves around the third axis which is then perpendicular to the orbital plane. The motion is basically controlled by an electrical motor with a mechanical reduction system, to be able to approximate different apparent velocities of a satellite. The angular velocity may be continuously changed from 0.1° to 0.8° per second and the instrument moves in an angular range of 120° deg. The observer only makes the corrections when some imprecise tracking occurs. Corrections are also made around the fourth axis which corrects the deviation of the apparent orbit from a proper plane. Those corrections can be made in a range from −5° to +12°.

The mount supports a laser transmitter with an energy of 1.8−2.0 J. The ruby laser gives a pulse of 18−20 ns width at a wavelength of 6943 Å. The transmitter gives 6 pulses per minute with an output angle of 5 mrad. From the transmitter optics goes a pulse within an angle smaller than 0.6 mrad. The laser transmitter may be triggered automatically. The whole system is water-cooled with a closed circuit.

The main data on the laser system are summarized in Table 1.

Table 1

Energy (J)	1.8−2.0
Pulse width (ns)	18−20
Full beam divergence (mrad)	0.6
Time counter resolution (ns)	4
Receiver aperture (mm)	320
Clock type	Quartz
Synchronization	OMA 50
Expected time accuracy (ms)	0.1
Tracking	semi-automatic visual

The receiver is a Cassegrain mirror system and the main reflecting mirror has an aperture of 320 mm. The focal length of the system is 1000 mm. The back of the main mirror is coated. The transmission of the optical system is about 60%. The receiver optics directs the light signal through a filter to a type EMI 9558 B photomultiplier. The filter transmission is better than 50% with a bandwidth of 20 Å.

The time interval between the transmitted and received signal is measured by a counter with a resolution of 4 ns. The main quartz clock receives the OMA 50 time signal and it gives the absolute time measurements with a precision of 0.1 ms. The time interval measurements and the absolute time of the laser shot are printed by separate printers.

The experimental measurements proved the good qualities of the whole system and its ability to give satisfactory range data. The instrument might be further improved to provide even higher accuracy of measurements.

The author of the paper served as a coordinator of the international work during the process of completion of the instrument; the full description of it with all the details will be published soon.

Space Research XIII — Akademie-Verlag, Berlin 1973

UTILISATION SCIENTIFIQUE POSSIBLE DE LONGUES BASES LASER

M. Lefebvre et J. Gaignebet

Centre National d'Etudes Spatiales, Département Géodésie Spatiale, Brétigny, France

Recent and future technological developments of very stable clocks, requirements for accurate synchronization on a world-wide basis for very long base interferometry, relativistic experiments, fundamental physics and new laser ranging systems lead to the development of a new method of subnanosecond and permanent synchronization and to the establishment of very long laser baselines using one-way measurements for the precise determination of probe or satellite trajectories in space.

1. Introduction

La synchronisation d'horloges à grandes distances en permanance, a de nombreux intérêts tant scientifiques que pratiques. Du point de vue scientifique, la définition d'étalons de fréquence ou d'horloges très précises, permet de rapporter tous les phénomènes à une même échelle de fréquence (pour la physique) ou de temps (pour l'astronomie), et partant, de pouvoir analyser des phénomènes en différents points du globe avec une même référence. Rappelons par exemple, que c'est l'apparition d'horloges à quartz précises et comparées qui a permis d'analyser les variations de rotation de la terre. Du point de vue pratique, une synchronisation précise globale peut servir de base à l'établissement de procédés de navigation précis.

2. Principe de l'experience

Le but de l'expérience est de synchroniser de façon permanente des horloges situées à grande distance avec une précision de l'ordre de la nanoseconde en utilisant des impulsions laser très brèves émises de stations terrestres vers un satellite stationnaire et d'un dispositif de chronométrie.

Soit un satellite stationnaire S muni de réflecteurs laser, d'un détecteur optique et d'un dispositif de chronométrie. Soit deux stations laser L_1 et L_2 équipées de deux horloges très précises H_1 et H_2. A l'instant t_1 dans l'échelle H_1 la station laser L_1 émet une impulsion très brève qui revient en L_1 après réflection sur S et on mesure $\tau_1 + \tau_1'$ le temps d'aller—retour. L'impulsion est détectée par S et déclenche le départ de la chronométrie à bord du satellite.

A l'instant t_2 dans l'échelle H_2 la station laser L_2 émet une impulsion, on mesure de même $\tau_2 + \tau_2'$ et l'impulsion détectée à bord du satellite arrête la chronométrie.

Soit Δt la valeur mesurée par le chronomètre et retransmise par télémesure.

Après correction des temps d'aberration dus aux mouvements relatifs de S, L_1 et L_2 on peut déterminer les temps de propagation moyens réduits au même instant soient τ_{1m} et τ_{2m}. On a alors:

$$\Delta t = (t_2 + \tau_{2m}) - (t_1 + \tau_{1m}).$$

On obtient donc directement les différences d'échelle de temps:

$$t_1 - t_2 = \Delta t - (\tau_{2m} - \tau_{1m}). \quad (1)$$

La précision de synchronisation dépendra de la précision de mesure de chacune des quantités de l'équation (1).

3. Dispositif expérimental

L'implantation des réflecteurs laser n'est pas un problème critique. Il s'agit surtout d'avoir un panneau concentré par exemple sur une demi-sphère de façon à pouvoir rapporter les mesures faites par les deux stations (qui n'interviennent que par leur différence) à un même point géométrique.

La définition des caractéristiques du détecteur nécessite le calcul du bilan de liaison en trajet aller et le calcul du bruit reçu de la terre éclairée par le soleil.

On caractérisera ces deux problèmes par deux critères: la surface minimale du collecteur nécessaire à la détection des éclairs laser vus par le satellite; la surface maximale S_M au-delà de laquelle les impulsions détectées ne sont plus discriminables du bruit de fond (fond du ciel, clair de terre).

Soient T_a coefficient de transmission atmosphérique, T_r transmission de l'optique de réception, T_f coefficient de transmission du filtre interférentiel (0,4 pour $\Delta\lambda = 5$ Å), D distance satellite station ($\sim$40 megamètres), ϱ rendement quantique du photodétecteur, G le gain du photodétecteur, n le nombre de photoélectrons nécessaires à la réception (dans le cas d'une coïncidence simple, $n = 2$), W_e l'énergie émise en joules, θ_e la divergence du faisceau émis, W_r l'énergie reçue $= n/p$.

La surface minimale S_m s'exprime par:

$$S_m = \frac{n}{p W_e} \frac{D^2 \Theta_e^2}{T_a T_r T_f} \times \frac{\Pi}{4}.$$

Le tableau 1 permet de résumer les résultats obtenus pour S_m et le rayon correspondant r selon le type de station et de détecteur.

La surface maximale de détection est définie par le champ nécessaire pour englober toute la terre soit 0.3 radian, et par le bruit de la terre éclairée par le soleil; pour un filtre de bande passante de 5 Å, on obtient les niveaux suivants en photons cm^{-2} sr^{-1} s^{-1}:

$\lambda = 0{,}53$	$2{,}5 \cdot 10^{10}$
$\lambda = 0{,}69$	$3 \cdot 10^{10}$
$\lambda = 1{,}06$	$5 \cdot 10^{10}$

Si on suppose que la synchronisation des stations est déjà connue à moins de 10 μs et qu'on veut une probabilité de fausse détection de 1% ceci nous permet de différencier les impulsions laser dans un bruit de fond égal à $4 \cdot 10^5$ électrons primaires/sec pour les stations actuelles et $6 \cdot 10^5$ pour les stations 2ème génération.

Tableau 1

S_m et le rayon correspondant pour divers types de stations

Type de station	2è. génération		5 joules, $\theta_e = 2 \cdot 10^{-4}$, 3 ns				1è. génér. 1 joule, $5 \cdot 10^{-4}$, 15 ns	
photo détecteur	$\lambda = 0{,}53$		$\lambda = 0{,}69$		$\lambda = 1{,}06$		$\lambda = 0{,}69$	
	S_m^2	r_{mm}	S_m^2	r_{mm}	S_m^2	r_{mm}	S_m^2	r_{mm}
PM S 20	$3 \cdot 10^{-6}$	10^{-2}	$8 \cdot 10^{-10}$	$2{,}5 \cdot 10^{-2}$	$2 \cdot 10^{-8}$	$8 \cdot 10^{-2}$	$2{,}5 \cdot 10^{-8}$	$9 \cdot 10^{-2}$
photodiode à avalanche	$2 \cdot 10^{-10}$	$8 \cdot 10^{-3}$	$2 \cdot 10^{-10}$	10^{-2}	$2 \cdot 10^{-10}$	$8 \cdot 10^{-3}$	$6{,}5 \cdot 10^{-9}$	$4{,}5 \cdot 10^{-2}$
PM	$1{,}5 \cdot 10^{-10}$	$7 \cdot 10^{-3}$	$1{,}6 \cdot 10^{-10}$	$8 \cdot 10^{-3}$	$5 \cdot 10^{-10}$	$1{,}6 \cdot 10^{-2}$	$5 \cdot 10^{-9}$	$4 \cdot 10^{-2}$

D'ou le tableau 2. Le rapport S_M/S_m est donc grand. On peut encore améliorer la détection avec une discrimination d'amplitude (seuil). Ce gain est alors très important, on profite de la grande concentration temporelle de l'énergie émise par le laser. (NB. L'utilisation des systèmes à coïncidence augmente également le gain de détection mais fait perdre en précision.)

Enfin la chronométrie à 0,1 nanoseconde ne pose pas de problèmes et étant donné les durées de comptage (quelques microsecondes), l'utilisation d'un simple quartz comme oscillateur de comptage, suffit.

Tableau 2

S_m pour divers types de stations

Type de station	2ème génération			lère génération
λ	0,53	0,69	1,06	0,69
Détecteurs				
PM S 20	60	60	60	1,2
Photodiode	60	60	60	1,2
PM	60	60	60	1,2

4. Precisions

La précision sur Δt ne dépend quede la précision de chronométrie (erreur inférieure à 0,1 ns) et des retards dans les circuits de détection, mais ceux-ci ne jouent qu'en différentiel les impulsions venant de L_1 et L_2 empruntant les mêmes chemins optiques et électroniques et à des instants très voisins. Les fronts de montée des détecteurs peuvent varier mais les erreurs introduites auront l'allure de bruits.

Mesure des temps de propagation (aller et retour). Il s'agit de mesures du temps aller et retour par laser. La précision d' 1 ns est déjà atteinte par exemple pour les mesures effectuées sur les réflecteurs Apollo à l'Observatoire MacDonald. Une précision de 0,1 ns est possible dans les dix ans à venir: diminution de la largeur d'impulsion, échantillonnage des impulsions aller et retour, amélioration des techniques de calibration, amélioration de la correction troposphérique, connaissance de la fonction de transfert géométrique du panneau de réflecteurs. (NB. La simultanéité des tirs n'est pas absolument nécessaire. Notons toutefois que l'intervalle de temps entre l'arrivée des signaux des deux stations — peut-être rendu aussi petit que l'on veut — au besoin par approximations successives, ce qui permet de rendre complètement négligeables les mouvements de translation et de rotation du satellite pendant cet intervalle.)

Il faut évidemment que les instants de déclenchements des impulsions t_1 et t_2 soient connus avec une grande précision dans l'échelle des horloges. Ceci peut être réalisé avec une précision inférieure à 100 picosecondes et doit pouvoir être amélioré rapidement.

En conclusion, on peut affirmer qu'une précision de 1 à 2 ns est d'ores et déjà possible avec la technologie existante et qu'elle pourra probablement atteindre 100 ps dans les dix années à venir.

5. Utilisation des longues bases laser

Le premier objectif, la synchronisation précise d'horloges très stables éloignées, étant réalisé, on peut atteindre de nombreux objectifs:

VLBI. L'enregistrement des émissions d'une même source à l'aide de radiotélescopes à grande distance l'un de l'autre en utilisant des horloges très stables, comme bases de temps, permet par des techniques de cohérence maximale, de trouver les franges d'interférence et la vitesse de défilement de ces franges. On peut retrouver la direction de la source et les mouvements apparents des stations par rapport à la source avec une grande précision.

La synchronisation entre les horloges peut également être prise comme inconnue. Si toutefois on veut déterminer des directions absolues, il faut utiliser plusieurs bases. Par ailleurs, une synchronisation externe simplifie considérablement les problèmes de dépouillement.

Si on utilise les méthodes VLBI, à partir de sources artificielles, la méthode de synchronisation laser apporte un contrôle externe et permet de déceler certaines erreurs systématiques du VLBI en particulier, les erreurs de propagation ionosphériques et troposphériques (composante humide), qui sont inéxistantes au niveau des longueurs d'ondes optiques.

Experiences relativistes. En utilisant comme oscillateur de comptage, une horloge embarquée (par exemple $5 \cdot 10^{-13}$ pour un rubidium) on peut déclencher le démarrage et l'arrêt du comptage par des impulsions laser émises de la même station. On mesure "l'effet horloge" avec une précision de 10^{-3} sur un jour de mesure.

Certaines expériences relativistes permettent de mesurer les coefficients β et γ de la relativité générale et également l'aplatissement dynamique.

L'expérience prévoit le lancement d'une sonde en orbite solaire équipée d'un système drag-free très précis (pour éliminer les forces non gravitationnelles) d'une horloge de grande stabilité et d'un système de détection optique et de comptage pour mesurer la distance laser entre la terre et la sonde.

Les variations de distance au voisinage de la conjonction (un coronographe permettant d'éliminer la lumière solaire), permettent de mesurer le coefficient γ par l'effet de déviation des rayons lumineux. L'étude de la trajectoire permet par contre de mesurer l'effet relativiste sur le mouvement du périhélie ainsi que l'effet dû à l'aplatissement dynamique du soleil.

Si nous utilisons deux stations laser préalablement synchronisées, il nous suffit de tirs laser aller, déjà prévus par l'expérience et de compter la différence de temps entre les impulsions recues. On mesure cette fois la différence des courbures des rayons au voisinage du soleil. Sur une mesure individuelle, on peut espérer obtenir une précision de 0,002″, la déviation du rayon lumineux étant de 1,7″; on peut espérer une précision entre 10^{-3} et 10^{-4} et d'autre part comme le fait remarquer Shapiro, la précision nécessaire pour le drag-free est moins contraignante.

Il s'agit donc d'une expérience complémentaire qui ne nécessite aucun dispositif supplémentaire.

Trajectographie de sondes ou satellites lointains. Le même dispositif (détecteur et chonométrie), dont le poids est peu élevé et qui ne consomme que très peu d'énergie, placé à bord de satellites proches ou lointains, permet une trajectographie très précise (directions à quelques millièmes de seconde d'arcs).

Lune. Le dispositif pourrait également être placé sur la lune et permettrait des mesures de direction précises qui seraient un complément idéal aux mesures de distances et plus faciles à réaliser puisqu'elles ne demandent que des mesures laser aller et qu'elles peuvent être réalisées de façon continue. La durée de vie dépend de l'électronique de bord (oscillateur et chronométrie), qui ne seonblent pas poser de problèmes particuliers.

Comparaisons à distance d'étalons très précis. Le développement rapide d'horloges à la fois stables, fiables et reproductibles rend plus que jamais nécessaire leur comparaison avec une très grande précision. Les transports sont coûteux, peu fréquents et posent des problèmes technologiques très importants.

Navigation aérienne. L'augmentation du trafic aérien en particulier sur l'atlantique nord, rend nécessaire de nouveaux moyens de navigation et de contrôle. L'utilisation d'horloges précises synchronisées en permanence apporte un élément de réponse.

Tirs laser simultanés. Des tirs laser simultanés à l'échelle intercontinentale sur un satellite haut, doivent permettre non seulement des rattachements entre stations mais également de détecter les mouvements entre plaques tectoniques. Cependant, il est probable qu'il sera difficile d'obtenir des réseaux étendus de stations suffisamment puissantes pour des tirs aller et retour sur satellites très hauts. Des stations simplifiées du type des stations existantes pourraient servir de réseau du second ordre avec des mesures de différences de distance seulement. On obtiendrait donc des renseignements non plus sur les mouvements relatifs de quelques points isolés, mais sur des mouvements entre plaques ou sous-plaques.

Mesures de la réfraction par utilisation de différentes longueur d'ondes. On sait qu'une des possibilités pour mesurer la correction troposphérique est de mesurer la différence de trajet entre deux rayons émis en même temps, mais de longueur d'onde différente. Un montage analogue à notre dispositif est possible, si toutefois la précision des détecteurs (temps de montée) est améliorée.

6. Conclusion

Telle qu'elle est, la mission présente donc un grand intérêt et doit pouvoir être embarquée sans complication excessive sur un satellite stationnaire d'autant que certains satellites météorologiques sont déjà équipés d'une partie des sous-systèmes nécessaires.

On peut envisager trois niveaux de missions:

(a) Embarquement des réflecteurs et d'un détecteur avec dispositif de chronomètre pilotée par un simple oscillateur, sur un satellite stationnaire: On peut alors synchroniser deux horloges très éloignées avec une précision de l'ordre de la nanoseconde (2 à 3 actuellement, 0,2 dans les dix ans à venir) en utilisant des stations laser 2ème génération et des horloges associées.

(b) On réalise la même expérience, mais une horloge très stable est embarquée et pilote la chronométrie. On peut alors mesurer l'effet horloge relativiste avec une précision de 10^{-3} sur une journée.

(c) Si on suppose des horloges synchronisées par notre méthode, on peut alors utiliser des impulsions laser pour faire des mesures aller de deux ou plusieurs stations sur un détecteur-chronomètre placé sur n'importe quel objet spatial, par exemple la lune, une sonde spatiale, un satellite lointain. La résolution est alors de quelques millièmes de secondes d'arcs, sur la détermination des directions ce qui permet l'étude fine des trajectoires: effets relativistes ou gravitationnels (differents corps du système solaire, champs de potentiel des planètes, mouvements de rotation des planètes autour de leur centre de gravité).

Space Research XIII — Akademie-Verlag, Berlin 1973

STELLAR TRIANGULATION MEASURED WITH BALLOON-BORNE BEACONS, ITS SCALE AND POSITION

T. J. Kukkamäki

Finnish Geodetic Institute, Helsinki, Finland

A general account is given of the measurement of the stellar triangulation net of Finland with sides 200 km in length. The net consists of 14 stations; five of these have already been used for observation. The balloon-borne beacon is described in more detail. The scale of the net will be determined with the aid of a precision traverse measured with a laser-geodimeter. The position of the net will be determined through observation of passive satellites on both ends of the precision traverse. Here a time micrometer with no moving parts is used and it is described in detail.

1. Introduction

In 1946 Väisälä [1] suggested that, instead of measuring angles between directions from the observation point to target points on solid ground, it is possible to determine directions to beacons elevated in free air by photographing these beacons against the fixed star background. In the same year experiments were carried out in Finland with balloon-borne beacons between Turku and Helsinki over a distance of 153 km. Due to post-war difficulties the work was interrupted and postponed. It was not taken up again until 1959. Photographic flashbulbs lifted by balloon and triggered with a clock mechanism were then photographed from Turku and Helsinki. From 26 simultaneous pairs of photographs the direction from Turku to Helsinki was computed. An accuracy of $\pm 1.05''$ in the hour angle and $\pm 0.62''$ in the declination was obtained. This result deviated from the classical triangulation -0.6 m in azimuth and $+0.2$ m in elevation at this distance of 153 km.

2. Stellar Triangulation of Finland

For stellar triangulation over the whole of Finland a method was selected in which flashes given by a flying device between two observation stations are photographed from these stations. The pairs of directions from the observation stations to each flash are determined with the aid of the fixed stars in the background and then the direction from one station to the other is computed. These directions are combined in a three-dimensional network.

The flash device, a Geosonde GS 13, consists of a battery, dc to dc converter and stabilizer, radio receiver and trigger circuits. The geosonde is carried by balloon to an altitude of 35—38 km. The flashes are released by impulses from a radio transmitter of 40 W power and 153 MHz frequency. The transmitter forwards manually selected second impulses from a Hewlett—Packard quartz clock and so an accuracy better than 1 ms is obtained.

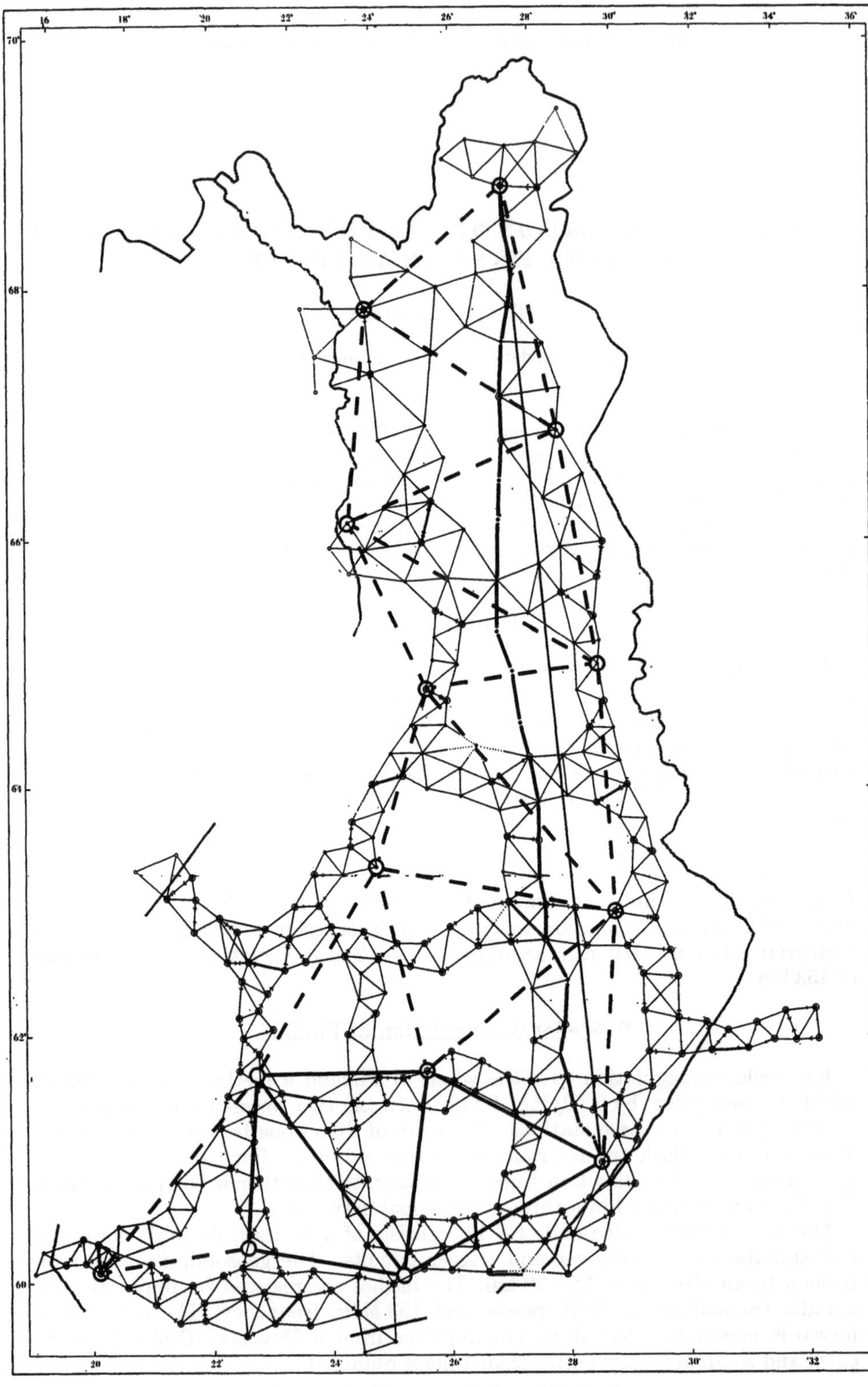

Fig. 1. Stellar triangulation of Finland.

To photograph the flashes three Schmidt—Väisälä telescopes are used, with a focal length of 1031 mm and a 340 mm aperture, the field of view being $5° \times 5°$ and the size of the plate 9 cm $\times$ 9 cm. The telescopes are set equatorially and they are driven by clock. They are transportable and can be set up or dismounted in one day. The flash energy is 50 W s and a flash photographed with the above telescope from a distance of 100—200 km appears on the plate as a star of sixth magnitude.

With the above equipment the longest side length of triangles is about 300 km. In order to make observation possible in somewhat poor circumstances a network with 200 km side lengths is being planned. All stations with the exception of the three first stations are placed in the immediate neighbourhood of a first-order triangulation station. The observations are carried out on the three triangles in the south. The progress of the work depends on the weather, so it is not easy to estimate the time needed for the whole net, but 8 years seems likely.

3. Geodimeter Traverse

The scale of the stellar triangulation net is derived from the classical triangulation net, which is derived from the Nummela Standard Baseline measured with a Väisälä light-interference comparator. In order to determine the scale of the stellar triangulation net independently from the classical triangulation a geodimeter traverse will be measured through the whole network.

A laser geodimeter model 8 will be used. The instrument, No. 80093, to be used here, is a modification of the common one so that only one of the frequencies is used and the three others are locked out. This arrangement provides better frequency stability. Further, in this instrument the demodulation is improved with an additional crystal. The instrument has been calibrated and tested on the Niinisalo 22.2 km calibration baseline and among others on a 41.1 km first-order triangulation side. The internal accuracy of one observation at these distances was found to be ± 9 and ± 18 mm, respectively.

The length of the calibration baseline derived with invar wires from the Nummela Standard Baseline is 22219848.3 mm and the length computed from 138 measurements with the geodimeter 22219849.5 mm. The deviation of 1.2 mm is much lower than the internal errors. Thus a real accuracy of 0.4 ppm for one measurement may be expected.

The traverse, as shown in the attached map, is provisionally planned along the first-order triangulation sides from the northernmost stellar triangulation station to one of the southernmost stellar triangulation stations. In this zone of the first-order triangulation net the average side length is 37 km. The quadratic mean of the deviation angles between individual sides and the diagonal between the terminal stellar triangulation stations is 14.2°. The accuracy of the azimuth of one side is $\pm 0.3''$ when derived from the general adjustment of the Finnish triangulation net. The arithmetical mean of the elevation differences between end points of the triangulation sides is 63 m, the quadratic means of the errors in these elevation differences ± 1.7 m and in the absolute elevations of the triangulation points ± 1.2 m.

We can estimate the accuracy to be expected on the basis of the above data in the following way.

(i) The error of one geodimeter measurement of an average side is ± 15 mm. This is computed with the value of the error of one measurement as ± 0.4 ppm. As we do not know how independent different geodimeter series really are, this may also be the proper error for the whole observation series at one station.

(ii) The error of the projection on the diagonal caused by the error in azimuth as ± 14 mm.

(iii) The error caused by the error in tilting of the side as ± 3 mm.

(iv) The error in projecting to the ellipsoid caused by the error in absolute elevations of the triangulation points as ± 7 mm.

These four errors make a total error of the projection of one side of ± 22 mm. The planned traverse consists of 26 sides and so the total error of the 890 km traverse diagonal is ± 112 mm.

4. Time Micrometer

The principle of the device, a time micrometer, for observing passive satellites has already been reported [2]. This instrument has been developed now in working stage and tested. A description of the instrument and the test results are given in a new report by the same authors [3]. Thus the following information about it can be given here.

A grid of parallel lines is placed in front of the photographic plate or film. The grid consists of two systems of non-transparent lines. The 0.1 mm thick coordinating lines are spaced 1.0 mm from each other. In model A, for fast satellites, the 0.1 mm thick timing lines are at 0.1 mm intervals, and in model B, for slow satellites, the 0.04 mm thick lines are at 0.04 mm intervals. The timing lines are in the centre of the plate and photographic emulsion is removed from under the grid area covered by timing lines. When using films a hole is made under the timing lines. A satellite moving across the coordinating lines exposes on the plate a trail cut in sections by the coordinating lines of the grid. The coordinates of the sections of the chopped trail are measured with a plate comparator in the same way as the coordinates of fixed stars. The coordinates of the crossing points of the satellite trail and timing lines are interpolated between the above trail sections as the mutual positions of the coordinating lines, and the timing lines are determined with the plate comparator. The corresponding times at which a satellite crosses the timing lines are determined with a photomultiplier placed behind the hole in the film. The photomultiplier impulses are recorded by the quartz clock impulses. The recording is made with a modulator, a normal magnetic tape recorder, a demodulator and a pen recorder. For reading the time from a frequency of 1 kHz of the quartz clock a time code generator has been built in which marks of 1 ms, 10 ms, 100 ms, 1 s, 10 s, and 1 min differ in amplitude and length. The recorded magnetic tape is read with a pen recorder.

Considering the apparatus described here for observing passive satellites there are several promising features. There are no moving mechanical parts, and so no vibration is to be anticipated. The trail of the satellite on the photographic plate is chopped into sections sharply, and the beginning and end of a section are completely similar. Thus in the coordinates of a section no systematical error due to possible asymmetry in the shape of the sections is to be anticipated.

The tests with the time micrometer, when it was attached to the telescope mentioned here, have shown the following accuracy.

The interpolation of the coordinates of the crossing points of the satellite trail with the timing lines is affected by irregularities in the coordinating lines of the grid, by inaccuracies in the plate comparator used for coordinate measurement, by rapid refractive motion of the image of the satellite and by irregular deformation in the film. All these sources of error cause an error of $\pm 0.52''$ along the trail and of $\pm 1.26''$ perpendicular to it in the position of one crossing. As there are 26 lines the final errors are $\pm 0.10''$ and $\pm 0.25''$ respectively. To the above errors we have still to add the effect of irregularities in the timing lines of the grid and the long-period refractive wandering of the image. Thus the total errors along the trail are $\pm 0.40''$, and perpendicular to it $\pm 0.39''$.

In determining the time at which a satellite crosses the timing lines the non-linearity of the registration apparatus and photomultiplier, the instability of the tape recorder speed, the delay of the tape recorder and some minor errors cause a total error of around ± 0.13 ms for a fast satellite and ± 0.23 ms for a slow satellite.

By observing satellites with a time micrometer at both ends of the precision traverse the direction position of the Finnish stellar triangulation net will be determined.

References

[1] Y. Väisälä, Sitz.-Ber. Finn. Akad. Wissenschaften 1946, p. 99 (1947).
[2] J. Kakkuri and K. Kalliomäki, Information Bulletin of Central Bureau for Satellite Geodesy No. 2 (1970).
[3] J. Kakkuri and K. Kalliomäki, Publ. Finnish Geodet. Inst. No. 74 (1972).

Space Research XIII — Akademie-Verlag, Berlin 1973

USES OF GEOCEIVER AS A GEODETIC INSTRUMENT

B. B. Holland

The Johns Hopkins University, Applied Physics Laboratory, Silver Spring, Md., USA

The precision capabilities and accuracy experience of the Geoceiver passive Doppler receiver are reported. Geoceiver receives either of two frequency pairs 150/400 or 162/324 MHz and operates with 120 watts. Paired frequencies are used for first order ionospheric refraction correction. The instrument is portable (45.6 kg) and is commercially available. Error sources are enumerated and quantified to provide an overall positioning error figure. This figure depends upon the accuracy with which the gravity field is known, and it is limited by current knowledge to 20 meters on a single transit of a satellite. Using multiple transits, the errors approach those imposed by the receiver. These errors, and thus the ultimate positioning capability, are less than 3 meters. The operating environment is discussed thus providing a measure of logistic and computing support requirements.

1. Introduction

The Johns Hopkins University Applied Physics Laboratory has had an interest in the development and use of artificial earth satellites since SPUTNIK I successfully achieved orbit in 1957. An analysis of the radio signals being emitted by SPUTNIK led to the discovery by Drs. Guier and Weiffenbach [1] that such satellites could be tracked using observations of the Doppler shift on these signals. Dr. F. T. McClure (also of this Laboratory) shortly thereafter, reasoned that the inverse was also true, i.e. that a ground observer knowing the satellite's orbit and using the received Doppler signal could locate himself relative to the earth.

Since that time the art of satellite tracking and navigation has been refined by two orders of magnitude. Geoceiver, which was originally developed at JHU/APL by Stansell et al. [2], is a product of this technological growth.

The Geoceiver is a small, lightweight radio receiver which will automatically lock on to and track either of two frequency pairs 162/324, 150/400 MHz. The receiver is phase-locked to the signal and an associated recorder outputs a punched paper (or mylar) tape which contains time, cycle count, and ionospheric refraction data. Information concerning the station, weather, and satellite are punched as a header and trailer to the measurement data.

The measurements are intended to be transmitted to a computing facility where they can be combined with the satellite orbital data to find the position of the Geoceiver's antenna. To attain the highest possible accuracy, the measurements must also be corrected for ionospheric and tropospheric refraction effects and the geoceiver's clock must be calibrated. Ionospheric refraction corrections are a part of the measurement data and tropospheric refraction can be calculated as a function of the submitted weather data. Clock calibrations are based on the receipt of navigation satellite timing marks which are used to produce epoch corrections to be applied throughout the data span. Accurate positioning cannot be attained without the use of these (or comparably equipped) satellites. The accuracy of the positioning also depends directly on the quality of the orbital information, which in turn depends on the quality and completeness of the physical model used for tracking.

A satellite system, currently in existence, which can meet the necessary criteria is the US Navy Navigation Satellite System. The satellites in this system are tracked regularly (daily), have stable oscillators, emit one of the receivable radio-frequency pairs, and transmit timing signals needed to calibrate the geoceiver clock.

2. Instrument

The electronic structure of the instrument is well documented [3] and will not be repeated here. Physically, Geoceiver is packaged in five travel cases: (1) receiver, (2) antenna, RF amplifier and cables, (3) recorder, (4) antenna tripod, (5) field spares. The shipping weight and volume are 93.4 kg and 0.261 m^3, respectively. When removed from the shipping cases the Geoceiver weighs 45.6 kg. The power supply will operate from any one of: 22—32 V dc, 103—127 V ac (47.5—420 Hz) single phase, 206—254 V ac (47.5—420 Hz) single phase, and requires an average of 120 watts of power.

A single wideband antenna is used for the reception of all four frequencies. The RF amplifier is mounted directly on the base of this antenna (to assure the lowest possible system noise figure) and connects to the receiver with a 100 ft coaxial cable.

Processing of the signals from the two dual-frequency receivers (162/324 MHz, 150/400 MHz) is under control of the data processing unit which, in addition to data recovery and output (to the recorder or associated computer), controls the signals for message synchronization, receiver tuning, navigation-satellite timing, and internal timing.

3. Operation

The mode switch (receiver front panel) is used to select one of the two receivable frequency pairs. Independently of which pair is being received, the higher frequency (of the pair) is locked on to and tracked with the narrowband phase-locked receiver. This signal is mixed with a locally derived signal (near the same frequency) to obtain the Doppler measurement. The lower frequency of the pair is processed through a second narrowband receiver which has been slaved to the primary phase-locked receiver. Its output frequency is thus proportional to the lack of phase coherence of the two received signals. Since the signals were coherent when transmitted, this output frequency provides a measurement of the ionospheric refraction.

The Doppler is counted by sensing positive zero crossings of the signal. The initiation and termination of the count is triggered by time marks derived from the satellite, if it is the navigation variety, or the geoceiver clock if it is not. The satellite memory meters out doublets ($^1/_2$ bits) such that it takes exactly two minutes by the satellite's clock to transmit 12206 doublets (= 6103 bits). Each time a navigation satellite's clock is to arrive at an even minute, the event is marked with a fixed bit pattern in the carrier frequency modulation. Geoceiver identifies this (time) mark to synchronize with the satellite emitted bit string. Both the synchronization word itself and bit counts between words can serve as time marks. The Geoceiver partitions each 2-minute interval into four nearly equal time segments by counting the received bits thus producing the triggering time marks. The intervals are exactly: 1638 bits $\cong$ 32.2071 s, 1404 bits $\cong$ 27.6061 s 1638 bits $\cong$ 32.2071 s, 1423 bits $\cong$ 27.9797 s, from beginning to end of the two-minute interval, respectively.

The limits of integration are the ground clock readings taken at the first positive zero crossing of the Doppler signal following the arrival of the time mark (regardless of its origin). The Doppler count is thus always an integer and continuous, i.e. the end time for one interval is the begin time for the next. The refraction count is integrated over the same time limits as the Doppler. Since these times are based on the Doppler zero crossings they will, in general, not coincide with zero crossings of the refraction signal, thus producing a truncation error. The receiver is actually counting an ionospheric refraction signal which is nine times the correction at 324 MHz (55/6 times the correction at 400 MHz). The measured refraction count must, therefore, be divided by 9 or 55/6 (depending on the frequency) prior to applying the ionospheric correction to the Doppler.

Geoceiver has ten different operating modes which are selectable through a switch on the front panel of the receiver section. These modes enable test and calibration of the instrument in addition to satellite tracking. They fall into three categories: (i) Attended operation: These modes require an operator's presence, e.g. to set the geoceiver clock. (ii) Test modes. For running checkout and calibration tests on the instrument. (iii) Unattended operation: These modes allow satellite pass data to be taken without requiring the operator's presence. Passes taken this way must later be identified by the computing center since a single satellite number is associated with many passes. Such identification is complicated but can be accomplished on about 95% of the passes.

Geoceiver requires almost nothing in the way of logistical support. Aside from life support for the operator(s) and electrical power (within one of the specified ranges), the instrument is self-supporting. The instrument's self-test capability and "*post-facto*" calibration of the clock means that no additional support facilities and/or manpower are needed at the geodetic site.

Site preparation consists of a permanent geodetic marker at or near the point where the antenna is to be located. This (recoverable) location must have an uncluttered radio horizon and be free of RF interference.

Prior to taking a pass of data, the data tape is prepunched with a "Header" which contains information about the pass to be taken. This information consists of: year, day number, satellite number. It is punched three times to assure future transmission success.

The pass data follow and consist of: time in hours, minutes, seconds (decimal) and fractional seconds (octal) to a precision of $\pm 2\,\mu s$; Doppler cycle count (octal); refraction cycle count (octal).

At the completion of the pass, a data "Trailer" is punched on the tape and consists of: geodetic marker (or station) number, $\pm$ temperature (°C), pressure (mb, leading 0 or 1 assumed), relative humidity (%), format number, operation mode, comment number.

4. Reduction

To obtain the geodetic position of the Geoceiver's antenna one must first obtain a reasonable number of passes. The experimental evidence available would indicate that at least 20 high-quality passes should be used. These should be spread uniformly east and west of the station and north and south rising. The best quality data are those passes which have no data points missing, i.e. the data are continuous from rise to set. If one is utilizing the Navy Navigation Satellite System the site should be occupied about three days. Currently there are five satellites in this system and one can acquire an average of four passes per day from each satellite. One day should be allowed for setup, oscillator stabilization, instrument checkout, etc., and an additional day for backup data.

A single three-dimensional, least-squares, solution using all the acceptable data probably yields the best result. Single-pass solutions give larger errors but, when grouped together, generally have a mean which coincides with the single, multiple-pass solution. The general approach is to form the theoretical expression of the measurement and to fit the difference of this and the measurement by least-squares (after correction for first order ionospheric refraction).

5. Error Sources

Geoceiver error sources and values are enumerated in Table 1. These include all sources of error which limit one's ability to determine the position of the Geoceiver's antenna.

To calibrate the geoceiver clock one must relate geoceiver clock time to navigation satellite clock time. The satellite time marks are used to trigger the reading of the geoceiver clock hence one can determine their epoch difference by accounting for the time delay between time mark emission and geoceiver clock reading. The time mark modulation can be applied to the carrier to an absolute accuracy of $\pm$ 50 μs rms. Any single time mark could, therefore, be biased by this amount.

The signal transmission delay time is large (7 ms) but can be calculated to within $\pm$ 5 μs. (This assumes a range error of 2 km; by looping on the range calculation one could reduce the range error to the order of meters and hence this error to nanoseconds.) The receiver delay time is calibrated to the order of $\pm$ 30 μs. An operation mode [3] is provided in Geoceiver whereby calibration measurements can be submitted as data to the computing center. Finally, there is the random time delay which occurs between the recognition of the (time signal) trigger and the next positive zero crossing of the Doppler signal. The computing center can calculate the Doppler frequency f_D and apply a time correction equal to $1/(2 f_D)$. The minimum Doppler frequency of a navigation satellite is about 24 kHz. Thus the maximum time delay ranges from 0 to 42 μs. The rms error suffered in assuming an error of $1/(2 f_D)$ (21 μs) is $\pm$ 12 μs.

The satellite carrier is phase-modulated by data in doublet form where the doublets are paired to form bits. To synchronize the receiver bit slots with the satellite bit slots, the cosine component of the phase modulation (doublet rate) is tracked in a separate phase-locked loop. This is the time recovery loop whose error (due primarily to jitter) should not exceed $\pm$ 40 μs rms.

Table 1

Uncorrected Geoceiver errors

	Peak bias	RMS random
A. Clock calibration	(μs)	(μs)
Satellite time signal accuracy	$\pm$ 50	
Signal propagation delay	$\pm$ 5	
Receiver delay	$\pm$ 30	
Time mark recover		$\pm$ 40
Wait time for zero crossing of Doppler signal		$\pm$ 12
Total (μs)		0.46
RMS (meters)	0.32	
	150/400	162/324
B. Refraction and Doppler count	(cycles rms)	(cycles rms)
Reference oscillator stability (short-term)	0.096	0.078
Clock readout resolution	0.052	0.042
Receiver phase shifts (Doppler)	0.136	0.136
Receiver phase shifts (refraction)	0.059	0.091
Refraction count round-off	0.045	0.045
Total (cycles)	0.190	0.191
RMS (meters)	0.143	0.177
C. Position determination	μs	Meters
Epoch correction (from A)	46.0	0.32
Instrumentation (from B)		0.216
Reference oscillator stability (long-term)	0.06	0
Uncorrected tropospheric refraction		3
Uncorrected ionospheric refraction		3
Neglected gravity harmonics		15—20
Satellite oscillator stability ($3/10^{11}$)	0.04	
Total (meters)		15.6—20.5

The clock epoch error can be calculated by solving the following equation for $\Delta t_{e\mu}$:

$$t_\mu = T_\mu + \frac{\varrho_\mu}{c} + \Delta t_r + \Delta t_c + \Delta t_{D\mu} + \Delta t_{e\mu}$$

where t_μ = Geoceiver clock time (measurement); T_μ = known time on satellite clock ($\pm$ 50 μs); ϱ_μ/c = signal transmission delay time ($\pm$ 5 μs); Δt_r = receiver time delay ($\pm$ 30 μs); Δt_c = time mark recovery delay ($\pm$ 40 μs); Δt_D = wait time for positive zero crossing of Doppler ($\pm$ 12 μs); Δt_e = epoch difference between geoceiver and satellite clocks. Epoch corrections for times between navigation satellite passes can be obtained to sufficient accuracy by linear interpolation on these values. The random errors give an rms of 33 μs or less for a pass with ten more data points. The bias errors have a uniform probability of attaining any value between $\pm$ 55 μs. This distribution gives an rms value of 32 μs. Thus the overall clock calibration error is 46 μs rms.

The reference oscillator is designed to maintain a short-term stability of 8 parts in 10^{12} rms. For a 30-second counting interval this error accumulates to a Doppler count error of 0.096 cycle at 400 MHz and 0.078 cycle at 324 MHz.

The digital clock readout error is $\pm$ 2 μs. This affects the time interval estimate and, in effect, reduces the measurement accuracy. The equivalent Doppler count error ranges from 0.042 cycle at 324 MHz to 0.052 cycle at 400 MHz rms.

Receiver phase shift variation errors occur in both Doppler and refraction signals. These can arise from temperature variations, signal strength variations, or reponse of the phase-locked loop to changes in the Doppler frequency. The maximum phase shift variation allowed by the receiver is $\pm$ 60°, which gives an rms error of 49° or 0.136 cycle rms for the Doppler signal and 0.059 and 0.091 cycle rms for the 150/400 and 162/324 MHz refraction signals respectively.

The refraction of the signal as it passes through the troposphere induces a positioning error which can be as large as 120 m depending on the maximum elevation attained. This error is primarily a function of elevation rate and can be reduced by removing data points near the horizons (i.e. near rise and set). If one deletes points with elevation less than 5° the uncorrected error is reduced to a maximum of about 25 m. The tropospheric effects have been modeled by Hopfield [4—6]. Using this model and weather conditions (of which it is a function), the unresolved error in the positioning due to tropospheric refraction of the signal can be reduced to 3 m. A convenient computational algorithm has been published by Yionoulis [7].

The RF signal is also refracted by the ionosphere. The effect of this refraction on the received frequency is inversely proportional to the frequency of the signal [8]. The dominant error may be removed by using the two coherent frequencies transmitted by the satellite. The refraction counts recorded by the Geoceiver are analog estimates of this dominant refraction error. The computing facility, by applying these measurements as corrections to the data, can reduce the remaining ionospheric refraction error to the order of 1.4—7 m rms. The variability of this error results from the fact that the state of the ionosphere varies with solar activity. There are indications from current investigations in the field that the remaining error effects can be reduced by 50% or more by ionospheric modeling [9]. Weiffenbach [10] estimated 3 m when averaging over a number of passes and the error is not likely to be less than this in the absence of such modeling.

As noted earlier the refraction count is truncated, producing an rms error of 0.408 cycle. Since the signal being counted is either 9 or 55/6 times the actual refraction contribution [2], the Doppler error is 0.045 cycle rms.

The earth's gravity field has been expanded in spherical harmonics by almost everyone. The fitted coefficients take account primarily of the terms of which they are coefficients but further reflect a part of the missing terms. As a result, each model (coefficient set) is both unique and better than theory would indicate. It is additionally true that multiple pass solutions statistically reduce the errors still further. SAO, APL, and NWL have all published coefficient sets which purport to yield rms errors of about 15—20 m. The experimental results presented in the next section speak for themselves.

The total expected error given in Table 1 is for single-pass positioning. If several passes are used (say 20 or more) then this error is statistically reduced to the 3—4 m level.

6. Experimental Results

Several experiments have been performed using the Geoceiver over the past two years. Two of these, which were well controlled, will be described and their results summarized. The first was a co-location (R. R. Newton, private communication) of two Geoceivers which were set up at APL and attached to a common antenna. Ten passes were taken under each of two conditions: first with both Geoceivers using the same reference oscillator and second using their individual oscillators. The results are summarized in Table 2. Half the passes taken had no

Table 2

Calculated antenna position using co-located Geoceivers

	Number of passes	Calculated position differences (cm) Along track	Slant range	Resultant
Different oscillators				
Passes with no points deleted	5	−11	+30	32
Passes with at least one point deleted	5	−104	−143	177
All passes	10	−58	−57	81
Same oscillators				
Passes with no points deleted	5	0	15	15
Passes with at least one point deleted	5	68	−13	69
All passes	10	43	− 3	43

Table 3

Geoceiver—satellite positions minus Geodimeter survey positions using satellites 30130 and 30140

Station No.	δ latitude	(differences in meters) δ longitude	δ height	Resultant
10003	−1	0	0	1
10006	1	−1	−1	1.7
10018	0	−3	−1	3.2
10019	0	−1	−2	2.2
10020	0	0	1	1
10022	0	0	−1	1
10023	−2	1	−2	2.2
20000	0	0	2	2
20001	0	0	1	1
20015	−2	.0	1	1.7
20016	−2	0	1	1.7
30025	0	0	0	0
30026	0	1	1	1.4
30027	−1	0	0	1
30028	0	−1	−1	1.4

Standard deviation about surveyed position = 1.67

points missing, i.e. the data were continuous throughout the transit of the satellite; and half had one or more points missing. There is improvement of at least a factor of 4 when the data are continuous. The entries in the tables are differences in the along-track and slant-range corrections to the antenna location as determined by the two Geoceivers. That is, the location and geodesy errors are subtracted from the results. Thus the tabulated values give a rather good measure of the instrumentation errors. These are 15 cm and 32 cm depending on which oscillator is used and even at its worst (32 cm) is quite small.

The second experiment [11] consists of several hundred satellite passes of two navigation satellite (30130 and 30140) taken with Geoceivers located at 15 sites on a high precision Geodimeter surveyed network over the south-eastern United States [12]. At least 20 passes were taken at each site. These sites were then positioned using the satellite data and the resultant position compared with the surveyed position. The overall results are presented in Table 3 [11].

In summary, Geoceiver instrument errors are of the order of 0.5 m and with sufficient passes and proper data selection and reduction, geodetic positions can be determined to within 3 m.

Acknowledgment

This work was supported by the Navy Space Project Office, under Task T of Contract No. N00017-72-C-4401.

References

[1] W. H. Guier and G. C. Weiffenbach, Proc. Inst. Radio Engrs **48**, 507 (1960).
[2] T. A. Stansell, W. A. Bourguignon, C. R. Karl and P. C. Marth, APL/JHU Rep. TG-710 (Rev.), 1968.
[3] Magnavox Research Laboratories, Navy NAVAIR 16-30 PRR 14-1, 2, 3, (three volumes) 1971.
[4] H. S. Hopfield, J. Geophys. Res. **68**, 5157 (1963).
[5] H. S. Hopfield, J. Geophys. Res. **74**, 4487 (1969).
[6] H. S. Hopfield, Space Research XII, 581 (1972).
[7] S. M. Yionoulis, J. Geophys. Res. **75**, 7636 (1970).
[8] W. H. Guier, Proc. Inst. Radio Engrs, **49**, 1680 (1961).
[9] M. M. Feen, private communication.
[10] G. C. Weiffenbach, in: Electromagnetic Distance Measurement, Hilger and Watts Ltd., London 1967 (p. 339).
[11] L. K. Beuglass, DMA Report 0001, July 1972.
[12] B. K. Meade, paper presented at XIV General Assembly of IUGG, Lucerne 1967.

Space Research XIII — Akademie-Verlag, Berlin 1973

SOME CHARACTERISTICS OF AN OPERATIONAL SYSTEM FOR MEASURING UT 1 USING VERY LONG BASELINE INTERFEROMETRY

J. M. Moran

Smithsonian Astrophysical Observatory, Cambridge, Mass., USA

The demonstrated fractional frequency stability of very long baseline interferometers is 1.0×10^{-13}. This stability is sufficient for determining UT 1 to an accuracy of 2 milliseconds per day on a 4000 km baseline without corroborative atmospheric measurements and 1 millisecond with ground-based atmospheric measurements. Some specifications are given for using galactic H_2O sources at 22 GHz or quasars at 8 GHz in an operational system for measuring UT 1.

1. Introduction

Microwave interferometers with baselines up to 9000 km have been used over the past five years in the range from 1 to 75 cm to measure the angular sizes of radio sources down to about 10^{-4} arcsec [1—4]. For these measurements only the magnitude of the cross correlation function is determined and phase coherence is required only for a time long enough to detect the signal. This is typically a few minutes. Phase coherence is generally maintained by the use of atomic frequency standards at each station to control the phases of the local oscillators. The use of very long baseline interferometers (VLBI) for geometric measurements of baseline or source coordinates through the analysis of the fringe phase information places much more stringent requirements on the phase stability of the instrument. The possible use of VLBI techniques for making geometric measurements of interest in geophysical research, in particular measurement of the rotation rate of the earth, was suggested very early (1967) by Gold [5] and MacDonald [6]. In this paper some requirements for the measurement of universal time with VLBI methods are considered.

Fluctuations in UT or sidereal time can be measured by noting changes in apparent right ascensions (and declinations) of stars or radio sources. Only the accuracy of the raw time measurements (UT 0) without the corrections necessary to obtain UT 1 and UT 2 will be considered here. Optical techniques utilizing zenith tubes give accuracies of about 0.075 arcsec (5 ms of time) averaged over one day [7]. These measurements are limited by atmospheric seeing which produces stellar images of about 1 arcsec in diameter. Short baseline radio interferometers with baselines less than a few kilometers operated by such groups as the National Radio Astronomy Observatory, Green Bank, W. Va. [8], the Royal Radar Establishment, Malvern, England [9] and the Mullard Radio Astronomy Observatory, Cambridge, England [10], have produced lists of positions of radio sources with accuracies of about 0.5 arcsec. Phase fluctuations induced by the atmosphere limit the accuracy of these measurements [11]. The phase fluctuations increase approximately linearly with the baseline length up to a certain limit when the two propagation paths through the atmosphere become independent [12]. In principle, then, the atmospheric problem should be alleviated by the use of extremely long baselines. For example, the entire atmospheric excess propagation path of about 2.5 meters (at zenith) represents an angular displacement of 0.1″ on a 4000 km baseline. The measurement of absolute source positions has proceeded slowly using VLBI, principally because of inadequate atomic frequency standards. Cohen and Shaffer [13] used rubidium standards on a 10000 km baseline at 13 cm to derive positions for 35 sources with typical accuracies of 3″. Hinteregger et al. [14] used hydrogen maser standards on an 845 km baseline and gave the positions of seven radio sources at 4 cm wavelength to an accuracy of about 0.5″. This paper will show the method of data analysis, the accuracy to which UT can be measured with existing techniques and equipment.

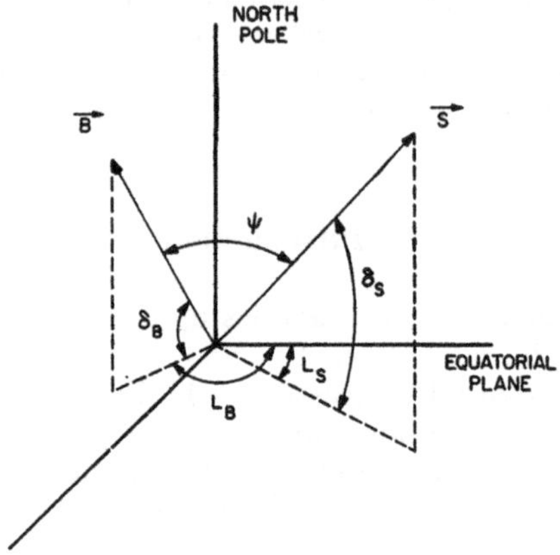

Fig. 1. Geometry of an interferometer.

2. Data Analysis

The geometry of a long baseline interferometer is shown in Fig. 1. The relative propagation time or geometric delay is, neglecting small effects,

$$\tau_g = \frac{1}{c} \boldsymbol{B} \cdot \boldsymbol{S} \tag{1}$$

where $\boldsymbol{B}$ = baseline vector, $\boldsymbol{S}$ = unit vector in direction of radio source, c = velocity of propagation. The fringe phase and fringe rate are defined by

$$\phi = 2\pi f \tau_g$$
$$f_R = d\phi/dt \tag{2}$$

where f is the signal frequency. The signals received at each antenna are broad band gaussian random variables which are characterized by an incident flux S. The power in the receiver is generally calibrated in terms of antenna temperature which is related to the flux by the equation $SA/2 = kT$ where A is the collecting area of the antenna and k is Boltzmann's constant. The received signals are converted to a video band by mixing with local oscillator signals controlled by atomic frequency standards and then recorded on magnetic tape. The tapes are brought together and played into a processor which (i) aligns the signals and removes the geometric delay τ_g, (ii) calculates the cross spectral function (or the cross correlation function which is then Fourier transformed), (iii) multiplies by $\exp(j\,\phi_{est})$, where ϕ_{est} is the best *a priori* estimate of the fringe phase, to "stop" the fringes which can have frequencies up to 50 kHz, and (iv) averages the data over the desired interval. Only the analysis of the fringe rate information will be discussed here although the delay information can also be used. The total phase rate or fringe rate is

$$f_R = -\frac{f B_H \Omega}{c} \cos\delta_s \sin(L_s - L_b) + f_{prop} + f_{std} \tag{3}$$

where $L_s = T - \alpha_s$. B_H is the length of the equatorial component of the baseline, Ω is the rotation frequency of the earth, f is the local oscillator frequency, δ_s is the source declination, L_s and L_b are the source and baseline local hour angles respectively, α_s is the source right ascension and T is the sidereal time. f_{prop} is the induced fringe rate due to the changing differential excess propagation path which can be written as

$$f_{prop} = \frac{d}{dt}\left[\frac{f}{c}\int (n-1)\,ds_1 - \frac{f}{c}\int (n-1)\,ds_2\right] \tag{4}$$

where n is the index of refraction and s_1 and s_2 are the propagation paths. f_{std} is the difference in local oscillator frequencies due to errors in the atomic frequency standards.

After the best estimate of the geometric fringe rate has been removed the residual fringe rate is

$$\Delta f_R = A \sin(L_s - L_b) + B \cos(L_s - L_b) + f_{prop} + f_{std} \tag{5}$$

where

$$A = \frac{f B_H \Omega}{c} \cos\delta_s \,(\Delta\alpha_s + \Delta T)$$

$$B = \frac{f B_H \Omega}{c}\left[\sin\delta_s\,\Delta\delta_s - \cos\delta_s \frac{\Delta B_H}{B_H}\right].$$

$\Delta\alpha_s$, ΔT, $\Delta\delta_s$, ΔB_H are errors in assumed parameters, and ΔT is equivalent to ΔUT. These errors cause the residual fringe rate to have a sinusoidal dependence with 24 hour period. If a source is tracked for a substantial portion of a day then Eq. (5) can be used in a least mean squares analysis to determine two variables, either $\Delta\alpha_s$ or ΔT and either $\Delta\delta_s$ or ΔB_H. The accuracy is limited by the signal to noise ratio and the adequacy of the models for f_{prop} and f_{std}. Note that since an initial epoch for the measurements is assumed it is not necessary to determine the actual baseline parameters accurately.

The uncertainty in the measurement of phase due to finite signal to noise ratio is approximately

$$\sigma(\phi) \simeq 2 T_R T_A^{-1} B^{-1/2} \tau^{-1/2} \quad \text{(radians)} \tag{6}$$

when $\sigma(\phi) < 1$, where T_R and T_A are the receiver and antenna temperatures, B is the receiver bandwidth and τ is the integration period. $B\tau$ represents the number of independent signal samples averaged. The uncertainty in the estimate of fringe rate is therefore approximately

$$\sigma(f_R) \simeq T_R T_A^{-1} B^{-1/2} \tau^{-3/2} \text{ (hertz)}. \tag{7}$$

3. Measurement Accuracy

The overall accuracy achievable with fringe rate measurements has been investigated empirically by a VLB group with members from the Massachusetts Institute of Technology, Naval Research Laboratory, and Smithsonian Astrophysical Observatory [15] using galactic H_2O sources at 1.35 cm. A typical set of residuals on the source W 49 after 10 hours of observation is shown in Fig. 2. Eq. (5) was fit to the data with four parameters: A, B, one atmospheric parameter (the zenith delay), and a constant frequency offset due to the frequency standards. The zenith delay estimate was 2.3 ± 0.1 m and the clock offset was 2.1 ± 4 mHz. The coordinate errors were $\sigma(\alpha_s) = 0.09$ arcsec and $\sigma(\delta_s) = 1.4$ arcsec. The standard deviation of the data was 2.4 mHz giving $\sigma(f_R)/f = 1.1 \times 10^{-13}$. This value for $\sigma(f_R)/f$ is the same as that measured by Hinteregger et al. [14] at 8 GHz. It is the most important characteristic of the performance of a VLBI system. The value for $\sigma(f)/f$ for the maser frequency standards alone is about 10^{-14} [16].

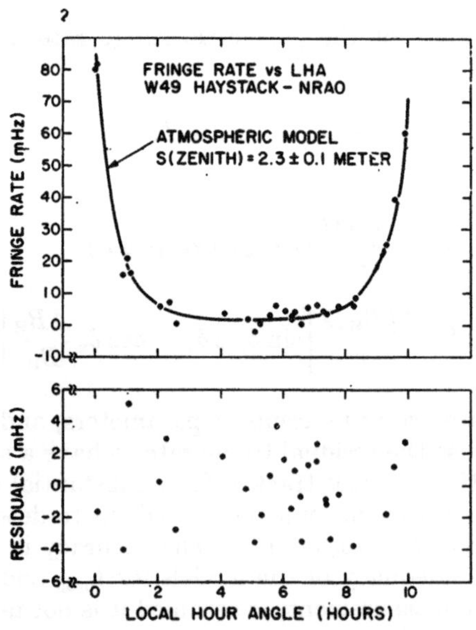

Fig. 2. Fringe rate residuals on an 845 km baseline using the 140 ft antenna of the National Radio Astronomy Observatory and the 120 ft antenna of the Haystack Observatory at $\lambda = 1.35$ cm (from [15]). Top: Without the atmospheric effects removed. Bottom: Residuals with effects of plane parallel atmosphere with zenith path length equal to 2.3 m removed.

4. Galactic Water Vapor Sources

Consider the use of galactic H_2O sources for measuring UT. Very strong emissions from galactic clouds of H_2O were discovered in 1969 [17] between excited rotational levels ($6_{16} \rightarrow 5_{23}$) at 22235.080 MHz (1.35 cm). About 50 sources [18] have been found so far down to a flux limit of about 10^{-25} W m^{-2} Hz^{-1}. Because of the tremendous power output and small angular sizes of these sources maser amplification is thought to be involved in the emission process [19]. The advantages of using these sources are:

(i) Because of the very high frequency the effects of propagation through the ionosphere are very small. The ionosphere, because of its variability and irregularity especially near sunset and sunrise is very difficult to model [20]. The excess propagation path is proportional to the square of the wavelength and during the daytime at the zenith is about 20 m at 1 GHz, 35 cm at 8 GHz and 4 cm at 22 GHz. The index of refraction due to atmospheric water vapor at 22 GHz is essentially independent of frequency. Hence the modelling of the tropospheric phase delay is frequency independent throughout the microwave region.

(ii) The receiving equipment can be used to make measurements of atmospheric brightness temperature in order to obtain the water vapor content of the atmosphere [21].

(iii) The sources are very strong (see Table 1) having fluxes up to 3×10^{-22} W m^{-2} Hz^{-1} or 3 orders of magnitude greater than the largest usable flux of a quasar at 8 GHz.

The disadvantages of using these sources:

(i) The distance of the sources is between 1 and 15 kiloparsec. Hence proper motions might be expected between 0.001″ and 0.01″ per year. These could presumably, however, be calibrated with respect to more distant sources.

(ii) The sources are variable in intensity [22, 23]. Intensities can vary by an order of magnitude on the time scale of months. Hence measurements of many sources are necessary to ensure continuity in the program.

(iii) The line widths are narrow ($\sim 10^5$ Hz). Hence the signal to noise ratio cannot be improved by increasing the receiver bandwidth and no effective measurement of the geometric delay can be made to estimate parameters.

(iv) The distribution over the sky is non-uniform. The sources are confined to the galactic plane which is inclined by 65° to the celestial equator.

Only (i) might present a serious problem because of the unknown nature of the proper or apparent motions.

5. Parameters for an Operational System

The basic design parameter is the achievable fringe rate noise level which has been found to be 2.4 mHz in the absence of supplemental tropospheric measurements at 22 GHz. Consider a baseline of length 3930 km and declination −7° such as the one formed by connecting the Haystack Observatory in Westford, Massachusetts, and the Owens Valley Radio Observatory in Big Pine, California ($B_H = 3903$ km). The antennas must be large enough to allow the fringe rate to

Table 1

The strongest galactic H_2O sources

Source name	Position[a] RA (hrs)	(1950 epoch) Decl. (°)	Flux[b] (1972) (10^{-26} W M^{-2} Hz^{-1})	T_A[c] (°K)	Bandwidth[d] (kHz)	Velocity[e] (km s^{-1})	LST (hours)
W 3 (C)	2.365	62.87	300	30	250	−36	19.2—16.1
W 3 (OH)	2.3880	61.648	1300	120	500	−49	19.2—16.1
Orion	5.5464	−5.405	2300	200	2000	4	04.2—10.2
VY Canis Majoris	7.3485	−25.670	600	50	1000	18	07.1—10.3
M 17	18.291	−16.23	1700	150	100	10	17.2—22.2
W 49 N	19.1305	9.022	30000	2500	2200	−2	16.6—00.6
W 51	19.357	14.41	700	60	1500	59	16.6—01.3
W 75 S	20.6208	42.203	600	50	500	1	16.1—04.6
ON 1	20.134	31.37	200	20	100	14	16.5—03.5
ON 2	20.331	37.27	200	20	1000	10	16.2—03.8

a. Most positions are known to an accuracy of 2 arcsec.
b. Variable on the time scale of months.
c. Antenna temperature on 25 m antenna with 40% efficiency or a 37 m antenna with 18% efficiency.
d. Bandwidth required to receive all major features. Individual features have bandwidths of about 50 kHz.
e. Velocity with respect to the local standard of rest of the main feature.
f. Local sidereal times (Haystack) when sources are available on the Haystack — OVRO baseline.

be measured to, say, 1 mHz within the coherence time of the system. With $\sigma(f)/f = 10^{-13}$ the coherence limit is 500 seconds. Assuming an integration time of 150 seconds, bandwidth of 50 kHz and receiver temperature of 2000 °K, and using Eqs. (6) and (7) the minimum value of T_A is 3 °K if the noise in the receiver is to be smaller than the atmospheric and frequency standard effects. With a minimum flux of 200×10^{-26} W m^{-2} Hz^{-1} the collecting area of the antennas must be 70 m^2 each, e.g. antennas of diameter 14 m at 50% efficiency or 25 m at 15% efficiency. Table 1 lists the strongest known sources and their characteristics. The source availability versus declination is shown in Fig. 3.

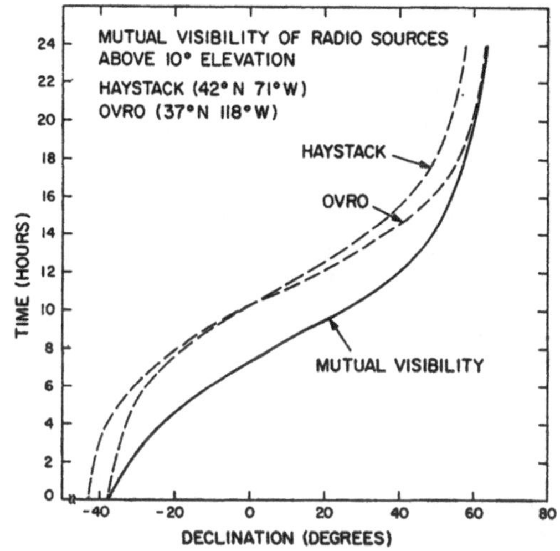

Fig. 3. The availability of sources for elevation angles greater than 10° plotted against declination on a 3900 km east—west baseline.

A covariance analysis was performed for sources at various declinations with five parameters in the model: ΔT, $\Delta \delta_s$, TZ1 (zenith path length 1), TZ2 (zenith path length 2) and Δf_{std} (local oscillator frequency offset). The measurements were assumed to be made every hour during the source availability time with an accuracy of 2.5 mHz. The results, shown in Table 2, indicate that the error in ΔT is about 2 ms (0.3″, 0.9 m in length at the equator) for sources between 10° and 60° declination. The constancy over the declination range is due to the fact that the increased observation time available on high declination sources offsets the decreased sensitivity due to the cos δ_s term in Eq. (5). The uncertainty in the clock frequency Δf_{std} is generally below 10 mHz so that the stations remain synchronized to better than 50 nanoseconds per day. A three parameter model with parameters ΔT, $\Delta \delta_s$ and Δf_s was also analysed and gave uncertainties in ΔT of about 1 millisecond (see Table 2). This model is appropriate when the tropospheric path length can be independently measured to better than 5 cm.

Table 3 shows a list of suitable extragalactic radio sources which could be used for measurement of UT drawn from the work of Cohen et al. [24]. These sources, quasars and radio galaxies are far enough away for proper motions to be negligible. However, many of the sources are variable in intensity and show differential

Table 2

Specification for UT measuring system

1. Accuracy: $\sigma(f_R) = 2.5$ mHz, $\sigma(T) = 2$ ms
2. Frequency: 22235 MHz
3. Equatorial baseline: 3900 km (nominal sensitivity = 100 mHz/arcsec)
4. Antenna size: 14 m at 50% efficiency
5. Receiver temperature: 2000°K
6. Recording bandwidth: 50 kHz
7. Frequency standard: hydrogen maser oscillator
8. Covariance analysis:

A. 3 parameters

δ_s (°)	$\sigma(T)$ (ms)	$\sigma(\delta_s)$ (arcsec)	$\sigma(\Delta f_{std})$ (mHz)
−10	1.3	4.8	7.5
10	1.0	2.6	3.7
30	0.8	0.6	2.1
50	0.9	0.2	0.9
70	1.4	0.1	0.5

B. 5 parameters

δ_s (°)	$\sigma(T)$ (ms)	$\sigma(\delta_s)$ (arcsec)	$\sigma(\Delta f_s)$ (mHz)	σ(TZ 1) (cm)	σ(TZ 2) (cm)
−10	3.1	16.2	28.0	9	5
10	1.9	6.7	11.4	10	4
30	1.6	1.3	6.5	6	4
50	1.7	0.4	3.1	7	7
70	3.2	0.1	1.0	41	33

Table 3

Strong extragalactic radio sources of small angular size

Source	Size[a]	Position RA (hrs)	Decl. (°)	Correlated flux (10^{-26} W m^{-2} Hz^{-1})
PKS 0106 + 01	U	1.10222	1.3217	8
3C84	S	3.27488	41.3311	15
3C120	S	4.50878	5.2499	6
OJ287	U	8.86591	20.2995	4
4C 39.25	U	9.39871	39.2568	9
PKS 1127 − 14	P	11.45990	−14.549	2
3C273	S	12.44258	2.3283	16
3C279	S	12.89328	−5.5188	3
3C345	S	16.68821	39.9029	10
PKS 2134 + 00	P	21.56803	0.47422	8
VRO 42.2201	P	22.01092	42.0357	8
3C454.3	P	22.85819	15.8816	7

a. 8 GHz, $D/\lambda = 10^8$, U = unresolved, P = partly resolved, i.e. fringe visibility between 1.0 and 0.5, S = strongly resolved. Data from Cohen et al. [24].

internal motion of the order of 0.003″ per year. The achievement of 1 mHz accuracy at 8000 MHz for 2 ms time measurement requires antennas 40 m in diameter if the receiver temperature is 200 °K and the bandwidth is 50 kHz. An advantage of using the quasars is that the bandwidth can be increased greatly and is limited only by the tape recorders.

6. Improvements in Measurement Accuracy

An improvement of about a factor of 2 can be achieved by doubling the baseline length (e.g. Spain to west coast USA). Beyond this, the accuracy does not improve because of the decrease in mutual visibility of sources.

An improvement in the system by one order of magnitude ($\Delta T = 0.2$ ms) should be possible within five years by (i) increasing the signal to noise ratio, (ii) improving local oscillator phase stability, and (iii) simultaneous independent measurement of the tropospheric path length.

The signal to noise ratio can be improved by upgrading the receivers from 2000 °K to 200 °K noise temperature with either cooled mixers or maser amplifiers which will be available in a few years. The phase stability of the interferometer can be improved by incorporating the maser frequency standards directly into the local oscillator system, that is, by deriving the local oscillator signal from the 1420.405 MHz maser signal rather than a phase locked 5.0 MHz signal. It is hoped that a stability of 10^{-14} can be achieved in this way.

The determination of the tropospheric path length presents a challenging problem. The part due to dry air is about 230 cm at zenith and varies by only 5 cm. Hopfield has shown [25, 26] that the dry component is proportional to surface pressure and can be measured to an accuracy of several millimeters from the ground. The wet component, due to tropospheric water vapor, varies from 0 to 30 cm. It is not well correlated with any surface parameters such as relative humidity. Schaper, Staelin and Waters [21] show that the water vapor density as a function of height can be deduced from ground-based measurements of the brightness temperature profile at 22 GHz of the thermal emission from water vapor. With these measurements they find that the error for the total tropospheric path is 1.4 cm at zenith. Effort in developing this technique will probably be very fruitful and important in allowing VLBI techniques to reach their full potential.

Acknowledgments

I thank the M. I. T. — N. R. L. — S. A. O. VLBI group for the use of their data in Fig. 2 [15].

References

[1] M. H. Cohen, D. L. Jauncey, K. I. Kellermann and B. G. Clark, Science **162**, 88 (1968).

[2] N. W. Broten, R. W. Clarke, T. H. Legg, J. L. Locke, J. A. Galt, J. L. Yen and R. M. Chisholm, Mon. Not. Roy. Astron. Soc. **146**, 313 (1969).

[3] K. I. Kellermann, D. L. Jauncey, M. H. Cohen, D. B. Shaffer, B. G. Clark, J. Broderick, B. Ronnang, O. E. H. Rydbeck, L. Matveyenko, I. Moiseyev, V. Vitkevitch, B. F. C. Cooper and R. Batchelor, Astrophys. J. **169**, 1 (1971).

[4] J. M. MORAN, B. F. BURKE, A. H. BARRETT, A. E. E. ROGERS, J. C. CARTER, J. A. BALL and D. D. CUDABACK, Astrophys. J. **152**, L 97 (1968).
[5] T. GOLD, Science **157**, 302 (1967).
[6] G. J. F. MACDONALD, Science **157**, 304 (1967).
[7] W. MARKOWITZ, Astron. J. **60**, 171 (1955).
[8] C. M. WADE, Astrophys. J. **162**, 381 (1970).
[9] H. GENT, R. L. ADGIE and J. H. CROWTHER, Nature, Lond. **228**, 146 (1970).
[10] J. W. SMITH, Nature Phys. Sci. **232**, 150 (1971).
[11] D. K. BARTON, Proc. Inst. Elect. Electron. Engrs **51**, 626 (1963).
[12] J. W. M. BAARS, Proc. Inst. Elect. Electron. Engrs Trans. Ant. and Prop. AP **15**, 582 (1967).
[13] M. H. COHEN and D. B. SHAFFER, Astron. J. **76**, 91 (1971).
[14] H. F. HINTEREGGER, R. ERGAS, C. A. KNIGHT, D. S. ROBERTSON, I. I. SHAPIRO, A. R. WHITNEY, A. E. E. ROGERS and T. A. CLARK, submitted to Science.
[15] J. M. MORAN, G. D. PAPADAPOULOS, B. F. BURKE, K. Y. LO, P. R. SCHWARTZ, K. J. JOHNSTON, S. H. KNOWLES, I. I. SHAPIRO and C. A. REISZ, submitted to Astrophys. J.
[16] M. W. LEVINE and R. F. C. VESSOT, Radio Sci. **5**, 1287 (1970).
[17] A. C. CHEUNG, D. M. RANK, C. H. TOWNES, D. D. THORNTON and W. J. WELCH, Nature, Lond. **221**, 626 (1969).
[18] M. L. MEEKS, J. C. CARTER, A. H. BARRETT, P. R. SCHWARTZ, J. W. WATERS and W. E. BROWN, Science **165**, 180 (1969).
[19] B. E. TURNER, Roy. Astron. Soc. Can. J. **64**, 221 (1971).
[20] N. C. MATHUR, M. D. GROSSI and M. R. PEARLMAN, Radio Sci. **5**, 1253 (1970).
[21] L. W. SCHAPER, D. H. STAELIN and J. W. WATERS, Proc. Inst. Elect. Electron. Engrs **58**, 272 (1970).
[22] W. T. SULLIVAN, Astrophys. J. **166**, 321 (1971).
[23] D. BUHL, L. E. SNYDER, P. R. SCHWARTZ and A. H. BARRETT, Astrophys. J. **158**, L 97 (1969).
[24] M. H. COHEN, W. CANNON, G. H. PURCELL, D. B. SHAFFER, J. J. BRODERICK, K. I. KELLERMANN and D. L. JAUNCEY, Astrophys. J. **170**, 207 (1971).
[25] H. S. HOPFIELD, Radio Sci. **6**, 357 (1971).
[26] H. S. HOPFIELD, J. Geophys. Res. **68**, 5157 (1963).

Space Research XIII — Akademie-Verlag, Berlin 1973

TESTING RELATIVISTIC GRAVITY THEORIES USING RADIO TRACKING DATA FROM PLANETARY ORBITING SPACECRAFT

J. F. JORDAN, W. G. MELBOURNE and J. D. ANDERSON

Jet Propulsion Laboratory, Pasadena, Calif., USA

We present an analysis of a computational method for determining the numerical values of the relativity and other related dynamical parameters using two-way Doppler and ranging data from planetary orbiting spacecraft. The computational method consists of the determination of earth—planet distance values to 100 ns accuracy from the tracking data and the treatment of these values, along with planetary radar and meridian circle data, as raw data in a regression analysis to solve for the significant solar system parameters. The main conclusions of the paper are that a 5% determination of the metric parameter β is attainable using a time base that spans both Mariner 9 and Viking '75 and includes ground-based radar tracking of Mars. Over the same span γ is determined to at least 1%.

1. Introduction

The testing of general relativity by classical astronomical techniques, although well understood, is not sufficiently accurate to discriminate between competing metric* theories of gravity. Of the three classical tests, only the precession of the perihelion of Mercury is capable of yielding accuracies of the order of $\pm 1\%$, and even that is degraded to a level of about $\pm 10\%$ by uncertainties in the amount of precession introduced by the solar oblateness. Further, an appeal to theoretical arguments to limit the number of viable theories, as done by Ni [1], for example, still leaves a fairly large class of theories which must be treated experimentally.

Precise determinations of the metric of general relativity are only possible with measurements that depend on space experiments and their associated technology. Radio tracking of spacecraft promises to determine the exact form of the relativistic space—time metric. The purpose of this paper is to examine the expected accuracy of relativistic tests which rely both on tracking data from planetary orbiters, such as Mariner 9 in 1972 or Viking in 1976, and on several years of radar ranging to the inner planets.

The application of spacecraft tracking data to the determination of the metric of general relativity was noted by Anderson and Lorell almost a decade ago [2], but it is only with the advent of planetary orbiters in the 1970's that this has become a reality. Recently, measurements of the delay in ranging signals to the Mariner 6 and 7 spacecraft at superior conjunction have been determined to 4% which yield the first order term in the space part of the metric to $\pm 8\%$ [3], but the second order term in the time part of the metric and a really precise test of

* The term "metric" is defined by Eq. (1). It describes the geometrical nature of four-dimensional space—time.

the first order term is inaccessible in the Mariner 6 and 7 data. Non-gravitational forces on the spacecraft corrupt the fit to the many months of data that would be needed for a complete determination of the metric. This difficulty is completely avoided for planetary orbiters. From the motion of the orbiter over one revolution it is possible to use the spacecraft tracking data to obtain a precise range measurement to the center of mass of the planet. Thus, it is the planet that is being tracked in the orbiter case, and non-gravitational forces are negligible. Tracking over the lifetime of the orbiting spacecraft can be used to determine the nature of the space—time metric which governs the motion of the planets.

Long arcs of planetary radar ranging are available at the present time. However, the analysis of these data is complicated by the fact that radar measurements are made to the sub-earth point on the planetary surface; the ranging accuracies to the center of the planet are corrupted by uncertainties in mean radius and topography. Nevertheless, by analyzing most of the existing radar data, Shapiro [4] has obtained a measurement of the time-delay effect at superior conjunction to $\pm 5\%$. When the 1971 superior conjunctions data for Venus are added to this radar solution, Shapiro predicts that an accuracy of from 2% to 3% will be achieved in the time delay measurement.

Any experimental determination of the space—time geometry can be accomplished by a parameterization of the metric and by a subsequent adjustment of the numerical values of the parameters to fit best the data in question. Post-Newtonian parameterizations have been studied in detail by Will [5], and by Nordtvedt [6]. Will has shown that if a metric theory of gravity possesses integral conservation laws for energy, momentum angular momentum, and center-of-mass motion, then there are only two unconstrained parameters in the metric. Further, the metric can be written in the Eddington form, which is [7]

$$ds^2 = \left[1 - 2\frac{m}{r} + 2\beta\left(\frac{m}{r}\right)^2 + \cdots\right] c^2\, dt^2 - \left[1 + 2\gamma\frac{m}{r} + \cdots\right](dx^2 + dy^2 + dz^2). \tag{1}$$

In this metric the two unconstrained parameters are β and γ, and m is equal to one-half the gravitational radius of the central body. For the sun,

$$m = \frac{GM_\odot}{c^2} = 1.4766\,\mathrm{km}. \tag{2}$$

In Einstein's theory of general relativity, the parameters β and γ are both unity. A determination of values for either β or γ which differed significantly from unity, would indicate that a metric theory of gravity other than the Einstein theory would be required. At this time no reliable experiment indicates values of β and γ that are different from one. However, no significant measurement of β has yet been achieved, and the $\pm 8\%$ superior conjunction measurement of γ is not sufficiently precise to definitely rule out a value which differs from Einstein.

A procedure for obtaining values of β and γ from the radar and orbiter data involves the prediction of the observed range, or round-trip light time to the planet, in the space—time geometry given by Eq. (1), and then comparison of the predicted range data with observations. A fundamental formula in this procedure is the parameterized total one-way light time (coordinate time) between the observer and the planet, which is:

$$c\,\Delta t = r_{01} + (1+\gamma)m \ln \left[\frac{r_0 + r_1 + r_{01}}{r_0 + r_1 - r_{01}}\right] \tag{3}$$

where r_{01}, r_0, and r_1 are coordinate distances representing, respectively, the observer—planet distance, the sun—observer distance, and the sun—planet distance. All distances are computed from orbits (geodesics) defined by the metric. As such they depend implicitly on both β and γ. Orbits are computed by numerical integration of the ordinary Newtonian heliocentric equations of motion for each planet, but with an added perturbative post-Newtonian acceleration of the form

$$\ddot{\boldsymbol{r}}_{\text{rel}} = \frac{m}{r^3}\left\{\left[2(\beta+\gamma)\frac{GM_\odot}{r} - \gamma v^2\right]\boldsymbol{r} + 2(1+\gamma)(\boldsymbol{r}\dot{\boldsymbol{r}})\,\dot{\boldsymbol{r}}\right\}. \tag{4}$$

The details of the computation of the observables are given in [3].

2. Planetary Satellite Orbit Determination and the Measurement of the Earth—Mars Range

Two Mars orbiter missions are available for use in the next five years for high accuracy determination of earth—Mars range. Mariner 9 is currently orbiting Mars, having been injected into orbit on 14 November 1971. The spacecraft has an orbital period of just less than 12 hours, and the orbit is inclined to the Martian equator by 64.3°. The Viking mission project has planned two orbiting spacecraft, the first of which is scheduled to arrive at Mars on 23 June 1976. The orbital periods of the probes are planned to be 24.6 hours. with respective inclinations of 33.2° and 79.3°.

The earth-based data used for tracking the spacecraft consist of two-way frequency difference (Doppler) measurements that are proportional to the tracking station—spacecraft range rate, and signal round-trip time-delay measurements that are proportional to the station—spacecraft range. Doppler measurements have been taken of the Mariner 9 spacecraft, typically at one minute sample times, from DSS-12 and 14 at Goldstone, California, DSS-41 at Woomera, Australia, and DSS-62 at Madrid, Spain. Range measurements have been obtained from Goldstone on a several-times-weekly basis. Individual Doppler measurements are accurate to less than 0.015 Hz, or 1 mm s^{-1}, and individual range measurements are accurate to less than 0.1 μs, or 15 meters.

Ideally, the total combined data set of range and range rate measurements could be processed simultaneously via regression analysis to estimate the parameter affecting the motion of Mars and the earth, and the radio signal. However, analysis of the Mariner 9 tracking data has led to the conclusion that Mars is gravitationally very rough, with second order tesseral harmonics 4 to 5 times larger than expected before flight [8]. Specification of a gravity model with a finite number of parameters accurate enough to integrate the probe motion over hundreds of revolutions may be impossible, even if the parameters are estimated directly from the tracking data. In addition, the computer expense of such regression analyses would be enormous. Hence, a data compression scheme has been devised to alleviate the accuracy and the cost difficulties of the direct approach. The compression scheme consists of using the Doppler data to relate the spacecraft to the center of mass of Mars, and converting the station—spacecraft range measurements to direct observations of the earth—Mars distance. These earth—Mars pseudo range measurements, called normal points, contain virtually all of the information in the individual range measurements and the local Doppler data pertinent to a detailed knowledge of the earth—Mars motion. The earth—Mars angular information content of the Doppler is weak because of errors in the location of the tracking stations on the earth, and can be neglected.

The earth—Mars pseudo range measurement depends on the magnitude of the coordinate vector ϱ as defined in the earth—Mars geometry diagram presented in Fig. 1, and as given by the exact expression

$$\varrho = |\boldsymbol{R} + \boldsymbol{R}_{s/c} - \boldsymbol{R}_s|\,. \tag{5}$$

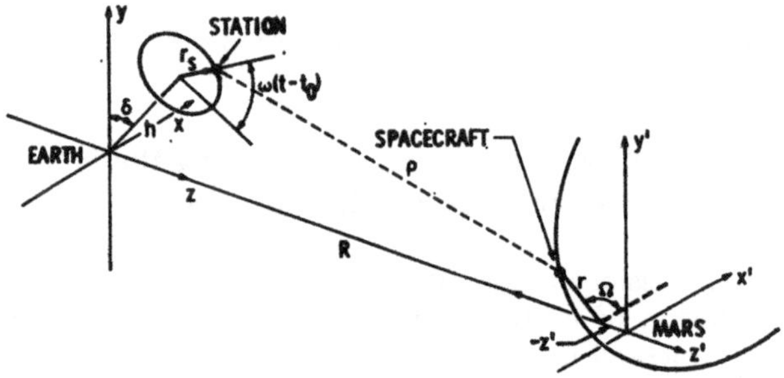

Fig. 1. Earth, station, Mars, spacecraft configuration.

The quantity $\boldsymbol{R}_{s/c}$ is the planet-centered coordinate vector of the spacecraft which is determined from Doppler tracking of the orbiter. The quantity $\boldsymbol{R}_s$ is the earth-centered coordinate vector of the station. Eq. (5) can be written in the following form, accurate to 0.1 m,

$$\varrho = R + z' - h_z + \frac{1}{2R}\{(r\sin\Omega - h_y)^2 + (r\cos\Omega - h_x)^2\} \tag{6}$$

where z', r, and Ω are cylindrical coordinates defining the position of the spacecraft relative to Mars at the time of range observation, and where h_x, h_y, and h_z are components of the station in the earth-centered cartesian coordinate system defined in Fig. 1, and are given by the expressions

$$h_x = r_s \sin\omega t$$

$$h_y = h\cos\delta - r_s \sin\delta\cos\omega t$$

$$h_z = h\sin\delta + r_s\cos\delta\cos\omega t$$

and where h = distance from station to earth equator along the earth spin axis, δ = the geocentric declination of Mars, ω = the angular rate of the earth, t = time from station meridian passage of Mars, r_s = station distance from the spin axis. Eq. (6) can be solved directly for the earth—Mars distance R from a measured value of the station—spacecraft range, using Eq. (3) to convert to coordinate "range" ϱ. In practice, an initial value of R is chosen, R_0, and the difference between the measured and predicted values of ϱ is then related to a differential correction in R. The correction is then added to R_0 yielding a pseudo measured value of R at the time of the spacecraft ranging. Actually, Eq. (3) is then used again to convert R into a relativistically consistent pseudo observable $c\,\Delta t$.

The error in R can be related to the error in the range determination $\Delta \varrho$, the error in the station locations (Δh_x, Δh_y, Δh_z), and the error in the Doppler determined spacecraft position relative to Mars by the following approximate expression, which is accurate to 0.1 m,

$$\Delta R = \Delta \varrho + \Delta h_z - \Delta z' - \frac{1}{R_0}\{(r - h_x \cos \Omega - h_y \sin \Omega)\,\Delta r + (h_x \sin \Omega - h_y \cos \Omega) r\,\Delta \Omega\}. \tag{7}$$

Since the distance of the station from the earth's spin axis and the station longitude are known to less than 10 m from analysis of data acquired during planetary encounter on past Mariner missions, the error in the distance of the station from the earth's equator is the only term which contributes more than 10 m to Δh_z. Hence h_z can be expressed as $\Delta h \sin \delta$, plus terms which contribute less than 10 m errors.

The remaining errors in Eq. (7) are due to the uncertainty in the Doppler determined spacecraft position relative to Mars. The coefficients of the errors of the spacecraft position in the earth's plane-of-the-sky, Δr and $r\,\Delta\Omega$, are 10^{-4}, hence Eq. (7) can be written in the form

$$\Delta R = \Delta \varrho + \Delta h \sin \delta - \Delta z' - (10^{-4})\Delta r - (10^{-4}) r\,\Delta \Omega + (< 10\ \text{m terms}). \tag{8}$$

The accuracies to which z', r, and Ω are known are related to the properties of earth-based Mars satellite orbit determination, which have been discussed in several references [9—11]. Covariance analyses and experiments with both real Mariner 9 data and simulated data before the Mars arrival of Mariner 9, have led to the following conclusions regarding the characteristics of the satellite orbital-determination problem:

(i) The major error source in locating the position of a planetary orbiting spacecraft relative to the center of mass of the planet consists of the accelerations due to an incomplete or erroneous mathematical model of the Mars gravity field, e.g., in the classical sense, uncertainties in the zonal and tesseral coefficients in a spherical harmonic gravity potential function. Other error sources such as Mars ephemeris errors, station location errors, and non-gravitational forces contribute errors in the spacecraft position an order of magnitude less than those due to the gravity uncertainties.

(ii) An optimal data arc for locating the position of an orbiting spacecraft is a single revolution of data, omitting the data within an hour of periapsis, which is particularly sensitive to errors in the gravity model.

(iii) The best determined orbital elements from Doppler data tracking of each revolution are the orbital period and the local times of periapsis passage, which are seen to less than 0.1 second or 1 part in 10^6.

(iv) The worst determined orbit elements is the longitude of node, Ω, of the orbit in the earth's plane-of-the-sky, the plane perpendicular to the earth—Mars line of sight, which is seen to 0.1° or 10^{-3} radian.

(v) Gravity models determined from multi-revolution data fits, if used in the spacecraft trajectory integration, can significantly reduce the uncertainty of the determined spacecraft position.

The 1σ uncertainties of the components of position, r, Ω, and z are shown in Table 1, along with their respective contributions to the error in R. Pre-flight, actual pre-gravity sensing, and post-gravity sensing results are shown in the table.

Table 1

1σ uncertainties in Mariner 9 position when ranging is taken

Pre-flight		Actual pre-gravity sensing	Post-gravity sensing
$r\Delta\Omega$	30 km	120 km	10 km
Δr	100 m	400 m	20 m
$\Delta z'$	50 m	200 m	20 m
		Effect on earth—Mars measurement	
$r\Delta\Omega$	3 m	12 m	1 m
Δr	0.01 m	0.04 m	0.002 m
$\Delta z'$	50 m	200 m	20 m

Contributions from $r\Delta\Omega$ and Δr are seen to be negligible leaving the 20 m contribution from Δz the only spacecraft position error of importance. Eq. (7) can now be expressed in the form

$$\Delta R = \Delta\varrho + \Delta h \sin\delta - \Delta z' + (< 10 \text{ m terms}) \quad (9)$$

where neither the contributions from measurement error $\Delta\varrho$ and spacecraft position error $\Delta z'$ are highly correlated in time. The station location error Δh is a constant.

From the data fits made over the first few months of the Mariner 9 orbiter mission, values of r, Ω, and z' have been determined at the times at which range measurements have been obtained. These position values have been used to predict values of ϱ. Observed residuals in ϱ have been related to values in R at all of the times at which range measurements have been obtained.

Residuals of the normal points during the first month of the Mariner 9 orbit are given in Fig. 2, where the elements of the orbits of both earth and Mars have been fit with the data. The sampled rms of the points obtained during the first month of Mariner 9 is 0.092 s or 14 m. The effect of an error in h has undoubtedly

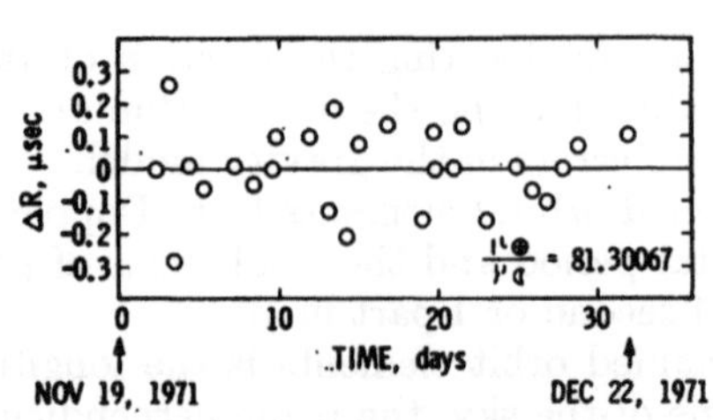

Fig. 2. ΔR residuals for the first month of Mariner 9.

been smoothed away by some of the earth and Mars orbital element corrections for such a short data span, and it is clear that h should be estimated along with the orbital elements.

3. Expected Accuracy of the Relativity Parameters

At what time in the future will the values of β and γ be known to a precision that will satisfy the astrophysicist and cosmologist? Will this occur when both parameters are known to a fraction of one per cent, or will further experimental tests at the level of a hundredth or a thousand of one per cent lead to profoundly deeper understanding of gravity, perhaps with unimagined theoretical and practical consequences? So much depends on the future course of space experiments, their results, and their interactions.

This paper considers the expected improvement in the relativistic parameters as a result of the analysis of data from the Mariner '71 and Viking satellites of Mars. These missions will yield useful data, and, as demonstrated by the analysis of the preceding section, the error sources in the reduced range measurements to the center of mass of Mars are well understood. If the success of the Martian missions continues, and if range and Doppler data can be obtained until at least several weeks beyond the respective superior conjunctions, then the expected uncertainties in β and γ presented here from the orbiter data from Mariner '71 and Viking, along with radar data from future Martian oppositions, should be a reliable and conservative estimate of what to expect in early 1977.

The covariance analysis is based on simulated range data to Mars for four intervals of time. These intervals are listed in Table 2. Although the rms residuals on the Mariner Mars '71 normal points is about ± 14 m, the uncertainty has been increased to ± 40 m to allow for the possibility of systematic errors or correlations in the data. The uncertainty of ± 450 m on the radar data is a realistic estimate of the degree to which the topography of Mars corrupts the reduction of the basic radar measurements (currently accurate to about 100 m) to the center of the planet.

In addition to the nominal standard deviation on each range measurement, the error in the simulated S band data is increased further near the sun in order to account for plasma irregularities or scintillations [12]. The value of σ that is assigned to each range point is computed from the formula

$$\sigma^2 = \sigma^2_{\text{nominal}} + \sigma^2_{\text{corona}}$$

where $\sigma_{\text{corona}} = 4900 p^{-3/2}$ (m) and where p is the distance of closest approach radii of the ray path to the center of the sun.

Table 2
Data intervals and accuracy

Data interval	Source	Assumed accuracy in one-way range
14 Nov. 1971 to 14 Nov. 1972	Mariner Mars '71	$\pm$ 40 m
1 July 1973 to 1 Jan. 1974	Opposition radar	$\pm$450 m
1 Sept. 1975 to 1 Feb. 1976	Opposition radar	$\pm$450 m
23 June 1976 to 1 April 1977	Viking orbiter	$\pm$ 40 m

The assumed sample interval for both the radar data and the normal points is once every five days, except for a region between 5° and 17° of the sun where the sample interval is once per day. The planet Mars remains within 17° of the sun for a period of about ± 1 month on either side of superior conjunction. This is a reasonable interval of time to expect daily normal points, but the increased cost and scheduling problems associated with daily tracking outside of $\pm 17°$ probably could not be justified on the basis of better accuracy in β and γ.

The parameters which are included in the covariance analysis are listed in Table 3 along with their assumed *a priori* errors.

Table 3

Estimated parameters

Parameter	*a priori* standard deviation
1. Relativity parameter β	0.2
2. Relativity parameter γ	0.2
3. Solar oblateness coefficient J_2	10^{-5}
4. Corona parameter A	0.9×10^{8} el cm^{-3}
5. Corona parameter B	0.7×10^{6} el cm^{-3}
6. Corona parameter ε	0.3
7. Variable gravitational constant $\dot{G}/G$	10^{-12}/day
8. Astronomical unit	3 km
9. Mass of Jupiter $\Delta M/M$	2×10^{-6}
10. Station coordinate parallel to earth's spin axis h	50 m
11.—16. Six orbital elements for earth	$\sim 10^{-3}$ s of arc
17.—22. Six orbital elements of Mars	$\sim 10^{-2}$ s of arc

The solar corona parameters A, B, and ε represent a steady state model of the interplanetary medium. The application of this model to the spacecraft tracking problem has been discussed in detail by Muhleman et al. [12]. The basic assumption of the model is that the electron density N_e at a distance r from the sun can be represented by

$$N_e = \frac{A}{r^6} + \frac{B}{r^{2+\varepsilon}}$$

where r is expressed in units of solar radii. The two parameters B and ε must be very highly correlated to assume a narrow range of values of N_e at the earth's distance from the sun. In the error studies presented here, a correlation coefficient of 0.999 is assumed. The *a priori* error estimates on the corona parameters represent an error in N at the earth of about ± 5 electrons cm^{-3}.

The *a priori* errors on the other parameters in the analysis are conservative estimates of uncertainties during the 1971 and 1977 time period. The mass of Jupiter is assumed known from Pioneer F/G to 2 parts per million, while the solar oblateness and variable gravitational constant are assumed bounded at the level of their *a priori* errors by past radar and optical data and by the Mercury radar data that will be obtained in the next few years.

For purposes of performing the error covariance analysis, the simulated data are processed sequentially. In this way it is possible to display the covariance matrix of the weighted least squares method as a function of time. Just such a plot of errors on β and γ is shown in Fig. 3. The *a priori* errors in the ephemerides for this plot are represented by a complete covariance matrix from the reduction of real radar and optical data (up to 1971). The diagonal elements of the matrix correspond to errors in the elements of the earth of about 10^{-3} second of arc and in the elements of Mars of about 10^{-2} second of arc. An exception to this is that the percentage error in the semi-major axes of the two planets, when expressed in seconds of arc, is about one order of magnitude smaller than the error in the other five elements to account for the accuracy in the mean motion values.

Two plots for the error in γ are shown in Fig. 3. The upper curve, which represents the maximum error on γ, is based on the extremely pessimistic assumption that all data from both Mariner '71 and Viking are lost within a region of $\pm 10°$

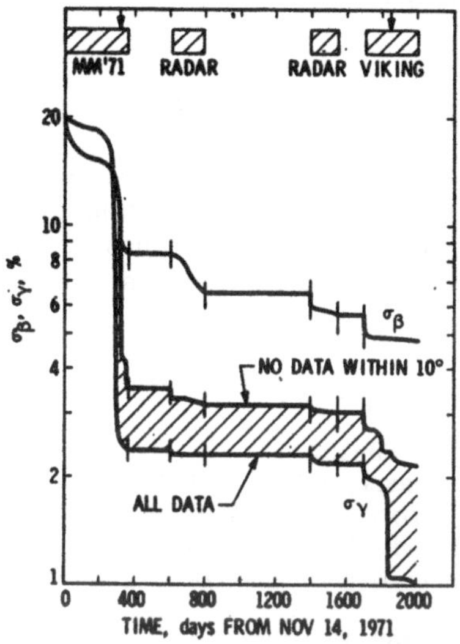

Fig. 3. Time history of 1σ uncertainty in β, γ.

of the sun. The lower, and more realistic curve, assumes that data are available within $\pm 10°$ and that the corona contributes no more error than that derived from the assumptions of this analysis. The purpose of showing the upper curve with over five weeks of data deleted is to demonstrate that the solar corona can at very most limit the measurement accuracy on γ to $\pm 2.2\%$ after Viking and to $\pm 3.5\%$ after Mariner '71. More realistically, the error will be limited by the corona to $\pm 1.0\%$ and $\pm 2.4\%$ at those dates.

With respect to the parameters of the study, a sensitivity analysis for each constant reveals that only the corona parameters have a significant effect on γ and that J_2 contributes most of the error on β. The parameters $\dot{G}/G$, AU, and h are negligible error sources. The mass of Jupiter could contribute about a 1% error in β if its present uncertainty of 40 ppm were not improved by the Pioneer F/G Jupiter flybys to the level of 2 ppm as assumed here.

The accuracy of γ is relatively insensitive to ephemeris errors, but the accuracy on β is extremely sensitive. For example, an improvement in the ephemerides of the earth and Mars from a level of $\pm 10^{-3}$ to $\pm 10^{-4}$ second of arc, with data other than those assumed for this study, could reduce the error on β from $\pm 5.0\%$ to

2.0% after Viking, while a complete lack of *a priori* data on the ephemerides, with a complete reliance on the orbiter and radar opposition data, would limit the determination of β to $\pm 8.3\%$ after Viking.

The accuracies presented here compare favorably with respective values of 16% and 3% determined by Cain [13] from analysis of the Viking lander. It is certainly reasonable to expect that the extensive radar tracking of Mercury, plus data from an extended Mariner Venus-Mercury '73 mission, when taken in combination with the Mariner '71 and Viking data, will reduce the uncertainty in both β and J_2 below the values given here. Also, the anticipated dual frequency capability at S and X bands on the Mariner '73 and Viking missions will yield new information on the solar corona and will bring about a corresponding reduction of the uncertainty in γ to 0.5%.

Acknowledgments

This paper presents one phase of research carried out at the Jet Propulsion Laboratory, California Institute of Technology, under NASA Contract NAS 7-100. The authors wish to acknowledge the assistance of Mrs. Eunice Lau and Miss Nancy Hamata for some valuable computational support.

References

[1] W. NI, Astrophys. J. **176**, 769 (1972).
[2] J. D. ANDERSON and J. LORELL, AIIA J. **1**, 1372 (1963).
[3] J. D. ANDERSON, P. B. ESPOSITO, W. L. MARTIN and D. O. MUHLEMAN, Proc. Conf. on Experimental Tests of Gravitation Theories, JPL TM 33-499 (1971), p. 111.
[4] I. I. SHAPIRO, Proc. Conf. on Experimental Tests of Gravitation Theories, JPL TM 33-499 (1971), p. 136.
[5] C. M. WILL, Astrophys. J. **169**, 125 (1971).
[6] K. NORDTVEDT, JR., Astrophys. J. **161**, 1059 (1970).
[7] A. S. EDDINGTON, The Mathematical Theory of Relativity, Cambridge University Press, London 1922.
[8] J. LORELL et al., Science **175**, 317 (1972).
[9] G. W. REYNOLDS, R. K. RUSSELL and J. F. JORDAN, paper presented at AIAA 9th Aerospace Sciences Meeting, New York 1971.
[10] J. W. ZIELENBACH and J. F. JORDAN, paper presented at AAS/AIAA Astrodynamics Specialists Conf. Fort Lauderdale, Fla 1971.
[11] J. F. JORDAN et al., paper presented at ION National Space Meeting, Orlando, Fla 1972.
[12] D. O. MUHLEMAN, J. D. ANDERSON, P. B. ESPOSITO and W. L. MARTIN, Proc. Conf. on Experimental Tests of Gravitation Theories, JPL TM 33-499 (1971), p. 92.
[13] D. L. CAIN, Proc. Conf. on Experimental Tests of Gravitation Theories, JPL TM 33-499 (1971), p. 245.

Space Research XIII — Akademie-Verlag, Berlin 1973

THE USE OF RADIO BEACONS IN GEOPHYSICS AND THEIR APPLICATIONS

R. Leitinger

Institut für Meteorologie und Geophysik, University, Graz, Austria

A review is given on the present situation of radio beacon observations, including methods and techniques of observations, cooperative programs and future experiments. The importance of satellite beacon measurements in geophysics is briefly demonstrated. Some aspects of the application of beacon observations in engineering problems are treated.

1. Introduction

Satellite beacon observations have been carried out since 1958. It has been proved that these measurements can give valuable data on the upper atmosphere. The results are not restricted to the ionosphere. In addition to their geophysical importance, radio beacon measurements are needed for application purposes where satellites are involved (among others in communication, satellite ranging, position finding, navigation). The interest of both geophysicists and application engineers in satellite beacon data is increasing. The material needed could be listed as follows: suitably reduced data gained over longer periods of time; intensive observation during major solar-terrestrial events; statistics on the spatial and temporal behaviour of parameters related to beacon observations; definition, probability, prediction of disturbed and undisturbed conditions for application systems.

2. Methods and Techniques of Observations; Data Handling

It is the primary purpose of satellite beacon observations to measure the influence of the medium between the satellite and the ground. The methods which are applied have changed considerably in recent years. Cheap electronic components allow measurement of phase differences at low costs. Digital readout systems and direct processing with electronic computers will be used by a great number of stations engaged in radio observations of geostationary satellites or of low orbiting navigation satellites: analogue outputs and scaling by hand are not adequate to manage the very large data fluxes which arise from observations on a continuous basis.

In the future amplitude measurements will be restricted to the observation of scintillations and of absorption. The observation of all other effects caused by the interlying medium can be reduced to measurement of phase differences after suitable processing of the received signals in a multichannel receiver; for example, the Faraday effect can be reduced to the phase path length differences between the two circularly polarized components of one radio wave, the group delay can be reduced to the differences of modulation phase if the same modulation is applied to two coherent carriers [1, 2]. The phase path length differences between two coherent beacons have been used systematically for over 13 years to measure the ionospheric electron content by the so-called differential Doppler method [3].

The influence of the interlying medium on the angle of arrival can also be measured by phase methods. In this case two or more phase-locked one-channel receiving systems can be used if the true zenith angle of the satellite is accurately known as a function of time [4]. To overcome the need for accurate orbital data one could use two coherent beacons with widely spaced frequencies and measure the difference in angles of arrival. To our knowledge this has not yet been done on a systematic basis.

Much work has been done to improve the methods of evaluation, e.g. to solve ambiguity problems, to apply corrections and to study the observable effects in second order approximations. Several systems for digital data processing have been tried and future programs will be planned with automated equipment. The data reduction is simple compared with ionogram evaluation and allows the use of fully digitized systems.

Future measurements will give large quantities of data: for example, a fully equipped receiving station for the ATS-F beacons could have more than ten channels; a redundant readout system is advisable which doubles the number of values which are read at a given time; readouts are made at intervals of 5 minutes, (for special purposes the intervals will even be shorter); this gives 5760 values a day. Each value is a 3-digit decimal number. Therefore the use of an electronic computer is absolutely necessary to reduce this amount of data. The quantity and quality of the results will justify the costs for computer time.

A review on scintillation studies was given during the 1971 COSPAR meeting [5]. Therefore it is not necessary to stress again the great importance of these phenomena which are often considered as a trouble in radio transmissions but can give valuable insight into processes in the upper atmosphere. The theory of scintillations has made considerable progress but the results are still far from satisfying. New experiments have to be made to give the necessary input for further theoretical investigations. On the ground it will be necessary to use spaced receiver techniques extensively to explore the spatial coherence of scintillations. Such measurements are under way but one should encourage their use on a larger scale. One other important problem is the frequency dependence of scintillations. To improve existing estimates it would be necessary to have beacons which radiate at least ten coherent waves with frequencies suitably distributed between about 40 MHz and a few GHz. As we know, one experiment of this type is firmly planned [6].

3. Existing Observing Stations, Participation in Cooperative Programmes

Recent inquiries on satellite beacon observations show that there are 89 stations, 72 in the northern hemisphere and 17 in the southern hemisphere. These are distributed as follows:

Europe (including the whole of the USSR)	33
Asia	13
Africa	5

N. America	21
S. America	7
Australia and New Zealand	8
in the Pacific	2
High latitudes ($\phi > 60°$)	10
Middle latitudes	69
Low latitudes ($\phi < 20°$)	10

Satellites which are observed or have been observed during the last two years: Low orbiting satellites BE-C (1965—032A), NNSS-satellites (1967-034A, 1967-048A, 1967-092A, 1968-012A, 1970-067A), Cosmos, Intercosmos, Electron satellites; a few observations have been made on telemetry carriers of several other satellites.
Geostationary satellites: ATS-1, ATS-3, ATS-5, Intelsat IIF2, Intelsat IIF3.

During the last few years cooperative programmes have been in operation within the Joint Satellite Studies Group and within the Intercosmos Group. After the end of the S-66 program no special beacon satellite has been available and no cooperation on a larger scale has been possible. In the meantime great efforts have been made to prepare for the ATS-F project. Nearly worldwide cooperation will be possible. These plans were discussed at the Symposium on satellite beacon measurements at Graz [13].

Strongly related to cooperation in the field of satellite beacon observations are the following problems on which agreements should be reached: rules for evaluation methods, standardization of terms and symbols, standardization of data formats, systems for data exchange on a multilateral basis, use of a World Data Center for collection and distribution of data banks (e.g. regular bulletins) [13]. These problems will also be considered by the "COSPAR Beacon Satellite Group".

4. Future Experiments

4.1. The ATS-F Project

The geostationary satellite ATS-F will be equipped with a multifrequency radio beacon. Launch is planned for December 1973. For about one year the satellite will be located at 95° W. It will be shifted to 35° E around December 1974. The frequencies of the beacon will be 40 MHz, 140 MHz, 360 MHz, all modulated with 1 MHz (amplitude modulation to give carrier and side frequencies of comparable transmitting power). The 40 MHz and the 360 MHz carriers will also be modulated with 100 kHz. Among others the following methods to measure atmospheric properties will be possible: Faraday effect methods (simultaneously on several frequencies to overcome the ambiguity problem), phase path length methods (dispersive Doppler) between 40, 140 and 360 MHz waves, group delay (modulation phase) methods. The group delay measurements will add an important new tool for accurate determinations of columnar electron content.

A very important opportunity of the ATS-F project is the combination of different methods to extend the range of useful results: for example, the Faraday effect is sensitive only to the ionosphere below about 1000 km whereas electron content values obtained by other methods (group delay, phase path length differences) give the total content from the ground to the height of the satellite. A suitable combination could give information on the exospheric electron content. It should also be possible to combine ATS-F data with data from low orbiting satellites (e.g. NNSS satellites or the Intasat satellite). For more detailed information see [2] and newsletters of Dr. K. Davies, NOAA, Boulder, Colorado, USA (available on request). Colleagues interested in ATS-F observations when positioned at 35° E should contact Dr. G. Hartmann, Max-Planck-Institut für Aeronomie, Lindau/Harz, W. Germany.

4.2. The Intasat Project

A Spanish low orbiting beacon satellite will be launched in 1974 under the designation Intasat. The beacon frequencies will be 40.010 and 41.010 25 MHz allowing Faraday effect measurements with the equipment which has been used in the S-66 program (the same frequencies were used on BE-B and BE-C). A circular orbit with an altitude of 1460 km and an inclination of 101.56° is planned. For details contact Instituto Nacional de Técnica Aerospacial (INTA), Madrid (Project Scientist: L. Sánchez Muniosguren).

4.3. Polar Orbiting Wideband Beacon of the Stanford Research Laboratory

(Beacon payload DNA-002)

Under this designation a 10-frequency beacon will be launched in September 1973. A circular orbit with an altitude of 740 km and an inclination of 98.5° is planned. Ten coherent signals are provided. Frequencies: 138.1 MHz, 7 equally spaced signals between 379.8 and 448.9 MHz, one signal at 1243.1 MHz, one, at 2900.5 MHz. The beacon is intended for observation of amplitude scintillation at all frequencies and of phase scintillation at the lower frequencies using the 2900 MHz signal as a reference. Electron content measurements are also planned using the second difference of phase among equally spaced frequency triads [6—8]. For details contact Dr. E. J. Fremouw, Radio Physics Laboratory, Stanford Research Institute, Menlo Park, Calif., USA.

5. Geophysical Phenomena to be studied with Satellite Beacon Observations

At the beginning of satellite beacon observation for geophysical purposes the "ionospheric electron content" was the only quantity to be evaluated from the raw data material (the electron content is the number of free electrons to be found in a column between the ground and the height of the satellite. The column has a cross section of 1 m^2). The electron content, its spatial and temporal behaviour, its correlation with other ionospheric parameters, with solar flux and geomagnetic parameters, is now only one aim among others. Satellite beacon measurement can give valuable insight into dynamic phenomena (travelling ionospheric disturbances, gravity waves, large scale irregularities, disturbed conditions, magnetic storm effects, small scale irregularities, etc.).

A review on published papers on satellite beacon measurements can give a good picture of the present use of these observations in geophysics. For the following summary restricted to 1970 to March 1972, the Proceedings of two Symposia [7, 9] have been used and the journals *Annls. Géophys.*, *Geomagn. and Aeron.*, *J. Atmos. Terr. Phys.*, *J. Geophys. Res.*, *Planet. Space Sci.*, *Radio Sci.* In four other journals related to the physics of the upper atmosphere no relevant articles were published. Here we give only a list of the most important topics and the number of papers in each category.

(i) Theory of radio wave propagation through the ionosphere; errors; evaluation methods.

(Theory in general; second order effects; Faraday rotation under unusual conditions; Faraday rotation across the transverse region; papers dealing with the $n\pi$ ambiguity in Faraday effect methods; guided propagation of HF waves; improvements in the closely spaced frequencies method; comparisons of reduction methods; application of ray tracing methods; estimation of errors; problems in differential Doppler interpretation; use of transfer functions).

24 papers.

(ii) Results in terms of columnar electron content and geophysical implications of these results. Other ionospheric parameters.

(Spatial and temporal changes of electron content in different regions of the earth, diurnal and seasonal changes, latitudinal variation; global surveys of electron content behaviour; electron content and topside plasma temperatures; electron content, slab thickness and their relation to other ionospheric parameters; influence of solar activity and of geomagnetic parameters).

26 papers.

(iii) Scintillations in amplitude and phase.

(Theoretical studies, statistics, relation to other phenomena; frequency dependence of scintillations; scintillation boundary; regions of high scintillation occurrence: polar scintillation region and low latitude scintillation region; dependence of scintillations on zenith angle and on elevation; nature of the irregularities which could produce scintillations; estimates of the height of the irregularities).

25 papers.

(iv) Studies of large scale irregularities by means of satellite beacon methods.

(Isolated ionospheric irregularities; non-periodic irregularities; the inhomogeneous structure of the outer ionosphere).

4 papers.

(v) Studies of dynamic phenomena by means of satellite beacon methods.

(Neutral winds; gravity waves; travelling ionospheric disturbances; dynamic effects induced by solar eclipses; ionospheric and magnetic storms).

19 papers of which 9 are related to gravity waves.

(vi) Proposals for new experiments, data on existing beacons.

8 papers.

(vii) Equipment, data handling.

6 papers.

(Each paper is counted only once in the category which seemed to fit best).

We would estimate that the above review covers about 70% of the total "output" in generally available form (not including "reports" etc.) and related to geophysics. It should be noted that most of the papers on the behaviour of the electron content, on scintillations and on disturbances of the ionosphere can be related directly to satellite ranging, communication, navigation and position finding problems. There is a strong tendency to publish papers on the application of satellite beacon measurements in engineering problems as "reports". Some important aspects of radio beacon observations are found in the review articles [5, 10—12] where detailed references are given.

6. The Application of Satellite Beacon Observations in Engineering

As mentioned above satellite beacon measurements give data which are used to define system parameters in satellite ranging, satellite communication, navigation and position finding. It is not necessary to stress the importance of these applications. It is sufficient to describe in a few words the sort of data banks and statements which are required:

The application engineer is not interested in electron content values but in correction factors. Conversion is easily done from electron content values into correction factors. Corrections are useless under disturbed conditions. Therefore the definition and recognition of disturbed conditions is necessary. Under disturbed conditions large errors in satellite ranging and in related applications may occur, errors which can not be reduced by correction methods. Disturbed and undisturbed conditions are independent of scintillations. The latter are the main source of limitation to some communication systems.

In connection with application needs it would be important to multiply the observations of absorption, especially at low elevation angles.

It will be one purpose of international cooperation of radio beacon observers to prepare data banks which are useful in application problems on a world-wide scale.

References

[1] O. K. GARRIOTT, A. V. DA ROSA and W. J. ROSS, J. Atmos. Terr. Phys. **32**, 705 (1970).
[2] J. K. HARGREAVES, in: Proc. Symp. on Future Application of Satellite Beacon Experiments, Max-Planck-Inst. f. Aeronomie, Lindau/Harz, FRG 1970.
[3] W. J. ROSS, J. Geophys. Res. **65**, 2601 (1960).
[4] B. W. REINISCH, Lowell Tech. Inst. Res. Found. J. **3**/2 (1970).
[5] G. K. HARTMANN, Space Research XII, 1221 (1972).
[6] E. J. FREMOUW and A. A. BURNS, Radio Sci. **5**, 917 (1970).
[7] Proc. Symp. on Future Application of Satellite Beacon Experiments, Max-Planck-Inst. f. Aeronomie, Lindau/Harz, FRG 1970.
[8] A. A. BURNS and E. J. FREMOUW, in: Proc. Symp. on Future Application of Satellite Beacon Experiments, Max-Planck-Inst. f. Aeronomie, Lindau/Harz, FRG 1970.
[9] Papers in: Radio Sci. **5**, No. 6 (1970) (special issue: papers presented at Symp. on Application of Atmospheric Studies to Satellite Transmissions, Boston, Mass., September 1969).
[10] K. RAWER, Space Sci. Rev. **3**, 380 (1964).
[11] YA. L. AL'PERT, Space Sci. Rev. **4**, 5 (1965).
[12] K. H. SCHMELOVSKY, Space Sci. Rev. **8**, 74 (1968).
[13] Proc. Symp. on Future Application of Satellite Beacon Measurements, Inst. f. Meteorologie u. Geophysik, Universität Graz, Austria 1972.

Remote Sensing of the Earth

Space Research XIII — Akademie-Verlag, Berlin 1973

A STUDY OF LINEAMENTS FROM A ZOND 5 PHOTOGRAPH OF NORTHERN AFRICA

A. A. GRIGORYEV

Laboratory of Space Methods for Earth Sciences, University of Leningrad, Leningrad, USSR

By interpreting the global photograph from the automated interplanetary station Zond 5, detection was made of an entire network of lineaments on a planetary scale crossing the northern part of the African continent and having an extent of 150 to several hundred kilometres (about 30 of them were revealed most reliably). These lineaments, evidently, constitute large broken tectonic destructions most of which were unknown earlier.

Space remote sensing occupies an important place among the new methods of studying the geomorphological structure of the earth. For this purpose photographs from manned space vehicles have yielded appreciable results; television pictures from meteorological satellites are not so extensively used. Pictures taken from the automated interplanetary stations, such as Zond 5 and Zond 7, make it possible to recognize macro-features of land relief, and to distinguish the wind-induced dust-sand flows and dust storms [1, 4]. But so far they have not been used for studying relatively stable features of the earth's relief.

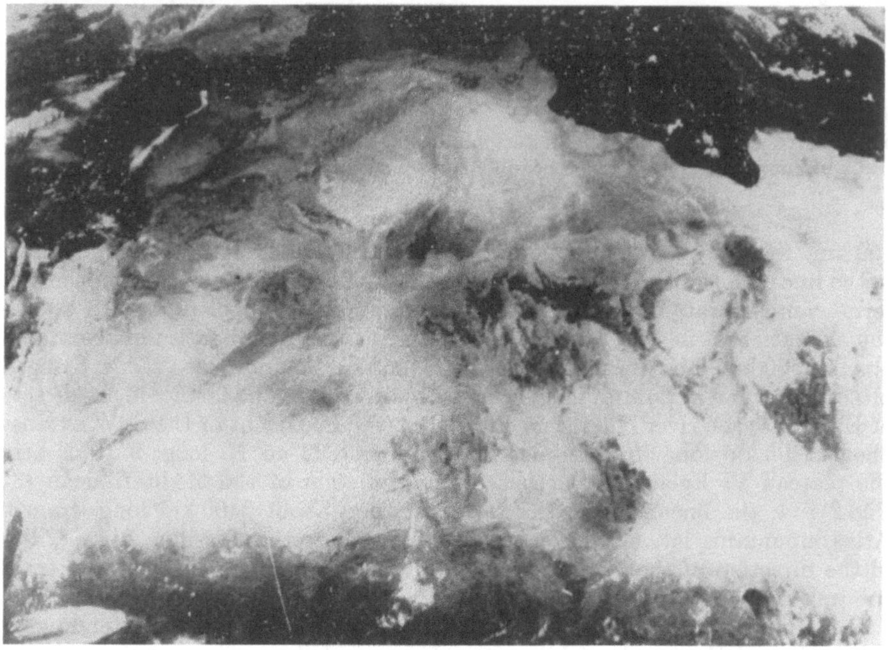

Fig. 1. A fragment of a global picture of the African continent as obtained on 21 September 1968 from the automated interplanetary station Zond 5 at a height of about 90000 km.

Analysis of a photograph of North Africa from Zond 5 (Fig. 1) has permitted the detection of several tens of lineaments. Lineaments are characterized by the linear elongation of a photograph of sharply outlined ridges, ledges, junction zones of sand, clay and stone plains and plateaux. Only one third of the lineaments (the smallest in extent) can be reliably recognized earlier. Corresponding to these lineaments in the small-scale geological maps are the broken tectonic destructions [2, 3].

Among the 30 lineaments shown in Fig. 2 are those known earlier including the Atlas break (labelled 6 in Fig. 2) smaller breaks within the elevation Ahaggar and a number previously unknown. They extend over great distances and cross different landscapes. The length of each lineament shown in the map exceeds

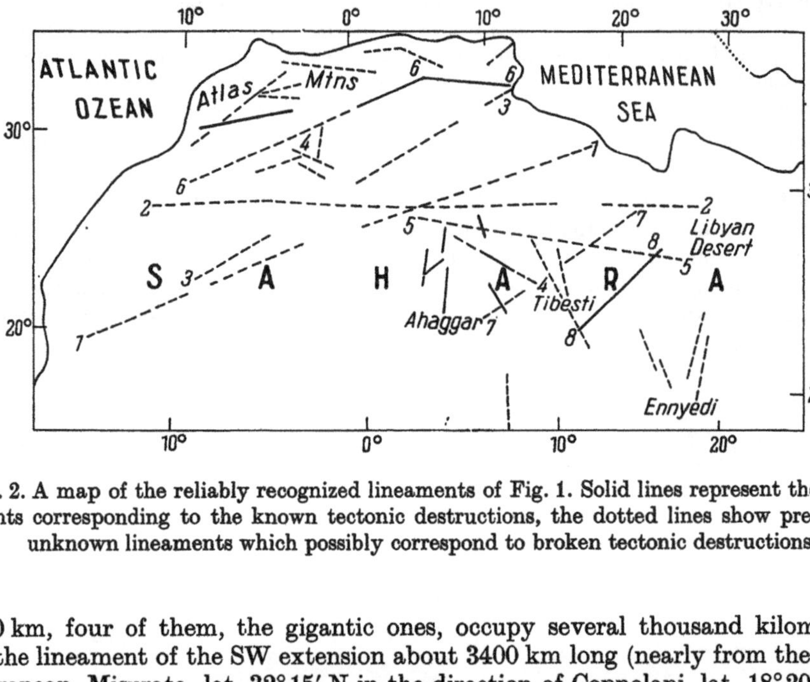

Fig. 2. A map of the reliably recognized lineaments of Fig. 1. Solid lines represent the lineaments corresponding to the known tectonic destructions, the dotted lines show previously unknown lineaments which possibly correspond to broken tectonic destructions.

150 km, four of them, the gigantic ones, occupy several thousand kilometres: 1, the lineament of the SW extension about 3400 km long (nearly from the Mediterranean, Misurata, lat. 32° 15′ N in the direction of Coppolani, lat. 18° 20′ N on the Atlantic) (1 in Fig. 2); 2, the lineament of the sub-latitudinal extension about 3500 km long (nearly from the Atlantic, Bojador, lat. 26° N along the plateaux Seguiet el Hamra, Tinghert, Tademait, es Soda and farther in the direction of El Minya, the river Nile) (2 in Fig. 2); 3, the lineament of the SSW extension about 2300 km long (from Aguet Lgoumair lat. 22° 45′ N, long. 9° 15′ E across the plateau El Eglab and farther in the direction of the Mediterranean (3 in Fig. 2); 4, the lineament of the NW extension about 3400 km long (from the Atlas mountains lat. 32° 15′ N long. 2° 45′ E along the Tassilin, Ajjer, Tibesti, in the direction of the plateau Dorfur (4 in Fig. 2), for whose south-east tracing use was made of a Zond 7 photograph.

It is impossible to notice such linearly oriented land features extending over hundreds and thousands of kilometres from airphotos and, even more so, from ground observations. This is explained, not only by their extensive length, but also by their considerable width, from several kilometres to several tens of kilometres. In many cases the revealed lineaments, those of large extent in particular, can not be observed in earlier photographs from manned space vehicles. The difficulty of detecting these lineaments is due to their becoming "lost" in desert regions. It is only in near-global photographs that one can detect the signs of their intersections with sands (as from a Zond 5 photo of the Erg Chech sand areas and the desert Macteir).

References

[1] B. V. Vinogradov, Global Photography of the Earth, Zemlya i Vselennaya No. 1, 1970.
[2] Carta tectonique international de l'Afrique. 1 : 5000000. Paris, 1968.
[3] Geological map of Africa. 1 : 5000000, Paris, 1963.
[4] A. A. Grigoryev, V. B. Lipatov and B. V. Vinogradov, Space Research XI, 727 (1971).

Space Research XIII — Akademie-Verlag, Berlin 1973

REMOTE SOUNDING OF WATER SURFACE CONDITIONS FROM ABOARD ARTIFICIAL SATELLITES

V. V. MELENTYEV and YU. I. RABINOVICH

Voyeykov Main Geophysical Observatory, Leningrad, USSR

Calculations of radiative coefficients for smooth surfaces of fresh and sea water at different temperatures and salinities for a number of wavelengths in the centimetre band have been performed and an expression obtained for the microwave brightness temperature of the system "rough sea — cloudless atmosphere".

A technique for laboratory measurements of the emissivity of different natural surfaces in the microwave region has been worked out.

Multichannel aircraft measurements of the rough sea surface microwave radiation with varying viewing angle show that sea roughness can be observed from aboard satellites.

A temperature chart of the Caspian Sea has been constructed according to the data on microwave radiation at 3.2 cm wavelength.

At the present time planetary investigations of meteorological characteristics of the atmosphere and the underlying surface by means of measuring the electromagnetic radiative field in different parts of the electromagnetic spectrum from aboard artificial earth satellites acquires more and more importance. For the interpretation of results of such investigations one needs data on the radiative and reflective properties of surfaces over which the measurements are performed. Theoretical and experimental research have shown that in the centimetre and millimetre regions the radiative ability of natural surfaces including that of the planetary ocean is a strongly variable characteristic.

Calculations of radiative coefficients for smooth surfaces of fresh water and sea water at different temperatures and salinities for a number of wavelengths in the centimetre band were performed. It appeared that the radiation of fresh water and sea water is mainly influenced by its temperature, while the influence of salinity shows itself in the long wave region of the band in question.

The results of calculations on the dependence of the rough sea surface microwave radiation for the characteristics actually observed are given. (The calculations, carried out earlier, of the rough sea surface microwave radiation assuming exclusive dependence on the wind speed over the sea, can not describe the phenomenon completely, because of the unrepresentativeness of wind.)

In the calculating scheme it is assumed that the surface of the rough sea can be described as the sum of an infinite number of elementary plates, oriented differently. As far as the microwave brightness temperature of the surface depends on the angle of vision, the differently sloping plates produce different inputs to the total radiation of the surface. Thus the dependence of the sea surface radiation is reduced to the dependence of radiation on the sloping angles of the infinite number of the surface elements, in other words on the function of the distribution of the rough sea surface slopes. An expression for calculating the radiative ability of the rough water surface was found. It is determined by the degree of development of sea roughness (the function of distribution of slopes), the wavelength of microwave radiation, the type of polarization, the dielectric constants of water surface at this wavelength, and hence the sea temperature and salinity, and the angle of vision.

When the sea is rough there appears a difference between the microwave radiation viewed vertically at horizontal and vertical polarizations, this difference growing with the roughness of the sea. The correctness of the model chosen for the radiative transfer in the centimetre band over the rough sea surface is proved by the fact that the calculated values of the microwave brightness temperatures appeared to be close to the results of aircraft measurements.

A technique for laboratory measurements of the radiative ability of different natural surfaces in the microwave region has been worked out, these values being necessary for the interpretation of the satellite data.

The measurements of the dependence of the radiative coefficients of a smooth water surface on the water temperature and salinity at 3.2, 1.6 and 0.8 cm wavelengths showed good agreement with the theoretical calculations.

The influence which the formation of foam or thin films of organic matter has on the microwave radiation of the sea surface has been considered. Then it is possible to determine the parameters of wind-induced waves and also to evaluate the degree of water surface pollution. In order to estimate the influence of foam and organic films on the microwave radiation, laboratory measurements of foam, kerosene and oil film samples on a water surface were performed. The increase of the microwave brightness temperature of a water surface covered with kerosene and oil films are 11 and 29°K respectively, if compared with the radiation of fresh water having the same temperature at the wavelength of 3.2 cm. For the samples of egg and soap foam at the same wavelength this increase was 143 and 152°K respectively.

Multichannel aircraft measurements of the rough sea surface microwave radiation at different viewing angles showed that at all the wavelengths, with horizontal polarization an increase of the rough sea surface microwave brightness temperature is observed. Such a dependence remained the same for all the observed cases of sea roughness. The rise of the wave height up to 1.5—2 m increases the microwave brightness temperature of the sea at the 1.6 cm wavelength with horizontal polarization by 8—10°K.

It is concluded that it is possible to evaluate the development of sea roughness from aboard artificial satellites by means of microwave radiation measurements.

Recommendations have been given for the choice of wavelength for the temperature survey of the sea surface, and a temperature chart of the Caspian Sea constructed according to the data on microwave radiation at 3.2 cm wavelength is available.

Space Research XIII — Akademie-Verlag, Berlin 1973

EXPERIMENTAL VERIFICATION OF PROPOSED SKYLAB ALTIMETER MEASUREMENTS

B. S. YAPLEE, A. SHAPIRO, E. A. ULIANA, D. L. HAMMOND and K. J. CRAIG

E. O. Hulburt Center for Space Research Naval Research Laboratory,
Washington, D. C., USA

The S-193 experiment in Skylab 1 includes a radar altimeter for the remote sensing of the earth's surface features. The expected performance of the altimeter will permit the resolution of ocean wave heights and the profiling of the mean sea surface. In order to evaluate the potential capability of the Skylab altimeter as a height sensor, a series of experiments has been conducted using nanosecond radars on tower and aircraft platforms to simulate the satellite altimeter performance.

The results of preliminary measurements indicate that the Skylab altimeter, assuming no additional instrumental range noise, should be able to measure geoidal and oceanographic height variations of the order of 10 cm over 10 km paths.

1. Introduction

The first radar altimeter for remote sensing of the earth's geometric surface properties will be aboard the Skylab satellite in the first quarter of 1973. The high range resolution mode of the altimeter will be used over the ocean for the direct determination of ocean topography and sea state, with an expected height measurement precision of the order of centimeters. However, the altimeter range precision is affected by the sea surface roughness, and thus an understanding of the effects of ocean waves on the radar return is needed.

In 1969 NRL initiated analytical studies [1] and a series of radar experiments to provide basic information on the interaction of narrow radar pulses with the rough sea surface. This paper will describe the NRL nanosecond radar ocean measurements from both fixed [2, 3] and moving platforms and the results will be used to predict the Skylab altimeter performance.

2. Preliminary Results

An analysis of pulse limited satellite altimeter operation was made in [1] which showed that

(i) the radar return of a narrow pulse altimeter will contain information on the electromagnetic distribution of the observed sea surface (this is defined as the radar energy that is reflected from the sea surface as a function of the wave height when a narrow radar pulse impinges at normal incidence on the water surface and resolves the vertical wave structure),

(ii) data processing methods for extracting the electromagnetic centroid and the rms width of the electromagnetic distribution need to be developed,

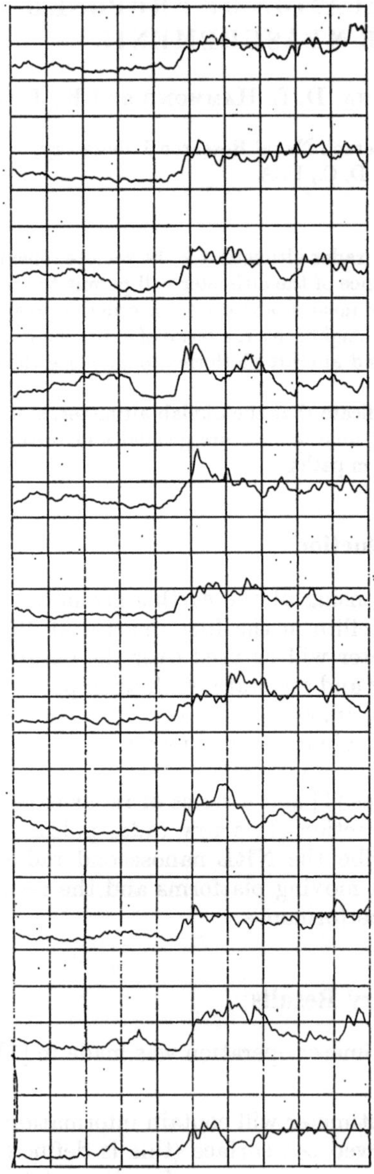

Fig. 1. Individual radar returns.

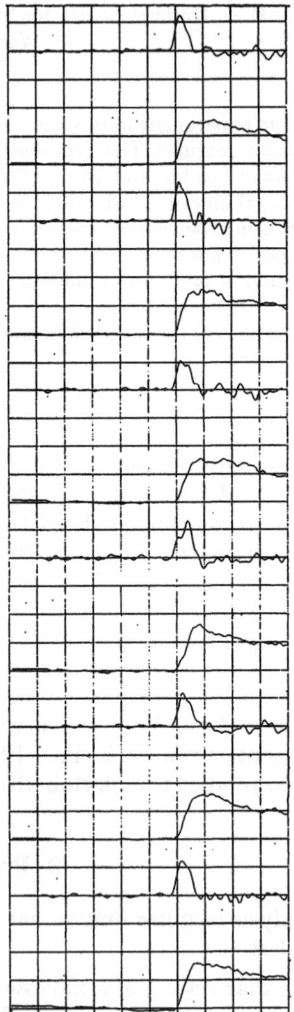

Fig. 2. Averages of 60 radar returns and their derivatives.

(iii) in order to relate the above parameters with the geometric surface parameters of height and sea state, measurements of the electromagnetic distribution for different sea states are required, and

(iv) experimental confirmation of assumed statistical and geometrical reflection properties is needed to verify the expected altimeter performance.

The electromagnetic distributions of the sea surface were determined from 1-ns radar tower measurements for different sea states. The tower measurements [2, 3] indicated that the geometric and electromagnetic distribution can be approximated by similar normal distributions for moderate sea states, but that the electromagnetic centroid was slightly shifted toward the troughs (about 5% of significant wave height). Thus relative height measurements can be made with high precision, as long as sea conditions remain uniform and moderate sea states are encountered.

3. Aircraft Altimeter Observations

Aircraft 1-ns radar measurements, made at an altitude of about 12 000 feet, simulated at least partially the geometric and instrumental conditions of the Skylab altimeter operations. The operating parameters of the two systems are compared in Table 1. It is seen that the combination of platform height and antenna

Table 1

Aircraft and Skylab altimeter parameters

Parameter	Aircraft	Skylab
Radio wavelength	3.1 cm	2.16 cm
Pulse width	1 ns	10 ns
Platform height h	3.7 km	370 km
Pulse width half angle	10^{-2} rad	3.2×10^{-3} rad
Antenna beamwidth half angle	6.5×10^{-2} rad	6.5×10^{-3} rad
Time delay for pulse width-antenna beamwidth match (3 dB point)	40 ns	40 ns
Typical spot size	200 m	3.4 km
Pulse repetition rate	90 pulses/sec	100 pulses/sec

beamwidth matches the two systems at the 40 ns delay relative to the initial contact with the water waves. For moderate sea states, where the dominant water wavelengths will be considerably less than the 200 m spot size, the basic statistical description of the sea, as seen from the aircraft, will approximate the satellite altimeter operation over larger spot sizes.

A series of individual 1-ns radar returns is shown in Fig. 1. The general shape of the radar return conforms with the pulse limited geometry, but the fluctuations appear to be smaller than expected from a purely random sea clutter process.

4. Data Processing

The data processing methods developed in [1] for obtaining the electromagnetic centroid require an *a priori* estimate of the position and width of the rise time to bracket the ramp. The Skylab altimeter obtains this information from a range tracker. For the aircraft measurements, the initial position estimate was obtained from the derivative of averaged radar returns (Fig. 2) where the averaging periods were small enough to assume linear vertical aircraft motion, but long enough to reduce amplitude fluctuations for reliable computer processing. The derivative method reduces the uncertainties of where the plateau region begins.

The centroid values obtained from the derivative method were used in the direct method as described in [1] to obtain the electromagnetic centroid for each individual radar return. The resultant height values were used in a linear least squares to provide the slope and the initial, middle and final delay of the electromagnetic centroid over 1-second periods.

5. Discussion of Results

The internal consistency of the 1-ns radar measurement for a relatively low sea state (4-foot significant wave height (SWH)) is indicated in Fig. 3 by the scatter of the points at 1-sec intervals, for groups of 90 radar returns. The circle represents the initial value of the linear least squares fit to 90 range values and the dot the final value of the linear least squares fit to the preceding 90 range values. The computed rms error at the end points of 8 cm is conservative, since it depends on

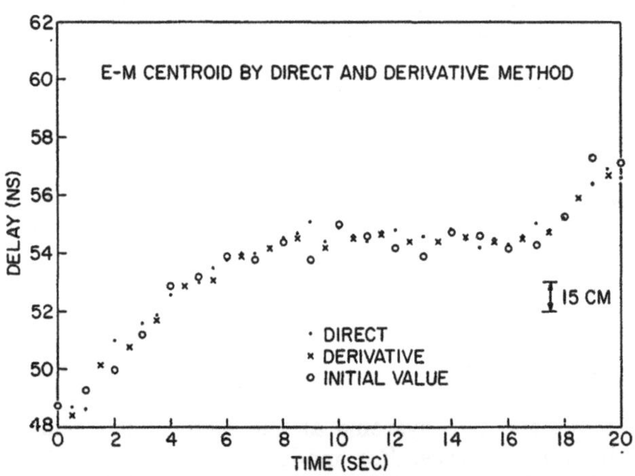

Fig. 3. Scatter of heigth measurements of aircraft motion.

the error of the computed slope. The long-term variations in delay show that the dominant periods of the vertical aircraft motion are longer than 10 seconds, and confirm the linear assumption over the 1-second averaging period. The figure also shows the values of the electromagnetic centroid as obtained from the derivative of the radar return (indicated by the crosses) at the half-second positions. The differences between the derivative and direct method values are small compared with the scatter of the measurements.

The 10-ns capability of the Skylab altimeter was simulated by integrating the 1-ns radar returns over 10-ns time intervals and computing the electromagnetic centroid as before, but utilizing the smaller number of effective sampling gates.

The 10-ns intervals were also shifted by 5 ns to determine the effect, if any, of the particular sampling position on the height precision.

The altimetry measurement precision for different conditions of sea state and pulse width are summarized in Table 2. The first column gives the sea state in terms of the radar significant wave height as obtained from the derivative method and shows an rms error of about 10% for 1-second samples. No direct ground truth measurements were available but visual observations from the tower in the general vicinity of the observed area agree with the measured difference in significant wave height.

The third column gives the measured rms height error of individual radar returns for different conditions of sea state and pulse width and column 4 provides

Table 2

Precision of aircraft altimeter measurements

Radar SWH (ft)	Pulse width (ns)	$\sigma_{hm/pulse}$ (cm)	$\sigma_{hc/pulse}$ SNR $\gg 1$ (cm)	σ_{hm} (90 pulses) (cm)	σ_{hc} (90 pulses) SNR $\gg 1$ (cm)	$\sigma_{(Der\text{-}Dir)}$ (90 pulses) (cm)
4.1 ± 0.5	1	27	27	8	6	2
	10	48	120	13	25	5
	10 (shift)	50	120	20	25	7
6.7 ± 0.6	1	39	34	12	7	4
	10	52	120	17	25	7
	10 (shift)	48	120	18	25	6

σ_{hm} = "height measured" uncertainty; σ_{hc} = theoretically derived "height computed" uncertainty; $\sigma_{(Der\text{-}Dir)}$ = height difference error obtained by processing the data in two different manners and comparing their derived heights. The Der applies to processing the derivative of the radar return and Dir applies to the processing of the direct radar return.

the corresponding predicted accuracy based on the scattering model used in [1]. The results show that for 1-ns pulse widths, the measured precision for one radar return is comparable with the expected precision of 27 cm. The 10-ns precision is considerably better than predicted. This is due to the smoothing effect of simulating 10-ns pulse returns by integration of 1-ns pulses within a 10-ns range window. Finally the error increases with increasing water wave heights for the 1-ns pulse width, but remains essentially constant for the 10-ns pulse width since the water wave heights are barely resolved.

Columns 5 and 6 show the corresponding precision at the end points of 1-second intervals. The expected increase of precision by about a factor of 5 is not realized, and is more like a factor of 3. However, in spite of the degradation, the effective precision still exceeds the theoretically predicted values from [1] for 10-ns pulse widths and shows potential errors of about 15 cm for 90 pulse averages. This will be improved by a factor of 2, if *a priori* vertical motion of the altimeter platform is available.

In the last column the rms deviation between the 1-second range values obtained from the derivative and direct method is given. While the error is a function of pulse width and sea state, it is small compared with the measured precision. This confirms the validity of the normal distribution model discussed in [3].

6. Conclusions

The preliminary results of the NRL altimetry program for a limited number of sea states, when extrapolated to the 10-ns Skylab altimeter, can be summarized as follows:

1. A range precision of about 50 cm for each radar return appears feasible.
2. Range variations of the order of 20 cm are discernible over 1-second intervals (about 7 km distance) and should be reduced to 10 cm, if *a priori* vertical motion of satellite is entered into the signal processing.
3. A normal electromagnetic distribution of sea states of up to 2 m of significant wave height can be assumed.

Based on the above studies, and assuming no additional instrumental range noise, the Skylab altimeter should be able to measure mean ocean surface slopes such as those exhibited by the Puerto Rican trench or the Gulf Stream.

Acknowledgment

The authors wish to express their appreciation to Mr. J. McGoogan of NASA-Wallops Station and his staff who provided fiscal and logistic support, which made the aircraft and tower measurements possible.

References

[1] A. Shapiro and B. S. Yaplee, NRL Rep. No. 7018 (1970).
[2] B. S. Yaplee, A. Shapiro, D. L. Hammond and E. A. Uliana, Symp. on Remote Sensing of Environment, Willow Run Laboratories, Ann Arbor, Michigan (1971) Vol. III, p. 1879.
[3] A. Shapiro, E. A. Uliana and B. S. Yaplee, Joint NOAA/NASA/NAVY Conf. "Sea Surface Topography from Space," Key Biscayne, Florida (1971) NOAA Tech. Rep. ERL 228-AOML 7, p. 11—1.

Space Research XIII — Akademie-Verlag, Berlin 1973

MULTI-STAGE ACQUISITION OF FOREST INFORMATION FROM SPACE AND AIRCRAFT IMAGERY AND GROUND SAMPLING

K. KUUSELA and S. POSO

Forest Research Institute, Helsinki, Finland

To develop a forest-inventory technique based on multi-stage acquisition of information, two experiments are reported. Pictures taken by the meteorological satellite Essa 8 on 21 March 1969 from an altitude of 1500 km and on a scale of 1 : 13 million were used in an experiment covering 17 million hectares. A highly significant regression was found between the darkness grades on the picture and the mean volume of the forest growing stock per area unit for the aggregate of land and inland waters. In the second experiment, a systematic sample of photo-interpretation plots was located on aircraft pictures taken from an altitude of 9.5 km at 1 : 60000 on an area of 2.8 million hectares. The photo plots were combined into groups which were homogeneous in regard to the interpreted forest characteristics. A random plot from each group was selected and measured in the field. Photo plots were furnished with data from the ground plot in each particular group and final forest-resource estimates were calculated for areas formed from any set of photo plots.

1. Introduction

The possibility of recording electromagnetic radiation with devices carried by aircraft and space vehicles has led to a new phase in the development of methods for the inventory of natural resources. The imagery from space vehicles and aircraft makes it possible to achieve results which otherwise were impossible or very expensive. It also allows methods which can in part substitute expensive field work by imagery.

In the countries where regular forestry has been practised longest, the inventory of forest resources is generally carried out by sampling based solely on field observations. In countries where regular forest policy started after World War II, the method based on two-phase sampling with photo and field estimations (double sampling) has often been applied.

Photos taken from spacecraft enable forest resources to be stratified. Then estimations are made based on interpretation of photographs taken from aircraft. The more important the forest area, the lower is the flying altitude which becomes feasible. Observations are then made in the field. Most weight is given again here to the most valuable forest areas. The objective of applying the sampling is to obtain the desired data with the minimum total cost, thus multi-stage sampling is especially suitable for the inventory of large forest areas which are not under commercial forestry. Most of these areas are found in Canada, the Soviet Union, South America and Africa.

2. Regression of Forest Resources from Space Imagery

The pictures used in the experiment were taken by the meteorological satellite ESSA 8 on 21 March 1969 (Fig. 1). The scale of the pictures was roughly 1:13 000 000 and the altitude of the spacecraft was 1500 km. At the moment of recording, the skies were clear. The depth of snow all over Finland exceeded 20 cm and the trees were free of snow. Thus, forests dominated mainly by coniferous trees made a clear, dark contrast against the white, snow-covered ground.

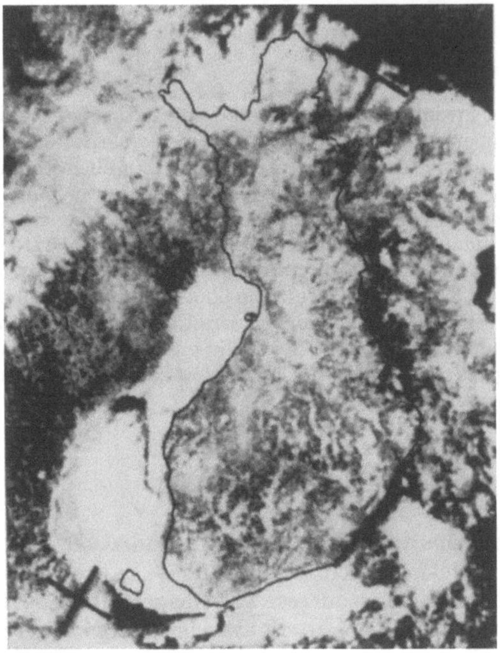

Fig. 1. Satellite picture from ESSA 8 on 21 March 1969. The length of Finland measured on the original picture was 85 mm.

In south Finland was an area of 17 million hectares, divided into 46 regions, where the mean volume of the forest growing stock per hectare for the aggregate of land and inland waters was known on the basis of a ground sample measured in the period 1964—1968. The total area was divided into a grid of squares each covering 10 000 hectares. The darkness on each square was analysed on six pictures in two grades and a mean darkness estimate was calculated for each of the sub-regions mentioned above. Several models of regression of the mean growing stock volume on the mean darkness and distance of the sub-region from the photo centre were tested. The best model gave the coefficient of multiple correlation a highly significant value of 0.84.

The experiment justified a conclusion that regions having varying mean volumes of growing stock can be delineated on the basis of space imagery. The regression equations obtained from the test area were applicable for estimating volumes of growing stock in adjoining regions.

3. Combining Air Photo and Field Sampling by a "Grouping Method" for Forest Inventory Purposes

The grouping method can be regarded as a special case of double sampling for stratification. The method was experimented with in northern Finland in an area consisting of 2.8 million hectares (about 7 million acres). The photos applied were taken with black-and-white panchromatic film from a flying altitude of 9.5 km on a scale of 1:60000.

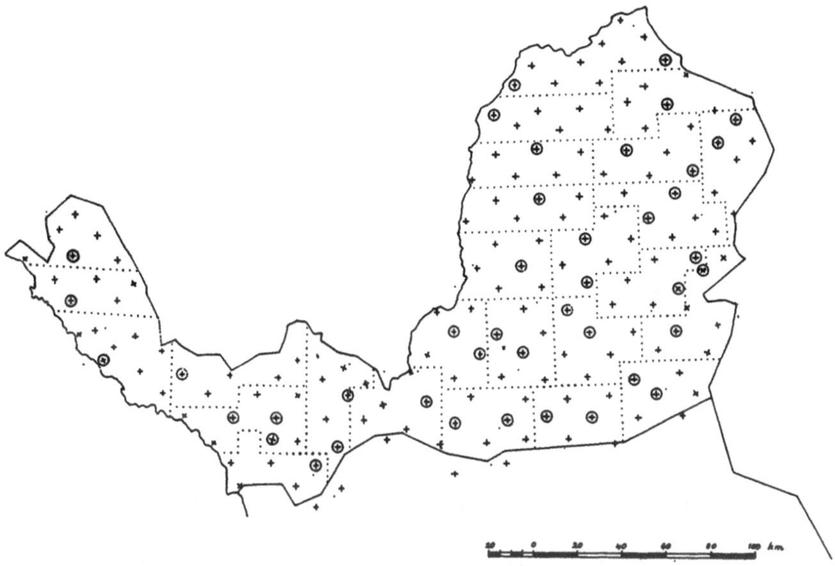

Fig. 2. The area subject to inventory. The crosses indicate the centre points of the aerial photographs on which the sample plots were photographed. A circle denotes a photograph which has been selected for a field check.

A systematic sample of 15441 points was placed on aerial photos. Each point defined a variable plot (relascope plot where the diameter of tree at breast height was at least 1:50 of the distance of the tree from the centre point). For each plot, the desirable characteristics describing the soil and growing stock were interpreted from stereoscopic pairs. The photo plots were then grouped into homogeneous groups in respect of all photo-interpreted characteristics. From each group, one plot was drawn by random selection for measurement in the field. Consequently, the number of groups was 659 and, according to the principles of the grouping method, this equalled the number of field sample plots.

The plots selected for field measurements were located in the field from photos. The location proved to be fairly easy. It was found that a photo point could be placed in the field in a few minutes with an accuracy of about 5 metres. All the necessary parameters were measured independently for each selected sample plot in the field.

The data of a field plot was then extended to all photo sample plots belonging to the respective group. Accordingly, each photo sample plot was supplied with as complete data as measured for field plots. After the extension, the new data of a photo sample plot could be used as an independent record. The new records of photo sample plots were inserted on to magnetic tapes for computer processing. The necessary estimates of parameters of forest resources could be easily calculated on the basis of photo sample plots for all desirable sub-areas as well as for the total area subject to inventory (Fig. 3).

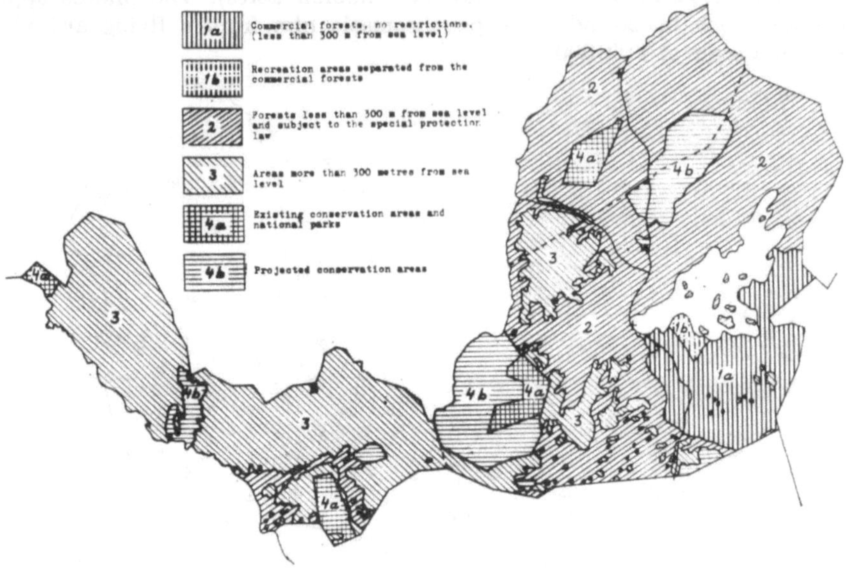

Fig. 3. The sub-areas for which the independent inventory results were calculated.

The two-phase grouping method described above is especially advantageous because of its good applicability to various conditions. The area subject to inventory can be divided before or after the data collection into desirable subdistricts. Their minimum size is restricted only by the precision required for the results. The precision, again, is largely dependent on the number of photo sample plots. Thus, it is possible to increase the precision of sub-areas to a certain level at relatively small expense by increasing the number of photo sample plots.

4. Outline for Multi-stage Sampling

On the basis of the two methods described above, it is possible to construct a multi-stage sampling method which is applicable to forest inventories over large areas where the resources are not well known. In the first stage, the satellite pictures (or other kind of radiation registers) should be analysed in respect of those characteristics whose interpretation results correlate distinctly with the true measures, which, again, are regarded as important for estimation. It is recommended that the sampling units applied are recorded according to a coordinate system on the ground. If correlation between remote recorders and field measures was known on the basis of earlier studies, it would be possible to calculate rough estimates of forest resources. In any case, it is possible to get information on the regional distribution of forest resources.

Aerial photography, photo sampling and its interpretation is planned and performed according to information concerning the regional distribution of forest resources. The most valuable regions are given the highest weight. The photo sample units (photo plots) are formed into homogeneous groups. The exact data for a photo plot of a group will be obtained from a plot measured in the field as described above. As every photo sample plot is supplied with horizontal and perhaps also with vertical coordinates, it is easy to form geographical sub-areas. For these sub-areas, or strata based on other criteria, it is possible to calculate results and to plot them on maps in the same way as has been done on the basis of systematic field sampling in the Finnish national forest inventories.

The application of the data records obtainable repeatedly through spacecraft sensors is especially useful for following the changes in forest resources in developed forestry areas. It is possible to discover the regional changes caused, for instance, by cutting, planting and seeding, growth and damage. After homogeneous areas in respect of the changes have been outlined in the first phase, the exact estimates will be derived by aerial-photo interpretation and field observation.

References

[1] K. Kuusela and S. Poso, Photogrammetric J. Finland **4**, 1 (1970).

[2] P. Langley in: Proc. Sixth Int. Symp. on Remote Sensing of Environment 1969, Ann Arbor, Mich. (p. 1179).

[3] S. Poso and M. Kujala, Folia Forestalia 132, Helsinki 1971.

[4] S. Poso, Comm. Inst. Forestalis Fenniae **76**, 1 (1972).

Space Research XIII — Akademie-Verlag, Berlin 1973

USE IN ISRAEL OF REMOTE SENSING FROM SATELLITES

J. H. JOSEPH

Department of Geophysical Environmental Sciences, Tel-Aviv University, Ramat Aviv, Israel

An outline is presented of the research and other activities carried out in Israel in connection with the utilization of the ERTS and Skylab-EREP Satellite data. The decisive factors in the choice of topics were the existence of active research and of expertise in research subjects in which the large-scale surveys to be provided by the earth resources satellites can further scientific knowledge and in which the experts are willing to incorporate the new surveying tool in their work. The national coordination of the research will be carried out by an organizing committee, the members of which are versed in the scientific, economic and other aspects of the planned surveys. One single center is being set up for data acquirement, handling, storage and dissemination.

1. Introduction

The future results of application of the technology and science embodied in the NASA Earth Resources Program can be put to varied and numerous uses [1—3]. The user's problem is the judicious and intelligent choice among the many possible applications. It is of interest to discuss here the organization set up in Israel to serve as the ultimate receiving antenna — the local non-NASA user — its rationale and the ways in which the applications most useful to Israel were chosen.

2. Choice of Research Topics

It was decided to design experiments only in fields in which there existed active local expertise, and in which active ground-based or airborne pertinent research has been in progress for some time. The active experts were asked: "Are you willing to allot part of your time to testing and using this new tool and apply it to your research? And if so, what are the problems in your field of active interest to which you would want to apply this new tool?" A number of specific areas of application could be listed in this way. The applications break up into two main families: a) applications of direct and immediate regional or national usefulness, where the ER satellites could add to knowledge of local conditions; b) applications that would help detemine the feasibility of using ER satellites as additional aids in several of the many disciplines proposed by NASA.

3. Outline of the Set of ERTS-A Proposals submitted to NASA

A common deminator of all proposals should be emphasized here: In the Israeli ERTS-A proposal, it is proposed that the multispectral scanner system be operated over Israel and the imagery provided to Tel-Aviv University for analysis. The analysis, and especially the utilization, will in most cases be jointly with other institutions.

(a) *Hydrology* in arid and semi-arid regions.

Important benefits are expected from mapping of the decrease in reflectivity due to moisture in the soil over areas in the arid and semi-arid Negev as a result of rainfalls or subsurface water. In order to delineate these potential water sources it is advisable to have information about the channel reaches where high infiltration is likely to take place. A possible technique, already implemented for obtaining this information is by taking records of rainfalls and the extent of the resultant flood flows. Owing to the aridity of the area, the available data collected on the surface are insufficient for a good hydrological evaluation of the water sources beneath the river channels. These data may be much improved by a series of ERTS-A pictures.

It seems that the present research project may lead toward the development of a method for obtaining the necessary data by the identification of the wet soil.

(b) *Agriculture* — Citrus and Wheat

The multispectral scanner imagery, with its very important reflective infrared band, will be used to map the extent of agricultural species, with attempts for vigour and health assessment, yield prediction and infestation detection. Primary attention will be given to two important crops: citrus and wheat.

A laboratory study into the effect of agricultural practice on the spectral reflectivity of citrus leaves started at the Volcani Institute of Agricultural Research in 1969. The effects of leaf pigment and water content on reflectivity are being measured and structural details are also being studied [4—7]. Later field studies will examine the effect of different cultivation practices and diseases on the spectral reflectivity of individual tree crowns and small areas of plantation canopy. The measurements might be regarded as establishment of keys or signatures for ERTS-A imagery.

Israel plans to increase considerably home wheat production, both by expanding the area currently devoted to the crop (1000 km^2) and increasing productivity. The vast majority of current production is on non-irrigated land, but much research has been devoted to the problem of profitable expansion to more arid areas where irrigation will be needed. Remote sensing could be of use in assessing nutrient and water requirements and predicting yields. The latter application is important to the operation of the drought insurance scheme by the Israeli Government.

In 1969 observations on the solar reflectivity of wheat crops were made, investigating the effect of age, variety, fertilizer and irrigation treatment on reflection in the visible and emissivity in the infrared spectral regions [5, 6]. The radiometric measurements were supplemented by aerial photography, assessing various film-filter combinations densitometrically. It is planned to supplement these observations with narrow band studies in both the field and laboratory.

(c) *Forestry and Arid Regions Vegetation*

Israel since its inception has been conducting an intense afforestation program. By now, the young forests cover hundreds of square kilometers. Monitoring the growth of these young forests from satellite altitudes will prove very helpful in further planning of the afforestation program. The study of desert vegetation is of scientific interest and at the same time has practical application, since it can help land use recommendations and uncover potential sites for planting special "desert crops" such as date palms. The intensity of desert vegetation when the canopy covers a measurable fraction of the ground can be assessed by reflectivity in the near infrared region. Ground-truth observations of some chosen sites will be carried out by the Hebrew University, at the Department of Botany.

d) Oceanography

The fishing industry in the Eastern Mediterranean plays a significant part in supplying protein to the Israeli diet. The industry is faced with considerable problems, one of which is the changes in the flow and nutrient contents of the Nile water after the Aswan Dam was put into use. It is proposed to attempt to map the ocean colour in the Eastern Mediterranean, and to correlate these maps with reports of catch by the fishing fleet.

It is expected that this activity will be closely correlated with mapping of upwellings and currents through thermal infrared scans by Skylab-EREP in late 1972—early 1973. Since those scans will be regrettably infrequent, thermal infrared maps of Eastern Mediterranean by Nimbus, though of poorer resolution, will also be analysed.

e) Limnology—Pollution in Lake Kinneret

The 170 km^2 lake forms the major reservoir in the National Water Carrier scheme and supplies one third of the national water supply. It is also an important tourist attraction and supplies a significant part of the fish production. Intensive agriculture in its watershed area has led to a sharp increase in the nutrient content of the water and algae frequently develop, accompanied by a marked visual change in color and decreases in water quality. Spectral observations might provide a useful method of assessing these changes and also indicate their areal distribution. Very intensive measurements of water quality by various standard limnological methods are routinely available. A few spot observations of broad-band reflectivity have been made and it is hoped to supplement these with more intensive narrow band measurements in the future.

f) Atmospheric Effects

An important part of all investigations will be the assessment of the distortion of satellite measurements by atmospheric interference. Theoretical and experimental studies are being carried out on problems of the mathematical inversion of spectral data and on the effects of clouds and aerosols on the satellite spectra [8—12]. Research on other related problems is also being carried out [13—15].

4. Organization

A National Coordination Committee has been set up, acting as a sub-committee of the National Committee for Space Research. Members of the Coordination Committee are representatives of the Space Committee, participating scientists and participating organizations and institutions. This body will act to control, coordinate and expedite efficient analysis, use and circulation of the data to come from the ERTS and Skylab satellites and from local sources. A national data handling and storage center is being organized and will be affiliated with the Department of Environmental Sciences at Tel-Aviv University [15].

Necessary instrumentation, not available in the country, like a multispectral color projector/viewer for "eyeball" analysis has been ordered. A scanning microdensitometer at the Weizmann Institute will be available on a fractional time basis to digitize imagery for computer data processing. In general, extensive use will be made of computer techniques. The necessary computer routines have been, or are being, developed.

Finally, in order to extend the data analysis and use regionally, common topics of research have been identified and submitted to NASA as a joint proposal by Ethiopian and Israeli scientists.

5. Summary

The important points in the choice of research fields in Israel into which use of the future ERTS results can be profitably integrated are the existence of active research and expertise in subjects in which the large-scale surveys to be provided by the Earth Resources Satellites can further scientific knowledge, and the willingness of experts to incorporate the new surveying tool in their work. The basic concepts of the national coordination of research are an organizing committee that has responsibilities and interests in both scientific and economic aspects and one single centre for data acquirement handling, storage and dissemination.

References

[1] NASA Goddard Space Flight Center. Ref. S-701-P-3 (1969).
[2] Data Users Handbook; NASA ERTS Document No. 71 SD 4249, NASA Goddard Space Flight Center (1971).
[3] G. E. Doc. No. 70SD4226 m (Final Report) (1970).
[4] G. Stanhill, G. J. Hofstede and J. D. Kalma, Quart. J. Roy. Met. Soc. **92**, 128 (1966).
[5] G. Stanhill, J. T. H. Cox and S. Moreshet, J. Appl. Ecol. **5**, 707 (1968).
[6] J. D. Kalma and G. Stanhill, Solar Energy, **12**, 491 (1969).
[7] G. D. Stanhill, Solar Energy, **13**, 59 (1970).
[8] J. H. Joseph, Israel J. Earth-Sciences, **19**, 51 (1970).
[9] J. H. Joseph, Israel J. Earth-Sciences, **20**, 7 (1971).
[10] J. H. Joseph and A. Manes, J. Appl. Met., **10**, 453 (1971).
[11] J. H. Joseph, A. Manes and D. Ashbel, submitted for publication 1972.
[12] J. H. Joseph, D. Guetta and G. Ohring, Israel Met. Service, Dept. of Envir. Sci., Tel-Aviv Univ., Interim Rep. Contr. No. 8309000 p. 44 (1972).
[13] J. Otterman, J. Brit. Interplan. Soc. **23**, 349 (1970).
[14] J. Otterman and R. Shilo, Israel J. Earth-Sciences **20**, 89 (1971).
[15] J. Otterman, Proc. XIX Rassegna Int. Elettron. Nucl. ed Aerosp. Rome, III-6, p. 269 (1972).

Space Research XIII — Akademie-Verlag, Berlin 1973

UK ACTIVITY IN EARTH SURVEYS USING REMOTE SENSING TECHNIQUES

D. D. Clark

Space Division, Department of Trade and Industry, London, UK

The Space Division of the Department of Trade and Industry is responsible for coordinating activity in UK on remote sensing for earth resources from spacecraft. As the result of a questionnaire it was established that some 17 University groups, 16 Government laboratories and 13 industrial or commercial firms employ remote sensing techniques for earth surveys using aircraft, balloons and in one case Skylark sounding rockets. The Space Division also coordinates UK participation in the NASA ERTS and Skylab programmes. Eight groups are involved in the ERTS-A and EREP programmes and some 20 groups have already declared their interest in participating in the ERTS-B programme. The possibilities of instrumenting an aircraft to operate in conjunction with the ERTS-B satellite, and of equipping one of the existing space ground stations in UK to receive data relayed via ERTS-B from remotely located automatic observing stations, are being investigated.

1. General

Remote sensing methods are employed in the United Kingdom for a variety of purposes and by a diversity of users. For example, conventional and multispectral photography from aircraft are used by the Department of the Environment for planning the use of land and water resources, by the Ministry of Agriculture to identify plant disease and mesoclimatic features and by the Central Electricity Generating Board to monitor thermal pollution of the rivers. Some new investigations involving remote sensing are encouraged by research grants from the Natural Environment Research Council, and, where there is a space implication, support may be given by the Space Division of the Department of Trade and Industry. In an attempt to identify all that is going on and to help in defining future requirements, a questionnaire on remote sensing was circulated to all those in UK likely to be interested. Among the 50 replies received, 17 were from Universities, 16 from Government departments and laboratories and 13 from industry and commerce. The distribution of interest of these groups is shown in Fig. 1. As some groups are pursuing several aims they may therefore appear under more than one heading. The replies also showed that, in addition to remote sensing in UK, work is also in progress on behalf of other governments or international organizations in ten other countries. Some of the work is concerned with mineral resources and is sponsored by industry. In one case there is cooperation between the Fisheries Laboratory at Lowestoft and the French Eole project to fix the position of drifting buoys to measure deep ocean currents. Post graduate courses in the use of remote sensing for earth surveys are available at the Universities of Sheffield, Reading and Aston (Birmingham).

2. Research

In the National Physical Laboratory a sub-millimetre interferometer spectrometer has been constructed which can be used to detect the amounts of pollutant gases present in the atmosphere, such as nitric oxide, nitric acid and sulphur dioxide, to 10 parts in 10^9. It can also be used to detect water vapour and ozone. In its present form this instrument can be mounted on a balloon or in an aircraft and has already undergone flight trials. Other work is being done in Government laboratories, Universities and in industry on instrument development including improved infrared detectors, infra-red line scanning equipment, infra-red sensitive image tubes, infra-red and sub-millimetre bandpass filters, sideways-looking radar and gamma-ray detectors. Research into the problems of data handling and interpretation is being done by the Experimental Cartography Unit of the Royal College of Arts, the Universities of Bristol, East Anglia and Aston and in industry by the Plessey Company.

EASAMS, a member of GEC-Marconi Electronics Ltd is carrying out studies for ESRO on sideways-looking radar and on an aircraft programme for earth resources surveys.

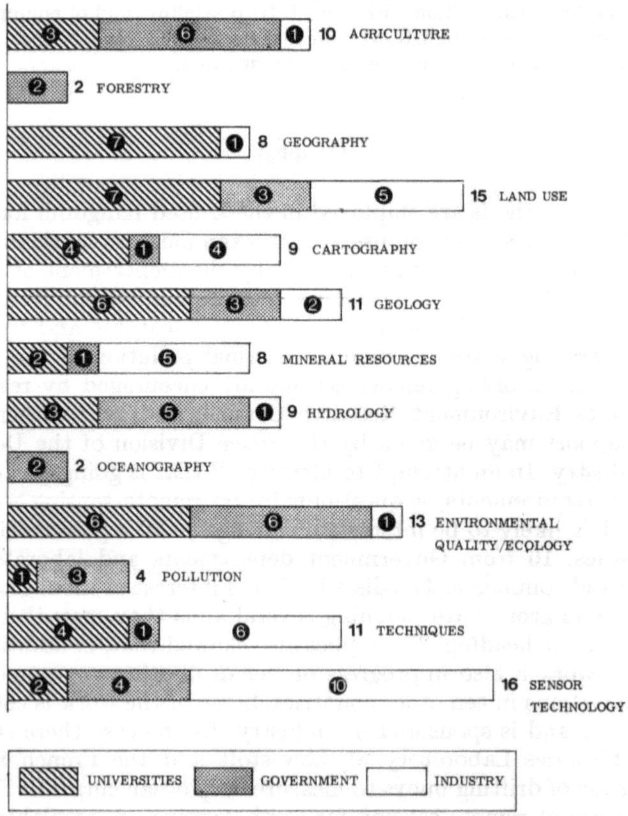

Fig. 1. Distribution of interest in UK in remote sensing methods.

3. Space Applications

In order to catalogue the methods and devices being used in connection with orbital remote sensings of the earth, a contract was given to EMI Electronics Ltd to compile a compendium on the subject which would feature current techniques, details of sensors, test facilities, truth sites, data acquisition and data handling.

Active involvement in space programmes for earth resources consists for the moment in participation in the NASA ERTS and EREP programmes and in the UK national technological sounding rocket programme.

3.1. Satellites

Table 1 lists the UK groups participating in ERTS-A and EREP programmes. Many more groups however have shown interest and so far we have received about 20 preliminary proposals for submission for ERTS-B. In anticipation of a probable UK participation in ERTS-B, we are studying the possibility of equipping an aircraft as a back-up facility for UK ERTS-B participants to obtain spectral reflectivity data from truth sites for comparison with ERTS-B data. In addition preliminary enquiries are being made on the possibility of adapting one of the UK satellite ground stations to receive transmissions from data collection platforms via the ERTS-B satellite when it is launched, should there be a sufficient demand.

3.2. Sounding Rockets

As a means to fill the gap represented by the long lead time of the satellite programmes and because it was felt that there were inherent advantages in the use of sounding rockets for remote sensing, the Royal Aircraft Establishment in 1971 started work on an earth resources sounding rocket programme, based on the Skylark rocket, to incorporate a horizontally stabilized platform. The first flight in this programme using a Skylark with a solid propellant Goldfinch boost motor and a Raven sustainer motor took place successfully from Woomera on 27 March 1972 and reached 297 km altitude. Intended primarily to prove the system, it carried a payload of two cameras selected from those readily available, of which more details are given in Table 2. In order to make most use of the altitude, the cameras were obliquely offset to cover only a segment of the total desired target area; the whole area was covered by causing the rocket body to perform five 60° distinct turns about the vertical axis using the gas jets. In each position sets of photographs were taken.

This programme was carried out in cooperation with CSIRO and the Department of Mines of South Australia and Reading University to obtain data on rock, soil, water, form, plant, land use and spectral reflectivity for comparison with those obtained from aerial survey and ERTS-A. Other campaigns are being studied using different film, filters and lens combinations or other sensing equipment. In particular a cooperative programme with the Argentine Authorities is under discussion whereby two earth resources Skylarks would be launched in Argentina later this year for the purposes of crop census and land use.

Table 1

UK Proposals for participation in ERTS-A and EREP

Institution	Experiment	Site	Discipline
Centre for Overseas Pest Research	Detection of potential locust breeding sites (ERTS-A)	South West Saudi-Arabia	Agriculture Ecology
Experimental Cartography Unit Royal College of Art (Fairey Surveys Ltd)	Experimental cartography using ERTS data (ERTS-A)	South/Central England	Cartography Geography
Hunting Surveys Ltd (George Philip & Son) (Bartholomew & Son)	Cartographic research programme for small scale mapping at 1:1000000 (ERTS-A)	UK	Cartography Interpretation Techniques
School of Environmental Sciences University of East Anglia	Development of computer-based system of generalizing remotely sensed data for unit areas of the earth's surface (ERTS-A)	UK South West USA	Interpretation Techniques Data processing
University of Leeds (Hunting Geology & Geophysics Ltd)	Regional structural and lithological mapping (ERTS-A and EREP)	Ethiopia	Geology
University of Sheffield	Integrated surveys of natural resources in a developing area (EREP)	South East Spain	Geology Geography
Hunting Surveys Ltd (George Philip & Son) (Bartholomew & Son)	Cartographic research programme for small scale mapping (EREP)	Nepal	Cartography Interpretation Techniques

Table 2

Skylark SL 1081 remote sensing equipment

Camera type	Lens	Focal length	Film	Passbands	Ground coverage and resolution
F 24	Ross EMI f/4	127 mm	Kodak plus X Aerographic type 2648	0.41 – 0.57 μm 0.48 – 0.60 μm 0.59 – 0.70 μm	88000 km²/frame 100 m
Hasselblad 500 EL/70 M	Zeiss f/2.8	80 mm	Kodak Aerochrome Infra-red Type 2443 False colour reversal	0.35 – 0.90 μm	40000 km²/frame 150 m

4. Conclusion

Activity in remote sensing has grown, albeit in a somewhat spontaneous manner according to the need, but now with the extension of remote sensing to spacecraft some coordination of effort is taking place. Geo-scientists have held meetings among themselves to discuss common aims and the pattern of future cooperative action. It is furthermore evident that a major effort will be required in the sphere of data reduction and analysis in anticipation of greatly increased flow rates of data associated with the space activities.

Space Research XIII — Akademie-Verlag, Berlin 1973

SATELLITE TECHNIQUES FOR AUTOMATIC PLATFORMS LOCATION AND DATA RELAY

PIERRE MOREL

University of Paris, Paris, France

The EOLE experiment, conducted in late 1971 and 1972, highlights the development of a new investigation technique based on using a large number of small, possibly drifting, instrumented automatic platforms and a satellite for location, data collection and relay. The EOLE experiment involved 500 constant-level balloon sondes floating at the nominal 200 mb pressure altitude and distributed around the southern hemisphere. These balloons are tracers of the general circulation of the atmosphere at 200 mb. Their trajectories, determined by 2 to 6 successive fixes every day, provide far-reaching information on the daily variation of the upper air flow patterns, the eddy kinetic energy spectrum and eddy diffusion due to two-dimensional turbulence, the mean meridional circulation and the horizontal flow divergence at various scales. It is the first time that such a wide Lagrangian observing system has been deployed in the atmosphere.

1. Historical Background

In spite of the exciting cloud cover images obtained with the first Tiros meteorological satellite, and the dramatic successes of the remote probing of planetary atmosphere exemplified by the Mariner 4 Mars occultation experiment (1965), there was considerable doubt in these early days that remote sensing of the earth's atmosphere could provide the kind of very precise *quantitative* data needed for dynamical weather forecasting. In fact, the excellent performances of the infrared spectrometers or radiometers on Nimbus 4 have demonstrated the feasibility of obtaining atmospheric temperature and moisture data applicable to operational weather prediction computations. It remains true nevertheless that some atmospheric parameters cannot be inferred conveniently from remote observation alone and this justifies the need for *in situ* measurements of the wind velocity field and surface pressure, mainly. Also, the more sophisticated coupled ocean-atmosphere models will eventually require sub-surface data such as sea temperature profiles which cannot be determined from a satellite by any conceivable extension of present technology. Now, the deployment and operation of a worldwide network of automatic stations would pose staggering logistics problems if it were not for the concurrent development of constant level balloons which could carry a small automatic weather station around the globe in 10—15 days [1,2] and satellite navigation and communication techniques for locating the stations and collecting their observations.

This idea was first developed by Lally [3] and Giles and Angell [4] who proposed a meteorological observation system based on horizontal sounding with constant level balloons: balloon displacements determined from successive fixes made by an appropriate navigation satellite, would provide exceedingly precise wind velocity determinations, while other meteorological parameters would be measured by the balloon sensors. Considerable attention was devoted to this concept in the United States, culminating with the STROBE study (for Satellite Tracking of Balloons and Emergencies) under the leadership of the NASA/Weather Bureau Joint Meteorological Satellite Advisory Committee (Goddard Space Flight Center, 1962).

The STROBE project was not in fact implemented in the United States but the development of the Interrogation, Recording and Location System (IRLS) was undertaken instead, for limited engineering tests on Nimbus 3 and 4 [5]. Simultaneously, the then newly formed Centre National d'Etudes Spatiales (CNES) of France took up the same concept of a meteorological sounding system for a large scale experiment, to be implemented in the southern hemisphere under the code name of EOLE (unlike other project headings, EOLE stands for nothing but the French name of the mythical god of winds Aeolus). The EOLE experiment consisted in launching, interrogating and tracking 500 constant-level balloons approximately, flying at the nominal flight level 200 mb (about 12000 m altitude) in

Fig. 1. EOLE satellite.

the southern hemisphere lower stratosphere. The location and communication functions are accomplished by the specially built EOLE (Fig. 1) satellite, launched from Wallops Island, Va, USA, on 16 August 1971 as a joint CNES/NASA undertaking. The spacecraft, about 50 remaining balloons and many other ground-based platforms are operating on the day of this presentation.

Another large experiment of this nature, the Tropical Wind, Energy Conversion and Reference Level Experiment (TWERLE) is being prepared by a team comprising the NASA Goddard Space Flight Center, the National Center for Atmospheric Research (NCAR) and the University of Wisconsin in the USA. This experiment is planned to be implemented during the GARP Atlantic Tropical Experiment in summer 1974. It will consist in launching 300 constant-level balloons from near equatorial sites for providing large scale wind data at a 150—200 mb level in the tropics in support of the Tropical Experiment. Each balloon will also carry a very precise radio-altimeter and a barometer, thus providing measurements of the pressure gradient force and the acceleration along the tracer trajectory, for a direct estimate of the potential to kinetic energy conversion process. Finally, as the balloons will hopefully drift to more southerly latitudes, they will provide reference level (pressure versus altitude) data against which remotely measured temperature profiles could be calibrated. The TWERLE experiment design is based on a new random access Doppler navigation and Data Collection System, to be carried by the Nimbus-F spacecraft [6].

Finally, the current planning for the First GARP Global Experiment (1976—1977?) calls for a mixed global observation system which would associate conventional meteorological soundings and remote observations from meteorological satellites with *in situ* measurements by automatic stations carried aloft by balloons or on the sea surface, by drifting buoys (The First GARP Global Experiment, Objectives and Plans, WMO Document, April 1972).

Table 1
Evolution of platform location and data relay systems

Year	68	69	70	71	74	76
System	ATS-1 OPLE	Nimbus 3 IRLS	Nimbus 4 IRLS	EOLE	Nimbus-F Radem	Tiros-N Rad
Number of platforms	10	20	30	500	300+	1000+
Payload weight (kg)	50	45	5	3	1	1
Peak radiated power (Watts)	25	25	6	5	1	2
Cost per platform (US$)	35000	50000	20000	10000	2000	2000

By this time, it has been realized that the satellite data collection technique has many other applications in addition to automatic meteorological stations on balloons or buoys, most appropriately for Tsunami warning, tide measurements, hydrological flow gauges, automated rain gauges, oceanographic measurements by instrumented buoys, data relay from merchant vessels, and so on. This is the reason why several satellite data collection programmes are now underway to be implemented in the 1970's: from geostationary satellites such as GOES/SMS (USA), the Japanese meteorological satellite project and the European Meteosat; and from low altitude orbiting satellites like ERTS A and B (USA) and the operational ESSA satellite series (starting with Tiros N in 1976) under a joint undertaking of the United States and France.

2. Principle of Platform Location Systems

Locating automatic stations from space is an altogether different problem from the satellite navigation techniques applicable to ships: for example, the transponder or beacon on an automatic station is by necessity fairly simple and cannot perform complex functions as expected from a Transit ground processing unit. Thus, one must rely on the spacecraft to accomplish the more sophisticated functions and some data reduction and storage. A corresponding reduction of navigational accuracy is to be expected then.

A prerequisite of any location system is naturally the establishment of a one-way (random-access Doppler only) or more conveniently, two-way communication link between a platform on or near the ground and the spacecraft. The UHF (400—460 MHz) band is used to minimize ionospheric refraction while retaining a reasonable propagation loss with omnidirectional antennae. On account of the quite small signal bandwidth necessary for location and the low rate data transmission, a comfortable signal/noise ratio obtains with moderate transmitted power. For platforms communicating with low altitude satellites like Nimbus 3 (1000 km) and EOLE (600—800 km), a few watts peak radiated power is adequate: IRLS (Nimbus) $\sim$ 5 watts (peak); EOLE $\sim$ 3 watts (peak.)

Using even less powerful signals is contemplated for future systems (TWERLE $\sim$ 0.6 watt). This will therefore appear to be the first striking difference between surface communications (requiring tens of watts) and ground-to-space relay. Next, one of the following three location methods must be chosen: range r only, range rate $\dot{r}$ only, range and range rate (r, $\dot{r}$).

Range r only

A fix is obtained in combining several (at least two) successive range measurements from different successive positions of the satellite during the period of visibility of a particular platform (Fig. 2). This method was chosen for the IRLS implemented on Nimbus 3 and 4. The range measurements were derived from pulse-code modulated signals transmitted from the satellite to the platform and back, on a two-way UHF link. The sequence used for ranging consists of 192 bits or 16 words of 12 bits transmitted at the basic rate of 12.5 kilobits/sec, corresponding to a bit period of 80 microseconds or a distance increment of 12000 m on the two-way propagation path. If the relative phase difference of the outgoing and incoming digital sequences can be measured within 5% of one revolution, the basic instrumental uncertainty is 0.6 km only.

It would be misleading however to consider this range measurement accuracy as the order of magnitude of the location accuracy because the successive fixed-range loci generally intersect at small angle thereby multiplying the range error

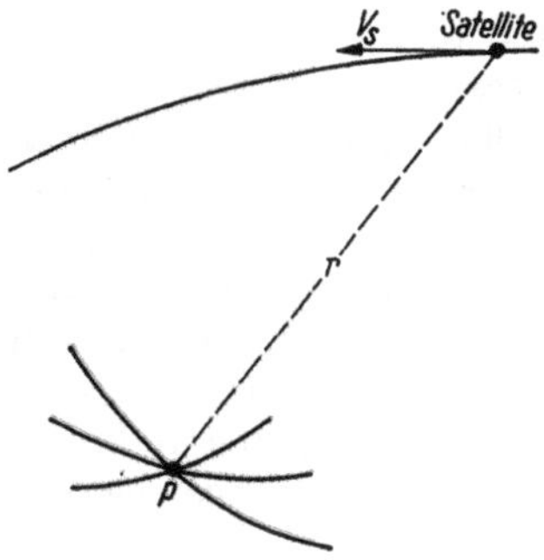

Fig. 2. Platform location by range r only.

by a factor of some $1/\sin \alpha$. Furthermore, any uncompensated displacement of the platform between successive ranging could result in a corresponding (often bigger) location error.

Range Rate $\dot{r}$ only

A perfectly good fix could also be derived from several (at least two) successive range rate or range increment measurements taken on different successive positions of the satellite (Fig. 3). The range rate or relative radial velocity $\dot{r}$ is naturally proportional to the Doppler radio-frequency shift of the satellite-to-platform link. It can be measured very accurately either on the ground (Transit navigational satellites) or on the spacecraft (Eole; Nimbus-F/TWERLE). The accuracy is thus

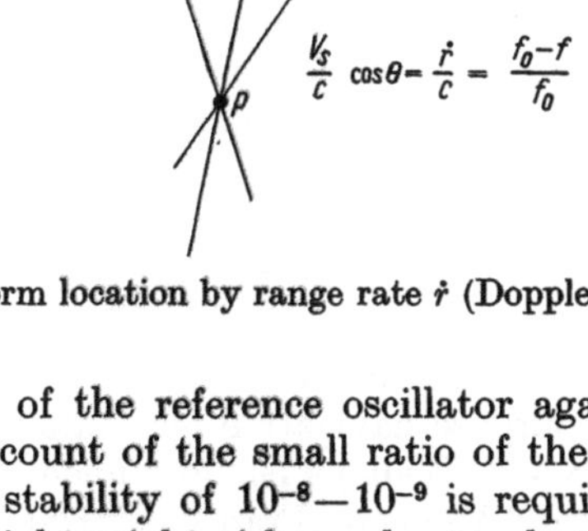

Fig. 3. Platform location by range rate $\dot{r}$ (Doppler shift) only.

as good as the stability of the reference oscillator against which the Doppler shift is measured. On account of the small ratio of the satellite velocity to the light velocity, oscillator stability of $10^{-8}-10^{-9}$ is required. This is not easy to come by on a small-size lightweight airborne beacon but it is nevertheless within the state of the art, as one expects to demonstrate in the TWERLE experiment (random-access one-way link from platforms to the satellite). In the Eole experiment, Doppler shifts were measured on a two-way phase-coherent radio link, thus eliminating the platform oscillator stability problem.

The same difficulty related to the shallow intersection of successive constant Doppler shift curves on the earth applies here: the basic location uncertainty can be 5 to 10 times more than the single range-rate measurement error. The remark relative to the effect of uncompensated platform displacements during the satellite passage applied also with one aggravating circumstance, i.e. the sensitivity to the platform instantaneous motion as well as to the platform displacement between successive $\dot{r}$ measurements. Indeed, if the uncompensated platform (drifting balloon) velocity is 40 m s^{-1} or about 1/200 the satellite velocity, a 1/200 radian angle error would result, corresponding to a 10 km error at the typical range of 2000 km. Clearly, the platform velocity must be compensated for and this does complicate data processing.

Range and Range Rate (r, $\dot{r}$)

The range and range rate location technique has been used for the EOLE experiment thereby providing an ultimate location accuracy of the same order as the basic range or range-rate measurement errors, i.e. about 300 m; this is a conse-

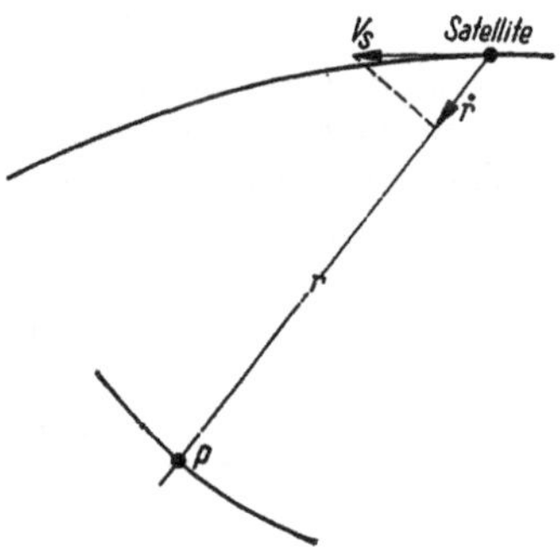

Fig. 4. Platform location by range and range rate.

quence of the fact that one fix is derived from two *simultaneous* measurements of r and $\dot{r}$ (Fig. 4) and that the r = constant sphere centred at the satellite position S intersects at *right angles* the $\dot{r}$ = constant hyperboloid of centre S. This does not eliminate the necessity of compensating the Doppler shift for platform motion however (which could otherwise easily induce a 10—50 km error).

3. The Eole Experiment

The EOLE experiment implementation began on 16 August 1971 with the launching of the special purpose navigation and data collection satellite EOLE 1 by a Scout rocket from NASA Wallops Island launching facility under an agreement of the United States (NASA) and France (CNES) *. The satellite was placed on a slightly elliptical orbit of inclination $i = 50°$ over the equator. This orbit allows the satellite to cover most of the earth's surface from 75°S to 75°N and inter-

* The EOLE programme is a joint undertaking of France and the United States also known under the name of CAS 1 (Cooperative Application Satellite No. 1).

rogate each platform several times in succession (up to six successive times for platforms located in the latitude band 30—60° S.

Further, 480 constant-level balloons were released from three launching sites located in Argentina, under an agreement of France and the Argentinian Republic. The launching schedule was as follows:

Mendoza (33° S, 69° W)	: 100 balloons
Neuquen (39° S, 68° W)	: 200 balloons
Lago Fagnano (55.5° S, 67° W)	: 180 balloons

The launching operations were conducted during the southern hemisphere spring, from late August to mid-December. Each balloon carried a solar generator and storage battery power supply, an EOLE transponder, two meteorological sensors (p, T) and a destruction device. The destruction capability was meant to eliminate, under a special command from the satellite, those balloons which would perchance stray into the northern hemisphere. The sure-fire efficiency of this device was amply demonstrated by a serious incident which occurred in the early

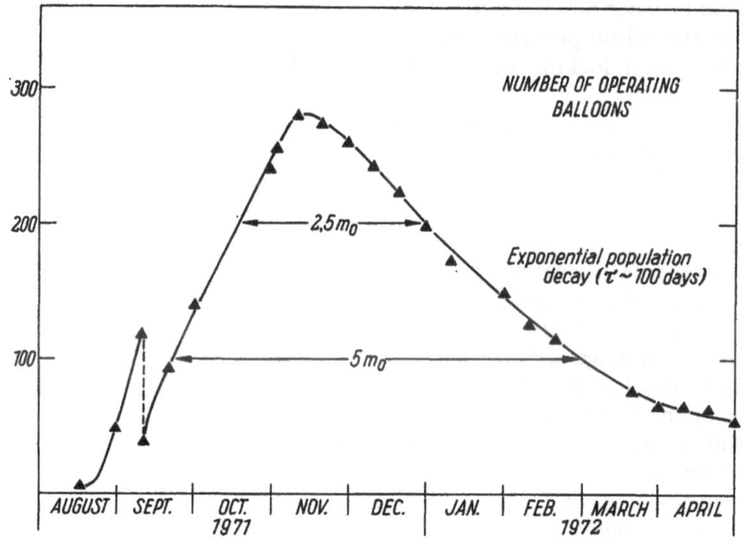

Fig. 5. Number of operating balloons.

phase of the launching programme: an erroneous command sent to the spacecraft resulted in turningon the destruction order for all balloons in sight and the loss of about 80 tracers. The build-up of the balloon network was resumed, however, to reach eventually a maximum number of 280 platforms in operation during the month of November 1971 (Fig. 5). With the end of the launching programme in December this number started to decrease, following an essentially exponential (random failure) law with an effective mean life of 100 days.

The detailed operation of the EOLE interrogation, location and data collection system is summarized below. The basic addressing, interrogation and measurement cycle extends over 500 milliseconds with a 250 millisecond stepping rate. Thus, platform address N is called during the first half cycle, while the data from platform N are received and address $N + 1$ is sent during the next half cycle. These functions are performed either sequentially: $1, 2 \ldots N - 1, N, N + 1 \ldots$ or individually, according to a schedule loaded from the ground into the satellite programme memory. The satellite interrogation signal includes a continuous carrier at 464.486 MHz and phase modulation corresponding to frequency shift keying at two frequencies $f_0 = 2304$ Hz (bit 0) and $f_1 = 2688$ Hz (bit 1) at the basic data rate of 48 bits per second. This modulation is used both for range measurement and platform address encoding.

The balloon transponders operate in each of the three following modes:

Mode I

Pulsed stand-by reception only, when no satellite signal (carrier + modulation) is present. When the EOLE satellite signal is detected the platform changes to Mode II.

Mode II

Logics and sensor circuits activated.

Receiver turned on permanently.

Acquisition and locking of the phase-lock loop on the satellite carrier frequency.

Bit rate acquisition and synchronization.

Address recognition. When the proper address is recognized, the platform changes to Mode III.

Mode III

Multiply the satellite carrier frequency by the rational factor 32/37 and transmit a 401.718 MHz return signal preserving the phase coherence with the satellite signal (phase locked bandwith 100 Hz).

Transmit a sequence of bits 1 and 0 for range tracking, i.e. preserving the phase coherence of the bit sequence (received from the satellite) as well as the phase coherence of the two frequencies f_0 and f_1.

Transmit frequency modulated video signals corresponding to four time-multiplexed information channels.

If balloon No. N is within interrogation range of the satellite (i.e. about 3000 km) its answer is received by the satellite system, demodulated, analysed for range (propagation delay), range rate (Doppler carrier frequency shift) and information channels, and recorded in either one of two 100000-bit random-access magnetic core storage memories. Format is as follows:

Satellite time
Platform number
Doppler frequency shift
Coarse (bit rate) phase shift
f_0 phase shift
f_1 phase shift
Information channel 1
Information channel 2
Information channel 3
Information channel 4

This information can be read out by telemetry upon command, from Command and Data Acquisition stations in Brétigny, Canary Islands, Ouagadougou, Brazzaville and Pretoria. Data are telemetered in digital form and temporarily stored on tape. Transmission to the Brétigny Space Centre proceeds immediately after the satellite passage via commercial telex lines. It is easy to see that EOLE data processing requires:

Establishing very precise ephemerides from the spacecraft orbital elements.

Establishing the precise synchronization of the satellite internal clock with universal time.

Reducing the Doppler shift to a range rate determination.

Reducing the propagation phase shifts of the 48 bps digital sequence, the subcarrier frequencies difference $f_1 - f_0 = 384$ Hz and the subcarrier frequency $f_1 = 2688$ Hz, to establish the propagation delay and determine the range. Note that the two lower frequencies serve only to lift the ambiguity of the f_1 subcarrier phase shift measured to 1% approximately (300 m random range error).

Calibrating and formating the four information channels (pressure, temperature, over-pressure, battery voltage).

Further processing consists in computing the location of each platform from the (known) satellite ephemeris and the values of r and $\dot{r}$. It can readily be seen that two locations, essentially symmetrical with respect to the satellite ground track, could fit (almost) equally well the same values $r, \dot{r}$. It was found feasible, under favourable geometrical conditions, to discriminate the spurious location using several interrogations of the same platform during one satellite passage. But this cannot be done reliably in all cases and it was found necessary to apply a trajectory continuity criterion for selecting the proper balloon locations from a set of interrogations on several successive orbits of the satellite.

Finally, the first estimate of the mean wind so derived from the raw range and Doppler data is applied to a differential wind-correction matrix computed at each individual balloon location, for compensating the Doppler shift error resulting from the balloon motion.

It can be guessed from this brief outline that the operation of a satellite location and data collection system is not a straightforward proposition. Keeping this background in mind, one could analyse the actual accuracy of the system under normal operating conditions by considering the results of interrogating nine EOLE transponders temporarily located in the same spot on Victoria Island, British Columbia (prior to sailing with nine drifting buoys). The actual buoy locations which resulted from the above were spread in a loose cluster 1 km across and separated from the true location by a distance of two kilometres or so (Fig. 6). By comparing with the fixes obtained on the next day under a similar platform-satellite geometrical configuration, one can see that the largest portion of the location error is systematic and could be corrected by further, more refined, processing. From this example, one would conclude then the EOLE location system accuracy in the northern hemisphere is limited by the following factors:

A systematic discrepancy between the satellite internal time and universal time, amounting to 0.25 second or a 2 km displacement along the satellite track. (Note that this orbital error is much less in the southern hemisphere where the CNES satellite tracking facilities are located.)

A systematic calibration error of the sub-carrier (f_0, f_1) modulation delay in the various transponders involved in this experiment, which amounts to a 1 km dispersion of the fixes.

A random electronic measurement (noise or ionospheric refraction) error of the order of 300 m.

We can therefore conclude that the ultimate precision of Eole and similar location systems on satellites is of the order of 300 metres, given very precise orbital elements and *a posteriori* calibration of the various transponders. Note

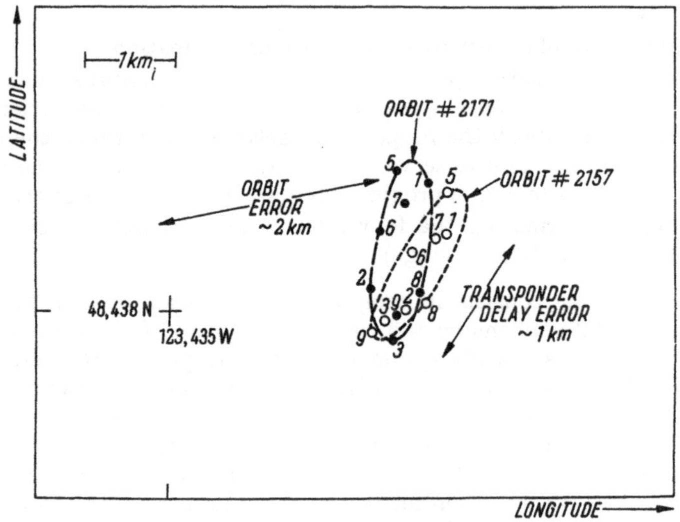

Fig. 6. Buoy locations.

that even without these refinements, the probable 2—3 km error (3σ) in the southern hemisphere is more than adequate to provide excellent mean wind velocity determinations (0.7 m s^{-1}, 3σ-error) over the 6000 seconds interval between two successive passages of the satellite.

4. The Constant Level Balloons

Constant-level balloons are essentially constant-volume floating crafts, not unlike drifting buoys, with the difference that they float at one nominal density level of the atmosphere instead of the density discontinuity at the sea surface. The Eole constant-level balloons were designed to float at the $\varrho = 0.32$ kg m^{-3} density level, corresponding to $p = 200$ mb in the standard atmosphere or an altitude of 12000 m approximately.

Constant-level balloon envelopes are made of inextensible mylar* or polyethylene-terephthalate, a high tensile strength plastic fabric. Such balloons are launched partially inflated with helium gas and become fully inflated, even overpressurized as they reach their nominal flight level. The over-pressure is not

* DuPont trademark.

much, usually 10% of the nominal outside pressure, but it does nevertheless produce a very significant stress on the very thin mylar skin (50 μm). Only mylar has been found to have sufficient tensile strength, of the order of 10 kg mm^{-2} (steel: 100 kg mm^{-2}) to bear this stress without yielding too much. This particular property of mylar obtains because this polymer is not actually a plastic material at room temperature but has the chemical structure and some of the brittleness of glass. This is the reason why empty or partially inflated mylar envelopes must be handled with great care, lest they be crumpled and develop pin-hole leaks or incipient rupture points. Special launching hangars were built at the three stations of Mendoza, Neuquen and Lago Fagnano in Argentina, large enough to accommodate four craft for final assembly and envelope inflation with complete protection from the outside environment. With this technique, the balloon is never exposed to aerodynamic buffeting even during the very quick launching operation: the balloon must be released as soon as it comes out of the preparation hangar.

Another technical problem was that of designing electronic equipments and power supply capable of operating at very low ambient temperature (−65 to −70 °C), yet packaged in a way which would be safe in case of collision with an aircraft. This aircraft safety requirement led to stringent design specifications [7] based on an extensive impact test programme on airframe mock-ups. The resulting design is a very elongated payload, including a solar generator, a nickel-cadmium battery operating at −65 °C, sensor and electronic packages, UHF antenna and radar reflector, spread over a height of 10 metres or so. The total payload mass is less than 3 kg.

Further, elementary hydrostatics show that the free lift A of the craft at launch is converted into over-pressure Δp when the envelope becomes fully inflated at the nominal flight level. Barring temperature differences between the lifting gas and the outside atmosphere, one gets

$$\frac{A}{M} = \frac{\Delta p}{p}$$

where M is the total mass of the craft, A the free lift (kg) and $\Delta p/p$ the relative over-pressure. Since $\Delta p/p$ cannot much exceed 0.1 and M is 8.5 kg, the nominal Eole balloon free lift is well under 1 kg. If any accretion of mass in excess of 850 grammes occurs, the free lift vanishes altogether and the balloon descends towards the ground. This could happen at night by the process of inverse sublimation of ice when the radiatively cooled envelope skin is colder than the local frostpoint. A very fast, and generally catastrophic, deposition of ice particles also occurs when the craft floats inside dense cirrus clouds, such as those produced by the powerful convective towers of the tropical zone (which can reach altitudes much higher than 12000 m). The tropical belt is thus a sink for the relatively low altitude Eole balloons: the probability of their being destroyed there (by icing) is much greater than in the mid-latitudes or polar stratosphere [8].

Fig. 7 shows the ascent and beginning of level flight of two constant-level balloons launched one shortly after the other and tracked by a high-accuracy missile-tracking radar. In addition to the slow upward drift due to the envelope approaching its final shape, one can see the more or less continuous vertical oscillations of the craft about their long-term equilibrium level. The oscillation frequency ω reflects the vertical stability of the craft for a given thermal stratification of the atmosphere:

$$\omega^2 = N^2 + \frac{g}{\gamma H}$$

where N is the Brunt-Vaïsala frequency of adiabatic vertical oscillations of an air parcel, $\gamma = C_p/C_v$ and H is the scale height. Normally, $N^2 \sim 10^{-4}\,s^{-2}$ and $g/H \sim 10^{-3}\,s^{-2}$ so that the period of constant-level balloon oscillations is 3—5 minutes. The vertical extent of these oscillations (about 100 m) is large enough for the tracer to sample the airflow at slightly different levels and therefore smooth the temporary vertical wind shears arising from the layered structure of the atmosphere.

Note finally that two different balloons reach essentially the same flight level (maybe somewhat above the nominal level). It is estimated, from (a) the accuracy

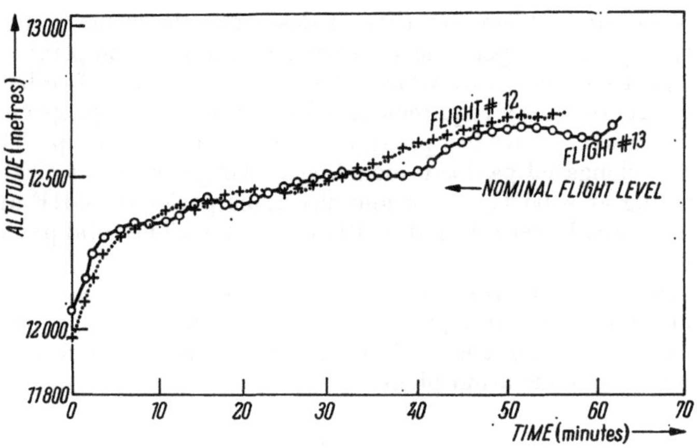

Fig. 7. Ascent and beginning of level-flight of two balloons.

of prelaunch measurements of the envelope volume and total balloon mass M and (b) the dispersion of flight-level densities from in-flight p and T data, that the vertical dispersion about the mean standard altitude is 50 metres (1σ).

5. The Preliminary Results

It may appear strange to space scientists apt to be probing far-away planetary atmospheres, that any new scientific knowledge about our own planet atmosphere could be gained by such a simple-minded experiment as tracking a few hundred balloons floating in it. But this would be overlooking the fact that atmospheric dynamics are those of a fully turbulent flow, unstable with respect even to small-scale perturbations which make its evolution unpredictable after some time. It is therefore of great theoretical and practical importance to investigate the dynamic effects of smaller scale motions of the atmosphere which escape notice between the discrete soundings of the operational meteorological network. Furthermore, even such planetary-scale phenomena as the mean meridional circulation are not well known, because the very small average velocity is masked by the very large wavelike perturbations of the general circulation. In the first case, discrete vertical soundings do not provide short enough sampling intervals. In the second, a stable estimate of a very small mean cannot be obtained from wildly fluctuating, independent discrete values.

A constant level balloon, on the other hand, follows essentially the horizontal motion of the atmospheric fluid and provides, accordingly, a *Lagrangian sounding* along the path of one air parcel. A tracer trajectory is sensitive to all components of the motion including the smallest scales, and tracer displacements reflect the only proper average of these various scales, i.e. the motion of the fluid particles themselves. Consequently, Lagrangian tracers are a very powerful tool for investigating slow long-term drifts like the mean meridional circulation as well as probing into the spectral characteristics of atmospheric turbulence. The Eole experiment is the first full-scale implementation of this investigation technique applied to a planetary flow, i.e. the general circulation of the southern hemisphere atmosphere.

First, let us ask where do these balloons go? Do they concentrate in permanent clusters along favoured zones or over the polar cap? Will they, on the contrary, cross the equator and spill over the northern hemisphere? Such questions could not in fact receive a definite answer before the experiment, on the strength of existing observational or theoretical meteorological knowledge, because soundings have been too scarce and because a spectral truncation is necessarily involved in any numerical modelling of turbulent geophysical flows. The Eole experiment has shown that the upper atmosphere winds *do not generally cross over* from the southern into the northern hemisphere: among 480 tracers released in the southern hemisphere and residing there more than three months on the average, only three cases of a balloon crossing into the northern hemisphere (and being destroyed on command) were observed.

Next, even artificially made balloon clusters could not persist indefinitely, as eddy diffusion by the smaller scale perturbations of the flow acted to separate the tracers more and more. This eddy diffusion process was investigated specifically by launching the balloons by groups of 4 to 6 within a short period (typically 2 hours) from one site, and studying their subsequent dispersion (Fig. 8). Analysis shows that the mean dispersion rate $d\bar{x}^2/dt$ increases as the separation d or $(x^2 + y^2)^{1/2}$ between the tracers grows, because larger scales of motion become effective in separating the tracers further. Inversely, it is possible to deduce the kinetic energy spectral law $E(k) \sim A k^{-\alpha}$ from the relation between $d\bar{x}^2/dt$ and the separation X^2. The preliminary analysis of the high quality processed Eole balloon location data is compatible with an exponential eddy diffusion law, i.e.

$$\frac{d\bar{x}^2}{dt} \sim x^2$$

$$x^2 \sim \exp (t/\tau)$$

with τ of the order of 2 days. These data are therefore evidence of a k^{-3} dependence of the kinetic energy spectrum, characteristic of a vorticity cascade in two-dimensional turbulent flow (rather than the familiar $k^{-5/3}$ spectrum characteristic of the energy cascade in three-dimensional turbulence). This k^{-3} spectral law for kinetic energy of tropospheric and lower stratospheric flow is also well based on sound theoretical grounds [9].

The fact is, anyway, that eddy diffusion is a very efficient process indeed in atmospheric flow so that the tracers diffuse very quickly from their initial mean latitude (launch site) to be distributed quite evenly over the whole hemisphere outside the tropical belt (Fig. 9). Note the almost homogeneous distribution of

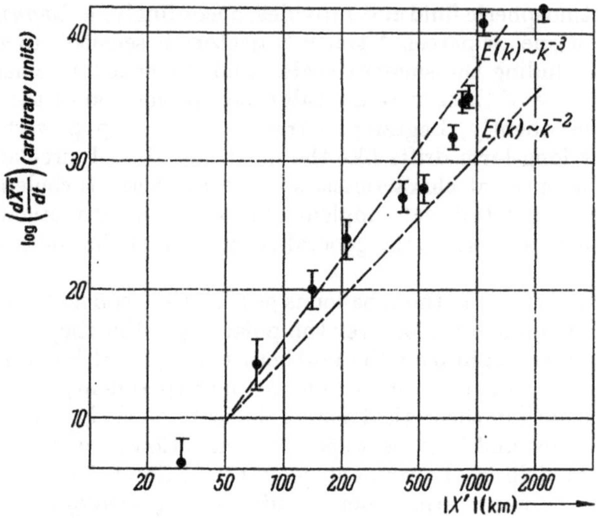

Fig. 8. Tracer dispersion in large scale atmospheric turbulence.

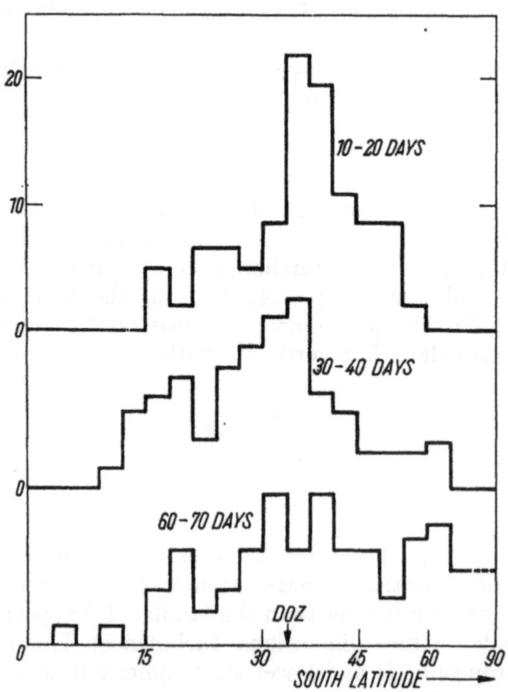

Fig. 9. Balloons launched from Mendoza (% population per equal area zone).

the balloons launched from Mendoza over the whole latitude range of the westerly circulation, achieved after about two months' flights. This fact has a very important impact on the logistics of any large horizontal sounding system of the same nature: only a few, probably two, balloon launching stations would be sufficient to maintain an optimal distribution of platforms over the whole hemisphere, a very advantageous situation indeed compared with the deployment of surface automatic stations. Thus, the eddy meridional transport at this altitude is much larger than the mean meridional velocity, an unfortunate circumstance for determining the latter. We cannot be sure, then, of the statistical stability of our current short-term estimates based on the mean latitude drift of balloon ensembles

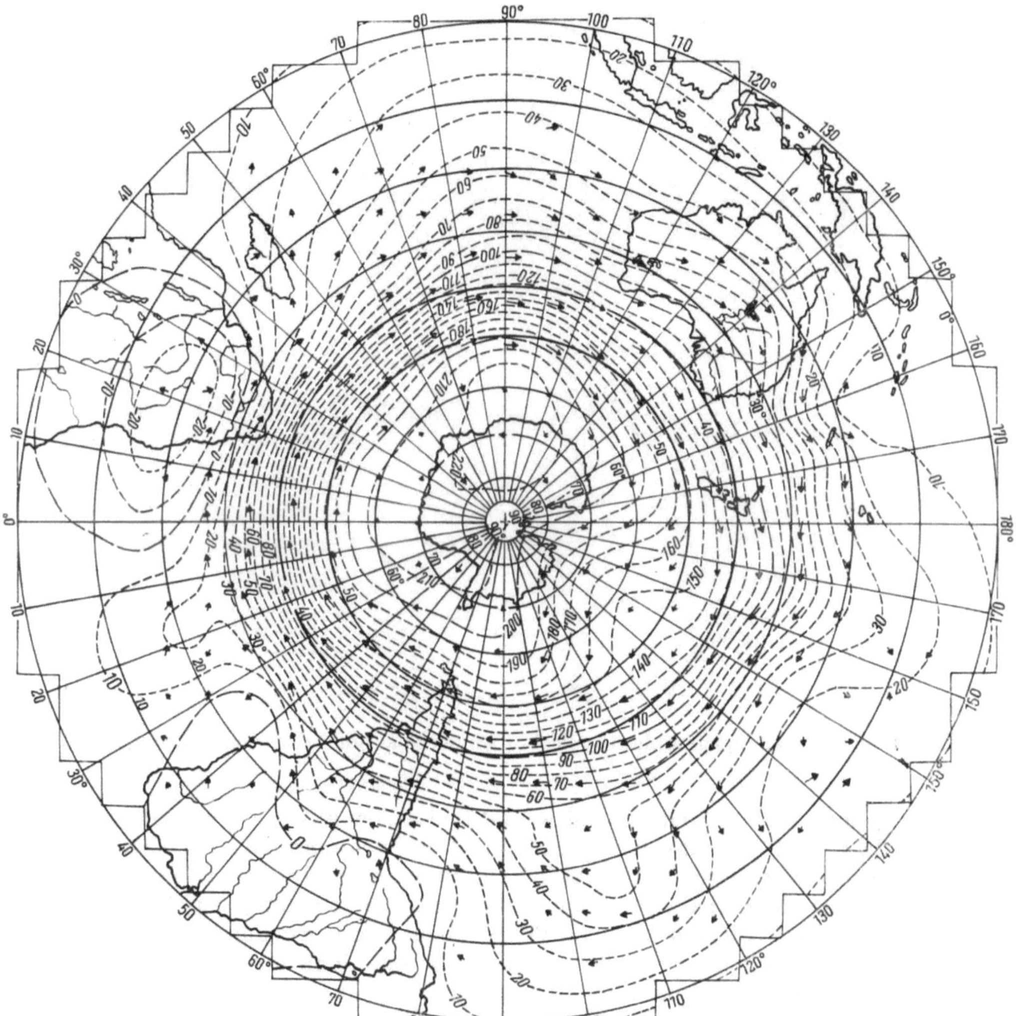

Fig. 10. Mean stream function chart showing temporary blocking of transient planetary waves by high pressure area (January 1972).

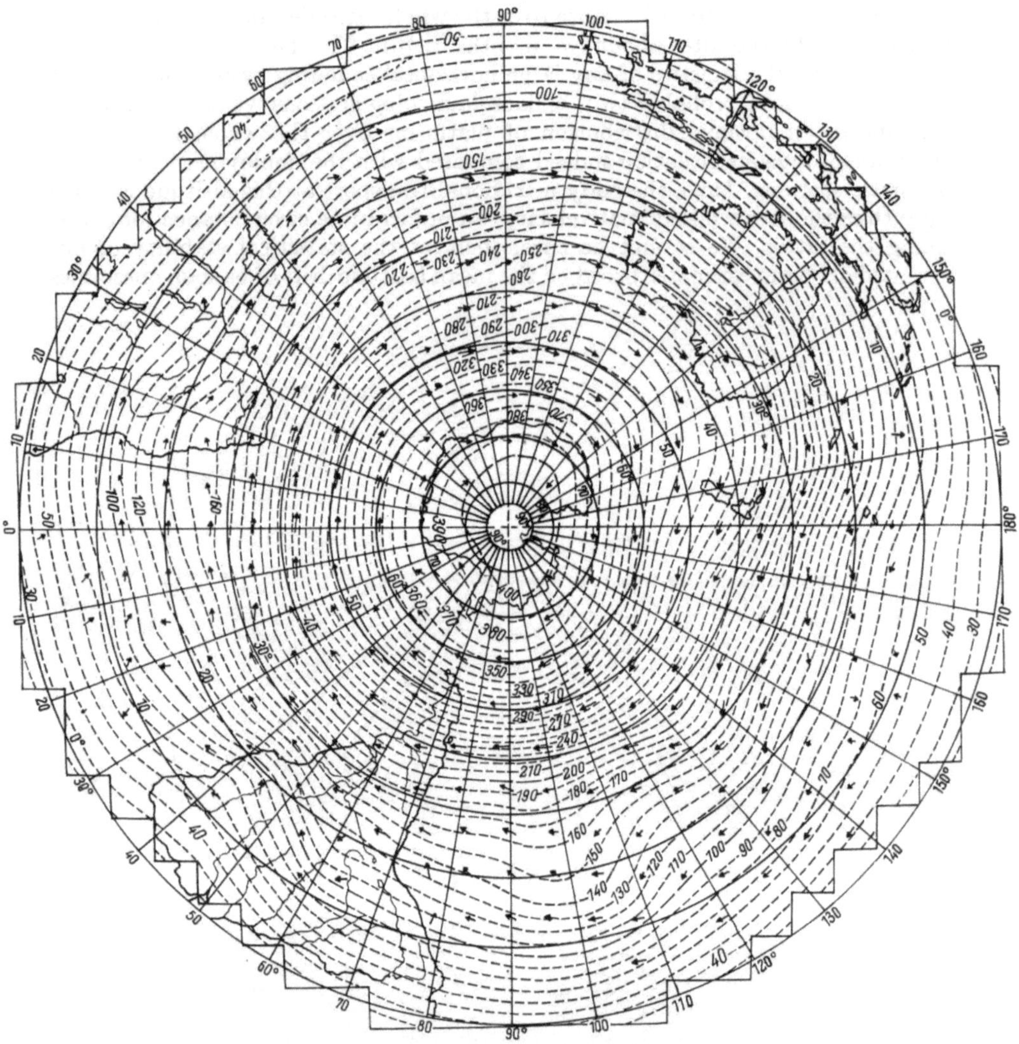

Fig. 11. Mean stream function chart typical of winter or autumn situations (November 1971).

over periods of 10 to 30 days. But we did find a slow (0.1 m s^{-1}) systematic poleward drift of all balloons during spring and early summer. We expect now this drift to reverse towards the equator in autumn and early winter in response to the changing meridional circulation of the stratosphere and mesosphere.

Two hundred balloons, each providing about two non-redundant wind determinations per 24 hours, are fully equivalent (at one single level, of course) to the northern hemisphere upper air sounding network and even better on account of their essentially homogeneous distribution over land and sea alike. It is a straightforward matter then to deduce the (daily) flow patterns of the general circulation at 200 mb from these wind data. We have chosen to update twice daily the stream function $\Psi(\lambda, \phi)$ which represents the non-divergent part of the wind velocity field (Fig. 10). Monthly mean stream function charts (Fig. 11) drawn from the same data base, show a temporary but very dramatic blocking of the transient planetary waves by the high-pressure area over Brazil and tropical Africa at the height of the southern hemisphere summer. This blocking produces a striking wavenumber $n = 4$ stationary perturbation of the normal zonal flow near the tropics.

Reliable estimates of the horizontal flow divergence cannot usually be derived from local vertical wind soundings because (i) radio wind or even radar wind measurement errors, taken as derivatives over a small time interval, are fairly large (1—3 m s^{-1}), (ii) instantaneous discrete wind velocity measurements are not truly representative of the air mass (1 m s^{-1}), (iii) the divergence is quite small but the flow vorticity is large so that $(\partial u/\partial x + \partial v/\partial y)$ appears often as the (very small) difference of two large Eulerian derivatives.

Nothing of the sort happens when Lagrangian data are available, because the horizontal flow divergence is computed directly from the variations of the horizontal area A of an air mass marked by the balloon tracers. Indeed:

$$\operatorname{div} \boldsymbol{V} = \frac{1}{A}\frac{dA}{dt}$$

where the area increase rate can be estimated very accurately on a period of two to four hours (Fig. 12). Accordingly, the EOLE experiment produced the best measurements of the horizontal divergence of atmospheric flow ever obtained in the history of meteorology. Now, one must visualize that the divergence has great

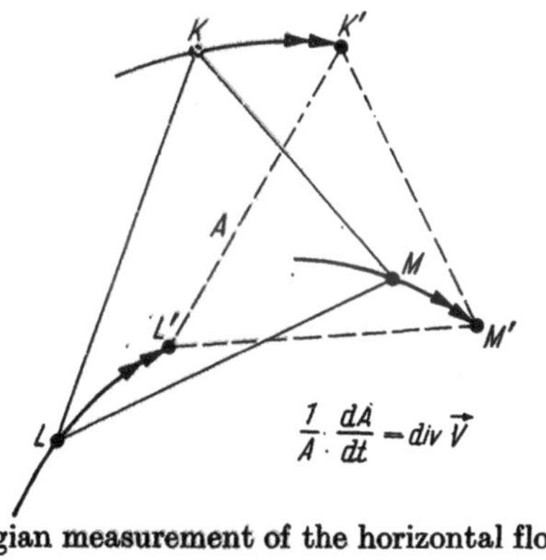

Fig. 12. Lagrangian measurement of the horizontal flow divergence.

dynamic significance, even though it is small, because it gives a cue to the nature of the perturbations we see in the atmospheric flow: quasi-geostrophic waves or gravity waves or practically discontinuous shocks (fronts)? A preliminary analysis of the spectral dependence of rms (div $\boldsymbol{V}$) versus scale L shows at least three scale ranges (Fig. 13): a large scale limit ($L > 1000$ km) where (div $\boldsymbol{V}$) drops rather rapidly as one goes through the peak of kinetic energy spectral density (energy production range for planetary disturbances); an inertial sub-range ($300 < L < 1000$ km) where rms (div $\boldsymbol{V}$) is essentially independent of scale, as one would expect for two-dimensional turbulence with inertial transfer of energy and vorticity; a mesoscale range ($L < 300$ km) where again rms (div $\boldsymbol{V}$) is a decreasing function of scale:

$$\text{rms}\,(\operatorname{div} \boldsymbol{V}) \sim L^{-1/2}$$

compatible with a k^{-2} kinetic energy spectral law, characteristic of an ensemble of discontinuous fronts.

Finally, one should note that the daily charts of the 200 mb horizontal divergence field show a striking correlation of large positive divergence areas with developing meteorological perturbations and cloudiness, indicating that the lower stratosphere flow patterns are strongly coupled with the tropospheric circulation: from the standpoint of general circulation, the troposphere extends to 150 mb

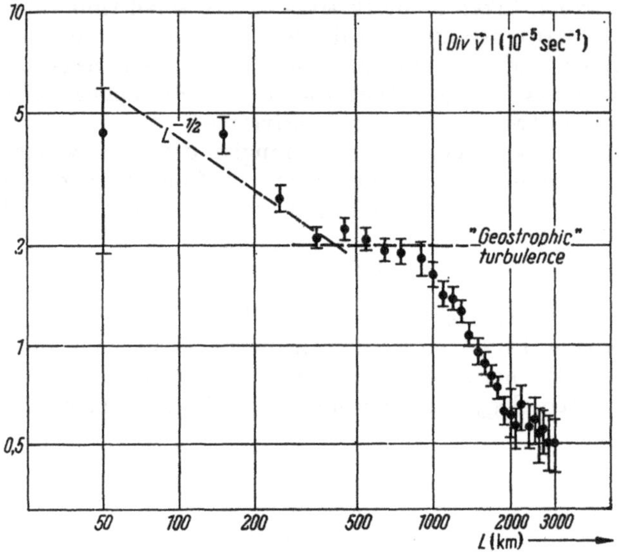

Fig. 13. Dependence of div $\boldsymbol{V}$ on scale L.

or maybe higher! Also, this correlation justifies the idea of using the 200 mb divergence data for estimating the vertical velocity at the quasi-non-divergent level (500 mb) and for diagnosing the occurrence of fronts and mesoscale phenomena in the lower atmosphere. But much more is yet to be done to extract the full meaning of EOLE data in this respect.

References

[1] A. D. ANDERSON and H. J. MASTENBROOK, Bull. Am. Met. Soc. **37**, 342 (1956).
[2] T. O. HAIG and V. E. LALLY, Bull. Am. Met. Soc. **39**, 401 (1958).
[3] V. E. LALLY, Bull. Am. Met. Soc. **41**, 429 (1960).
[4] K. C. GILES and J. K. ANGELL, Bull. Am. Met. Soc. **44**, 687 (1963).
[5] G. D. HOGAN and M. R. TOWNSEND, paper presented at WMO Tech. Conf. on Automated Weather Stations, Geneva 1966.
[6] J. E. MASTERSON, Bull. Am. Met. Soc. **51**, 222 (1970).
[7] P. MOREL, Constant level balloon impact on aircraft structures, COSPAR Working Group VI Report, February 1970.
[8] P. MOREL, J. FOURRIER and P. SITBON, J. Appl. Met. **7**, 626 (1968).
[9] J. CHARNEY, J. Atmos. Sci. **28**, 1087 (1971).

Space Research XIII — Akademie-Verlag, Berlin 1973

USE OF METEOROLOGICAL ROCKETSONDE AND SATELLITE RADIATION DATA FOR CONSTANT-PRESSURE ANALYSES AT LEVELS BETWEEN 5 AND 0.4 mb

F. G. Finger and M. E. Gelman

National Meteorological Center, NWS, NOAA, Hillcrest Heights, Md, USA

Meteorological rocketsonde observations in the northern hemisphere allow the definition of large-scale circulation features to about 0.4 mb (55 km). Confidence in the accuracy of analysis is limited since rocketsonde observations are sparse compared with lower-level coverage by the international radiosonde network.

The Satellite Infrared Spectrometers (SIRS) on board Nimbus 3 and 4 satellites have provided large amounts of radiance information from which, through the inverse solution of the radiative transfer equation, temperature and geopotential height can be derived for the troposphere and stratosphere to the 10-mb level (30 km). The Selective Chopper Radiometer (SCR) is able to yield information to higher levels.

A method based on regression of measured satellite radiances with a selected group of rocketsonde temperature profiles has also been developed to specify thickness in various layers between 100 mb (16 km) and 0.4 mb. Heights of the upper-level pressure surfaces are obtained by adding the specified thicknesses to the height field of lower-level charts.

A method is given for analyzing charts of the 5-, 2-, and 0.4-mb levels using all available satellite data for a "first approximation" field and rocketsonde data in finalizing this field. Relative compatibility of the two sets of data is discussed, using a chart series depicting the northern hemisphere winter circulation in early 1972.

1. Introduction

In the past, data from meteorological rocketsondes have been the sole source of information for synoptic constant-pressure analysis in the upper stratosphere (up to 0.4 mb, or about 55 km) [1]. Charts constructed from these data have necessarily presented only the gross picture of the high-level circulation, since the number of rocketsonde observations is relatively small. Even so, much information regarding the behavior of the circulation of the upper stratosphere and lower mesosphere has emerged from studies based on such analyses.

Perhaps the foremost problem facing the analyst of high-level charts has been to determine the overall circulation features in the sometimes vast areas between the rocket stations. Heretofore, these features were determined only by the most judicious use of the rocket data, for example with the aid of time and cross-section analyses, thermal wind considerations, etc. Even then, in many cases parts of the analyses could only be considered as educated guesses. It is, therefore, easy to understand that the stratospheric meteorologist was eagerly awaiting satellite radiation information (such as the Satellite Infrared Spectrometer, SIRS and Selective Chopper Radiometer, SCR) with the expectation that these new data would significantly aid in the determination of high-level circulation patterns.

2. Derivation of Data from Satellite Observations

Several methods to gain temperature and height information at the higher stratosphere levels from the SIRS and SCR radiation data are now available. The most well-known retrieval schemes [2, 3] yield temperature information to 0.01 mb. Heights are then computed using the hydrostatic relationship. However, due to decreased information content of the weighting functions at very high stratospheric levels, temperature values derived for levels above about 10 mb become progressively unreliable with height.

A relatively simple scheme to derive height fields directly from either SIRS or SCR radiance observations has recently been developed [4]. This scheme basically regresses observed satellite radiance values against height thickness values, the latter computed from combined rawinsonde—rocketsonde soundings taken in the same area and at the same time as the satellite observations. The method takes advantage of the fact that satellite observed radiance represents the weighted mean temperature over a thick layer of the atmosphere. For example, approximately 80% of the energy sensed by SIRS channel 8 originates from the layer between 100 and 2 mb. A regression relationship for the thickness between 100 and 2 mb is shown in Fig. 1. The rms height error of the observations is well within the acceptable range for a 2-mb analysis.

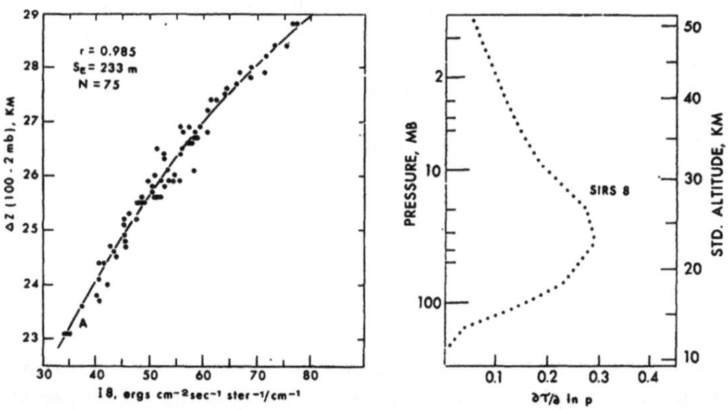

Fig. 1. Left, relationship between SIRS channel 8 radiance and thickness of a layer 100—2 mb. Right, transmittance weighting function for SIRS channel 8 (669 cm^{-1}).

Presently there are two operating stratospheric channels on SIRS and three on SCR. Thus, a series of optimum regression relationships may be derived for the major portions of the layers sensed by each channel. For example, SIRS channel 7 can be used to derive thickness between 100 and 5 mb and SCR channel A radiance was found to be a good indicator of thickness for the 10—0.4 mb layer.

3. Height Field Analysis

The SIRS channel 8 analyzed radiance field for 9 February 1972 is shown in Fig. 2. The dots on the chart represent the position of usable satellite data gathered along the various paths during a 24-hour period.

With the aid of the regression relationship which was presented in Fig. 1 it is an easy matter to relabel the radiance analysis in terms of height thickness for the layer from 100 to 2 mb. As shown in Fig. 2, the thickness values at the cold or low radiance center is 24 km, while that in the warm or high radiance center over the Aleutians is 26.5 km. The isolines between these extreme values can be labeled for intermediate thickness values.

The analyzed radiances converted to thickness can be added to a given base map for the derivation of the "first guess" height field at some higher level. This can be accomplished by simple graphical addition. The basic prerequisites are the radiance field of the particular channel and the analyzed lower level chart. This first guess field can then be adjusted to conform with the rocketsonde height and wind observations.

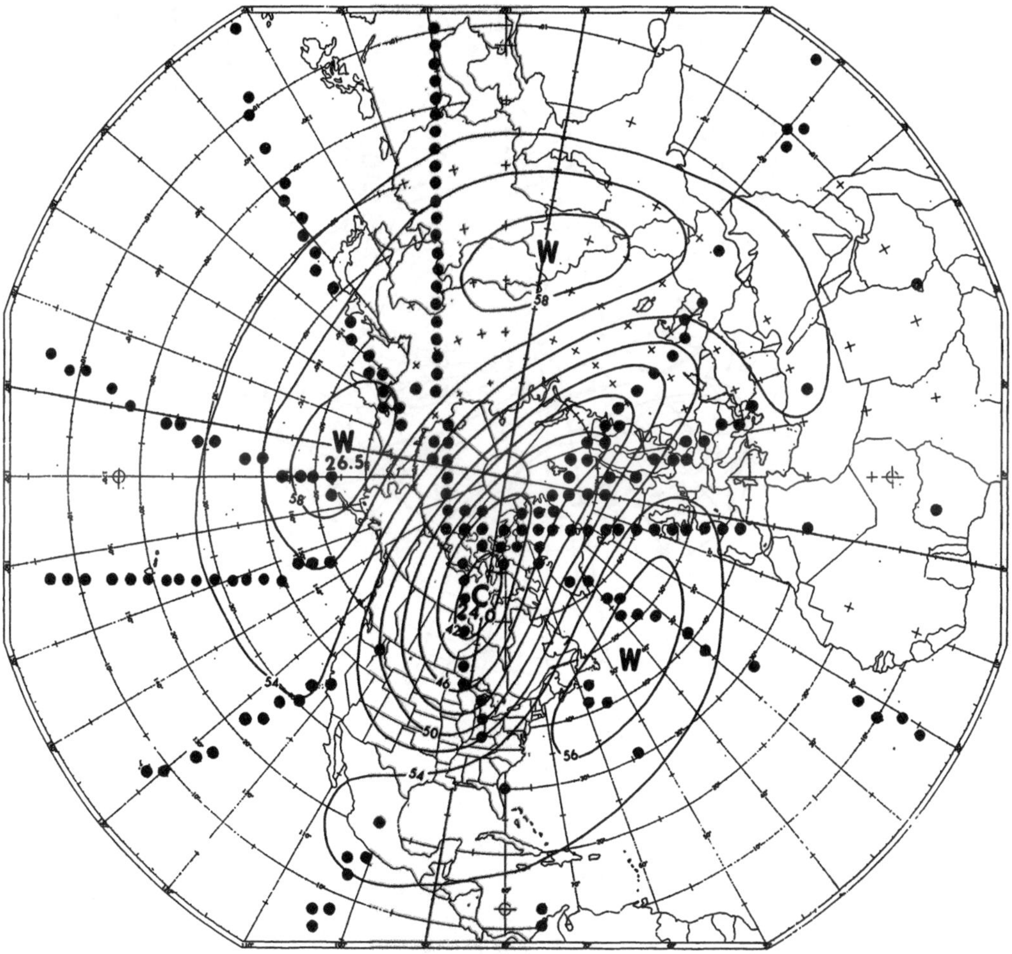

Fig. 2. Map of radiance for SIRS channel 8 on 9 February 1972 (contours labelled in ergs cm^{-2} s^{-1} sr^{-1} cm^{-1}). Corresponding values for thickness (km) between 100 and 2 mb shown at coldest and warmest centers. Distribution of SIRS subsatellite observations shown by filled circles.

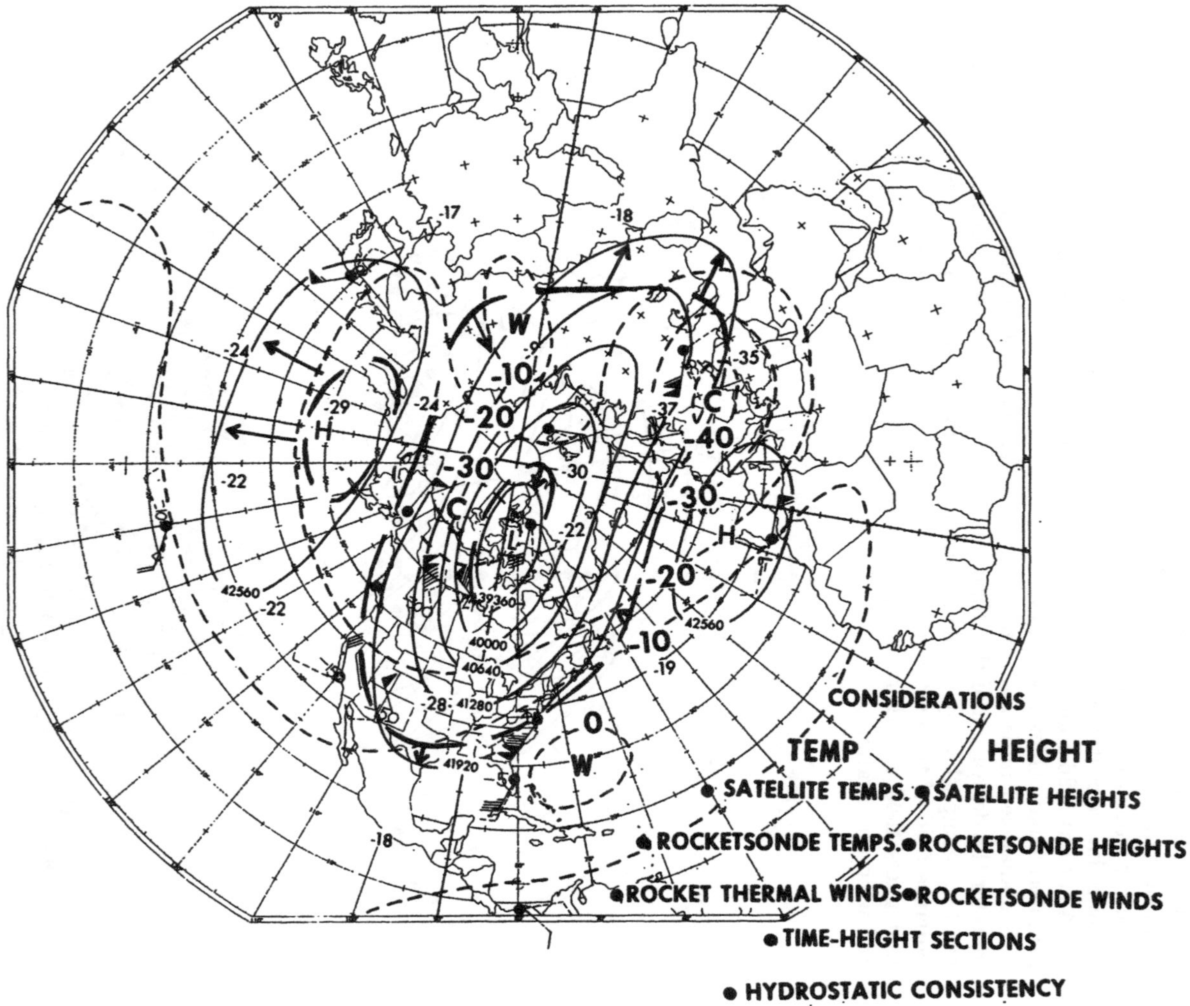

Fig.3. Map of 2-mb height (solid lines, in geopotential meters) and temperature (short dashed lines, °C) on 9 February 1972. Long dashed lines represent height first guess (as explained in text) which was adjusted according to various considerations in direction of arrows to obtain final contours. Selected satellite retrieval temperatures as well as rocketsonde observed temperatures are shown by small numbers. Wind shown at rocketsonde stations (flag 50 knots, full barb 10 kt). Direction of thermal wind derived from vertical shear of rocketsonde wind observation, is also shown (dashed wind bars).

Rocketsonde observations during this period were available from nine stations in the western hemisphere as well as two USSR sites and Thumba, India. The final height analysis (Fig. 3) is shown by the solid lines, while parts of the first guess field are outlined by the long dashed lines. Although differences of up to 600 meters exist, the locations of the centers of the cyclone and anticyclones are very close, and the height gradients are quite compatible. The 2-mb height field on 9 February is suggestive of a disturbed stratosphere, with an elongated vortex in the polar region, and with well developed anticyclones at lower latitudes.

4. Temperature Field Analysis

The temperature for a particular level derived from remote soundings is more uncertain than a mean layer temperature or thickness. Retrieval schemes using multiple channel radiance data together with sophisticated reduction techniques present more reliable temperature information than can be obtained by simple regressions. However, single channel relationships may at times be quite useful and have the advantage of being very simple to apply. In practice we find it convenient to derive the first guess temperature field by using information from both sources. Finalization of the temperature field consists of adjustment of first guess temperatures to observed rocketsonde temperatures and thermal winds.

The temperature pattern for 9 February (Fig. 3, short dashed lines) indicates the beginning of a perturbed situation. The most significant feature is the relatively warm air over northern Eurasia. This is associated with a southward movement and splitting of the cold center normally situated in the polar area.

A weekly series of hemispheric charts for the 5-, 2-, and 0.4-mb levels has been produced for the period from January through March using the methods outlined above. This series indicates that although there were a series of weak to moderate warmings during the period, a stratospheric warming and circulation breakdown of major proportions did not occur. A major event can be described as one where the temperature changes at least 50 °C over a large area, within a brief period, and the wintertime polar cyclone is destroyed in a deep layer of the stratosphere. In these cases the polar cyclone is replaced by a warm anticyclone.

The 2-mb chart for 16 February 1972 (Fig. 4) shows a significant change in the height and temperature fields (cf. Fig. 3). For example, temperature over Siberia increased during the week by more than 25 °C and the polar low filled by nearly 1 km in response to the warming. This represented the "high point" in the stratospheric variability during the winter season of 1971—1972.

5. Concluding Remarks

The major purpose of this note was to present a most recently developed method of analyzing constant-pressure charts for upper stratospheric levels. Great improvement in the accuracy of such analyses has come from use of the large mass of radiation data which is now available from SIRS and SCR. However, it is important to stress that the satellite data complement the *in situ* rocketsonde observations rather than replacing them. It is only through judicious merging of both data sets that we can realize the improvements in analysis.

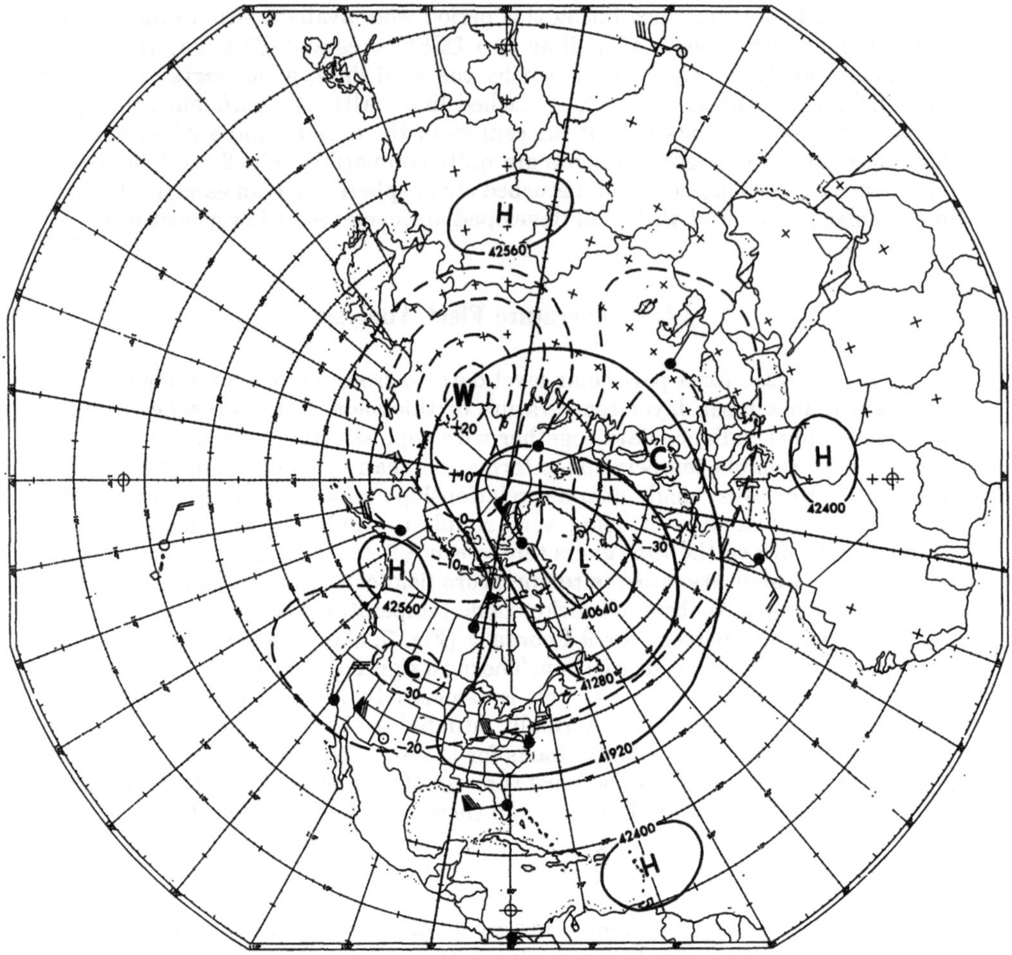

Fig. 4. Height (solid, gpm) and temperature (dashed, °C) contours for 16 February 1972.

References

[1] Staff, Upper Air Branch, National Meteorological Center, National Oceanic and Atmospheric Administration Technical Reports, National Weather Service, Numbers 2, 3, 9, 12, 14 (1967–1971).
[2] W. L. SMITH, H. M. WOOLF and H. E. FLEMING, J. Appl. Met., **11**, 113 (1972).
[3] C. D. RODGERS, Quart. J. Roy. Met. Soc., **96**, 654 (1970).
[4] R. S. QUIROZ and M. E. GELMAN, Mon. Weath. Rev. **100**, 11 (1972).

Space Research XIII — Akademie-Verlag, Berlin 1973

SOME RESULTS OF LASER RADAR STUDIES OF CLOUDS AND UNDERLYING SURFACES

A. I. HERMAN, V. M. ZAKHAROV, O. K. KOSTKO, V. E. ROKOTYAN and E. A. CHAYANOVA

USSR

Some results of aircraft laser radar experiments on the determination of an upper cloud boundary and reflectivities of underlying surfaces are given.

Pulsed laser radars operating at 0.6 943 μm and 1.06 μm were used. Experiments were conducted in various meteorological conditions. Preliminary results suggest that it is possible to use laser radars for the determination of a cloud's topography and phase state, as well as the type and characteristics of an underlying surface.

In satellite laser studies it is necessary to know atmospheric transparency for the determination of the topography of the upper cloud boundary and solution of other meteorological problems. It is known that the troposphere plays the main role in such measurements. In connection with this a series of aircraft measurements of atmospheric backscattering coefficients under various meteorological conditions were conducted. Experiments were made at a 0.5—10 km height for ruby (0.6943 μm) and neodymium (1.06 μm) laser wavelengths over different types of underlying surfaces. The volume scattering coefficient σ and backscattering coefficient σ_π depend on an incident radiation wavelength, as well as on the number, size, form and refraction properties of particles per unit volume. The highest values of these coefficients correspond to precipitation, clouds and fogs, and fall with increase of atmospheric transparency to the values corresponding to atmospheric gases.

For the atmospheric conditions under which our experiments were conducted the relation between σ and σ_π can only be obtained from the literature [2, 3]. Evidently, it is an approximate relation. Most of the measured σ_π values vary significantly not only with height but even at one and the same height in different portions of the flight path. Since these variations are beyond the limits of errors of measurement, it can surely be said that in the atmosphere there exist rather thin layers of increased turbidity and thus the transition from a vertical atmospheric transparency to a sloping transparency can not be expressed in terms of a simple dependence on an atmospheric mass increase. In most cases of vertical atmospheric soundings there is observed an increased turbidity layer in the height range around 3 km, a transparency increase up to 3.5—4.5 km, and its subsequent decrease in the region of 5—6 km height. Thus, a 3—4 km layer is evidently rather "pure" and its contribution to the attenuation of radiation is negligible. It was also found that the atmospheric backscattering coefficient at the neodymium laser wavelength (1.06 μm) is less variable with height than the same coefficient for $\lambda = 0.6943$ μm.

There was not found any significant relation between values of backscattering coefficient and type of an underlying surface, except that in general σ_π values for deserts are several times as high as the values obtained for steppe and sea surface. This effect is more strongly pronounced for ruby than for a neodymium laser.

With measuring conditions on board an aircraft one cannot obtain a vertical atmospheric profile below 500 m over the surface, i.e. where the strongest attenuation takes place. In order to avoid this disadvantage scattering coefficient measurements were conducted in the ground layer in various meteorological conditions. Ruby laser returns were received from two identical diffusely reflecting surfaces located on a horizontal path at distances $Z_1 = 200$ m and $Z_2 = 525$ m, and averaged over a series of five measurements.

Measurements conducted under conditions when the meteorological visibility range (MVR) was about 2 km, which corresponds to a dense haze, gave $\sigma = 1.64 \times 10^{-3}$ m^{-1}; calculations according to the Ångström formula

$$\mathrm{MVR} = \frac{3.91}{\sigma} \left(\frac{0.55}{0.694}\right)^q,$$

where $q = 1$ for our case, give a value MVR = 1.889 km which indirectly confirms a satisfactory accuracy of the determination of σ. Processing of the atmospheric backscattering oscillograms obtained simultaneously at a distance of 60–90 m from the instrument gives $\sigma_\pi = (1.0 + 1.2) \times 10^{-4}$ m^{-1}. The measurements of backscattering coefficients in stratocumuli give $\sigma_\pi = 1.1 + 2.8 \times 10^{-5}$ m^{-1} at 0.694 μm wavelength. Since for cloudiness the backscattering coefficient $\varkappa_{180} = 0.01$ the scattering coefficient $\sigma = (3 + 10) \times 10^{-3}$ m^{-1}.

In processing the results attenuation in the atmospheric layer between the aircraft and underlying surface depending on the given scattering coefficient was taken into account. Practically in all measurements reflectivity varies significantly even for one and the same type of underlying surface, which is evidently caused by relatively small (of the order of several metres) inhomogeneities in the surface properties, since the diameter of a radiation spot is about 3 m from 1 km height.

Taking this into account we present mean (for example, arithmetic mean) reflectivities of the main types of underlying surfaces in different seasons. The Table gives some of the measured reflectivities of various underlying surfaces.

In conclusion we shall note that reflectivities for $\lambda = 1.06$ μm are lower than those for $\lambda = 0.6943$ μm [1].

Type of underlying surface	$\lambda = 1.06$ μm		
	mean	max.	min.
Ploughed fields	0.013	0.019	0.003
Green fields	0.043	0.049	0.034
Kizil-Kum in spring (April)	0.009	0.013	0.005
Kizil-Kum in summer (June)	0.022	0.024	0.021
Water (the sea)	0.002	0.003	0.0002
Snowed ice (the sea)	0.002	0.0024	0.0015

References

[1] V. M. Zakharov, O. K. Kostko, E. A. Chayanova and V. E. Rokotyan, Space Research XII, 499 (1972).

The Neutral Atmosphere

Space Research XIII — Akademie-Verlag, Berlin 1973

PRELIMINARY ANALYSIS OF EOLE METEOROLOGICAL DATA

M. Desbois

Laboratoire de Météorologie Dynamique du CNRS, Paris, France

Eole balloons provide good coverage of the southern hemisphere from 30° S to 70° S. Mean zonal velocity, mean meridional velocity and mean temperature versus latitude are obtained from statistical estimates. Time variations of these parameters show in particular that an important slowing down of zonal circulation, a southward meridional velocity and a rapid heating of low stratosphere in high latitudes occur at the onset of the Spring season. Charts of mean monthly streamlines show the circulation evolution in greater detail.

1. Introduction

The most direct results obtained from Eole data are mean characteristics of atmospheric flow at 200 mb in the southern hemisphere. Mean zonal velocity versus latitude can be obtained by conventional methods, but evaluation of mean meridional velocity is more difficult. However, a constant-level balloon network is quite appropriate for this evaluation, as it allows the study of mean latitudinal variations of every balloon. Moreover, despite the lack of precision of individual temperature measurements, their number is sufficient to establish accurately the meridional variations of this parameter.

Mean monthly charts of streamlines can also be drawn to visualize the general circulation. To know the accuracy of these statistics, it is necessary to examine the spatial distribution of data, hence of balloons.

2. Spatial Distribution of Balloons

Several factors can influence meridional distribution of balloon density: launching station location, mean meridional velocity and turbulent dispersion. Also, the probability of balloon loss is greater in low latitudes than in higher ones (because of icing phenomena) which influences latitudinal distribution.

The distribution of balloons, plotted every five days, shows that their density is large only from 25° S to the southward limit of satellite visibility (73° S). Inside this zone, the distribution is either approximately flat, or shows a peak to the south.

The small number of balloons north of 25° S can be explained on the one hand by a southward meridional circulation in these latitudes (Hadley cell) and on the other hand by the greater icing probability in tropical regions. Most balloons reaching these regions are carried into some restricted areas by permanent deformations of the general circulation (Central Pacific, for example). Therefore, they are not well distributed with respect to longitude and, north of 25° S, statistics on distribution cannot be representative of the whole hemisphere.

3. Mean Zonal Velocity

"Instantaneous" wind velocity is determined by balloon displacement during a time corresponding to one or two satellite orbits (100 or 200 minutes). Mean zonal velocity is calculated over periods ranging from 3 to 7 days according to the density of data. Their duration is chosen so as to yield a total number of data between 1000 and 2000. For one latitude zone of five degrees width, values are first computed in 16 equal areas bounded by meridians, and then the mean is taken over the whole zone. This reduces the influence of areas for which we have a large number of data.

The error of this zonal wind velocity is estimated to be no greater than 3 m s^{-1} between 30° S and 73° S. Equatorward of this zone, the error first increases because of the scarcity of data and then, north of 20° S, the effect described in § 2 causes a systematic overestimation of eastward velocity.

Fig. 1 presents the results for October and November. During this period all the results south of 20° S have an accuracy of 5 m s^{-1} or higher. The modification of mean circulation is striking: there is a rapid slowing down of zonal velocities which induces first two maxima at very different latitudes; subsequent evolution (up to January) is still marked by a decrease of zonal velocities, except in mid-latitudes between 40° S and 50° S. These results do not quite coincide with those

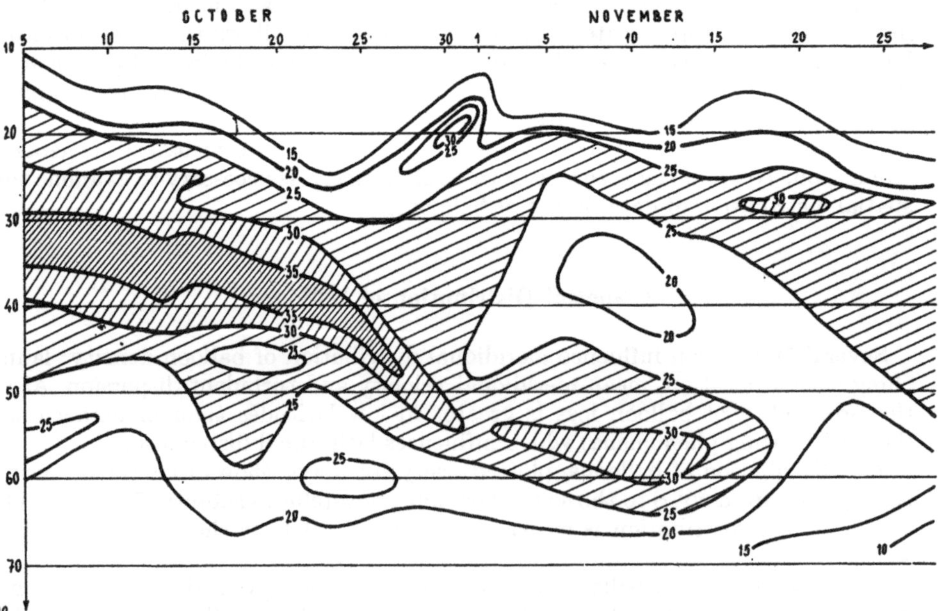

Fig. 1. The evolution of mean zonal velocity (m s^{-1}) with respect to latitude for October and November 1971.

obtained by Solot and Angell [1] from the GHOST test flights, possibly because they had only nine balloons and the bias due to distribution was greater, especially in low latitudes.

4. Mean Meridional Velocity

Unlike zonal velocity, it is very difficult to determine meridional velocity by a conventional method [2, 3]. Indeed, we have to measure mean velocities less than 1 m s^{-1} and the standard deviation to this mean reaches 15 m s^{-1}. Considering the density of the data, we should take a time average over several months to get an error less than 0.1 m s^{-1}.

For this reason, a different method has been chosen: the latitudinal displacements of balloons with respect to time are smoothed by running averages over periods of ten days at least; this eliminates fluctuations of periods smaller than ten days due to large atmospheric waves. Meridional velocities are then inferred from mean latitudinal displacement of balloons over more or less large periods. The chief objection to this sort of method has been pointed out by Solot and Angell [4]. Balloons would be entrained into the jet-stream core and their meridional displacements would be more representative of jet-stream position than of the meridional component of the wind. This argument seems to be questioned by the EOLE results, where there is no evident correlation between the latitudes where the balloon density is maximum and those of zonal wind maximum. Indeed, the zone where balloon density is maximum is often located in light wind regions, for example from 60° S to 70° S.

Preliminary results show that, for the monthly means, meridional winds do not exceed 0.5 m s^{-1} south of 30° S. In November, the velocity is northward north

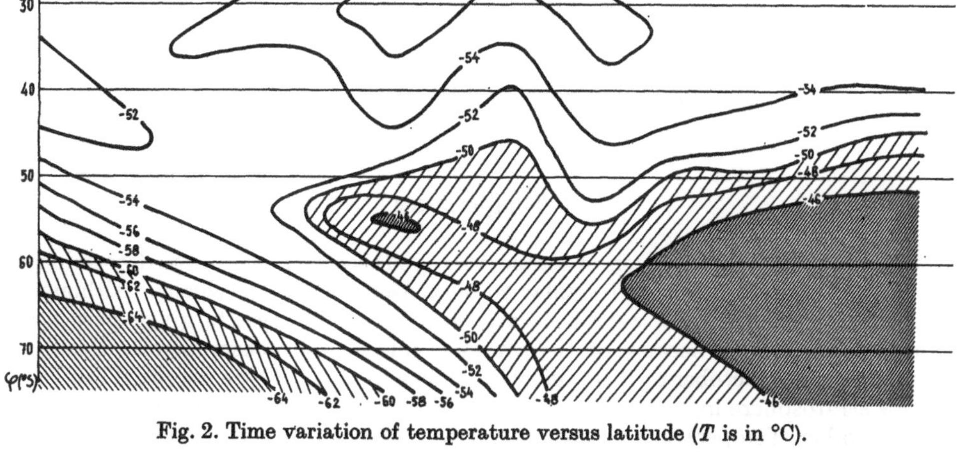

Fig. 2. Time variation of temperature versus latitude (T is in °C).

of 45° S, and southward south of 45° S. In December, it is very small at every latitude. In January and February, it is equatorward. This last circulation is similar to that of a Ferrel cell. These results require further confirmation, for we must study the influence of the chosen method, particularly of the time averaging duration.

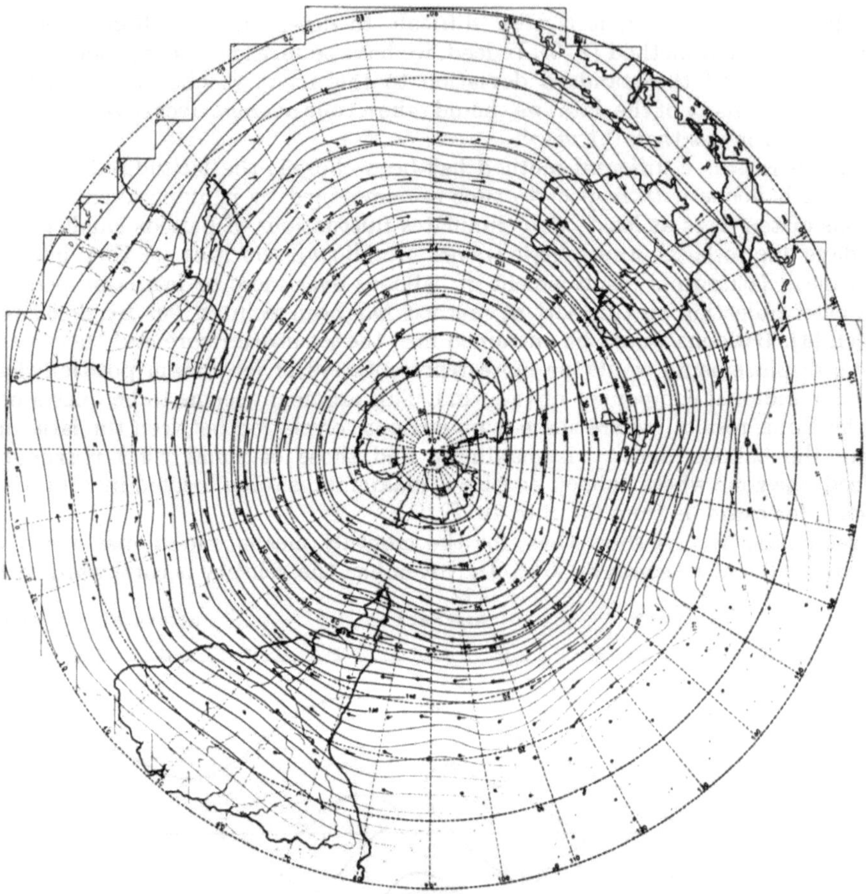

Fig. 3. Streamlines, L.M.D./Eole Oct. 1971.

5. Temperature

The latitudinal distribution of temperature reduced to 200 mb was determined every ten days by a method similar to that used for zonal velocity. Although individual measurements were accurate only to $\pm 2\,°C$, results given here are accurate to $\pm 1\,°C$, because of the available number of measurements. Results obtained from 1 October to 10 January are shown in Fig. 2. There is a rapid heating of low stratosphere in high latitudes during the end of October and the beginning of November. Temperature variations in lower latitude regions are much smaller.

6. Mean Monthly Charts of Streamlines

To know more precisely the seasonal variation of general circulation, mean monthly charts of the streamlines were drawn. Figs. 3 and 4 show two examples of these charts, for October 1971 and January 1972. In October and November South America and Africa do not produce any notable deviation of streamlines,

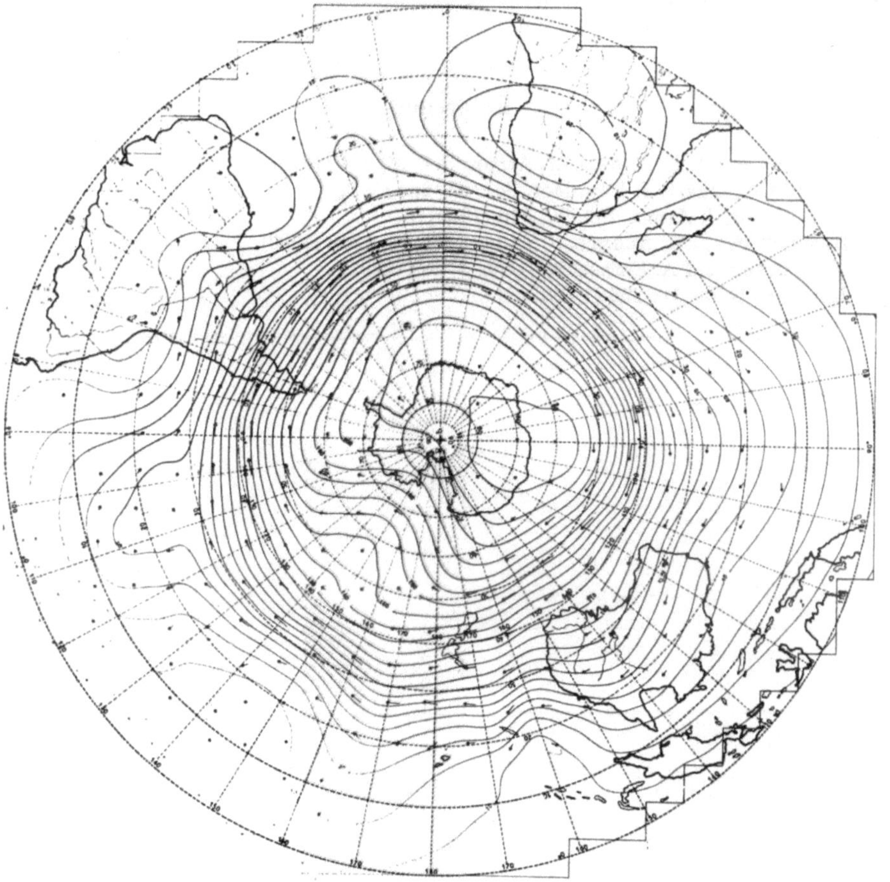

Fig. 4. Streamlines, L.M.D./EOLE Jan. 1972.

but from December on, their influence becomes important. Therefore continental deviations of streamlines seem to have to be chiefly ascribed to the contrast between heat sources and heat sinks. This contrast between oceans and continents, which was very small during the spring, becomes greater during the summer. Continental elevation influence seems to be weak at this level, contrary to the assertions of Charney and Eliassen [5] and Eliasen [6]. Indeed, such an effect would have no reason for changing with seasons.

7. Conclusion

The Eole experiment is especially suitable for the study of the mean characteristics of the atmospheric flow at 200 mb. Indeed, balloons provide a sufficiently good coverage of the southern hemisphere to provide reliable statistical treatment. The most striking feature is the transition between the winter and summer circulations, the latter being less intense and less regular. In high latitudes, this transition is accompanied by a heating of the low stratosphere and a mean poleward meridional velocity.

References

[1] S. B. Solot and J. K. Angell, J. Atmos. Sci. **26**, 574 (1970).
[2] G. B. Tucker, Quart. J. Roy. Met. Soc. **85**, 209 (1959).
[3] G. O. P. Obasi, J. Atmos. Sci. **20**, 516 (1963).
[4] S. B. Solot and J. K. Angell, J. Atmos. Sci. **26**, 1299 (1969).
[5] J. G. Charney and A. Eliassen, Tellus **1**, 38 (1949).
[6] E. Eliasen, Tellus **10**, 206 (1958).

Space Research XIII — Akademie-Verlag, Berlin 1973

ESTIMATION DU CHAMP DE DIVERGENCE A PARTIR DE TRACEURS LAGRANGIENS

G. Necco*

Laboratoire de Météorologie Dynamique du CNRS, Paris, France

L'expérience Eole a permis des mesures directes du champ de divergence du vent à partir des variations relatives avec le temps des aires de triangles formés par des triades de ballons plafonnants.

En introduisant une échelle caractéristique (proportionnelle à la racine carrée de la surface moyenne des triangles considérés dans chaque mesure) on obtient expérimentalement une relation entre la magnitude de la divergence et cette échelle qui montre une variation différente dans les différents domaines du spectre de mouvement: la magnitude de la divergence est inversement proportionnelle à l'échelle pour des mouvements mésosynoptiques et inversement proportionnelle à l'échelle pour des mouvements de dimensions synoptiques ou planétaires avec un possible plateau de transition entre ces deux domaines.

On a developpé aussi une méthode d'analyse spatiale pour obtenir le champ de divergence planétaire dans un intervalle de temps donné. On présente les résultats fournis par cette méthode pour une période de 24 heures et on les compare avec les systèmes nuageux obtenus pour la même journée par le satellite ESSA 9. On observe une coïncidence entre les centres de divergence et de convergence et les systèmes synoptiques principaux qui montre l'applicabilité de la méthode si l'on admet le mécanisme de compensation dite de Dynes.

1. Introduction

Il est bien connu que la divergence du champ de vitesse d'un écoulement fluide bidimensionnel put être exprimée par la dérivée lagrangienne de la surface A (en projection horizontale) d'une parcelle infinitésimale de fluide:

$$\operatorname{div} \boldsymbol{V} = \frac{1}{A}\frac{dA}{dt}.$$

De même, en calculant la variation dans le temps d'une surface limitée par une courbe matérielle, c'est-à-dire composée par les mêmes particules du fluide, on peut estimer la divergence moyenne spatio-temporelle sur l'aire considérée.

Dans les cas de l'atmosphère terrestre, on peut déduire cette dérivée lagrangienne du déplacement de traceurs, des ballons à plafond constant qui suivent l'écoulement atmosphérique. On peut donc, à partir des positions d'un groupe de ballons à différents instants, calculer les variations des surfaces des polygones formés par les traceurs et obtenir ainsi une estimation de la divergence moyenne. Cette méthode directe, lagrangienne, évite les problèmes des précisions posés par le calcul de la divergence à partir d'observations eulériennes. La première application

* détaché du "Servicio Meteorologico Nacional" d'Argentine.

de cette méthode à la mesure de la divergence (et d'autres caractéristiques cinématiques de l'écoulement atmosphérique) a été réalisée par Angell [1] en utilisant les trajectoires d'un groupe de ballons transosondes. L'expérience Eole (décrite dans la communication de P. Morel [2]) fournit un outil particulièrement adapté à l'obtention de la divergence par la méthode citée.

2. Estimation de la divergence

2.1. Représentativité du ballon plafonnant comme traceur lagrangien

Il faut discuter l'aptitude des ballons plafonnants du projet Eole à suivre exactement le mouvement horizontal de l'air dans lequel ils se trouvent, c'est-à-dire la qualité des ballons comme traceurs.

Les résultats de Greif [3] montrent que les ballons de l'expérience Eole effectuent un filtrage inertiel pour les fréquences supérieures à 0.01 hertz. De plus, ils oscillent verticalement autour de leur niveau d'équilibre avec une période propre de 400 secondes environ. On est ainsi assuré que les ballons suivent les mouvements atmosphériques d'échelles mésosynoptiques et supérieures; c'est le domaine dans lequel on pourra mesurer la divergence.

En ce qui concerne la dispersion due au cisaillement vertical du vent, de l'ordre de 1 m s^{-1} par 100 m dans sa structure fine (voir p. e. Levi [4]), et si l'on admet des fluctuations typiques du niveau de vol de 20 m, l'aire d'indétermination de localisation d'un ballon après deux heures (borne supérieure de l'intervalle de temps entre deux orbites successives) sera un cercle d'environ 1.5 km de diamètre.

Par ailleurs, les effets de la dispersion horizontale par diffusion turbulente sont inférieurs d'un ordre de grandeur, au moins, à la magnitude de la divergence, étant donné le court intervalle de temps considéré (110 minutes).

2.2. Calcul de la divergence

On s'est limité à l'étude des variations d'aire entre deux orbites successives, des triangles formés par trois ballons. Les aires ont été calculées directement sur le domaine sphérique. Dans l'écoulement fluide l'arc de grand cercle matériel joignant les sommets initiaux quelconques va être déformé. En admettant que le champ du vent est relativement continu et régulier, cet effet de deuxième ordre a été négligé sur l'intervalle de temps considéré.

Les ballons étant interrogés séquentiellement par le satellite, l'instant précis de la localisation des différents sommets d'un triangle est différent pour une orbite donnée: on définit le "temps de localisation d'un triangle" comme la moyenne des temps de localisation des trois sommets. Une correction différentielle (interpolation) est nécessaire pour ramener les mesures de position brutes des trois ballons à un temps unique. Pour éviter des corrections trop importantes on s'est limité aux triangles dont les sommets sont localisés dans un intervalle de six minutes au maximum. On élimine aussi dans la sélection des triangles ceux qui par leur forme ne donnent pas une estimation correcte de la variation relative d'aire (voir plus bas). Finalement, la divergence moyenne est estimée par la formule:

$$\operatorname{div} \boldsymbol{V} = 2(A_f - A_i)\,\{A_f + A_i)\,(T_f - T_i)\}^{-1}$$

où: A_i, A_f sont les aires sphériques initiale et finale évaluées à partir des positions corrigées, et T_i, T_f sont les temps moyens de localisation des triangles initial et final.

Les valeurs ainsi calculées sont considérées comme représentatives de la valeur moyenne de la divergence sur l'intervalle de temps considéré et sur un domaine spatial centré au barycentre des six sommets et d'aire égale à l'aire moyenne des deux triangles. Bien que les fichiers de position (préliminaire) utilisés ne contiennent pas encore une estimation individuelle des erreurs de localisation, on a déduit une erreur moyenne de 2 km (1σ) pour les positions brutes.

3. Relation entre la magnitude de la divergence et l'échelle de mouvement

Les données de l'expérience ont permis l'évaluation d'un très grand nombre des valeurs individuelles de la divergence à partir desquelles on a effectué une étude statistique de l'ordre de grandeur de la div en fonction de l'échelle des triangles initial et final. La décroissance de la magnitude de la divergence horizontale du champ de vents avec l'échelle est un fait bien établi en météorologie depuis longtemps. Par exemple, Panofsky [5] propose une loi de variation de la divergence inversement proportionelle à l'échelle.

On a éliminé les triangles à forme très allongée par le critère conventionnel suivant (appliqué aux triangles initiaux et finaux): si A, B et 1 sont les rapports entre la longueur des côtés du triangle et la longueur du côté minimum, on

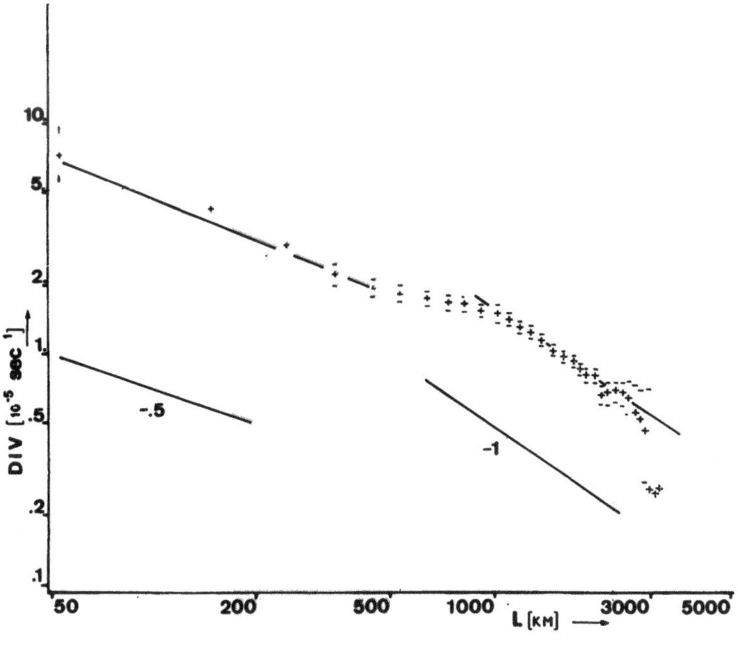

Fig. 1. Divergence (unité $10^{-5}\,s^{-1}$) en fonction de l'échelle (en km). Moyenne sur tous les échantillons.

rejette tout triangle tel que $|A - B| \geqq 0.9$. On rejette aussi les cas où l'aire finale change de signe par rapport à l'aire initiale.

Les différents échantillons analysés (10 jours régulièrement répartis entre le 15 octobre 1971 et le 15 février 1972) montrent la variation de la magnitude de la divergence qui ne peut être interprétée par une loi unique (div $\sim L^{\alpha}$) pour toutes les échelles. En effet, les valeurs correspondant à des triangles d'échelles caractéristiques inférieures à 400 km sont bien représentées par une loi de pente $\alpha = -0{,}5$ tandis que celles qui correspondent à des échelles supérieures à 1000 km sont représentées par une pente $\alpha = -1$ (Fig. 1) (la brusque chute des valeurs à l'extrémité supérieure du spectre est due en partie au fait qu'on arrive à la limite d'applicabilité de la méthode; de plus, on ne dispose que d'un nombre de cas très réduits dans ces intervalles ce qui en restreint la signification statistique). Tout cela semble indiquer un comportement spectral différent pour les processus d'échelles mésosynoptiques et pour les processus d'échelles synoptiques et supérieures, avec un possible plateau de transition (quasi-constant?) entre ces échelles.

On notera finalement que les résultats sont bien comparables aux valeurs estimées par différents auteurs pour la divergence à des niveaux proches de 200 mb: Landers [6] calcule à 8000 m, environ, une moyenne de l'ordre de $1{,}5 \cdot 10^{-5}\,s^{-1}$ pour des échelles synoptiques; Lateef [7] dans les tropiques, sur des aires carrées de 4 degrés de côté obtient à 200 mb des maxima de l'ordre de $0{,}6 \cdot 10^{-5}\,s^{-1}$; Fankhauser [8] sur des aires de 200 km $\times$ 200 km à 225 mb obtient des valeurs de l'ordre de $0{,}5 \cdot 10^{-4}\,s^{-1}$; Krishnamurti et Rodgers [8] sur des cartes planétaires moyennes à 200 mb observent des valeurs de l'ordre de $2-4 \cdot 10^{-6}\,s^{-1}$.

4. Champs moyens spatiaux

Dans le but d'obtenir un champ hémisphérique moyen sur une période donnée (disons 12 ou 24 heures) on a essayé une méthode d'analyse adaptée au type de mesure utilisé pour la divergence. On considère un réseau sphérique uniforme dans le plan $[\lambda, \phi]$; λ: longitude, ϕ: latitude, formé de points séparés de 5 degrés en latitude et en longitude. Autour de chaque point du réseau, on considère une "aire de référence" carrée de 555,55 km de côté.

Pour calculer la divergence en un point du réseau, on prend toutes les mesures comprises dans la période choisie et on moyenne les valeurs correspondant à des triangles d'échelle L comprise entre 500 et 2000 km et dont les aires caractéristiques (L^2) chevauchent ou interceptent l'aire de référence du point envisagé en pondérant chaque mesure avec un poids proportionnel à la surface d'intersection relativement à l'aire de référence. Comme le spectre spatial est limité approximativement au domaine synoptique, la magnitude de la divergence sera de l'ordre de $10^{-5}\,s^{-1}$; donc, étant donné les intervalles de temps sur lesquels on l'estime ($\sim$2 heures), la surface des triangles ne variera au plus que de 15%, permettant ainsi un critère d'élimination plus serré à partir de la forme des triangles.

On a appliqué la méthode aux ballons localisés dans la journée du 25 janvier 1972. Le résultat est le champ de Fig. 2 obtenu à partir de la moyenne de 4418 mesures de divergence. La comparaison du champ ainsi obtenu avec les systèmes nuageux synoptiques obtenus à partir d'une mosaïque de photos prises par le satellite ESSA 9 pour le même jour (Fig. 3) montre une coïncidence notable entre les systèmes cycloniques et anticycloniques et les maxima de divergence et convergence. On note par exemple que le train de systèmes cycloniques de centres situés aux points 30° E 60° S, 90° E 50° S, 135° E 55° S et 175° E 60° S est surmonté par des zones de divergence et que de plus la position des anticyclones situés entre ces cyclones coïncide avec les centres de convergence situés à 85° E 45° S, 120° E 45° S et 160° E 55° S qui ont d'ailleurs à peu près la même intensité. Les centres de divergence sur le continent australien et sur l'Est de la Nouvelle-Zélande (ce dernier relativement intense) surmontent des tourbillons cycloniques.

La vaste zone de divergence située entre les longitudes 120° W et 130° W coïncide avec un système cyclonique multiple assez developpé et il est intéressant de noter que la langue de convergence située à 135° E 45° S correspond à la zone de ciel clair située entre deux ensembles nuageux. Cependant, il existe des centres de convergence et divergence qui ne surmontent pas un système donné dans la troposphère et qui peuvent correspondre à des régions de grandes advections de vorticité, comme par exemple dans un "high level trough"; mais néanmoins sur presque chaque système cyclonique—anticyclonique ou maximum de vorticité identifiable par la configuration nuageuse il existe un centre de divergence ou

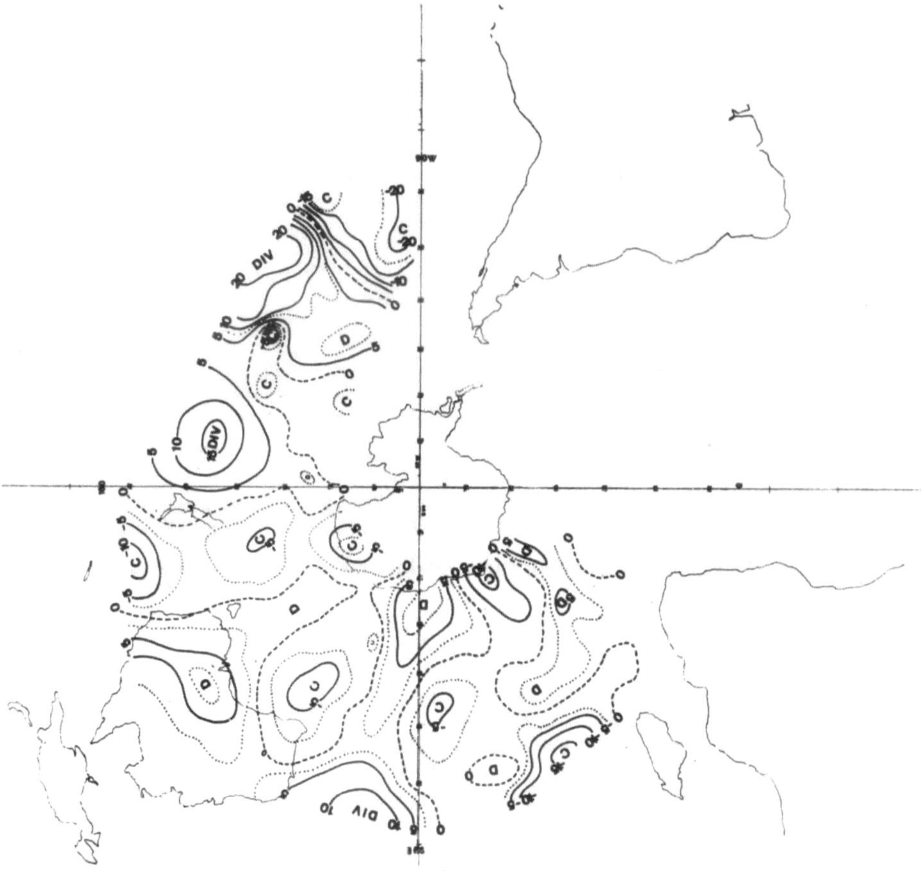

Fig. 2. Champ de divergence moyen sur 24 heures le 25 janvier 1972. Unités 10^{-6} s^{-1}.

convergence à 200 mb. Donc, bien que les ballons se trouvent pratiquement dans l'écoulement stratosphérique à partir de 40°—50° S, on retrouve le mécanisme de compensation de Dines.

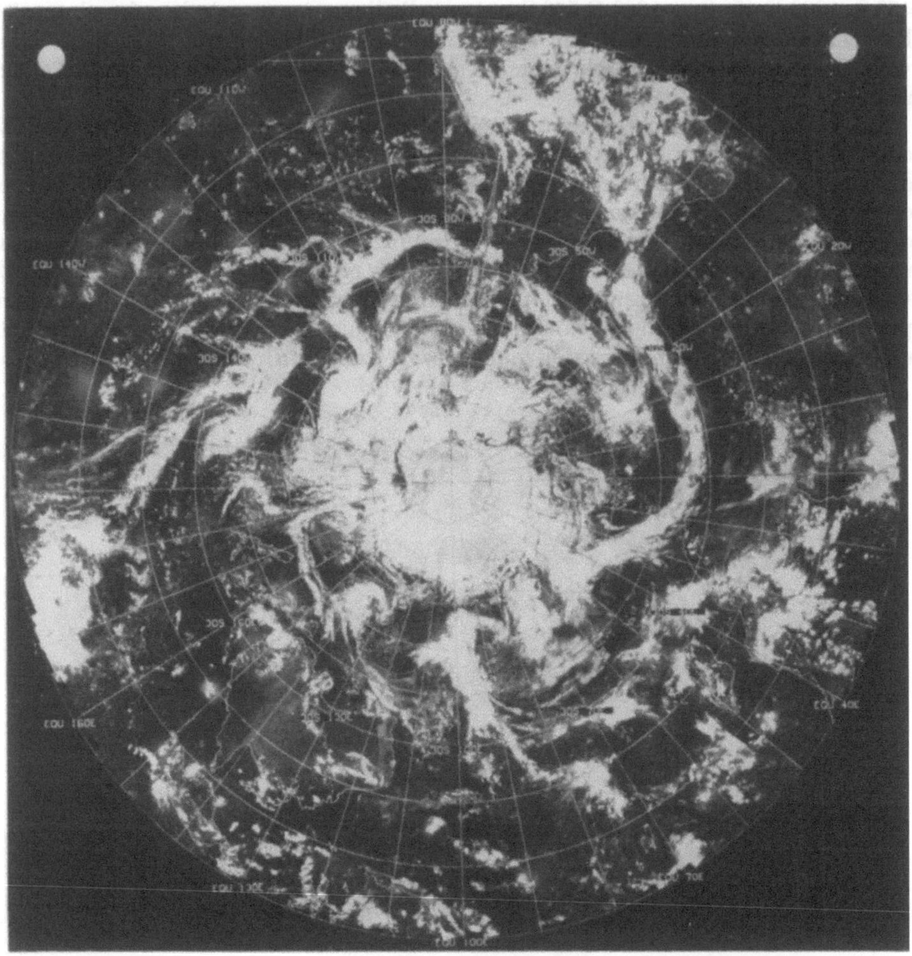

Fig. 3. Champ nuageux obtenu par le satellite ESSA 9, le 25 janvier 1972.

References

[1] J. K. ANGELL, Adv. in Geophys. 8, 137 (1961).
[2] P. MOREL, Space Research XIII, 129 (1973).
[3] M. GREIF, Thèse de 3ème cycle, Fac. Sciences, Université de Paris, (1970).
[4] D. LEVI, Thèse de 3ème cycle, Fac. Sciences, Université de Paris, (1970).
[5] H. H. PANOFSKY in: Compendium of Meteorology, Ed. T. Malone, (p. 639) (1951).
[6] H. LANDERS, J. Met. 12, 415 (1955).
[7] M. A. LATEEF, Mon. Weath. Rev. 95, 778 (1967).
[8] J. C. FANKHAUSER, J. Appl. Met., 8, 778 (1969).
[9] T. N. KRISHNAMURTI et E. B. RODGERS, Flo. State Univ.; Report N° 70—2 (Dec. 1970).

Space Research XIII — Akademie-Verlag, Berlin 1973

TURBULENT DISPERSION — EOLE EXPERIMENT

M. Larcheveque

Laboratoire de Météorologie Dynamique du CNRS, Paris, France

The results of the Eole experiment on the dispersion of Lagrangian tracers in a quasi-stationary twodimensional turbulent flow, show that the rms distance of paired balloons follows an exponential law versus time. Dimensional arguments based upon the specific behaviour of a two-dimensional flow justify this law.

1. Introduction

For the first time, a great many data about dispersion of constant level balloon flights are provided by an experiment (Eole experiment). This paper presents the analysis of the results obtained. The main conclusion is that a power law, in $t^{3/2}$, predicted by Batchelor [1] for the rms distance of two tracers versus time in homogeneous three-dimensional flow, or in $t^{5/2}$, proposed by Morel [2], to take the influence of a shear flow into consideration, does not explain the numerical results computed. Dimensional arguments justify these laws for a three-dimensional flow. Thus this failure can be explained by the two-dimensional behaviour of the atmospheric flow for the scales considered. Lin [3] has been the first to prove that, in this case, an exponential law is derived from dimensional arguments. In this paper, slightly different reasoning leads to the same law which fits well the experimental results obtained.

2. Theoretical Considerations

The intermediate scale of meteorological flow can be considered as two-dimensional and incompressible, and it is well known that, in this case, enstrophy (half mean squared vorticity) as well as kinetic energy integrated over all the fluid are constants. If homogeneity is assumed, this first invariant has profound effects on the inertial energy transfer [4, 5], and an inertial enstrophy cascading range exists with $E(k) \sim \eta^{2/3} k^{-3}$ where η is the cascade rate of enstrophy,

the kinetic energy is $\int_0^{+\infty} E(k)\, dk$

and the enstrophy is $\int_0^{+\infty} k^2 E(k)\, dk$.

Now let us consider pairs of Lagrangian particles with mean square distance at time t equal to $\overline{D^2}(t)$ in such a way that $(\overline{D^2})^{-1/2}$ belongs to the inertial enstrophy cascading range. We may suppose that the dispersion rate is a function of the enstrophy spectrum $k^2 E(k)$ and of k only, integrated over all the k of this inertial range greater than or equal to $(D^2)^{-1/2}$: indeed, vorticity is the curl of velocity and takes into consideration velocity gradients which are efficient for particle separation and further, the transfer of enstrophy is non-local in the inertial range considered above [5]. Consequently, we can write:

$$\frac{d\overline{D^2}}{dt} = \int_{k_0}^{k_*} f(k^2 E(k), k)\, dk \quad \text{with} \quad k_0 = [\overline{D^2}]^{-1/2} \quad k_* \gg k_0 .$$

Dimensional arguments show that

$$\frac{d\overline{D^2}}{dt} = A \int_{k_0}^{k_*} E^{1/2} k^{-3/2}\, dk \qquad (A \text{ constant}).$$

Consequently with $E(k) \sim \eta^{2/3} k^{-3}$ we obtain, neglecting the term k_*^{-2}

$$\overline{D^2} \sim \exp(Bt) \tag{1}$$

$$B \sim {}^1/_2 A \eta^{1/3}.$$

Therefore, in a two-dimensional, incompressible, turbulent, homogeneous and stationary flow, we expect an exponential law for $\overline{D^2}$ versus time.

But the atmospheric flow is not exactly homogeneous. A first approach of shear flow is developed by Morel [2] who shows that a mean shear induces:

$$\overline{x^2} \sim t^2 \left(\frac{\partial \overline{u}}{\partial y}\right)^2 \overline{y^2} \tag{2}$$

where $\overline{u}$ is the mean zonal velocity (in the x direction), y the meridional direction and the $\overline{y^2}(t)$ law the same as in the homogeneous case.

3. Data Analysis

483 constant level balloons (200 mb) have been launched from three stations in Argentina: Mendoza (33° S 291° E; 100 balloons), Neuquen (39° S 292° E; 202 balloons) and Lago Fagnano (54° S 293° E); 181 balloons), during September, October and November 1971. They are located by the satellite EOLE which covers the southern hemisphere in twelve hours with a period of 100 minutes. The fact that the satellite cannot cover all the hemisphere at the same time does not allow synoptic interrogations of two balloons separated by a distance greater than three or four thousand kilometres. Consequently, the dispersion results are obtained for balloon distances increasing from 50 kilometres to 3000 kilometres.

3.1. Means Computation

In a first step the clusters used for computation are composed of two balloons launched from the same site, at almost the same time (the difference between launching times is less than forty minutes). The distance $\overrightarrow{B_1 B_2}$ between the two balloons of a cluster is split into two components x and y. Two kinds of coordinates are used: first along the zonal (x) and the meridional (y) directions, and then along directions moving with the middle M of $B_1 B_2$; more precisely in this second case, the trajectory of M between two successive localizations $\overrightarrow{B_1 B_2}(t_1)$ and $\overrightarrow{B_1 B_2}(t_2)$, ($t_2 - t_1 \leqq 10$ hours) is defined as the great circle running between $M(t_1)$ and $M(t_2)$; the x direction is tangent to this circle and the y direction normal to the x direction. In each case, derivatives are calculated by a linear method from two successive localizations. (The small difference $t_2 - t_1$ justifies these linear approximations.)

The means are computed in the following way: the results obtained for a two-balloon distance belonging to an interval $[D_n, D_{n+1}]$ are averaged and provide $\overline{x^2}$, $\overline{y^2}$, $\overline{dx^2}/dt$, $\overline{dy^2}/dt$. But with this paired balloons choice, we have only three data groups corresponding to the three launching stations located in the same region. Consequently, to enlarge the field of results, the same computations are done with other groups: we chose all the groups of balloons composed of two balloons flying 250 kilometres apart at most, over a region defined by $\lambda_1 \leqq \lambda \leqq \lambda_2$ and $\phi_1 \leqq \phi \leqq \phi_2$ at a time t ($t_1 \leqq t \leqq t_2$), (λ longitude, ϕ latitude). Generally, we have taken $\lambda_1 + 50°$ S, $\lambda_2 = 30°$ S, $\phi_2 - \phi_1 - 20°$ and $t_2 - t_1 = 30$ days, to ensure statistical significance.

3.2. Results

The results for groups launched from one station (Lago Fagnano, 118 groups) are summarized in Fig. 1 and those given for groups chosen through the second method ($\phi_1 = 50°$ S, $\phi_2 = 30°$ S, $\lambda_1 = 80$ °E, $\lambda_2 = 100°$ E, during October 1971, 100 groups), by Fig. 2. The obvious conclusion is that these results fit well an exponential law in accordance with (1). Fig. 3 shows a very interesting result: for distances less than 600 kilometres $(\overline{x^2})^{1/2}$ is nearly equal to $1.4(\overline{y^2})^{1/2}$.

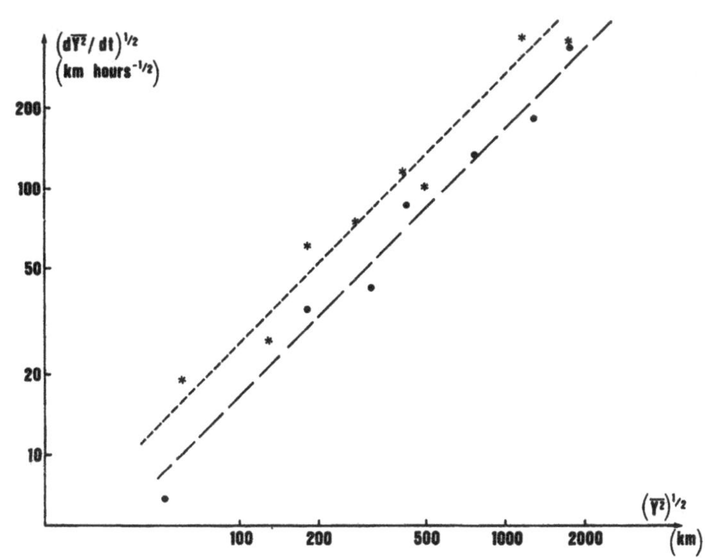

Fig. 1. Time means: • longitude—latitude frame, * "moving frame".

But for greater distances this equality becomes: $(\overline{x^2})^{1/2} \sim (\overline{y^2})^{1/2} \log \overline{D^2}$. Relation (1) provides $\log \overline{D^2} \sim t$. Thus relation (2) is satisfied for distances greater than 600 kilometres and in this range of distances the influence of shear flow is important.

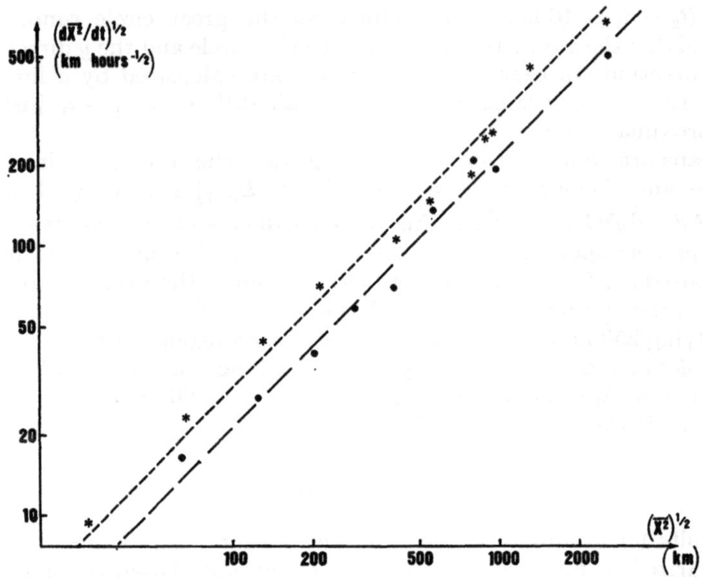

Fig. 2. Space means: • longitude—latitude frame, * "moving frame".

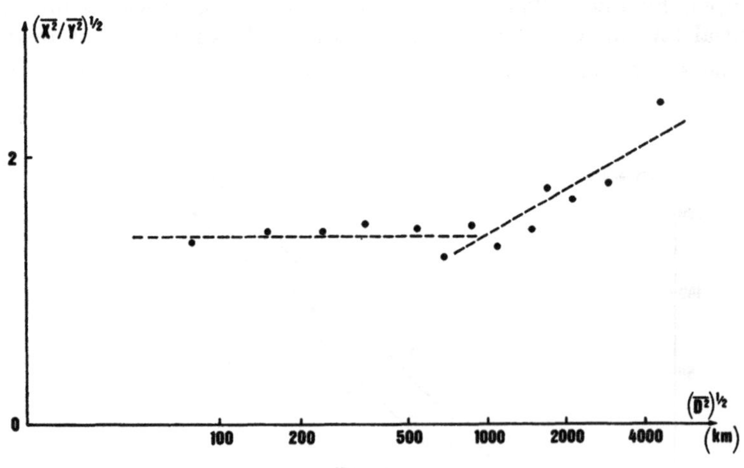

Fig. 3.

References

[1] G. K. BATCHELOR, Quart. J. Roy. Met. Soc., **76**, 133 (1950).
[2] P. MOREL, Annls Géophys., **26**, 815 (1970).
[3] J. T. LIN, Private communication, NCAR, Boulder, Col., (1971).
[4] G. K. BATCHELOR, Phys. Fluids, **12**, 233 (1969).
[5] R. H. KRAICHNAN, Phys. Fluids, **10**, 1417 (1967).

Space Research XIII — Akademie-Verlag, Berlin 1973

TEMPERATURE AND WIND IN THE UPPER ATMOSPHERE OVER THE ANTARCTIC

S. S. Gaigerov and Yu. P. Koshelkov

USSR

Seasonal temperature and wind variations in the Antarctic stratosphere and lower mesosphere are considered on the basis of meteorological rocket firings at the Molodezhnaya station (68° S 46° E). Wind speeds in the upper stratosphere in summer are close to those known for the northern hemisphere, and wind direction is nearly zonal. The period with easterly winds is shorter in the Antarctic than in the Arctic. In winter, westerly stratospheric flow is definitely more intense in the southern hemisphere than in the northern.

The temperature regime of the Antarctic stratosphere is characterized by a great seasonal range. Temperature maximum and minimum in the annual cycle are shifted to late spring and late autumn respectively.

Since the middle of 1969 soundings of the upper atmosphere by means of M-100 meteorological rocket have been made at the Molodezhnaya Station (68° S 46° E) About 120 successful flights took place during the period from July 1969 to February 1972, approximately from 2 to 5 per month.

Monthly mean values of meteorological parameters have been calculated that enabled time sections to be constructed and seasonal variations considered. The winter temperature minimum in the stratosphere occurs in different months at different levels. Both in 1970 and in 1971 in the 30—45 km layer the minimum was recorded in May and June and shifted to the autumn above 45 km. The same phase change, though less pronounced, is notable over the Heiss Island (80.5°N 58° E) [1], but almost absent in the new reference atmosphere models [2] in the polar region.

The time of the spring—summer temperature maximum was also variable with altitude, changing from December at 20 km to November (1969) or October (1970 and 1971) from 30 to 40 km. At 45—50 km the peak values were delayed until December or January, though a relatively high value had been recorded in the upper stratosphere already in spring. No similar phase variation with height has been reported for the Arctic [1, 2].

A reversal of sign of seasonal temperature variation was observed over Molodezhnaya station at about 60—65 km which is in agreement with the northern hemisphere data.

Absolute temperatures in the lower half of the Antarctic winter stratosphere are known to be lower than those in the Arctic. As the comparison of Molodezhnaya and the northern hemisphere rocket data shows (Fig. 1), in the upper stratosphere this discrepancy weakens and can be traced only in the early winter. On the other hand, in July and August the rapid increase of temperature with height is evident in the upper stratosphere over Molodezhnaya station which results in higher values over about 35 km compared with the Arctic data for January and February.

Summer temperature records in the Antarctic stratosphere up to the height of 40—45 km proved to be several degrees higher than in the Arctic, in accordance with earlier findings [3—5].

The range of seasonal temperature variations in the polar stratosphere of the southern hemisphere is greater than that of the northern hemisphere. Also, the level of maximum range in the Antarctic is located lower in the stratosphere than

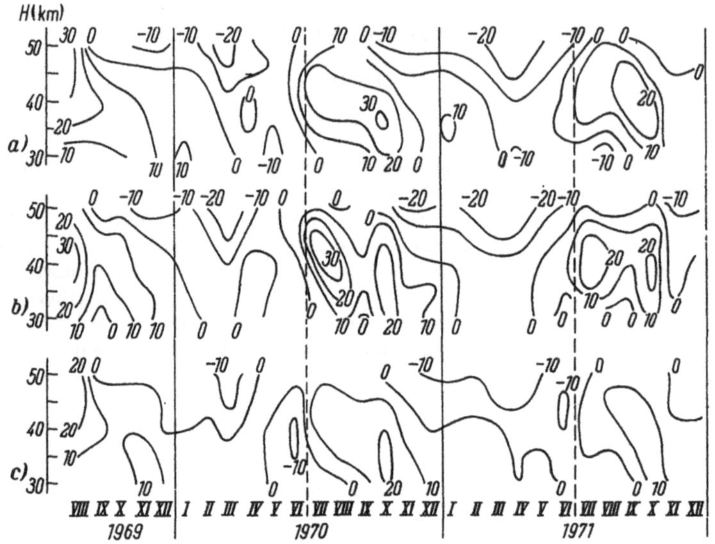

Fig. 1. Deviations of monthly mean temperatures at Molodezhnaya Station from corresponding values for the northern hemisphere: a) from [2], 70°; b) from Thule, 16° N; c) from Heiss Is., 80° N.

in the Arctic. Over the Molodezhnaya station the range amounts to 40—45° at 30—40 km and decreases with altitude. This is in agreement with McMurdo and research vessel data [6—8] and is about 10° higher than obtained at appropriate latitudes in the Arctic.

It is difficult now to make a reliable estimate of the temperature range over the Molodezhnaya Station for the levels above 40 km. Smaller values of the range are characteristic for these higher levels in comparison with lower ones. The values of the range may become comparable in the two hemispheres at these levels.

As for the wind regime of the Antarctic middle stratosphere, it is characteristic that the spring reversal of zonal wind direction initially takes place in the middle latitudes and reaches the Antarctic continent from 2 to 3 weeks later as noted in [4]. In 1970 at the 35 km level easterly winds appeared over Kerguelen Is. on 26—27 October and over Molodezhnaya [9] in mid-November. The vernal reversal of circulation over both stations is first noted in the upper stratosphere and then spreads into the lower layers, which corresponds to the McMurdo results [6]. This agrees with the fact that in the middle stratosphere the warm region advancing in the southward direction during spring [4] is located farther to the south than the high pressure belt and is responsible for the southward tilt of the latter. This specific feature of the vernal processes in the southern hemisphere may reflect the influence of the lower atmospheric layers with their rapid decrease of pressure towards high southern latitudes.

The vernal change of dominant wind direction over Molodezhnaya Station in the upper stratosphere usually takes place in November, while the date of change may vary from year to year. This time variability, however, is far less marked than is known for the northern hemisphere.

In spring 1970 northerly wind components were prevailing in the stratosphere over the southern part of Indian Ocean showing a pressure increase towards the Australian sector which might be the region of high pressure development. A similar pressure distribution is known to have occurred in previous years. In 1969 and 1971, however, meridional components in the upper stratosphere over Molodezhnaya Station in spring were not so stable, probably reflecting considerable wave activity in these years.

The autumn transfer from easterly to westerly winds in 1970 progressed somewhat earlier at the Molodezhnaya Station than over Kerguelen Is., depicting a rapid retreat of the high pressure belt from polar to middle latitudes. Since in February 1970 the meridional temperature gradient in the Kerguelen—Molodezhnaya region at 20—35 km was still of a summer character, the high pressure belt was tilted to the polar region. As a result the easterly winds over Molodezhnaya Station persisted for a longer period at the height of 40—45 km rather than at 20—30 km. In early March 1970 when the wind reversal was observed over Kerguelen Is. in the middle latitudes it was nearly simultaneous in the height region 25—43 km [9].

According to preliminary data in the Antarctic stratosphere the summer period with easterly circulation grows in duration with height more rapidly than in the Arctic, the difference accounting for the fading away of the tropospheric influences on stratospheric circulation with increasing height.

Similar effects are displayed in the character of the vertical profile of zonal wind in summer. In the middle stratosphere (30 mb level) the easterly wind speed is somewhat lower in the Antarctic than in the Arctic [4]. Judging by the December—January firings at Molodezhnaya Station wind speed values above 28—30 km may be in excess of those typical for 70° N (determined by interpolation of Heiss—Volgograd data, or taken from [10]).

In winter mean speeds of westerly wind over Molodezhnaya Station as well as over Kerguelen Is. are considerably higher than those in the northern hemisphere as supposed previously for the South Indian Ocean on the basis of research vessel soundings [11]. It is of interest to note the maximum of westerly wind speed reported at the height of 40 km at Molodezhnaya corresponding to the similar maximum in McMurdo data. This fact supposes the reversal of meridional temperature gradient in the uppermost stratospheric layers, the central polar region being warmer than the periphery of the continent. Further data are needed to prove or disprove this, as well as the other findings outlined here.

References

[1] S. S. Gaigerov, L. M. Kolomiitseva and D. A. Tarasenko, Space Research XII, 637 (1972).
[2] G. V. Groves, paper presented at COSPAR XII Plenary Meeting, Leningrad 1970.
[3] W. L. Godson, Meteor. Abhandl. Fr. Univ. Berlin, 36 (1963).

[4] YU. P. KOSHELKOV, Trudy CAO, No. 99 (1971).
[5] H. WEXSLER, Quart. J. Roy. Met. Soc. 85, (1959).
[6] R. S. BRIGGS, J. Appl. Met. 4, No. 2 (1965).
[7] S. S. GAIGEROV, YU. P. KOSHELKOV and M. YA. KALIKHMAN, Trudy CAO, No. 93 (1970).
[8] S. S. GAIGEROV, Polar Meteorology, Tech. Notes, No. 87 (1967), WMO No. 211, TP. III.
[9] YU. P. KOSHELKOV, Meteorol. i Gidrol. No. 4 (1972).
[10] G. V. GROVES, Space Research X, 137 (1970).
[11] YU. P. KOSHELKOV, Meteorol. i Gidrol. No. 9 (1969).

Space Research XIII — Akademie-Verlag, Berlin 1973

THE ESTABLISHMENT OF THE WINTER POLAR VORTEX IN MIDDLE LATITUDES IN 1971

A. Azcárraga and J. M. Cisneros

Instituto Nacional de Téchnica Aeroespacial "Esteban Terradas", Madrid, Spain

Several rocket launches, to evaluate the summer to winter change of stratospheric winds from easterlies to westerlies, and thus the establishment of the polar vortex in the northern hemisphere, have been made from various stations in middle latitudes on both sides of the Atlantic. The change in the stratospheric circulation index is measured and a comparison is made with the circulation over the southern hemisphere at the same time as the change happens.

1. Introduction

Taking advantage of the cooperative programme between Spain and the US to study the circulation and behaviour of the upper atmosphere on both sides of the Atlantic Ocean, plus the existence of the Exametnet network, an analysis has been made of the establishment of the winter circulation pattern in the northern hemisphere and its cooperation with the breakdown of that in the southern hemisphere.

The data came from three stations: Wallops Island in the US (37.8° N 75.5° W), Arenosillo in Spain (37.1° N 6.7° W) and Mar Chiquita in Argentina (37.8° S 57.4° W). All stations are thus conveniently located at the same latitude (change of hemisphere notwithstanding) and latitudinal variations do not need to be taken into account. The rockets used were Loki instrumented meteorological systems in both Mar Chiquita and Arenosillo, and Loki and Arcas type at Wallops Island. Due to the consistent failures of the WOX 1-A temperature sensors at Arenosillo only wind comparisons are available.

2. Presentation of Data and Comments

Figs. 1, 2, and 3 represent the derived zonal component of the winds. Fig. 4 shows the stratospheric circulation indexes (SCI) of Arenosillo and Wallops stations for September 1971. First it may be noticed that the Mar Chiquita data show that change to easterly winds, i.e. from winter to summer conditions, takes place in October, whereas the Arenosillo and Wallops data show that the change occurs in September. This means that the break-up of the polar vortex in one hemisphere is not symmetrical with its build-up in the opposite hemisphere. In other words, the "winter situation" lasts longer than the "summer situation" in both hemispheres. Secondly, in Arenosillo some observations indicate no wind;

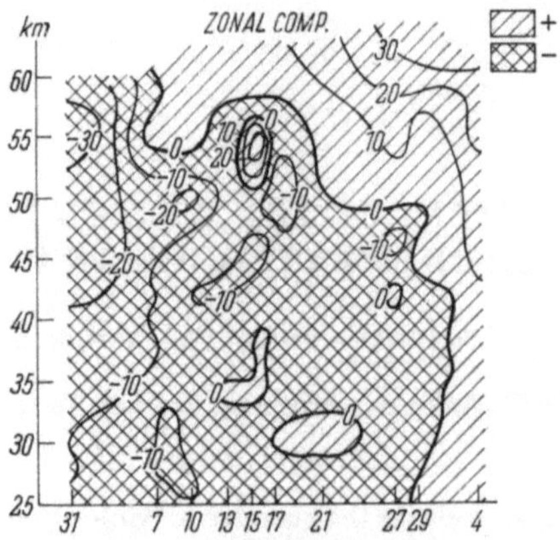

Fig. 1. Zonal components of winds on Wallops. September 1971; + westerly (winter), — easterly (summer). The velocity contours are given in units of m s^{-1}.

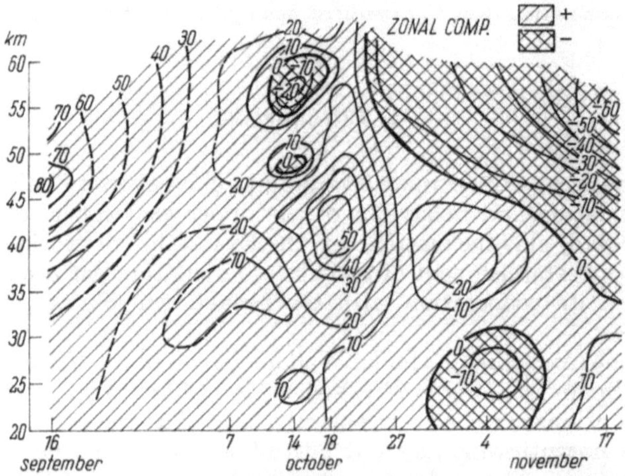

Fig. 2. Zonal components of winds on Mar Chiquita. September, October, November 1971. Rey and contours as in Fig. 1.

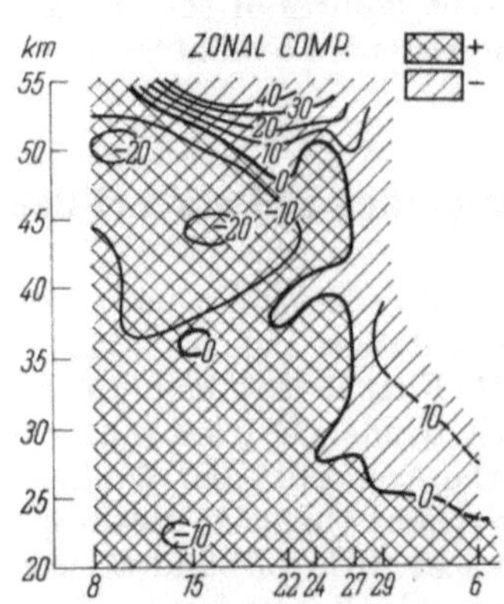

Fig. 3. Zonal components of winds on Arenosillo. September 1971. Rey and contours as in Fig. 1.

thus, the crucial point of the change of wind system was actually recorded. Thirdly, the lack of any points on the Arenosillo data between 15 and 22 September makes it difficult to compare the SCI's in Fig. 4.

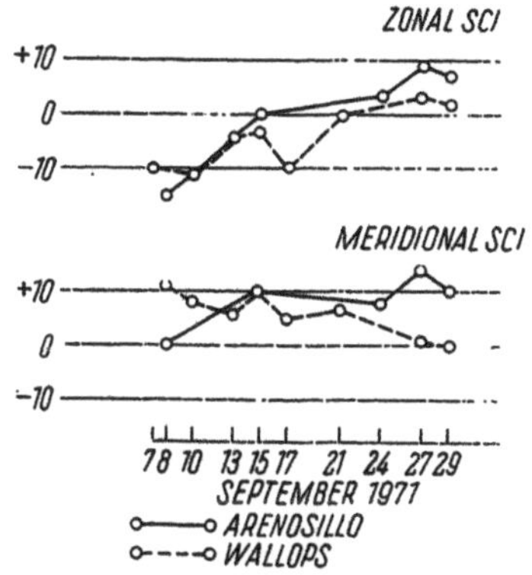

Fig. 4. Zonal and meridional SCI on Arenosillo and Wallops. September 1971. Westerly +, easterly —. Velocity in m s^{-1}.

3. Data Analysis

Before the winter circulation is established there is a consistent southerly flow. Such a flow is not apparent in Mar Chiquita, but there the change is the opposite, from winter circulation to summer circulation. Looking at Fig. 4, this meridional flow is clearly indicated by the SCI. This feature was not noticed in earlier measurements [1].

It is evident that at both northern stations the change follows exactly the pattern given by the CIRA models [2]. The only significant difference is that, as in past years [3], western winds over Arenosillo are established approximately one week earlier than over Wallops Island. At both stations, however, the change-over is rapid and moves from the upper layers to lower layers with some very minor discrepancies (Figs. 1 and 3).

In September at Mar Chiquita the wind has its winter circulation, and the change occurs one month later, in October (Fig. 2). Also here the change goes swiftly from the upper layers keeping always in mind that this is the opposite change, i.e. from winter to summer.

It is evident that at the latitude level of 37°, irrespective of hemisphere, the winter polar vortex lasts more than half a year, of course not taking into account sudden warmings with their possible break-up of the circulation pattern.

References

[1] A. Azcárraga and L. Sánchez Muniosguren, in: Progress in Astronautics and Aeronautics, 22, 522 (1969).

[2] CIRA 1965 (COSPAR International Atmosphere 1965), North-Holland Publ. Co., Amsterdam, 1965.

[3] Data Report Meteorological Rocket Network Firings. September 1968. Vol. V No. 9; September 1969. Vol. VI No. 9 World Data Center A — Meteorology.

Space Research XIII — Akademie-Verlag, Berlin 1973

THE STRATOSPHERIC CIRCULATION OVER MIDDLE ATLANTIC LATITUDES IN THE 1966—1971 PERIOD

J. M. Cisneros

Instituto Nacional de Técnica Aeroespacial "Esteban Terrada", Madrid, Spain

Meteorological sounding rockets have been launched from Arenosillo (37°6′ N and 6°44′ W) in cooperation with NASA Wallops Station. Both the stratospheric circulation and the stratospheric winds measured during sudden warmings which took place in January 1969, January 1970 and February 1972, have been compared.

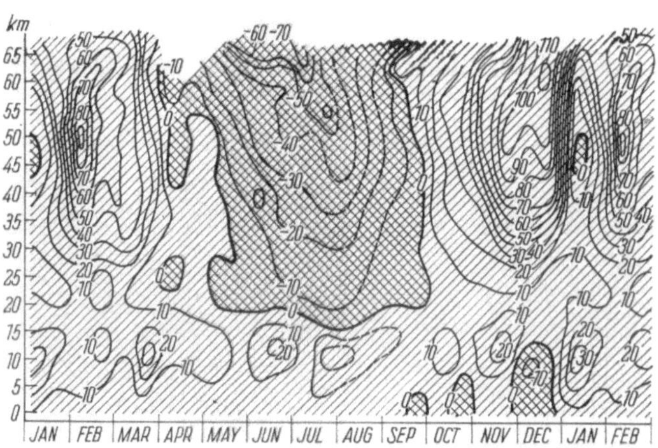

Fig. 1. Arenosillo 1966—1971 mean zonal components of winds; velocities (m s^{-1}): + towards west, — towards east.

Fig. 1 shows the smoothed variation of zonal winds from surface up to beyond 60 km. Two maximum values are shown:

(a) One positive (W) during the first fortnight of February, of over 90 m s^{-1} centred at 50 km altitude,

(b) One negative (E) during the first fortnight of July of over 70 m s^{-1} beyond 60 km.

Fig. 2 gives the zonal stratospheric circulation index (SCI) distribution [1] in Arenosillo during a year (average 1966—1971). At Arenosillo small (even zero) values of the stratospheric circulation index during January differ from values of this index ($\sim +50$ m s^{-1}) on Wallops during the same month. This is because the

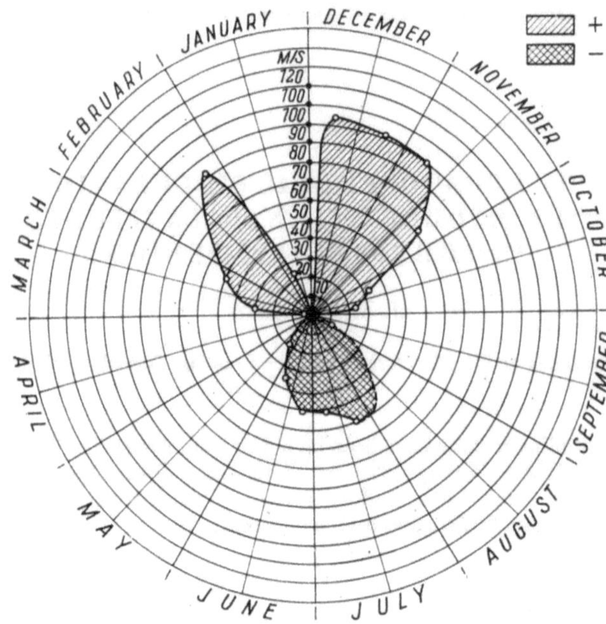

Fig. 2. Arenosillo zonal SCI. Mean values 1966—1971; winds blowing from the west (westerly) are +, from the east —.

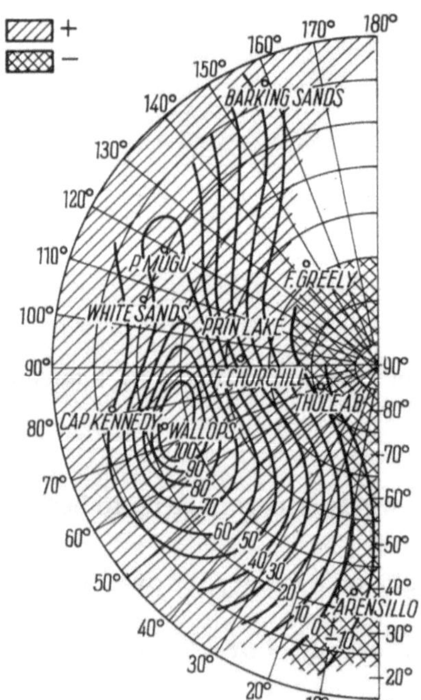

Fig. 3. Western part of the northern hemisphere zonal SCI.
19, 20, 21 January 1970.

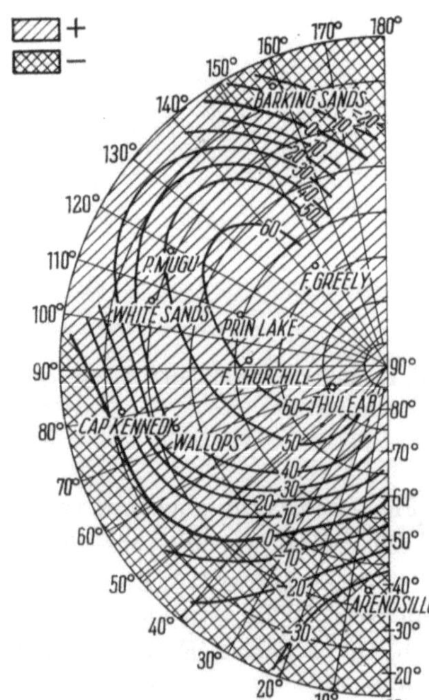

Fig. 4. Western part of the northern hemisphere zonal SCI.
28, 29, 30 January 1970.

Winds blowing from the west are +, from the east —

measurements in Arenosillo were made in a month when there was a sudden warming.

The first sudden warming was detected on Arenosillo in January 1969; another took place in January 1970. Easterly zonal winds were observed at altitudes between 40 and 60 km. Figs. 3 and 4 show the evolution of this sudden warming in the northern hemisphere. Data have been taken from soundings carried out at the stations shown and on the stated days.

A "midwinter warming" [2] took place in February 1972 and was registered in Arenosillo, as Figs. 5 and 6 show. It was preceded by an intense cyclonic activity in the troposphere [3]. Zonal winds at a level of 45 km show a variation with respect to values on the previous days of about 100 m s^{-1}. Temperatures increase by about 40 °C at 55 km altitude. The simultaneous occurrence of the temperature maximum in Fig. 6, with the maximum component of east wind, make it difficult to confirm that stratospheric circulation changes follow temperature changes [4].

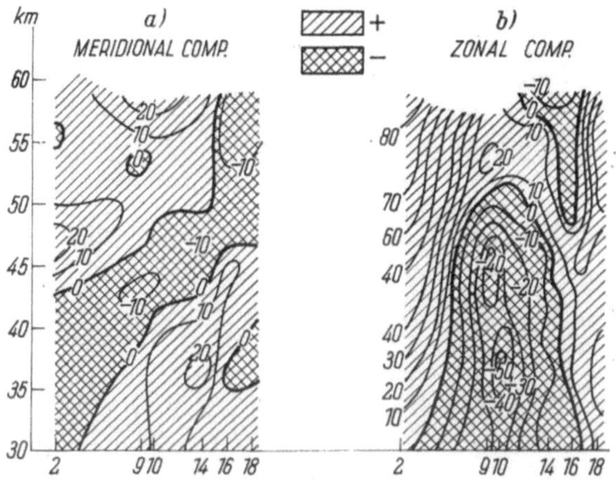

Fig. 5. Winds on Arenosillo, sudden warming February 1972.

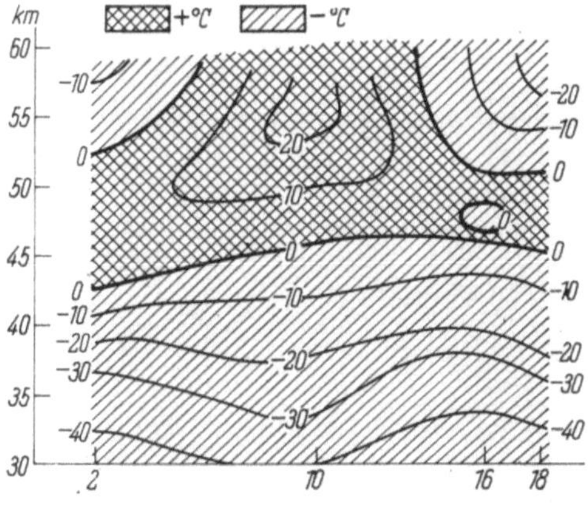

Fig. 6. Temperatures on Arenosillo, sudden warming February 1972.

References

[1] W. L. Webb, Structure of the Stratosphere and Mesosphere, Academic Press, New York, London 1969.
[2] G. V. Groves, Space Research X, 137 (1970).
[3] K. Labitzke, J. Appl. Met. 4, 91 (1965).
[4] B. H. Williams and B. T. Miers, in: Stratospheric Circulation, Academic Press, New York, London 1969.

Space Research XIII — Akademie-Verlag, Berlin 1973

DEVIATIONS OF ATMOSPHERIC PARAMETERS FROM MODEL VALUES IN THE STRATOSPHERE AND MESOSPHERE AT HIGH LATITUDES

S. S. Gaigerov, V. G. Kidiyarova, L. M. Kolomiitseva and D. A. Tarasenko

USSR

The deviations of the data obtained from individual temperature and density measurements from the mean season models in the 30—80 km layer for winter and summer periods are presented. The statistical asymmetry and excess characteristics for the temperature and density distribution curves are estimated. The calculations of the changeability of thermodynamic parameters are compared with the available information on the activity of atmospheric processes.

1. Introduction

The probable deviations of thermodynamic parameters of the atmosphere from model values are a supplement to the International Standard Atmosphere. In order to obtain and analyse the probable deviations of temperature, as well as the information used for establishing the standard atmospheres we drew on the results of rocket soundings at Heiss Island during 1970. The number of observations is given in Table 1.

Table 1

Temperature observations at Heiss Island

Month/Altitude (km)	25	30	35	40	45	50	55	60	65	70	75	80
Dec.—Jan.	98	98	94	87	65	48	44	39	42	37	35	37
Jun.—Jul.	83	74	80	77	63	53	43	42	25	39	33	16

Individual rocket observations over Heiss Island, in which the density profile ϱ was calculated from the temperature profile obtained by M-100 rocket and the pressure measured at a 20 km altitude, have been used. Earlier standard density profiles ϱ_{st} calculated from mean temperature distributions have been compared with the individual observations at Heiss Island. Throughout the 25—80 km layer, the density deviations are within $\pm$ 10% of the standard values.

In accordance with Cole's calculations for other latitudes [1], the estimates of the probability characteristics of temperature and density deviations for the 80° latitude were obtained using the statistical method described by Kimball [2].

2. Probable Deviations in Temperature and Density for 80°N

In Fig. 1 one can see some cases of frequency distribution of temperature (T °C) and relative density $(\varrho - \varrho_{st})/\varrho_{st}$ %. The distribution curves indicate that the variability increases with altitude, and the range of deviations is greater in winter. From such curves constructed for 5 km altitude intervals we took the probable

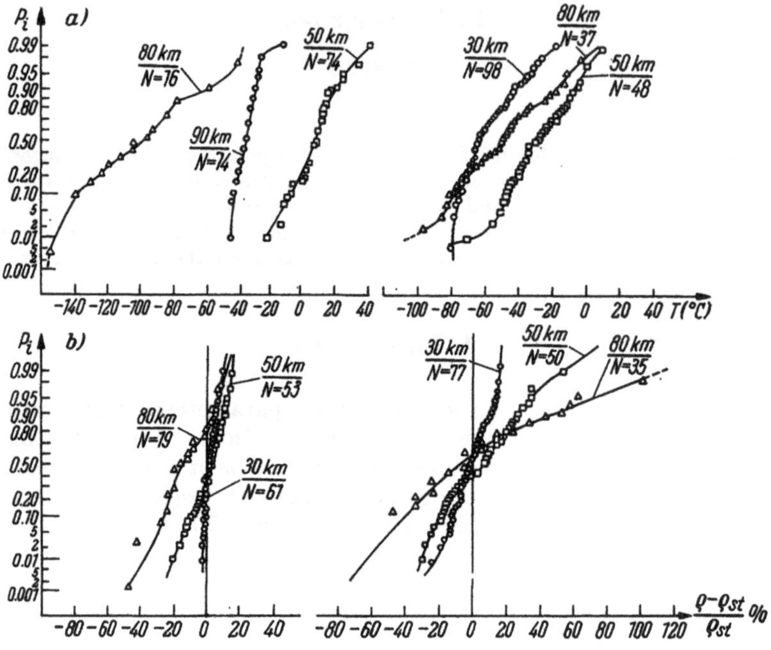

Fig. 1. Cases of frequency distribution of temperature (a) and density deviation (b) over Heiss Island: left, summer; right, winter.

deviations with frequencies of 1, 5, 10, 20, 50, 80, 90, 95 and 99%. These values were used to plot the altitude distributions of probable deviations of temperature and density for winter and summer seasons (Fig. 2a and b). A comparison of the characteristics obtained with the corresponding values presented by Cole [1] for other latitudes has shown a much greater scatter of values for high latitudes.

3. Theoretical Estimations of the Deviation Limits of Temperature and Density

The values of temperature and density deviations from the mean state were theoretically estimated supposing that the distributions of these parameters are normal in winter and summer seasons. We considered the deviations corresponding to 1, 5 and 10% probabilities as well as the extreme deviations corresponding to a 0.03% probability (Fig. 3). Only observations which are continuous throughout the whole range of altitudes involved (30—80 km) have been used. In order to have some more cases, for Heiss Island observations during three-month seasons

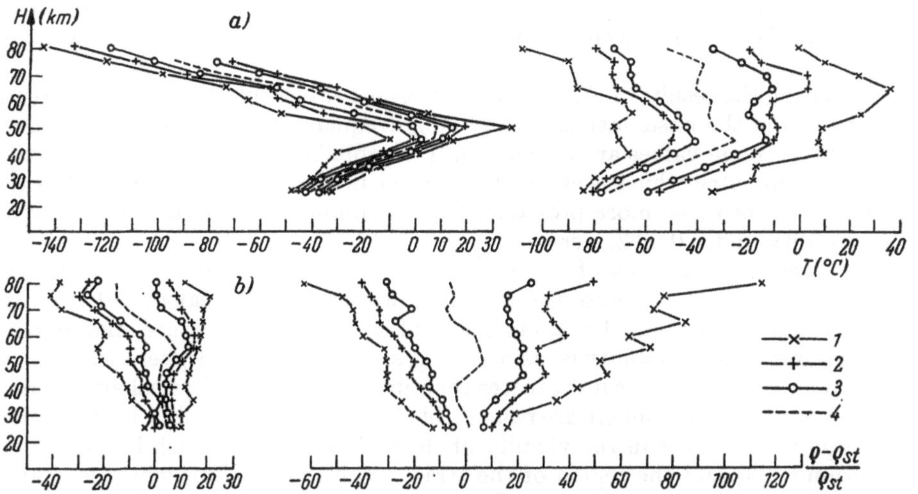

Fig. 2. Probability limits of changeability of temperature (a) and density (b) over Heiss Island: left, summer; right, winter. 1, frequency limits for 1 and 99% probabilities; 2, frequency limits for 10 and 90% probabilities; 3, frequency limits for 20 and 80% probabilities; 4, median distribution.

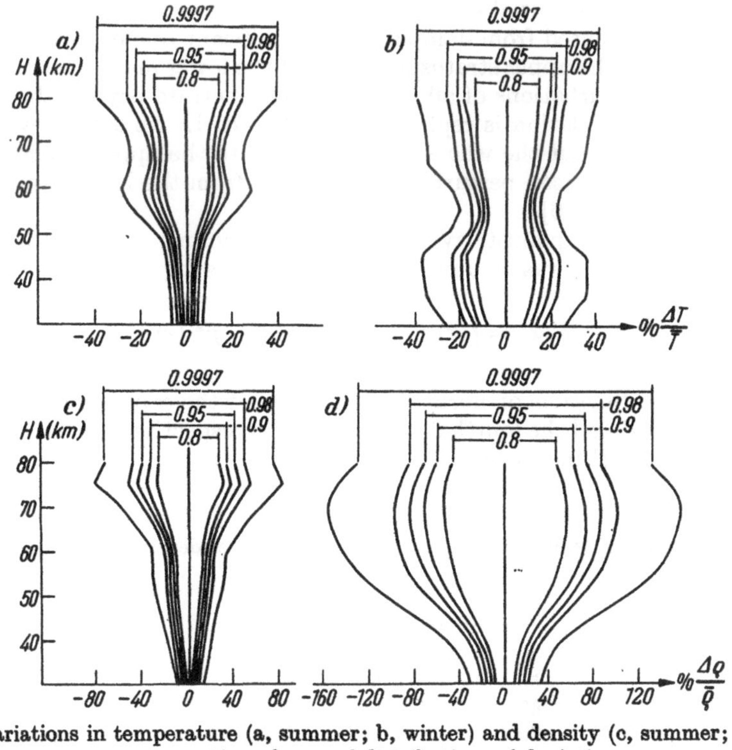

Fig. 3. Variations in temperature (a, summer; b, winter) and density (c, summer; d, winter) on assumption of normal distributions of deviations.

were taken. We had thirty continuous winter (Dec.—Feb.) observations in the 30—80 km layer and twenty seven summer (Jun.—Aug.) continuous observations.

Comparing the results of Fig. 3 with the extreme values (plotted in Fig. 2) directly obtained from the observations (1 and 99% frequency), it is evident that sometimes the extreme values are beyond the 1% probability in Fig 3. Moreover they involve asymmetry which is especially distinct in the vicinity of 60 km. Positive deviations seem to be more probable. In this connection, the deviations having a probability under 0.03% were calculated.

Fig. 3 also gives an idea of the altitude and season dependence of variability ($\sigma_T/\overline{T}$ % and $\sigma_\varrho/\bar{\varrho}$ %) because the probable deviations were calculated from the standard deviations on the assumption of normal temperature and density distributions. The variability is, as a rule, greater in the mesosphere than in the stratosphere and in winter it is greater than in summer. The physical and thermal conditions in the stratosphere are responsible for the special stability in this region. The largest stability is in the vicinity of the 20 km altitude which is close to the "stratonull" [3] and the region of the stratopause (45—55 km) which is characterized by a stable thermal regime. Between these extreme regions the variability increases. The static stability in the mesosphere is much less due to its thermal structure. Because of its low density wave phenomena in the mesosphere must become stronger. As some authors show, perturbations of various scales can penetrate from the troposphere into the mesosphere. In this case energy of vertical motions can transform into energy of turbulent eddies [4—6]. As there is no sufficient reserve of potential energy in the tenuous mesosphere, random turbulent eddies are prevented from damping. The significant role of turbulence in the thermal balance of the upper stratosphere and of the lower mesosphere is shown in [7]. In summer perturbations of all kinds do not penetrate beyond the boundary where westward circulation is replaced by eastward. In winter westward circulation occurs throughout the whole atmosphere up to the upper boundary of the mesosphere. In this season perturbations can pass from the lower atmosphere to the upper atmosphere [6].

In order to consider quantitatively the effect of asymmetry upon the deviation distribution, the latter was assumed to obey the following law:

$$F(x) = f(x) + \frac{K}{6} f^{(3)}(x) + \frac{E}{24} f^{(4)}(x)$$

where $f(x)$ is the function of normal distribution of the x parameter, $f^{(3)}(x)$ and $f^{(4)}(x)$ are the third and the fourth derivative of $f(x)$ respectively, K is the asymmetry parameter and E the excess parameter.

Table 2

Parameters of asymmetry K and excess E for winter density distributions

H (km)	K	E
40	−0.70	−1.7
50	0.40	−1.8
60	0.46	−0.8
70	1.05	−2.1
80	0.67	−0.4

Table 2 contains values of asymmetry and excess parameters for density distribution calculated for the winter season. It follows from Table 2 that the distribution curves of density have positive asymmetry and appear flatter than the normal distribution curves.

It is important to note that the quantitative estimations of the asymmetry and excess parameters are quite reliable provided the number of continuous observations exceeds 70.

References

[1] A. E. Cole, Space Research XI, 813 (1971).
[2] B. F. Kimball, J. Am. Statist. Ass., 55, 546 (1960).
[3] W. L. Webb, Structure of the Stratosphere and Mesosphere, Academic Press, New York, London 1966.
[4] B. N. Trubnikov and I. A. Shcherba, Trudy CAO, No. 76, 44 (1967).
[5] C. O. Hines and C. A. Reddy, J. Geophys. Res. 72, 1015 (1967).
[6] B. W. Boville, Space Research VII, 20 (1967).
[7] A. I. Ivanovsky and A. F. Kivganov, Trudy CAO, No. 96, 54 (1970).

Space Research XIII — Akademie-Verlag, Berlin 1973

RESULTS OF SIMULTANEOUS WIND MEASUREMENTS IN THE STRATOSPHERE, MESOSPHERE AND LOW THERMOSPHERE

L. A. Andreeva, B. O. Vugmeister, Yu. D. Ilyichev, E. S. Kazimirovsky, L. A. Katasev, V. D. Kokourov, N. S. Lifshitz, S. V. Pahomov and D. B. Uvarov

USSR

Results of simultaneous wind measurements in the height interval up to 150 km are presented from rocket experiments carried out in July 1970 and September 1971. The methods used include radar tracking of a cloud of dipoles and the head of a rocket attached to a parachute, optical tracking of luminous sodium and trimethylaluminium clouds, radar studies of meteors and horizontal drift measurements of ionization irregularities in the ionosphere by the close-spaced antenna method. It is found that the results of rocket measurements and those of ground-based methods are in good agreement. It is concluded that such simultaneous wind measurements made by different methods are important for increasing the height interval under investigation and for inter-calibration between the various methods.

Rocket experiments have been carried out in Volgograd in July 1970 and September 1971 using the following techniques for measurement of winds: (i) radar tracking of the head of a geophysical rocket attached to a tripped parachute (to 50 km); (ii) radar tracking of a cloud of dipoles released from a rocket (from 50 to 75—80 km); (iii) radar meteor technique; (iv) photography of artificial luminous clouds (80—150 km); (v) horizontal drift velocity measurements of ionization irregularities by the vertical ionosonde technique and by the close-spaced antenna method (100—400 km).

Wind measurements by the radar tracking of a cloud of dipoles and of the head of a rocket were carried out at Central Aerological Observatory (CAO) during the launchings of rocket M-100. The technique and equipment are described in [1, 2].

Radar meteor observations were carried out by the Institute of Experimental Meteorology (IEM) by means of special coherent-pulse radar working at a frequency of 24 MHz. The equipment and measurement procedures do not differ from those described earlier [3, 4]. The reflecting region heights of meteor trails were determined according to the technique suggested in [5]; it was concluded that there is a dependence between the height h and the slant range of the meteor trail reflecting point:

$$h(R) = \frac{\int_0^{2\pi} h(R, \Psi) P_1(R/\Psi) G(\Psi)\, d\Psi}{\int_0^{2\pi} P_1(R/\Psi) G(\Psi)\, d\Psi}$$

where $P_1(R/\Psi)$ is the conventional probability of the meteor trail reflecting point appearing with a range R at an angle Ψ, $G(\Psi)$ is the resultant coefficient of directional antenna effect of the "reception—transmission" system and $h(R, \Psi)$

is the height of the reflecting meteor trail region, defined as a function of R and Ψ:

$$h(R, \Psi) = \left[\frac{R^2}{D} + R \sin \Psi\right]$$

where D is the diameter of the earth.

Two mutually perpendicular wind velocity components were measured for every hour of observation. The measurements were conducted alternately during the half-hour intervals in the zonal (U) and meridional (V) directions.

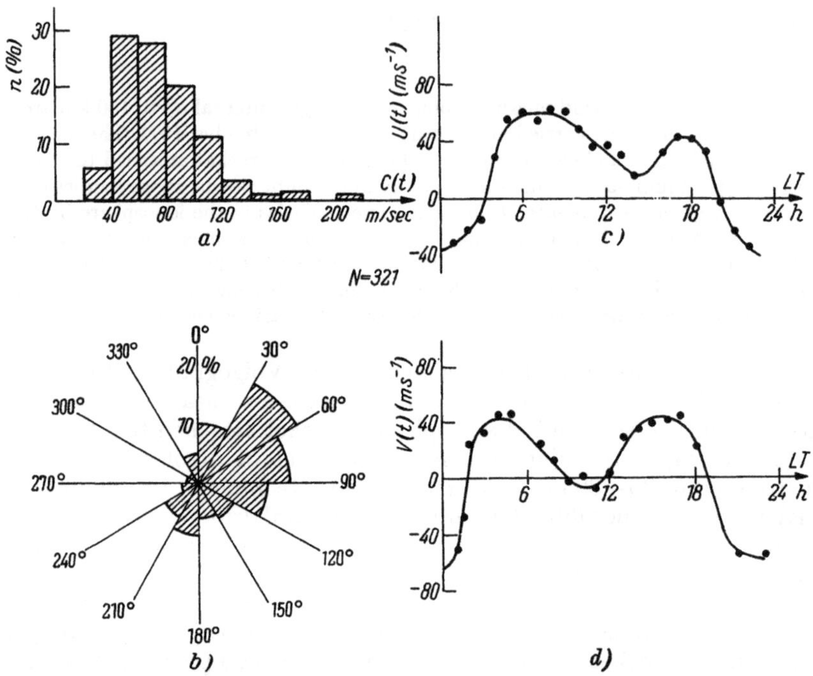

Fig. 1. Investigation results of the wind regime in lower ionosphere by the close-spaced antenna method.

a) histogram of the velocity vector modulus
b) histogram of the azimuth motion
c) daily variation of the zonal wind
d) daily variation of the meridional wind.

Artificial luminous clouds were formed as a result of the release from MR-12 rockets of different chemicals into the atmosphere at heights greater than 80 km. In experiments carried out in IEM, spherical lithium clouds on a sodium trail background and trimethylaluminium trails were used. The sodium cloud was formed in the evening twilight on 6 July 1970 at 2050 and the trimethylaluminium clouds on 7 July 1970 at 0243 and on 15 October 1971 at 2228 (Moscow time). To determine the spatial coordinates of features of the clouds they were photographed simultaneously from three points. Coordinates of separate cloud points were calculated using star coordinates found on the photograph [6].

The horizontal drift measurements of small-scale ionization irregularities were conducted using the close-spaced antenna method. The measurement technique and data treatment are published in [7]. The drift was observed every day and every hour from 7 June 1970 to 7 July 1970. Radio sounding was carried out at a frequency of 2.27 MHz, the radio-echo from the Es and E layers in the daytime and from the Es and F regions at night were used. The wind regime investigation

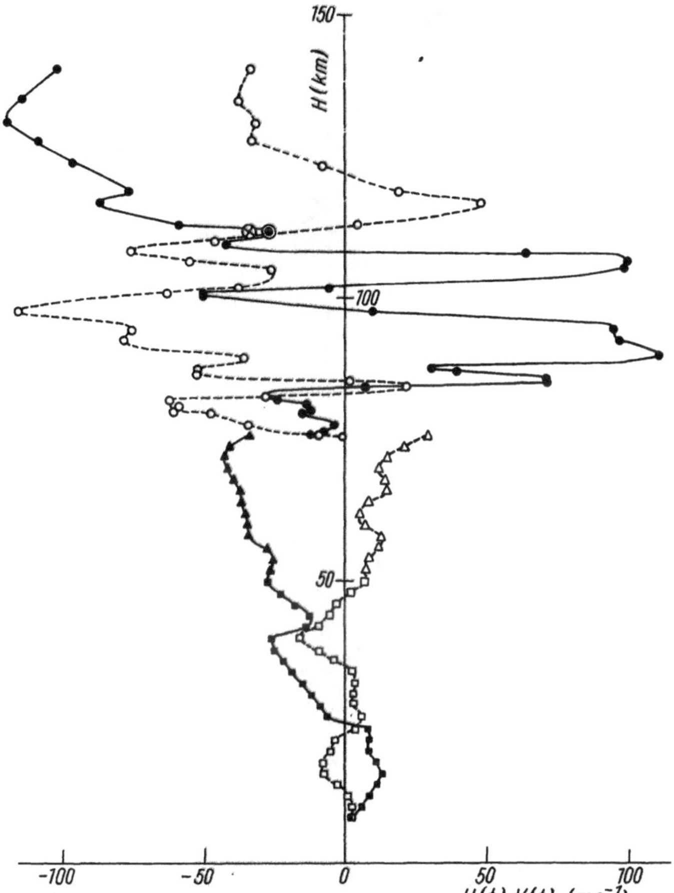

Fig. 2. Wind profile of 6 July 1970: ■ U, □ V, tracking of the rocket head; ▲ U, △ V, tracking of cloud of dipoles; ⊙ U, ⊗ V, close-spaced method (Es layer); • U, ∘ V, artificial clouds; —— zonal wind velocity component; ------ meridional wind velocity component.

results for the lower ionosphere (E and Es layers) are given in Fig. 1. Distribution of the velocity vector modulus makes it possible to conclude that during the investigation period the wind was directed mainly to the east. In the average daily variations of the zonal and meridional velocity components of motion the 12-hour periodicity prevails. These results are in good agreement with the general atmospheric circulation at ionospheric levels for mid-latitudes in the northern hemisphere in the period of summer solstice and high solar activity [7, 8]. Irregularity motions in the lower ionosphere measured by the close-spaced antenna method may be identified with neutral wind in the 130—140 km altitude region [7, 9].

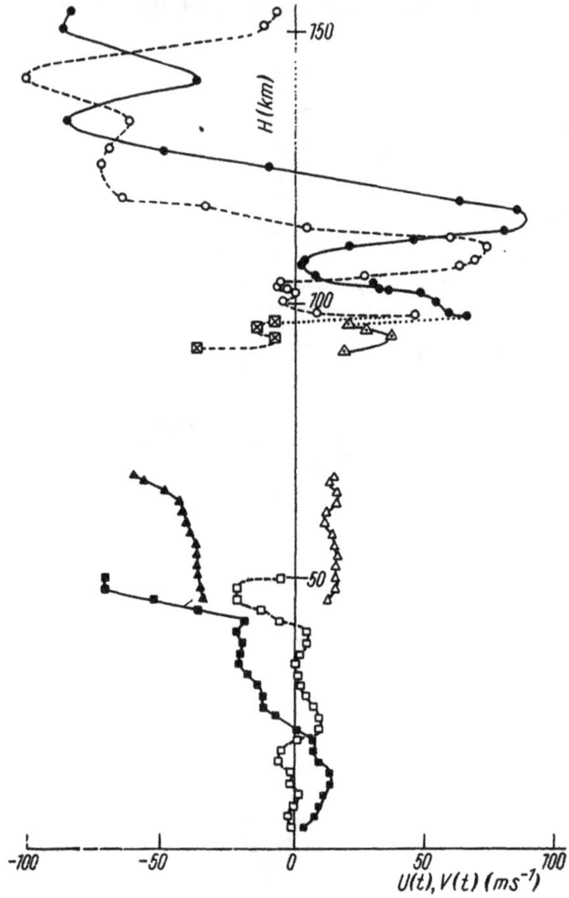

Fig. 3. Wind profile of 7 July 1970: ▪ U, □ V, tracking of the rocket head; ▲ U, △ V, tracking of cloud of dipoles; △ U, ⊠ V, radar meteor technique; • U, ∘ V, artificial clouds; —— zonal wind velocity component; ----- meridional wind velocity component.

The wind profiles of particular interest for 6 July 1970, 7 July 1970 and 15 October 1971 when the wind measurements were carried out by all techniques are given in Figs. 2, 3 and 4.

The most effective was the experiment on 6 July 1970 when we succeeded in obtaining an almost continuous wind profile from 7 to 140 km. Wind values calculated according to the artificial luminous cloud displacement are referred to the average time moment of measurements mentioned above. Data obtained when tracking the head of a rocket and dipole reflectors are true for 2122 and the irregularity drift data in the Es layer for 2059.

In profiles for 7 July 1970 wind values were not obtained in the 68—90 km height region and there is some disagreement at heights of 50 km which at present it is difficult to explain. The heights where the wind data were obtained on 7 July 1970 according to the artificial luminous clouds and meteor trails are not overlapping but it can be supposed that these profiles may be combined (a dotted line in Fig. 3) because in this altitude range significant vertical gradients of the

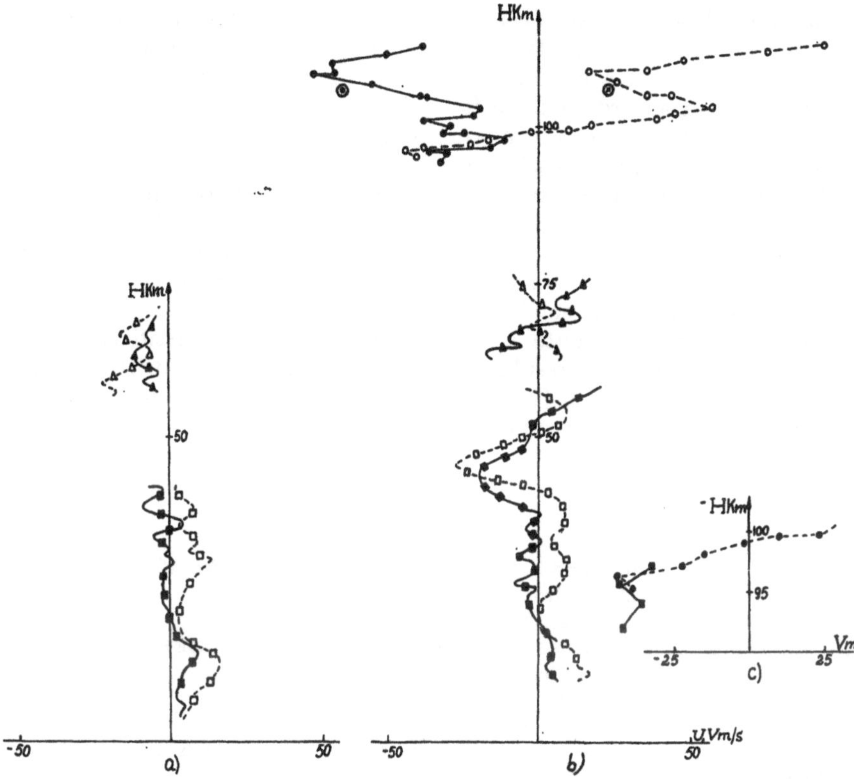

Fig. 4. Wind profile of 15 October 1971 (a, 2100; b, 2300): ■ U, □ V, tracking of the rocket head; ▲ U, △ V, tracing of cloud of dipoles; ⊙ U, ⊗ V, close-spaced method (Es layer); ⊙ U ∘ V, artificial luminous clouds; —— zonal wind velocity component; ----- meridional wind velocity component. c, longitudinal wind velocity component: • artificial luminous cloud method, ■ radar meteor technique.

horizontal wind can occur. Wind values obtained when tracking the meteor trails and artificial luminous clouds refer to 0250 and those obtained when tracking the head of a rocket and a cloud of dipoles to 0303.

In the experiment conducted on 15 October 1971 the results of tracking the head of a rocket and a cloud of dipoles are related to 2100 and 2300. Fig. 4 shows that within two hours the wind direction and its value in some regions of the 10—75 km height interval change significantly. Wind data (Fig. 4c) according to artificial luminous clouds were obtained 30 minutes earlier than these times.

Despite the lack of data at some height intervals the profiles show the existence of quasi-periodic wind velocity variations in the whole region, the amplitude of these variations increasing with height. Above 80 km significant wind velocity gradients are observed in the zonal as well as in the meridional components.

Drift measurements in the Es layer on 7 July 1970 were conducted from 2054 to 2104 and on 15 October 1971 from 2234 to 2250. For 6 July 1970 the result was related to the middle of time interval when the reflecting height from the Es layer was 111.8 km; and on 15 October 1971 the Es layer height was 106 km. It is seen that in both cases the wind velocity measured by the radio technique was close to that measured according to the displacement of the luminous cloud within the limits of experimental error of both techniques. The wind measurement error according to artificial luminous clouds was not greater than ± 3 m s^{-1} and according to drift in the Es layer it was ± 6 m s^{-1}. The azimuth measuring error in both techniques was $\pm 4-6°$ and the height determination error was ± 0.5 km and ± 2 km respectively.

Good agreement is also observed in the meridional wind component values obtained on 15 October 1971 according to the artificial luminous clouds and by the radar meteor technique when areas of observation overlapped. These data were both obtained at 2228. The zonal wind component according to meteor trails was not obtained due to the small number of meteors.

The good agreement of data obtained according to the artificial luminous clouds with the observation results of ionospheric irregularity drift and meteor trails confirms the conclusion [7, 9] that using the radio technique it is possible to measure neutral winds in the lower ionosphere.

References

[1] M. N. IZAKOV, G. A. KOKIN and S. P. PEROV, Meteorol. i Gidrol. No. 12, 70 (1967).
[2] S. V. PAHOMOV, Meteorol. i Gidrol. No. 3, 103 (1969).
[3] I. A. LYSENKO, Astronom. J. **41**, No. 1 (1963).
[4] A. V. AREFYEVA, V. M. KORPUSOV, I. A. LYSENKO, A. D. ORLYANSKY, A. N. RYABCHIKOV and N. F. SHUVARIKOVA, Geomagn. i Aeronom. **6**, 703 (1966).
[5] YU. D. ILYICHEV, Proc. Inst. Exp. Meteorol. **24** (1970).
[6] D. B. UVAROV, Proc. Inst. Appl. Geophys. **15** (1969).
[7] V. M. POLYAKOV, L. A. SHCHEPKIN, E. S. KAZIMIROVSKY and V. D. KOKOUROV, Ionospheric Processes, Izd. Nauka, Moscow 1968.
[8] E. S. KASIMIROVSKY, Geomagn. i Aeronom. **3**, 463 (1963).
[9] J. M. WRIGHT, J. Atmos. Terr. Phys. **30**, 919 (1969).

Space Research XIII — Akademie-Verlag, Berlin 1973

VARIABILITY IN DENSITY AND TEMPERATURE MEASUREMENTS AT WHITE SANDS DURING THE 1971 WINTER

A. C. Faire, K. S. W. Champion and E. A. Murphy

Air Force Cambridge Research Laboratories, L. G. Hanscom Field, Bedford, Mass., USA

Density and temperature results are presented which were obtained with rocketborne 7-inch rigid spheres launched in January and March 1971 at White Sands Missile Range, New Mexico. Two rockets were fired on 8 January at 1441 and 1531 hours MST, respectively, and one on 11 March at 2212 hours MST. Excellent data were obtained from all three rockets within the altitude range 36—108 km. The January results, obtained during the winter storm period, show large density excursions which at 95 km are almost 70% relative to the 30°N January model from the US Standard Atmosphere Supplements, 1966. The density departures observed are in the form of a wave or oscillatory pattern with amplitude increasing with altitude. The results also indicate that on this occasion the absolute density variability was 20% in two measurements made 50 minutes apart. The March density values are much closer to the models, but again show an oscillating pattern. Temperature profiles have been deduced from the density results. The January measurements show a decrease in the stratopause altitude with a slightly enhanced temperature and a mesopause raised by 14 km with temperatures observed in the mesosphere that are as much as 45°K higher than normal. Departures of the March temperature results from the model values are smaller than those of January.

1. Introduction

Density and temperature results are presented that were obtained at White Sands Missile Range (WSMR), New Mexico, during January and March 1971 using AFCRL rocketborne 7-inch falling spheres. The purposes of the measurements was to investigate the nature and extent of short-term density and temperature variations during the winter storm season when upper atmospheric variability appears to be greatest. Measurements made by WSMR Meteorological Arcasonde and Rawinsonde Rockets [1, 2] near sphere launch times overlap the falling sphere data below 60 km.

2. Experimental Approach

The AFCRL-7-inch rigid falling sphere, instrumented with an omnidirectional accelerometer, is used to determine atmospheric density at high altitudes [3]. The sphere is ejected from a rocket at an altitude of 50—60 km during rocket vehicle ascent. The accelerometer senses atmospheric drag on the sphere during its free flight. Density is determined from the drag data within the altitude range of approximately 30—120 km [4], the upper level being dependent upon the rocket vehicle velocity and the sensitivity of the accelerometer employed. Temperature results are computed from the density data. At the highest levels of the measurements the system error is less than 5% and becomes even smaller at lower altitudes, until about 35 km where it increases again due to uncertainties in the sphere drag coefficient.

Launch data are summarized in Table 1 for falling sphere rockets and Arcasondes.

Table 1

Launch summary of falling sphere rockets and Arcasondes
White Sands Missile Range, New Mexico, Jan.—Mar. 1971

Launch date	Launch time (hours, MST)	Payload type	Vehicle type	Altitude range of data (km)	Quality of data
8 Jan.	1230	Arcasonde	ARCAS	23—60	Excellent
8 Jan.	1441	Falling Sphere	NIRO	36—108	Excellent
8 Jan.	1531	Falling Sphere	NIRO	36—104	Excellent
8 Jan.	1547	Arcasonde	ARCAS	23—60	Excellent
11 Mar.	1930	Arcasonde	ARCAS	23—60	Very good
11 Mar.	2212	Falling sphere	NIRO	36—102	Very good

3. Results

Measured densities obtained from AFCRL falling spheres and WSMR meteorological rocketsondes are plotted as percentage departures relative to annual mean values obtained from the US Standard Atmosphere Supplements, 1962 (USSA-62) [5]. In addition, appropriate seasonal models from the US Standard Atmosphere Supplements, 1966 (USSAS-66) [6] are plotted for comparison with the *in situ* data. Temperature height profiles determined from sphere density results are also compared with model and meteorological rocket temperatures in the altitude regions where the data overlap.

3.1. Density Results

Density results obtained on the two falling sphere rockets launched 50 minutes apart on 8 January are shown in Fig. 1. Because of an uncertainty of approximately 0.4 second in sphere ejection time on the first sphere rocket, launched at 1441 hours, the measured densities were adjusted by raising the altitudes 0.8 km to normalize the data with those obtained from the Arcasonde launched at 1547 hours. This small correction can be justified by the fact that the short-term variability between successive falling sphere measurements did not exceed 5% in the altitude range 44—60 km, but was observed to increase slowly with height until it reached a maximum of 20% at 100 km. In the range 36—60 km the sphere and MRN data overlap and the measurements agree within 5%. It should be noted that density departures are plotted using USSA-62 as a base representing 100%. The maximum difference between the measurements was observed at 100 km where the first and second measurements, ϱ_1 and ϱ_2, gave density departures relative to USSA-62 of 40% and 71%, respectively. It is then apparent that the density departure of ϱ_2, *relative to USSA-62*, was 31% greater than that of ϱ_1. However, the absolute density variability or the percentage difference between the two measurements is 20%, obtained by substituting appropriate numbers from the graph in the ratio of the change in density to the average density between measurements. The actual values of density at 100 km are $\varrho_1 = 6.97 \times 10^{-10}$ g cm^{-3} and $\varrho_2 = 8.55 \times 10^{-10}$ g cm^{-3}.

Data obtained from the first sphere show a maximum negative deviation of 18% with regard to USSA-62 while those of the second are about 12%. The differences are even less relative to the January model which is approximately 4% smaller than USSA-62 at this altitude. However, above 73 km the data of the

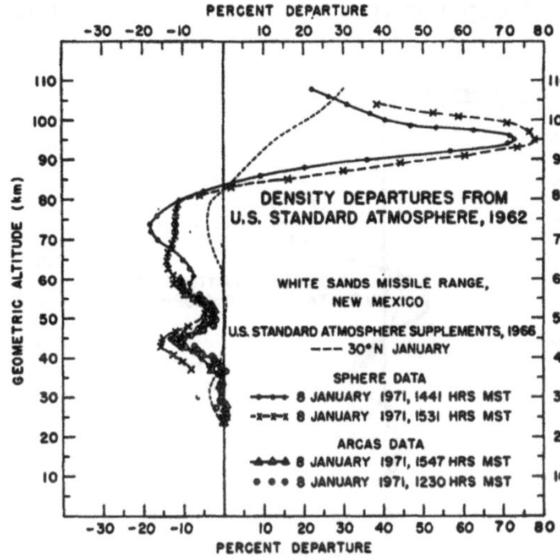

Fig. 1. Density departures from the US Standard Atmosphere, 1962, 8 January 1971, 1441 and 1531 MST.

first and second falling spheres are described by large positive-going oscillations that have maxima at 95 km and density departures of 73% and 78% respectively. The oscillations decrease in amplitude above 95 km giving departure values of 22% at 108 km for the first rocket and 38% at 104 km for the second. It is interesting to note that the curve of the first rocket has a path that is almost normal to the direction of the curve of the January model at 105 km where the two curves intersect. If one assumes a linear extrapolation of about 3 km to the same altitude of intersection a similar comment can be made with regard to the second sphere curve which shows a decreasing departure relative to USSA-62 as it approaches the turbopause level from 95 km.

The large density perturbations shown in these data indicate disturbed conditions in the upper atmosphere and they are consistent with previous observations made at WSMR during the winter storm season in 1967 [7, 8]. Furthermore, unusually large pressure gradients were recorded at altitudes of 40 and 50 km over WSMR during the 8 January 1971 measurements [9]. Such strong pressure gradients in the stratosphere appear to be associated with highly disturbed conditions in the upper atmosphere and these could have given rise to the upward propagating waves observed in the data.

In Fig. 2 is shown a plot similar to that of Fig. 1 for sphere density departure data obtained on 11 March at 2212 MST. Arcasonde data are also shown for a 1930 launch on the same date. Both the 30° N January and the mid-latitude spring/fall models from USSAS-66 are plotted for comparison. Although a maximum departure of 45% relative to USSA-62 occurs at approximately 100 km the sphere data are within 20% of the seasonal model values throughout the range of the measurements.

The basic similarity between the data of 11 March and those of 8 January lies in the fact that both exhibit wave-like structures which increase in amplitude with height. However, the sphere curve of Fig. 2 shows a well defined oscillatory pattern which is almost sinusoidal up to about 78 km. Above this level the wave becomes irregular as it expands to a departure of 28% relative to USSA-62 in

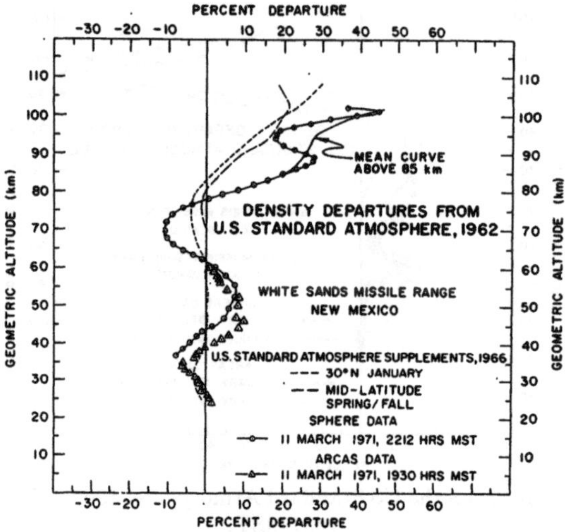

Fig. 2. Density departures from the US Standard Atmosphere, 1962, 11 March 1971, 2212 MST.

the positive direction. The wave pattern then reverses itself and forms shorter wavelengths until the terminal altitude of the measurements is reached at about 102 km. Because of some scatter in the data a mean curve was calculated for the region above 85 km which tends to damp out the oscillations contained in the raw data. The agreement between sphere and Arcasonde data tends to support the notion of upward propagating waves due to lower atmospheric dynamics.

3.2. Temperature Results

Molecular-scale temperatures computed from the density results of 8 January are shown in Fig. 3. For computational details the reader is referred to a previous paper by Faire and Champion [4]. Also plotted for comparison are selected models and meteorological rocketsonde data. Both sphere temperature profiles display some interesting features. The sphere and Arcasonde data show excellent agreement up to 55 km and indicate a warm stratopause positioned about 4 km lower than that of the models. Both sphere curves show the mesopause raised by 15 km from model values. The data indicate significant cooling at the mesopause when compared with the seasonal model. Also, in the upper mesosphere temperatures

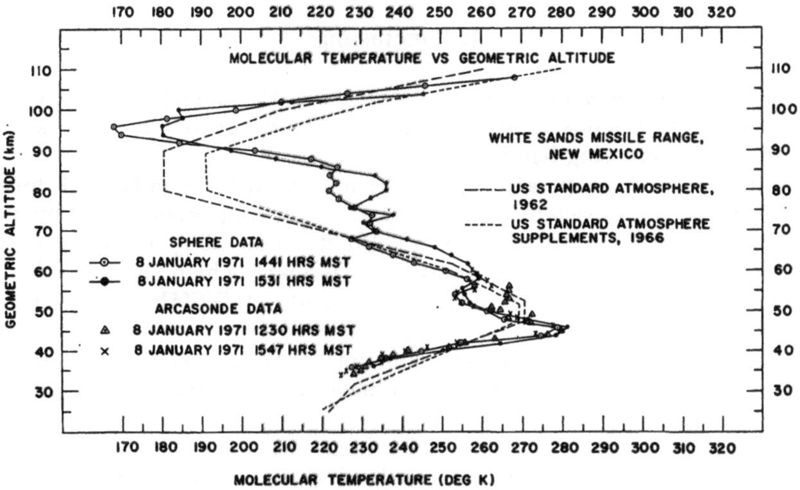

Fig. 3. Molecular-scale temperatures, 8 January 1971, 1441 and 1531 MST.

are observed that are as much as 45°K higher than model values. Both sphere curves are characterized by considerable structure between 70 and 80 km which strongly suggests atmospheric instability.

Sphere temperatures obtained on 11 March are shown in Fig. 4. The agreement between sphere and Arcasonde data is moderately good but there are some differences in fine structure. The differences no doubt are primarily due to the time

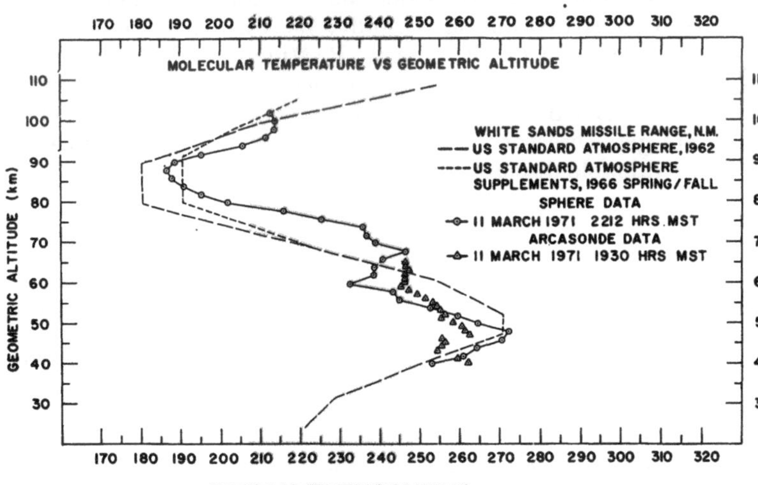

Fig. 4. Molecular-scale temperatures, 11 March 1971, 2212 MST.

difference of 2 hours 42 minutes between the measurements. The structure observed in the sphere data between 60 and 70 km suggests atmospheric instability as in the data shown in Fig. 3. However, the March data are closer to the mean seasonal model values than the data from the January rounds.

4. Conclusions

Density and temperature perturbations exist throughout the range of the measurements and are observed as wave-like patterns or oscillations that grow in amplitude with increasing height.

These density and temperature perturbations seem to be enhanced during the winter storm season, especially at altitudes above 60 km, as supported by previous observations. The absolute density variability was 20% between successive measurements made 50 minutes apart on 8 January which supports the results of previous short-term density observations made at WSMR during the winter season.

While all of the density and temperature height profiles exhibit some similarities there are significant differences which can most likely be attributed to day to day variations.

Although there is much observational evidence to support the notion of upward propagation of energy from the lower levels of the atmosphere to the higher levels, an unambiguous interpretation of results such as those presented here requires simultaneous measurements of parameters that are known to be capable of inducing changes in thermodynamic structure of the neutral atmosphere. These include all mass transport properties and independent temperature measurements.

5. Acknowledgments

We are indebted to personnel of WSMR and Wentworth Institute, Boston, Massachusetts. Without their excellent support and cooperation this project could not have been carried out successfully. We wish to acknowledge, in particular, the contributions of G. Stromberg, P. Mundis and M. Baratz and their associates at Wentworth Institute who provided falling sphere instrumentation, payload integration and field support. Also we wish to express our appreciation to Dr. C. G. Stergis, Director of the Aeronomy Laboratory, AFCRL, for his interest and for valuable discussions.

References

[1] G. L. Dunaway, Atmos. Sciences Lab., WSMR, N. Mex., Meteorological Data Report, DR-616 (1971).
[2] G. L. Dunaway, Atmos. Sciences Lab., WSMR, N. Mex., Meteorological Data Report, DR-597 (1971).
[3] A. C. Faire, in: Proc. of 1966 Natl. Telemetering Conf., Boston, Mass. 1966 (p. 70).
[4] A. C. Faire and K. S. W. Champion, Space Research V, 1039 (1965).
[5] COESA, US Standard Atmosphere, 1962, Washington, DC.
[6] COESA, US Standard Atmosphere Supplements, 1966, Washington, DC.
[7] A. C. Faire, E. A. Murphy and O. W. Thiele, Space Research X, 188 (1970).
[8] A. C. Faire, K. S. W. Champion and S. P. Zimmerman, Space Research X, 198 (1970).
[9] J. Sharp, Atmos. Sciences Lab., WSMR, N. Mex., Synoptic Pressure Maps for the 40 to 50 km levels at WSMR (1971).

Space Research XIII — Akademie-Verlag, Berlin 1973

THE MEASUREMENT OF TURBULENT SPECTRA AND DIFFUSION COEFFICIENTS IN THE ALTITUDE REGION 95 to ~110 km

S. P. ZIMMERMAN[a] and C. A. TROWBRIDGE[b]

[a]Air Force Cambridge Research Laboratories, L. G. Hanscom Field, Bedford, Mass., USA
[b]PhotoMetrics, Inc., Lexington, Mass., USA

Four rocket-borne chemical trails were launched from Eglin AFB on 20 November 1970. They covered the altitude interval 85—140 km, and were separated in the horizontal direction by ~50 km at ~100 km. These trails are analyzed for turbulent atmospheric coefficients over the altitude range 95—100 km, where turbulence was indicated by intense fluctuations in the scalar trail density distributions. This is accomplished by determining the power spectra of the density fluctuations and the meaned density distributions. The former shows evidence of an "inertial" subrange which allows the determination of the rates of viscous dissipation. In the latter case, the halfwidth of the spectrum of the roughly Gaussian density distribution is simply related to the halfwidth in configuration space by $h_f^2 = {}^1/_2\,\pi^2 h_x^2$, where h_x is the halfwidth of the density distribution in configuration space and h_f is the halfwidth in Fourier space. Thus by measuring the time rate of change of h_f the effect of density fluctuations upon the determination of h_x is minimized, and accurate determination of turbulent diffusion coefficients is made.

1. Introduction

The utilization of spectral analysis as a tool to measure turbulent diffusion coefficients and turbulent energy spectra from photographs of chemical releases in the upper atmosphere was demonstrated to be feasible by Zimmerman et al. [1]. In this paper we are continuing the usage of this technique, which eliminates the errors generated by the density fluctuations in the determination of the mean density distribution, resulting in more accurate determination of turbulent diffusion coefficients. We shall also show in this paper that an independent measurement of the gradient of the scalar tracer fluctuations $n'(x)$, when applied to fluctuation spectra measurements, will allow the independent determination of the rate of viscous dissipation of turbulent kinetic energy ε. This quantity ε may also be determined from the rate of growth of the tracer material when the growth rate is cubic in time t^3 measured from release. As will be shown, the rates of viscous dissipation, though separately measured by different techniques, compare well.

2. Turbulent Spectra Theory

Zimmerman et al. have demonstrated theoretically that the spectra of the passive scalar fluctuations, in the presence of a gradient of the mean scalar distribution, may be described by

$$\mathscr{E}_\gamma(k) = \alpha \left(\frac{\partial \bar{n}}{\partial x}\right)^2 k^{-3} + \beta\,\varepsilon^{-1/3}\varepsilon_\gamma k^{-5/3}. \tag{1}$$

The underlying assumption here is that the velocity field, which produces the fluctuations by interactions with the gradient of the mean density, is "inertial", i.e. $\mathscr{E}(k) = \alpha' \varepsilon^{2/3} k^{-5/3}$ where $\mathscr{E}_\gamma(k)$ is the scalar spectrum defined by

$$\overline{n'^2} = \int_0^\infty \mathscr{E}_\gamma(k)\, dk$$

$\mathscr{E}(k)$ is the energy spectrum of the velocity field, defined by

$$\overline{v^2} = \int_0^\infty \mathscr{E}(k)\, dk$$

$\bar{n}$ is the meaned density, $\partial \bar{n}/\partial x$ is the local gradient of the meaned density, n' is the scalar density fluctuation, ε_γ is the rate of decay of the scalar fluctuations due molecular diffusion, ε is the rate of viscous dissipation of the turbulent kinetic energy, k is wavenumber, $\overline{v^2}$ is the turbulent kinetic energy per unit mass, and α and β are constants of the order of unity which will be more fully discussed.

The comparison of this theory against the measurements showed conclusively that the observed spectra could be completely explained by the gradient of the mean scalar density and inertial transfer mechanisms. Also the observed time rate of decay of the scalar "inertial" subrange was paralleled by the time rate of decay of the mean squared scalar density gradient, supporting the hypothesis of the mean scalar density gradient being acted upon by the turbulent velocity field thus creating the scalar fluctuations. This also shows that the explicit time dependence $\partial \mathscr{E}_\gamma(k, t)/\partial t$ of the scalar fluctuation spectrum is in essence zero, resulting in a steady state condition of the scalar energy spectrum and supporting the assumption of a time independent atmospheric turbulent velocity field, at least over the time of experimental measurements. Thus, given an "inertial" energy field, there is exhibited an "inertial" passive scalar field [2, 3], given by the second term on the right side of Eq. (1). From this relation, a one-dimensional scalar spectrum is derived from

$$\mathscr{E}_{\gamma_1}(k_1) = \int_{k_1}^\infty \frac{\mathscr{E}_\gamma(k)}{k}\, dk \tag{2}$$

which for the inertial range is

$$\mathscr{E}_{\gamma_1}(k_1) = \frac{3}{5}\, \beta \varepsilon^{-1/3} \varepsilon_\gamma k_1^{-5/3}. \tag{3}$$

The value of this one-dimensional coefficient $3\beta/5$ is a matter of some controversy, with values reported ranging from ~ 0.31 [4] to 1.17 [5]. Reasons hypothesized to explain these larger values were anisotropy of the temperature fine structure, and very large intermittency of the velocity dissipation. A resolution or explanation of these differences is not in the scope of this paper but, as shown below, the measured amplitude variations of this constant causes extremely large changes in the determination of the rate of turbulent dissipation ε, from the measured spectra.

From Eq. (3)

$$\varepsilon = \frac{A \varepsilon_\gamma{}^3}{\mathscr{E}^3_{\gamma_1}(k_1) k_1{}^5} \tag{4}$$

where A is $(3\beta/5)^3$, and ε_γ, the dissipation of the scalar field, can then be estimated from

$$\varepsilon_\gamma = 6D \left\langle \left(\frac{\partial n}{\partial x}\right)^2 \right\rangle \approx 6D \left\langle \left(\frac{\partial n'}{\partial x}\right)^2 \right\rangle. \tag{5}$$

It is quite apparent that as ε is proportional to the constant A the measured variations of β can cause ε to vary by a factor of ~ 50.

3. Data

The data that were analyzed for the turbulent scalar spectra were obtained by the photography of chemical releases in the altitude region 95—125 km. There was evidence of turbulent or unstable layers [6] at 95 to $\sim$ 100 km, from 102 to 104 km, 107 to 109, and 109 to 112 km. These layers were all analyzed, but the one centered at 103 km was selected for this report since the diffusion measurements, discussed later, show evidence of a cubic growth rate of the density variance. Thus, we can derive an independent measurement of ε [7] from these same data.

Fig. 1 shows the chemical trail analyzed, and the highly evident turbulent region is the layer around 103 km. Fig. 2 shows the resultant scalar fluctuation spectrum of 45000 data points as a result of 50 scans in the longitudinal direction.

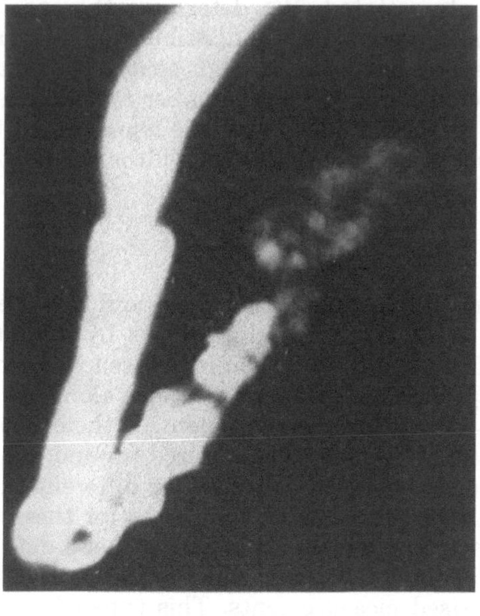

Fig. 1 Chemical trail (Hilda) showing the relative altitude regions of turbulence, instability and stability. The lowest observable turbulent region (right-hand side) is at $\sim$103 km.

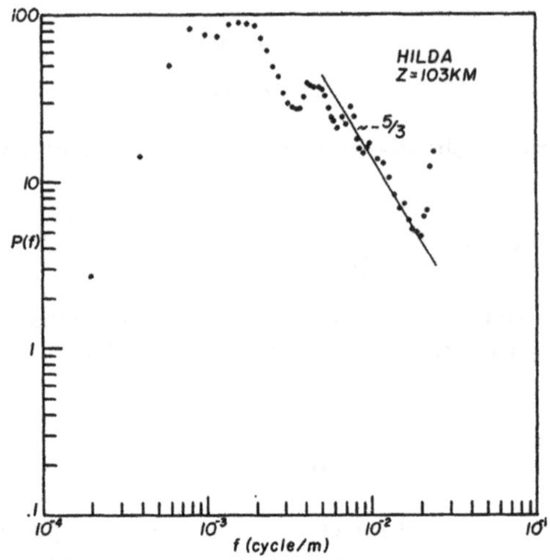

Fig. 2. Turbulent passive scalar spectrum of Hilda analyzed at the altitude of 103 km. The $-{}^5/_3$ section is that of the "inertia like" or "inertial" subrange.

The application of Eq. (4), with the independent determination of ε_γ, results in values of ε of 7.3×10^5 cm^2 s^{-3} for a value of $3\beta/5$ of $\sim$1.00, and 2.7×10^4 cm^2 s^{-3} when $3\beta/5$ has the value 0.33.

4. Diffusion Analysis

The theory is simply restated. We determine the spectral transform of the autocorrelation of the observed density distribution. By this process the fluctuating components are presented at the high spatial wavenumber region leaving the mean distribution in the lower wavenumber region. Then, a simple Gaussian density distribution in configuration space is transformed to a simple Gaussian distribution in Fourier space, where the halfwidths are related by

$$h_x^2 = \frac{1}{2\pi^2 h_\mathfrak{f}^2}$$

where h_x is the halfwidth in configuration space and $h_\mathfrak{f}$ is the halfwidth in Fourier space of dimension cycle/meter. The halfwidths by definition are the position where the density of spectral distribution is reduced to 1/e of the peak.

The diffusion analysis was at different positions along the trail, and the section chosen as representative of these measurements is the obviously turbulent region in the lower right-hand corner which is centered around 103 km. Fig. 3 shows the spectral distributions of the density profiles for different times after release, and in Fig. 4 are the transverse measured halfwidths for these times plotted against time from release. As observed, there is at early times a definite cubic time regime with a flattening at about 180 seconds and a gradual disappearance at about 300 seconds, where we ceased measurements. This type of dispersion behavior is to be expected since we are in a high shear region and are measuring transversely to the wind direction.

The rates of turbulent dissipation ε may be determined from the observed growth rate using Batchelor's theory [7] of the relative dispersion of passive tracers

$$\overline{r^2}(t) \sim C\varepsilon t^3 \tag{6}$$

where $\overline{r^2}(t)$ is the variance or the square of the Gaussian halfwidth, and C is a constant of the order unity. A recent calculation by Krasnoff and Peskin [8] shows this coefficient C to be 4/9. Thus application of Batchelor's theory as amended by Krasnoff and Peskin results in $\varepsilon = 1.31 \times 10^4$ cm^2 s^{-3}. This compares

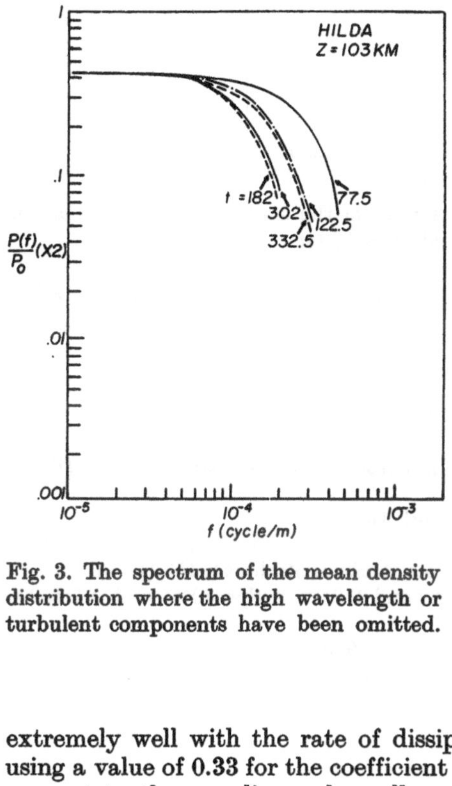

Fig. 3. The spectrum of the mean density distribution where the high wavelength or turbulent components have been omitted.

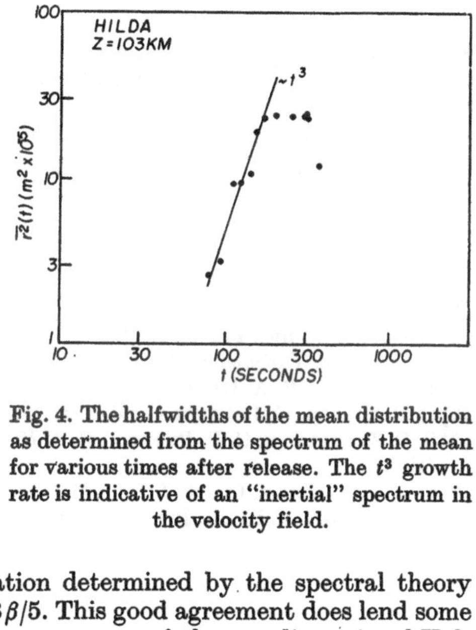

Fig. 4. The halfwidths of the mean distribution as determined from the spectrum of the mean for various times after release. The t^3 growth rate is indicative of an "inertial" spectrum in the velocity field.

extremely well with the rate of dissipation determined by the spectral theory using a value of 0.33 for the coefficient $3\beta/5$. This good agreement does lend some support to these earlier and smaller measurements of the one-dimensional Kolmogoroff constant in the "inertial" subrange determined from temperature data.

The measured dispersion of tracers in the other turbulent layers is shown in Fig. 5, where only the 110 km measurement shows a definite t^3 regime. This results in a value for ε of 4.5×10^2 cm^2 s^{-3}, much smaller than the value determined at 103 km. The other measured altitudes were 95 and 108 km. These showed respectively a linear growth rate where $K = 3.27 \times 10^7$ cm^2 s^{-1}, a value much larger than molecular, and a t^2 regime, which at present has not been analyzed for dissipation rates.

In conclusion, we have shown that the measurement of the spectrum of the passive scalar fluctuations can reveal information about the turbulent velocity field if there exists an "inertial" or "inertia like" spectral subrange. The value of the rate of turbulent energy dissipation so determined is supported by the

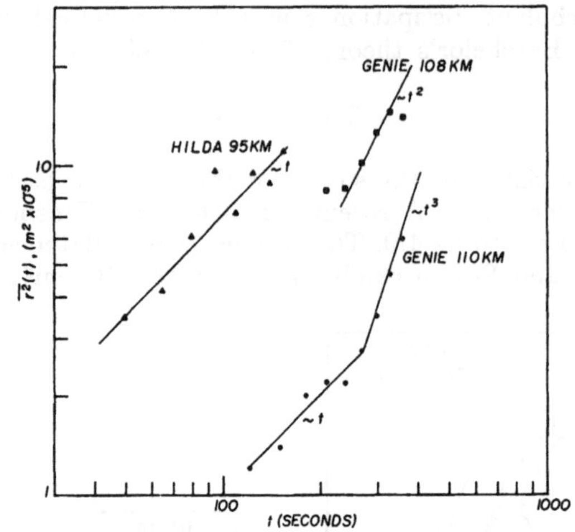

Fig. 5. The time-dependent halfwidths of other sections of the trail "Hilda" as well as the trail "Genie" separated by ~50 km.

simultaneous determination of ε from the diffusion studies with the further assumption that the Grant et al. measurements of the Kolmogoroff scalar constant are applicable.

References

[1] S. P. Zimmerman, C. A. Trowbridge and I. L. Kofsky, Space Research XI, 907 (1971).
[2] A. Corrsin, Phys. Fluids. 7, 1156 (1964).
[3] J. O. Hinze, Turbulence, McGraw-Hill, New York 1959.
[4] H. L. Grant, B. A. Hughes, W. L. Vogel and A. Moilliet, J. Fluid Mech. 34, 423 (1968).
[5] C. H. Gibson, G. R. Stegen and R. B. Williams, J. Fluid Mech., 41, 153 (1970).
[6] S. P. Zimmerman, G. P. Pereira, E. A. Murphy and J. Theon, Space Research XIII, 209 (1973).
[7] G. K. Batchelor, Quart. J. Roy. Met. Soc. 76, 113 (1950).
[8] E. Krasnoff and R. L. Peskin, Geophys. Fluid Dyn., 2, 123 (1971).

Space Research XIII — Akademie-Verlag, Berlin 1973

INTERNAL GRAVITY WAVES AND TURBULENCE IN SIMULTANEOUS UPPER ATMOSPHERE TEMPERATURE AND WIND MEASUREMENTS

S. P. Zimmerman[a], G. P. Pereira[a*], E. A. Murphy[a] and J. Theon[b]

[a]Air Force Cambridge Research Laboratories, L. G. Hanscom Field, Bedford, Mass., USA
[b]NASA, Goddard Space Flight Center, Greenbelt, Md, USA

Rocket-borne pitot tube and chemical puff releases, over the altitude interval 85—125 km, results in simultaneous temperature, wind and turbulence measurements. Comparison of the wind and temperature oscillations shows conclusively that the temperature oscillations are not simply related to the wind oscillations by a constant phase shift as predicted by linear internal gravity wave theory. Turbulence, as observed in the density fluctuations of the lithium trail, occurs where the Richardson number (Ri) varies from values of $\sim$10 to 0.23. For a critical Ri of 1/4 or 1, this observed turbulence could not be simply attributed to atmospheric processes. However with the inclusion of Coriolis forces f, horizontal shear $\partial U/\partial y$ and kinematic viscosity ν, a new critical Richardson number (Ri_C) is introduced where instability will occur for

$$Ri \leqq Ri_C = f^2 \left[\nu^2 m^4 - f\left(\frac{\partial U}{\partial y} - f\right)\right],$$

where m is the vertical wave number. This results in a better correspondence between observed turbulence or instability and the criterion of instability. The theory will be briefly examined, and comparison with experimental measurements will be presented.

1. Introduction

The emphasis upon this section of the Aladdin experiment is the determination of the transport characteristics of the atmospheric region 85—125 km. This includes such parameters as the Richardson number, a measure of atmospheric instability, turbulent diffusion coefficients, and the amplitudes and phases of the wind and temperature oscillations due to tidal and/or internal gravity waves. The measured values of these parameters are then compared with theoretical calculations, with agreement and disagreement being emphasized.

2. Wind and Temperature Modes

The east—west and north—south wind amplitudes for the experiment Genie are shown in Fig. 1. This shows the north—south and east—west winds as out of phase from $\sim$ 85 to $\sim$ 100 km. For a single mode oscillation, the zero crossings are separated by a quarter wavelength; the measured value of the wind gives a value of $\sim$ 4 km which would indicate a wavelength of $\sim$ 16 km, not inconsistent

* deceased

with the wavelength of the primary diurnal mode [1]. The amplitude and hour of maximum westerly wind from Lindzen and Blake (L-B) [2] are shown in Figs. 2 and 3 (Figs. 2 and 3 of [2]), with the data also shown. Although the L-B calculations are done for equatorial conditions, the amplitude and, particularly, the phase are not much different from those of a 30° latitude. As is evident, the amplitude is not in great disagreement, and the phase, or hour of maximum westerly, is in very good agreement at ~ 102 and ~ 130 km.

The northerlies, as mentioned above, show what appears to be a phase shift consistent with the concept that the major element in the wind system is the single propagating mode. That is until ~ 100 km, where there is a sudden reversal.

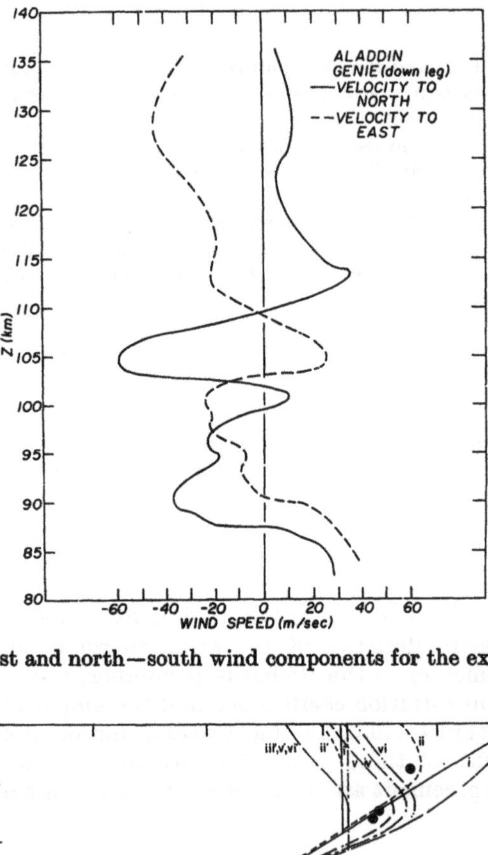

Fig. 1. East—west and north—south wind components for the experiment Genie.

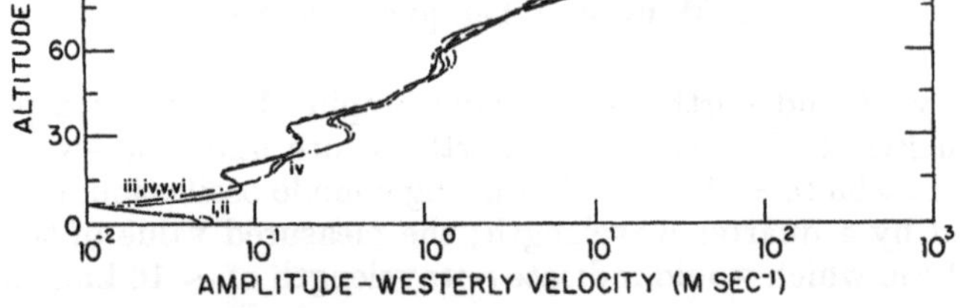

Fig. 2. The Lindzen and Blake theoretical amplitude of the westerly component of the diurnal solar tidal oscillations, with the data gathered from the trail Genie included [2, Fig. 2].

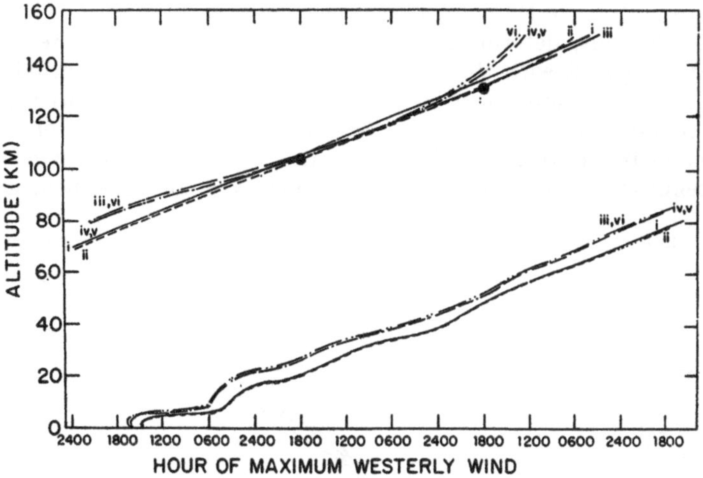

Fig. 3. The hour of the maximum westerly component of the diurnal solar tide, with the data determined from this experiment included, as above [2, Fig. 3].

so that the E—W and N—S winds are in opposition at 105 km. This anomalous event is not quantitatively or qualitatively explicable from the existent theories of tidal or internal gravity waves.

The temperature oscillations [3] superimposed upon the wind measurements are shown in Fig. 4. These are determined by subtracting from the measured

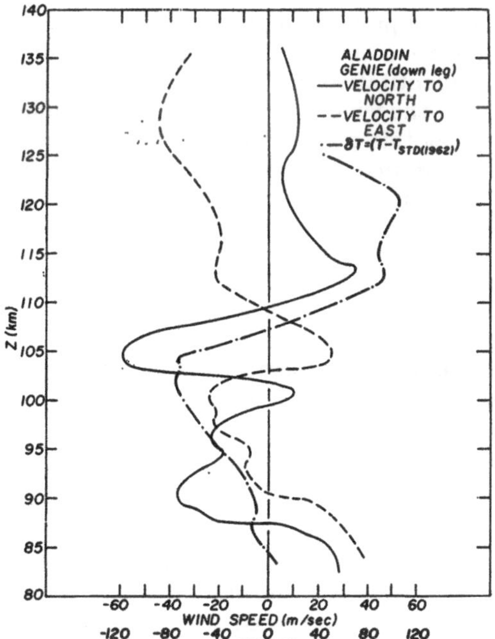

Fig. 4. The superposed wind and temperature oscillations showing some agreement in phase, particularly between the temperature and the easterly wind in the region below 105 km.

temperature profile the 1962 Standard Atmosphere temperature. The amplitudes of these oscillations, determined from the measured maxima are displayed in Fig. 5 and compared with theory of the major propagating diurnal mode (this is Fig. 4 of [2]). The agreement in the upper regions (~ 100 km) is fairly good, but unfortunately we do not have the theoretical calculation of temperature phase for comparison with the L-B theory. Also, in comparing the wind oscillations with the temperature oscillations, there is observed some agreement in the phase

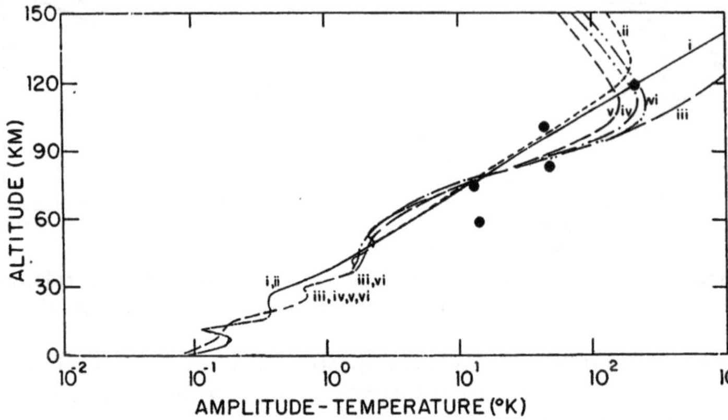

Fig. 5. The theoretical amplitude of the diurnal temperature oscillation of Lindzen and Blake, with the experimental data included [2, Fig. 4].

relation between the wind profiles and the temperature profile up to ~ 100 km, with a divergence of phase occurring above this altitude until about 107 km.

Thus there is some quantitative support, up to ~ 100 km, for the identification of the major features of the wind system as being the major propagating diurnal mode. What happens at this point, reflection from a turbulent layer or superposition of tidal and internal gravity wave modes, is shear conjecture, since further time experiments to study the time-dependent phase shifts in wind, temperature and turbulence were not performed.

3. Turbulence

The simultaneous wind and temperature observations allow for the first time the calculation of the Richardson number

$$Ri = \frac{g}{T}\left[\frac{\partial T}{\partial z} + \Gamma\right] \Big/ \left(\frac{\partial U}{\partial z}\right)^2$$

in the altitude region 85—125 km, where g is the acceleration of gravity, T is the kinetic temperature, $\partial T/\partial z$ is the vertical temperature gradient, Γ is the adiabatic lapse rate $= 10\,°\mathrm{K\ km}^{-1}$ and $\partial U/\partial z$ is the vertical shear of the horizontal wind. These values are compared wich the chemical trail release photographs, where the density fluctuations are indications of a turbulent, or unstable condition, and the smooth Gaussian density distributions are indications of a lack of turbulence or unstable conditions. If one accepts the critical values of

Ri of $^1/_4$ or 1, where $Ri_C < 1$ or $^1/_4$ is an indication of instability, then Fig. 6 shows that we should have instabilities evident in the photographs at altitudes from 85 km (where there is a superadiabatic lapse rate) to 86, 87 to approximately 91 or 92 km, and 101 to 104 km. However, the photographs show turbulence from 94 to ~ 99 km, with a rapid necking of the turbulently diffuse trail at ~ 99 km, indicative of a local cessation of turbulence. Then turbulence again from 102 to ~ 104, turbulence to ~ 107 km, a small necking there, and indications of a larger scale size instability to the altitude where turbulence is finally observed to cease, at approximately 111 km for the trail Genie and 109 for Hilda.

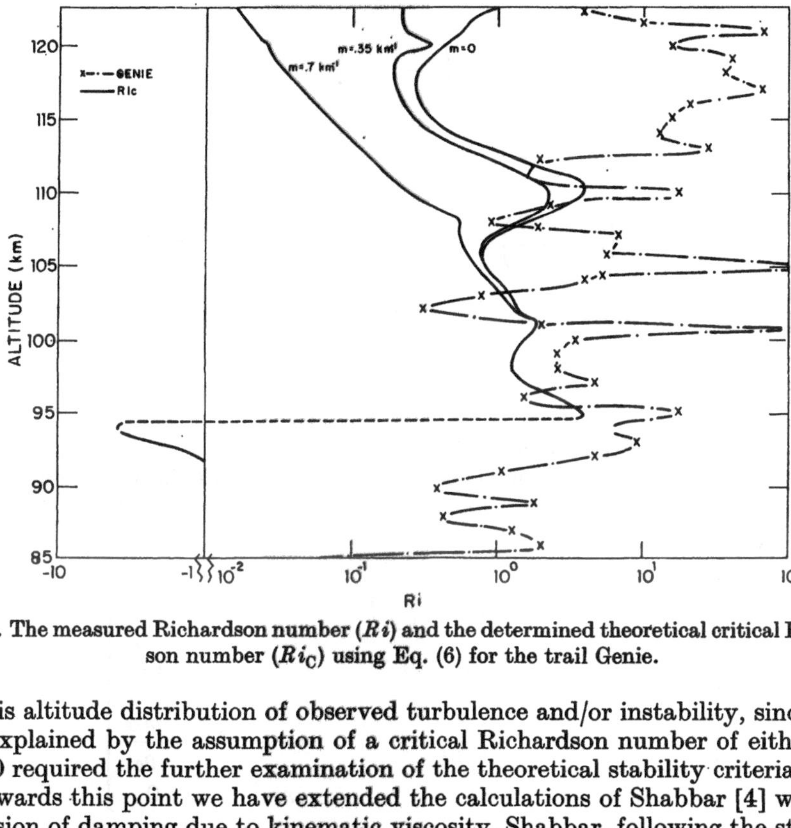

Fig. 6. The measured Richardson number (Ri) and the determined theoretical critical Richardson number (Ri_C) using Eq. (6) for the trail Genie.

This altitude distribution of observed turbulence and/or instability, since it is not explained by the assumption of a critical Richardson number of either 0.25 or 1.0 required the further examination of the theoretical stability criteria.

Towards this point we have extended the calculations of Shabbar [4] with the inclusion of damping due to kinematic viscosity. Shabbar, following the stability analysis of Wilson [5] included vertical shear in the equations of continuity and momentum where Wilson had considered only the instability created by the transverse shear of a steady zonal wind and the Coriolis forces.

Starting with the equations of continuity and momentum

$$\frac{D}{Dt}(\varrho_0 + \varrho) = 0 \tag{1}$$

$$\frac{D}{Dt}(U_0 + u, v, w) = -\frac{1}{\varrho_0 + \varrho}\nabla(P_0 + P) - g + (U_0 + u, v, w) \times f - \nu m^2 (u, v, w) \tag{2}$$

where ϱ_0, P_0 and U_0 are the undisturbed density, pressure and zonal velocity and ϱ, P, u, v, w are the perturbation components. The Coriolis parameter f, and g, the acceleration of gravity, are in the vertical plane, ν is the kinematic viscosity and D/Dt is the convected time derivative

$$\frac{D}{Dt} \equiv \frac{\partial}{\partial t} + (U + u, v, w) \cdot \nabla .$$

Then considering time and space perturbation variations as $\exp\{i(\sigma t + lx + mz)\}$ and meridional variation of U, the rearranged equations of motion and continuity become

$$-(i\sigma + \nu m^2)u - \left(f - \frac{\partial U}{\partial y}\right) v + \frac{\partial U}{\partial z} w = 0$$

$$fu - (i\sigma + \nu m^2)v + \frac{ilP}{\varrho_0} = 0$$

$$-f\left(f - \frac{\partial U}{\partial y}\right) v + (g\beta - \sigma^2)w + \frac{m\sigma}{\varrho_0} P = 0 \tag{3}$$

$$ilv + imw = 0$$

where
$$\beta = \frac{g}{T}\left(\frac{\partial T}{\partial z} + \Gamma\right),$$

and Γ is the adiabatic lapse rate.

This is Shabbar's set of equations with the addition of the viscous dissipation terms. This set has a unique solution when the determinant of the coefficients vanishes, i.e.

$$\begin{vmatrix} -i\sigma + \nu m^2 & -\left(f - \frac{\partial U}{\partial y}\right) & \frac{\partial U}{\partial z} & 0 \\ f & -i\sigma - \nu m^2 & 0 & i\frac{l}{\varrho_0} \\ 0 & +f\frac{\partial U}{\partial z} & g\beta - \sigma^2 & \frac{m\sigma}{\varrho_0} \\ 0 & -il & im & 0 \end{vmatrix} \equiv 0. \tag{4}$$

Equating the real component to zero, we have

$$\sigma^2\left(1 + \frac{l^2}{m^2}\right) = g\beta\frac{l^2}{m^2} + 2\frac{l}{m} f \frac{\partial U}{\partial z} + f\left(f - \frac{\partial U}{\partial y}\right) + \nu^2 m^4. \tag{5}$$

For the conditions of σ marginally unstable (gravity waves), and l/m real and $\ll 1$, we arrive at the following stability criterion

$$Ri \equiv g\beta \Big/ \left(\frac{\partial U}{\partial z}\right)^2 \leq f^2 \Big/ \left[\nu^2 m^4 + f\left(f - \frac{\partial U}{\partial y}\right)\right]. \tag{6}$$

It should be noted here that this result is in essence identical to Kuo's [6] Eq. (13b).

Utilizing the measured horizontal shear, vertical shear, temperature and density we calculated Ri and Ri_C for two wind trails and for three values of m. An example is shown in Fig. 6; and considering the uncertainties in the measurements and the quasi two dimensionality of the theory, the correspondence between the observed regions of turbulence and/or instability and the altitude regions where the theory predicts instabilities is quite good.

In conclusion, this agreement between the experimental observations of turbulence and instabilities and the derived theoretical criterion for instability shows that, for the atmospheric condition where strong horizontal shears are present over large distances the simple Richardson stability criterion of $^1/_4$ or 1 is not sufficient to establish the presence or absence of turbulence and/or instabilities.

References

[1] R. S. Lindzen, Quart. J. Roy. Met. Soc. **93**, 18 (1967).
[2] R. S. Lindzen and D. Blake, Geophys. Fluid Dynam. **2**, 31 (1970).
[3] J. Theon, Trans. AGU, **53**, 463 (1972).
[4] M. Shabbar, Tech. Mem. No. TEC 713, Dept. of Transport, Met. Branch, Toronto, Canada.
[5] S. D. R. Wilson, Quart J. Roy. Met. Soc. **91**, 132 (1965).
[6] H. L. Kuo, J. Met. **13**, 561 (1956).

Space Research XIII — Akademie-Verlag, Berlin 1973

LA POLARISATION DES MARÉES ET DES ONDES DE GRAVITÉ

H. Teitelbaum

Service d'Aéronomie du CNRS, Verrières-le-Buisson, France

We show that the non-linear interaction between two gravity waves can give secondary waves of opposite polarization, that is, the horizontal velocity vector turns anticlockwise in the northern hemisphere and clockwise in the southern.
Analysis of experimental results confirms this theoretical standpoint.

Il est couramment admis que les marées et les ondes de gravité doivent être polarisées à gauche dans l'hémisphère nord et à droite dans l'hémisphère sud. L'expérience montre qu'il n'en est pas toujours ainsi. Sur un hodographe instantané, si l'on admet que le flux vertical d'énergie est positif, le vecteur de vitesse horizontal devrait tourner dans le sens des aiguilles d'une montre dans l'hémisphère nord, en sens contraire dans l'hémisphère sud. Or, il existe des hodographes où le sens de rotation est inversé.

Nous avons déjà prouvé théoriquement [1] qu'il existe des modes de marée dont la polarisation est différente dans des zones de latitude que nous avons pu fixer avec certitude. Pour les ondes de gravité de courte période [2], inférieure à 30 minutes, la force de Coriolis est négligeable et il n'y a pas de rotation. Pour étudier le cas des ondes de gravité de période plus longue, on peut introduire la force de Coriolis dans la théorie classique, tout en la considérant localement constante. Dans ces conditions on trouve pour la moyenne du flux vertical d'énergie par cycle

$$E_{\mathrm{v}} = \gamma^2 p_0(0) \left\{2\sigma \left(\frac{1}{h} - \frac{\sigma^2}{g}\right)\right\}^{-1} \mathrm{Im} \left(y^* \frac{dy}{d\xi}\right). \tag{1}$$

Sauf l'hauteur reduite

$$\xi = \int \frac{dz}{H}$$

notre écriture et celle de Siebert [3].

Et pour la variation avec l'altitude de l'angle du vecteur de vitesse horizontal avec l'axe N—S

$$\frac{d\alpha}{d\xi} = -\frac{1}{u^2 + v^2} \frac{\cos\theta}{f} (k_x^2 + k_y^2) \left\{k_\xi^2 + \left(\frac{\sigma^2 H}{g} - \frac{1}{2}\right)^2\right\} \mathrm{Im} \left(y^* \frac{dy}{d\xi}\right) \tag{2}$$

u, v = composantes N—S, O—E du vent, p = pression, H = hauteur d'échelle, k_x, k_y = nombres d'onde horizontaux, vers le sud et vers l'est respectivement, σ = fréquence de l'onde, θ = co-latitude (0 au pôle nord), γ = constante adiabatique. On voit que les ondes de gravité sont à polarisation gauche dans l'hémisphère nord et droite dans l'hémisphère sud.

Le sens de rotation dépend de la différence de phase entre les composantes horizontales du vent. La tangente de la différence de phase est

$$-\frac{\cos\theta}{f}(k_x^2+k_y^2)\left\{\left(1-\frac{\cos^2\theta}{f^2}\right)k_xk_y\right\}^{-1}. \tag{3}$$

Cette expression montre que à l'équateur la différence de phase approche de 0° ou 180°, c'est-à-dire que l'onde est polarisée dans un plan.

Nous allons prouver que l'interaction non linéaire de deux ondes peut produire des ondes secondaires dont la polarisation est inversée et donner des exemples prouvant l'existence de ce phénomène. Pour étudier l'interaction non linéaire nous avons employé la méthode mise au point par Spizzichino [4]. Le calcul que nous avons fait porte sur l'interaction de la marée diurne avec différentes ondes de gravité; en effet, dans les mesures que nous avons analysées, nous avons constaté la présence de cette marée. La figure 1 montrent quelques résultats. Le sens de rotation n'est pas suffisant pour déterminer l'allure d'un hodographe;

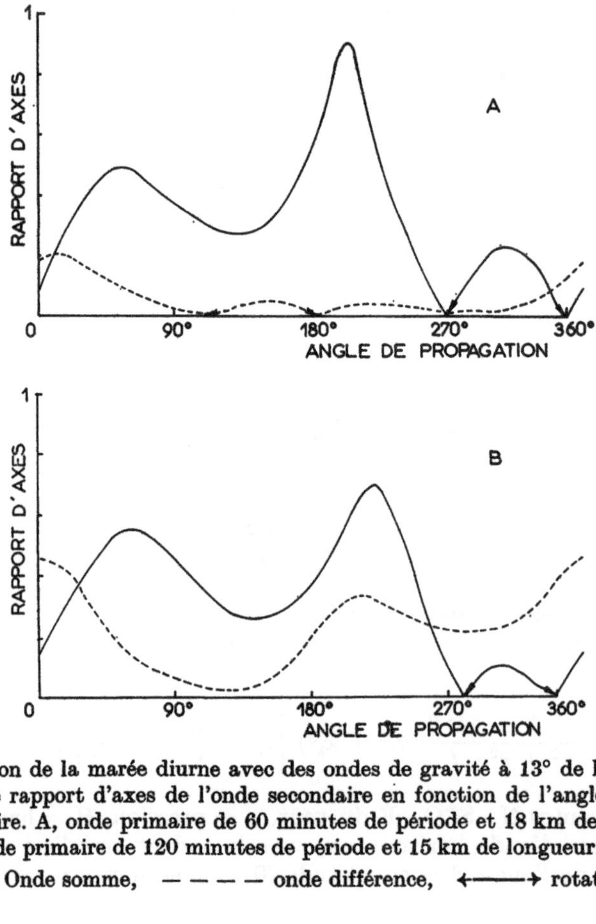

Fig. 1. Interaction de la marée diurne avec des ondes de gravité à 13° de latitude nord. La courbe donne le rapport d'axes de l'onde secondaire en fonction de l'angle de propagation de l'onde primaire. A, onde primaire de 60 minutes de période et 18 km de longueur d'onde verticale; B, onde primaire de 120 minutes de période et 15 km de longueur d'onde verticale. ——— Onde somme, — — — — onde différence, ←——→ rotation directe.

c'est pourquoi nous avons préféré donner le rapport d'axes de l'ellipse de polarisation en fonction de la direction de propagation de l'onde de gravité primaire sur le plan horizontal (0° dans la direction N—S). On constate que même des ondes de faible période peuvent donner après interaction avec la marée des ondes secondaires avec une ellipse de polarisation dont le rapport d'axes est proche de l'unité. Pour chaque onde de gravité on a deux ondes secondaires: l'onde somme et l'onde différence selon que la fréquence résultante est la somme ou la différence des fréquences. L'onde de gravité est de période plus courte que la marée. Si la longueur d'onde verticale est également plus courte les deux ondes secondaires auront la fréquence et le nombre d'onde du même signe, c'est-à-dire que le flux vertical d'énergie sera positif. Si la longueur d'onde verticale de l'onde de gravité est supérieure à celle de la marée, le flux vertical d'énergie de l'onde différence sera négatif.

Nous allons faire l'analyse des mesures faites à 13° de latitude la nuit du 19 au 20 septembre 1966 et dont les hodographes ont été publiés par Murphy [5]. Sur les deux premiers hodographes (2055 et 2224 TL) on voit que la rotation est inversée à partir de 105 km. Dans les profils de vent correspondant à ces hodographes on remarque la présence d'une onde dont la différence de phase des composants ne correspond pas à un mode de marée et dont la phase elle-même se déplace vers le bas; il s'agit bien d'une onde secondaire produite par l'interaction de deux autres. Sa longueur d'onde est de 30 km et la période, d'après le décalage de la phase, de 7 heures. L'interaction commence a être importante à partir d'une altitude qui dépend de l'amplitude des mouvements. On doit trouver à une certaine altitude les deux ondes indépendantes superposées, puis au-dessus, une zone de transition où commence l'interaction et où les mouvements sont compliqués et finalement, les ondes secondaires qui ont absorbé l'énergie des ondes primaires. C'est ce qui se passe pour l'hodographe que nous avons analysé.

A 0010 et 0123 TL dans la partie inférieure des profils O—E on remarque une onde d'environ 10 km de longueur d'onde. La phase est descendue d'environ 2 km, en 73 minutes, ce qui donne une période de 6 heures.

L'onde de gravité qu'on voit à partir de 100 km peut être expliquée par l'interaction de la marée diurne avec l'onde de 10 km de longueur d'onde visible au-dessous. Ce résultat implique que la longueur d'onde verticale de la marée diurne soit de 16 km à 90 km d'altitude. Nous allons voir qu'il s'agit de la marée diurne et que sa longueur d'onde verticale est de 16 km. Si on admet à partir de 90 km d'altitude un gradient de température constant et positif, l'équation radiale des marées admet une solution analytique et on peut essayer de décomposer le profil de vent tout en ajustant la température à la limite inférieure et le gradient de température [6]. Quand il s'agit de plusieurs profils qui se suivent dans le temps, la méthode devient plus exacte. Nous avons employé cette méthode dans les quatre derniers profils N—S. Le résultat qu'on voit sur la figure 2 a été obtenu avec la marée diurne seulement. La température qui a donné une bonne approximation est de 140° à 90 km avec un gradient de 8°/km. L'augmentation d'amplitude avec l'altitude montre que la dissipation visqueuse n'est pas négligeable c'est pourquoi nous avons ajouté un facteur empirique de forme exponentielle $\exp(-\alpha z/H)$. Le fait que pendant quelques heures la marée diurne puisse être remarquée apparait également sur les hodographes.

La figures 3 [7] montre deux hodographes qui correspondent aux mesures faites à 117 minutes d'intervalle. Le premier montre une rotation inverse à partir de 105 km d'altitude. Le second montre une structure très compliquée à partir de

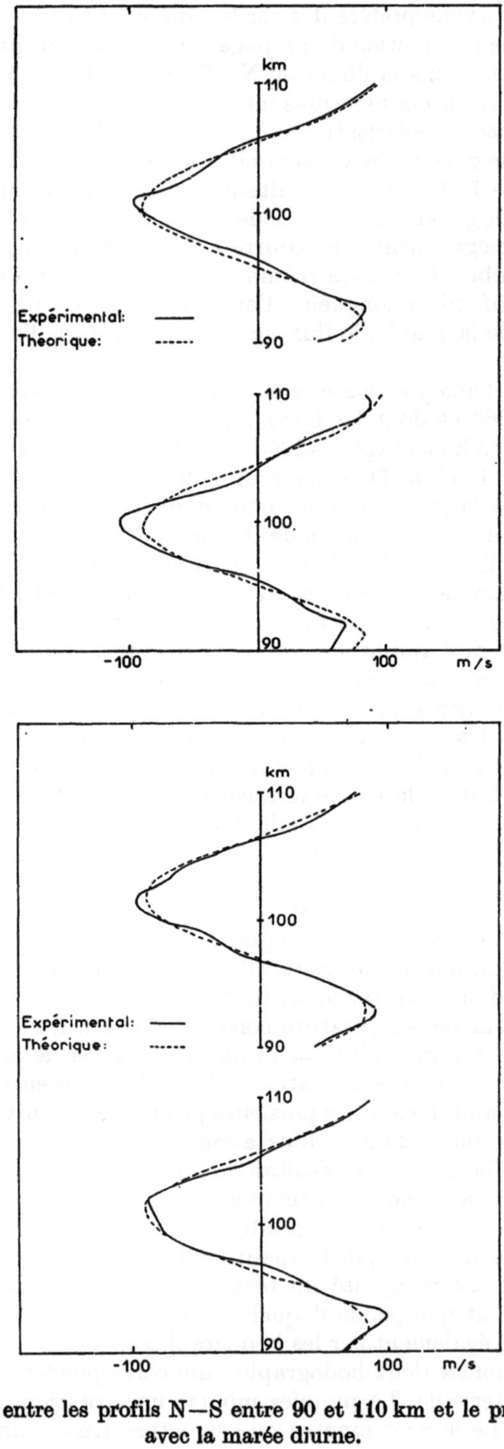

Fig. 2. Comparaison entre les profils N—S entre 90 et 110 km et le profil théorique calculé avec la marée diurne.

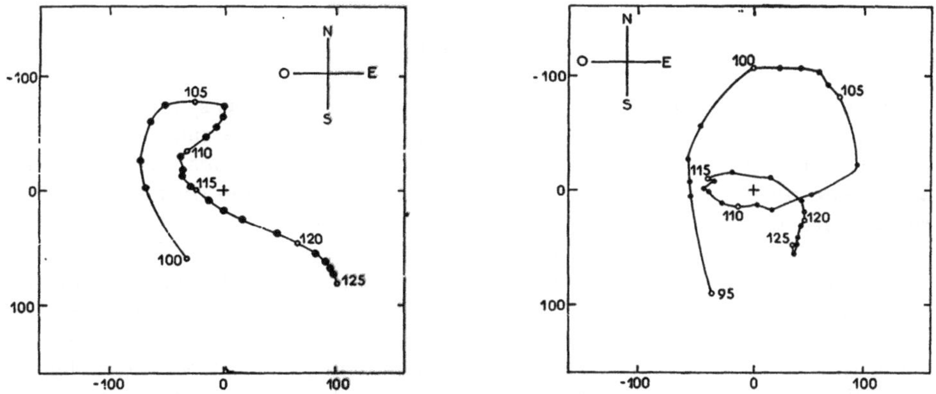

Fig. 3. Hodographes correspondant aux mesures de vent faites à 13° de latitude nord (Barbados) dans la nuit du 22 au 23 novembre 1965. Les heures locales sont 0131 et 0328 respectivement [7].

la même altitude mais de 95 à 105 km la forme de l'hodographe correspond à la marée diurne. Murphy [5] a remarqué le phénomène d'apparition de la rotation classique à certaines heures de la nuit. Nous l'avons retrouvé dans plusieurs données expérimentales. Nous proposons d'expliquer ce phénomène par les variations que subit l'onde de gravité quand elle traverse le milieu modifié par la marée. Quand une onde pénètre dans uns couche d'air qui se déplace avec une vitesse $\boldsymbol{U}$, sa fréquence dans le milieu mobile prend la valeur [8, 9]:

$$\sigma' = \sigma - \boldsymbol{k}\boldsymbol{U} \tag{4}$$

où σ est la fréquence dans le milieu fixe et k le nombre d'onde horizontale de l'onde de gravité. La relation de dispersion est aussi valable dans le milieu mobile ce qui implique que lorsque la fréquence diminue le nombre d'onde verticale augmente, le nombre d'onde horizontale étant constant. L'annulation de la fréquence σ' ne peut pas être atteinte parce que la longueur d'onde verticale aura diminué avant, jusqu'à produire une dissipation visqueuse. A une altitude où le module de la vitesse horizontale induite par la marée serait capable d'annuler la fréquence σ', l'onde sera dissipée quand $\boldsymbol{k}$ et $\boldsymbol{U}$ sont parallèles.

On admet que la plupart des ondes de gravité ont leur origine dans les zones polaires. A basse latitude on trouvera des ondes qui se propagent presque horizontalement, c'est-à-dire de longue période. La direction de propagation doit être proche de la direction N—S. Sa basse fréquence explique que l'onde puisse être réfléchie facilement. A l'heure où le vecteur vitesse horizontale induite par la marée pointe vers le sud à l'altitude où $\sigma' \simeq 0$, l'onde de gravité est dissipée. Le reste du temps l'onde continue à se propager vers le haut, augmente d'amplitude et les phénomènes non linéaires prennent place. Il peut y avoir aussi réflexion partielle avec atténuation de l'onde de gravité de façon que le dissipation visqueuse intervienne avant que les phénomènes non linéaires soient importants. Dans les mesures que nous avons analysées la fréquence de l'onde de gravité primaire est $\sigma = 3 \times 10^{-4}\ \mathrm{s}^{-1}$. Avec la longueur d'onde de 10 km on a un nombre d'onde horizontale $k = 9 \times 10^{-6}\ \mathrm{m}^{-1}$. Il en résulte une vitesse de vent nécessaire pour produire la dissipation de $30\ \mathrm{m\ s}^{-1}$. D'après la vitesse qu'on trouve au-dessus de 90 km on peut avoir une vitesse de $30\ \mathrm{m\ s}^{-1}$ à 80 km.

Conclusion

La polarisation des marées et des ondes de gravité n'est pas systématiquement gauche dans l'hémisphère nord et droite dans l'hémisphère sud. A l'existence pour certains modes de marée de zones de latitude que nous avons appelées à rotation inverse, on peut ajouter la possibilité d'apparition du même type de rotation par interaction de deux ondes.

Les données expérimentales provenant des mesures faites à basse latitude montrent ce genre de phénomène. On constate aussi l'arrêt de l'interaction avec apparition de la marée à l'état presque pure pendant quelques heures. Ceci s'explique par la dissipation de l'onde de gravité à plus basse altitude lorsque les modifications du milieu induites par la marée augmentent considérablement le nombre d'onde verticale. Les données expérimentales que nous avons utilisées correspondent à des mesures faites à 13° de latitude nord. Pour des mesures faites à 33° [7] on constate également une rotation inverse et l'arrêt de l'interaction pendant quelques heures ce qui prouve qu'il ne s'agit pas d'un phénomène fortuit mais assez systématique au moins aux basses latitudes.

Remerciements

Je suis heureux de remercier le Professeur J. E. Blamont qui par de fructueux commentaires m'a aidé à mener à bien ce travail.

References

[1] J. E. Blamont et H. Teitelbaum, Annls Géophys. **24**, 387 (1968).
[2] C. O. Hines, Canad. J. Phys. **38**, 1441 (1960).
[3] M. Siebert, in: Advanc. Geophys. **7**, Academic Press, New York 1961 (p. 108).
[4] A. Spizzichino, Annls Géophys. **25**, 755 (1969).
[5] C. H. Murphy, J. Geophys. Res. **74**, 339 (1969).
[6] H. Teitelbaum, Space Research X, 180 (1969).
[7] Data Report of the Meteorological Rocket Network, Superintendent of Documents Washington D.C. Nov. 1965.
[8] C. O. Hines and C. A. Reddy, J. Geophys. Res. **72**, 1015 (1967).
[9] D. H. Cowling, H. D. Webb and K. C. Yeh, J. Geophys. Res. **76**, 213 (1971).

Space Research XIII — Akademie-Verlag, Berlin 1973

ATMOSPHERIC DENSITY VALUES FROM ORBITAL DRAG ON THE CANNON BALL 2 AND MUSKET BALL SATELLITES

K. S. W. Champion, F. A. Marcos and R. A. Schweinfurth

Air Force Cambridge Research Laboratories, Bedford, Mass., USA

Two satellites Cannon Ball 2 (1971-67C) and Musket Ball (1971-67D) were launched at 0011 GMT, 7 Aug. 1971 into elliptical, near polar orbits. Cannon Ball 2 had an initial perigee of 132 km and apogee of 1968 km and Musket Ball an initial perigee of 139 km and apogee of 870 km. The density of the upper atmosphere between 132 km and 144 km has been determined at intervals of approximately 24 hours from analysis of orbital changes induced by aerodynamic drag. Data are presented for the time period 7—24 Aug. 1971. This time period has been divided into three intervals corresponding to when the perigee of Musket Ball was at low, middle, or high northern latitudes. The perigee latitude of Cannon Ball 2 was approximately 10° less than that of Musket Ball during each interval. The average Musket Ball perigee density was 8% less than the Jacchia 1971 model at low latitudes. At middle latitudes the average density was equal to the model value, and at high latitudes it was 6% greater than the model value. The Cannon Ball 2 average densities agreed with the Musket Ball data within 3% for all three intervals. Considerable variability was observed in the density data, particularly at high latitudes. Any difference in the drag coefficient of the two satellites was less than 3%.

1. Introduction

Two low altitude density research satellites, Cannon Ball 2 and Musket Ball, were successfully launched into near polar orbits from Vandenberg AFB at 1711 PDT on 6 Aug. 1971. This paper describes perigee densities obtained from analysis of the orbital drag effects on these satellites. Cannon Ball 2, a 363.6 kg 66 cm diameter sphere (see Fig. 1) was injected into a 92° inclination orbit with an exceptionally low perigee of 132 km and apogee of 1968 km. Lifetime was 178 days. Perigee initially was at 6° N and precessed northward at 3.0 degrees/day. One primary objective of this satellite was to obtain high resolution density profiles in the lower thermosphere with an ultrasensitive triaxial accelerometer. These results and comparison with orbital decay data have been described elsewhere [1, 2]. The satellite contained a minitrack beacon and a C-band tracking beacon which could be turned on by ground command. The NASA STADAN Network had responsibility for command and control of Cannon Ball 2. Since the STADAN did not have C-band tracking equipment they turned the beacon on and tracking was accomplished by manned space flight network stations.

Musket Ball was designed as a very inexpensive means to measure density at low perigee altitudes by the orbital drag technique. The satellite was a 61.4 kg 31.3 cm diameter sphere containing a C-band tracking beacon, batteries and a timer which activated the C-band beacon for a four-hour period each day for a two-week period. The stations that tracked the satellite beacon included those of the Eastern Test Range and the Bermuda and Wallops Island stations. Initial orbital parameters were: perigee 139 km, apogee 870 km, inclination 87°, initial northward precession 3.83 degrees/day. Lifetime was 43.5 days.

Fig. 1. Photograph of the Cannon Ball 2 satellite.

2. Data Analysis

The average density at perigee was deduced from analysis of beacon and skin track observations of the two satellites. Data are presented for the period 7—24 Aug. 1971. The method used consists of a numerical integration scheme combined with a differential orbit correction procedure where the discrepancies between computed and observed satellite position are reconciled by adjusting the assumed atmospheric model. Orbital drag densities for individual perigee passes were determined by employing the method of sliding fit spans.

3. Results

The ratios of the orbital decay densities obtained at perigee to the appropriate Jacchia 1971 models [3] have been calculated. Fig. 2 shows these results as a function of time for the Musket Ball satellite. Also shown on the figure are the 3-hourly *Kp* values plotted at the midpoint of their interval and the value of the

10.7 cm solar flux. During the period 7 Aug. (day 41170) to 24 Aug. (day 41187) perigee altitude varied from 136 km to 144 km and perigee latitude varied from 15° N to 82° N. These results have been separated into three intervals according to whether perigee was at low, middle or high latitudes: 7—11 Aug. when perigee latitude was between 15 and 31° N (low latitudes), 12—18 Aug. when perigee latitude was between 32 and 54° N (middle latitudes) and 19—23 Aug. when perigee latitude was between 55 and 78° N (high latitudes). For each interval the

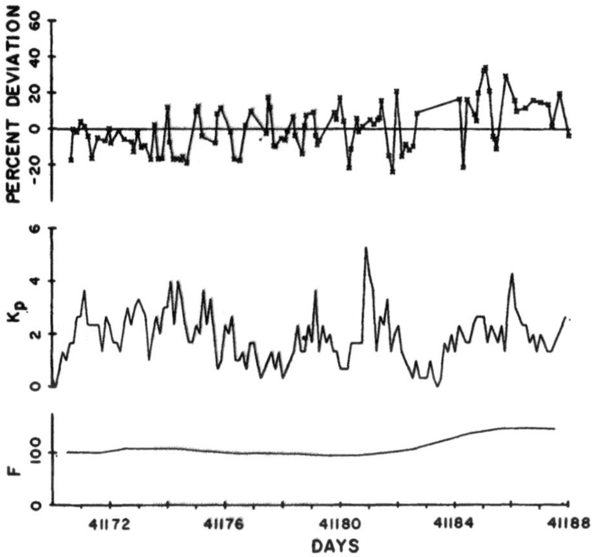

Fig. 2. Orbital drag density deviations from the Jacchia 1971 models for Musket Ball, *Kp*, and $F_{10.7}$ for the period 7—24 Aug. 1971.

average ratio was calculated. The low latitude data are an average of 8% less than the model. The average daily sum of *Kp* had a low value of 18 and the average solar flux was 105. The average middle latitude data showed no deviation from the Jacchia 1971 models. For this period the average daily *Kp* was 14 and the average solar flux was 99.5. At the high latitudes the average density was 6% higher than the models. The average daily sum of *Kp* was 13+, the lowest value for these three intervals. However, the solar flux increased markedly on 19 Aug. and the average value for this period was 140.2. The average density ratios show very good agreement with the Jacchia 1971 models.

Fig. 3 shows Cannon Ball 2 density ratios plotted as a function of time, together with *Kp* and solar flux, as in Fig. 2. These results have been separated into the same three time periods used for the Musket Ball data. Because the perigee precession was slower, the Cannon Ball 2 data do not cover as wide a range of latitude. The corresponding latitudes are: 7—11 Aug., 9—21° N; 12—18 Aug., 22—42° N; 19—23 Aug., 43—55° N. For the 7—11 Aug. data the average density ratio is 11% less than the model; from 12—18 Aug. it is 3% less and from 19—23 Aug. it is 2.5% greater.

The Cannon Ball 2 average densities agree very well with those from Musket Ball for each interval, being from 3—3.5% higher. The Musket Ball data show an increase in the density ratio of 14% between the 7—11 Aug. data and the 19—23

Aug. data. The Cannon Ball 2 data show an increase of 13.5% for the same time interval. These results are summarized in Table 1. With the amount of data presently available the model improvements for solar and/or geophysical relations to density variations cannot be specified. Variations with solar flux, latitude, local time and geomagnetic activity are observed for this set of data. The solar flux was approximately constant during the first two time intervals, during

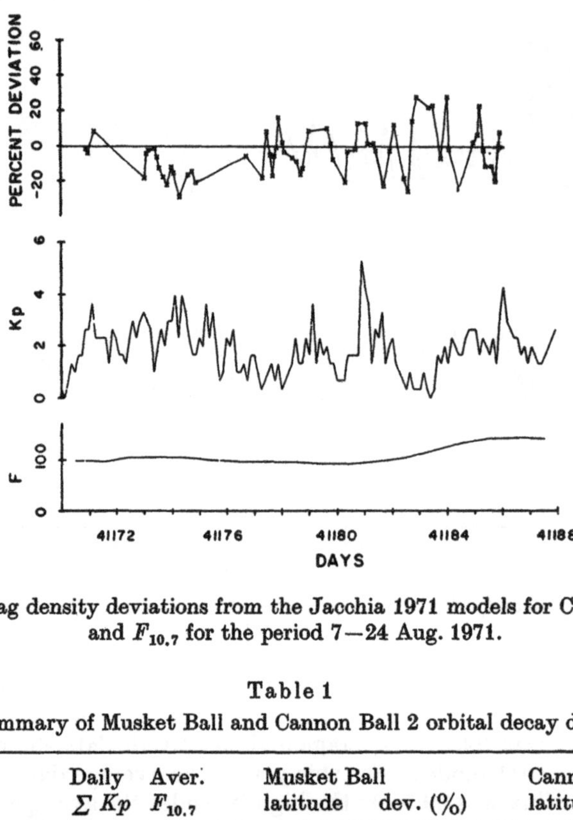

Fig. 3. Orbital drag density deviations from the Jacchia 1971 models for Cannon Ball 2, Kp, and $F_{10.7}$ for the period 7—24 Aug. 1971.

Table 1

Summary of Musket Ball and Cannon Ball 2 orbital decay data

Date	Daily ΣKp	Aver. $F_{10.7}$	Musket Ball latitude	Musket Ball dev. (%)	Cannon Ball 2 latitude	Cannon Ball 2 dev. (%)
7—11 Aug. 1971	18	104.6	15—31°N	−8	9—21° N	−11
12—18 Aug. 1971	14	99.6	32—54°N	0	22—42° N	−3
19—23 Aug. 1971	13+	140.2	55—78°N	+6	43—55° N	+2.5

which period the average ratio increased by 8%, but when the solar flux increased by some 40 units, the average ratio increase was only 6%. The increase in the ratios is probably at least partly latitudinal. Local time varied from 1600 hrs on 7 Aug. to 1330 hrs on 24 Aug. Some of the variation between the measured data and the model may be due to local time effects. The average daily sum Kp decreased from 18 to 13+ during the measurement interval. This corresponds to a temperature difference of approximately 20°K. The model density variation for a 20°K difference at 135 km, a typical average perigee altitude for the two satellites, is less than 1%. Thus the increase in density ratios with time does not appear to be related to geomagnetic activity modeling.

4. Conclusions

The density at perigee has been determined from orbital decay analysis of the Cannon Ball 2 and Musket Ball satellites for the time period 7—24 Aug. 1971. Results from the two satellites show that the Jacchia 1971 models provide a very good representation of the atmosphere between 132 and 144 km for the solar and geophysical conditions that existed. A systematic increase with time in the average ratio of the measured density to model density is observed. Some of the long term variations may be due to model errors which vary with latitude or local time, although geomagnetic activity and solar flux are also possible factors. Only a portion of the available data are reduced at present. Analysis of the complete results will be made to determine improvements to be recommended in the present models. Another important conclusion is that the drag coefficients of the two satellites (one with a brass outer shell and the other almost entirely covered with solar cells) are the same in this altitude regime within one or two per cent.

References

[1] F. A. Marcos and K. S. W. Champion, First Internat. Conf. on Aerospace and Aeronautical Meteorology, May 1972 (p. 173).
[2] K. S. W. Champion and F. A. Marcos, Space Research XIII, 229 (1973).
[3] L. G. Jacchia, Smithson. Astrophys. Obs. Spec. Rep. No. 332 (1971).

Space Research XIII — Akademie-Verlag, Berlin 1973

LOWER THERMOSPHERE DENSITY VARIATIONS DETERMINED FROM ACCELEROMETERS ON THE CANNON BALL 2 SATELLITE

K. S. W. Champion and F. A. Marcos

Air Force Cambridge Research Laboratories, Bedford, Mass., USA

Cannon Ball 2 (1971-67C) carried as its primary experiment an ultrasensitive triaxial electrostatic force rebalance accelerometer for measurement of aerodynamic drag. Profiles of the atmospheric neutral density were obtained in the altitude region 132—200 km. The latitudinal span of each profile is 30° on either side of perigee with a local time of the data of approximately 1700 hours. Results are presented for the period 9—12 Aug. 1971, during which time geomagnetic activity was low. Perigee precessed from 13° N to 22° N during this period. Eight orbits of data were obtained for each day. The results are compared with the Jacchia 1971 models and are correlated with variations in latitude and solar and geomagnetic parameters. The measured density values are generally within ±10% of the Jacchia models. The average density at perigee is 7% less than the model value, in excellent agreement with orbital decay data from this satellite. The scale heights are larger than the model values below about 160 km and less than the model values above about 180 km.

1. Introduction

The Cannon Ball 2 (1971-67C) satellite was specifically designed for *in situ* density measurements at very low satellite altitudes. The spacecraft was a 363.6 kg 66 cm diameter sphere. The resulting high mass-to-area ratio of 106 g cm^{-2} was achieved mainly by its one inch thick brass outer shell weighing 308 kg. An ultrasensitive triaxial accelerometer comprised the primary payload. This experiment determines atmospheric density from direct measurement of aerodynamic drag. Other significant components of the spacecraft were a solar eye magnetometer aspect system, tape recorder, and C-band and minitrack tracking beacons. Solar cells provided the electrical power. The satellite was successfully launched as part of Air Force STP Flight 70-2 from Vandenberg AFB at 1711 PDT on 6 Aug. 1971. An exceptionally low perigee of 132 km was achieved with an apogee of 1968 km. Lifetime of the satellite was 178 days. The inclination was 92°. Data were recorded either for one 90-minute period or for three 30-minute periods centered on perigee during consecutive orbits, before dumping the tape recorder.

2. Flight Data

The scale factor for each accelerometer was 1 pps $= 1.5 \times 10^{-7}$ g. The sample time was one second. Output from the three accelerometers is shown in Fig. 1. These results were obtained on revolution number 76 on 12 Aug. 1971. The spin rate of the spacecraft had increased to 1.1 rev min^{-1}, from its initial 0.3 rev min^{-1} owing to aerodynamic effects. The figure shows accelerometer count rate as a function of GMT in seconds. The data sample starts at 500 km altitude at 51130 seconds and ends at 500 km at 52800 seconds. Perigee occurs at 51960 seconds. For each axis accelerations due to vehicle dynamics only are seen at the higher altitudes. Air drag modulates the accelerometer output at the spin frequency at the lower altitudes, between 51700 seconds and 52300 seconds. Aerodynamic drag

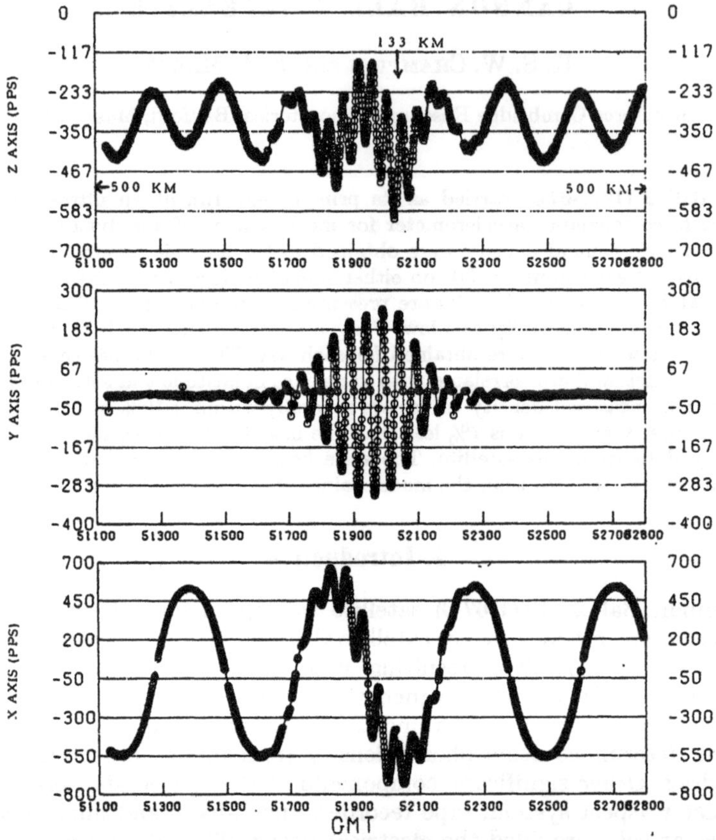

Fig. 1. Accelerations measured by each uniaxial accelerometer unit on Cannon Ball 2 for revolution No. 76 on 12 Aug. 1971 at altitudes between 500 km and 132.2 km.

accelerations are mathematically extracted from the total acceleration data. Total drag a_D is:

$$a_D = (a_x^2 + a_y^2 + a_z^2)^{1/2}$$

where a_x, a_y and a_z are the drag accelerations sensed at the same time by the X, Y and Z accelerometers. The density ϱ is determined from

$$\varrho = \frac{2 m a_D}{C_D A V^2}$$

where m, C_D, A and V are the satellite's mass, drag coefficient, area and velocity respectively.

3. Density Results

The density profile obtained from rev. 76 is shown in Fig. 2. Open circles represent data taken as the satellite approached perigee (downleg), solid circles represent data taken as the satellite moved away from perigee (upleg). The measured data start on the downleg at 200 km at a latitude of 47° N and end at 200 km upleg at 3° S. Perigee is at 22° N. Longitude is 5° W and local time is 1600 hours. The subsolar point was at 15.5° N latitude. The data were obtained during a period of quiet solar and geomagnetic conditions with $F_{10.7} = 99$ units and $Kp = 4^-$, assuming a six-hour time lag. Good agreement is found between upleg

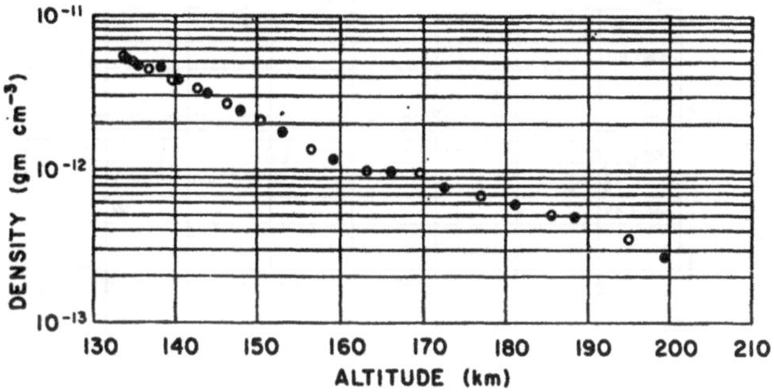

Fig. 2. Density data obtained from the triaxial accelerometer arrangement on Cannon Ball 2 for revolution No. 76 on 12 Aug. 1971: ∘ downleg; • upleg; perigee 22°N lat., 5° W long.

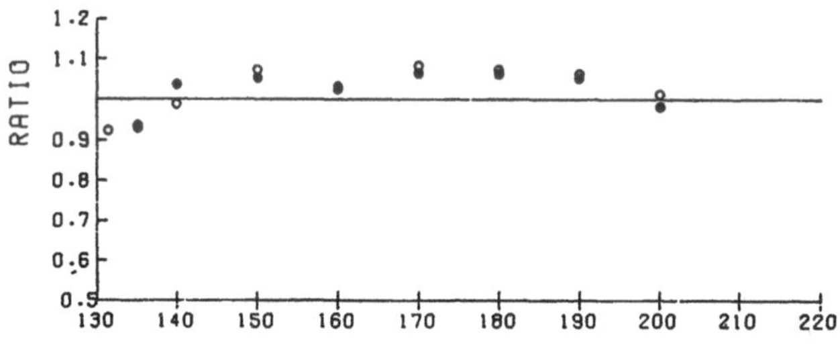

Fig. 3. Ratios of revolution No. 76 density data to the Jacchia 1971 model density values: ∘ downleg; • upleg.

and downleg data. No significant effect due to the diurnal variation is expected since perigee is close to the sub-solar point. These density data are compared with the Jacchia 1971 models [1] in Fig. 3. This figure shows the ratio of the measured density to the model density as a function of altitude from perigee, 132 km, to 200 km. The measured densities fall within $\pm 9\%$ of the models with the lowest ratios occurring near perigee. The models show excellent agreement below 200 km with these data measured at low and middle latitudes during a quiet solar and geomagnetic period. The ratios increase up to about 175 km indicating larger scale heights, then begin to decrease toward 1.0 at 200 km.

Density results automatically recorded on three consecutive orbits are shown in Fig. 4. These data are from revs. 50, 51 and 52 on 10 Aug. 1971. The $F_{10.7}$ cm solar flux was 107 units for this day. The applicable average 3-hourly Kp was 1 for rev. 50 and 2 for revs. 51 and 52. Local time for these three orbits was approximately 1600 hours. The rev. 51 data are compared with the Jacchia 1971 model in Fig. 5. Again the agreement between the measured data and the models is very good. The shape of the ratio vs altitude plot is similar to that for rev. 76 shown in Fig. 3, with lower ratios found near perigee, maximum ratios near 175 km and an average ratio of about 1.0 at 200 km.

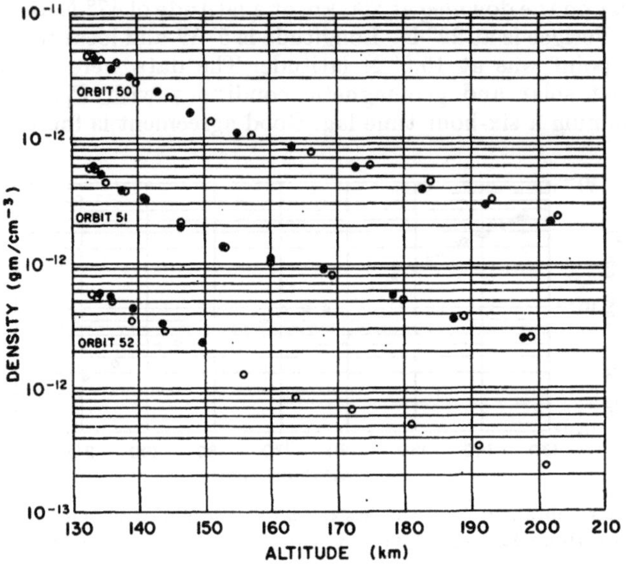

Fig. 4. Density data obtained from the triaxial accelerometer on Cannon Ball 2 for revolutions 50, 51 and 52 on 10 Aug. 1971: ∘ downleg; • upleg; perigee 16° N lat.

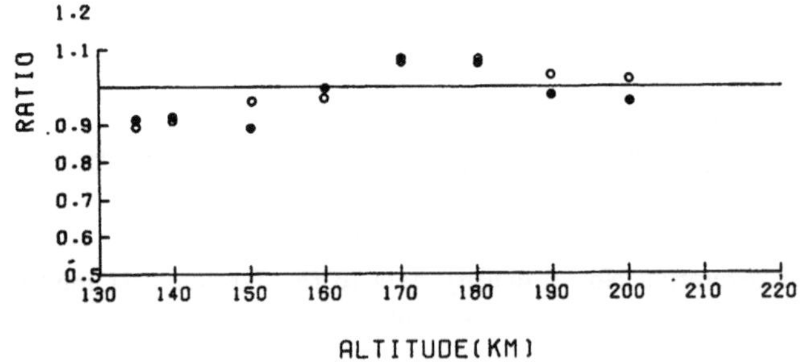

Fig. 5. Ratios of revolution No. 51 density data to the Jacchia 1971 model density values: ∘ downleg; • upleg.

Table 1 lists orbital, solar and geomagnetic parameters for data from seven revolutions. The average density at 135 km and 175 km for each revolution was compared with the Jacchia 1971 models. This ratio is plotted against revolution number in Fig. 6. Solid circles represent the average density at 135 km for each revolution; open circles represent the average density at 175 km for each revolution. In all cases lower ratios are found at the lower altitudes, a typical difference

Table 1

Orbital, solar, and geomagnetic parameters for data plotted in Fig. 6

Rev. No.	Date	Local time	GMT	Perigee alt. (km)	Perigee lat. (°N)	Perigee long. (°E)	$F_{10.7}$	Kp (6 hr. lag)
31	9 Aug. 1971	1557	0853	132.1	12.7	106.4	109	2+
46	10 Aug. 1971	1555	1120	132.7	15.8	95.3	107	3
50	10 Aug. 1971	1554	1823	132.5	16.7	322.8	107	2
51	10 Aug. 1971	1554	2009	132.5	16.9	296.3	107	2
52	10 Aug. 1971	1554	2155	132.8	17.1	269.8	107	3−
69	12 Aug. 1971	1551	0352	132.5	20.7	179.8	99	2−
76	12 Aug. 1971	1550	1426	133.0	21.9	355	99	4−

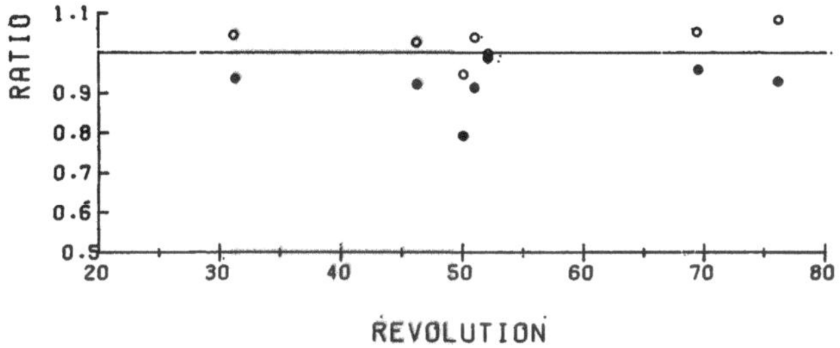

Fig. 6. Ratios of Cannon Ball 2 density data for seven revolutions at 135 km and 175 km to the Jacchia 1971 model density values: • average 135 km; ∘ average 175 km.

being 10%. Consistently good agreement is found between the measured data and model data, the largest deviation being a ratio of 0.8 for the rev. 50 135 km data. The cause of the lower ratios for rev. 50, both at 135 km and 175 km, is not known.

The agreement between the accelerometer data and orbital decay data is quite good. The average perigee density for these seven orbits is 9% less than the model. Orbital decay densities from this satellite for the time period 7—11 Aug. were an average of 11% less than the model [2].

4. Conclusions

Data from seven orbits obtained from the Cannon Ball 2 satellite have been presented. Measurements were acquired to the low altitude of 132 km. The data are for low and middle latitudes and for a period of relatively low solar and geomagnetic activity. The results show excellent agreement with orbital decay data and with the Jacchia 1971 models. Most of the data fall within $\pm 10\%$ of the models. Larger scale heights are indicated for the altitude region 135 km to 175 km. Further results will be analyzed to determine the response of the neutral atmosphere to a wide variation of solar and geophysical conditions.

References

[1] L. G. Jacchia, Smithson. Astrophys. Obs. Spec. Rep. No. 332 (1971).
[2] K. S. W. Champion, F. A. Marcos and R. A. Schweinfurth, Space Research XIII, 223 (1973).

Space Research XIII — Akademie-Verlag, Berlin 1973

STUDY OF NEUTRAL COMPOSITION OF LOWER THERMOSPHERE AT FORT CHURCHILL

A. O. NIER and D. R. HICKMAN

University of Minnesota, Minneapolis, Minnesota, USA

On 4 and 6 February 1969 and 11 May 1970, Aerobee rockets carrying neutral mass spectrometers were flown at Fort Churchill, Canada (59° N 94° W) during conditions of low geomagnetic activity. As in earlier flights at White Sands, New Mexico, each rocket carried both "open" and "closed" ion source instruments. Vertical profiles of N_2, O_2, O, Ar and He were measured. Results obtained were essentially the same as observed at White Sands except that for the winter flights helium appeared to be in diffusive equilibrium.

1. Introduction

During the period 1965—1967, the Minnesota group studied the neutral composition of the lower thermosphere in a number of rocket flights [1—3] launched at the White Sands, New Mexico Missile Range. In each of these flights the payload consisted of three magnetic deflection mass spectrometers. Spectrometer 1 had an "open" ion source to facilitate the measurement of reactive species such as atomic oxygen. Spectrometers 2 and 3 had their ion sources in a common chamber connected to the ambient atmosphere by a knife-edged hole in a wall of the chamber extending beyond the skin of the rocket. In flight 4.322 UA the ion source as well as the chamber were gold-plated. In 4.272 UA and 4.273 UA, as well as in earlier flights, the surfaces were principally of stainless steel. Because of the carefully defined geometry, a quantitative relation could be established between particle density as measured by the instruments and that in the ambient atmosphere. Spectrometers 1 and 2 scan mass spectra from 4 to 44 amu with a period of approximately 2.5 seconds whereas spectrometer 3 scans only the mass 28 peak, and is thus essentially a continuous N_2 monitor.

In 1969 the study was extended to the polar regions with flights 4.272 UA launched at 0835 CST, 4 February 1969, 4.273 UA at 0835 CST, 6 February 1969 and 4.322 UA at 1342 CST, 11 May 1970. A report on the 4.272 UA flight has been given [4]. The present paper summarizes the results of all three polar flights.

As described by Hedin and Nier [1], the mass spectrometers are mounted with their openings to the atmosphere facing out from the skin of the cylindrical part of the Aerobee rocket. The vacuum sealed instruments are turned on prior to launch, and when an altitude of approximately 100 km is reached, protective caps [5] which sealed the instruments are removed so that the ion sources are exposed to the ambient atmosphere. Data are taken on both the upleg and downleg of the trajectory. In a typical case, the rocket spins with a period of about 4 seconds. The axis of the rocket also undergoes a precessional motion. Since the speed of the rocket is comparable with average molecular velocities, the combination of the several motions of the rocket results in a rather intense modulation of the measured signal. Fig. 1 shows how the mass 28 peak in the No. 3 instrument varied with time in flight 4.322 UA. Hedin et al. [6] showed how a curve of this sort could be reduced to a smooth monotonic one giving the relation between altitude and ambient particle density. Although the procedure followed in the analysis of 4.273 UA and 4.322 UA was somewhat different from that described by Hedin and Nier [1] and Hedin et al. [6], the same relation between ambient and ion source particle densities was assumed.

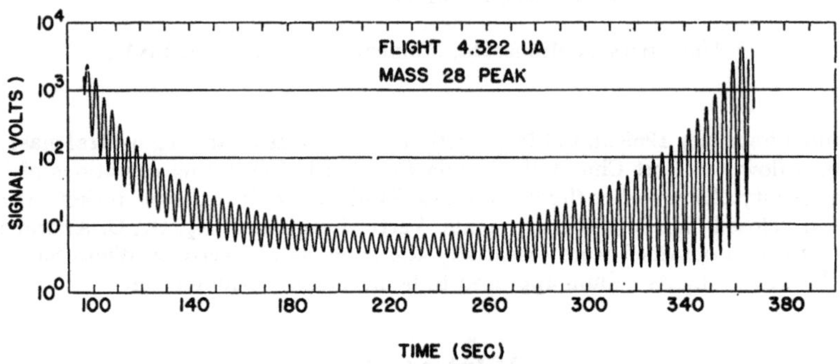

Fig. 1. Output signal in equivalent volts of the No. 3 instrument on flight 4.322 UA showing the variation of the mass 28 peak with time. Times 109 and 354 seconds correspond to altitude 120 km; time 231 seconds corresponds to 190 km, the apex of the flight. A one volt output signal corresponds to an ion source density of approximately 3×10^9 N_2 molecules cm^{-3}. The rapid modulation is due to the alternate ram and wake conditions caused by the rocket spin. The greater amount of modulation toward the end of the flight results from an increased angle of attack for the rocket.

2. Results

2.1. Molecular Nitrogen, Molecular Oxygen, Argon

Fig. 2 gives the variation with altitude of the N_2, O_2 and Ar number densities as determined in the three flights. Also shown for comparison are the Jacchia 1971 model values [7]. In the case of 4.272 UA, the method of reduction of raw data to final results is discussed by Hickman and Nier [4].

Because of a rocket motor malfunction, flight 4.273 UA reached an altitude of only 133 km. Also, spectrometer 2 suffered a systematic loss in sensitivity between laboratory calibration and flight, so the N_2 results are the weighted average of those for the No. 1 and No. 3 instruments while the O_2 and Ar results are those found from instrument No. 1 alone. In the case of flight 4.322 UA, the No. 1 instrument failed to function properly, so the N_2 values are the weighted average of the No. 2 and No. 3 instruments while the O_2 values are those given by the No. 2 instrument alone. The Ar results did not appear as reliable as usual and hence are not reported for this flight.

The N_2 values are slightly higher than found in previous flights at White Sands, New Mexico [1—3]. At 150 km they agree well with the N_2 values 3.06×10^{10} cm^{-3} and 3.41×10^{10} cm^{-3} given for a 12 Dec. 1966 Fort Churchill flight by von Zahn and Gross [8] and Müller and Hartmann [9] respectively. The 4.272 UA and

4.273 UA N_2 number densities agree well with the 1100 °K exosphere Jacchia 1971 values [7] while the 4.322 UA curve falls off less rapidly with altitude. As in the case of N_2, the O_2 densities are slightly higher than observed at White Sands, New Mexico, and for flights 4.272 UA and 4.273 UA are in excellent agreement with the 1100 °K exosphere model, Jacchia 1971 [7]. While the agreement for 4.322 UA is also very good at low altitudes, as in the case of N_2, the

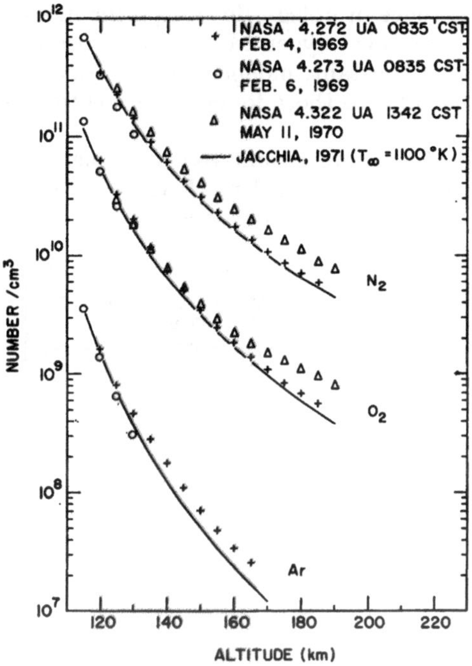

Fig. 2. Number densities of N_2, O_2 and Ar as a function of altitude. Shown for comparison are the Jacchia 1971 model values for an exospheric temperature of 1100 °K.

measured values at higher altitudes are greater than those given by the model. At 150 km the present values fall considerably below the figure 5.74×10^9 cm^{-3} given by Müller and Hartmann [9].

The failure of the N_2 and O_2 curves of flight 4.322 UA to fall off with altitude as rapidly as the corresponding curves in previous flights is puzzling. While it would be tempting to attribute this to some geophysical effect such as an abnormally high atmospheric temperature, this seems unlikely. It is probably due to some instrumental difficulty not readily explicable. It is not likely that the shape of the N_2 curve can be attributed to CO, present as an impurity formed from atomic oxygen. The N_2 abundance computed from the 14 peak in the No. 2 instrument follows the 28 peak very well. This would not be the case if the 28 peak contained any appreciable amount of CO. The similar departures from expected values of the N_2 and O_2 curves suggest a common cause. The 4.272 UA and 4.273 UA argon values agree well with the Jacchia 1971 model [7] but like the N_2 and O_2 results are slightly higher than corresponding values found at White Sands, New Mexico.

2.2. Atomic Oxygen

Fig. 3 shows the measured atomic oxygen as determined by the open source spectrometers, uncorrected for losses, for flights 4.272 UA and 4.273 UA. Also shown for comparison are the Jacchia 1971 model values [7]. While the measured densities fall below the model values, the curves differ by constant factors and this, along with the absence of upleg—downleg differences in the data, indicates that if there is loss of atomic oxygen in the mass spectrometer ion source, the fraction lost is probably constant, independent of altitude or time. The present results are thus very similar to those found with identical instruments in earlier flights.

The determination of absolute atomic oxygen densities by rocketborne mass spectrometers continues to be a problem. Total mass densities computed from mass spectrometrically-determined constituent number densities have usually

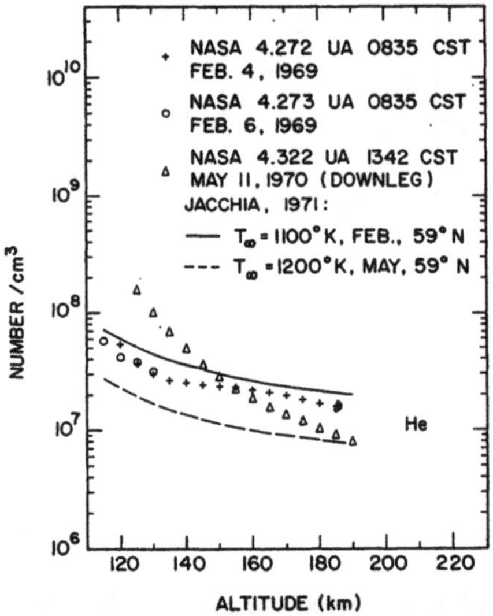

Fig. 3. Number density of O as a function of altitude for flights 4.272 UA and 4.273 UA. The values given are calculated on the basis of no atomic oxygen loss in the ion source and hence represent minimum values. Shown for comparison are the Jacchia 1971 model values for an exospheric temperature of 1100°K.

been lower than those found from satellite drag measurements. From a comprehensive comparison of composition-determined mass densities at 150 km with those computed from satellite drag data, von Zahn [10] concluded that the results could be reconciled if one increased the mass spectrometrically-determined atomic oxygen densities by a substantial factor, perhaps 3 to 5, and also decreased satellite determined densities by 10%. In support of higher atomic oxygen densities Offermann and von Zahn [11] presented results determined with a rocketborne mass spectrometer having a helium-cooled ion source.

On the other hand, Nier [12] directly compared mass density scale heights in the altitude range 130—180 km found by the low altitude satellites OV1-15 and

OV 1-16 with ones found from mass spectrometric composition studies. He concluded that an arbitrary increase of the measured atomic oxygen densities by large factors destroyed the relatively good agreement which existed in the scale heights. A comparison is shown in Fig. 4 where mass densities computed from the number densities given in Fig. 2 and 3 are plotted for flight 4.272 UA. The several curves correspond to doubling, tripling, etc. the measured atomic oxygen number densities. The points correspond to the OV 1-15 and OV 1-16 results as presented

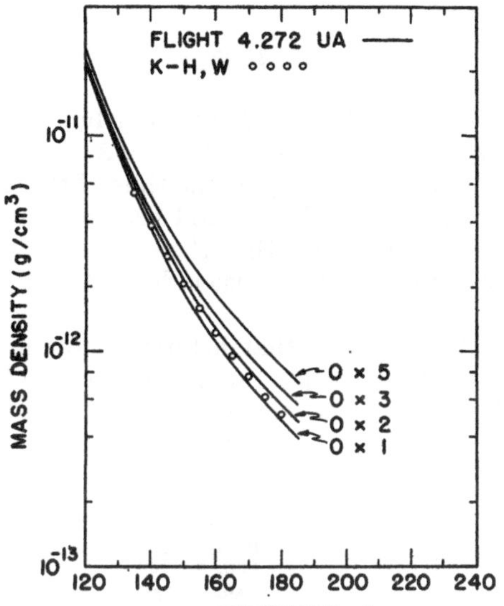

Fig. 4. Comparison between mass density versus altitude as given for satellites OV 1-15 and OV 1-16 by King-Hele and Walker [13] and as computed from composition found in rocket flight 4.272 UA.

by King-Hele and Walker [13]. It is seen that while a doubling of the measured oxygen number densities might be in order, an increase by a larger factor would clearly destroy the scale height agreement which existed.

In flights such as discussed here, and earlier [1—3], the instruments were always opened to the ambient atmosphere at an altitude below 100 km on the upleg of the trajectory. The ion source surfaces were thus exposed to a relatively large amount of atomic oxygen, as well as other atmospheric gases, before data were actually taken. This may have "conditioned" the surfaces in such a way as to reduce atomic oxygen loss, a fortuitous situation. That such an effect may have taken place seems likely in view of the experience of Mauersberger [14], who showed that in a flight where the instrument was opened to the ambient atmosphere at apogee, 132 km, the initial atomic oxygen measured was abnormally low. By the time the rocket had dropped below 110 km, the amount of atomic oxygen indicated by the instrument was much more reasonable.

Nier et al. [15] have reported on a laboratory study in which an atomic oxygen beam was directed at a mass spectrometer ion source similar to that employed in the present flights. The results suggest that the atomic oxygen loss in "open" source mass spectrometers is not a serious factor. On the other hand, the authors caution that the experiments were conducted with an oil-pumped vacuum facility which was not as "clean" as is generally used in preparing mass spectrometers for rocket or space flights. It may be this lack of cleanliness which inhibits the loss of atomic oxygen on surfaces!

These experiments have been repeated [16] with a facility which was somewhat cleaner than that employed in the Nier et al. [15] study. The results indicate some loss and suggest that in flights such as discussed here the atomic oxygen measured may be low by a factor lying between 2 and 3.

2.3. Helium

Fig. 5 shows the variation with altitude of the helium in the three flights. The 4.273 UA results agree well with those of 4.272 UA and support the conclusion that helium was in diffusive equilibrium at the time of the flights, February 1969. This agrees with the conclusions of Müller and Hartmann [9] resulting from their 1966 winter flight at Fort Churchill. Bitterberg et al. [17], reporting on a pair of flights from Sardinia in October 1967, also indicated that helium was in diffusive equilibrium.

The curve for 4.322 UA suggests that helium is not in diffusive equilibrium. The result should, however, be accepted with some reservation. During the upleg trajectory the mass 4 peak suffered an abrupt drop in intensity by a factor of almost 10 at an altitude of approximately 180 km. The mass 7 and 8 peaks,

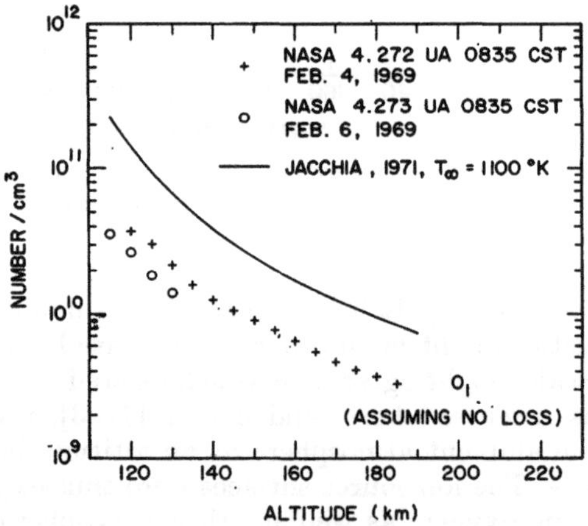

Fig. 5. Number density of helium as a function of altitude.

measured with the same collector and electron multiplier tube, behaved in a perfectly normal manner, ruling out the possibility of some electronic malfunction. During the remainder of the trajectory all of the peaks were normal. A possible explanation is that one of the valves sealing the fuel or oxidizer tank of the rocket failed to close properly after powered flight. Since helium is used for pressurization in these tanks, one is tempted to attribute the initial high helium concentration to helium leakage. For this reason only downleg data are plotted for the flight. It is hard to see how a malfunction such as hypothesized could affect the data on the downleg in such a way as to indicate a departure from diffusive equilibrium of the kind observed here. While the No. 1 instrument suffered a serious loss in sensitivity during powered flight, as well as some other difficulties, and hence data from it are not included in this paper, the helium values obtained from it tend to support non-diffusive equilibrium for helium rather than diffusive equilibrium.

Acknowledgments

This research was supported from grants NGL-24-005-009 and NGR-24-005-196 from the National Aeronautics and Space Administration. We are indebted to Jeffrey Hayden and Herbert Ballmann for their assistance during the flights and David Olson for his help in reducing the data.

References

[1] A. E. Hedin and A. O. Nier, J. Geophys. Res. **71**, 4121 (1966).
[2] W. T. Kasprzak, D. Krankowsky and A. O. Nier, J. Geophys. Res. **73**, 6765 (1968).
[3] D. Krankowsky, W. T. Kasprzak and A. O. Nier, J. Geophys. Res. **73**, 7291 (1968).
[4] D. R. Hickman and A. O. Nier, J. Geophys. Res. **77**, 2880 (1972).
[5] R. B. Thorness and A. O. Nier, Rev. Sci. Instrum. **33**, 1005 (1962).
[6] A. E. Hedin, C. P. Avery and C. D. Tschetter, J. Geophys. Res. **69**, 4637 (1964).
[7] L. G. Jacchia, Smithson. Astrophys. Obs. Spec. Rep. No. 332 (1971).
[8] U. von Zahn and J. Gross, J. Geophys. Res. **74**, 4055 (1969).
[9] D. Müller and G. Hartmann, J. Geophys. Res. **74**, 1287 (1969).
[10] U. von Zahn, J. Geophys. Res. **75**, 5517 (1970).
[11] D. Offermann and U. von Zahn, J. Geophys. Res. **76**, 2520 (1971).
[12] A. O. Nier, Space Research XII, 881 (1972).
[13] D. G. King-Hele and D. M. C. Walker, Planet. Space Sci. **17**, 2027 (1969).
[14] K. Mauersberger, Trans. Am. Geophys. Un. **52**, 876 (1971).
[15] A. O. Nier, J. L. Hayden, J. B. French and N. M. Reid, J. Geophys. Res. **77**, 1987 (1972).
[16] L. R. Lake, Ph. D. Thesis, University of Minnesota (1972).
[17] W. Bitterberg, K. Bruchhausen, D. Offermann and U. von Zahn, J. Geophys. Res. **75**, 5528 (1970).

Space Research XIII — Akademie-Verlag, Berlin 1973

MOLECULAR OXYGEN DENSITY IN THE LOWER THERMOSPHERE FROM MAY TO NOVEMBER 1967

B. R. May

Meteorological Office, Bracknell, Berkshire, England

Measurements of molecular oxygen density at heights from 150 to 240 km were made from May to November 1967 by a solar grazing-ray absorption experiment flown on the Ariel 3 satellite.

The density was observed to be very variable with an extreme range of 7:1 at 180 km. The greatest densities were observed at medium and high latitudes and always in association with magnetic disturbances. At magnetic latitudes greater than 40° a change in *Kp* from 1— to 4— was accompanied by an average increase of 46% in the density at 180 km, after a mean delay of about 4.5 hours: at lower latitudes the increase was 11% after 10.5 hours. Greater changes of density occurred in the morning hours of local solar magnetic time than in the afternoon hours during a magnetic disturbance in October 1967.

During quiet magnetic conditions the density at 180 km at very high latitudes, within 10° of the magnetic north pole, was 45% greater than at low latitudes. At low latitudes the density at 180 km was observed to be 16% less, on average, at sunset than sunrise.

The observations indicate that the molecular oxygen density at 180 km given by Jacchia's 1971 model atmosphere is too great by about 50% on average and that the model does not correctly represent the sunrise—sunset difference at low latitudes.

1. Introduction

On 5 May 1967 the Ariel 3 satellite was put into orbit about the earth, carrying, among others, an experiment designed and built by the staff of the Meteorological Office, Bracknell, to measure the density of molecular oxygen in the lower thermosphere using the solar grazing-ray absorption technique. The principle and design of the equipment and its behaviour in orbit have been described by Wildman et al. [1] and Stewart and Wildman [2]. The equipment continued to work much longer than expected and gave useful measurements of density up to November 1967. The main results of a study of these densities is given in this paper.

2. Calculation of Molecular Oxygen Densities from Absorption Measurements

The measurement of solar radiation intensity was made over a wavelength range from 1425 to 1490 Å in which the radiation is heavily absorbed by oxygen molecules in the atmosphere. This bandwidth, of the ionization chambers used to measure the intensity, was narrow enough for the logarithmic extinction law to be used directly to estimate the total number of molecules in a unit cross-sectional area beam from the satellite to the sun. A value of the cross-sectional area of oxygen molecules of 1.40×10^{-17} cm^2 was used [3]. Assuming spherical symmetry of the atmosphere the density at heights from 150 to 240 km could then be deduced from the intensity measurements made during each eclipse passage of the satellite using a method suggested by May [4]. Densities were obtained from 656 passages of the satellite and of these 521 yielded the density at 180 km, the height at which it is measured with the least error. Two measurements were made in each orbit (though telemetry limitations prevented more than 20—30% of these being received); successive orbits were separated by 24 degrees in longitude and 95 minutes in time. The instrumental standard deviations associated with each density were also estimated and these averaged 9%, 6% and 35% at 150, 180 and 240 km respectively. In this paper the variation of density with height is expressed in terms of the number density and scale height at 180 km, n_{180} and H_{180}, and the gradient of scale height β, with the standard deviations given in brackets.

3. The Variation of Molecular Oxygen Density

3.1. Variation of Density with Magnetic Activity and Latitude

The daily maximum molecular oxygen density observed at 180 km is plotted against time in Fig. 1. The most conspicuous feature is the occurrence of large increases of density up to 11.1×10^8 cm^{-3}, superimposed upon a mean undisturbed

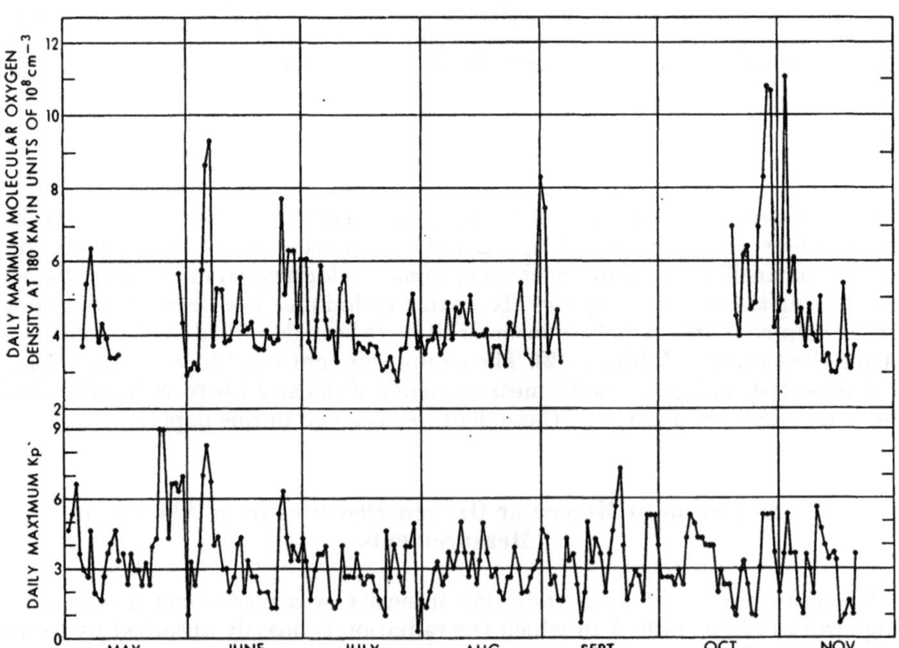

Fig. 1. Variations of daily maximum molecular oxygen density at 180 km and Kp from May to November 1967.

density of about 3.2×10^8 cm^{-3}. These increases are invariably associated with magnetic storms as can be seen by a comparison with the accompanying plot of the daily maximum 3-hourly range activity index Kp. They occur at all heights from 150 to 240 km, increasing in amplitude with height. This behaviour is similar to that of the total air density in the lower thermosphere deduced from drag measurements on artificial earth satellites. It is also apparent from Fig. 1 that the response of n_{180} to magnetic activity is variable, being smaller from mid-July to mid-August during which time the observations were made at low latitudes. In order to study the difference in response of the density at different latitudes to magnetic activity, the observations of n_{180} were divided into two groups to represent the low (0°—40° N and S magnetic) and the mid to high (40°—80°) latitude regions. In Fig. 2 the correlation coefficients r of n_{180} with the

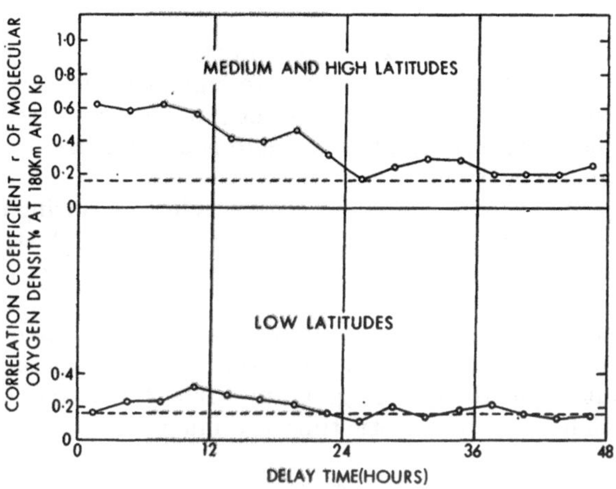

Fig. 2. Variation with delay time of correlation coefficient of molecular oxygen density at 180 km and Kp. The dashed lines represent the 5% significance level.

magnetic activity 1.5, 4.5, 7.5 ... hours earlier (denoted by $Kp_{1.5}$, $Kp_{4.5}$ etc.) are plotted against the delay time; the dashed line is the 5% significance level. At low latitudes r is generally small but significant and reaches a definite maximum value of 0.33 for a delay of 10.5 hours. In contrast, at mid to high latitudes r is generally much larger with values of 0.6 for up to 7.5 hours delay and then decreasing; the centre of the peak in r is estimated to occur for a 4.5 hour delay. These times clearly show that the response of oxygen density to magnetic activity is later at the equator than at higher latitudes and they agree fairly well with the corresponding mean values of the delay for air density quoted by Jacchia et al. [5] of 5.8 hours at 65° and 7.2 hours at 25° geographic latitude. The mean delay times of 4.5 hours at mid to high latitudes and 10.5 hours at low latitudes have been used in the rest of the analysis.

Values of density were combined together to produce reliable means for the two levels of magnetic activity $Kp = 1-$ and $4-$ in these two latitude regions and also for $Kp = 1-$ at even higher latitudes, within 10° of the magnetic north pole. These mean densities are accurately represented over the height range 150 to 210 km by the values of n_{180}, H_{180} and β in Table 1.

Table 1

Variation of molecular oxygen density with magnetic activity and latitude

Magnetic latitude range		n_{180} (units of 10^8 cm^{-3})	H_{180} (km)	β (km km^{-1})	ratio of n_{180}
Polar 80–90°	$\bar{K}p_{4.5} = 1-$	4.74 (0.30)	21.8 (1.7)	0.11 (0.25)	—
	$\bar{K}p_{4.5} = 4-$	—	—	—	
Mid to high lat. 40–80°	$\bar{K}p_{4.5} = 1-$	3.66 (0.10)	19.6 (0.8)	0.13 (0.05)	1.46
	$\bar{K}p_{4.5} = 4-$	5.36 (0.50)	21.6 (1.1)	0.06 (0.09)	
Low latitude 0–40°	$\bar{K}p_{10.5} = 1-$	3.29 (0.07)	19.0 (0.5)	0.11 (0.04)	1.11
	$\bar{K}p_{10.5} = 4-$	3.65 (0.15)	18.7 (0.7)	0.08 (0.06)	

A change in Kp from 1— to 4— is accompanied by an 11% increase in n_{180} at low latitudes and a 46% increase at mid to high latitudes, after an appropriate delay. At mid to high latitudes the scale height also increases by about 2 km. Even in very quiet magnetic conditions there is an increase of density and scale height with magnetic latitude.

Table 1 represents average behaviour, but the changes during individual storms were sometimes much larger or smaller than the average. A striking illustration of the non-uniform response of density to magnetic disturbance is provided by the observations of n_{180} from two successive orbits of the satellite during a magnetic storm on 30 October. At this time the two observations made on each orbit were on opposite sides of the pole, at nearly the same magnetic latitudes and separated by only 14 minutes; the details are given in Table 2. On both orbits the density measured near 16 hours geomagnetic time was much less than that near 08 hours.

Table 2

Observations of molecular oxygen density at 180 km from orbits 2672 and 2673

Orbit	Time (UT, 30 Oct 1967)	n_{180} (units of 10^8 cm^{-3})	Geomag. lat. (°)	Geomag. time (h)
2672	0253	2.52 (0.28)	49 N	15.9
	0307	7.30 (0.73)	48 N	8.3
2673	0428	3.12 (0.35)	46 N	16.2
	0442	7.53 (0.75)	52 N	8.5

It is of interest to note that Hartz and Brice [6] have suggested that several high latitude phenomena have a diurnal minimum of activity near 16 hours geomagnetic time and maxima near to 22 and 08 hours indicating that the latter are the preferred times for the influx of charged particles into the upper atmosphere. According to Cole [7] these particles could be a powerful source of thermospheric heating in the auroral zone causing the increases in air and molecular oxygen density observed at high latitudes during magnetic storms.

However this mechanism cannot be directly responsible for the increases in density observed at low latitudes but it is possible that energy is transferred to these regions by gravity waves. The delay time of 10.5 hours already quoted for the effects of magnetic storms on oxygen densities at low latitudes would correspond to a velocity of about 200 m s^{-1}. This speed agrees favourably with the observed speeds of "travelling ionospheric disturbances" (which are believed to be caused by gravity waves propagating through the F region) of 180 ms^{-1} reported by Du Castel and Faynot [8] and a median value of 150 m s^{-1} by Hunsucker and Tveten [9]. The smaller response of the molecular oxygen density to changes in Kp at low latitudes compared with high latitudes is also consistent with the transport of energy by waves as these would be considerably attenuated by the time they reach the equator.

3.2. Variations of Molecular Oxygen Density observed at Low Latitudes

At high latitudes the effects of magnetic storms are sufficiently large to obscure other effects, but at low latitudes they are much less and so other variations could be searched for.

There is a difference in the density of molecular oxygen in the lower thermosphere at low latitudes between sunrise and sunset. This is apparent from the values of n_{180}, H_{180} and β given in Table 3 which represent the mean densities from 150 to 210 km for three groups of sunrise and sunset observations.

Table 3

Difference in molecular oxygen density at low latitude between sunrise and sunset

Epoch (1967)	n_{180} (units of 10^8 cm^{-3}) Observed	Jacchia 1971 model	H_{180} (km) Observed	Jacchia 1971 model	β (km km^{-1}) Observed	Jacchia 1971 model
			Sunrise			
28 May—4 June	3.52 (0.16)	4.81	18.7 (0.3)	19.5	0.12 (0.02)	0.13
10—27 Aug.	3.94 (0.06)	4.21	19.6 (0.5)	19.1	0.12 (0.02)	0.13
01—21 Nov.	3.60 (0.10)	4.89	19.2 (0.4)	18.5	0.10 (0.06)	0.12
			Sunset			
5—14 May	3.21 (0.04)	5.78	18.5 (0.7)	20.7	0.10 (0.06)	0.20
17 July—10 Aug.	3.10 (0.06)	5.24	18.6 (0.4)	20.8	0.07 (0.03)	0.12
19—27 Oct.	2.92 (0.12)	5.66	20.2 (1.1)	19.7	0.13 (0.12)	0.14

The density at 180 km at sunset is consistently less than at sunrise, by about 16% on average, in contrast with the air density which is observed to be greater at sunset than sunrise. This decrease in density is probably the result of the dissociation of oxygen molecules by solar ultra-violet radiation during the daytime. A comparison with Jacchia's 1971 model densities [10] reveals that the model overestimates the O_2 density at low latitudes by a factor of 1.3 at sunrise and 1.5 at sunset.

There is no clear indication of a semi-annual variation of molecular oxygen density in the lower thermosphere which is a prominent feature in the air density. The density at 180 km at sunrise is certainly higher during August than in May or November (opposite to that of the air) but this pattern is not present at sunset. The values of H_{180} and β in Table 3 reveal no systematic sunrise—sunset difference or semi-annual variation. The presence of a 27-day variation of molecular oxygen density associated with solar activity is not clear in Fig. 1; if this variation exists then its amplitude is less than 10% at 180 km.

Acknowledgments

The author would like to acknowledge useful discussions on this topic with Dr. K. H. Stewart and the assistance of Mrs. M. Broadhurst and Mr. R. D. Carter in the numerical work.

This paper is published by permission of the Director-General of the Meteorological Office.

References

[1] P. J. L. WILDMAN, G. P. CARRUTHERS and C. V. ELSE, Radio Electron. Engr **35**, 97 (1965).
[2] K. H. STEWART and P. J. L. WILDMAN, Proc. Roy. Soc. A **311**, 591 (1969).
[3] A. J. BLAKE, J. H. CARVER and G. N. HADDAD, J. Quant. Spectrosc. Rad. Transf. **61**, 451 (1966).
[4] B. R. MAY, Planet. Space Sci. **19**, 27 (1971).
[5] L. G. JACCHIA, J. SLOWEY and F. VERNIANI, J. Geophys. Res. **72**, 1423 (1967).
[6] T. R. HARTZ and N. BRICE, Planet. Space Sci. **15**, 301 (1967).
[7] K. D. COLE, Aust. J. Phys. **15**, 223 (1962).
[8] F. DU CASTEL and J.-M. FAYNOT, Nature, Lond. **204**, 984 (1964).
[9] R. D. HUNSUCKER and L. H. TVETEN, J. Atmos. Terr. Phys. **29**, 909 (1967).
[10] L. G. JACCHIA, Smithson. Astrophys. Obs. Spec. Rep. No. 332 (1971).

Space Research XIII — Akademie-Verlag, Berlin 1973

ATOMIC OXYGEN PROFILES IN THE LOWER THERMOSPHERE

R. E. Good and D. Golomb

Air Force Cambridge Research Laboratories, Bedford, Mass., USA

A recent release of nitric oxide on 18 May 71, 2141 LT, at Eglin, Florida, yielded an atomic oxygen profile that peaked at 99 km with $[O] \simeq 8 \times 10^{11}$ cm^{-3}. The number density decreased monotonically to 1.2×10^{10} cm^{-3} at 150 km. This was not the case in two previous flights at Churchill (1968) and Hawaii (1969) where a layered structure of atomic oxygen was observed. A compilation is made of all reported nitric oxide releases and mass-spectrometric measurements that covered the altitude span where peak atomic oxygen densities occur. Maximum densities are found at a mean altitude of 98.6 ± 1 km, with not enough statistical evidence as to the height variation with latitude, season or solar effects.

1. Introduction

In the previous COSPAR Symposium, we reported on a novel technique of releasing nitric oxide from a rocket to determine atomic oxygen number densities from the ensuing NO/O headglow [1]. The technique consists of releasing a jet of NO from a pressurized container through a sonic orifice into the forward direction of the flight. A hemispherical mixing zone is generated, bounded by an inner and bow shock. In the mixing zone, NO and atomic oxygen of the air interact to give the characteristic NO/O chemiluminescence. A photometer is situated near the sonic orifice, viewing the headglow from its center, and monitoring and telemetering the radiance during flight. From wind-tunnel experiments it was determined that for the release conditions the radiance is nearly independent of the NO partial pressure in the mixing zone, provided a strong jet is maintained in which part of the expanding gas forms dimers, molecular clusters and perhaps particulate condensate. The radiance is linearly dependent, however, on the atomic oxygen flux impinging on the bow shock:

$$B = F_e q [O] v. \tag{1}$$

Here B is radiance in rayleighs (photons cm^{-2} s^{-1}), $[O]$ is number density (atom cm^{-3}), v is vehicle velocity (cm s^{-1}). The factor q is the quantum yield, expressing the conversion efficiency of an impinging oxygen atom into a photon. F_e is the entrainment factor which takes into account that at lower altitudes where the Reynolds number is of the order of 10^2—10^3, a portion of the incident air/oxygen flux, after passing the bow shock, flows as an *unmixed* film along the edges of the mixing zone into the wake, and therefore is not effective in producing photons. This factor becomes unity at 120 km, where the mixing zone in considered to be completely viscid and well-mixed. The uncertainty of q is $\pm 100\%$ (a factor of two), of F_e $\pm 50\%$, resulting in a combined error of $\pm 150\%$ below 120 km, and $\pm 100\%$ above 120 km. The measurement errors of B and v are negligible in comparison with these uncertainties.

2. New Results

The recent NO release rocket was launched on 17 May 1971, 2141 LT, at Eglin, Fla (30° N). The experiment was similar to the previous flights [1] except that the gas was released continuously and not intermittently. The solenoid valves turned out to be erratic and costly; therefore, it was decided to forgo their use. To conserve gas over a wider altitude span, a narrower orifice diameter was used,

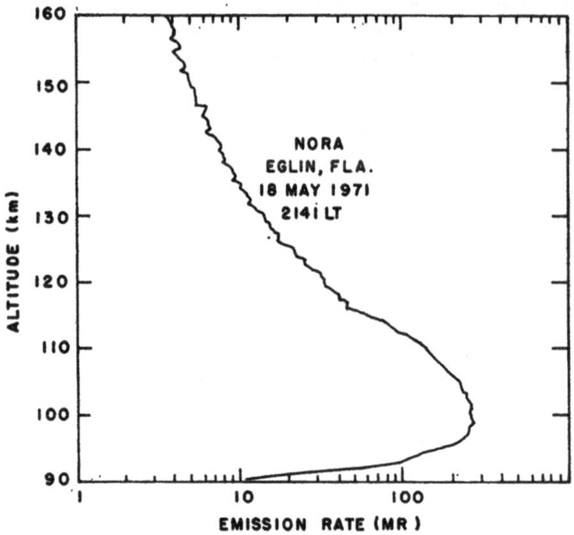

Fig. 1. Total headglow radiance B vs. altitude.

$d^* = 0.355$ cm, compared with 0.7 cm used before. This resulted in smaller headglow diameters, but since the radiance was measured in flight, and not from ground photography, the headglow diameter was of no major consequence.

A further important change was to use a mixture of 75% N_2—25% NO for the release gas. The NO is preloaded into the payload tank at 35 atm, and in the field 100 atm of N_2 is added. Wind-tunnel measurements indicated that the headglow brightness decreased by only 20% when such a mixture was used [2]. This is interpreted as being due to mixed clusters, such as NO—N_2, which are nearly as efficient in producing the chemiluminous reaction with O as the pure NO clusters. The quantum yield for this mixture is $q = 0.004 \pm 100\%$.

The total radiance vs. altitude of this flight is shown in Fig. 1. The total radiance was obtained by correcting the measured radiance by the ratio of the NO/O spectrum given by Fontijn et al. [3] to the portion of the spectrum detected by the photometer. The [O] values derived from Eq. (1) using the flight parameters from Table 1 are given in Fig. 2. The spectral emission rate during flight shown in

Table 1
Flight parameters

Alt. (km)	90	100	110	120	130	140	150
v (cm s^{-1})	1.50	1.42	1.35	1.27	1.19	1.10	1.01
F_e	0.34	0.58	0.94	1	1	1	1
q			0.004 ± 100%				

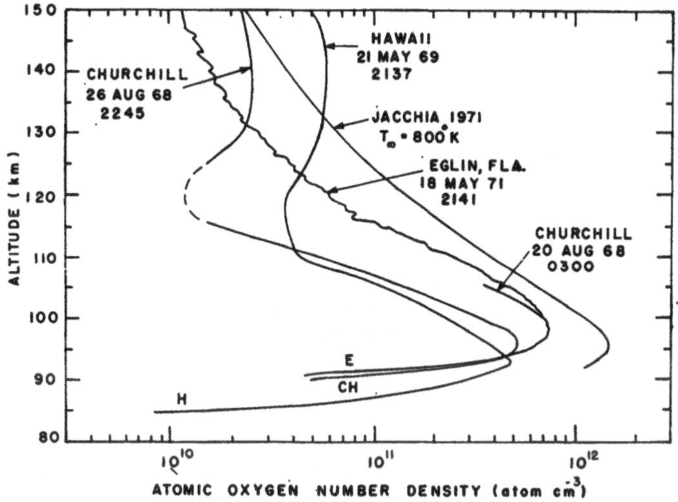

Fig. 2. Atomic oxygen number densities vs. altitude.

Fig. 3 was monitored with five narrow bandpass filters. The spectrum of the headglow follows closely the laboratory NO/O chemiluminescent spectrum at all altitudes.

The present data show a "normal" profile with peak $[O] \simeq 8 \times 10^{11}$ cm^{-3} at 99 km, decreasing to 1.2×10^{10} cm^{-3} at 150 km. Also shown in Fig. 2 are the data from two previous flights in Churchill (1968) and Hawaii (1969) where "anomalous" profiles have been obtained with major peaks at 96 and 93 km, respectively, a trough in the 115–120 km region, and a secondary broad peak at 140 km. The present profile is parallel to Jacchia's 71 model [4] for $T_{exo} = 800$ °K, which closely corresponds to $F_{10.7} = 117$, prevailing at the time of the experiment. However, the model number densities are larger by a factor of about two, stemming from the fact that the model ties into von Zahn's recommended ratio $[O]/[O_2] = 9.2$ at 150 km [5]. This difference is within the stated experimental error, but it may be real.

3. Definition of Peak [O] Layer

A compilation was made of all reported nitric oxide releases and mass-spectrometric measurements that covered the 90–110 km region in which maximum O densities are found, to determine the mean height of $[O]_{max}$. This is presented in Table 2. The arithmetic mean of all observations is 98.6 ± 1 km with a confidence level of 95% to find the layer within ± 9.5 km of that altitude. Not enough

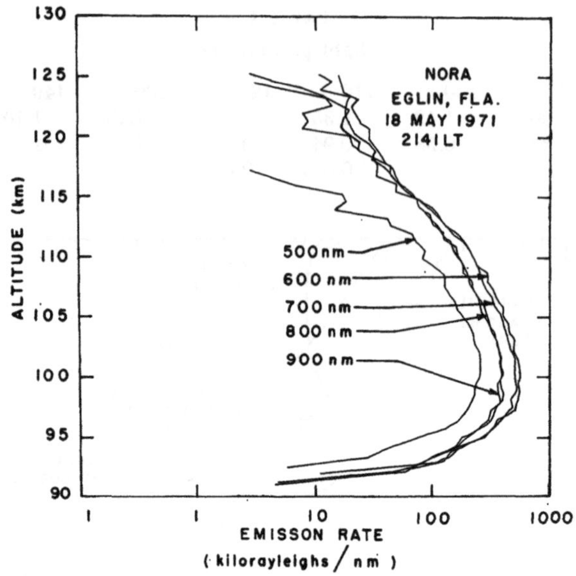

Fig. 3.

Table 2
Peak [O] layer altitudes

Date	Time (LT)	Coord.	$F_{10.7}$ (10^{-22} W $m^{-2}Hz^{-1}$)	Alt. (km)	Method	Ref.
16 Nov. 62	1907	36° N 75° W	80	101	NO	[7]
27 Nov. 62	1800	30° N 87° W	80	102	NO	[8]
3 Dec. 62	2245	30° N 87° W	80	105	NO	[8]
6 Dec. 62	0421	30° N 87° W	80	108	NO	[8]
24 Mar. 63	2118	59° N 94° W	75	95	NO	[7]
18 Oct. 63	0320	59° N 94° W	75	95	NO	[7]
28 Oct. 63	2053	59° N 94° W	75	96	NO	[7]
26 Aug. 68	2245	59° N 94° W	125	96*	NO	[1]
21 May 69	2137	22° N 160° W	125	93*	NO	[1]
20 Nov. 70	1723	30° N 87° W	180	97*	MS	[9]
18 May 71	2141	30° N 87° W	117	99	NO	Present
5 Oct. 71	1833	36° N 75° W	106	99	MS	[9]
5 Oct. 71	2045	36° N 75° W	106	97	MS	[10]

* Layered profile observed, the given altitude is for peak [O].

statistical evidence is present to determine the height variation with latitude, time or solar flux. It is noted that peak emission rates of the [OI] 5577 Å green line are also found around 100 km [6].

It is probable that the $[O]_{max}$ height is dependent on the turbulent structure of the atmosphere; it appears that $[O]_{max}$ occurs just below the cessation of intense turbulence. However, recently it became evident that the turbopause is not a sharply defined boundary; less intense turbulent eddies and cells can be traced on chemical trails to an altitude of 120 km. Shears as high as 30 m s^{-1} km^{-1} are found in 10% of wind shears measured in the 110—120 km region [11].

Acknowledgments

We are indebted to Mr. W. K. Vickery of AFCRL for the engineering supervision of the payload construction and launch logistics. Also acknowledged is the devoted effort of Mr. G. F. Harpell of AFCRL in designing and building of the photometer.

References

[1] D. Golomb and R. E. Good, Space Science XII, 675 (1972).
[2] D. Golomb and R. E. Good, J. Chem. Phys. **49**, 4176 (1968).
[3] A. Fontijn et al., J. Chem. Phys. **40**, 64 (1964).
[4] L. G. Jacchia, Smithson. Astrophys. Obs. Spec. Rep. No. 332 (1971).
[5] U. von Zahn, J. Geophys. Res. **75**, 5517 (1970).
[6] B. S. Dandekar and J. P. Turtle, Planet. Space Sci. **19**, 949 (1971).
[7] G. B. Spindler, Planet. Space Sci. **14**, 53 (1966).
[8] D. Golomb et al., J. Geophys. Res. **70**, 1155 (1965).
[9] C. R. Philbrick et al., Space Research XIII, 255 (1973).
[10] C. R. Philbrick, unpublished results.
[11] N. W. Rosenberg and S. P. Zimmerman, Radio Sci. **7**, 377 (1972).

Space Research XIII — Akademie-Verlag, Berlin 1973

ROCKET MEASUREMENTS OF MESOSPHERIC AND LOWER THERMOSPHERIC COMPOSITION

C. R. Philbrick, G. A. Faucher and E. Trzcinski

Air Force Cambridge Research Laboratories, Bedford, Mass., USA

Measurements of the major and some of the minor neutral atmospheric species have been obtained from six rocket flights in the 65—120 km altitude region. The instruments were launched on 7 March 1970 (solar eclipse), 20 November 1970 ($\chi = 98°$), 5 October 1971 ($\chi = 90°$ and $\chi = 115°$), and 12 April 1972 (2000 and 2016 LST). The instrument used to obtain these measurements is a recently developed RF quadrupole mass spectrometer which employs a liquid nitrogen cooled cryosorption pump. The species measured include N_2, O, O_2, Ar, CO_2, O_3, and other minor constituents. The instrument uses two ionizing energies on alternate one-second spectral scans. The lower ionizing energy of about 20 eV allows extension of the atomic oxygen measurement down to about 85 km. Profiles of the individual constituents obtained from the 20 November flight are presented and comparison is made with the results of one of the flights on 5 October 1971. The individual atomic oxygen profiles exhibit variations in the 85—120 km region which show strong effects from dynamic and transport processes in this region.

1. Description of the Experiment

The experiment, referred to as NACS (Neutral Atmospheric Composition Spectrometer) is a quadrupole analyzer utilizing rods of 0.381 cm diameter and 7.6 cm length (Fig. 1). The ion source is mounted directly below the entrance orifice plate which also serves as the anode for the ion source. The filament and accelerator electrode are cylindrically symmetrical. The objective of this design is to provide a high density of ionizing electrons in the region where the gas sample effuses through the orifice. This configuration enhances the signal for the incoming sample and minimizes the background signal from unpumped gases in the instrument and/or from particles reflected in the ion source region. Since several species of interest are chemically reactive, reflected particles may not give a true measure of the atmospheric species. Another advantage of this type of source is that its symmetry minimizes effects associated with the vehicle motion. The electron ionizing energy was switched between two values on alternate mass spectra. This capability allows a better measurement of some of the species (particularly atomic oxygen) because the fraction of molecules which are dissociatively ionized in the ion source is considerably less for 20 eV ionizing electrons than for 100 eV ionizing electrons (less than 0.5% for 20 eV electrons compared with about 10% for 100 eV electrons).

The cryosorption pump removes the incoming gas flux and maintains a low background pressure in the region of the analyzer. The cryosorption pump was made by bonding a 5 Å molecular sieve onto aluminum plates which have been

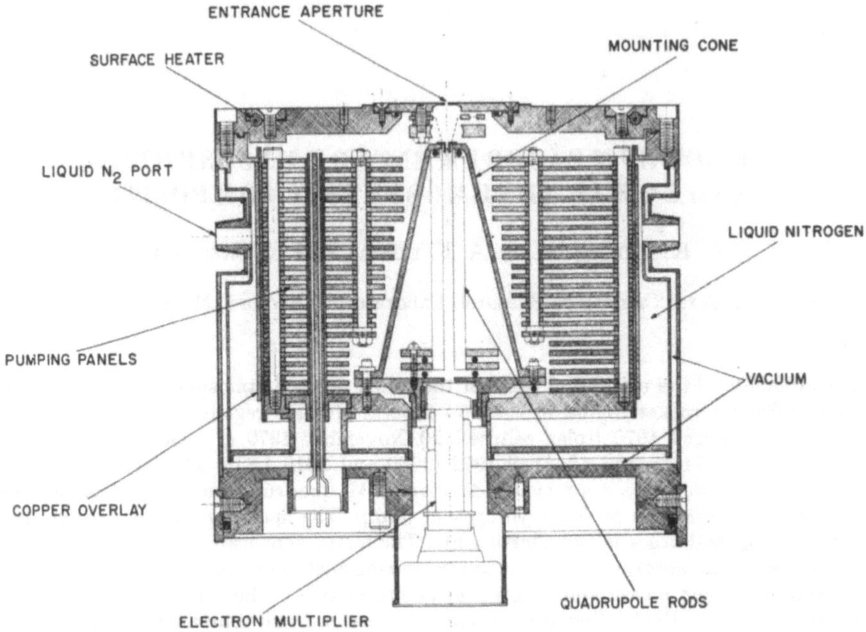

Fig. 1. Schematic drawing of the NACS instrument.

cross-mill grooved. These plates are sandwiched into a dewar which is cooled with liquid nitrogen. High pumping speeds are obtained for most atmospheric species (about 100 liter s^{-1} for nitrogen). The gas capacity of the pump is many times greater than the gas load presented in a single rocket flight. The holding time after removal of liquid nitrogen is considerably longer than the typical 6-minute rocket flight.

2. Flight Data and Results

The results presented were obtained by the following procedure. The current for each mass of interest was divided by the N_2 current in that spectrum. When the nitrogen mass peak was off scale, its value was obtained by taking the proper ratio of the 14-amu peak which comes from dissociative ionization of N_2. The current ratio was then divided by the ratio of ionization cross sections of the mass of interest and molecular nitrogen. The laboratory calibrations had shown that the only measurable mass discrimination in this instrument was due to the differences in ionization cross sections.

Fig. 2 shows the ratio of the currents measured at the 16 and 32 amu peaks as a function of altitude. The advantage of using two electron ionization energies is apparent in this figure. The background level due to dissociative ionization of molecular oxygen is considerably less for the lower ionizing energy and allows the atomic oxygen profile to be extended to significantly lower altitude. The higher ionization energy is used to determine the atomic oxygen density for the upper part of the profile and the curve for the lower ionization energy is normalized to it.

Altitude profiles for several of the constituents measured on 20 November 1970 are shown in Fig. 3. The mass density, measured by Theon and Horvath [1], was used to determine the species densities from the current ratios of each of the species to molecular nitrogen, the ionization cross section, and the mean molecular mass determined from the mass spectrometer measurement. The ionization cross sections used for N_2, O_2, Ar, and CO_2 were those of Rapp and Golden [2]

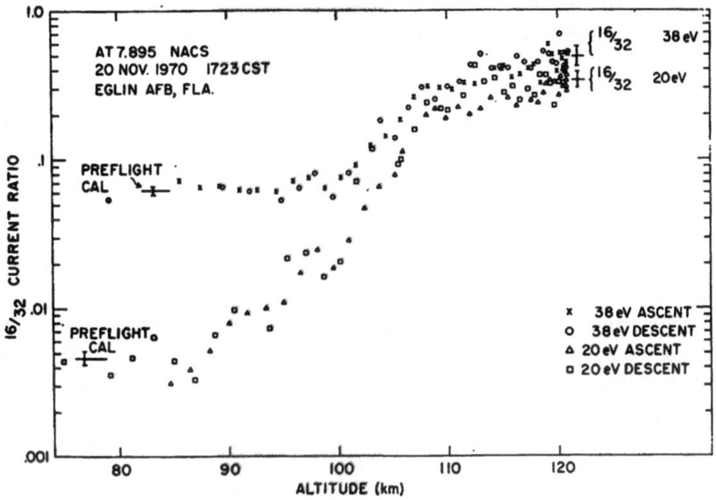

Fig. 2. Current ratio of the 16 to 32 amu peaks for two ionizing energies for 20 November 1970 flight.

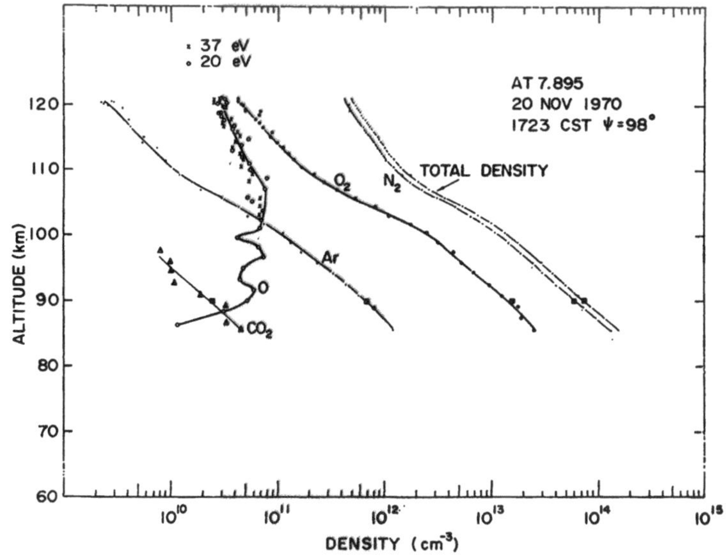

Fig. 3. Altitude profiles for major constituents measured on the 20 November 1970 flight.

and the O cross section used was from the work of Fite and Brackmann [3]. A background signal from residual gas which entered at low altitude was removed from the O_2 and Ar signals. The vehicle developed a very large coning angle which caused large modulations of the incoming gas flux. This modulation was used to remove the background signals of the residual gas. The modulation of the N_2 signal agreed well with that expected for the $F(s)$ sampling function. The departure of the O_2 and Ar signal modulation from that expected from the sampling theory was used to determine the magnitude of the background signal. At the 90 km level, the squares shown represent the ground level mixing ratios of each of the constituents. The N_2, O_2, Ar, and CO_2 measurements agree well with the ground level concentrations at lower altitudes. The O profile shows very complicated structure with a peak density slightly less than 10^{11} cm^{-3} in the vicinity of 107 km. The structure in the atomic oxygen profile was observed on both the ascent and descent portions of the flight. Turbulent layers were observed on chemical trails up to an altitude of 112 km during the same time period as this measurement. A comparison of this result with a theoretical calculation exhibits a similar profile with even an indication of the structure in the O profile presented here [4].

In Fig. 4 the measured ratio of the Ar/N_2 densities is shown as a function of altitude. Because of the scatter in the data when this ratio is determined, the curves shown follow the limits on the experimental points. The ratio indicates a mixed atmosphere to an altitude in the vicinity of 105 — 110 km. Chemical trail measurements showed turbulent and laminar layers up to 112 km.

In Fig. 5, the N_2 and O profiles from the flight on 20 November 1970 are compared with a measurement on 5 October 1971. A large difference in the atomic oxygen profile shape and density between the two flights was observed. The

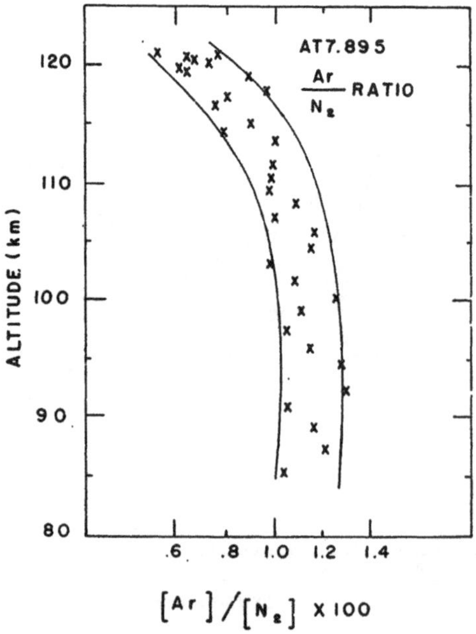

Fig. 4. Measured Ar/N_2 density ratio for the 20 November 1970 flight.

5 October 1971 flight shows an atomic oxygen profile more typical of that expected for a lower altitude turbopause. The measurements show that large variations in the atomic oxygen profile do occur and that atmospheric dynamics is quite important in establishing the atomic oxygen profile. The theoretical predictions of Keneshea and Zimmerman [5] agree well with these measurements.

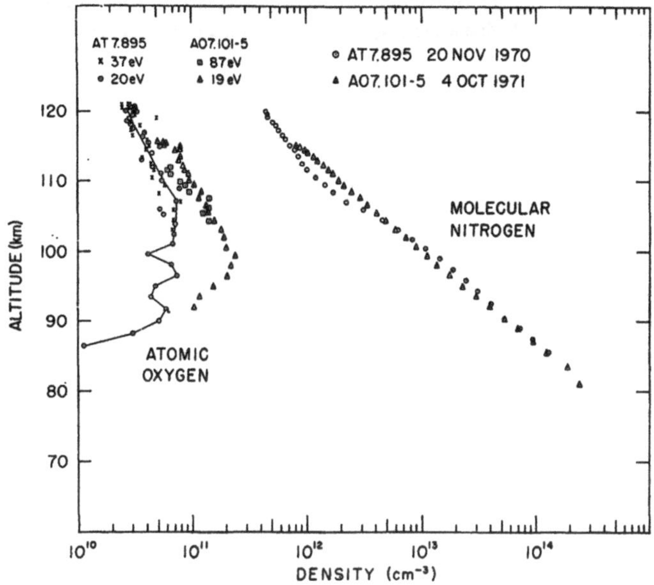

Fig. 5. Comparison of the O and N_2 profiles for flights on 20 November 1970 and 5 October 1971.

The atomic oxygen measurements should be regarded as a lower limit at the present time. Wind tunnel simulations and further measurements of atomic oxygen should clarify its actual density in this region.

Acknowledgments

The authors wish to acknowledge and express appreciation to Mr. R. E. McInerney for providing the gyro reduction for the vehicle altitude, Mr. G. S. Federico and Capt. Dalton Baker for their contributions to the mechanical design and development, Mr. R. A. Wlodyka and Mr. R. B. E. Moren for their efforts in the electrical design and development of the instrument. Mr. B. D. Donovan and Miss M. E. Gardner for their efforts in the reduction of the data. The efforts of the following individuals of Northeastern University; Mr. R. D. Anderson, Mr. R. L. Morin, Mr. R. P. Healey, Mr. R. A. Fonteyn, Mr. F. J. Tracy, and Mr. J. J. Mc Elhinney who have contributed significantly together with Mr. E. F. McKenna and Mr. J. F. Geary to the payload design and the success of these flights.

References

[1] J. S. Theon and J. J. Horvath, Trans Am. Geophys. Un. **53**, 463 (1972).
[2] D. Rapp and P. Englander-Golden, J. Chem. Phys. **43**, 1464 (1965).
[3] W. L. Fite and R. T. Brackmann, Phys. Rev. **113**, 815 (1959).
[4] C. R. Philbrick, R. S. Narcisi, R. E. Good, H. S. Hoffman, T. J. Keneshea, M. A. MacLeod, S. P. Zimmerman and B. W. Reinisch, Space Research XIII, 441 (1973).
[5] T. J. Keneshea and S. P. Zimmerman, J. Atmos. Sci. **27**, 831 (1970).

Space Reesarch XIII — Akademie-Verlag, Berlin 1973

MINOR CONSTITUENTS OF THE ARCTIC THERMOSPHERE DURING DISTURBANCES

G. M. MARTYNKEVICH and E. D. BYURO

USSR

Neutral composition variations measured by mass spectrometers during periods of disturbances including aurorae are described. The experiments were performed over Heiss Island in 1966—1971. Particular emphasis is laid on such minor constituents as N, NO, OH and H. The data obtained show sharp simultaneous increases in concentrations of these components during periods of disturbances. In one experiment nitric oxide forms a layer with a maximum at 96 km, the number density reaching a value of $(6-9) \times 10^{11}$ cm^{-3} in the layer maximum. The data obtained are briefly analysed and compared with earlier published data.

The Table lists some information on seven experiments performed aboard the MP-12 rockets during periods of various disturbances by means of radio frequency [1] and time-of-flight [2] mass spectrometers. Some appreciable and even radical changes in the neutral composition of the lower thermosphere have been observed during these disturbances.

A description of geophysical conditions under which the launchings were made is given in the table. The most typical qualitative changes in the composition and the phenomena associated with these changes will first be described. Among the latter there are: (i) fine structure (or microstructure) of the neutral composition; (ii) appreciable heating.

The changes in the composition, at least for the disturbances under study, always begin with an increase by one or two orders of magnitude in the number density n_H of the hydrogen atoms although both smaller and greater increases in n_H may take place. The growth of n_H is associated with an increase in number densities of N, O, of NO, OH, H_2O and occasionally of some other components.

Fig. 1 is a vivid presentation of both the changes and the pattern of microstructure initiation. It should be emphasized that the duration of the processes resulting in the composition changes under consideration, particularly during the aurorae, is often comparable with the duration of the rocket flight [3]. Consequently, thermospheric composition changes observed during periods of disturbances are not so much space as space—time ones, and sometimes essentially time changes. This can be confirmed by the fact that the changes in the composition start, according to the experiments listed in the table and described in [4], at any part (upleg, apogee, downleg) of the rocket trajectory.

Let us consider in greater detail the results of measurements of each of the above components.

Atomic hydrogen. In Flight 2—7 the ratio of number densities n_H/n_{N_2} reaches values of the order of between 1—20. The smallest value was recorded during Flight 3 near the trajectory peak. The lower boundary where an appreciable and even large increase in n_H was still observed is situated below 90—110 km.

Table 1
Data on MP-12 rocket experiments

Flight	Date	Local time	Locality	Apogee (km)	Characteristics of disturbance	Situation
1.	1 Oct. 1966	1259	Heiss Island	174.2	Ionograms typical for aurora, auroral Es layer.	a
2.	15 Jan. 1969	0006	Heiss Island	176.3	Non-zenith, aurora, ray arcs, 2-3 IBS	b
3.	25 Jul. 1969	0427	p. Volgograd	157.4	Local minimum of radio wave absorption	c
4.	22 Dec. 1970	2217	Heiss Island	184	Non-zenith aurora	
5.	28 Dec. 1970	1752	Heiss Island	181.6	Non-zenith aurora	
6.	15 Mar. 1971	2209	Heiss Island	187	Aurora in zenith	
7.	8 Oct. 1971	1359	p. Volgograd	167	Solar flare 2^+ 3 Oct. 1971	d

Ionospheric and geomagnetic situation; corpuscular fluxes:

a f_0Es = 6.5 MHz at 1252, Es disappearance at 1310, diffusion of reflection from F2, magnetic storm of 1 hour duration 17 hours before the flight

b Diffusion of reflection from F2 layer Δf_0F2 up to 60% in 1 hour interval after flight

c Radio wave absorption < lower limit of registration from 0407 to 0412

d Shielding Es layer; f_0Es = 8 MHz; magnetic storm with sudden commencement ΔD and ΔH up to 40—50 γ; fluxes of electrons and protons on boundary of atmosphere (cm^{-2} s^{-1}) 7 Oct. 1971, 1706 $N_e(E > 25 \text{ keV}) = 2.67 \times 10^2 \pm 5$, $N_e(E > 40 \text{ keV}) = 8.8 \times 10^1 \pm 8$, $N_p(24 < E < 27 \text{ MeV}) = 2.6 \times 10^{-1} \pm 0.06$; 8 Oct. 1971, 1705, $N_e(E > 25 \text{ keV}) = 4.85 \times 10^2 \pm 7$, $N_e(E > 40 \text{ keV}) = 1.29 \times 10^2 \pm 5$, $N_p(1.5 < E < 2.5 \text{ MeV}) = 1.7 \times 10^{-1} \pm 0.05$, $N_p(24 < E < 25.6 \text{ MeV}) = 2.6 \times 10^{-1} \pm 0.07$.

Nitric oxide. Unlike the atoms of H, the molecules of NO were not registered during all the flights. They were, undoubtedly, recorded during the 1st, 5th and 6th launchings. The results of the flight 1 briefly described before are of greatest interest. Fig. 2 shows a vertical distribution of the NO molecules number density (n_{NO}, cm^{-3}), measured during flight 1. The NO molecules form a layer with its maximum lying at the level of about 96 km. The number density n_{NO} in the maximum of the layer is as high as 8×10^{11} cm^{-3}, the accuracy being within a factor of 2. In flights 5 and 6, n_{NO}/n_{N_2} does not exceed 2×10^{-3}, the accuracy being within a factor of 2, over a wide range of altitudes, i.e. n_{NO} lies within 5×10^7 to 1×10^{10} cm^{-3}. Distinct layers of NO are not formed.

Atomic nitrogen. Atmospheric nitrogen atoms were observed during six out of seven launchings, and during the 8th one, described in [4]. The curve n_N/N_{N_2} given in Fig. 3 shows the behaviour of the number density ratio n_N/n_{N_2} inside the ion source as a function of altitude for flight 5.

Atomic oxygen. Relative concentrations of n_O/n_{O_2}, higher than those under normal conditions, calculated only from the ion current $I(O^+)/I(O_2^+)$ ratio were observed in six out of seven launchings. The n_O/n_{O_2} ratio showed the maximal excess ($\sim$ three times) over the values typical for normal conditions during flight 7, i.e. five days after a solar flare of the 2^+ class and during considerable geomagnetic

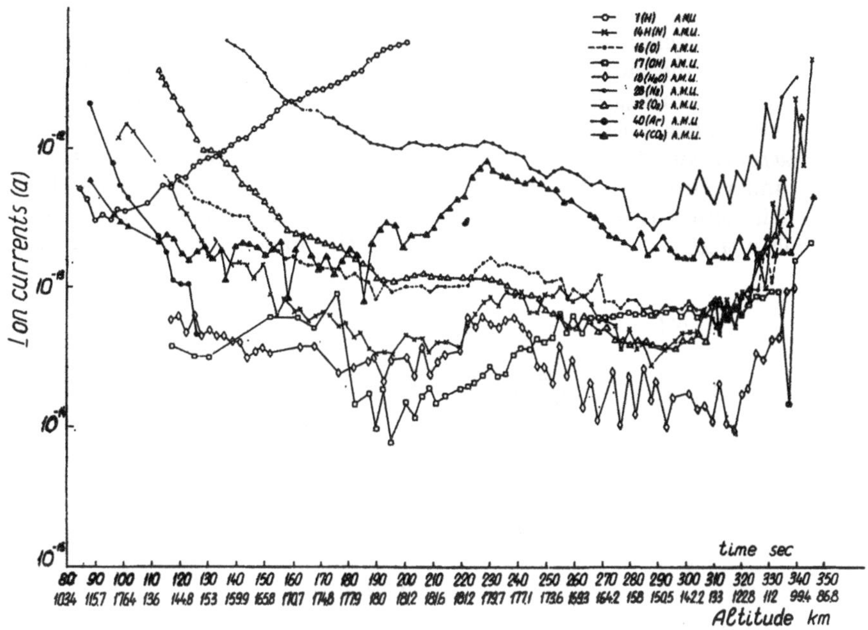

Fig. 1. Output signals (ion currents) of some constituents versus altitude and time of flight in flight 5. Ion current $I(H^+)$ exceeded the upper limit of registration 6×10^{-12} ampere after 200 sec. of time of flight.

disturbances (ΔD and $\Delta H \sim 40-50\gamma$) with a sudden commencement recorded on 6 and 8 October 1971. The excess in all other launchings was less pronounced.

Hydroxyl. The increase in hydroxyl concentration during the periods of disturbances is of much interest, especially the simultaneous n_H and n_{OH} increase observed during the 2nd and 5th launchings. This suggests a rapid oxidation of the precipitated energetic hydrogen atoms [5—7]. The increase was observed in five out of seven launchings (1st, 2nd, 4th, 5th, 6th), though it was not significant during flight 1. In the 2nd and 5th flights, however, an increase in n_{OH}/n_{H_2O}

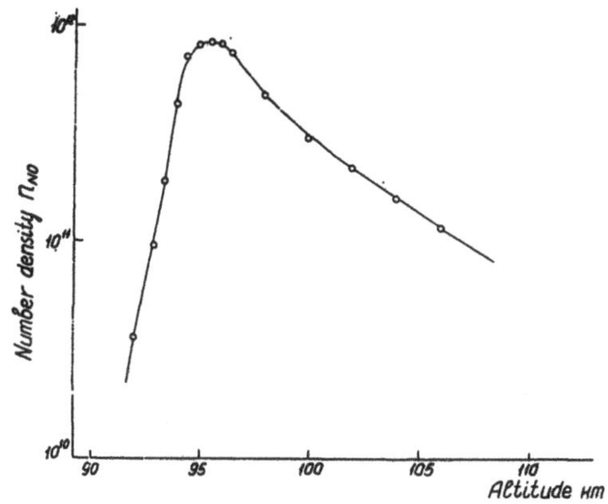

Fig. 2. Vertical distribution of number density of NO molecules in flight 1.

(by one order of magnitude and more), n_{OH}/n_O and n_{OH}/n_{N_2} ratios was so pronounced that a new mechanism of hydroxyl formation was proposed [5—8]. Fig. 3 shows these ratios measured in the ion source, the dependence on altitude in flight 5 where an increase in n_{OH} is most distinct. For comparison, the upper limit of the n_{OH}/n_{H_2O} ratio under normal conditions is about 3.6×10^{-1}.

Water vapour. An adequately detailed description of the behaviour of the H_2O molecule under normal and disturbed conditions is given in [9] so we shall not dwell on it in detail. However, we should like to consider the correlation between the atomic nitrogen, molecular nitrogen and the molecules of H_2O previously

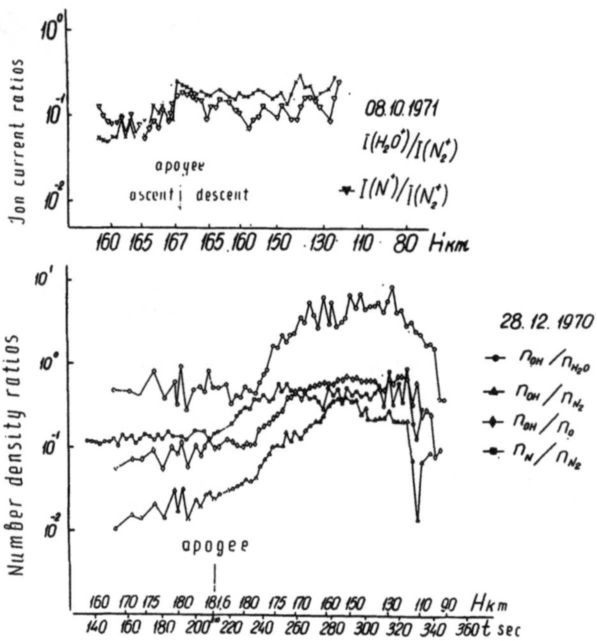

Fig. 3. Number density ratios n_N/n_{N_2}, n_{OH}/n_{H_2O}, n_{OH}/n_O, n_{OH}/n_{N_2} in flight 5 and ion current ratios $I(N^+)/I(N_2^+)$, $I(H_2O^+)/I(N_2^+)$ in flight 7 as functions of altitude and time of flight.

described [10]. We mean a close correlation between the output signals of the lines $14^+(N)$, $18^+(H_2O)$ and $28^+(N_2)$ during flight 7. The ratios of ion currents $I(N^+)/I(N_2^+)$ and $I(H_2O^+)/I(N_2^+)$ are given for illustration in Fig. 3, as well as the ratios of number densities n_N/n_{N_2}, n_{OH}/n_{H_2O}, n_{OH}/n_O and n_{OH}/n_{N_2}. It is obvious that the two ion curves can be correlated, and in the interval between 155 km and 167 km not only can they be correlated but they are of principally similar form which goes to prove that the majority of the H_2O molecules recorded during flight 7 are formed from the precipitated hydrogen atoms. This conclusion is in full accordance with those drawn before [5—7, 9—11].

Of interest is also the fact that though the aurora during the 2nd launching was due to electrons, it was, nevertheless, associated with a considerable increase in n_H.

In conclusion, significant changes in the composition of the lower thermosphere occur, in all probability, quite often and not only in high (Heiss Island) but also in middle (p. Volgograd) latitudes, as the disturbances causing these changes occur quite often.

Acknowledgments

The authors consider it their duty to thank Ya. I. Feldstein to whom they are indebted for the interpretation of the data on the aurora obtained with the all-sky camera during flight 2.

References

[1] B. I. Zharkhin, V. A. Pavlenko, A. E. Rafalson and M. E. Slutsky, Kosm. Issled. 4, 457 (1966).
[2] G. M. Martynkevich, G. N. Kudryavtsev, V. T. Nenarokomova, V. A. Pavlenko, A. E. Rafalson, E. G. Shvidkovsky and A. P. Zhukov, Space Research XI, 869 (1970).
[3] S.-I. Akasofu, Polar and Magnetospheric Substorms, "Mir", Moscow 1971.
[4] G. M. Martynkevich and E. D. Byuro, Space Research XII, 691 (1972).
[5] G. M. Martynkevich, Meteorol. i Gidrol. No. 12, 44 (1971).
[6] G. M. Martynkevich, preprint presented to the XVth Plenary Meeting of IUGG, Moscow, August 1971.
[7] G. M. Martynkevich, Space Research XII, 1311 (1972).
[8] G. M. Martynkevich, E. D. Byuro and G. V. Malyarova, Space Research VIII, 947 (1968).
[9] G. M. Martynkevich, paper presented at the XVth Plenary Meeting of COSPAR, Madrid, May 1972.
[10] G. M. Martynkevich and E. G. Shvidkovsky, Space Research XI, 865 (1971).
[11] G. M. Martynkevich, Meteorol. i Gidrol. No. 12, 3 (1969).

Space Research XIII — Akademie-Verlag, Berlin 1973

MEASUREMENTS OF THERMOSPHERIC TEMPERATURES BY INCOHERENT SCATTER RADAR

J. E. SALAH[a] and J. V. EVANS[b]

[a]Department of Meteorology and Lincoln Laboratory, Massachusetts Institute of Technology, Cambridge, Mass. USA

[b]Lincoln Laboratory, Massachusetts Institute of Technology, Lexington, Mass., USA

Exospheric temperatures have been obtained from Millstone Hill incoherent scatter radar measurements of ion and electron temperatures and electron density by calculations based on the ion thermal energy balance equation. The observed variations of the exospheric temperature for 48 days, spanning a two-year period, have been compared with predictions from atmospheric models based on other radar observations, and on satellite drag analysis. In general, the measurements confirm the structure of the daily variation deduced from earlier incoherent scatter results, but exhibit a larger seasonal variation in the temperature extrema and no variation in the day-to-night ratio. In order to compare the measurements with recent tidal theory predictions, the observed daily variations were represented by Fourier time series. In conflict with theoretical predictions, the observed semidiurnal components in the exospheric temperature are less than 50°K amplitude and appear to be generated *in situ*. In contrast, incoherent scatter measurements made between 110 and 130 km, where thermal equilibrium is thought to prevail, indicate the presence of marked semidiurnal tides propagating from below. We observe that these tides dissipate below 150 km and appear to influence the spatial and temporal variations of the temperature in the thermosphere.

1. Introduction

Incoherent scatter radar measurements of the plasma properties of the ionosphere have provided a new direct source for the study of the thermal structure of the earth's thermosphere. Based upon the assumption of ion thermal energy balance, Carru et al. [1] and Nisbet [2] have demonstrated the application of the measurements to the calculation of neutral temperatures above 250 km. The method has been successfully employed at Saint-Santin, France (45° N) [3], Arecibo, Puerto Rico (18° N) [4] and Jicamarca, Peru (12° S) [5] to study the diurnal and longer term variations of the exospheric (or thermopause) temperature, and has resulted in a new global temperature model [6]. Based upon the existence of thermal equilibrium between the ionized and neutral species below 130 km, Carru et al. [1] and Wand and Perkins [7] have derived the neutral temperature and density variations in the lower thermosphere. Studies of the interrelation between the thermospheric temperature gradient and the exospheric temperature and neutral density have also been conducted [8, 9]. The results from the various incoherent scatter experiments have been compared with rocket and satellite measurements, and have complemented the latter in improving our understanding and modeling of the structure of the thermosphere. Morever, various discrepancies and problem areas emerged when different measurement techniques and theoretical predictions were compared, particularly in the determination of the shape of

Table 1

Incoherent scatter observations

Date	Period (GMT)	$F_{10.7}$	$\bar{F}_{10.7}$	ΣKp
25–26 March 1969	2130–2130	190	159	22^+
9–10 April	1430–1430	146	159	19^+
23–24 April	0320–0320	142	158	16^-
6–7 May	1300–1300	151	156	16^-
5–6 June	1330–1400	155	152	9
23–24 June	1320–1320	129	150	15^+
1–2 July	1300–1315	132	149	25
9–10 July	1140–0945	160	148	13^-
29–30 July	2300–2400	133	144	11
14–15 August	1315–1345	120	143	10^-
26–27 August	2030–2030	152	142	27^+
9–10 September	1415–1410	122	143	20
23–24 September	1300–1300	140	145	15^+
30 September–1 October	0300–1400	140	145	50^+
14–15 October	2200–2200	125	147	9^+
3–4 November	1900–2200	141	149	20
20–21 November	1330–1500	190	152	4^-
8–9 December	1745–1745	124	154	17^-
6–7 January 1970	1400–1400	131	158	8
20–21 January	1630–1700	181	160	15^-
17–18 February	1430–1430	198	163	21
23–24 February	2000–2000	187	164	13
17–18 March	1530–1530	140	164	11^+
23–24 March	2100–2045	152	164	5^-
14–15 April	1400–1400	184	164	6
28–29 April	1300–1300	136	162	9^+
12–13 May	2000–2000	173	160	21
18–19 May	2000–2000	193	160	10^-
10–11 June	0700–0700	134	156	13^-
23–24 June	2200–2000	145	154	8^+
7–8 July	1800–1800	180	154	11
18–19 July	1300–1600	122	151	10^-
24–25 August	1900–1900	143	147	10^+
31 August–1 September	1900–1930	146	146	21^-
16–17 September	1420–1420	117	146	14^+
28–29 September	2000–2000	149	148	9^+
5–6 October	2000–2000	137	148	12
13–14 October	1900–1950	136	149	19^-
31 October–2 November	1220–0030	178	152	7^-
7–8 November	1240–1240	153	154	34
8–9 November	1240–1300	154	154	12
21–22 December	2037–2100	161	154	9
28–29 December	2040–2100	124	152	22^-
11–12 January 1971	2120–2130	157	148	10^-
20–21 January	1730–1730	171	146	25
19–20 February	0030–0030	134	137	19^+
8–9 March	2040–2050	107	131	19
30–31 March	2100–2100	104	124	29^-

the diurnal variations of density and temperature, the tidal influence in the lower thermosphere, the specification of the spectral characteristics of the time response of the thermosphere to the energy input, and the dynamic influence of neutral air motions on the thermal structure.

The purpose of this paper is to present thermospheric temperatures deduced from incoherent scatter radar measurements at Millstone Hill, Mass. (42.6° N 71.5° W).

2. Exospheric Temperature Variations

2.1. Incoherent Scatter Single-pulse Experiment

The vertically directed incoherent scatter radar system at Millstone Hill operates at a frequency of 440 MHz and, in the mode used to collect the data described in this section, is capable of transmitting single pulses of lengths 0.1, 0.5 or 1.0 ms and yields ion and electron temperatures in the height region 225—900 km and electron densities in the range 150—900 km. This mode of operation has been described in detail [10]. The theory of deducing the ion temperature T_i the electron temperature T_e and electron density N as well as other ionospheric parameters from incoherent scatter measurements has been reviewed by Evans [11].

At Millstone Hill, the backscattered signals are applied to a bank of filters whose frequency response is matched to that of the power distribution of the pulse. The filter outputs are sampled and integrated in a digital computer over several minutes. In the experiments discussed here, combinations of the various pulse lengths were transmitted for 4—8 minutes, resulting in complete height profiles of plasma temperatures and density with a time resolution of about 30 minutes. The height resolution of the temperature profiles is 75 km and that of the electron density is 15 km. Observation periods of 24 hours duration have been routinely carried out at Millstone in this manner about twice a month since March 1969. Our investigations used data collected from March 1969 to March 1971, listed in Table 1. On two occasions, the periods starting on 30 September 1969 and 7 November 1970, geomagnetic storms were in progress and the results from these periods merit a detailed discussion which has not been included in this paper. The remaining days are generally characterized with a 24-hour sum of the geomagnetic index Kp less than 25 and moderate solar radiation flux indicated by an average 10.7 cm intensity $\bar{F} \simeq 150 \times 10^{-22}$ W m^{-2} Hz^{-1} (Table 1).

2.2. Theory for inferring the Neutral Temperature

It has been established [12] that local energy deposition from photoionization, thermal conduction and transport processes largely determine the observed neutral temperature structure in the thermosphere. When ionospheric excitation mechanisms become ineffective at low energies, the photoelectron kinetic energy is dissipated in the ambient electron gas thereby raising the temperature of the electrons T_e. Elastic collisions between electrons and ions transfer some of this energy to the ions causing T_i also to rise above the temperature T_n of the neutral gas. Below about 250 km the ions are rapidly cooled by neutrals. Thus the ions provide the thermal coupling between the electrons and neutrals. The study of the neutral atmosphere from incoherent scatter measurements is based upon consideration of the ion thermal energy balance equation, which is written as

$$4.82 \times 10^7\, N \frac{T_e - T_i}{T_e^{3/2}} = \{6.6\, n(N_2) + 5.8\, n(O_2) + 0.21\, n(O)(T_i + T_n)^{1/2}\}\,(T_i - T_n). \tag{1}$$

At 250 km, $n(N_2)/n(O) \simeq 0.2$ and $n(O_2)/n(O) \simeq 0.015$; these ratios decrease with altitude. It is worth noting that Eq. (1) correctly predicts the observed behavior of the temperature, i.e. at low altitudes $N/n(O)$ becomes small and T_i approaches T_n while at high altitudes $N/n(O)$ increases and T_i tends towards T_e.

Using measurements of T_e, T_i and N at heights of 300, 375 and 450 km and given the neutral species number densities from the Jacchia 1971 model [13] one may easily calculate T_n. Our measurements and calculations show that in most cases $T_i - T_n$ rarely exceeds 100°K in the daytime and decreases to values within the statistical accuracy of the measurements at night. Above 300 km the neutral temperatures are approximately constant within statistical errors and can be related to the exospheric temperature T_∞ by the expression [14]

$$T_\infty = [T_n - T_0 \exp\{-s(z - z_0)\}]/[1 - \exp\{-s(z - z_0)\}] \tag{2}$$

where T_0 is assumed constant (355°K) at $z_0 = 120$ km and s, the thermospheric shape parameter, is also assumed constant ($= 0.02\ \text{km}^{-1}$). Since the measurements above 300 km are already in the range where the neutral temperature gradient is

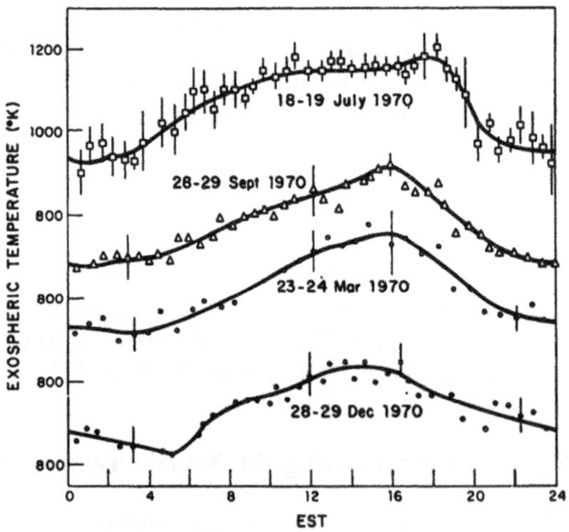

Fig. 1. Sample daily variations of the exospheric temperature measured at Millstone Hill.

small, an error in T_0 by $\pm$ 100°K introduces less than $\pm$ 1% error in T_∞, and a variation of s in the range 0.015—0.03 km^{-1} introduces less than $\pm$ 2% error in T_∞. In applying Eq. (2) and for purposes of finding the neutral number densities from the Jacchia 1971 model, the center height of the scattering volume rather than the nominal height is used. The distinction arises because the pulses illuminate a large altitude interval (75 km for 0.5 ms pulses and 150 km for 1 ms pulses). Thus, account must be taken of the variation of $N(z)/z^2$ in the scattered power in this height interval, together with a triangular weighting imposed jointly by the finite length of the pulse and width of the filters. The 'nominal' height of 300 km can be changed by $\pm$ 10 km depending upon whether the peak of the $N(z)$ profile is below or above 300 km. At 450 km, the actual height is decreased to about 410 km. In combining $T_n(z)$ to give T_∞ the temperatures are weighted by the signal-to-noise ratio at each height.

Fig. 1 shows sample results of the diurnal variation of T_∞ as calculated for a summer day, a winter day and at the equinox. The error bars indicate the standard deviation in T_∞ from combining the measurements at various heights. The uncertainty in T_∞ is estimated to be about $\pm$ 50°K, or about $\pm$ 5%. At night, the uncertainty may be slightly higher because of empirical corrections imposed and the possibility of operative non-solar heat sources that have been neglected in the theory. In view of the simplified theory, the use of experimental momentum transfer cross sections, and the inherent accuracy of the measured parameters, the error denoted above is a reasonable representative estimate of the total uncertainty in the technique.

2.3. Results

2.3.1. Daily Variations — General Features

The exospheric temperature variations with local time shown in Fig. 1 exemplify the observations recorded for most days. The major feature discussed widely in the literature and observed at all other incoherent scatter stations, is that the time of maximum temperature t_{max} lags the time of maximum heat input. We find

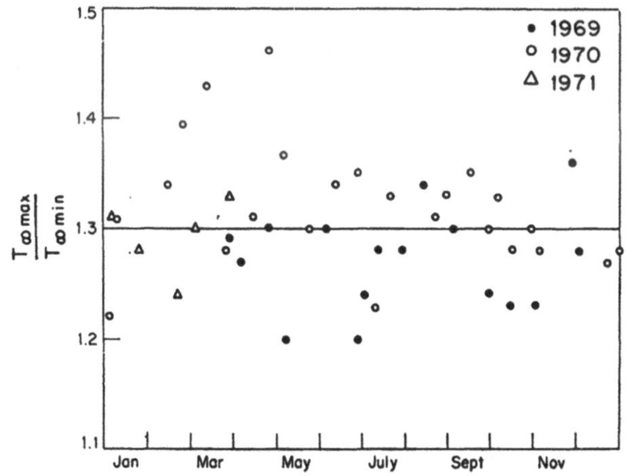

Fig. 2. Amplitude of the daily exospheric temperature oscillation at Millstone Hill.

t_{max} occurs earlier in the winter than in the summer. For the purpose of statistical averages we define the "summer" season as the months of May to August, "winter" as November to February and "equinox" as March, April, September and October. According to this grouping, t_{max} occurs at about 1530 LT in winter, 1600 LT at the equinox, and 1700 LT in summer, with a standard deviation of $\pm$ 1 hour. The annual average is close to the equinoctial value. The time of minimum t_{min} is generally less well-defined, and occurs at about 0030 LT in summer, 0200 LT at the equinox and 0400 LT in winter. It is noted that t_{max} occurs about 3 hours before thermospheric sunset (300 km) and t_{min} occurs about 2 hours before thermospheric sunrise. Particularly evident at the solstices are the temperature gradients at sunrise and sunset which exceed those during the day and night. The shape of the temperature variation with season thus reflects the length of the day. The actual rates observed are of course influenced by short-term effects such as geomagnetic activity variations.

The second general feature of the daily temperature oscillation is its amplitude $R = T_{max}/T_{min}$, shown in Fig. 2 for the days of observations. The summer, equinox and winter averages are respectively, 1.29, 1.31 and 1.29. The annual average is about 1.3 ± 0.05. It must be noted that the value of R calculated on a particular day is contaminated by the geomagnetic activity at t_{max} and t_{min}. Since the temperatures are known with an accuracy of $\pm 5\%$, the accuracy with which R can be determined is about $\pm 10\%$ and is perhaps reduced to ± 5 to $\pm 6\%$ by the smoothing introduced in finding the temperature extrema. Within this uncertainty, no seasonal or other long-term variations are evident in R.

On some days night-time fluctuations that are larger than the statistical uncertainty in the temperature have been observed after midnight. These are of amplitude about 50 °K and a period of 2—3 hours. In some cases the oscillation begins just prior to sunrise and is lost in the sunrise temperature increase. Waves in the neutral gas suggest a plausible explanation. This hypothesis can be checked by examining whether the horizontal neutral wind derived from vertical ion drifts measured simultaneously on the days in question [15] possesses similar oscillations. The magnitude of the neutral wind thus derived may also confirm whether the neglect of frictional heating which leads to an overestimation of the neutral temperature is justified. The electron heat input from conjugate photoelectrons transferred to the ions by Coulomb collisions may also contribute to the predawn increase but must await detailed investigation.

2.3.2. *Daily Variations — Harmonic Analysis*

In an effort to define the daily exospheric temperature variation in a mathematical model and to compare this with recent theoretical calculations, the observed temperature oscillations have been represented in terms of a Fourier time series. While the temporal resolution afforded by the incoherent scatter experiments is most satisfactory, the frequency resolution is limited to 1 cycle/day since our observations extend over 24 hours only. Spectral content cannot thus be rigorously defined or proved. In addition the underlying natural non-linear time-variant response of the thermosphere would render the interpretation difficult and objectionable. One must, therefore, assume the presence of certain frequency harmonics as is customary in tidal analysis and solve for their amplitudes and phases. The exospheric temperatures can then be expressed as

$$\begin{aligned} T_\infty(t) = \overline{T}_\infty + T_1 \cos(t - \tau_1) + T_2 \cos 2(t - \tau_2) \\ + T_3 \cos 3(t - \tau_3) + - - - \end{aligned} \tag{3}$$

where $\overline{T}_\infty$ is the mean daily temperature; T_1, T_2, T_3 are the amplitudes of the 1 cycle/day component (diurnal), 2 cycle/day (semidiurnal), 3 cycle/day (terdiurnal) etc; $\tau_1, \tau_2, \tau_3, \ldots$ are the times of maximum for the diurnal, semidiurnal and terdiurnal components, etc. Using standard Fourier analysis techniques we

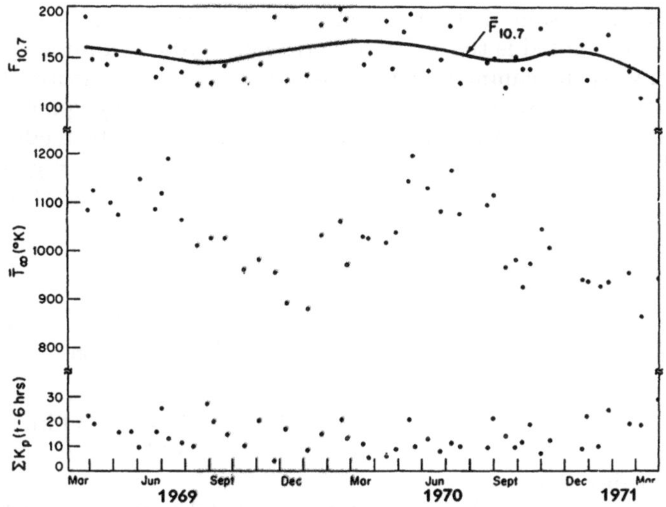

Fig. 3. Daily average of the exospheric temperatures measured at Millstone Hill and the solar flux and 24-hour geomagnetic indices for the two-year period of observations.

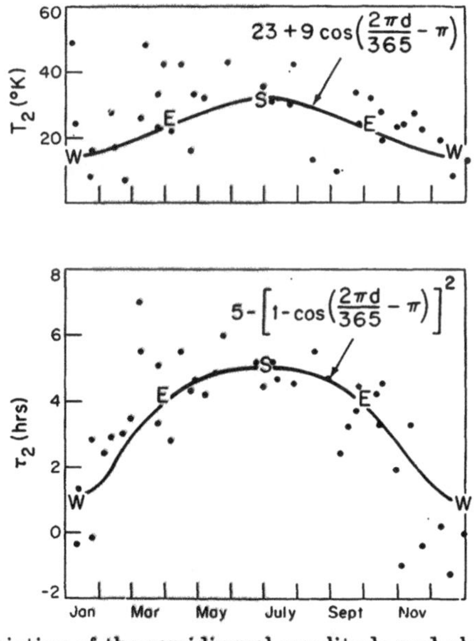

Fig. 4. Seasonal variation of the semidiurnal amplitude and phase of the exospheric temperature.

have solved for the mean and the sinusoidal components up to frequencies of 12 cycles/day.

The daily mean component $\overline{T}_\infty$ is shown for the days of observations in Fig. 3 together with $F_{10.7}$ the value of solar flux 10.7 cm (measured one day earlier at Ottawa), $\overline{F}_{10\cdot 7}$ the solar flux averaged over three solar rotations, and $\Sigma Kp(t-6)$.

the twenty-four hour sum of the geomagnetic indices (measured six hours earlier). The prime feature noted is the seasonal variation of 100°K amplitude with maximum temperature in summer. The two-year period of measurements studied is of moderate slowly varying solar activity and reasonably quiet geomagnetic conditions. Based on this analysis, the daily oscillation may be modeled to reflect the seasonal influence as illustrated for the semidiurnal mode in Fig. 4. Our model expresses the daily variation as the sum of the mean and the first three harmonics in Eq. (3) with

$$T_1 = 110\,°\mathrm{K} \qquad \tau_1 = 14 \text{ hrs.}$$

$$T_2 = 23 + 9 \cos\left(\frac{2\pi d}{365} - \pi\right) °\mathrm{K} \qquad \tau_2 = 5 - \left[1 - \cos\left(\frac{2\pi d}{365} - \pi\right)\right]^2 \text{ hrs.} \tag{4}$$

$$T_3 = 12\,°\mathrm{K} \qquad \tau_3 = 1 + \cos\left(\frac{2\pi d}{365} - \pi\right) \text{ hrs.}$$

where d is the day number.

In order to specify the daily variation completely, one must add the mean term $\bar{T}_\infty$. For this purpose an analytical function has been derived by least squares fitting to the data in Fig. 3 as a function of $F_{10.7}$ $(t - 1 \text{ day})$, $\bar{F}_{10.7}$, $\Sigma Kp(t - 6)$ modulated by an annual and a semi-annual variation. The latter variation was determined to be of the order of 1% with a statistical error of the same order as the calculated amplitude; it will therefore be neglected. We express our result in the format of the satellite drag models [13]

$$\bar{T}_\infty = (475 + 3.38\bar{F})\left[1 + 0.107 \sin\left(\frac{2\pi d}{365} - 1.58\right)\right]$$
$$+ 1.41\,(F - \bar{F}) + 2.95\,\Sigma Kp. \tag{5}$$

The coefficients have been determined with a statistical accuracy of ± 20 to ± 25% except for the annual variation coefficient whose amplitude is known

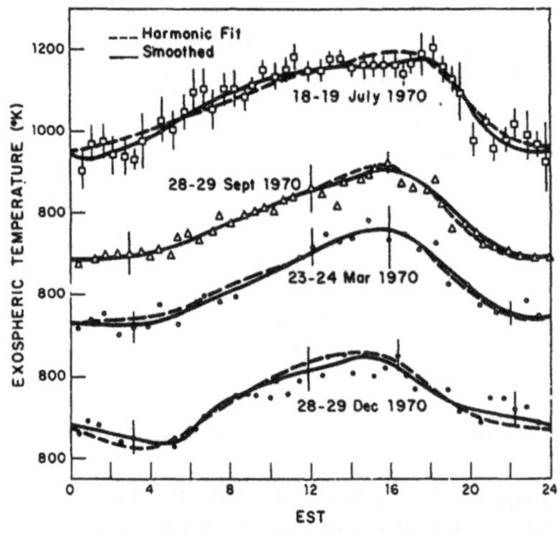

Fig. 5. Comparison of observations with the harmonic model for the daily exospheric temperature variation at Millstone.

within an uncertainty of ± 7% and its phase within ± 5%. This model fits the data with an rms deviation of about 30 °K.

Prediction of the daily variation using Eq. (3), (4) and (5) for the four days shown in Fig. 1 is compared with the observations in Fig. 5. The fits are seen to be reasonable except for a slight advance in t_{max} for the summer day and a slightly larger daytime temperature for the December day. To within 5 °K, the peak occurs at 1500 LT at the winter solstice, 1600 LT at the equinox and 1700 LT at summer solstice in good agreement with the averages obtained from the data.

2.3.3. Long and Short Term Variations

Eq. (5) describes the long and short term variations of the average daily temperature. A seasonal variation of ± 10% with a maximum in summer is seen to modulate the temperature. The semi-annual variation in the temperature was found to be negligible as mentioned previously. The largest component of the drive is the solar EUV flux followed by the geomagnetic effect and the 27-day solar rotation effect. Eq. (5) has been optimized for a time delay of 1 day in selecting $F_{10.7}$ and 6 hours in selecting Kp.

2.3.4. Discussion

The observations of the daily exospheric temperature variation at Millstone Hill support those made at the other incoherent scatter radars, particularly in the shape of the variation and time of temperature maxima. The most relevant comparisons would be with the results obtained at Saint-Santin (45° N) from which an exospheric model has been derived [6]. The prime discrepancy between the measurements is the lack of evidence of a seasonal variation in $R(= T_{max}/T_{min})$ which has been observed in France, but not at low latitudes or from satellite drag analysis. We have already commented on the accuracy of determining R in our measurements, and believe that a variation of 5% amplitude, if present, should be apparent. It is noted that the Millstone Hill measurements were carried out at a time when the solar cycle was varying relatively slowly compared with the time when the French results were gathered. Variations in R with solar cycle have been suggested from satellite data [16]. The annual average of $R (\simeq 1.3)$ is in close agreement at the two stations.

Comparison of the results at the two stations reveals a larger seasonal variation at Millstone in the daily average temperature, or the maximum temperature. A comparison of least-squares fix to T_{max} at Millstone Hill similar to Waldteufel's model, reveals a reasonable agreement in all parameters within the statistical uncertainty, except for the annual variation coefficient which is twice as large at Millstone Hill. The scatter between the observed $T_{\infty(max)}$ and that predicted by Waldteufel's model is illustrated in Fig. 6b and 7b; the latter confirms the seasonal discrepancy. A least squares fit in Fig. 6b can be described as

$$T_{\infty(max)}\ (\text{Millstone Hill}) = \underset{\pm\,0.11}{1.03}\, T_{\infty(max)}\ (\text{Waldteufel 1971}) - \underset{\pm\,130}{40}\ {}^\circ\text{K} \tag{6}$$

with an rms deviation of 55 °K and a correlation coefficient of 0.81.

Comparison with the Jacchia 1971 model yields in general better agreement with incoherent scatter measurements than earlier models based on satellite drag analysis. The model, however, does not accurately predict the daily temperature variation since the time of the temperature maximum is fixed at 1400 LT. The temperature oscillations in the Jacchia model have been synthesized as a diurnal and a semidiurnal component in phase at 1400 LT. The global temperature amplitude $R - 1$ for $Kp = 0$ is set in the model at 0.3 which appears to agree with our midlatitude measurements. However, the temperature ampli-

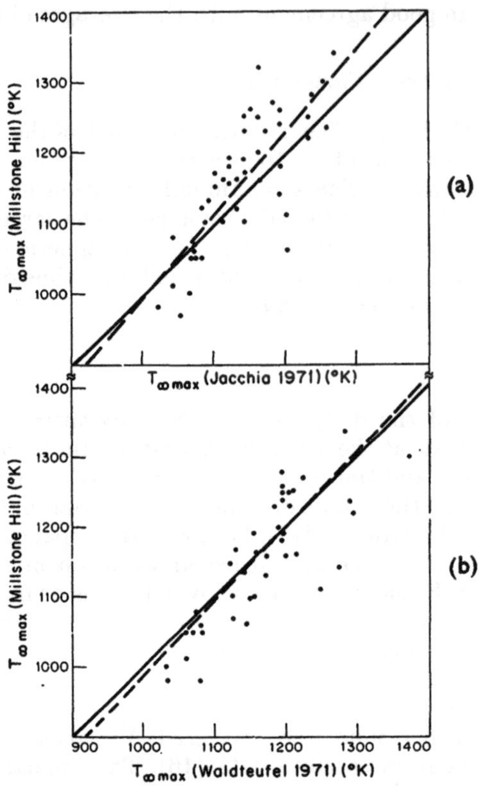

Fig. 6. Comparison of the daily maximum exospheric temperature at Millstone Hill with (a), above, the Jacchia 1971 model (b), below, the Waldteufel 1971 model. Dashed line is a least squares fit to the data.

tude of our extreme temperatures is about 0.6 indicating a discrepancy with the model. The model predicts an average amplitude of 0.22 at 42° N which is smaller than the measurements indicate. A scatter diagram between observations and model is shown in Fig. 6a and the deviation is interpreted from Fig. 7a as a difference in the seasonal variation. From this comparison one may write

$$T_{\infty(\max)}\text{ (Millstone Hill)} = \underset{\pm 0.14}{1.16}T_{\infty(\max)}\text{ (Jacchia 1971)} - \underset{\pm 150}{160°\text{K}} \tag{7}$$

with a correlation coefficient of 0.79 and an rms deviation of 57°K. Finally, the negligible semi-annual temperature variation observed in the Millstone Hill observations is consistent with Jacchia's reappraisal of this component from satellite drag data [17].

A three dimensional model of thermospheric dynamics has been recently developed [18] in terms of the eigen functions of the atmospheric system allowing tidal propagation through the lower thermospheric boundary from the mesosphere. A rigorous solution of the dynamic equations of the neutral atmosphere is compli-

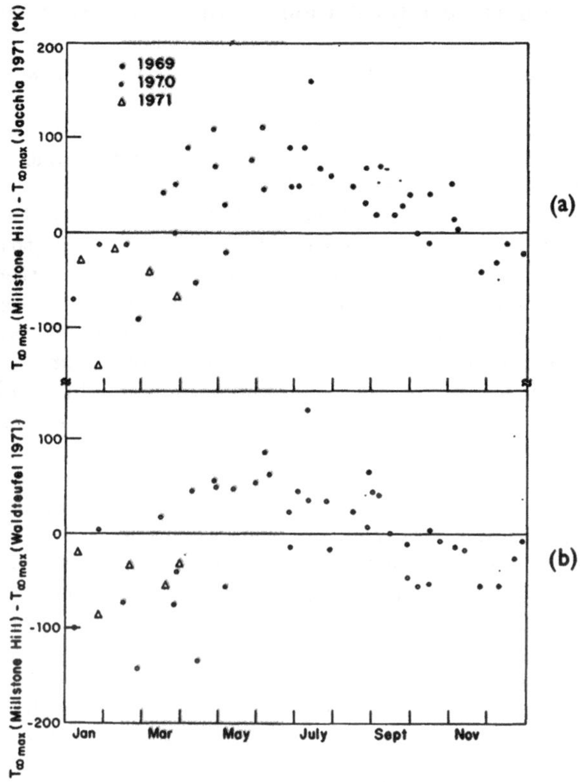

Fig. 7. Seasonal synthesis of the deviation between the Millstone Hill observations (a), above, the Jacchia 1971 model and (b), below, the Waldteufel 1971 model.

cated by the presence of viscosity, ion drag and Coriolis forces in the momentum equation, heat conduction and viscous dissipation in the energy equation, and complex boundary conditions that must allow for tidal propagation from the lower atmosphere. While the various studies have contributed substantially to the understanding of thermospheric structure, the theoretical problem is formidable because the equations, even when simplified, remain intractable. One application of the theory of atmospheric tides to a viscous heat-conducting atmosphere with ion drag [19, 20] resulted in solutions in which the main semidiurnal tide generated in the mesosphere by ozone absorption of insolation penetrated deeply into the equatorial thermosphere and gave rise to exospheric temperature oscillations that were equal to the diurnal oscillation driven by the solar EUV heating. The harmonic analysis conducted on the exospheric temperatures reported in § 2.3.2 shows that the observed semidiurnal component at midlatitudes is only a minor fraction of the diurnal oscillation. Comparison with the amplitude of theoretical *in situ* generated tides [19] and the strong seasonal influence observed on the semidiurnal component, leads us to conjecture that the observed tides above 300 km are primarily driven by the solar EUV source.

3. Temperature Oscillations in the Lower Thermosphere

3.1. Measurement and Analysis Procedure: Multi-pulse Experiment

In order to extend the incoherent scatter measurements below 200 km and to obtain better height resolution than is available from single-pulse observations, a scheme employing pairs of short pulses was implemented at Millstone Hill and successfully put into operation in early 1970 [21]. Briefly, the scheme involves the transmission of two short pulses (40 or 100 μs in length) separated by an interval which varies from 0 to 380 μs in 20 μs steps for scattering heights below about 300 km. The correlation is computed between the echoes from the altitude of interest by a procedure similar to that adopted by Wand and Perkins [7]. The multi-pulse technique allows measurements of plasma temperatures to be made in the altitude range 105—525 km with a height resolution of 6 km at the lower altitudes, 15 km in the middle range, and 30 km at the upper altitudes. Above about 225 km, where the ionic composition may be assumed to be totally O^+, the measured autocorrelation functions are matched in a least-squares sense to theoretical functions to derive T_e and T_i. Between about 130 km and 225 km, the ionic composition varies from predominant molecular ions (NO^+, O_2^+) at the lower boundary to atomic ions at the upper boundary. Accurate analysis of the measured correlations in this height interval to yield charged particle temperatures becomes difficult owing to the variability of the composition. Below about 130 km thermal equilibrium generally exists ($T_e = T_i = T_n$) and the composition may be assumed to be predominantly molecular (mass $\simeq$ 31 amu). However, the measured correlation for this region is now influenced by ion—neutral collisions. Then it is the ion—neutral collision frequency and the ion temperature T_i which may be derived from the observed autocorrelation function. Due to the observed experimental correlation between the temperature and collision frequency and general difficulties in accurate estimation of the latter, we will omit the collision frequency results in this paper and discuss only the thermal structure. The analysis programs for these measurements have only been recently completed and we present sample results collected for two days close to the 1971 vernal equinox.

3.2. Results

Due to the decrease of the electron density in the E region at night, backscattered power from heights below about 250 km is very weak. Measurements in the lower thermosphere are therefore limited to daytime hours. Fig. 8 illustrates the temperature results obtained on 25—26 March 1971 and 14 April 1971. The average accuracy in the temperature measurements is about ± 15 °K; representative statistical estimates of the error are shown in the graphs. The autocorrelation functions used to derive these results were obtained by integration of echo returns for about 12 minutes repeated every 30 minutes to allow measurements using longer pulses at higher altitudes. To decrease statistical noise in the measurements, a one-hour sliding average was performed to emphasize the major features in the results.

A large oscillation of amplitude 10—15% of the mean is clearly revealed in the temperatures at all heights. The absolute value of the peak-to-peak oscillation is close to 100 °K. The period of the oscillation is close to, but slightly less than,

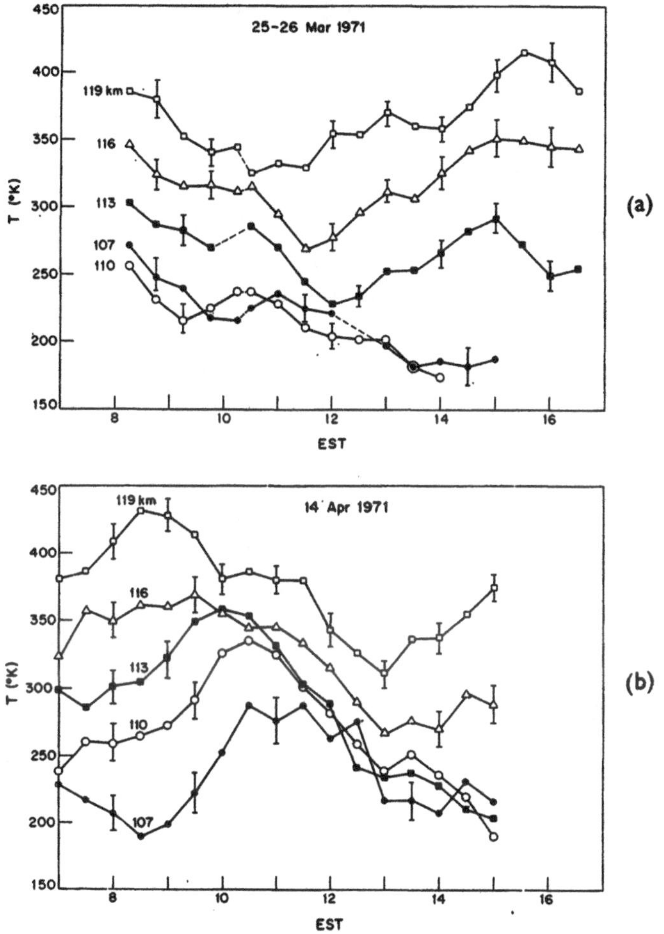

Fig. 8. Temperature oscillations in the E region for (a), above, 25—26 March 1971 and (b), below, 14 April 1971.

12 hours. At the lower heights, particularly on the April day, a shorter period closer to 6 hours is evident. The time of the minimum, clearly revealed at 116 and 119 km is seen to occur about 2 hours earlier on the March day. The most important feature observed is the change of phase of the oscillation with height. This downward phase velocity for gravity waves implies an upward group velocity which indicates that the energy of the oscillation is generated in the lower atmosphere. The temperature oscillations are primarily a manifestation of reversible heating due to adiabatic compression of the neutral gas by the wave motion.

Without measurements at various latitudes and longitudes, it is improper to employ the term "tidal" to observations at one location, according to COSPAR definitions [22]. However, the observed amplitude and phase of these measure-

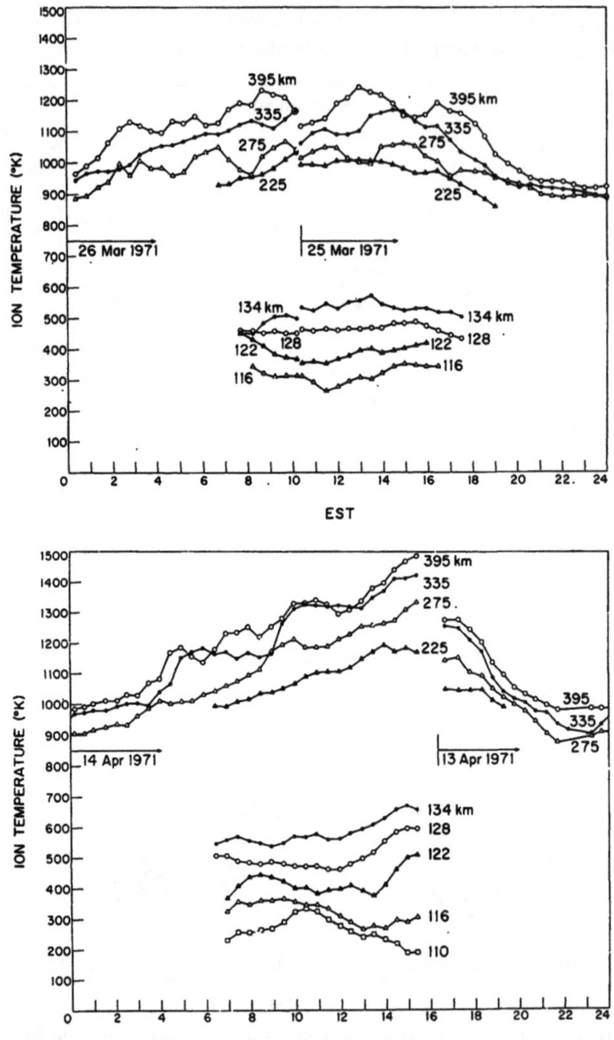

(a)

(b)

Fig. 9. Ion temperature measurements in the thermosphere for (a) 25—26 March 1971 and (b) 14 April 1971.

ments closely approximate theoretical tidal calculations [19] and general features of the semidiurnal tides near 120 km, so that the term "tidal" can be justified. The exact specification of the tidal component that we have observed is difficult since our measurements are limited to about 10 hours and little can be said about oscillations of larger periods. It is clear, however, that the observations show combinations of different spectral components. This may explain the phase variation between the two days. The ozone source drive in the stratosphere, the mesospheric thermal structure, and the background winds, all affect the propagation of the tides, and different components may reach different heights on different days.

If one assumes that the oscillations possess a 12-hour period and fits them with a 12-hour sinusoid, the vertical wavelength on the March day is found to increase from 30 km between 113 and 116 km to 70 km between 116 and 119 km. The amplitude increases from 20 °K at 113 km to a constant 38 °K at 116 and 119 km. On the April day, the vertical wavelength is about 45 km between 110 and 119 km, and the amplitude decreases from 73 °K at 113 km to 37 °K at 119 km. The vertical wavelengths compare satisfactorily with the 3-year average of 48 $\pm$ 9 km reported for Saint-Santin measurements [23]. The temperature—height gradient averaged over the time of observations is about 17 °K/km on both days.

Fig. 9a, b, shows the ion temperature daily variations at 6 km height intervals up to 134 km. The collision-influenced results are included at the lowest heights. The measurements at 122, 128 and 134 km were obtained by assuming a collisionless plasma with an ionic composition modeled according to earlier incoherent scatter results [24] which at these altitudes is at least 90% molecular ions. Above 225 km, where the composition is predominantly atomic ions, the ion temperature is slightly larger than the neutral temperature. Adjustment according to the ion heat balance used in § 1 would introduce nearly isothermal conditions at these heights. The region between 134 km and 225 km is ignored because of the uncertainty in the ionic compositions.

Concentrating on the behavior of the semi-diurnal tides (Fig. 9), it is clear that these have largely dissipated by 134 km since at that altitude the variations are dominated by a diurnal component typical of altitudes greater than 225 km and studied in § 1. Exact comparison with tidal predictions [19] adjusted for midlatitudes is difficult since we cannot separate exactly the tidal modes in the observations. Neither the amplitude nor the phase behavior, within experimental error, were found to be very decisive indices. The most important parameter to consider, perhaps, is the vertical wavelength λ_z which shows preference for the (2,4) mode as dominant in the observations rather than the (2,2) mode.

The present observations, similar to observations by Bernard and Spizzichino [23] and Wand and Perkins [25] do not reveal the presence of the (2,2) tide at heights of observation above about 110 km. In contrast, phase variations of the horizontal drift measurements at 130 km at Millstone Hill by Evans [26] indicate the likelihood of a (2,2) mode to explain the observed electric field-driven pattern. Theoretical calculations [19] predict the dominance of the (2,2) mode at these heights but some mechanisms have been suggested [27, 28] for damping these oscillations. Further study must await confirmation of the dominant semidiurnal tidal mode in the E region and definition of the influence of the mesospheric barrier and background winds on its propagation in the thermosphere at midlatitudes.

4. Thermospheric Temperature Structure

4.1. Theoretical Basis and Procedure

In the first two sections we examined the daily temperature variations at altitudes above the thermopause and at the base of the thermosphere independently. It is evidently desirable to determine the temperature variation in the middle thermosphere and to investigate the thermal structure of the entire thermosphere in order to understand the mutual influences and interrelations of the various parameters that characterize this region of the atmosphere. To this end, we have used the plasma temperature and electron density measurements from the multi-pulse technique at heights 110—400 km. Above 225 km, where O^+ predominates, one invokes the ion thermal balance hypothesis given by Eq. (1). This equation may be simplified by neglecting $n(N_2)$ and $n(O_2)$ and thus slightly underestimating $T_i - T_n$ at the lowest heights. Since $T_i - T_n$ is calculated to be less than 50°K at those heights in the daytime, the error introduced is within the experimental uncertainty and can be neglected. Eq. (1) is then rewritten as

$$n(\mathrm{O})\,(T_i + T_n)^{1/2}\,(T_i - T_n) = 2.3 \times 10^8\, N\, \frac{T_e - T_i}{T_e^{3/2}} \tag{8}$$

where the right-hand side of the equation is known from the experimental measurements. The atomic oxygen number density may be assumed to be governed by the diffusive equilibrium law and its height variation is therefore a function of $T_n(z)$ and a reference density, taken at 400 km,

$$n(\mathrm{O})_z = n\ (\mathrm{O})_{400}\, \frac{T_{n(400)}}{T_{n(z)}} \exp\left[-\left\{\int_{400}^{z} \frac{m_0 g(z)}{k T_n(z)}\, dz\right\}\right]. \tag{9}$$

In turn, $T_n(z)$ may be expressed by the simple model (Eq. (2)) which is the familiar formula due to Bates [14]

$$T_n(z) = T_\infty - (T_\infty - T_0) \exp\{-s(z - z_0)\}. \tag{10}$$

Below about 135 km, where the ion composition is predominantly molecular (O_2^+ and NO^+), our measurements yield $T_i = T_n$. The region between 140 km and 225 km where the ionic composition changes from molecular to atomic is then ignored because the variation of the composition is not well known. From the knowledge of T_i below 140 km and T_i, T_e, N above 225 km, it is adequate, as will be illustrated, to estimate the daily variability of T_∞, s, T_0 and $n(\mathrm{O})_{400}$.

In the presence of large waves with phase progression in altitude at the lower boundary, the simple exponential temperature law (Eq. 10) may become invalid [25]. For the measurements discussed here, the relatively small amplitude of the temperature oscillation and its dissipation below 135 km seems to justify the use of this relation. At night, the ionospheric echo is weak and only measurements above about 250 km can be obtained. This limits the applicability of the method to the daytime hours.

The accuracy with which the various parameters are calculated depends on statistical evaluations. The most difficult measurement is that of $n(\mathrm{O})_{400}$ and it is noted that this parameter cannot be calculated reliably at all times. The difficulty arises because the number density is used only in the thermal balance correction to T_i in order to obtain T_n. Since $T_i - T_n$ is small, as mentioned earlier, the minimization of the deviation depends only weakly on $n(\mathrm{O})$. In our investigations, whenever the error exceeded reasonable limits, the calculated $n(\mathrm{O})_{400}$ was replaced with that from the Jacchia 1971 model in order to allow the other parameters to be calculated more reliably. The sensitivity of these parameters to $n(\mathrm{O})$ is relatively weak and thus this procedure is adequate. Our presentation of $n(\mathrm{O})_{400}$ shows only those estimates of which we are confident. The uncertainty in these values is about $\pm$ 25% to $\pm$ 50% but could exceed these limits. The temperature at 120 km, T_0, is established reliably by data below and above that height and the uncertainty in its estimate is $\pm$ 5 to $\pm$ 10%. The value of s is somewhat difficult to determine since it represents the gradient in the thermospheric temperature profile and is most sensitive to the temperature in the omitted region 140—225 km. The uncertainty in s from our measurements is estimated to be $\pm$ 15 to $\pm$ 30%. The exospheric temperature T_∞ is the easiest parameter to obtain since it depends on measurements above 225 km of which there is an adequate number. The accuracy of T_∞ is estimated at $\pm$ 2 to $\pm$ 5%.

We have applied the above technique to the two equinox days discussed earlier, namely 25—26 March 1971 and 13—14 April 1971.

4.2. Results

Figs. 10 and 11 illustrate the estimates of the parameters T_∞, T_0, s and $n(\mathrm{O})_{400}$ for the two days. The dashed lines in parts a, b are the Jacchia 1971 model predictions; the model variability for T_0 and s is slow with time and almost constant

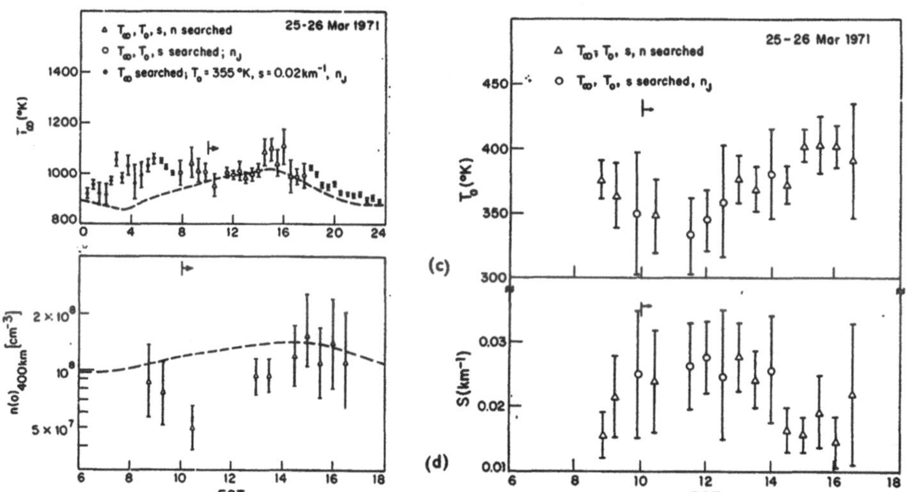

Fig. 10. Thermospheric structure calculated from 25—26 March 1971 incoherent scatter measurements: (a) exospheric temperature, (b) atomic oxygen density at 400 km, (c) neutral temperature at 120 km, and (d) thermospheric temperature profile shape parameter. Dashed line is the Jacchia 1971 model prediction (subscript J).

for the hours shown. The exospheric temperatures reveal a maximum at 1600 LT and are generally consistent with the equinoctial results in § 1. The slow post-sunrise rate of change on the April day is primarily a geomagnetic effect since $Kp\ (t - 6)$ on this day is 3^+ at 0300 LT and decreases to a value of 1 at 0900 LT thus suppressing the rate of increase. The temperature oscillation at 120 km is consistent in amplitude and phase with the tidal oscillation presented in § 2. The value of s appears to oscillate about a value close to 0.02 and on the March day indicates the same frequency but an out-of phase relationship with the T_0 oscillation.

Statistically, however, the correlation of s with T_∞ is stronger than with T_0. It is likely that T_0, as well as T_∞, contribute to the structure of s, since s is proportional to the temperature gradient at the boundary. The reversible heating effects at the lowest heights which are translated into irreversible dissipation at higher altitudes are apt to influence the thermospheric temperature structure. An oscillation frequency in s greater than 1 cycle/day is consistent with rocket measurements [29] and other incoherent scatter results [8, 9].

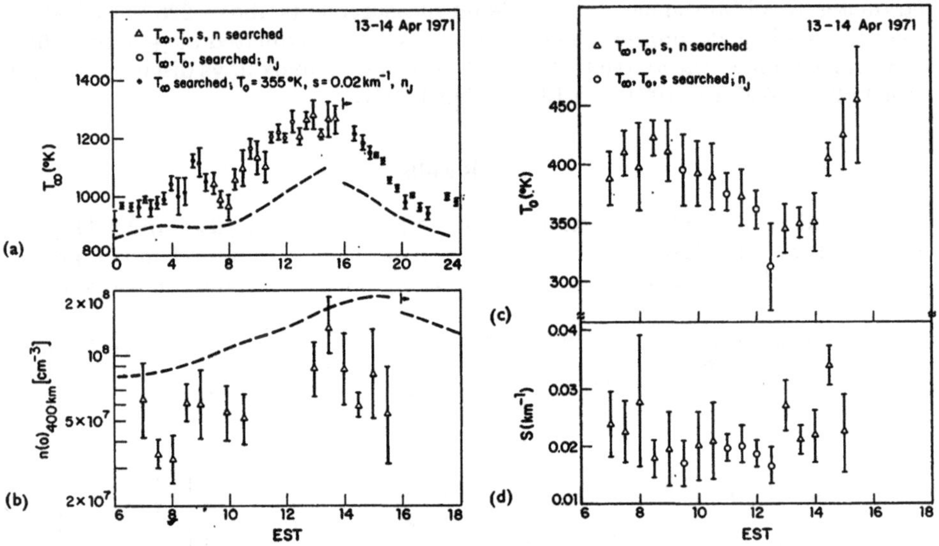

Fig. 11. Thermospheric structure calculated from 13—14 April 1971 incoherent scatter measurements: (a) exospheric temperature, (b) atomic oxygen density at 400 km, (c) neutral temperature at 120 km, and (d) thermospheric temperature profile shape parameter. Dashed line is the Jacchia 1971 model prediction (subscript J).

The agreement between the atomic oxygen results and the Jacchia model is satisfactory (within a factor of 2). The uncertainty in the measurements, however, makes it difficult to ascertain exactly the time of maximum of the density. The ability to measure simultaneously $n(O)$ and T_∞ would resolve the possible discrepancy between the time of density and temperature maxima or rather between satellite drag and incoherent scatter temperature measurements. Various explanations involving horizontal wind transport [30] and heat advection [31], adiabatic heating by vertical motion [32], and oscillations at the lower boundary [33] have been suggested. Our experimental results on 13—14 April tend to indicate a density maximum close to 1400 LT followed by a steep decrease as noted by Alcayde et al. [34]. On the other hand, the March day does not indicate a well-defined maximum, but suggests that the maximum occurs between 1400 and 1600 LT. It seems, however, that the density does not maximize in the same manner as the temperature.

5. Conclusion

Neutral temperatures at heights 110—400 km have been derived from incoherent scatter measurements at Millstone Hill. Exospheric temperatures for 46 days were used to characterize statistically the observed short and long term variations of the thermosphere. The daily temperature extrema follow a seasonal pattern that reflects the length of the day; the amplitude $(T_{max} - T_{min})/T_{min}$ is about 30% and shows no long-term variations. The maximum temperatures agree reasonably well with the Waldteufel and the Jacchia models except for a larger seasonal variability (± 10%) with a peak in the summer. The exospheric temperatures were synthesized in terms of frequency harmonics that indicate a semidiurnal component whose amplitude is only 30% of the diurnal, and whose phase undergoes a seasonal variability. The diurnal component is about 110 °K in amplitude and maximizes at about 1400 LT. A model for the mean daily exospheric temperature is derived from the data and the daily oscillation is represented as the sum of the first three harmonics.

A study of the temperature at the base of the thermosphere for two equinoctial days shows evidence of dominant semidiurnal oscillations (100 °K peak-to-peak) propagating from below and damping below 135 km. When the total temperature profile (110—400 km) is considered, these waves are shown to influence the thermospheric temperature gradient and presumably contribute to the temporal variability of the thermopause temperature. Within experimental uncertainties, the daily variations of the neutral density and the exospheric temperature are apparently out of phase. This is shown to be related to variations in the lower and middle thermosphere.

We intend to expand the data base for the study of the oscillations in the lower thermosphere, and to investigate possible seasonal variability at those altitudes. Further calculations of neutral temperature parameters will be carried out particularly for summer conditions where the phase difference between the neutral density and temperature will perhaps be more apparent. Deduction of horizontal neutral winds from ionospheric drifts measured simultaneously with the temperatures is presently underway and should complement this study in helping us assess the dynamic influence on the thermosphere structure.

Acknowledgments

We gratefully acknowledge the expert computer programming by J. M. Holt and R. F. Julian. We thank R. H. Wand and D. M. Cunnold for helpful discussions. One of us (JES) is indebted to Professors N. A. Phillips and R. E. Newell of the MIT Meteorology Department for their encouragement and advice.

The work was supported by the Department of the Army.

References

[1] H. M. Carru, M. Petit, G. Vasseur and P. Waldteufel, Annls Géophys. **23**, 455 (1967).
[2] J. S. Nisbet, J. Atmos. Sci. **5**, 586 (1967).
[3] H. Carru and P. Waldteufel, Annls Géophys. **25**, 485 (1969).

[4] P. WALDTEUFEL and L. COGGER, J. Geophys. Res. 76, 5322 (1971).
[5] J. P. McCLURE, J. Geophys. Res. 76, 3106 (1971).
[6] P. WALDTEUFEL, J. Geophys. Res. 76, 6990 (1971).
[7] R. H. WAND and F. W. PERKINS, J. Geophys. Res. 73, 6370 (1968).
[8] P. BAUER, P. WALDTEUFEL and D. ALCAYDE, J. Geophys. Res. 25, 4825 (1970).
[9] W. E. SWARTZ and J. S. NISBET, J. Geophys. Res. 76, 185 (1971).
[10] J. V. EVANS, R. F. JULIAN and W. A. REID, Lincoln Laboratory, Tech. Rep. No. 477 (1970), DDC AD-706863.
[11] J. V. EVANS, Proc. Inst. Elect. Electron. Engrs 57, 496 (1969).
[12] P. M. BANKS, Proc. Inst. Elect. Electron. Engrs 57, 258 (1969).
[13] L. G. JACCHIA, Smithson. Astrophys. Obs. Spec. Rep. No. 332 (1971).
[14] D. R. BATES, Proc. Roy. Soc. A 253, 451 (1958).
[15] J. V. EVANS, Radio Sci. 6, 609 (1971).
[16] L. G. JACCHIA, J. Geophys. Res. 75, 4347 (1970).
[17] L. G. JACCHIA, J. Geophys. Res. 76, 4602 (1971).
[18] H. VOLLAND and H. G. MAYR, Goddard Space Flight Center Rep. No. X-621-71-240 (1971).
[19] R. S. LINDZEN, Geophys. Fluid Dynam., 1, 303 (1970) and 2, 89 (1971).
[20] R. S. LINDZEN and D. BLAKE, Geophys. Fluid Dynam., 2, 31 (1971).
[21] J. V. EVANS, G. W. ARMISTEAD, W. A. REID, J. E. SALAH and R. H. WAND, Lincoln Lab, Tech. Rep., in preparation.
[22] COSPAR Information Bulletin, No. 56, 17 (1970).
[23] R. BERNARD and A. SPIZZICHINO, J. Atmos. Terr. Phys. 33, 1345 (1971).
[24] L. P. COX and J. V. EVANS, J. Geophys. Res. 75, 6271 (1970).
[25] R. H. WAND and F. W. PERKINS, J. Atmos. Terr. Phys. 32, 1921 (1970).
[26] J. V. EVANS, J. Geophys. Res. 77, 2341 (1972).
[27] R. E. DICKINSON, Space Research XII, 1015 (1972).
[28] H. VOLLAND and H. G. MAYR, Goddard Space Flight Center, Rep. No. X-621-71-331 (1971).
[29] G. R. SWENSON, J. Geophys. Res. 74, 4074 (1969).
[30] H. RISHBETH, Annls Géophys. 25, 495 (1969).
[31] H. VOLLAND and H. G. MAYR, Annls Géophys. 26, 907 (1970).
[32] R. E. DICKINSON, C. P. LAGOS and R. E. NEWELL, J. Geophys. Res. 73, 4299 (1968).
[33] S. CHANDRA and P. STUBBE, Planet. Space Sci. 18, 1021 (1970).
[34] D. ALCAYDE, P. BAUER, C. JAECK and J. FALIN, J. Geophys. Res. 76, 7814 (1971).

Space Research XIII — Akademie-Verlag, Berlin 1973

EQUATORIAL THERMOSPHERIC COMPOSITION AND ITS VARIATIONS

G. P. Newton, D. T. Pelz and W. T. Kasprzak

NASA/Goddard Space Flight Center, Greenbelt, Md., USA

The neutral atmospheric composition experiment on the San Marco 3 satellite has measured the composition of the equatorial atmosphere from 29 April to 29 November 1971. Preliminary results on the diurnal variation of atmospheric composition from 19 May to 30 June at 225 km altitude are presented. The diurnal variation of helium is seen to reach its maximum near 0800 hours and its minimum in the late afternoon in contrast to the behavior of molecular nitrogen and argon. The atomic oxygen densities show smaller variations than those of the other gases. The mass densities calculated from the composition data agree well with those determined from the *in situ* drag force measurements and from orbital decay measurements.

The Italian satellite San Marco 3 was launched into a three degree inclination orbit on 24 April 1971 with initial perigee and apogee of 213 and 713 km respectively. The satellite carried the Italian air density experiment (ADE) and two Goddard Space Flight Center instruments, namely, the neutral atmospheric temperature experiment (NATE) and the neutral atmospheric composition experiment (NACE). The satellite was spin stabilized with the spin axis maintained nearly normal to the orbit plane, and the spin period varied between six and ten seconds during the seven months of orbital life. Data were obtained from all experiments during the interval from approximately 29 April through re-entry of the satellite on 29 November 1971, and the structure parameters of the neutral equatorial atmosphere were measured over five complete cycles of local solar time.

This paper presents preliminary results from the NACE covering nearly 19 hours of local time of the diurnal cycle obtained during the period 19 May through 30 June (limited by the current status of data processing). The results are considered preliminary for three reasons. First, only 16 satellite interrogation intervals have been used which represents less than 20% of the data for this cycle. Second, no corrections have been made for physical-chemical effects (such as CO_2 and CO). These physical and chemical corrections are expected to be small, however, because of the agreement of the mass density from the NACE with those of the ADE and orbital decay techniques. Third, no adjustments have been made to the data for solar and geomagnetic activity effects.

The NACE mass spectrometer analyzer was a double focusing instrument of the Mattauch design and was a modification of that flown on Explorers 17 and 32 [1]. The ion source was of the gold plated, dual filament, enclosed variety. The thermalization chamber for the atmospheric particles, which entered through a knife edge orifice, was gold plated stainless steel. The experiment system allowed essentially simultaneous monitoring of five masses (4, 16, 28, 32, 40). The instrumentation could be tuned on command and measurements of masses 14, 18 and 44 were obtained periodically throughout the satellite lifetime.

Fig. 1 shows the atmospheric densities of atomic oxygen, molecular nitrogen. helium and argon measured near perigee on 20 May 1971. The total mass density calculated from the composition is compared with the mass densities determined from the ADE [2] and orbital decay [3] on this same data pass. The agreement between the three density data sets is representative of all passes examined to date. It is interesting to note that the NACE densities are slightly larger than the other densities.

The atomic oxygen densities for this paper were determined from the measured source densities at mass 32 by assuming that all atomic oxygen which enters the antechamber immediately recombines to form molecular oxygen, that no molecular or atomic oxygen is permanently lost on the interior surfaces, and that there is no molecular oxygen in the atmosphere.

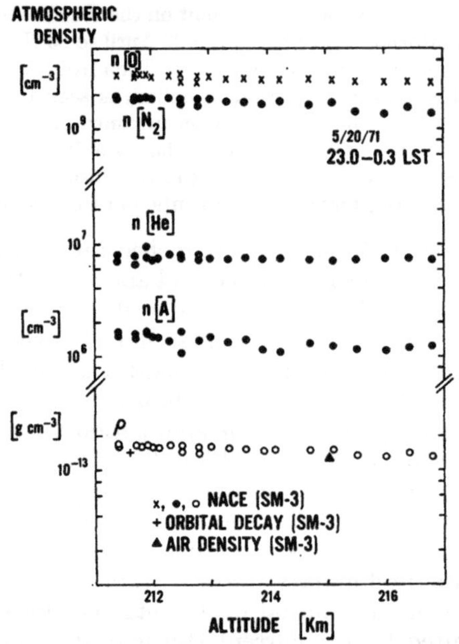

Fig. 1. Atmospheric composition and mass density measured near perigee by the NACE and a comparison with mass densities measured by the ADE and orbital decay.

Fig. 2 shows the atmospheric densities measured at 225 km altitude versus local time. The data have not been adjusted for the solar activity index $F_{10.7}$ and geomagnetic index Ap which are also shown in the figure. Although the activity adjustments will influence the diurnal variation amplitudes shown in Fig. 2, the nearly inverse diurnal variation of helium compared with those of molecular nitrogen and argon is noteworthy. The maximum in the helium variation appears to occur near 0800 hours with the minimum in the late afternoon. The oxygen densities show only small variations, but these may be influenced by solar activity. The agreement of the three mass density data sets is again good.

Fig. 3 shows the same NACE data as Fig. 2 but with the addition of model atmosphere results calculated for the average activity conditions of the data interval. The OGO 6 model [4], determined from composition data near 450 km

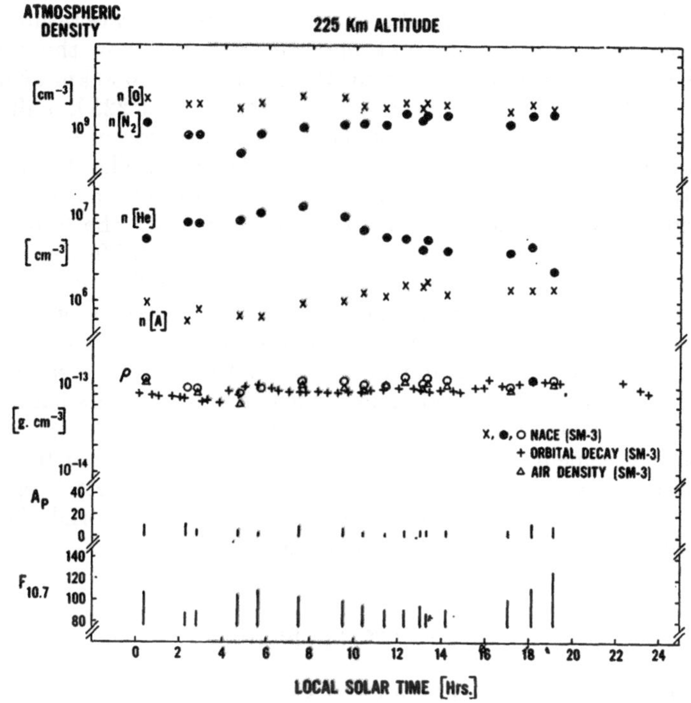

Fig. 2. Atmospheric compositions and mass densities at 225 km altitude measured by the NACE and mass densities measured by the ADE and orbital decay. The solar activity index $F_{10.7}$ and daily geomagnetic activity index correspond to the day on which the NACE measurements were made.

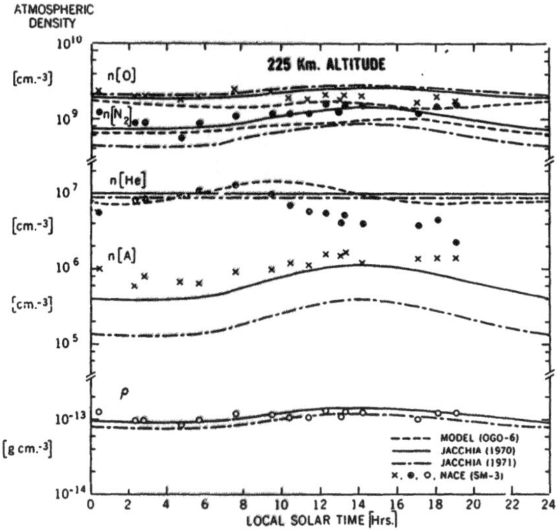

Fig. 3. The OGO 6, Jacchia 1970 and 1971 models compared with the NACE data of Fig. 2.

altitude, used static, diffusive equilibrium equations to generate results at 225 km altitude. It is interesting to note that the OGO 6 results show the maximum of the diurnal variation of helium to occur at 1000 hours at 450 km altitude while the San Marco results show the maximum at 0800 hours at 225 km altitude. This suggests that the time of maximum of the diurnal variation of helium may shift to earlier times at lower altitudes in the thermosphere. The Jacchia 1970 [5] and 1971 [6] models apparently underestimate the argon content of the equatorial thermosphere. However, the Jacchia 70 model represents the San Marco data for molecular nitrogen and argon better than the 71 model. Until activity effects are accounted for in detail and more data are examined, it is not possible to determine which model is more accurate for atomic oxygen or mass density because of the small differences between the two models at this altitude.

It can be concluded that helium in the lower thermosphere shows a diurnal variation maximum at 0800 local time, which is several hours earlier than the diurnal maximum time of the heavier gases molecular nitrogen and argon, that there is good agreement between total mass densities measured by the NACE, ADE and orbital decay, and that the Jacchia 1970 model is a better representation of the molecular nitrogen and argon densities in the equatorial thermosphere at 225 km altitude than is the Jacchia 1971 model.

Acknowledgments

This experiment was possible because of the dedicated contributions of a large number of people. We would particularly like to acknowledge the major part in the NACE development contributed by H. B. Benton, T. D. Clem and B. L. Johnson.

References

[1] C. A. Reber and M. Nicolet, Planet. Space Sci. **13**, 617 (1965).
[2] L. Broglio, paper presented at XIV COSPAR Plenary Meeting, Seattle 1971.
[3] L. H. Anderson and W. R. Burrowbridge, private communication.
[4] A. E. Hedin, H. G. Mayr, C. A. Reber, G. R. Carignan and N. W. Spencer, Space Research XIII, 315 (1973).
[5] L. G. Jacchia, Smithson. Astrophys. Obs. Spec. Rep. No. 313 (1970).
[6] L. G. Jacchia, Smithson. Astrophys. Obs. Spec. Rep. No. 332 (1971).

Space Research XIII — Akademie-Verlag, Berlin 1973

ON THE DIURNAL VARIATIONS IN THE THERMOSPHERE

M. N. Izakov, S. K. Morozov and I. A. Yashchenko

USSR

Data on the diurnal variations of structural parameters of the thermosphere obtained from observations of the orbital decay of satellites, from rocket measurements of vertical distributions of temperature and concentrations of the atmospheric components, and from incoherent scatter observations are summarized and compared with data obtained from the theoretical model. The comparison of these data allow us to conclude that the form of diurnal variations changes and depends on several determining factors. The theoretical model provides an explanation of the super-rotation of the atmosphere found from observations of satellites.

1. Diurnal variations of structural parameters of the earth's thermosphere have been studied using data obtained by different methods. From the data on density, measured by satellite drag (SD), it was concluded that the diurnal maximum of density lies at 14 hours [1—3] (the same value of time corresponds to maximum of temperature calculated from the density by means of static models). On the other hand, the diurnal maximum of temperature measured by the method of incoherent radiowave scatter (IS) lies one, or several, hours later [4—10].

There is also considerable difference in the amplitude of diurnal variations: the ratio of maximum to minimum temperatures in the global diurnal variation is 1.3 according to SD [3] and 1.4—1.6 according to IS, depending on the season [8]. There are also other differences in the global picture of the diurnal bulge (the global minimum during the equinox lies, according to IS, considerably farther from the equator: T_∞ isotherms are more stretched equatorward (cf. [3, 8, 9]).

To explain these divergent data it was suggested that a considerable phase difference between density and temperature variations exists (with the latter lagging) and some explanations for causes of such a phase difference have been proposed [11—13].

At the same time the clarification of the true character of diurnal variations is very important [14, 15] since it would clarify the energy sources and provide a complete and representative description of the thermosphere structure and dynamics.

In this paper recent experimental data on the diurnal variations are compared with our theoretical model and some conclusions are drawn which seem to be useful for the clarification of the problem under discussion.

2. The largest part of information about the thermospheric density and its variations was obtained from satellite drag data (SD). It must be emphasized, however, that the determination of diurnal (and, generally, short-term) variations by this method is difficult because (i) it yields the density averaged over a considerable angle near the perigee, and (ii) the perigee changes slowly (over months) with time angle and latitude.

Therefore, for the determination of diurnal variations one must exclude all other variations by means of some model [16].

The accuracy of determination of the maximum position is obviously not very high and is hardly better than 1—2 hours. Furthermore, not all SD data yield the density maximum at 14 hours. In [17—19] the maximum position obtained from Cosmos satellite drag data falls at 15—17 hours. True, the perigees of Cosmos satellites lie between 200—300 km, and in other works [3] satellites with higher perigees were mainly used; so that if the maximum would change its position with height, these data could be correlated. Data (though isolated ones) on the later maximum position (about 18 hours), were obtained using a satellite with perigee near 400 km [20].

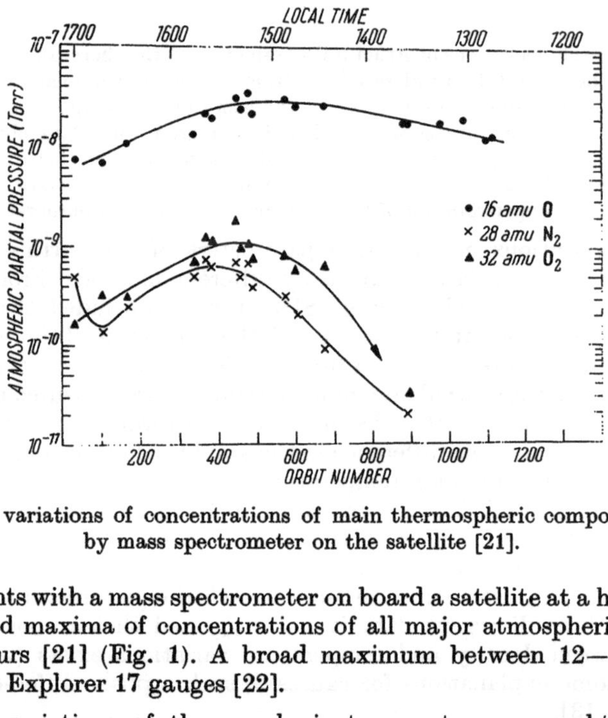

Fig. 1. Diurnal variations of concentrations of main thermospheric components measured by mass spectrometer on the satellite [21].

Measurements with a mass spectrometer on board a satellite at a height of about 400 km yielded maxima of concentrations of all major atmospheric components at 15—16 hours [21] (Fig. 1). A broad maximum between 12—15 hours was obtained with Explorer 17 gauges [22].

3. Diurnal variations of thermospheric temperature were obtained by the method of incoherent radiowave scatter (IS) [4—10, 13, 23] that measures directly ion temperature T_i, from which the neutral temperature T can be calculated with an accuracy from several to ten per cent [9, 24]. This method also meets with difficulties in the determination of diurnal variations (contrary to SD difficulties) associated with the fact that, due to high time resolution, short-term irregular variations are registered (so to speak, thermospheric "weather") which have to be smoothed out. From the first IS data it appeared that the temperature in the exosphere T_∞ was a maximum at sunset (18—19 hours). But subsequent data have changed this notion.

The histogram we have constructed from IS measurements in San-Santin (45° N) and Arecibo (18° N) [25] shows (Fig. 2) that, firstly, the most probable maximum position falls at 15—16 hours and, secondly, there exist deviations from this position by several hours to earlier and later times. In [25] it was also shown that there was high correlation of T_∞ with decimetric solar flux $F_{10.7}$ (with correlation coefficient 0.8—0.9) and that the temperature ratio in the global maximum and minimum lies in the limits $T_{\infty d}/T_{\infty n} = 1.2—1.7$ and is independent of $F_{10.7}$. The same conclusion about the maximum position follows from [9] that summarizes many IS data obtained at Arecibo in 1965—1970. The most probable maximum position is 16 hours in summer and 17 hours in winter and minimum 5 hours in winter and 4 hours in summer, with random deviations of 2—3 hours around this position to both earlier and later times.

We shall show below that the shift of the diurnal maximum can be a consequence of the influence of motions in the thermosphere on its thermal regime.

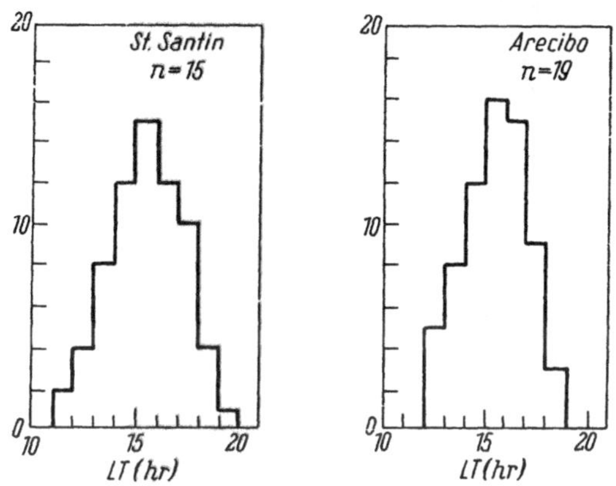

Fig. 2. Position of maximum of diurnal variations of temperature T_∞ from the incoherent scatter data [25].

Recently from IS measurements the data about atomic oxygen concentration n_0 in the daytime were obtained with an accuracy of about 20% [13, 23, 24]. The possibility of simultaneous measurements of T and n_0 is very valuable, because it is helpful for solving the problem of the existence of phase difference in the diurnal variations of T and ϱ.

Both T and n_0 maxima occur near 15 hours, with n_0 decreasing after the maximum the more rapidly. This latter fact has been confirmed [26] based on more SD data.

4. Diurnal variations of the equatorial thermosphere during the equinox were studied with the help of our (space) two-dimensional non-stationary theoretical model [27] by means of the integration of a system of hydrodynamic equations applicable to the thermosphere. New results presented here differ from those of [27], firstly by further improvement of the heat source (in agreement with the recent Hinteregger's data [28] which relate to solar activity corresponding to $F_{10.7} = 150 \times 10^{-22}$ W m^{-2} Hz^{-1}, and secondly by improvement of ion drag (in the same form as in [27], namely, the term describing ion drag in the equation of motion $^1/_2 \varrho_i \nu_{in} (v_i - v)$ and the term describing heat production due to this process in the energy equation $^1/_4 \varrho_i \nu_{in} (v - v_i)^2$, but with a more accurate value $\nu_{in} = 10^{-9}\, n$, according to [29] and [30] and with ion concentration approximately according to Nisbet's model [31]). Let us notice that in [27] ion drag effect was

greatly underestimated and the corresponding term practically had no influence on the solution.

Let us remind ourselves that the main simplified assumptions in the model are the following: neglect of meridional motions due to two-dimensionality, neglect of diurnal variations of composition, the lower boundary at 120 km, and neglect of the dynamic influence of the mesosphere.

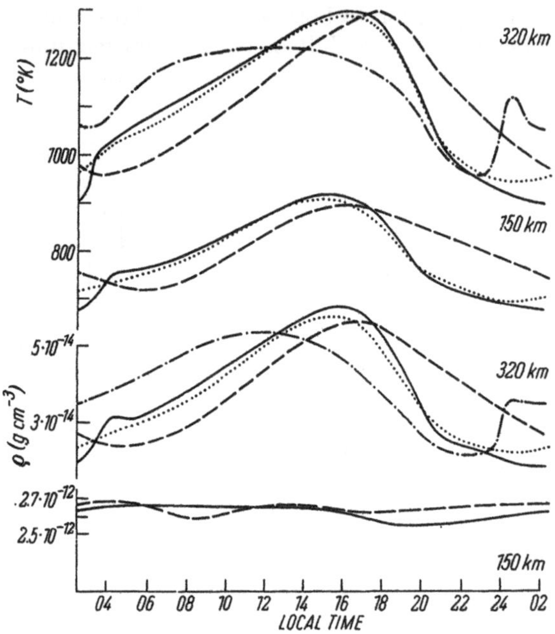

Fig. 3. Diurnal variations of thermospheric temperature and density from the theoretical model. ———— the main model, ·········· the main variant with turbulent viscosity ν_T, —·—·—· variant with the ion drag 10% of normal, — — — — variant without winds ($v_\varphi = 0$).

Diurnal variations of T and ϱ thus obtained are shown in Fig. 3 by solid lines, the case in the absence of horizontal winds ($v_\varphi = 0$) is shown by dotted lines, and chain lines show the case with the ion drag term decreased to 10 times compared with the normal one.

The main conclusion from our results is: the various motions, including ion drag, exert a great influence on the thermal regime of the thermosphere. Indeed, if without winds the thermosphere were heated practically up to sunset (dotted lines in Fig. 4), and with strong winds, developing with negligible ion drag, maxima of T and ϱ were in the morning (case in [27]), then under the most probable conditions (solid lines) the diurnal maxima of both temperature and density lie at 16—17 hours, close to the experimental data considered above.

The second important conclusion is a consequence of the first one: the shape of the diurnal variations, including the positions of the maxima, is variable. Actually variability of the winds and variability of ion concentration can affect the temperature and density variations. To evaluate such a possibility the case was calculated with the ion drag equal to 10% of normal (chain lines). One can see here far earlier T and ϱ maxima — near 13 hours — and noticeably different shapes of diurnal profile: a very broad, nearly symmetrical maximum of T and a sharper maximum of ϱ due to its more rapid decrease after maximum. We have seen already that such early maxima occur sometimes in the experimental data (see Fig. 2). One notices also that the shape of the diurnal profile corresponding to both the main case and the additional case (chain lines) is found in the experimental data.

Finally, the inverse deviation from the most probable diurnal profile is possible: for instance, as ion concentration increases, the maximum position will shift to a later time and the diurnal profile will approach the dotted line.

Interesting data on the diurnal variations of winds were also obtained in the model (Fig. 4): westward winds with velocities up to 30—150 m s^{-1}, prevailing

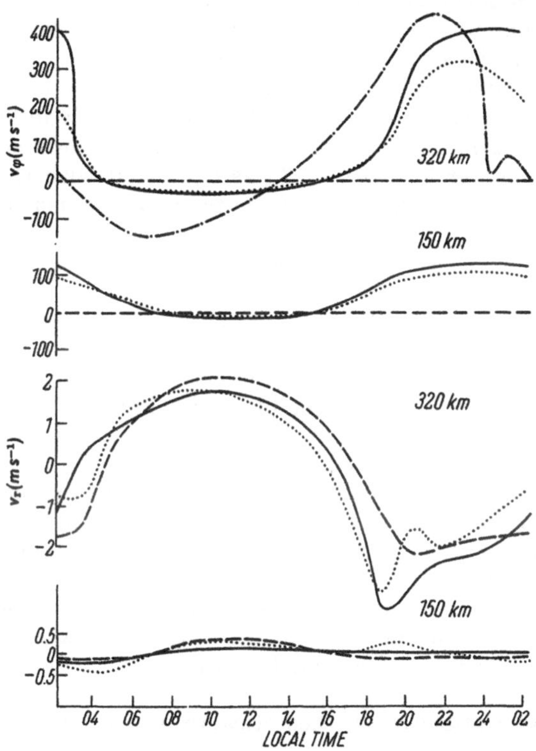

Fig. 4. Diurnal variations of thermospheric winds from the theoretical model: v_φ, horizontal wind component; v_r, vertical wind component. Notation as in Fig. 3.

in the morning and day times (before the diurnal T and ϱ maxima), change after the maximum to eastward winds with velocities up to 300—350 m s^{-1}. Hence, it is evident that the diurnal average velocity contains an eastward component — the phenomenon discovered by King-Hele and named super-rotation [32, 33]. It is also evident that its cause is the influence of diurnal variations of ion concentration through ion drag: the larger day-time ion concentration leads through increased ion drag to a decrease of wind velocities in the day time as compared with the night time. This conclusion is confirmed by the case [27] where neglect of the ion drag caused the absence of super-rotation. The ratio of the average diurnal velocity of the atmospheric rotation to the rotation velocity as a whole with the earth at the altitude of 320 km is equal in our model to $\Lambda = 1.15-1.20$ (compared with 1.3 by King-Hele [33]).

One more feature of the diurnal variations of winds is the shock wave formation in conditions where the Mach number $M > 1$ (velocity of gas relative to the heat source), as already mentioned in [27]. Such a shock wave can be seen on the wind profile $v_\varphi(t)$ near 0—2 hours. True, this shock wave disappears if we calculate the case with the inclusion of the horizontal turbulent viscosity coefficient $\nu_T = 5 \times 10^{12}$ cm² s⁻¹. Such a coefficient may be included if we interpret Blamont's results [34] (on the irregular temperature deviations of about $\pm 100°$ with a spatial scale of the order of 10^4 km) as a consequence of two-dimensional large-scale turbulence. In this case (which is shown as dotted curves in Fig. 3 and 4) the stationary regime was established, and in cases with zero ν_T the computation was performed until the shock wave appeared.

The vertical velocities are directed upwards in the day time and downwards at night and reach values up to 2 m s^{-1}. They considerably influence the thermal structure: the adiabatic heating due to the downward motion can be a reason for the somewhat slower decrease of T, as compared with that of ϱ, after the maximum and also for the beginning of the temperature increase after the minimum even at night.

In principle, our model allows us to take into account the dynamic influence of the mesosphere, which, according to some data, can be considerable [35—39] and, in particular, can cause a phase difference between ϱ and T [12]. For the present the influence of different boundary conditions ($\varrho_0 = \text{const}$, or $v_r = 0$) was tested, and it was found that this difference has a considerable effect at an altitude of 150 km and is almost negligible at 320 km. This is in agreement with the results of [39] which show very rapid decrease of semidiurnal velocity component within the height range of 120—140 km.

As far as the role of thermosphere heating due to the dissipation of energy of waves coming from below is concerned [35—39], let us note that, though the only heat source due to solar EUV is included in our model, the temperatures obtained are close to the real ones.

It should be emphasized once more that our model was constructed for the equatorial thermosphere during the equinox, so that the extrapolation of its data to high latitudes is unjustified. This points to the necessity of constructing a three-dimensional model.

More detailed descriptions of our theoretical results will be published in the near future.

5. It seems to us possible to make the following main conclusions on the thermospheric diurnal variations from the above mentioned experimental and theoretical data:

(i) The most probable position of the maximum of the temperature and density diurnal variations is at 15—17 hours, and of the minimum at 2—5 hours.

(ii) The shape of the diurnal profile varies with season and with changes in the motions, including the ion drag. The time of diurnal maximum can shift by several hours to earlier or later times.

(iii) A considerable phase difference (more than one hour) between temperature and density diurnal variations is hardly probable. If it still exists, its cause may be the dynamic influence of the mesosphere.

(iv) The influence of variations of the lower boundary is noticeable up to 150—200 km and is negligible at higher altitudes.

(v) The diurnal average of wind velocity contains an eastward component (super-rotation) which is a consequence of the influence of the diurnal variation of ion concentration through ion drag.

References

[1] W. Priester, H. A. Martin and K. Kramp, Nature, Lond. **188**, 202 (1960).
[2] L. G. Jacchia, Smithson. Astrophys. Obs. Spec. Rep. No. 313 (1970).
[3] L. G. Jacchia, Smithson. Astrophys. Obs. Spec. Rep. No. 332 (1971).
[4] H. Carru, M. Petit, G. Vasseur and P. Waldteufel, Annls Géophys. **23**, 455 (1967).
[5] J. S. Nisbet, J. Atmos. Sci. **24**, 586 (1967).
[6] J. P. McClure, J. Geophys. Res. **74**, 279 (1969).
[7] K. K. Mahajan, J. Atmos. Terr. Phys. **31**, 93 (1969).
[8] P. Waldteufel and J. P. McClure, Annls Géophys. **25**, 785 (1969).
[9] P. Waldteufel and L. Cogger, J. Geophys. Res. **76**, 5322 (1971).
[10] P. Waldteufel, Thèse de Doctorat D'état, Paris (1970).
[11] H. Rishbeth, Annls Géophys. **25**, 495 (1969).
[12] S. Chandra and P. Stubbe, Planet. Space Sci. **18**, 1021 (1970).
[13] W. E. Swartz and J. S. Nisbet, J. Geophys. Res. **76**, 185 (1971).
[14] P. Blum, I. Harris and W. Priester, in: CIRA 1972, Academie Verlag, Berlin 1972 (p. 397).
[15] M. N. Izakov, Space Sci. Rev. **12**, 261 (1971).
[16] S. P. Wyatt, Nature Lond. **184**, Suppl., 351 (1959).
[17] M. Ya. Marov, Space Research VIII, 811 (1968).
[18] M. Ya. Marov and A. M. Alpherov, Space Research IX, 487 (1969).
[19] M. Ya. Marov and A. M. Alpherov, Space Research X, 419 (1970).
[20] R. R. Allan, paper presented at COSPAR Meeting, Seattle 1971.
[21] C. R. Philbrick and J. P. McIsaac, Space Research XII, 743 (1972).
[22] G. P. Newton, R. Horowitz and W. Priester, Planet. Space Sci. **13**, 599 (1965).
[23] P. Bauer, P. Waldteufel and D. Alcayde, J. Geophys. Res. **75**, 4825 (1970).
[24] D. Alcaydé, P. Bauer, C. Jaeck and J. L. Falin, paper presented at COSPAR Meeting, Seattle 1971.
[25] M. N. Izakov and I. A. Yashchenko, Geomagn. i Aeron. **12**, 1037 (1972).
[26] D. Alcayde, P. Bauer, C. Jaeck and J. L. Falin, J. Geophys. Res. **76**, 7814 (1971).
[27] M. N. Izakov, S. K. Morozov and E. E. Shnoll, Space Research XII, 1029 (1972).
[28] H. E. Hinteregger, Annls Géophys. **26**, 547 (1970).
[29] P. Banks, Planet. Space Sci. **14**, 1105 (1966).
[30] P. Stubbe, J. Atmos. Terr. Phys. **30**, 1965 (1968).
[31] J. S. Nisbet, Pennsylvania State Univ. Spec. Rep. No. 355, p. 362 (E) (1970).
[32] D. G. King-Hele, Planet. Space Sci. **12**, 835 (1964).
[33] D. G. King-Hele, Nature, Lond. **233**, 325 (1971).
[34] J. E. Blamont and J. M. Luton, paper presented at COSPAR meeting, Leningrad 1970.
[35] C. O. Hines, J. Geophys. Res. **70**, 177 (1965).
[36] R. S. Lindzen and D. Blake, J. Geophys. Res. **75**, 6868 (1970).
[37] S. P. Zimmerman and N. W. Rosenberg, Space Research XII, 623 (1972).
[38] S. P. Zimmerman, A. C. Faire and E. A. Murphy, Space Research XII, 615 (1972).
[39] R. Bernard, paper presented at COSPAR meeting, Seattle 1971.

Space Research XIII — Akademie-Verlag Berlin 1973

NEUTRAL COMPOSITION AND ITS VARIATIONS IN THE LOWER THERMOSPHERE

K. K. Mahajan and S. Kumari

Radio Science Division, National Physical Laboratory, New Delhi, India

Number densities of N_2, O_2 and O, reported from various mass spectrometer and EUV extinction measurements are re-examined for an altitude of 150 km. The number densities of N_2 do not show any significant day to night change. The summer and equinox values of N_2, on the average, are about 30% higher than the winter values. The O_2 number densities, however, do not show any significant diurnal and seasonal change. The number densities of atomic oxygen obtained at 400 km from the incoherent-scatter measurements, are extrapolated down to 150 km by using the observed temperature profiles. The average daytime value is found to be 1.6×10^{10} cm^{-3} (at 150 km), close to the CIRA 1965 model value, but smaller than the figure suggested by von Zahn and used in the Jacchia 1971 model.

1. Introduction

Several measurements of the neutral composition in the lower thermosphere with the mass spectrometers [1—3] and with the EUV extinction technique (e.g. [4]) are now available. In this paper we shall use these measurements to study the diurnal and seasonal variations in the lower thermosphere. As the largest number of rocket data exists for an altitude of about 150 km, we shall select this height for our analysis (see also [5]).

Also the total atmospheric density obtained from the composition measurements has been found to be smaller by about 30% than that derived from the recent drag measurements with the low altitude satellites OV1-15 and OV1-16 [2, 5]. According to von Zahn [5], this discrepancy is, in part, due to losses at the ion source in the atomic oxygen measurements and in part due to possible 10—15% overestimation in the calculations of the atmospheric density. Nier [2], on the other hand, thinks that atomic oxygen values are as reliable as those of O_2 and N_2 and the discrepancy should be explained as due to the wrong value used for drag coefficient in the calculation of the atmospheric density. We shall, in this paper, check this discrepancy by using the N_2 and O_2 measurements from the mass spectrometers and the atomic oxygen measurements from the incoherent-scatter radar at St. Santin-Nançay. As the incoherent scatter measurements correspond to a height of 400 km, we shall extrapolate these down into the lower thermosphere, by using the observed exospheric temperature and the shape parameter of the temperature profile and assuming a diffusive equilibrium distribution.

2. Data and Results

The number density measurements of N_2, O_2 and O for a height of 150 km are very well compiled in Table 4 of von Zahn [5]. These include the mass spectrometer measurements of the university of Michigan and NASA—GSFC [3, 6—9], University of Minnesota [2, 10—13] and the University of Bonn [1, 14—17] and EUV extinction measurements of AFCRL [4, 18—22].

2.1. Molecular Nitrogen Number Densities

Of all the neutral constituents, there exists the largest number of measurements on the number density of N_2. We have plotted all these measurements against local time in Fig. 1. As pointed out by von Zahn [5], the N_2 number densities obtained by the University of Michigan and NASA—GSFC have been increased by a constant factor of 1.2 over the published values, on advice from D. R. Taeusch, due to an undetermined calibration error. One immediately notes a

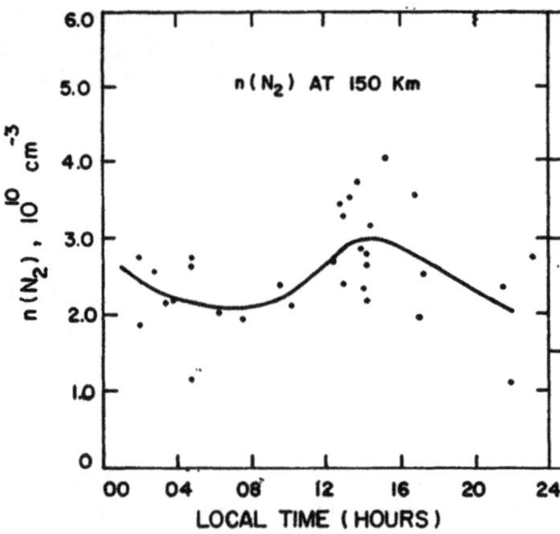

SEASON	AVERAGE $n(N_2)$ $10^{10}\ cm^{-3}$	
	DAY	NIGHT
SUMMER	3.03	2.75
WINTER	2.27	1.96
EQUINOX	3.09	2.69

Fig. 1. Diurnal variation of $n(N_2)$ at 150 km. The variation shown by the solid curve is, however, not real, because the day-to-night change vanishes when one divides the measurements into different seasons, as shown on the accompanying table.

day-to-night change of about 30% in the N_2 number densities. This diurnal change is, however, not real because it vanishes when one takes the day and night seasonal averages for the various measurements, as seen in the same figure. One, in fact, sees a seasonal change in N_2, with summer and equinox values about 30% above the winter values.

2.2. Molecular Oxygen Number Densities

The measured O_2 densities for all geographic latitudes up to 40° are listed in Table 1 and are divided into two categories — the day and the night measurements. As the total number of measurements is not very large, we have obtained the day and night averages without making any distinction about the seasons.

The day and night values appear to be about the same. Also, from the measurements there does not seem to be much seasonal change in O_2. But as the spread in the individual measurements is larger and the number of measurements smaller than the N_2 data, this result may not be statistically very significant.

Table 1

Number densities of O_2 at 150 km

Payload	Date	Time	Latitude	$n(O_2)$ 10^9 cm^{-3})	Ref.
	23 Aug. 1961	1004	33°N	4.20	[18]
	10 Jul. 1963	1004	33°N	3.30	[20]
	2 Mar. 1965	1713	32°N	2.10	[4]
	4 Mar. 1965	1715	32°N	2.40	[4]
	2 Mar. 1966	1259	38°N	2.10	[4]
NASA 4.180	2 Dec. 1966	1409	32°N	2.47	[12]
NASA 4.179	21 Jun. 1967	1249	32°N	2.28	[12]
NASA 4.211	20 Jul. 1967	1224	32°N	2.46	[13]
ESRO S19/1	4 Oct. 1967	1401	40°N	2.00	[1]
SL 407	1 Nov. 1967	1430	31°S	1.70	[22]
	Day time average 2.50				
	6 Mar. 1962	0600	33°N	3.10	[19]
	3 Mar. 1965	0719	32°N	2.70	[4]
NASA 4.127	15 Apr. 1965	0345	33°N	2.56	[11]
NASA 4.181	30 Nov. 1966	0445	32°N	2.87	[12]
NASA 4.212	20 Jul. 1967	0200	32°N	1.40	[13]
NC 3.115F	6 Jun. 1963	0730	33°N	2.05	[10]
ESRO CO7/2	11 Dec. 1965	0443	40°N	1.49	[14]
	Night-time average 2.31				

2.3. Atomic Oxygen Number Densities

While the average O_2 and N_2 values measured by the various groups have shown a remarkable agreement amongst themselves, the atomic oxygen measurements have shown large differences. AFCRL and the Minnesota measurements for example have shown up to a factor of 10 discrepancy between themselves. The average of the Minnesota value is, however, a factor of 3.5 smaller than the AFCRL, while the Bonn average is about in the middle (see also [5]). In view of this discrepancy, we have not used the mass spectrometer measurements of atomic oxygen in the our analysis, but have used the reported incoherent-scatter radar results for studying the diurnal changes in atomic oxygen.

Bauer et al. [23] have reported atomic oxygen number densities at a height of 400 km from the incoherent-scatter measurements at St. Santin-Nançay. In these measurements exospheric temperature and the shape parameter of the temperature profile are also determined simultaneously. By using the observed temperature and the shape parameter, we have extrapolated the 400 km values down to 150 km, by assuming oxygen in diffusive equilibrium. Fig. 2 shows the diurnal variations of $n(O)$ at 150 km for one of the days. We have also plotted the exospheric temperature T_∞, the shape parameters S, and $n(O)$ at 400 km, as reported by Bauer et al. [23] in Fig. 2. There is some evidence for the forenoon n(O) values being larger than the afternoon values at 150 km. No such effect is seen at 400 km, however.

Inherent in our calculations of n(O) at 150 km is the assumption of an approximately constant temperature of 350 °K at 120 km; an assumption made by Bauer

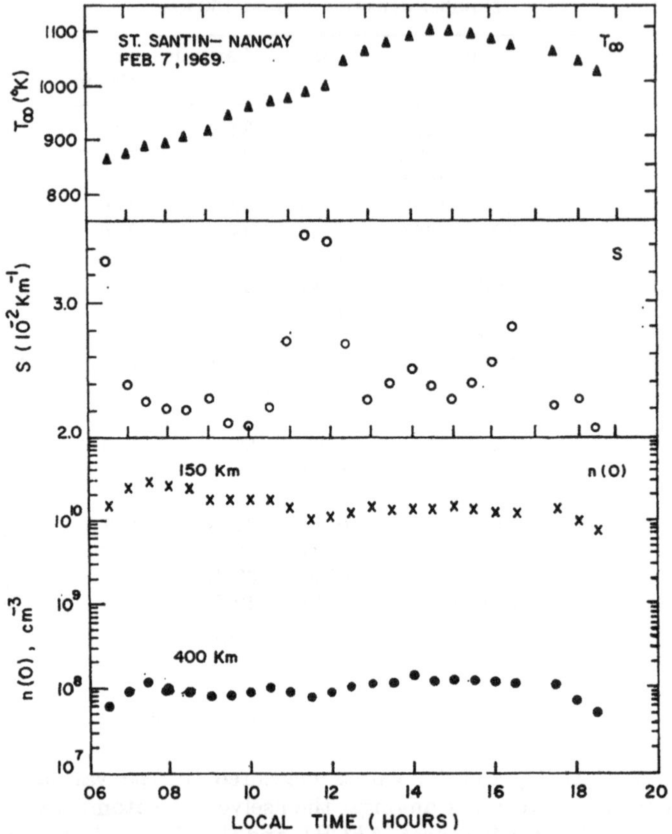

Fig. 2. Diurnal variations of n(O) at 150 km. The values of T_∞, S and n(O) at 400 km taken from [23] are also shown.

et al. [23] for obtaining n(O), T_∞ and S by performing a non-linear regression analysis of the dependence of the observed ion temperature on these parameters. The temperature at 120 km (T_{120}) has been observed to vary between 350 °K and 450 °K at St. Santin-Nançay [24]. According to Bauer et al. [23], it is rather difficult to predict the effect of such a variation of T_{120} on n(O) and S determinations. A numerical trial by these authors, however, showed that changing T_{120} from 355 °K to 380 °K had insignificant effect on the values obtained for S and n(O).

We are, however, aware that the accuracy of our calculations at 150 km very much depends upon the accuracy of the neutral temperature profile and thus upon the shape parameter S. Bauer et al. [23] have placed error bars on the parameter S in their measurements and we have used these in our calculations.

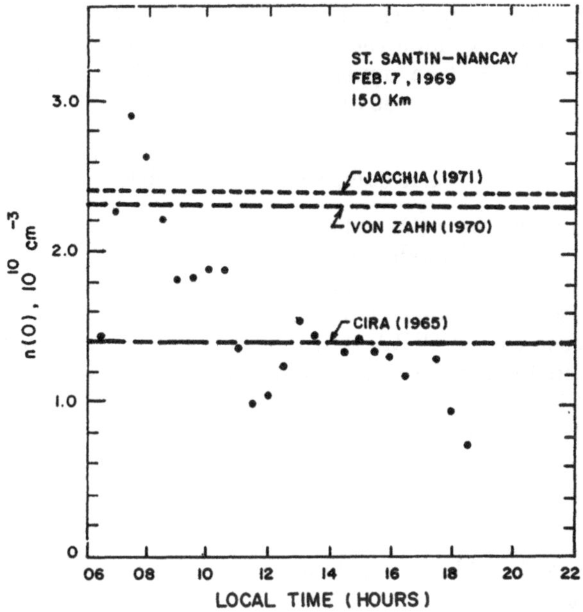

Fig. 3. Comparison of n(O) at 150 km for a winter day with CIRA 1965 and Jacchia 1971 models.

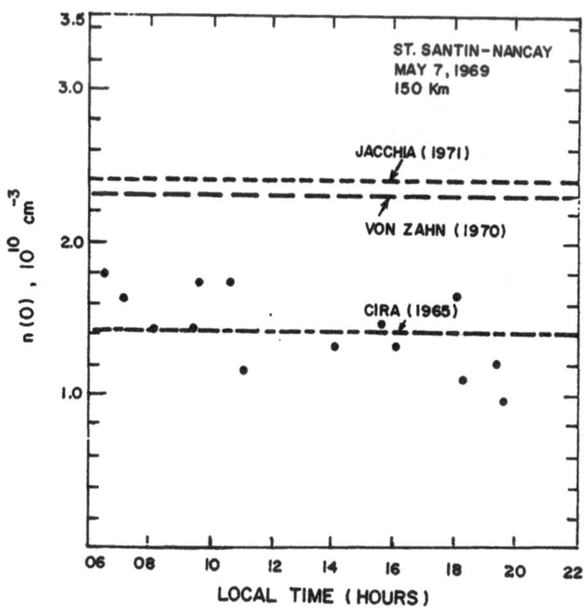

Fig. 4. Comparison of n(O) at 150 km for a summer day with CIRA 1965 and Jacchia 1971 models.

These error bars have very little effect on the extrapolated values at 150 km. We have also tried to examine the effect of change of T_{120} in our calculations. Changing T_{120} to 450 °K (the upper limit of neutral temperature observed at 120 km) decreased the n(O) value at 150 km by about 10%.

2.4. Comparison with Models

We have compared the n(O) values at 150 km for two days with the CIRA [25] and the Jacchia [26] models in Fig. 3 and 4. The values from the models correspond to the measured exospheric temperature for the particular local times. It can be seen that the incoherent-scatter-deduced atomic oxygen number densities are closer to the CIRA [25] model. Value of n(O) for an altitude of 150 km, suggested by von Zahn [5], is also shown. The average daytime value between 06—18 hours for the two days is 1.6×10^{10} cm^{-3}, while that suggested by von Zahn [5] and used in the Jacchia [26] model is 2.3×10^{10} cm^{-3} for a height of 150 km.

3. Discussion

Following von Zahn [5] one can take the following average number densities for N_2 and O_2 at 150 km:

$$n(N_2) = 26 \times 10^9 \text{ cm}^{-3}$$

$$n(O_2) = 2.5 \times 10^9 \text{ cm}^{-3}$$

and taking

$$n(O) = 16 \times 10^9 \text{ cm}^{-3}$$

as deduced from incoherent-scatter measurements, one gets the total atmospheric density at 150 km as:

$$\varrho_{150} = 1.78 \times 10^{-12} \text{ g cm}^{-3}.$$

The orbital decays of the low-altitude satellites OV 1—15, OV 1—16 and Cosmos 316 indicate an average density of 2.2×10^{-12} g cm^{-3} at a height of 150 km [27, 28]. This 20% discrepancy could probably be explained by revising the drag coefficient value used in deducing the atmospheric density from the drag data of the above mentioned satellites, as suggested earlier [2, 5].

In conclusion we suggest the following average values of O_2, N_2 and O at 150 km for the new COSPAR International Reference Atmosphere:

$$N_2 = 26 \times 10^9 \text{ cm}^{-3}$$

$$O_2 = 2.5 \times 10^9 \text{ cm}^{-3}$$

$$O = 16 \times 10^9 \text{ cm}^{-3}.$$

References

[1] W. BITTERBERG, K. BRUCHHAUSEN, D. OFFERMANN and U. VON ZAHN, J. Geophys. Res. 75, 5528 (1970).
[2] A. O. NIER, Space Research XII, 881 (1972).
[3] N. W. SPENCER, G. P. NEWTON, G. R. CARIGNAN and D. R. TAEUSCH, Space Research X, 389 (1970).

[4] L. A. Hall, C. W. Chagnon and H. E. Hinteregger, J. Geophys. Res. 72, 3425 (1967).
[5] U. von Zahn, J. Geophys. Res. 75, 5517 (1970).
[6] N. W. Spencer, L. H. Brace, G. R. Carignan, D. R. Taeusch and H. Niemann, J. Geophys. Res. 70, 2665 (1965).
[7] C. A. Reber and J. E. Cooley, Trans AGU 48, 75 (1967).
[8] D. R. Taeusch and G. R. Carignan, Sounding Rocket Flight Rep. 07065-6-R, Space Phys. Res. Lab. Univ. Michigan (1968).
[9] D. T. Pelz and G. P. Newton, J. Geophys. Res. 74, 267 (1969).
[10] A. E. Hedin, C. P. Avery and C. D. Tschetter, J. Geophys. Res. 69, 4637 (1964).
[11] A. E. Hedin and A. O. Nier, J. Geophys. Res. 71, 4121 (1966).
[12] W. T. Kasprzak, D. Krankowsky and A. O. Nier, J. Geophys. Res. 73, 6765 (1968).
[13] D. Krankowsky, W. T. Kasprzak and A. O. Nier, Geophys. Res. 73, 7291 (1968).
[14] K. Mauersberger, D. Müller, D. Offermann and U. von Zahn, J. Geophys. Res. 73, 1071 (1968).
[15] D. Müller and G. Hartmann, J. Geophys. Res. 74, 1287 (1969).
[16] U. von Zahn and J. Gross, J. Geophys. Res. 74, 4055 (1969).
[17] J. Gross, paper presented at 34th Physikertagung, Salzburg, Austria, 1969.
[18] H. E. Hinteregger, J. Atmos. Sci. 19, 351 (1962).
[19] A. S. Jursa, M. Nakamura and Y. Tanaka, J. Geophys. Res. 68, 6145 (1963).
[20] L. A. Hall, W. Schweizer and H. E. Hinteregger, J. Geophys. Res. 70, 105 (1965).
[21] L. H. Weeks and L. G. Smith, J. Geophys. Res. 73, 4835 (1968).
[22] P. J. L. Wildman, M. J. Kerley and M. S. Shaw, J. Atmos. Terr. Phys. 31, 951 (1969).
[23] P. Bauer, P. Waldteufel and D. Alcayde, J. Geophys. Res. 75, 4825 (1970).
[24] H. Carru, G. Pettit, P. Vasseur and P. Waldteufel, Annls Géophys. 23, 455 (1967).
[25] CIRA, COSPAR International Reference Atmosphere 1965, North-Holland Publ. Co., Amsterdam 1965.
[26] L. G. Jacchia, Smithson. Astrophys. Obs. Spec. Rep. No. 332 (1971).
[27] D. G. King-Hele and D. M. C. Walker, Planet. Space Sci. 17, 2027 (1969).
[28] D. G. King-Hele and D. M. C. Walker, paper presented at the 14th COSPAR meeting, Seattle, 1971.

Space Research XIII — Akademie-Verlag, Berlin 1973

ESTIMATES OF THERMOSPHERIC NEUTRAL CONSTITUENTS FROM ION COMPOSITION MEASUREMENTS

A. P. MITRA and P. CHAKRABARTY

Radio Science Division, National Physical Laboratory, New Delhi, India

Examination of O^+ concentration in fourteen rocket flights show that in most cases the required production rates are appreciably larger than would be given by CIRA 1965. The discrepancy is particularly pronounced during high solar activity. During low solar activity the discrepancy is only large below 200 km. Evidence is presented to show that the substantially larger atomic oxygen concentrations used in the Jacchia 1971 model is an over correction for low solar activity, but adequate for medium and high solar activity periods. There is also some evidence that the ratio $[O]/[O_2]$ obtained from simultaneous use of O^+, O_2^+ and NO^+ concentrations is larger in winter than in summer.

1. Introduction

In previous works dealing with the chemistry of O^+ ions we have noticed that the production rate of O^+ when computed with the CIRA 1965 model usually gives an O^+ concentration substantially lower than the observed concentration of O^+ ions at heights between 150 and 250 km. The implication that at these heights atomic oxygen concentration is larger than given in CIRA 1965 [1] has assumed urgency in view of the larger $[O]/[O_2]$ ratios derived by von Zahn and his colleagues [2] from mass-spectrometric measurements in West Germany, and the current controversy between these large values and the lower values of Nier and Hedin [3] on which the CIRA 1965 model was based.

In this work we examine the ion composition measurements made in fourteen flights in the USA and the USSR covering all periods of solar activity and different seasons.

2. Selection of Data

The ion composition measurements used are listed in Table 1, with the corresponding conditions of solar activity (in $F_{10.7}$ flux), the solar zenith angle χ at the time of the rocket flight, and the local time of measurement. Of the 14 cases chosen, 1—4 represent high solar activity conditions and 8—10 low solar activity conditions. When considering average conditions we take the averages of these two sets.

3. Neutral Composition from O^+ Measurements alone

The relevant equation relating O^+ concentration with neutral O concentration is given by:

$$[O^+] = \frac{R_1[O]}{k_1[N_2] + k_2[O_2]} \tag{1}$$

where

$$R_1 = \sum_{\lambda} \eta_\lambda(O)\, A_\lambda(O)\, Q_{\infty,\lambda} \exp(-\chi) \tag{1a}$$

and k_1 and k_2 are the reaction rates for the processes

$$O^+ + N_2 \xrightarrow{k_1} NO^+ + N$$

$$O^+ + O_2 \xrightarrow{k_2} O_2^+ + O$$

When N_2 is mostly in ground state, k_1 assumes a constant value of 5×10^{-13} cm³ s⁻¹ at $T > 600$ °K, and consequently in the absence of a sizable population of vibrationally excited N_2, a constant value of 5×10^{-13} $cm^3\,s^{-1}$ can be used for thermospheric levels for all conditions of solar activity. In [14] we have, however, shown, from the increase in the ratio $[NO^+]/[O^+]$ with solar activity, that the observed results are reproduced only if a variation in k_1 is introduced with solar activity and with height, at heights above 160 km, the values below 160 km being around the expected value of 5×10^{-13} $cm^3\,s^{-1}$. The requirements in k_1 are given in Fig. 1. The gradual increase in k_1 at greater heights would confirm the concept that k_1 does in fact consist of two parts, one depending on gas temperature, and decreasing to a low value around 5×10^{-13} $cm^3\,s^{-1}$ at temperatures of 600 °K and above, the other a function of the vibrational temperature. The net result is that k_1 can assume values as large as 25×10^{-13} s^{-1} at heights around 200 km and above during high solar activity. For k_2, laboratory measurements are consistent with the expression $1 \times 10^{-9}\, T^{-0.7}$ $cm^3\,s^{-1}$.

The use of Eq. (1) immediately shows that $q(O)$ computed from CIRA 1965 are always much too low. In fact, one obtains in most cases a *negative* value of k_1 if CIRA 1965 is used. On the other hand, use of Jacchia 1971 model tends to make k_1 too large during low solar activity. The k_1 values that one would obtain from the use of (a) CIRA 1965 (b) Jacchia 1970 [15] and (c) Jacchia 1971 [16] (hereafter

Table 1
Ion composition measurements

No.	Date of flight	$F_{10.7}$ ($\times 10^{-22}$ W m^{-2} Hz^{-1} s^{-1})	χ (°)	LT (hrs)	Ref.
1.	23 Mar. 58	268	58	1207	[4]
2.	2 Aug. 58	254	54	1000	[5]
3.	29 Apr. 60	153	28	1047	[6]
4.	15 Nov. 60	183	58	1141	[6]
5.	15 Jun. 60	166	75	0600	[5]
6.	13 Aug. 58	236	90	0500	[5]
7.	22 Jun. 59	178	90	0500	[5]
8.	15 Feb. 63	70	60	0936	[7]
9.	IQSY	70	60	0936	[8]
10.	23 Mar. 66	78	47	1259	[9]
11.	15 Aug. 66	78	78	1828	[10]
12.	7 Nov. 67	126	78	1132	[11]
13.	15 Mar. 68	127	49	1415	[12]
14.	6 Feb. 69	143	63	1030	[13]

referred to as J 70 and J 71) from the 14 ion composition measurements listed in Table 1 have been obtained individually; in Fig. 1 we show their averages for low and high solar activity conditions. Since we cannot accept negative values of k_1 we would conclude that CIRA 1965 O-concentrations are too low, that J 71 O-concentrations are too high and that J 70 values are roughly in conformity with the observed values.

To obtain the neutral model that would fit with the results best, we start with CIRA 1965 model, then increase the values for O, at the same time keeping the

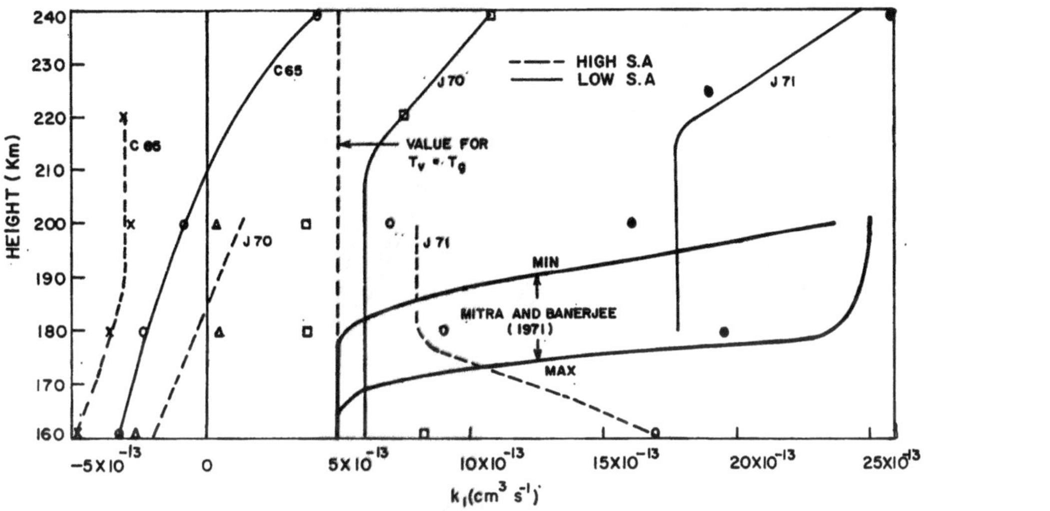

Fig. 1. Average values of k_1 obtained with CIRA 65, J 70 and J 71 models for rocket-borne ion composition measurements.

local concentration constant, until there is agreement with the observed O^+ concentration for a given value of k_1. To do this we transform Eq. (1) into the following

$$\frac{[O]}{[O_2]} = [O^+] \frac{k_2}{R_1} \left(1 + \frac{k_1}{k_2} \frac{[N_2]}{[O_2]}\right) \tag{2}$$

and assume that even when we alter atomic oxygen concentration in CIRA 1965 model, the ratio $[N_2]/[O_2]$ remains essentially the same as in CIRA model. Since the term (k_1/k_2) $([N_2]/[O_2])$ ranges between 0.5 and 2.5, any slight modification in the ratio $[N_2]/[O_2]$ has only a secondary importance.

The accuracy of the $[O]/[O_2]$ ratio so calculated depends on the reliability of our knowledge of changes of EUV flux with solar activity (which affects the value of R_1), apart from the observational accuracy on the estimates of O^+. In calculating R_1 we have taken the EUV flux list given by Hinteregger [27] over the spectral range 100—1300 °A for low solar activity ($F_{10.7} = 75$ units). For high solar activity we use a multiplying factor given by Tyagi et al. [18] from an examination of electron content measurements with satellite radio beacon transmissions. The latter show that the total EUV flux changes by a factor of about 2 as the $F_{10.7}$ flux changes from 75 to 250 units.

When the [O]/[O_2] ratio is obtained, the O_2 concentration may be obtained from the relation

$$[O_2]\left(1 + \frac{[O]}{[O_2]} + \frac{[N_2]}{[O_2]}\right) = \text{constant}$$

and finally N_2 concentration.

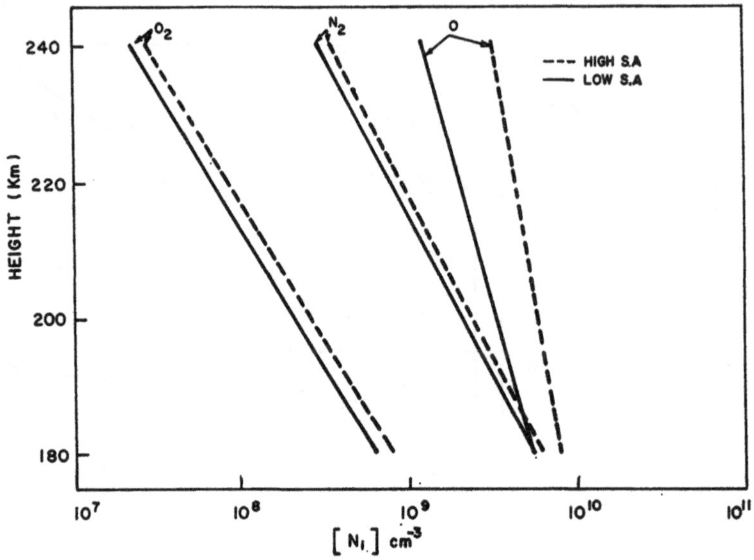

Fig. 2. Models of O, O_2 and N_2 for low and high solar activity conditions derived from the measurements of O^+ ions.

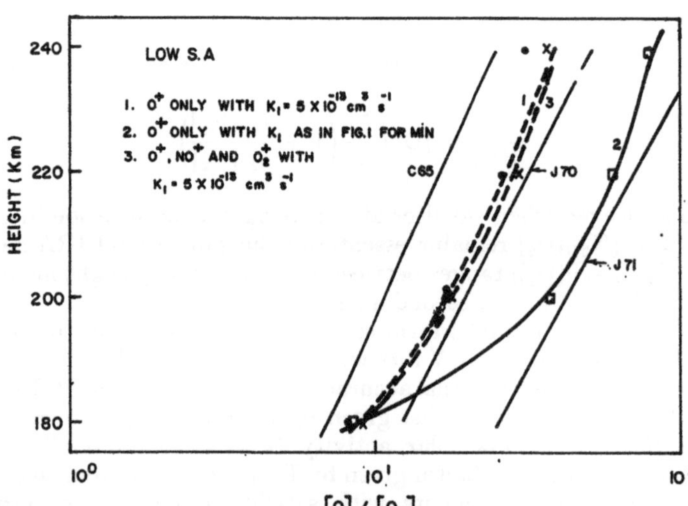

Fig. 3. Height distribution of O/O_2 ratio for low solar activity conditions derived from ion composition. For comparison CIRA 65, J 70 and J 71 values are also shown.

Models prepared in this way for both low and high solar activity conditions from averages of observed O^+ distribution are shown in Fig. 2. The ratios of $[O]/[O_2]$ are compared for low solar activity with those in CIRA 1965, J 70 and J 71 models as shown in Fig. 3. Similarly the ratios of $[O]/[N_2]$ are compared

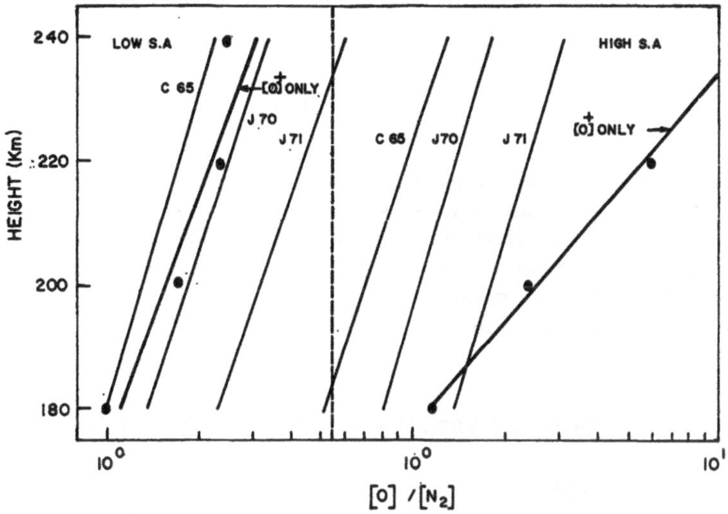

Fig. 4. Height distributions of O/N_2 ratio for low and high solar activity conditions derived from O^+ ions. For comparison CIRA 65, J 70 and J 71 values are also shown.

separately for low and high solar activity with CIRA 1965, J 70 and J 71 models as shown in Fig. 4. We notice that the modification required in the CIRA model to produce the observed O^+ concentration raises $[O]/[O_2]$ and $[O]/[N_2]$ ratio to levels of J 70 for low solar activity conditions and to levels of J 71 for high solar activity.

4. Neutral Composition from Simultaneous Use of O^+, NO^+, N_2^+, O_2^+ and N^+ Observations

The assumption of keeping the value of $[O] + [O_2] + [N_2]$ constant while modifying the individual concentrations can be removed if one uses simultaneously the observed distributions of all ions: O^+, O_2^+, NO^+, N_2^+ and N^+. There is, however, the possibility that this essentially minor advantage may be nullified by the accumulation of errors in the measurements of these ions and by inadequate consideration of the chemistry of the ions N_2^+ and N^+. We find, for example, that the USSR measurements of N_2^+ for high solar activity conditions are too low and yield unusually large values for the ratio $[O]/[N_2]$.

The equations that we have used, ignoring the measurements of N_2^+ which, we concluded, are not at present acceptable for this study, involve O^+, O_2^+ and NO^+ ions, and are

$$[O] = \frac{\alpha_D(O_2^+)\,[O_2^+]\,Ne}{\{R_3 + k_2[O^+]\}\,R_1}\left[k_2[O^+] + \frac{2.5\,\{[NO^+]/[O_2^+]\}\,\{R_3 + k_2\,[O^+]\}}{1 + R_2/k_1\,[O^+]}\right] \qquad (3)$$

$$[O_2] = \frac{\alpha_D(O_2^+)\,[O_2^+]\,Ne}{R_3 + k_2[O^+]} \tag{4}$$

$$[N_2] = \frac{1}{k_1}\left[\frac{R_1[O]}{[O^+]} - k_2[O_2]\right] \tag{5}$$

where
$$R_2 = \sum_\lambda \eta_\lambda(N_2)\,A_\lambda(N_2)\,Q_{\infty,\lambda}$$

$$R_3 = \sum_\lambda \eta_\lambda(O_2)\,A_\lambda(O_2)\,Q_{\infty,\lambda}$$

and $\alpha_D(O_2^+)$ is the dissociative recombination coefficient for O_2^+.

Eq. (3)—(5) are simultaneously solved, for each rocket flight listed in Table 1, for the ions O^+, O_2^+ and NO^+ and for the heights 180—260 km. The results are then averaged for low solar activity conditions. These average values are shown in Fig. 3. It is seen that the values of $[O]/[O_2]$ obtained in this way are not much different from estimates made from $[O^+]$ alone.

5. Seasonal Variation of Neutral Composition

Although the sample is inadequate, there is some indication that $[O]/[O_2]$ and $[O]/[N_2]$ derived from winter flights are somewhat larger than those from summer flights in both low and high solar activity conditions. The results of $[O]/[O_2]$ ratios are shown in Fig. 5. It is important to point out that in [18] in

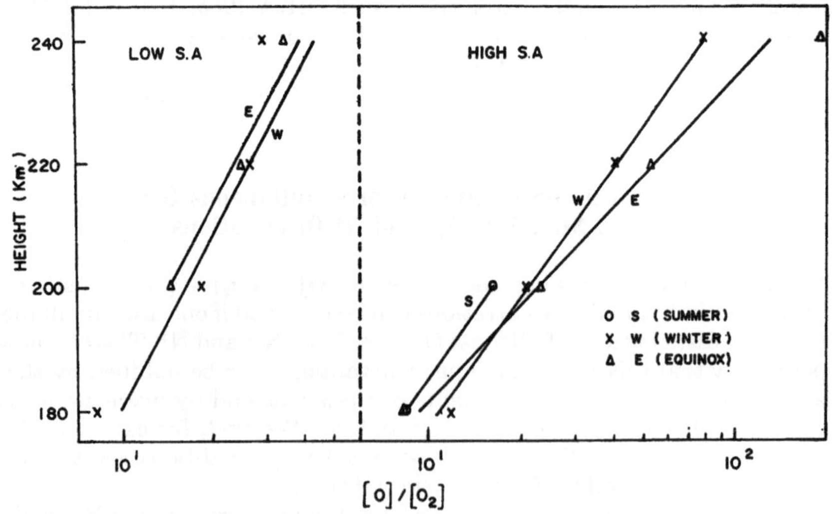

Fig. 5. Seasonal variation in O/O_2 ratios derived from O^+ ions.

which seasonal variation is more reliably determined, no appreciable seasonal variation exists for low solar activity, although the larger winter values are similar in both cases.

6. Conclusions

We conclude from the examination of the fourteen rocket-borne ion composition measurements that the atomic oxygen concentration is too low in CIRA 1965 and that J 71 is an over-correction for low solar activity, although it may be adequate for high solar activity conditions. A comparison of the model derived in this work with CIRA 1965, J 70 and J 71 for 200 km and for day time conditions is given in Table 2.

Table 2
Neutral composition at 200 km

Parameter	CIRA 1965	J 70	J 71	Present work
	a. Low solar activity ($T_\infty = 750$°K)			
[O]	2.8 + 9	3.47 + 9	4.5 + 9	3.2 + 9
[N_2]	2.3 + 9	2.05 + 9	1.5 + 9	1.9 + 9
[O_2]	2.3 + 8	1.77 + 8	1.1 + 8	1.9 + 8
[O]/[O_2]	1.2 + 1	2.0 + 1	4.09 + 1	1.7 + 1
	b. High solar activity ($T_\infty = 1281$°K)			
[O]	3.4 + 9	4.6 + 9	6.3 + 9	5.7 + 9
[N_2]	4.54 + 9	4.35 + 9	3.7 + 9	3.4 + 9
[O_2]	5.4 + 8	4.4 + 8	3.2 + 8	2.7 + 8
[O]/[O_2]	6.3 + 0	1.05 + 1	2.0 + 1	2.05 + 1

Acknowledgment

The authors wish to thank Mr. D. K. Chakrabarty for his assistance in the preparation of this paper.

References

[1] COSPAR International Reference Atmosphere, North-Holland Publ. Co., Amsterdam 1965.
[2] U. von Zahn, J. Geophys. Res. **72**, 5933 (1967).
[3] A. O. Nier and A. E. Hedin, J. Geophys. Res. **71**, 4121 (1966).
[4] C. Y. Johnson, Annls Géophys. **17**, 78 (1961).
[5] V. G. Istomin and A. A. Pokhunkov, Space Research III, 209 (1963).
[6] H. A. Taylor, Jr. and H. C. Brinton, J. Geophys. Res. **66**, 2587 (1961).
[7] J. C. Holmes, C. Y. Johnson and J. M. Young, Space Research V, 756 (1965).
[8] C. Y. Johnson in: Annals of IQSY, Vol. 5, Solar Terrestrial Physics, Terrestrial Aspects, MIT Press, Cambridge, Mass. 1969 (p. 197).
[9] H. C. Brinton, M. W. Pharo III, H. G. Mayr and H. A. Taylor, Jr., J. Geophys. Res. **274**, 2941 (1969).
[10] J. H. Hoffmann, C. Y. Johnson, J. C. Holmes and J. H. Young, J. Geophys. Res. **74**, 29 (1969).
[11] F. Arnold, W. Berthold, B. Betz, P. Lammerzaht and J. Zahringer, Space Research IX, 256 (1969).

[12] R. A. GOLDBERG and L. J. BLUMLE, J. Geophys. Res. **35**, 133 (1970).
[13] A. GIRAUD, G. SCIALOM, A. POKHUNKOV, S. POLOSKOV, G. TULINOV, Space Research XI, 1057 (1971).
[14] A. P. MITRA and P. BANERJEE, Space Research XII, 1315 (1972).
[15] L. G. JACCHIA, Smithson. Astrophys. Obs. Spec. Rep. No. 313 (1970).
[16] L. G. JACCHIA, Smithson. Astrophys. Obs. Spec. Rep. No. 332 (1971).
[17] L. A. HALL and H. E. HINTEREGGER, J. Geophys. Res. **75**, 6959 (1970).
[18] T. R. TYAGI, K. K. MITTAL, A. P. MITRA and S. PRAKASH, Space Research XIII, 461 (1973).

Space Research XIII — Akademie-Verlag, Berlin 1973

A GLOBAL EMPIRICAL MODEL OF THERMOSPHERIC COMPOSITION BASED ON OGO 6 MASS SPECTROMETER MEASUREMENTS

A. E. HEDIN[a], H. G. MAYR[a], C. A. REBER[a], G. R. CARIGNAN[b] and N. W. SPENCER[a]

[a]Goddard Space Flight Center, Greenbelt, Md., USA
[b]Space Physics Research Laboratory, University of Michigan, Ann Arbor, Mich., USA

A global empirical model for magnetically quiet conditions has been developed to describe longitudinally averaged OGO 6 N_2, O, and He densities (in the altitude range 400 to 600 km) by means of an expansion into spherical harmonics. The accuracy of the analytic description is of the order of the experimental error, thus providing an excellent representation of the satellite observations. Some of the results are summarized. (1) The annual asymmetric component in N_2 density corresponds to an increase of about 400°K between winter and summer poles, which is about a factor of three greater than in the Jacchia 1965 model. However, the corresponding oxygen component at satellite altitudes reflects an increase between winter and summer comparable with that in the Jacchia model; this implies that at 120 km atomic oxygen increases by more than a factor of two between summer and winter poles. (2) The amplitude of the symmetric semiannual component in N_2 is larger at the poles than at the equator. At the poles the N_2 variation corresponds to a temperature decrease of about 70°K from mid-April to mid-July. (3) At low latitudes the daily temperature maximum occurs close to 16 hours local time, nearly in agreement with radar backscatter observations.

1. Introduction and Data Analysis

Until recently, the bulk of the data available for determining the global behaviour of neutral density in the thermosphere has consisted of total density measurements obtained from satellite drag. Variations in composition can not be uniquely determined from these data and the atmospheric models [1, 2] based on satellite drag have, until recently, generally assumed diffusive equilibrium with fixed boundary densities at 120 km.

This paper reports on some aspects of the global behaviour of N_2, O, and He as determined from data obtained with the neutral mass spectrometer carried aboard the OGO 6 satellite. This satellite was launched on 5 June 1969 into an 82° inclination orbit with a 398 km perigee and 1100 km apogee. The data cover the time period 27 June 1969 to 22 June 1970 during which time data were obtained in the altitude range between perigee and 600 km (800 km for He). In order to study the behaviour of the thermosphere independent of longitude and universal time effects [3, 4] the globe was divided into 5° latitude bands (with a further subdivision near the poles into south and north bound passes) and data falling within each band were averaged over a 24 hour (universal time) period. Data from days which had an *Ap* index greater than 5 were excluded in order to minimize magnetic activity effects [5].

Two different schemes have been used to describe the data. First, the longitudinal averages have been expressed as a spherical harmonic expansion (in geographic latitude — local time coordinates) in which the composition is described in terms of density variations at 450 km with an altitude dependence assuming isothermal conditions at any given time. The expansion includes local time independent, diurnal, and semidiurnal terms and the coefficients have been written as the sums of terms representing the annual, semiannual, solar flux, and magnetic activity variations. The temperature used in the calculation was determined from the N_2 data in a manner to be described shortly, but it is in any case not critical to determining 450 km densities as this altitude is near the average altitude of the measurements. Thirty-three parameters were determined by a least squares fit to the data for each constituent. The average error of the fit was 20% for N_2 (about twice the experimental error) and 10% for He and O (about equal to the experimental error).

A second approach used to describe the measured densities was analogous to the method employed in the Jacchia J65 static diffusion model [1]. With the density of N_2 fixed at 120 km, the exospheric temperature necessary to calculate the measured densities was determined using analytic density profiles suggested by Walker [6]. The temperature derivation was based on the N_2 density, rather than O or He, because this species is theoretically closest to being in diffusive equilibrium [7]. However, the average temperature needed to fit the measurements is not arbitrary, as the measurements are made at different altitudes, and there is an average temperature which will fit the data best. Details of the global variation of temperature cannot, however, be obtained with much significance from the scale height information. The procedure followed was to fix the 120 km N_2 density, fit the measured densities using a spherical harmonic formula for exospheric temperature variation analogous to that used for the 450 km density variations, and allow the temperature gradient at 120 km to be an additional free parameter so that the average exospheric temperature and density would be consistent with the measurements. By using model temperatures determined from N_2, the average densities and variations at 120 km needed to reproduce the measurements of O and He were determined. It must be kept clearly in mind that although this model represents the measured data quite well, the densities at 120 km are only inferred on the basis of the diffusive equilibrium assumption. However, the variations in composition at 120 km are a convenient way of illustrating departures from fixed boundary diffuse equilibrium models and to the extent that diffusive equilibrium holds, the O to N_2 and He to N_2 ratios inferred at 120 km are relatively independent of errors in the derived exospheric temperature.

The global average densities at 450 km (average $F_{10.7} = 150$, $Ap = 3.5$), determined as described above, are 2.52×10^6 cm^{-3} for N_2, 6.42×10^7 cm^{-3} for O, and 3.55×10^6 cm^{-3} for He. Assuming an N_2 density at 120 km of 3×10^{11} cm^{-3} and a temperature of 355 °K the inferred densities at 120 km are 8.81×10^{10} cm^{-3} for O, and 3.20×10^7 cm^{-3} for He. The average exospheric temperature was 1116 °K. The S value in the temperature profile [6] was 0.021. In discussing particular density variations, the results will be given in terms of departures from the above averages.

2. Diurnal and Semidiurnal Components

The variation of density with local time at the equator for equinox conditions is shown in Fig. 1. At 450 km the N_2 density has a maximum at 15.6 hours, the O density at 14.9 hours and the He density at 10.2 hours. The phase of the O density maximum is not significantly later than that in the J 65 model. As compared with the J 65 model, the later temperature phase and larger maximum to minimum ratio is in agreement with incoherent scatter observations [8]. The amplitudes of the N_2 and O diurnal density variations are in agreement with the J 65 model and the small (statistically insignificant) inferred O density variation at 120 km indicates that the measured 450 km N_2 and O densities are reasonably

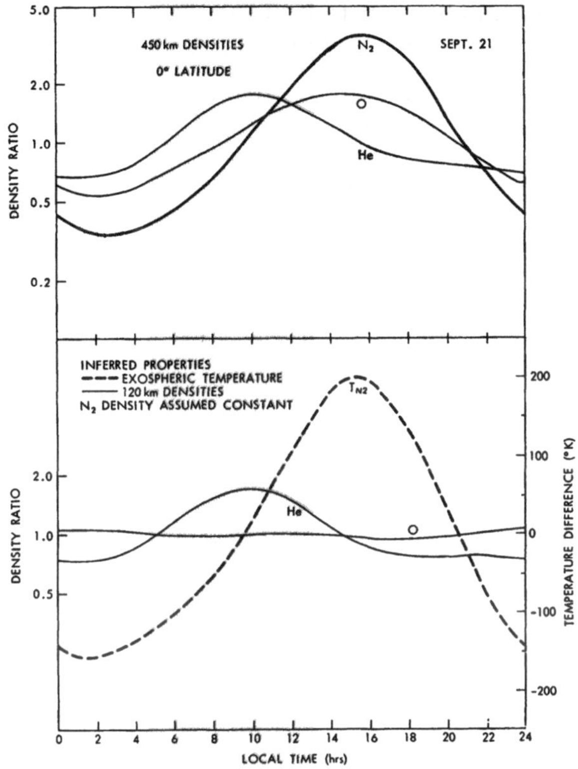

Fig. 1. Diurnal and semidiurnal components of density and temperature variation from the OGO 6 empirical density model for the equator at equinox.

consistent with a fixed boundary diffusive equilibrium model. The morning maximum for He densities and the systematic change in phase from N_2 to O to He, however, indicate that the thermosphere may not be in complete diffusive equilibrium and the measured trend is consistent with predictions of dynamic effects due to winds [7].

3. Asymmetrical Annual Components

The seasonal variations in density lead to the latitudinal density profiles seen in Fig. 2 for solstice conditions. The measured oxygen density variation at 450 km is comparable with that in the J 65 model, but the N_2 density variation is about three times larger. Thus, contrary to the situation for the diurnal variations, the relative variation of N_2 and O densities shows a striking departure from a fixed boundary diffusive equilibrium model and the inferred 120 km oxygen densities show a winter density enhancement of 2.4 times over summer values. The possibility of a winter oxygen enhancement has been inferred from ionospheric data [9, 10] and a similar trend has been recently deduced from a combination of drag and mass spectrometer data [11]. The greater exospheric temperature variation, from winter to summer pole, deduced from the N_2 densities is in the right direction to resolve a recent [12] objection to explaining the F-region seasonal anomaly by changes in the oxygen to nitrogen ratio.

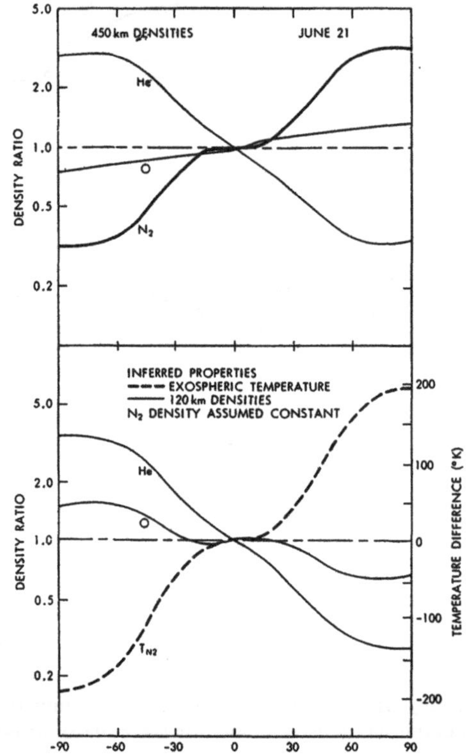

Fig. 2. Asymmetrical annual component of density and temperature variation from the OGO 6 empirical density model for solstice conditions.

The measured helium densities clearly reflect the winter helium bulge [13, 14]; however, the inferred helium density variation at 120 km is about double that given in a recent Jacchia model [2]. The latitude profile in the summer hemisphere is not as well defined by the OGO 6 measurements as in the winter hemisphere because of a lack of sensitivity, but the indicated gradient is consistent with measurements from Explorer 32 [15].

The latitudinal—seasonal variations in both helium and oxygen are consistent with a circulation pattern in the thermosphere which has air rising near the summer pole and transporting air rich in helium and oxygen to the winter pole [16—19].

4. Symmetrical Semiannual Components

The latitudinal profile of the amplitude of the semiannual variation is illustrated in Fig. 3. The amplitudes are quite small and are believed to be accurate to no more than 5 or 10%. Nevertheless, the fact that the phase of the variation was independently determined for each gas to be mid-April (or mid-October) leads to some confidence that the semiannual effect has indeed been detected. Furthermore, the latitudinal variations of the amplitudes show a systematic trend from nitrogen to oxygen to helium. The oxygen amplitude shows only a slight variation

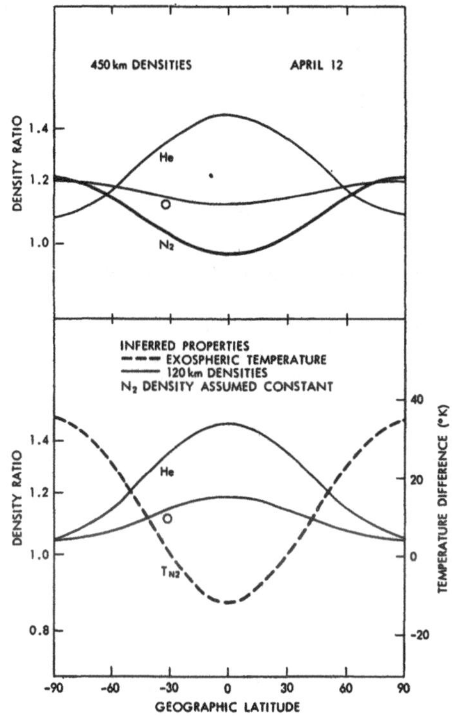

Fig. 3. Symmetrical semiannual component of density and temperature variation from the OGO 6 empirical density model.

with latitude, consistent with J65. The increase of N_2 densities toward the poles (and presumably temperature) combined with the relative depletion of oxygen and helium is qualitatively consistent with dynamical models [19]. The importance of the composition is also in line with the recently revised Jacchia model [20] which represents the semiannual effect in terms of density rather than temperature variations.

5. Conclusion

The analysis of composition measurements made with the neutral mass spectrometer aboard the OGO 6 satellite leads to the following conclusions: The measured atomic oxygen densities are generally in good agreement with those deduced from drag. The molecular nitrogen densities in the annual and semiannual variations depart significantly from those predicted by drag models and suggest similar departures for exospheric temperatures. The helium densities generally tend to vary in an inverse manner to the nitrogen densities. These composition changes are consistent with dynamical processes associated with the global circulation in the thermosphere.

Acknowledgments

We are indebted to the many people in the Space Physics Research Laboratory of the University of Michigan, particularly W. H. Pinkus and D. R. Taeusch, and in the Laboratory for Planetary Atmospheres of the Goddard Space Flight Center, particularly D. N. Harpold and J. E. Cooley, who contributed to the success of this experiment.

References

[1] L. G. JACCHIA, Smithson. Contr. Astrophys. 8, 215 (1965).
[2] L. G. JACCHIA, Smithson. Astrophys. Obs. Spec. Rep. No. 332 (1971).
[3] A. E. HEDIN and C. A. REBER, J. Geophys. Res. 77, 2871 (1972).
[4] A. E. HEDIN and H. G. MAYR, Trans. Am. Geophys. Un. 52, 872 (1971).
[5] D. R. TAEUSCH, G. R. CARIGNAN and C. A. REBER, J. Geophys. Res. 76, 8318 (1971).
[6] J. C. G. WALKER, J. Atmos. Sci. 22, 462 (1965).
[7] H. G. MAYR and H. VOLLAND, J. Geophys. Res. 77, 2359 (1972).
[8] P. WALDTEUFEL and L. COGGER, J. Geophys. Res. 76, 5322 (1971).
[9] H. RISHBETH and C. S. G. K. SETTY, J. Atmos. Terr. Phys. 20, 263 (1961).
[10] H. G. MAYR and K. K. MAHAJAN, J. Geophys. Res. 76, 1017 (1971).
[11] G. M. KEATING and E. J. PRIOR, Trans Am. Geophys. Un. 52, 872 (1971).
[12] S. CHANDRA and P. STUBBE, Planet. Space Sci. 19, 1014 (1971).
[13] C. A. REBER and M. NICOLET, Planet. Space Sci. 13, 617 (1965).
[14] G. M. KEATING and E. J. PRIOR, Space Research VIII, 982 (1968).
[15] C. A. REBER, A. E. HEDIN, J. E. COOLEY and D. N. HARPOLD, NASA Rep. No. TN D-7042 (1971).
[16] F. S. JOHNSON, in: Electron Density Distribution in the Ionosphere and Exosphere, North-Holland, Amsterdam 1964 (p. 81).
[17] G. A. M. KING, J. Atmos. Sci. 21, 231 (1964).
[18] C. A. REBER, H. G. MAYR and P. B. HAYS, Trans. Am. Geophys. Un. 51, 379 (1970).
[19] H. G. MAYR and H. VOLLAND, Annls Géophys. 27, 513 (1971).
[20] L. G. JACCHIA, J. Geophys. Res. 76, 4602 (1971).

Space Research XIII — Akademie-Verlag, Berlin 1973

SATELLITE MEASUREMENTS OF NEUTRAL COMPOSITION WITH A VELOCITY MASS SPECTROMETER

C. R. Philbrick, R. S. Narcisi, D. W. Baker, E. Trzcinski and M. E. Gardner

Air Force Cambridge Research Laboratories, Bedford, Mass., USA

A new instrument has been designed and successfully flown which is capable of measuring the atmospheric composition by energy analysis of an incoming stream of gas at satellite velocity. At satellite velocity the energies of several species are approximately H (0.3 eV), He (1.3 eV), O (5.3 eV), and N_2 (9.3 eV). The operating principle is based on the fact that electron bombardment ionization does not impart significant energy to the ion formed in the majority of the interactions and retarding energy analysis can be used to separate the ionized stream by mass. The ion source is a modification of the Pierce type which provides a uniform sheet of electrons through a field-free ionization region. The instrument was flown on the OV 1—21 satellite which was launched into a 780 km × 920 km polar orbit on 7 August 1971. Real time data were obtained from the experiment for a period of one month. Sample spectra obtained from the instrument are shown and some of the details of the experiment are discussed. Preliminary results from the experiment which obtained measurements of He and O are presented.

1. Introduction

One approach to the problem of measuring the composition of a uniform high velocity stream of particles such as that encountered by a satellite, is to perform an energy analysis to separate the components by mass. An instrument, referred to as a Velocity Mass Spectrometer (VMS), has been designed, developed, and successfully flown on a satellite wich can measure the atmospheric composition by an energy analysis. The technique of using a retarding potential analyzer (RPA) for satellite measurements of electron and ion energies has been accomplished by several investigators (cf. [1, 2]). Apparently there have been no previous attempts to perform an energy analysis of the neutral species of the upper atmosphere. The VMS uses an electron impact ionization source to ionize the neutral gas flux which is then analyzed by retardation in an electric field. The major requirement of this relatively simple instrument is to perform the ionization in a region free of electric fields with a well defined electron beam. At satellite velocities the energies of the atmospheric species are sufficiently large, compared with the mean thermal energy, that mass analysis by energy separation of the species is practicable. Gravitational separation of species in the upper atmosphere results generally in smaller concentrations of the heavier masses. This distribution allows retarding energy analysis that separates the larger concentration of lighter masses and still measures the lesser constituents because of their appearance at a greater retarding energy.

The technique of energy separation offers several advantages over the more conventional type of mass spectrometer. One of the chief concerns with conventional techniques is the question of chemical changes to the ambient composition caused by the instrument surfaces and the hot filament of the ion source. Measurements of O, O_2, and H suffer to some degree due to chemical adsorption (recombination) on the instrument surfaces. It is generally agreed that to a greater or lesser degree all measurements to date show smaller O concentrations and larger O_2 concentrations than exist in the ambient atmosphere due to the surface recombination process. The detailed processes involved with surface reactions are not completely understood. However, it is likely that when a high velocity particle (8 km s^{-1}) undergoes chemical adsorption on an instrument surface, it will not leave the surface with an appreciable fraction of its initial velocity. Thus, those atoms which recombine would not contribute to the composition measured by an energy analysis. In fact, most of the atoms or molecules which strike any instrument surface would not contribute to the measurement, thus eliminating or at least minimizing effects associated with both chemical and physical adsorption. A second advantage is that the sampling characteristics of the instrument can be chosen by the design of the limiting aperture. Concern over the question of thermal accommodation is minimized because the reflected particles should not be detected in the VMS, whereas the reflected particles provide most of the measured signal in conventional open and semi-open ion source mass spectrometers. The experiment uses an electron impact source to ionize the neutral gas flux. The ion which results from stripping an electron from an atom or molecule by electron collision does not receive significant kinetic energy. Since the neutral species are ionized in an enclosed region where electric fields are negligible, the measurement in the retarding field is not influenced by the satellite potential. The simplicity of this experiment makes it attractive for satellite measurements since no rf or magnetic fields are required to obtain a mass analysis. A density analysis is relatively straightforward since the main factors which are needed are the ionization cross section of the species, the instrument geometry, grid transmission, and a detection system for which the efficiency is known.

2. Experiment Description

The instrument shown in Fig. 1 (left-hand side) was developed and flown on the OV1-21 satellite on 7 August 1971. The orbit was 780 km $\times$ 920 km with an inclination of 88°. Data were collected on an average of 6 to 10 real time passes per day over a period of four weeks. The instrument was mounted with its axis perpendicular to the spin axis and the spin axis was normal to the orbit plane. Thus, once each spin period the instrument axis pointed along the velocity vector.

The operation of the instrument is shown in an idealized case in Fig. 1 (right-hand side). The important part of the instrument is the ion source which has been patterned after the Pierce-type [3] source which is especially suited to provide a well-defined electron sheet through the field-free ionizing region. For example, as much as 80% of the electron current entering a slit on one side of the box was found to exit from a slit of the same size on the other side of the region 4 cm away. Fig. 1 also shows an idealized mass spectrum that would be obtained from the retarding energy analysis. The instrument is also capable of providing temperature information from an analysis of the step occurring at the energy corresponding to a particular species. Several factors contribute to a background which will tend to distort this idealized representation. If appreciable numbers of molecular species are present, some energetic ions will be formed through dissociative processes with a spectrum of energies decreasing as a function of energy [4—6]. Another factor which may provide some background contribution at low retarding voltages is the fact that the thermal accommodation coefficient is not unity and some fraction of the reflected species will have sufficient energy to penetrate the retarding field.

The ion source contained two filaments, included for redundancy, which were mounted opposite one another. The electron current could be switched by ground

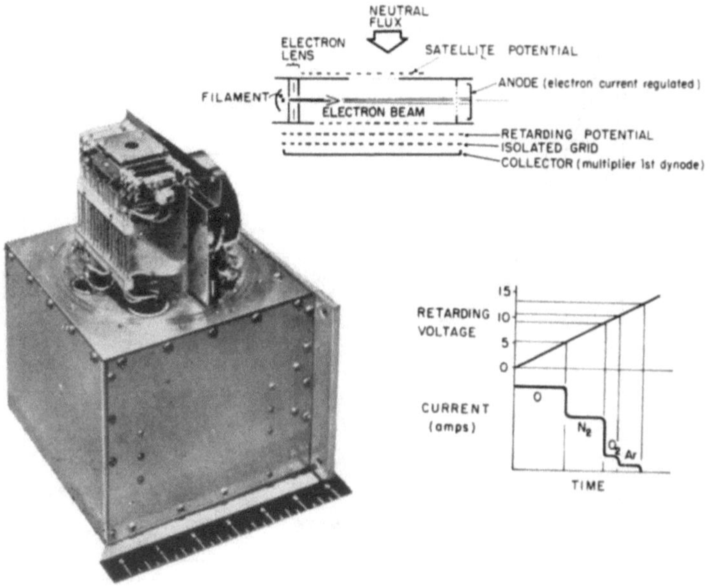

Fig. 1. A schematic drawing of the velocity mass spectrometer with an idealized representation of the retarding energy analysis spectrum together with a picture of the instrument with its vacuum housing removed.

command between levels of 10 μA or 100 μA. The 100 μA electron beam current allowed a large increase in sensitivity and the 10 μA level produced negligible distortion of the electric field in the source. The electron energy in the ionization region was 150 eV. The grid above the ionization region was used to repel ambient positive ions. A grid below the ionization region was used to isolate the high electric field of the secondary emission multiplier. The ion current was detected by a secondary emission multiplier and pulse counting techniques were used to measure the outputs. Two threshold levels were available in the pulse amplifier to allow determination of possible changes in detection system sensitivity. The retarding voltage was stepped through 48 levels each second. The spin period of the satellite was approximately 50 seconds and the geometry of the ion source allowed particles to enter the instrument without reflection when the instrument axis was within $\pm 20°$ of the velocity vector. Since the instrument program of operation was based on a 1 second period, a series of four retarding energy spectra was obtained during each spin cycle where atmospheric species could be measured. When the angle of attack is sufficiently small that the particles of the direct flux are not shadowed by the defining aperture, the particle flux will not change with angle of attack. The retarding energy associated with a particular mass will change as $\cos^2 \alpha$ (where α is the angle of attack between the velocity vector and the instrument axis).

During the first two weeks of operation, the instrument did not provide usable information on the atmospheric species because the background signal of the outgassing molecular constituents was high. The background was produced by dissociative ionization of species such as H_2O and CO. Dissociative ionization

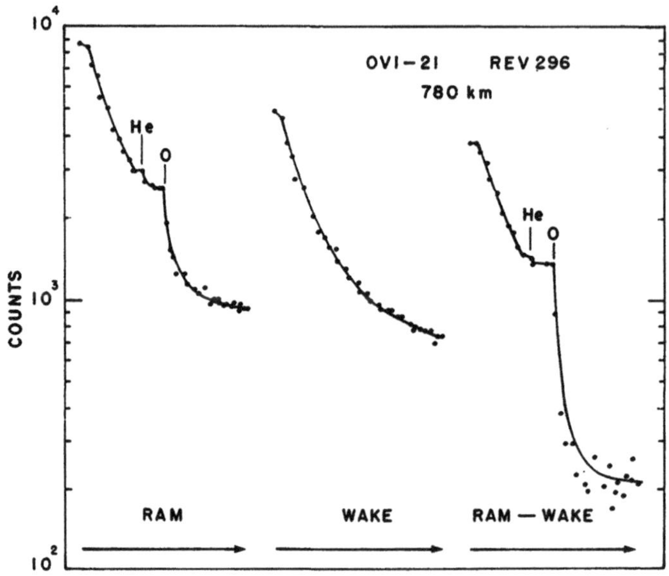

Fig. 2. A sample spectrum from orbit 296 on 27 August 1971. The conditions at minimum and maximum angle of attack are shown together with the difference between these two cases.

of these outgassing constituents can provide particles with energies up to tens of electrons volts [4—6]. Since the atmospheric density near 800 km is typically on the order of 10^6 cm^{-3}, the outgassing background signal can override the signal from the atmospheric species for a considerable period. The electron multiplier was mounted off-axis in order that effects associated with solar illumination of the multiplier and the effects of energetic particles would be minimized.

In Fig. 2, a sample spectrum is shown. The pulse count output from a secondary emission multiplier is shown as a function of retarding voltage for a spectrum near minimum angle of attack and for a spectrum in the vehicle wake. The difference between the ram and the wake, plotted at the right-hand side of the figure, is caused by the flux of gas intercepted by the satellite. The wake spectrum exhibits a gradual drop as a function of increased retarding voltage and is due to energy carried off by products of dissociative ionization processes in the ion source. Distinct drops are seen at energies corresponding to helium and atomic oxygen. Converting the count measured by making use of the entrance geometry, grid transmission factors, ionizing electron current, and ionization cross sections, He and O densities are roughly 3×10^5 cm^{-3} and 2×10^6 cm^{-3} respectively near 780 km. Preparation and analysis of the data are in preliminary stages. Approximately 130 satellite passes of about 10 minutes duration were obtained.

3. Discussion of Results

Although the data from the experiment are limited, several very interesting points can be realized from the preliminary examination of the results. First, the results have shown that this technique can provide measurements of the neutral species on a high velocity vehicle such as a satellite. In addition, charged particle fluxes have been detected in the polar regions moving both up and down the magnetic field lines. The two factors that most adversely affect the measurements were the contribution from the dissociative ionization of outgassing products and solar ultraviolet reflections on the secondary emission multiplier. The effect of the solar UV radiation was to provide a spurious signal when the instrument axis pointed in the general direction of the sun.

An instrument is presently being designed and constructed which will take advantage of the idea of energy separation used in the velocity mass spectrometer combined with the resolving power of a quadrupole mass spectrometer. The result of combining the two instruments is that the advantages of the velocity mass spectrometer can be maintained and the background contribution of dissociative ionization can be eliminated by tuning the quadrupole to a particular mass of interest and then performing a retarding energy analysis. The new combined instrument will be launched during the latter part of 1973.

Acknowledgments

The authors appreciate the efforts of Mr. Robert McInerney in providing the vehicle attitude information. The contributions of Mr. R. A. Wlodyka and Mr. R. B. E. Moren to the development of the electronics and Mr. G. S. Federico to the mechanical design are gratefully acknowledged.

References

[1] W. B. Hanson, S. Sanatani, D. Zuccaro and T. W. Flowerday, J. Geophys. Res. **75**, 5483 (1970).
[2] S. Chandra, B. E. Troy, Jr., J. L. Donley and R. E. Bourdeau, J. Geophys. Res. **75**, 3867 (1970).
[3] J. R. Pierce, Theory and Design of Electron Beams, Van Nostrand, New York City 1949.
[4] L. J. Kieffer and R. J. VonBrunt, J. Chem. Phys. **46**, 2728 (1967).
[5] G. H. Dunn and L. J. Kieffer, Phys. Rev. **132**, 2109 (1963).
[6] D. Rapp, P. Englander-Golden and D. D. Briglia, J. Chem. Phys. **42**, 4081 (1965).

Space Research XIII — Akademie-Verlag, Berlin 1973

NORTH—SOUTH ASYMMETRY OF THE NEUTRAL EXOSPHERE

G. M. Keating, D. S. McDougal, E. J. Prior and J. S. Levine

NASA Langley Research Center, Hampton, Va., USA

A systematic asymmetry in the neutral exosphere between the northern and southern hemispheres has been identified for the first time after correcting for all other known systematic variations. This phenomena was clearly detected from recent measurements of helium drag effects near 1000 km on the low mass-to-area ratio Explorer 19 satellite. From a re-analysis of all four NASA Langley Air Density Explorer satellites (Explorers 9, 19, 24 and 39) this systematic asymmetry was traced over an entire solar activity cycle. It was found that helium concentrations in local summer at high latitudes are generally more than 50% lower in the southern hemisphere than in the northern hemisphere. In addition, the increase in helium concentrations from summer to winter was found to be approximately 80% greater in the southern hemisphere than in the northern hemisphere.

It is generally assumed that the seasonal variations of helium are principally due to meridional flow of the major atmospheric constituents in the lower thermosphere. Thus the very low summer concentrations of helium in the southern hemisphere could result from outflow from the southern hemisphere in December stronger than the corresponding outflow from the northern hemisphere in June. In addition, a variation in turbopause altitude could also contribute to the asymmetry, since an increase in turbopause altitude resulting from greater turbulence causes a depletion in helium concentrations.

1. Introduction

From studies of the atmospheric drag effects on earth satellites a number of systematic variations in the neutral thermosphere and exosphere have been detected. The systematic variations identified to date include the following: variation with 11-year solar activity cycle, variation with solar rotation, variation with geomagnetic activity, altitudinal variation, the summer atomic oxygen bulge in the upper thermosphere, evidence of a winter enhancement of atomic oxygen in the lower thermosphere, the winter helium bulge, the diurnal variation, evidence of a high latitude nitrogen bulge in the lower thermosphere, and a global semi-annual variation. To date, no systematic asymmetry in the neutral exosphere has been detected between the northern and southern hemispheres. While there is slightly greater solar heat input in the southern hemisphere during local summer as the earth approaches perihelion, this effect is naturally accounted for as part of the variation associated with the solar flux incident upon the earth. The semi-annual variation is regarded as a global phenomenon occurring throughout the thermosphere and exosphere at the same time and therefore is not taken to exhibit a north—south asymmetry.

In the present study an investigation is performed to determine whether a north—south asymmetry exists in the lower exosphere. The most recent satellite drag data from Explorer 19 is found to give evidence for such an asymmetry.

In the light of this new information, an investigation is reported to determine whether such an asymmetry is a constant feature of the atmosphere occurring over the entire solar activity cycle. Only satellite drag data afford an opportunity for such a long term investigation and only the Air Density Explorer satellites are sensitive enough to detect clearly such an hemispheric asymmetry in the lower exosphere. After allowing for the effect of atomic oxygen drag, a remarkably consistent asymmetry in helium concentrations is discovered over the entire olar activity cycle between the northern and southern hemispheres.

2. Air Density Explorers

Isolating this new systematic variation required a comprehensive re-analysis of drag data from the whole series of Air Density Explorers: Explorer 9, Explorer 19, Explorer 24 and Explorer 39. These satellites were all designed and built at the NASA Langley Research Center specifically to obtain accurate measurements of the atmospheric density at satellite altitudes by the drag technique. All were rigid spheres 12.0 feet (3.66 m) in diameter with extremely low mass-to-area ratios and thus were highly sensitive to atmospheric drag.

The methods developed for determination of atmospheric densities from the orbital decay of Air Density Explorer satellites are described in [1—3].

3. North—South Asymmetry

For many years, the perigee precession of the Explorer 19 satellite was synchronized with the sun in such a fashion as to measure conditions in the northern hemisphere near the winter and summer solstices and conditions in the southern hemisphere near the spring and fall equinoxes. Thus the seasonal variation in the southern hemisphere at the measurement altitudes of Explorer 19 ($\simeq$ 1000 km) was not clearly delineated. However, drag measurements of the seasonal variation in the northern hemisphere resulted in the discovery of the winter helium bulge [1, 4, 5]. At these altitudes helium replaces atomic oxygen as the principal drag constituent due to the diffusive separation of atmospheric constituents. Helium concentrations were determined by subtracting from the total observed density the effect of other constituents, especially atomic oxygen, the only other significant constituent at these altitudes. The distribution of atomic oxygen is known independently from knowledge of the density at lower altitudes ($\simeq$ 500 km) where atomic oxygen is the overwhelming constituent. Later, with the launch of Explorer 39, measurements were obtained resulting in the discovery of the winter helium bulge in the southern hemisphere [6] which had been postulated by Keating and Prior [1]. Although Explorer 39 gave a clear picture of the increase of helium concentrations in local winter near 800 km, the summer concentrations of helium were more difficult to separate from the drag effects of atomic oxygen. This is because of the simultaneous effects of the summer atomic oxygen bulge and the winter helium bulge. As helium concentrations decrease in summer, atomic oxygen concentrations increase, resulting in a large correction that must be subtracted from the total measured density in order to determine the contribution of helium. Conversely, in winter when the atmosphere cools, helium concentrations increase and atomic oxygen concentrations decrease. Thus there is a much smaller atomic oxygen correction in order to determine winter helium concentrations.

Recently, however, Explorer 19 has become less synchronized with the sun and has obtained a number of measurements in the southern hemisphere summer at a high enough altitude that the helium drag effect ranged from 77 to 97% of the total drag. The results obtained are shown in Fig. 1 where $\log_{10}$ of the ratio

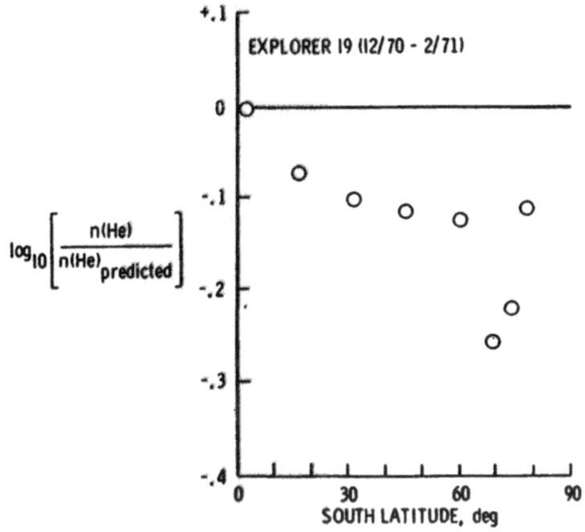

Fig. 1. Residuals between observations and predictions of helium concentrations in southern hemisphere summer (21 December 1970—8 February 1971) measured by Explorer 19.

of the observed helium density to the predicted helium density is shown as a function of southern latitude for the period 21 December 1970 through 8 February 1971 (summer in the southern hemisphere). The summer helium concentrations in the southern hemisphere are much lower than the predicted variation (indicated by the zero line) which is based on the summer helium concentrations observed in the northern hemisphere. Predictions of helium and atomic oxygen concentrations are based on the spring/fall model of the US Standard Atmosphere Supplements, 1966, with the following improvements [5—7], henceforth referred USSAS:

$$R = 0.4, \qquad T_0 = T_0' \left[\frac{1.28}{1.4}\right] \qquad \delta_B = \delta_S$$

$$\log_{10}[n(\mathrm{He})] = \log_{10}[n(\mathrm{He})_{\mathrm{mod}}] + \Delta \log_{10}[n(\mathrm{He})]$$

$$\Delta \log_{10}[n(\mathrm{He})] = K_1 + K_2\,\delta_P\delta_S + K_3\,|\delta_P\delta_S| \qquad (1)$$

where $R = (T_{\max} - T_0)/T_0$; $T_{\max}$ = maximum exospheric temperature (°K); T_0 = improved minimum exospheric temperature (°K); T_0' = minimum exospheric temperature excluding semi-annual effect in USSAS (°K); δ_B = latitude of $T_{\max}$ (radians); δ_S = declination or latitude of sun (radians); δ_P = latitude of measurements (radians); $n(\mathrm{He})$ = number density of helium (cm^{-3}); $n(\mathrm{Hs})_{\mathrm{mod}}$

= number density of helium in USSAS (cm^{-3}). It was assumed for Fig. 1 that $K_1 = +0.05$, $K_2 = -0.4$, and $K_3 = +0.2$ [7]. The semi-annual effect given in [8] was assumed.

Although previous helium drag measurements in the summer hemisphere generally required a larger correction for atomic oxygen, the earlier data were re-examined to establish whether the very low helium concentrations in the southern hemisphere summer were a regular systematic feature which occurred over the entire solar cycle or were just a short-term anomaly. First the simultaneous measurements of Explorers 19 and 39 since the launch of Explorer 39 in August 1968 were studied through September 1971. The coefficients K_2 and K_3 in Eq. (1), giving the minimum root mean square deviation of the data set from predictions, were determined separately for the northern and southern hemispheres. Coefficient K_1 was set equal to +0.1. Atomic oxygen concentrations were determined using the model described above. Then the helium density was determined by subtracting the atomic oxygen density from the total observed density. The resulting values of K_2 and K_3 for each hemisphere are tabulated in the first line of Table 1. The results correspond to the helium residuals shown in Fig. 2. It

Table 1

Coefficients describing north—south asymmetry of exosphere

$$\log_{10}\left[\frac{n(\mathrm{He})_{\mathrm{obs}}}{n(\mathrm{He})_{\mathrm{mod}}}\right] = 0.1 + K_2\delta_P\delta_S + K_3\,|\delta_P\delta_S|$$

Explorer satellites	Time period	Southern hemisphere K_2	K_3	Northern hemisphere K_2	K_3
19, 39	Aug. 68—Sept. 71	−0.6	−0.15	−0.4	−0.05
39	Aug. 68—Sept. 71	−0.5	−0.05	−0.3	−0.05
19	Dec. 63—Sept. 71	−0.6	−0.10	−0.4	+0.05
24	Dec. 64—Aug. 68	−0.5	+0.35	−0.3	+0.40
9	Feb. 61—Sept. 63	−0.7	+0.20	−0.5	+0.20
9, 19, 24, 39	Feb. 61—Sept. 71	−0.6	−0.05	−0.4	+0.10

may be seen (i) that the summer to winter seasonal variation in the southern hemisphere is considerably larger than the seasonal variation in the northern hemisphere, (ii) that helium concentrations are much lower in southern hemisphere summer than in northern hemisphere summer consistent with the recent measurements of Explorer 19, and (iii) that helium concentrations are apparently greater in southern hemisphere winter than in northern hemisphere winter. These three characteristics are also evident in the other data sets studied.

Performing a similar study of the most appropriate values of K_2 and K_3 in the northern and southern hemispheres for all the Explorer 39 data separately gives the results tabulated in the second line of Table 1. Again it is noted that the seasonal variation is greater in the southern hemisphere, and helium concentrations are lower in southern hemisphere local summer than northern hemisphere summer.

A similar study was performed for all the Explorer 19 date from December 1963 through September 1971. Thus data were included through the period of minimum solar activity. The results are tabulated in the third line of Table 1; again the larger seasonal variation in the southern hemisphere and the minimum in southern hemisphere summer is observed.

Next, Explorer 24 results were studied over the period December 1964 through August 1968. The corrections for atomic oxygen for this satellite were much greater than for the other satellites because of its lower altitude and thus the measurements were not nearly as sensitive to helium. The coefficients giving the lowest root mean square deviations are tabulated in Table 1. In Fig. 3, the corresponding seasonal variations are shown. Again, we see a much larger seasonal variation in the southern hemisphere and lowest concentrations in southern hemisphere summer.

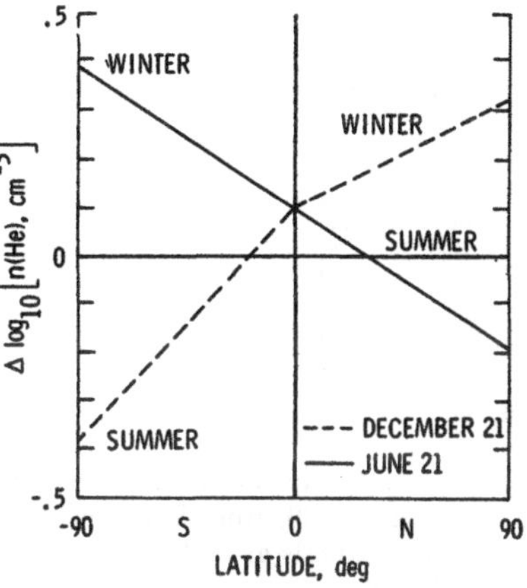

Fig. 2. Seasonal variation of exospheric helium in the northern (N) and southern (S) hemispheres based on Explorer 19 and 39 data from Aug. 1968 through Sept. 1971. Residuals are relative to invariant boundary conditions of USSAS.

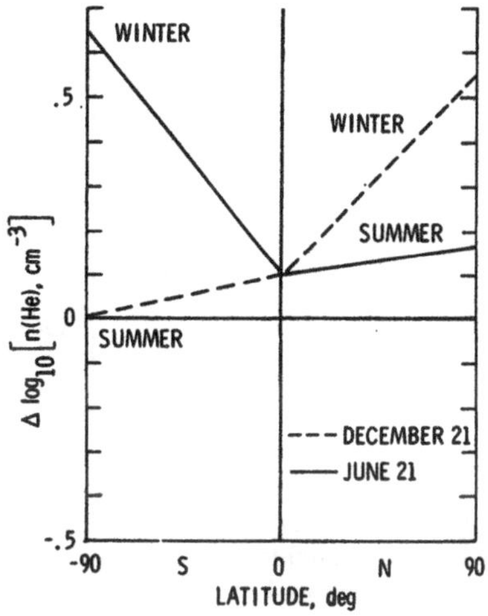

Fig. 3. Seasonal variation of exospheric helium in the northern (N) and southern (S) hemispheres based on Explorer 24 data from Dec. 1964 through Aug. 1968. Residuals are relative to invariant boundary conditions of USSAS.

Finally, Explorer 9 data were studied over the period before minimum solar activity, February 1961 through September 1963. Because of its low inclination, this satellite only obtained measurements to 39° latitude. The resulting coefficients are tabulated in Table 1.

Putting the data from all four satellites together, we obtain the results shown in the bottom of Table 1 for the period February 1961 through September 1971. Referring to Table 1, for each data set the absolute value of K_2 in the southern hemisphere is seen to be consistently 0.2 larger than in the northern hemisphere. This difference is a measure of the asymmetry of the seasonal variation in the two hemispheres and consistently corresponds to an 80% larger seasonal variation at the south pole than at the north pole.

The data were also divided into 10° increments in latitude from +70° to −70° latitude. Increments of 12° were used for +70° to +82° and for −70° to −82°. The results are shown in Fig. 4. Clearly, the seasonal variation is greater in the southern hemisphere and the minimum concentrations are found in the summer southern hemisphere. The individual data points also show that a straight line from zero degrees latitude is a good fit to the data over the whole range of latitudes. It may be seen that setting K_1 equal to 0.1 gives good agreement with observations

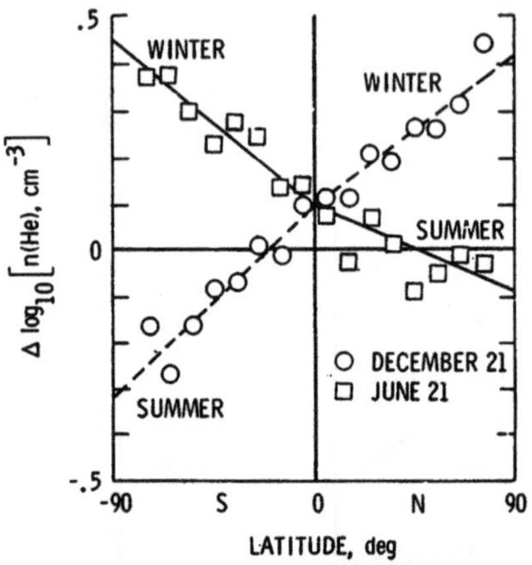

Fig. 4. Seasonal variation of exospheric helium in the northern (N) ans southern (S) hemispheres based on all available helium drag data (Explorers 9, 19, 24 and 39) from Feb. 1961 through Sept. 1971 evaluated in separate 10° latitude increments. Residuals are relative to invariant boundary conditions of USSAS.

This means equatorial helium concentrations exceed the USSAS model by about 23%.

Table 2 gives the corrections to apply to $\log_{10}$ of helium concentrations in USSAS for every month and every 10° of latitude in the northern and southern hemisphere to predict helium drag effects. The correction can be applied at any altitude from 200 km to 1000 km. The table is based on the following expressions for the northern and southern hemispheres:

$$\Delta \log_{10}[n(\mathrm{He})] = 0.1 - 0.4\delta_P\delta_S + 0.10\,|\delta_P\delta_S|, \quad (2)$$

(northern hemisphere)

$$\Delta \log_{10}[n(\mathrm{He})] = 0.1 - 0.6\,\delta_P\delta_S - 0.05\,|\delta_P\delta_S|. \quad (3)$$

(southern hemisphere)

4. Discussion

Referring to Fig. 4, the most prominent feature of the north—south asymmetry in the exosphere is the very low helium concentrations in the southern hemisphere during local summer. Two mechanisms have been proposed which may control the global distribution of helium in the exosphere. One mechanism is a reduction

Table 2

Global distribution of exospheric helium

$\Delta \log_{10} [n(\mathrm{He})]$

Date	Latitude																		
	Southern hemisphere										Northern hemisphere								
	–90	–80	–70	–60	–50	–40	–30	–20	–10	0	10	20	30	40	50	60	70	80	90
Jan. 21	–0.26	–0.22	–0.18	–0.14	–0.10	–0.06	–0.02	0.02	0.06	0.10	0.13	0.16	0.19	0.22	0.25	0.28	0.31	0.34	0.37
Feb. 21	–0.09	–0.07	–0.05	–0.03	–0.01	0.01	0.04	0.06	0.08	0.10	0.12	0.13	0.15	0.17	0.18	0.20	0.22	0.23	0.25
Mar. 21	0.10	0.10	0.10	0.10	0.10	0.10	0.10	0.10	0.10	0.10	0.10	0.10	0.10	0.10	0.10	0.10	0.10	0.10	0.10
Apr. 21	0.28	0.26	0.24	0.22	0.20	0.18	0.16	0.14	0.12	0.10	0.09	0.08	0.07	0.06	0.05	0.04	0.03	0.02	0.00
May 21	0.40	0.37	0.34	0.30	0.27	0.23	0.20	0.17	0.13	0.10	0.08	0.06	0.05	0.03	0.01	–0.01	–0.03	–0.05	–0.07
June 21	0.45	0.41	0.38	0.34	0.30	0.26	0.22	0.18	0.14	0.10	0.08	0.06	0.04	0.01	–0.01	–0.03	–0.05	–0.07	–0.09
July 21	0.41	0.38	0.34	0.31	0.27	0.24	0.20	0.17	0.14	0.10	0.08	0.06	0.04	0.03	0.01	–0.01	–0.03	–0.05	–0.07
Aug. 21	0.29	0.27	0.25	0.22	0.20	0.18	0.16	0.14	0.12	0.10	0.09	0.08	0.07	0.06	0.04	0.03	0.02	0.01	0.00
Sept. 21	0.12	0.11	0.11	0.11	0.11	0.11	0.11	0.10	0.10	0.10	0.10	0.10	0.10	0.10	0.10	0.10	0.09	0.09	0.09
Oct. 21	–0.09	–0.07	–0.05	–0.03	0.00	0.02	0.04	0.06	0.08	0.10	0.12	0.13	0.15	0.16	0.18	0.20	0.21	0.23	0.24
Nov. 21	–0.25	–0.21	–0.17	–0.14	–0.10	–0.06	–0.02	0.02	0.06	0.10	0.13	0.16	0.19	0.22	0.25	0.28	0.31	0.34	0.37
Dec. 21	–0.32	–0.27	–0.23	–0.18	–0.13	–0.08	–0.04	0.01	0.05	0.10	0.14	0.17	0.21	0.24	0.28	0.32	0.35	0.39	0.42

in the turbopause altitude which liberates helium [1, 4, 9] and the second is a seasonal variation of the meridional wind patterns in the lower thermosphere. It has been calculated that a reduction in turbopause altitude cannot by itself produce all of the helium observed in the winter thermosphere and exosphere and thus the meridional circulation mechanism had to be invoked [10—12]. However, a change in turbopause altitude may explain the difference between the summer minimum in the two hemispheres. The systematic difference could be explained if the mean turbopause altitude at high latitudes in the southern hemisphere summer were approximately 4 km higher than in the northern hemisphere summer, due to higher turbulence. If, on the other hand, the difference is assumed to occur because of meridional circulation, there must be greater outflow from the southern hemisphere during local summer than from the northern hemisphere during local summer six months later. Pressure gradients might develop producing such a flow if the average heat input, and hence temperatures, in the southern hemisphere exceeded the input to the northern hemisphere. Results of 6300 Å airglow [13] suggest higher temperatures in the southern hemisphere near the September equinox even though the heating of the two hemispheres due to extreme ultraviolet radiation should be the same. The asymmetry of the magnetic field of the earth may act to produce systematic north—south asymmetries in upper atmospheric heating. Of course, the sun approaches perihelion in January (southern hemisphere summer), but the difference in solar heat input is probably not large enough to produce the large observed effect.

With the lower concentrations of helium in the southern hemisphere summer (December), it would be expected, considering meridional circulation, that the northern hemisphere winter concentrations would be correspondingly greater. However, there is no evidence of this in the data. In fact, all the figures indicate that winter helium concentrations are apparently greater in June than in December. Since most of the helium is concentrated in the winter hemisphere, a decrease in the remaining helium in the summer hemisphere should not substantially increase the winter concentrations, and thus the increase in winter might be disguised in the scatter of the data. Nevertheless, the figures consistently show lower winter and summer concentrations in December than in June. In other words, the figures indicate that the net exospheric helium concentrations in December are less than those in June. This possible annual variation in helium concentrations may be related to the annual component of the semi-annual variation [8].

5. Conclusions

After accounting for all known systematic variations, including the global semi-annual variation, a large, remarkably systematic north—south asymmetry has been discovered in the exospheric helium concentrations. This asymmetry has been successfully traced by re-analysis of data of the four Air Density Explorer drag satellites over the period February 1961 through September 1971, and thus occurs throughout the solar activity cycle. The phenomenon may be indicative of a basic asymmetry in heating of the lower thermosphere causing a meridional outflow from the southern hemisphere in December (local summer) stronger than the corresponding outflow from the northern hemisphere in June (local summer). The asymmetry may also be related to systematic differences in the turbopause altitudes of the two hemispheres.

References

[1] G. M. Keating and E. J. Prior, Space Research VIII, 982 (1968).
[2] J. A. Mullins et al., NASA TN D-3432 (1966).
[3] G. M. Keating et al., NASA TN D-2895 (1965).
[4] G. M. Keating and E. J. Prior, Am. Geophys. Un. 48, No. 1, March 1967.
[6] G. M. Keating, J. A. Mullins and E. J. Prior, Space Research X, 439 (1970).
[6] G. M. Keating, J. A. Mullins and E. J. Prior, Space Research XI, 987 (1971).
[7] G. M. Keating et al., Space Research XII, 765 (1972).
[8] L. G. Jacchia, J. W. Slowey and I. G. Campbell, Smithson. Astrophys. Obs. Spec. Rep. No. 265 (1968).
[9] F. S. Johnson, in: Proc. Symp. on the Years of the Quiet Sun, National Academy of Science 104th Annual Meeting, National Academy of Sciences, Washington, D. C. (1967).
[10] F. S. Johnson, paper presented at 14th General Assembly of the International Union of Geodesy and Geophysics, Switzerland (1967).
[11] R. R. Hodges and F. S. Johnson, J. Geophys. Res. 73, 7307 (1968).
[12] F. S. Johnson and B. Gottlieb, Planet. Space. Sci. 18, 1707 (1970).
[13] J. E. Blamont and J. M. Luton, paper presented at XIVth COSPAR Meeting, Seattle, 1971.

Space Research XIII — Akademie-Verlag, Berlin 1973

DIURNAL VARIATIONS IN THE THERMOSPHERE

M. YA. MAROV and A. M. ALPHEROV
USSR

The results of the analysis of diurnal variations of thermospheric density from the Cosmos drag data for the period of 1960—1971 are discussed taking into account the influence of semi-annual variations. Some peculiarities in periodic density variations earlier revealed and verified by new experimental data require the development of more adequate models.

The diurnal bulge observed experimentally can be represented by means of a spherical harmonic expansion. The global distribution of density obtained for several seasons differs slightly from those recommended by the COSPAR reference atmospheric model, especially for the period of the solstices.

1. Introduction

The determination of density from the satellite drag data continues to remain the most effective method of investigating space—time variations of the upper atmosphere, and the values obtained are used with certain simplifying assumptions, for outlining models of the upper atmosphere [1, 2]. Diurnal and seasonal—latitudinal variations of density and temperature (the form and orientation of the diurnal bulge relative to the subsolar point) have been identified. Several conclusions made from the drag data [3—7] and the recently obtained results by means of a ground-based incoherent scatter radar [8] and satellite measurements of the Doppler broadening of the $\lambda = 6300$ Å line of atomic oxygen [9], indicate strong deviations of the observed diurnal effect (its amplitude, phase and seasonal—latitudinal change) compared with its model representation.

Errors in determining diurnal variations from satellite data can be due to averaging the atmospheric density in the vicinity of perigee [10] and can be introduced in the analytical form of the diurnal bulge. Other variations (semi-annual and those associated with solar and geomagnetic activity) lead to a more representative analytical expression for the diurnal bulge.

In this paper an attempt is made to present the diurnal variations deduced from satellite drag by expansion in spherical harmonics. In spite of some preliminary assumptions the model seems to deserve attention because it indicates a reasonable way of modifying the simpler empirical models of the diurnal effect.

2. Experimental data and Model Representation

Included in this paper are atmospheric density data resulting from processing of the orbital parameters of 12 new satellites launched during 1970—1971:

Cosmos 334 (1970-33 A), Cosmos 347 (1970-43 A),
Cosmos 357 (1970-63 A), Cosmos 362 (1970-73 A),
Cosmos 369 (1970-81 A), Cosmos 378 (1970-97 A),
Cosmos 388 (1970-112 A), Cosmos 393 (1971-7 A),
Cosmos 408 (1971-37 A), Cosmos 421 (1971-44 A),
Cosmos 423 (1971-47 A), Cosmos 435 (1971-72 A).

Excepting Cosmos 347 all satellites had an orbital inclination greater than 70°.

Calculation of density ϱ and fitting its values to the fixed height Z, was made by a routine procedure described earlier [3—5]. From experimental data the values of maximum day and minimum night densities were selected under following assumptions:

$$\varrho = \varrho_{\max} \text{ if } |\varphi - \delta_\odot| < 10^\circ, \qquad |t - 14^{\mathrm{h}}| < 2^{\mathrm{h}}$$
$$\varrho = \varrho_{\min} \text{ if } |\varphi + \delta_\odot| < 10^\circ, \qquad |t - 3^{\mathrm{h}}| < 3^{\mathrm{h}}$$

where φ is the geographical latitude, t local time, $\delta_\odot$ solar declination. When using condition (1) for the period 1960—1971 the maximum and minimum density distributions in the height range 180—300 km were approximated in the following relations:

$$\log \varrho_{\max} = \sum_{i=0}^{3} A_i(z - 270)^i + S(z, \gamma) + A_4(z - 270)/\Phi(\bar{F}, F, \Sigma Kp)$$
$$\log \varrho_{\min} = \sum_{i=0}^{3} A_i(z - 270)^i + S(z, \gamma) + A_4(z - 270)/f(\bar{F}, F, \Sigma Kp) \qquad (2)$$

where

$$\Phi(\bar{F}, F, \Sigma Kp) = A_5 + A_6(\bar{F} - 150) + A_7(F - \bar{F}) + Kp(A_8 + A_9 \sin|\varphi|)$$
$$f(\bar{F}, F, \Sigma Kp) = a_1 + a_2(\bar{F} - 150) + a_3(F - \bar{F}) + Kp(a_4 + a_5 \sin|\varphi|) \qquad (3)$$

$$S(z, \gamma) = g(\gamma) \sum_{i=10}^{12} A_i(z - 270)^{i-10}. \qquad (4)$$

F is the value of 10.7 cm solar flux (in units of 10^{-22} W m^{-2}H z^{-1}) one day before the day of measuring the density; $\bar{F}$ is the same flux averaged for the 3-month period; ΣKp the diurnal sum of planetary 3-hour Kp indices of geomagnetic disturbance; γ real (calendar) time; $S(z, \gamma)$ corrections to the density to allow for the semi-annual effect, where the function $g(\gamma)$, showing semi-annual variations at height $z = 270$ km was used according to the data published in [6]. The coefficients A_i ($i = 0, \ldots 12$), a_j ($j = 1, \ldots 5$) were defined by the method of least squares.

Assuming that at 200—300 km height semi-annual density variations are independent of latitude and local time, and a diurnal bulge has the same configuration, current values of density were approximated as:

$$\varrho(z, \varphi, t, \bar{F}, \Sigma Kp) = \varrho_{\min}(\varrho_{\max}/\varrho_{\min})^{q(t, \varphi)} \qquad (5)$$

The function $q(t, \varphi)$ was taken as a spherical harmonic expansion including fourth order terms

$$q(t, \varphi) = (\log_{10} \varrho - \log_{10} \varrho_{min})/(\log_{10} \varrho_{max} - \log_{10} \varrho_{min})$$

$$q(t, \varphi) = \sum_{n=0}^{4} \sum_{m=0}^{4} (g_n{}^m \cos mt + h_n{}^m \sin mt) \cdot \mathrm{P}_n{}^m (\sin \varphi) \tag{6}$$

where $\mathrm{P}_n{}^m (\sin \varphi)$ is Legendre's function.

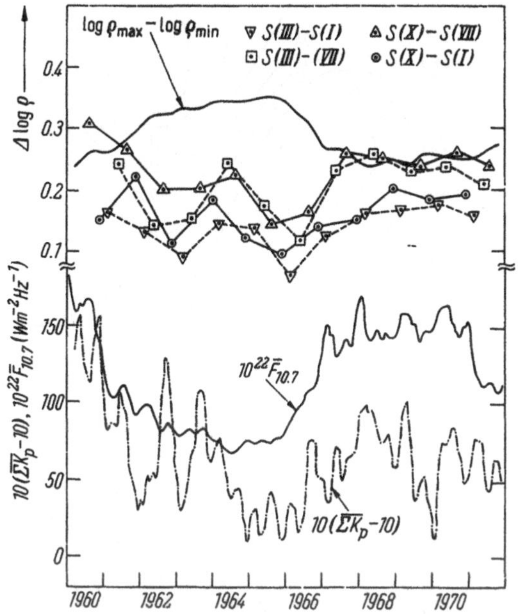

Fig. 1. Diurnal and semi-annual variations at 270 km height deduced from Cosmos drag data for the period of 1960—1971. The diurnal density ratio is presented as the solid line in the upper part and labelled $\log_{10} \rho_{max} - \log_{10} \rho_{min}$. Semi-annual variations are given as the difference of $S(\gamma)$ functions in neighbouring extrema, I to XII indicate the months. Average solar flux and Kp variations are also shown.

3. Discussion of Results

Fig. 1 shows changes of the diurnal and semi-annual variations observed in the period 1960—1971 at a height of 270 km. The amplitude of the diurnal variations is expressed as the difference of the logarithms of maximum day and minimum night densities according to the condition (1). Semi-annual variations are given as the difference of values of the $S(\gamma)$ function in neighbouring extrema. It follows from Fig. 1 that during the 11-year period at 270 km height the coefficient characterizing the amplitude of diurnal variations changed from 1.75 to 2.2. The model data [2] agree well with these values. The amplitude of semi-annual variation sfor this period changes from 1.2 to 1.8. If model values agree sufficiently with the mean coefficient of the amplitude of the semi-annual variation, 1.45, then the secular term observed is not exhibited in the model.

Coefficients of expansion (6) for the summer solstice are given in Table 1. Fig. 2a shows in a cylindrical Mercator projection the global density distribution for values of the function $q(t, \varphi)$ from $q = 0.0$ to $q = 1.0$ in 0.1 steps. For comparison, in Fig. 2b isolines of the function $q(t, \varphi)$ defined with help of model [2] are given. The distribution obtained differs from the model [2] by an asym-

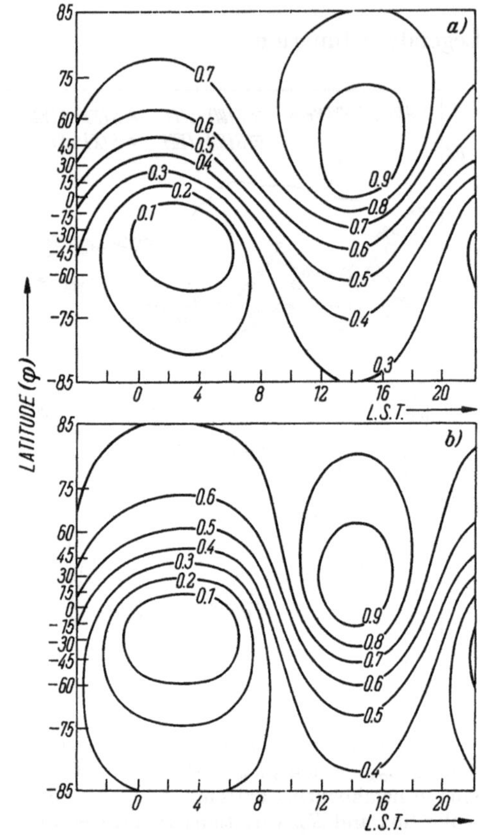

Fig. 2. Global density distribution for the time of summer solstice ($\delta_{\odot} = 23.44°$), in a cylindical Mercator projection: (a) according to the Cosmos drag data, (b) calculated from the Jacchia's 1971 model [2].

metry of 3 hours and 14.3 hours of local solar time. This leads to the idea that during the summer solstice the minimum density can be from 02 h at latitudes $\varphi > 70°$ to 05 h at latitudes $\varphi < -70°$. The time of maximum density varies from 16 h at latitudes $\varphi > 70°$ to 13 h at latitudes $\varphi < -70°$.

It should be noticed that the results obtained can depend slightly on our initial assumption (1). However, the data of incoherent scatter [9], excluding disagreement in phase, agree better with the given distribution.

Table 1

Coefficients of expansion of the $q(t, \varphi)$ in spherical functions for conditions of summer solstice ($\delta_{\odot} = 23.44°$)

n	m	g_n^m	h_n^m
0	0	5.071 E-1	0
1	0	2.857 E-1	0
1	1	−3.459 E-1	−1.693 E-1
2	0	1.139 E-2	0
2	1	4.781 E-3	1.158 E-2
2	2	1.254 E-3	−2.591 E-3
3	0	−5.820 E-2	0
3	1	1.316 E-3	−1.165 E-2
3	2	−8.672 E-3	4.255 E-3
3	3	1.277 E-3	−8.093 E-4
4	0	4.855 E-3	0
4	1	1.521 E-3	9.955 E-4
4	2	6.135 E-4	3.974 E-4
4	3	−3.220 E-5	−1.259 E-4
4	4	−7.891 E-5	4.186 E-5

4. Conclusion

The results obtained indicate that maximum and minimum atmospheric densities apparently depend on latitude and season. The model [2] does not show this effect, probably as a consequence of the initial assumptions.

References

[1] CIRA-1965, COSPAR International Reference Atmosphere, North-Holland Publ. Co., Amsterdam 1965.
[2] L. G. Jacchia, Smithson, Astrophys. Obs. Spec. Rep. No. 332 (1971).
[3] M. Ya. Marov, Space Research VIII, 811 (1968).
[4] M. Ya. Marov and A. M. Alpherov, Space Research IX, 497 (1969).
[5] M. Ya. Marov and A. M. Alpherov. Space Research X, 419 (1970).
[6] B. K. Ching, Space Research XII, 841 (1972).
[7] R. R. Allan, Royal Aircraft Establishment Tech. Rep. No. 71167 (1971).
[8] P. Waldteufel and L. Cogger, J. Geophys. Res. 76, 5322 (1971).
[9] J. E. Blamont and J. M. Luton, Paper presented at XIV COSPAR Meeting, Seattle 1971.
[10] M. I. Voiskovsky, B. V. Kugaenko, V. M. Synitsyn and P. E. Elyasberg, Space Research XI, 953 (1971).

Space Research XIII — Akademie-Verlag, Berlin 1973

A STUDY OF THE VARIATIONS IN THE THERMOSPHERE RELATED TO SOLAR ACTIVITY

L. G. JACCHIA and J. W. SLOWEY

Smithsonian Astrophysical Observatory, Cambridge, Mass., USA

Densities derived from the drag of five artificial satellites and covering the time interval from 1958 to 1971 have been used to study the variations in the thermosphere arising from changes in solar activity. Exospheric temperatures derived from the densities by means of atmospheric models have been related to the 10.7-cm solar flux F, in which the active-area component had been separated from the slow-varying clear-disk component by means of a relation between the flux and sunspot areas. When the day-to-day ("27-day") variation is suppressed, the remaining slow temperature variation appears to be linearly related to the clear-disk component F_0. If we also relate linearly the day-to-day temperature variations with the day-to-day variations of the flux, we find that the regression coefficient changes from 1 K (W m^{-2} Hz^{-1})$^{-1}$ at sunspot maximum to 2 K at low solar activity; a nearly constant coefficient is obtained by relating the temperature variations to the ratio of the active-area component to the clear-disk component of the flux. A very satisfactory fit to the observations is obtained by an equation using only F and F_0 as independent variables.

The time lag of the atmospheric temperature variations behind those of the solar flux F is investigated and found to be 1.14 ± 0.06 days.

1. Solar XUV and Thermospheric Variations

It is well known that variations in the solar XUV radiation are the cause of large temperature variations in the thermosphere and exosphere. Intensity variations of selected regions and emission lines in the solar XUV spectrum integrated over the solar disk have occasionally been measured by means of satellite-borne spectrometers (e.g. on OSO 1 [1], on OSO 3 [2], on OSO 4 [3], and on OSO 6 [4]). Although essential for the interpretation of atmospheric variations, such measurements have, so far, been too selective and limited in time to establish meaningful correlations to be used in atmospheric models. If they are to be used for practical purposes, atmospheric models must make use of an index of solar activity that is monitored without interruption and is made readily available, provided this index shows a fair degree of correlation with atmospheric temperatures. This is why the solar flux at 2800 MHz (10.7 cm) has been widely used in atmospheric models. The recent publication by the Ionosphere Research Laboratory of the Pennsylvania State University of Ca II plage indexes going back to 1 January 1958 [5] makes it now possible to use this parameter also, which logically should correlate better with thermospheric temperatures.

The higher the excitation potential of a solar XUV line, the more restricted is the area of the solar disk above which it is formed. Lower excitation lines such as H I Lyman α (1216 Å), He I (584 Å), and He II (304 Å) originate over the entirety of the solar disk, with only a moderate enhancement in active areas. High-excitation lines such as Fe XV (284 Å) or Fe XVI (335 Å), on the other hand, are practically absent over quiet regions of the sun and show up strongly only in active areas. As a consequence, the variation during one solar rotation of the intensity of a particular line integrated over the whole disk of the sun is different according to the excitation potential of the line. As observed by the OSO 1 satellite in 1961, the intensity of He II varied typically by 25%, while that of Fe XV (284 Å) in the same interval changed by a factor of 4 or 5 [1]. Also, since the area and intensity of active regions varies greatly with the solar cycle, the relative contribution of individual XUV lines and continuum regions to the integrated XUV radiation is vastly different at different phases of the cycle. For these reasons, when we use a single solar-activity index, such as the 10.7-cm flux, to find correlations with exospheric temperatures, we must expect to find different correlations according to whether we analyze variations with the solar cycle or within a solar rotation — and even within a solar rotation we can expect the correlation to change with the phase of the solar cycle.

2. Correlations with the 10.7-cm Solar Flux

The global night-time minimum T_0 of the exospheric temperature can be related to the 10.7-cm solar flux F by means of a formula [6] of the type

$$T_0 = a + b\bar{F} + c(F - \bar{F}). \tag{1}$$

In this formula $\bar{F}$ is a smoothed value of the solar flux, in which the "27-day" oscillations of the daily values of the flux F have been suppressed. To obtain $\bar{F}$, we had resorted in the past to drawing by hand a smooth curve through the monthly means of F. To achieve a greater measure of objectivity, it has been variously recommended to take for $\bar{F}$ the mean of F over three, four, or five solar rotations. Recent comparisons have shown that our $\bar{F}$ curve is best approximated by the mean of F over *six* solar rotations; by taking a smaller number of rotations, undesirable short time fluctuations are occasionally still left in the curve.

The term $b\bar{F}$ In eq. (1) should allow for the slow variation connected with the solar cycle, while the last term, $c(F - \bar{F})$, represents the day-to-day variations related to changes in the structure of active areas and to solar rotation. It is questionable whether $\bar{F}$ is the best indicator of the slow, 11-year component. Off hand, it would seem more appropriate to use for this purpose the clear-disk component F_0, which is obtained by subtracting from F the contribution from active areas. Hachenberg [7] derived F_0 using correlations between F and the sunspot number R in half-year intervals and found it to vary by a factor of 2 during one solar cycle (1954—1964). We found that the relation between F and R is not quite linear and preferred to use the Greenwich sunspot areas A. The relation between F and A is quite linear, and we derived values of F_0 from overlapping 150-day intervals in the years 1957—1965. The relation between F and A is shown in Fig. 1 for three such intervals, at different phases of the sunspot cycle. We are currently re-determining F_0 using the Ca II plage indexes, which became available only after the present analysis had been done; we do not expect, however, that the results will be significantly different.

A plot of F_0 against $\bar{F}$ shows that the relation between these two quantities is linear for all practical purposes (Fig. 2). This is a very fortunate circumstance, because it makes it possible to compute F_0 directly from $\bar{F}$, which is by far easier to obtain from the observed data. A least-squares solution gives the following relation:

$$F_0 = 30.3 + 0.562\bar{F}. \tag{2}$$

Here F, and hence $\bar{F}$ and F_0, is expressed in units of 10^{-22} W m^{-2} Hz^{-1} bandwidth.

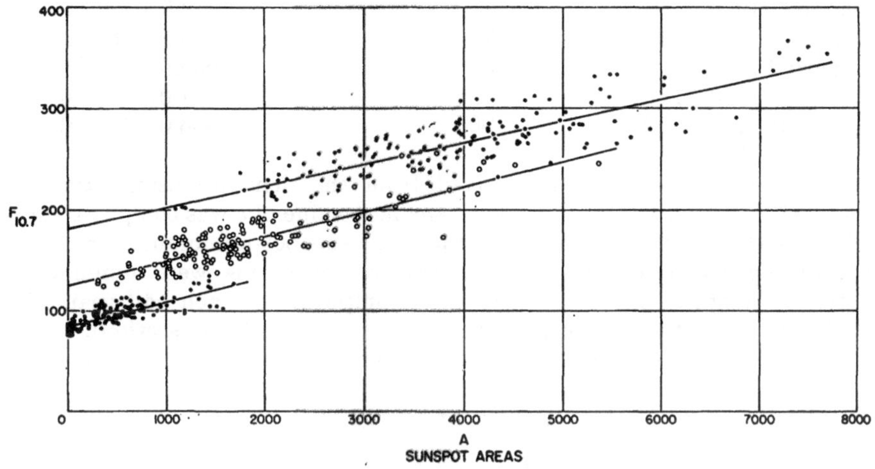

Fig. 1. Relation between the Greenwich sunspot areas A and the 10.7-cm solar flux F during three 150-day intervals at different phases of the solar cycle. The individual values of F have been corrected for the variation of $\bar{F}$ relative to the average value for the particular interval using preliminary values in the relationship between F_0 and F. Data are from 28 August 1957 through 25 January 1958 (upper plot), 4 April 1960 through 1 September 1960 (middle plot), and 17 August 1961 through 14 January 1962 (lower plot). The intercept of the regression lines determines the clear-disk component F_0 of the flux.

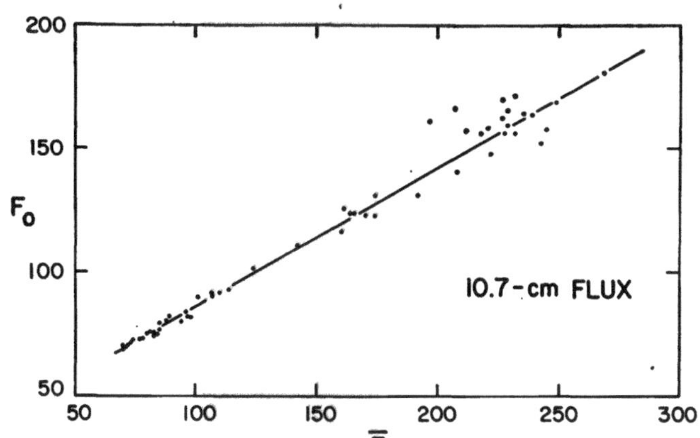

Fig. 2. Relation between $\bar{F}$, the 10.7-cm solar flux averaged over six solar rotations, and F_0, the clear-disk component of F.

Values ranging from 1.3° to 1.9° K(W m^{-2} Hz^{-1}) have been variously derived for the coefficient c in Eq. (1). To check on a possible variation of c with the solar cycle, we have determined its value individually within half-year intervals from density data derived from the drag of the following five satellites:

Satellite	Effective height (km)	Time interval
Explorer 1 (1958 α)	370	1958.5—1970.0
Vanguard 1 (1958 β2)	710	1958.5—1961.5
Vanguard 2 (1959 α1)	594	1959.5—1971.0
Explorer 8 (1960 ξ1)	455	1961.0—1971.0
Explorer 19 (1963-53A)	900	1966.0—1971.0

Exospheric temperatures were derived from the densities using atmospheric models [8]. For each half-year interval, the best value of c was computed for each satellite by the minimum-variance method; the sunspot-minimum years 1964 and 1965, during which day-to-day temperature variations were minimal, were omitted. A weighted mean of c was then computed for each half-year interval and plotted against the corresponding average $\bar{F}$ (Fig. 3, top section). From the plot it is evident that c is larger at sunspot minimum than at sunspot maximum:

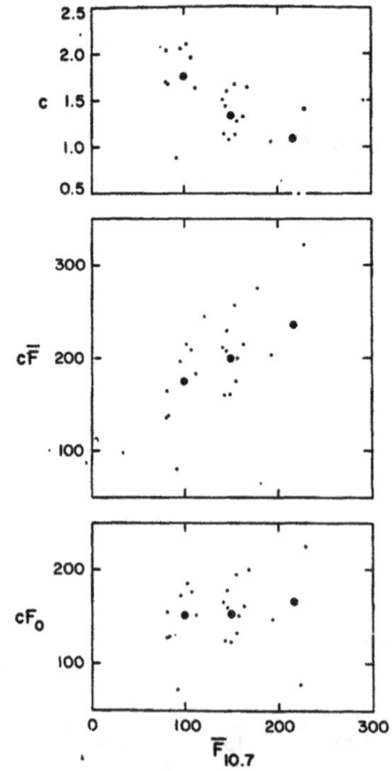

Fig. 3. Values of c, the coefficient of $F-\bar{F}$ in Eq. (1), are plotted against $\bar{F}$ in the top section. In the middle and bottom sections, the products $c\bar{F}$ and cF_0, respectively, are plotted against $\bar{F}$. The large filled circles represent normal points.

The range of the variation amounts roughly to a factor of 2 (the large filled circles in the plots represent normal points). In the bottom section of Fig. 3, we have plotted the product cF_0 against $\bar{F}$. As can be seen, this product seems to be constant; on the other hand, the product $c\bar{F}$ (middle section of Fig. 3) shows a dependence on $\bar{F}$. The weighted mean of cF_0 is 154 ± 6 (SD).

From the preceding considerations, it appears that the form of the last term of Eq. (1) should be changed to $c'(F - \bar{F})/F_0$. At this point, the question arises whether it would not be more convenient to drop $\bar{F}$ altogether in Eq. (1) in favor of F_0, to make it read

$$T_0 = a' + b'F_0 + c' \frac{F - F_0}{F_0}. \tag{3}$$

The last term of this equation represents the relative contribution of the sunspot (or active-area) component of F; since we always have $F \geqq F_0$, the term must always be positive and vanish only for $F = F_0$ — which could presumably occur only at the very minimum of solar activity. In this respect, this term behaves quite differently from the last term of Eq. (1), which can be both positive and negative and vanishes when $F = \bar{F}$.

Making $F = F$ we obtain $\bar{T}_0$, a temperature in which the 27-day fluctuation has been suppressed. From 88 values of $\bar{T}_0$ and $\bar{T}_M$ (the daytime maximum when $F = \bar{F}$) determined from the drag of three satellites, with the constraint $\bar{T}_M/\bar{T}_0 = \text{const} = 1 + R$, and assuming for c' the value found above, we obtained by least squares numerical values for a', b', and R, with the following result (degrees Kelvin, standard errors):

$$T_0 = \underset{\pm 10}{272} + \underset{\pm 0.10}{4.78}F_0 + \underset{\pm 6}{154} \frac{F - F_0}{F_0}; \qquad R = \underset{\pm 0.008}{0.315}. \tag{4}$$

Owing to the presence of an absolute term in the linear relation (Eq. (2)) between $\bar{F}$ and F_0, we find that $\bar{T}_0$, when computed from Eq. (4), is a nonlinear function of $\bar{F}$. This can be seen at a glance in Fig. 4, where we have drawn as solid lines the curves of $\bar{T}_0$ and $\bar{T}_M$ computed from Eq. (4); the data points are the observed values of T_0 and T_M. The straight broken lines represent $\bar{T}_0$ and $\bar{T}_M$ as computed from

$$\bar{T}_0 = \underset{\pm 7}{392} + \underset{\pm 0.05}{3.15}\bar{F}; \qquad R = \underset{\pm 0.008}{0.315}. \tag{5}$$

The numerical coefficients in Eq. (5) were obtained by least squares, again with the constraint $T_M/T_0 = 1 + R$. As can be seen, the solid curves fit the observations as well as, if not better than, the broken lines. The standard deviation of an observation is 24.1° from Eq. (4) and 24.5° from Eq. (5).

3. Time Lag

From the same observational material we used to determine the variations of the coefficient c, we have derived, again by the minimum-variance method, the time lag of the atmospheric variations behind those of F. Assuming that the

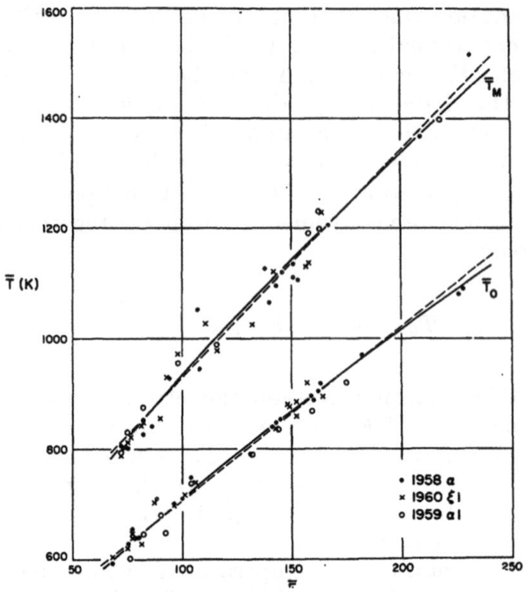

Fig. 4. Global nighttime exospheric-temperature minima $\overline{T}_0$ and daytime maxima $\overline{T}_M$ determined from three satellites, compared with Eq. (4) (solid curves) and Eq. (5) (broken straight lines).

average time of observation of the 2800-MHz solar flux at Ottawa is 1200 EST (= 1700 UT), we obtain for the lag the value

$$\Delta t = 1.14 \pm 0.06 \text{ (SD) days}.$$

This compares with the value $\Delta t = 1.0 \pm 0.12$ days determined by Roemer [9].

Acknowledgment

This work was supported in part by grant NGR 98-015-002 from the National Aeronautics and Space Administration.

References

[1] W. M. NEUPERT, W. E. BEHRING and J. C. LINDSAY, Space Research IV, 719 (1964).
[2] L. A. HALL and H. E. HINTEREGGER, J. Geophys. Res. 75, 6959 (1970).
[3] A. F. TIMOTHY and J. G. TIMOTHY, J. Geophys. Res. 75, 6950 (1970).
[4] B. E. WOODGATE, D. E. KNIGHT, R. URIBE, P. SHEATHER and J. BOWLES, Proc. Roy. Soc. 332, 291 (1973).
[5] W. E. SWARTZ and R. OVERBECK, Ionosphere Research Laboratory, The Pennsylvania State University, Sci. Rep. 373 (E) (1971).
[6] L. G. JACCHIA, Smithson. Contr. Astrophys. 8, 215 (1965).
[7] O. HACHENBERG, in: Solar System Radio Astronomy, Plenum Press, New York 1965 (p. 95).
[8] L. G. JACCHIA, Smithson. Astrophys. Obs. Spec. Rep. No. 332 (1971).
[9] M. ROEMER, Space Research VII, 1091 (1967).

Space Research XIII — Akademie-Verlag, Berlin 1973

STRUCTURE OF THE NEUTRAL ATMOSPHERE BETWEEN 150 AND 500 km

F. BARLIER[a], J. L. FALIN[b], M. ILL[c], and C. JAECK[a]

[a]Observatoire de Paris, Meudon, France
[b]Observatoire de Besançon, France
[c]Baja Observatory, Hungary, presently at University of Besançon, France

In the present paper we give preliminary results concerning the amplitude and the form of the semi-annual effect and a global analysis of the diurnal variation of density as a function of latitude. These results are obtained by a statistical method on the basis of 12000 values of density, deduced from drag data of 87 satellites for a time period ranging from 1967 to 1970.

1. Introduction

The aim of an atmospheric model is to give the best possible estimation of different parameters of the upper atmosphere as a function of time and space. However, a model must be simple with a limited number of tables permitting easy use. Consequently, a model is only a first approach to the real conditions and there exist residuals between the real atmosphere and a given model. Analysis of the densities given by the model Jacchia 71 shows that the residuals rarely exceed 20%, but they have not a completely random character [2]. The analysis of these residuals is possible on the basis of a large number of values, using statistical methods. In the present paper we give the first details of an analysis based on 12000 independent density values, deduced from the orbital analysis of 87 satellites for the time period ranging from 1967 to 1970. The results presented are the preliminary part of a more general analysis.

2. Method

2.1. Determination of Density

We used orbital elements published by GSFC, SAO and CNES. It has been demonstrated by different authors that the determination of density values is accurate to within about 10% [2, 3].

To determine the density we used a method based on the numerical integration of Lagrange's equation. Details have been published several times, now we recall only that we calculate from each set of orbital elements a density value at perigee, corresponding to a given date and to a point in space. These densities we call "observed". Then we calculate also the ratio f of the observed density ϱ_0 and the corresponding value of a model ϱ_J. In the present work we use the model J 71 [1].

2.2. Method of Analysis

Let us take into consideration the 12000 calculated values of f as a function of two parameters, e.g. local solar time and latitude of perigee. Let us divide the first parameter into n intervals, the second into m intervals. In this manner we obtain $n \times m$ "boxes". The division of each parameter into intervals is made in such a manner that we obtain 50 to 150 individual values of f in each box. The arithmetic mean of the values of f in any one box represents the correction to be applied to the model density corresponding to the box in question. It follows that each box gives one point in our analysis. The standard deviations of the points obtained in such a way vary between 2% and 6% depending on the number of individual values f in a box. With a large number of values in each box we can validly suppose that the distribution of the values of f is quite uniform in relation to the height of perigee and sufficiently homogeneous with respect to solar activity.

3. Results

3.1. The Semi-annual Effect

Different authors have established that the amplitude of the semi-annual effect is a function of height [1, 4, 5]. Therefore we show the semi-annual effect for five different heights (Fig. 1).

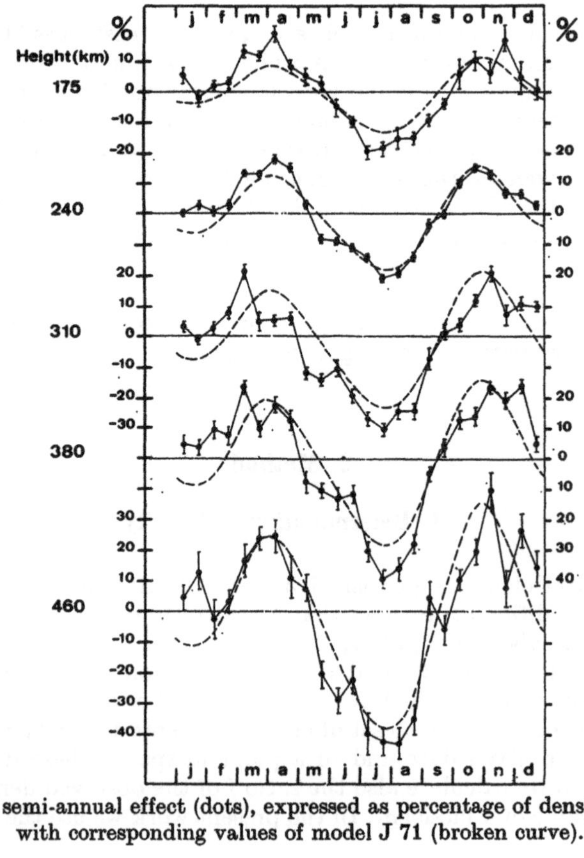

Fig. 1. Observed semi-annual effect (dots), expressed as percentage of density and compared with corresponding values of model J 71 (broken curve).

On each height we accepted as unity the density ϱ_j' given by J 71 without the semi-annual effect. In this case the semi-annual effect can be expressed as a percentage x_j or x_0 of this "unit density". In Fig. 1 the value x_j of the model is plotted as a broken line while the value x_0 of our study is represented by dots with standard deviations indicated by vertical bars. In our analysis the values of four years are mixed, thus the curve obtained is the mean effect of the years 1967—1970, i.e. it corresponds to a sufficiently stable 10.7 cm solar flux with a mean value of 150 (in units of 10^{-22} W m^{-1} Hz^{-1}).

On examining the five curves we find general accordance with J 71. Nevertheless, one can establish systematic differences reaching 15%, corresponding to about 40% of the semi-annual effect. Another feature is an asymmetry of the semi-annual effect, observed at each height: the minimum in January is less pronounced while the minimum of July is deeper than the corresponding value of J 71, but the dates of maxima and minima are in accordance with the model.

It is interesting to establish that the curves representing the semi-annual effect are not smoothed lines, but present secondary minima which perhaps can be considered as quasi-periodic variations with a period of about 2 months.

3.2. Diurnal and Latitudinal Variations

In a preliminary study of the ratios f as a function of local time and latitude for different heights we obtained curves of identical form and we present here only our analysis of the diurnal variation for a mean height of 280 km. We studied the effect for three periods of the year, centred around the following declinations of the sun: $\delta_\odot = -20°, 0°, +20°$.

To examine the diurnal variation we introduced the ratio B as the density at a given point divided by the night-time minimum density of J 71. In this manner we have:

$$B_{\text{obs}} = \text{density}_{\text{obs}}/\text{density}_{\text{min}\cdot J71}$$

and

$$B_{J71} = \text{density}_{J71}/\text{density}_{\text{min}\,J71}.$$

In our statistics we analysed the values of B in latitude intervals of 30° and in local time intervals of 2 hours. The obtained results are presented in the form of B contour maps as a function of latitude and local time, each map being the result of about 4000 values (Figs. 2, 3, 4).

Although it may be established that the model J 71 is a good representation of the real conditions, there are significant residuals. On analysing the observed maps we find the following general phenomena:

(i) The densities at the poles are generally 15% to 20% less than predicted by the model.

Let us examine the different polar values of B, given in the following table:

	Northern Hemisphere			Southern Hemisphere		
	Obs	J 71	$O-C$	Obs	J 71	$O-C$
$\delta = -20°$	1.10	1.26	−0.16 winter N	1.50	1.57	−0.07 summer S
$\delta = +20°$	1.30	1.57	−0.27 summer N	1.08	1.26	−0.18 winter S

It is an interesting feature that we have for the winter poles the same values of $O - C$ (observed — calculated), while the $O - C$ observed on the N pole is four times larger than the corresponding value on the S pole.

(ii) The time of maximum density lies between 13 h and 15 h local time.

(iii) The latitude of maximum density is about 0° for $\delta_\odot = 0°$. For the map corresponding to $\delta_\odot = +20°$, this latitude is slightly positive, but below $+10°$. In the case of $\delta_\odot = -20°$ the latitude of maximum density lies between $-10°$

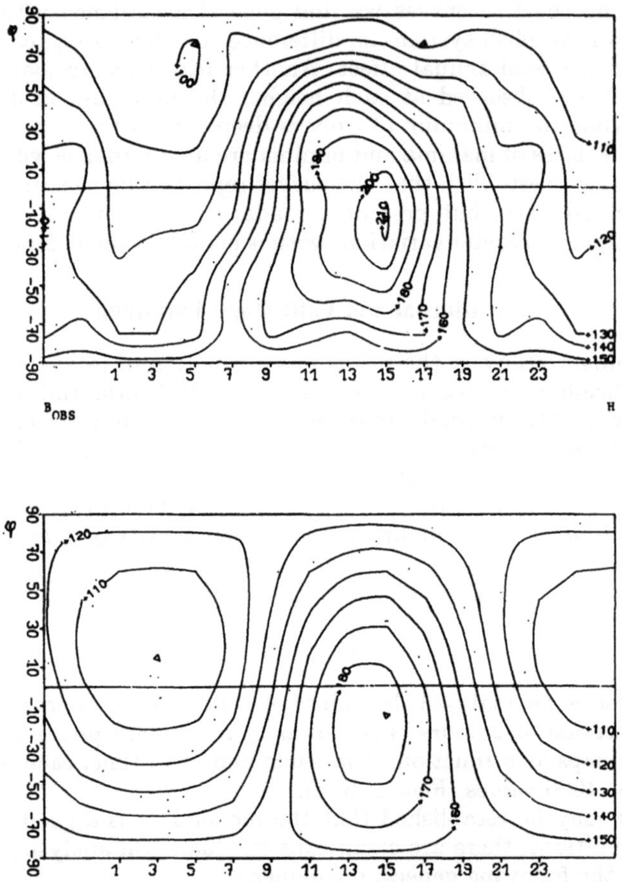

Fig. 2. Values of B (see text) as function of local solar time and latitude for the time period centred around $\delta_\odot = -20°$.

and $-20°$. Thus, we can establish that the migration in latitude of the point of maximum density has a smaller amplitude than predicted by the model and it is centred on a southern latitude.

(iv) On considering the area limited by the curves $B = 1.0$ and $B = 1.20$ as the region of minimum density, we can establish that the observed minimum region is much more extended in latitude than in J 71. In local time the minimum is centred on the same value as J 71 and it has the same width.

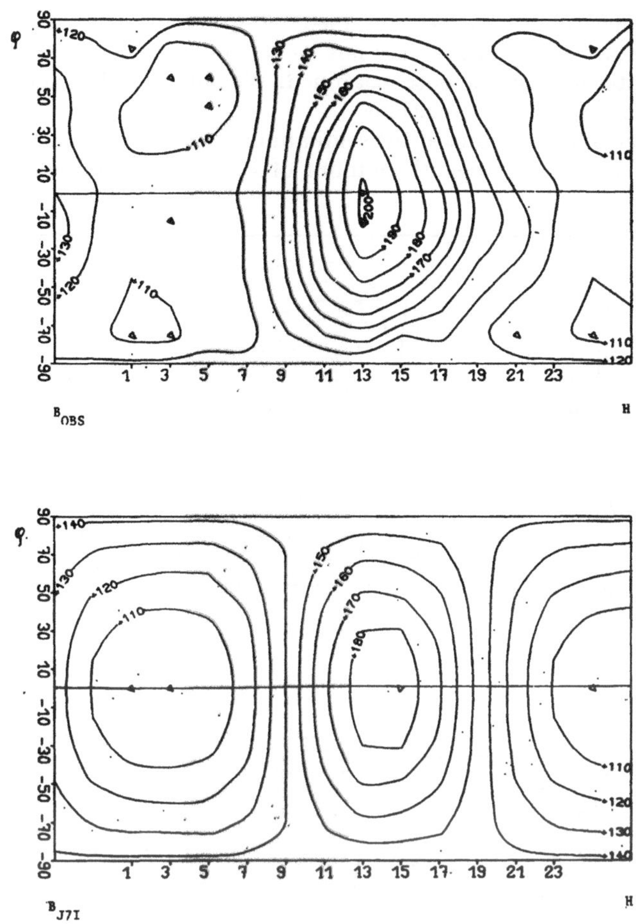

Fig. 3. B (as Fig. 2) for $\delta_{\odot} = 0°$.

(v) The model J 71 provides for a migration in latitude of the point of minimum density in opposition with the sun's declination, i.e. an amplitude of $\Delta \varphi = \pm 23°$, while we observed a much larger amplitude of at least $\Delta \varphi = \pm 50°$.

(vi) The observed amplitude of the diurnal variation is greatly the same as that of J 71, but we have generally a larger amplitude, the difference being about 10%.

4. Discussion and Conclusion

Some of the statistical studies have been repeated with 8000 values of f instead of 12000, but the results obtained were always within the range of standard deviation of the first analysis. Therefore, we concluded that with 12000 density values we have a significant number in the statistical sense and so we can suppose that the observed phenomena are real features of the atmosphere. On the other hand, we consider as arguments for the significance of our results their general accordance with the model, the regular character of the contour maps obtained, and the coherence of the curves obtained independently of each other. Nevertheless, the results of our study may not be generalized since they are valid only for the time intervals and heights examined. We consider that they describe the mean behaviour of density variations.

The residuals of the minima of the semi-annual effect underline the asymmetric character of this effect between the minima of July and the minima of January.

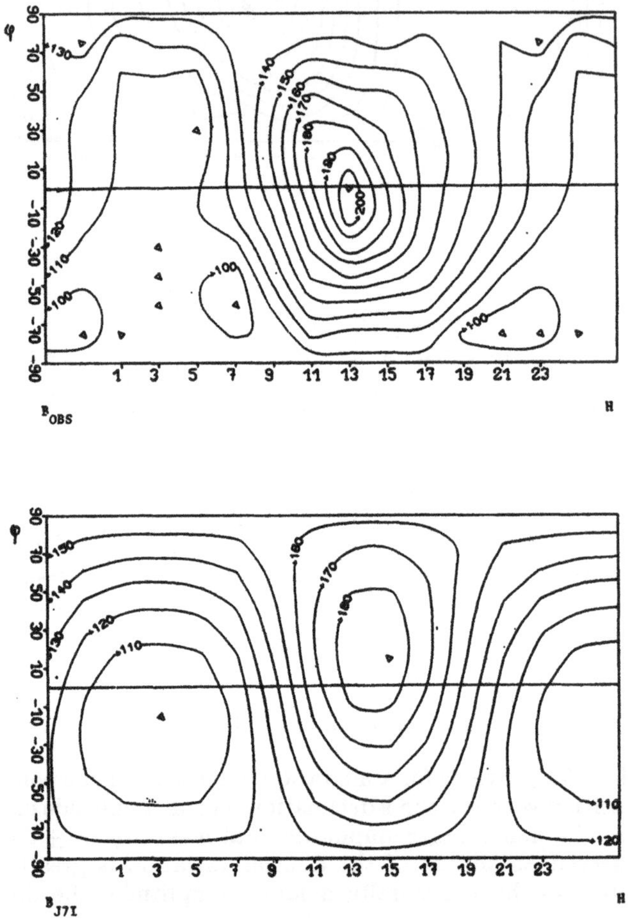

Fig. 4. B (as Fig. 2) for $\delta_{\odot} = +20°$.

We found secondary minima by a statistical method, using mean values of a large number of satellites. On the basis of completely different observational material and analysing one by one the curves of a few satellites, we have already remarked the existence of such secondary minima [6, 7], a part of them being identical with the secondary minima detected by King-Hele et al. [3, 8]. Unfortunately, at present we have no complete explanation for these phenomena, nor for the migration in latitude of the points of maximum and minimum density and the asymmetric behaviour at the summer poles.

It is perhaps surprising to find systematically much lower density values at the poles than those of J 71, but we remark that Marcos and Champion [9, 10], using micro-accelerometric measurements, obtained density values at the poles 20% lower than predicted by J 71 at a height of 220 km.

References

[1] L. G. Jacchia, Smithson. Astrophys. Obs. Spec. Rep. No. 332 (1971).
[2] J. P. Lespes et al., Space Research X, 505 (1970).
[3] D. G. King-Hele et al., Planet. Space Sci. **6**, 675 (1968).
[4] G. E. Cook, Annls Géophys. **25**, 451 (1969).
[5] C. Wulf-Mathies, Space Research XI, 981 (1971).
[6] M. Ill et al., Space Research X, 485 (1970).
[7] F. Barlier and M. Ill, C. R. Acad. Sci. Paris **268**, 437 (1969).
[8] D. G. King-Hele et al., Planet. Space Sci. **16**, 937 (1968).
[9] F. A. Marcos and K. S. W. Champion, paper presented at XIV COSPAR Plenary Meeting, Seattle 1971.
[10] F. A. Marcos et al., Space Research XI, 940 (1971).

Space Research XIII — Akademie-Verlag, Berlin 1973

FURTHER STUDIES OF UPPER ATMOSPHERE DENSITIES AT HIGH LATITUDES

R. R. ALLAN

Royal Aircraft Establishment, Farnborough, Hants., England

Analysis of the orbit of Molniya 1K (1968-85A) revealed very high air drag in October—November 1969 near 18 h local time at 400 km in the southern auroral zone, and it is natural to ask whether other such satellites show similar behaviour. Two of the five Molniya satellites launched in 1970, namely Molniya 1N (1970-13A) and Molniya 1Q (1970-77A), would be expected from their initial orbits to show detectable effects of air drag by now.

The orbital parameters for 1N are very similar to 1K but the satellite does not show the same high density in 1970 as did 1K in 1969. Molniya 1Q is decaying in the atmosphere and shows an even lower air density in 1971 than did 1N in 1970; this is consistent with the generally lower level of solar activity in 1971.

1. Introduction

The Russian Molniya communications satellites provide unique probes of the atmospheric density in a particularly interesting and important region — the southern auroral zones. Twenty of these satellites have now been launched; their orbits are inclined at $\sim 65°$ and are highly eccentric, with an initial perigee height between 450 and 700 km. The semi-major axis is ~ 26550 km giving a period of ~ 12 h, and is adjusted after injection so that the ground track is nearly fixed relative to the earth. Moreover the argument of perigee is near 270° so that apogee occurs in the northern hemisphere near the apex of the orbit; in fact the initial value is consistently 285° and is decreasing slowly at a rate between 5°/yr and 10°/yr depending on the satellite.

Because of the high eccentricity the atmosphere is sampled over only short arcs, and, since perigee remains near 65° S, each arc is at virtually a fixed latitude with the extension purely in longitude. Furthermore the atmosphere is sampled only near two (slowly changing) longitudes since the orbits are nearly 2:1 commensurable with the earth's rotation. The perigee height of Molniya satellites is, however, severely perturbed by luni-solar forces [1] owing to the high eccentricity. Essentially the height of perigee varies sinusoidally with the orientation of the orbital plane in space. Since the plane is regressing at a rate $\dot{\Omega} \simeq -50°/\text{yr}$, the period of oscillation is about seven years; the amplitude is of the order of ± 500 km. Usually perigee either ascends fairly rapidly so that the satellite will survive for at least about seven years, or else perigee descends into the denser atmosphere and the satellite decays within about two years.

At the time, the satellite Molniya IK (1968-85A), with an initial perigee height of 500 km and $\Omega_0 \simeq 60°$, was unique in that perigee fell to a minimum near 380 km about a year later and then rose again. This satellite apparently showed

[2, 3] very high air densities in October—November 1969 when perigee was near 420 km and the local time at perigee ~18 h. It is therefore important to see whether similar effects are shown by any later Molniya satellites.

Two of the five such satellites launched in 1970, namely Molniyas 1N and 1Q (1970-13A and 77A), will show detectable effects of air drag by now because of their initial orbits and the consequent variation of perigee height as shown in the lower part of Fig. 1. In fact Molniya 1N with perigee initially at 500 km and

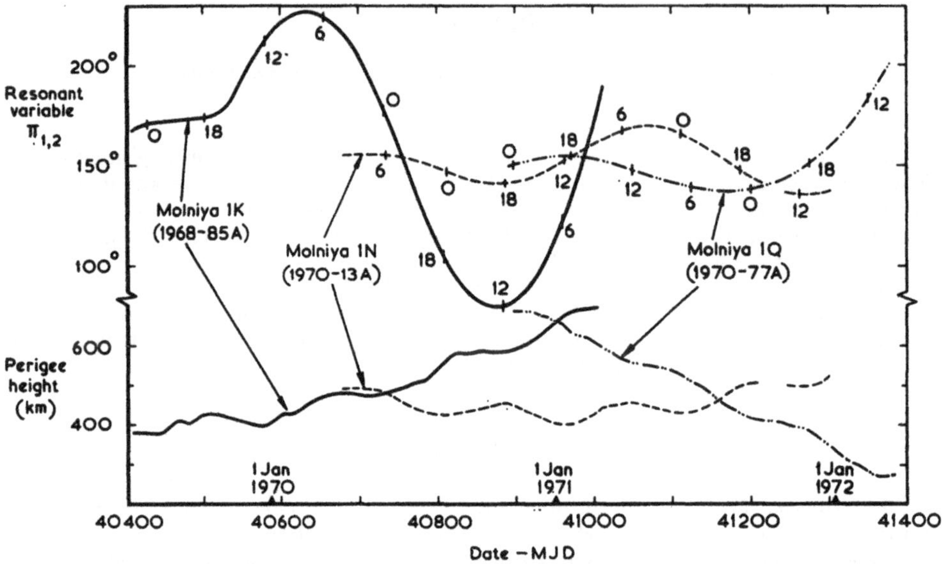

Fig. 1. Resonant variable and perigee height for Molniyas 1K, 1N and 1Q.

$\Omega_0 \simeq 52°$ is nearly as suitable as Molniya 1K, and sampled the air density near 18 h local time in October—November 1970 when perigee was near 450 km. On the other hand Molniya 1Q, with perigee initially near 700 km and $\Omega_0 \simeq 109°$, is now decaying in the atmosphere, with perigee below 300 km.

2. Method of Analysis

Basically the drag acceleration can be determined by taking first differences of the mean motion, or alternatively by taking second differences of the along-track position or its equivalent. The difficulty arises because the ground-track is nearly stationary relative to the earth, and the satellite is therefore in resonance with the longitude-dependent part of the earth's gravitation. It can be shown [4], however, that the effect on Molniya-type orbits can be described in terms of the resonant variable

$$\Pi = M + 2(\Omega - \nu), \tag{1}$$

where Ω and M are two of the usual set of osculating elliptic elements a, e, I, Ω, ω, M, and ν is the sidereal hour angle. The behaviour under resonance and drag combined is then given by

$$d^2\Pi/dt^2 = (dn/dt)_{\text{res}} + (dn/dt)_{\text{drag}}, \tag{2}$$

where the effect of the resonance forces is given by

$$\left(\frac{dn}{dt}\right)_{\text{res}} = -3n^2 \sum J_{lm} \left(\frac{R}{a}\right)^l \Re i\gamma F_{lmp}(I) G_{lpq}(e) \exp\{i(l-2p)\omega + i\gamma\Pi - im\phi_{lm}\}, \tag{3}$$

summed over all values of l, m, p, q which satisfy the inequalities

$$2 \leqq l \leqq \infty, \quad 1 \leqq m \leqq l, \quad 0 \leqq p \leqq l, \quad -\infty \leqq q \leqq \infty \tag{4}$$

and also the resonance conditions

$$l - 2p + q = \gamma, \quad m = 2\gamma, \quad \gamma = 1, 2, 3, \ldots \tag{5}$$

A detailed explanation may be found elsewhere [4], and would be out of place here.

The resonance effects are dominated by the term $(l, m, p, q) = (2, 2, 1, 1)$ with a smaller contribution from the term $(2, 2, 0, -1)$, both of which involve the gravity coefficient $J_{2,2}$ corresponding to the 'ellipticity of the equator'. It is desirable, however, to include further terms and this has now been done by a computer program. The opportunity has therefore been taken of comparing the resonant accelerations given by (i) the SAO Standard Earth 1969 [5] up to $l = m = 6$, (ii) Kaula's 'combined-adjusted' solution [6], also up to $l = m = 6$, (iii) the simple approximation of including only $J_{2,2}$ as previously [2, 3], taking Kaula's values. When applied to Molniya 1K these three approximations differ by at most 2×10^{-6} rev day^{-2}, and none of the conclusions needs to be modified on this count.

In the absence of any perturbing forces other than resonance, the variable Π would oscillate about some stable value which in practice lies between about 145° and 150°, depending on the value of ω, and which is changing slowly (of order 1°/yr) as ω changes. The effect of air drag is therefore manifest as a departure from a simple oscillation.

3. Results

The behaviour of the resonant variable and perigee height for Molniyas 1N and 1Q, deduced from USAF Spacetrack elements, are shown in Fig. 1 with also the same quantities for 1K from MJD 40400 for comparison. The local time at perigee, which is decreasing by 2 h about every 25 days mainly due to the annual motion of the earth (with corrections for $\dot{\Omega}$ and $\dot{\omega}$), is marked on the libration curves. Clearly 1N suffers some drag acceleration near the minimum at about MJD 40900, but the effect is less marked than for 1K, which was admittedly somewhat lower. Also 1Q is subject to increasing drag as perigee descends into the atmosphere, but even with perigee near 270 km the curvature is not as great as for Molniya 1K near MJD 40500.

The drag acceleration is found both from first differences of n and from second differences of Π, subtracting the resonant acceleration given by the SAO Standard Earth 1969 up to $l = m = 6$. To improve the statistics all values in each 2 h local time interval are grouped and averaged, with appropriate weighting and splitting between adjacent intervals.

3.1. Molniya 1 Q

There appears to be no significant drag before MJD 41200 (6 Sept 1971) when the local time is 24 h and perigee has fallen to 420 km. Up till then the mean drags from single and double differences are 0.7 and 0.4×10^{-6} rev day^{-2}, and the corresponding mean square values are 1.5 and 2.3×10^{-6} rev day^{-2} which can be taken as upper limits to the errors. This seems to be a convincing demonstration that the resonance effects are being calculated to adequate accuracy. The drag accelerations from MJD 41200 onwards are given in Table 1.

Table 1
Drag acceleration for Molniya 1 Q

Local time interval	Period	Mean perigee height (km)	Drag acceleration (10^{-6} rev day^{-2}) from n	from Π
24—22	6 Sept.—2 Oct. 1971	418	2.2	4.2
22—20	2 Oct.—27 Oct. 1971	404	4.6	2.4
20—18	27 Oct.—21 Nov. 1971	391	8.1	5.2
18—16	21 Nov.—15 Dec. 1971	365	14.8	18.5
16—14	15 Dec. 1971—10 Jan. 1972	325	26.3	23.4
14—12	10 Jan.—4 Feb. 1972	292	25.7	27.5
12—10	4 Feb.—28 Feb. 1972	272	37.9	39.7
10—8	28 Feb.—8 Mar. 1972	273	55.1	54.0

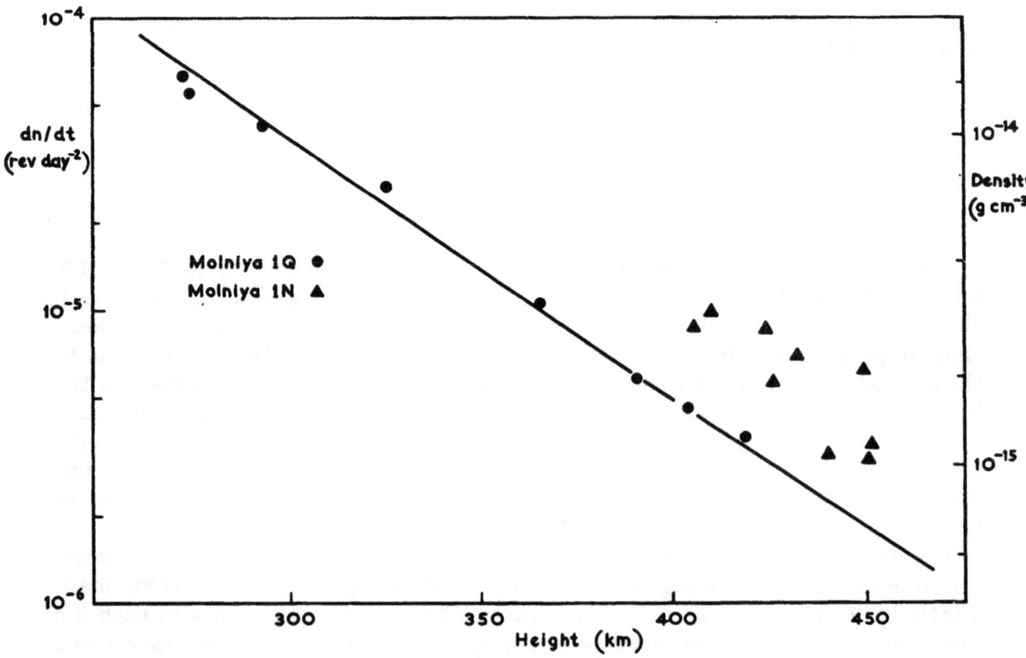

Fig. 2. Drag accelerations of Molniyas 1 Q and 1 N (adjusted to a mean cross section of 13.5 m²).

In operation the Molniya satellites are stabilized so that the large area of solar cells points at the sun; the satellite then presents nearly its maximum cross-sectional area ($\sim 20\ m^2$) to the atmosphere when perigee is at 06 h or 18 h local time, and nearly its minimum ($\sim 7\ m^2$) when perigee is at 12 h or 24 h local time. time. Assuming that the satellite remains sun-stabilized and that the cross-sectional area is varying sinusoidally between its maximum and minimum, the values in Table 1 derived from single differences of n have been adjusted to a mean cross section of $13.5\ m^2$ and plotted in Fig. 2. Although they refer to different local times and times of year, the points in Fig. 2 fit well and the best-fitting line gives a drag acceleration of $\sim 5 \times 10^{-6}$ rev day^{-2} at 400 km and a mean scale height of between 50 and 55 km. Taking $S = 13.5\ m^2$, $C_D = 2.2$, the mass as 1000 kg, and the scale height as 60 km, standard methods [7] give a density of $1.7 \times 10^{-15}\ g\ cm^{-3}$ at 400 km (the density scale on the right of Fig. 2 assumes the mass to be 1000 kg). If an alternative estimate of the mass as 1600 kg is used, the derived density rises to $2.7 \times 10^{-15}\ g\ cm^{-3}$. These values compare with a maximum equatorial density (at ~ 15 h) of $3.8 \times 10^{-15}\ g\ cm^{-3}$ at 400 km given by the CIRA 1965 model atmosphere, for $\bar{F}_{10.7} = 100$ which is typical of late 1971 (although the index is much higher in early 1972).

3.2. Molniya 1 N

The drag acceleration is significant only around the minimum near MJD 40900 and Table 2 shows the values derived for the period from 13 Aug 1970 to 27 Mar. 1971.

The drag acceleration from the single differences of n has also been adjusted to a mean cross-sectional area of $13.5\ m^2$ assuming the satellite is sun-stabilized as for 1 Q, and the resulting values are plotted as the triangles in Fig. 2. Although the points show a greater scatter than for 1 Q (the data seem to be intrinsically noisier with more 'wild' points), the density is clearly higher by a factor between 2.0 and 2.5 than for 1 Q in 1971, but lower than for 1 K in 1969 by a factor of about 10.

It is not known whether the satellites 1 N and 1 Q were sun-stabilized throughout. In either case, however, the alternative assumption that the satellite was

Table 2

Drag acceleration for Molniya 1 N

Local time interval	Period	Mean perigee height (km)	Drag acceleration (10^{-6} rev day^{-2}) from n	from Π
24–22	13 Aug.–7 Sept. 1970	433	4.2	4.4
22–20	7 Sept.–3 Oct. 1970	442	3.2	5.1
20–18	3 Oct.–27 Oct. 1970	452	4.9	5.5
18–16	27 Oct.–21 Nov. 1970	448	8.4	7.5
16–14	21 Nov.–17 Dec. 1970	425	8.8	7.4
14–12	17 Dec. 1970–11 Jan. 1971	406	5.2	3.0
12–10	11 Jan.–5 Feb. 1971	410	5.9	5.0
10–8	5 Feb.–2 Mar. 1971	427	5.8	8.5
8–6	2 Mar–27 Mar. 1971	451	4.3	5.3

completely unstabilized would superimpose on the densities plotted in Fig. 2 a regular variation with no apparent physical meaning. The assumption that the satellites remain sun-stabilized gives densities which are more self-consistent.

4. Conclusions

The very high drag shown by Molniya 1K in October—November 1969 is not repeated by 1N in 1970 although the characteristics of the orbits are very similar. For 1K, 18 h local time occurred on 9 Oct 1969 with perigee appearing at 15° E and 195° E and perigee height 420 km; while for 1N, 18 h local time occurred on 27 Oct 1970 with perigee appearing at 5° E and 185° E and perigee height 450 km. 1K also appeared to show detectably high drag in August—September 1970 when perigee height was between 550 and 600 km. There is no direct comparison with 1N, however, since the local times differ by about 6 h and perigee for 1K was then appearing about 40° W (approaching the South Atlantic geomagnetic anomaly) and 140° E (well inside the polar cap). 1Q decaying in the atmosphere shows a density in 1971 smaller by a factor of 2 than for 1N in 1970, and this is consistent with the generally lower level of solar activity in 1971.

It would be valuable to have further information on the dimensions and masses of the Molniya satellites, and to know when the orbit and attitude control systems may have been operating.

Acknowledgments

The writer is indebted to A. W. Odell for much assistance and advice on data handling and to M. R. Gordon for valuable programming assistance.

The paper is reproduced with the permission of the Controller, Her Majesty's Stationery Office.

References

[1] G. E. Cook and D. W. Scott, Planet. Space Sci., **15**, 1549 (1967).
[2] R. R. Allan, Space Research XII, 857 (1972).
[3] R. R. Allan, Nature, Lond. **235**, 100 (1972).
[4] R. R. Allan, Cel. Mech. **3**, 320 (1971).
[5] E. M. Gaposchkin and K. Lambeck, J. Geophys. Res. **76**, 4855 (1971).
[6] W. M. Kaula, J. Geophys. Res. **71**, 5303 (1966).
[7] D. G. King-Hele, Theory of Satellite Orbits in an Atmosphere, Butterworths, London 1964.

Space Research XIII — Akademie-Verlag, Berlin 1973

AN ANALYSIS OF THE ALTITUDE DEPENDENCE OF THE GEOMAGNETIC EFFECT BY MEANS OF "EQUIVALENT DURATIONS"

I. ALMÁR and E. ILLÉS-ALMÁR

Konkoly Observatory of the Hungarian Academy of Sciences,
Budapest, Hungary

Six geomagnetic storms were analysed by means of 35 equivalent duration (*D*) values determined from visual observations and orbital elements. Between 200 and 300 km the measured *D* values proved to be significantly larger than those calculated from the Jacchia 71 model, and an appropriate model correction is suggested. Above 300 km the coincidence with predictions of the model is satisfactory. In the case of two sudden atmospheric events, not correlated with strong geomagnetic activity, a single temperature increment can explain the observed density changes at all levels above 200 km.

1. Introduction

The equivalent duration D has been introduced [1] to characterize the total response of the atmosphere to corpuscular heating during geomagnetic storms. It is the integral of the relative density changes during a geomagnetic storm, i. e.

$$D = \int_{t_1}^{t_2} \frac{\varrho - \varrho_0}{\varrho_0} dt$$

a convenient measure of which is the length of time of a hypothetical disturbance having the same integral as the actual one, but giving rise to a constant 100% density increase for D days [1]. It can be determined either (i) by integrating directly the $\dot{P}$ or ϱ curves published in the literature (a very approximate method unless the time resolution of the data is extremely good), or (ii) by using the parallel displacement of a linear $P(t)$ curve or (iii) by the $O-C$ method described in detail in [1]. The second and third methods have been successfully applied to a whole series of orbital data derived from visually observed transits through the celestial equator by means of the PERLO programme [2], and complemented by published orbital elements (period, time of perigee passages) from the Current Gear Ratio Elements of NASA GSFC, as well as other sources [3—7]. The D values obtained by different methods are averaged; their estimated probable error is 40%.

Since D values obviously depend on the shape of the storm and on the altitude and position of the satellite's perigee, a model atmosphere must be used to provide D values for comparison. A computer programme based on the Jacchia 71 model [8] first calculates instantaneous density values ϱ every 3 hours, then selects an appropriate ϱ_0 corresponding to quiet atmospheric conditions and integrates the $(\varrho - \varrho_0)/\varrho_0$ curve in storm-time to obtain D_{J71}. The amplitude of the geomagnetic disturbance is defined as $A_{J71} = (\varrho_{max} - \varrho_0)/\varrho_0$.

2. Results concerning Six Geomagnetic Events

Six large and isolated geomagnetic storms were selected between 1966 and 1970 and equivalent durations calculated in 35 cases. The results, together with model values, are given in Table 1. Within the limits of error, all D/D_{J71} values fit on one curve in the altitude diagram, where h is the height at λH^* above perigee (Fig. 1). H^* is the best estimate of H_p (scale height at perigee), $\lambda = z^* - {}^1/_2 z^{*2}$

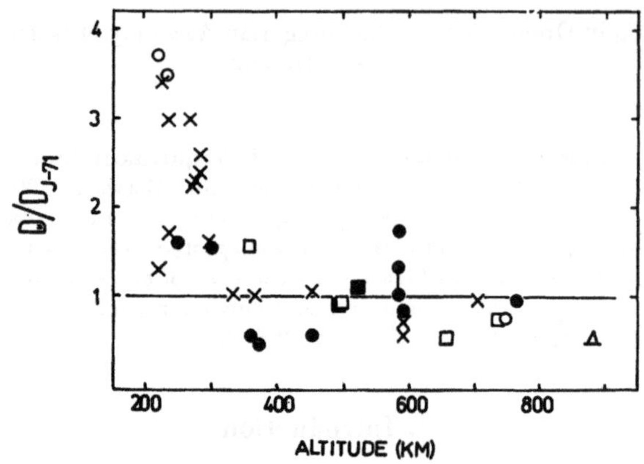

Fig. 1. Measured equivalent durations divided by the corresponding model values, as a function of altitude h. Different symbols represent different geomagnetic storms (see Table 1).

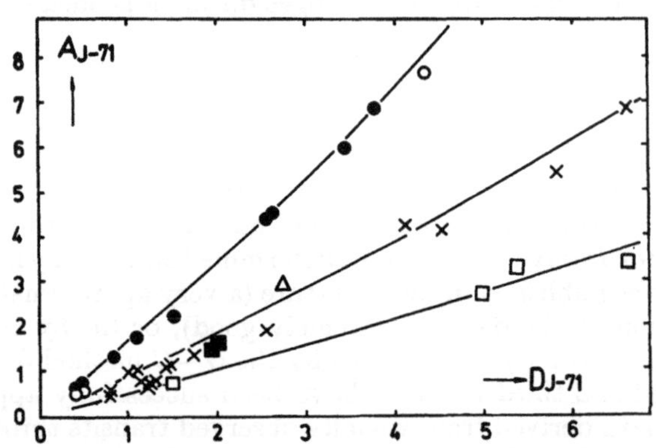

Fig. 2. Relation between amplitude and equivalent duration of different geomagnetic storms calculated by means of the Jacchia 71 model.

for $0 \leqq z^* \leqq 1$; $\lambda = {}^1/_2$ for $z^* > 1$, where $z^* = ae/H^*$ [9]. The average value of D/D_{J71} between 200 and 300 km is 2.41, and its deviation from unity is clearly significant ($\sigma = 0.28$). In order to determine the density correction D_{J71} is plotted against A_{J71}, i.e. $(\varrho_{max} - \varrho_0)/\varrho_0$ according to the J 71 model. The relation depends on the particular storm and is not linear (Fig. 2). A free-hand curve has been used to transform the measured D values into $A = (\varrho_{max} - \varrho_0)/\varrho_0$, a quantity, though often published, in most cases distorted by the limited time resolution of the actual observations. The ratio A/A_{J71} has been plotted against height (Fig. 3,

Table 1

Equivalent durations for 6 geomagnetic events

Storm MJD	*Ap* max		Ident.	*h* (km)	D_{obs}	Remarks	D_{J71}	A_{J71}
39369	112	×	58001 A	368	1.72	e, [3]	1.72	1.36
			59001 A	591	2.20	e, [3]	4.10	4.23
			63053 A	703	5.50	a,d	5.82	5.40
			64076 A	590	4.29	a,d	6.60	6.83
			65011 A	282	3.10	c,d	1.30	0.78
			65011 B	284	2.97	c,d	1.15	0.75
			65011 D	298	1.60	a,d	1.00	0.97
			65052 A	235	2.05	c,d	1.20	0.61
			65095 A	226	2.65	c,d	0.78	0.45
			65095 B	236	2.38	c,d	0.80	0.53
			65112 P	267	3.29	c,d	1.10	1.03
			66009 B	453	4.77	c	4.53	4.12
			66036 A	274	3.30	c,d	1.45	1.15
			66043 A	220	1.61	c,d	1.25	0.69
			66061 A	270	3.12	c,d	1.40	1.13
			66065 A	333	2.60	c	2.56	1.92
39636	146	●	58001 A	361	0.47	f, [4]	0.85	1.31
			59001 A	590	3.13	f, [4]	3.77	6.83
			60014 A	454	0.80	f, [4]	1.41	2.21
			62076 F	251	0.65	c, [5]	0.41	0.59
			63043 A	373	0.50	a	1.10	1.73
			63053 A	763	3.23	f, [4]	3.44	5.94
			64076 A	583	2.68	f, [4]	2.62	4.51
					3.41	a		
			65011 D	300	0.75	a	0.49	0.73
			65053 F	586	4.46	a, b	2.58	4.34
40018	103	■	61001 A	490	1.80	b	1.94	1.50
			64076 A	524	2.19	b	2.00	1.59
40161	120	△	63053 A	881	1.85	a, b, c	2.75	2.90
40356	131	□	64004 A	655	2.76	e, [6]	5.36	3.28
			64035 A	357	2.33	c	1.50	0.70
			65082 LP	734	4.83	c	6.66	3.43
			67042 A	496	4.63	c, [7]	4.97	2.70
40654	150	○	69082 DX	747	3.20	c	4.35	7.63
			69110 A	222	1.70	c	0.46	0.53
			70004 B	236	1.46	c	0.42	0.49

The remarks column is a one-letter code referring to the source and the method applied to determine D: a visual obs., $O-C$ method; b visual obs., P curve; c orbital elem., P curve; d orbital elem., $O-C$ method; e integral of P curve; f retransformation of ΔT into ρ and integral of ρ curve. Symbols refer to Fig. 1 and lower part of Fig. 3.

lower part), and a simple linear relation derived which might be used to correct model density values during strong magnetic storms up to 350 km. The factor, multiplying model $(\varrho - \varrho_0)/\varrho_0$ values between 200 and 350 km is 6.67−0.017 h. The correction yields larger D_{J71} values. A similar correction is already included in the Jacchia 71 model for $h < 200$ km; our procedure is, however, more complicated, because it involves the determination of ϱ_0, corresponding to "background" geomagnetic activity.

In any case, an obvious implication of the equivalent duration curve is that during magnetic storms a considerable amount of energy is deposited below 350 km, causing a positive deviation from model density profiles.

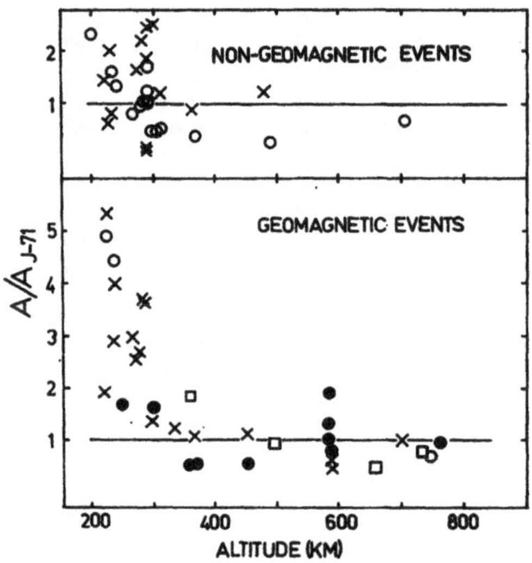

Fig. 3. Amplitude of the density changes, divided by the corresponding model values, during 6 geomagnetic storms (lower part) and during 2 non-geomagnetic events (upper part) as a function of altitude.

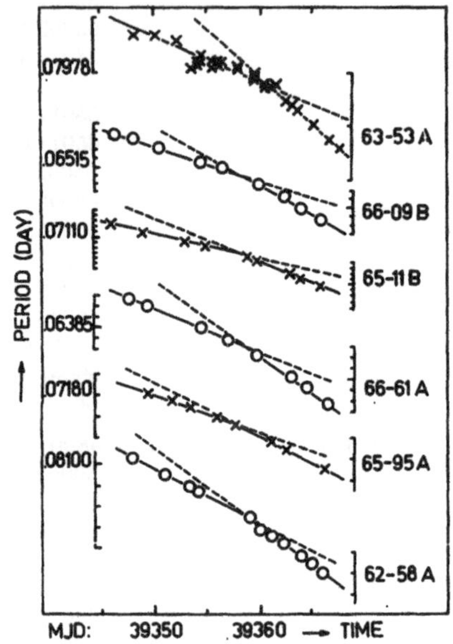

Fig. 4. A sample of period curves of different satellites around one of the "non-geomagnetic events".

3. Comparison: Analysis of two Non-geomagnetic Events

In order to investigate the possibility that this deviation is only a consequence of some systematic bias in our method, we carried out the analysis of two events of another type. Plotting P as a function of time, in August—October 1966 we observed quasi-simultaneous sudden changes in the slope of the $P(t)$ curve of many satellites (Fig. 4). In spite of the fact that these two events were not preceded by a geomagnetic storm or sudden increase in $F_{10.7}$, their reality seems to be unquestionable, since the effect can be traced on 33 diversely derived period curves of 19 satellites. These non-geomagnetic events and the usual geomagnetic storms are distinguished from each other by the fact that in the former case the atmosphere changes from one relatively stable situation to another.

Supposing that $\dot{P}_b/\dot{P}_a = \varrho_b/\varrho_a$ (where the indices b and a correspond to values before and after the event respectively) the amplitude of the density increase $A = (\varrho_a - \varrho_b)/\varrho_b$ can be calculated at different heights (Table 2). By means of these observed amplitudes adequate exospheric temperatures have been derived from model $d\varrho/dT$ curves. Accordingly the night-time minimum of the global exospheric temperature changed from 635 °K to 750 °K at MJD 39358 and from 750 °K to 815 °K at MJD 39387. Using these temperature values we calculated density amplitudes from the Jacchia 71 model and formed the A/A_{J71} ratios as previously. The result, as a function of height, is shown in the upper part of Fig. 3. The average amplitude ratio between 200 and 300 km is 1.30 ($\sigma = 0.40$), not differing significantly from unity.

Table 2

Acceleration changes for 2 non-geomagnetic events

Event		Ident.	h (km)	$\dot{P}_a/\dot{P}_b$	ϱ_a/ϱ_b J71
39358.7	○	62058 A	220	1.65	1.28
59.6		63053 A	706	2.12	2.67
60.0		64015 A	311	1.31:	1.58
57.8		65011 A	290	1.58	1.47
58.2		65011 B	291	1.79	1.47
59.5		65011 C	284	1.45	1.43
60 :		65020 A	289	1.80:	1.76
60.3 :		65020 B	290	1.72:	1.71
59.6		65020 C	286	1.67	1.65
59.6		65052 A	237	1.24	1.31
57.0		65095 A	238	1.41	1.32
55.6:		65095 B	234	1.47	1.30
57.8		65112 P	294	1.45	1.91
57.9		66009 B	491	1.64	3.34
58.3		66036 A	306	1.58	2.20
59.2		66061 A	278	1.73	1.77
58.9		66061 B	267	1.57	1.70
39360		66065 A	368	1.50	2.28
39385.6	×	62058 A	220	1.17	1.12
85.5		64015 A	308	1.35	1.29
90 :		65011 A	288	1.03:	1.23
87.0		65011 B	290	1.04	1.23
83.7		65020 A	288	1.47:	1.19
87.1		65020 C	282	1.40:	1.18
86.7		65052 A	233	1.10	1.12
87.3		65095 A	231	1.24	1.12
89.6		65095 B	234	1.16	1.12
85.9		65112 P	290	1.53	1.29
87.1		66009 B	479	1.84	1.70
87.8		66036 A	297	1.67:	1.27
89.3:		66043 A	225	1.12	1.19
87.6		66061 A	273	1.46	1.29
39385.5		66065 A	361	1.38	1.43

Symbols refer to upper part of Fig. 3.

4. Conclusions

Our results support Jacchia's conclusion that "since the characteristic time of geomagnetic disturbances is hours rather than days, static models cannot be expected to represent correctly both temperature and density variations" and a hybrid formula is needed to represent the geomagnetic variations in the thermosphere [8]. Recently Mayr and Volland also stated that "magnetic storm variations with a time scale about one day, comparable with the response time of the atmosphere, cannot be excited readily at all altitudes" [10]. On the other hand, we are unable to agree with Roemer's conclusion that "in the upper thermosphere the amplitude of the density variation during geomagnetic disturbances at various altitudes can be described by a single value of ΔT_∞ for a given storm" [11]. We are convinced that the distribution in height of the energy input (as characterized by the equivalent durations) during geomagnetic storms is different, at least below 300 km, from that of EUV absorption. Nevertheless more data and more careful correlation studies are needed before a final quantitative model correction can be made.

Acknowledgments

The authors are indebted to the Astronomical Council of the Soviet Academy of Sciences, to Dr. D. G. King-Hele, Dr. A. Manara, Prof. G. Chis, Dr. I. Curea, Dr. W. Baran and Mr. E. Otto for contributing orbital data and satellite observations.

References

[1] I. Almár and E. Illés-Almár, Space Research XI, 975 (1971).
[2] A. Horváth and E. Illés-Almár, in: Observations of Artificial Satellites of the Earth, IX, 277 (Warsaw 1970).
[3] L. G. Jacchia and J. W. Slowey, Smithson. Astron. Obs. Spec. Rep. No. 326 (1970).
[4] L. G. Jacchia, Space Research X, 367 (1970).
[5] B. R. May, Planet. Space Sci. 19, 691 (1971).
D. W. Scott, RAE Tech. Rep. No. 69255 (1969).
[6] I. Almár and E. Illés-Almár, in: Observations of Artificial Satellites of the Earth, X, 272 (Ulan Bator 1971).
[7] R. H. Gooding, RAE Tech. Rep. No. 69275 (1969).
[8] L. G. Jacchia, Smithson. Astron. Obs. Spec. Rep. No. 332 (1971).
[9] D. G. King-Hele, Theory of Satellite Orbits in an Atmosphere, Butterworths, London 1964 (p. 124).
[10] H. G. Mayr and H. Volland, Planet. Space Sci. 20, 379 (1972).
[11] M. Roemer, Space Research XI, 965 (1971).

Space Research XIII — Akademie-Verlag, Berlin 1973

THE GLOBAL WIND SYSTEM IN THE THERMOSPHERE

P. W. Blum and I. Harris

NASA/Goddard Space Flight Center, Greenbelt, Md., USA

The upper atmosphere wind distribution is calculated by solving the Navier—Stokes equations with models of upper atmospheric densities and ion distribution. All non-linear terms and viscous terms are included. It is found that the velocities from the non-linear solutions are smaller than those obtained from a linear treatment and that in the equatorial region the magnitude of the average meridional velocity is strongly dependent upon the latitudinal coupling. No appreciable super-rotation is obtained by the inclusion of the non-linear terms; the small value of the average zonal velocity is mainly due to the ion distribution.

1. Introduction

The wind system of the thermosphere has been the subject of a considerable number of observational and theoretical investigations during the last decade. A knowledge of this wind system is required for an understanding of the structure of the neutral atmosphere and its energy balance. It has been suggested that the solution of the phase anomaly of the neutral atmosphere is intimately connected with the wind system. Many ionospheric effects concerning the latitudinal distribution of charged particles, the daily variation of electron density and the maintenance of the night-time ionosphere cannot be fully explained without taking into account the effect of the atmospheric wind system.

Several computations of the thermospheric wind pattern have been made. All of them solve the horizontal equations of motion and assume hydrostatic equilibrium. There are two general approaches; a perturbation treatment of the full set of hydrodynamic equations [1, 2] or a solution of the equation of motion using a given model of atmospheric structure [3—5]. The various computations differ in the assumptions regarding the effects of ion drag and in the viscosity terms. All of them have in common that the non-linear terms of the convective derivative of the flow velocity were not fully included. Especially the main non-linear term, the meridional velocity gradients, have not been included by any of the investigators.

We have chosen the second approach, i.e. we have used in our calculations the atmospheric model of Jacchia [6] and the Penn State Ionospheric Model of Nisbet [7] and solved the full Navier—Stokes equations including all non-linear terms and the full expression for the viscous forces. None of the earlier work has included in the computations of the viscous terms the horizontal velocity gradients.

2. Equations of Motion

The Navier—Stokes equations in vector form are

$$\frac{D\boldsymbol{U}}{Dt} + 2\boldsymbol{\omega}\times\boldsymbol{U} + \boldsymbol{F}_{\text{ion}} = -\frac{\operatorname{grad} p}{\varrho} + \boldsymbol{g} + \nu \operatorname{div} \operatorname{grad} \boldsymbol{U}$$

where $\boldsymbol{U}$ is the velocity vector, ω the rotational velocity of the earth, $\boldsymbol{F}_{\text{ion}}$ the ion drag force, p the pressure, ρ the density and ν the kinematic viscosity.

We are concerned with the global wind pattern with periods of the order of one day and its harmonics. The appropriate coordinate system for this problem is spherical. As the vertical velocities are smaller by more than one order of magnitude than the horizontal velocities, we may solve the horizontal equations neglecting the vertical velocity terms that appear in the convective derivative, the Coriolis force and the viscous forces.

Let u be the meridional velocity, measured positive from north to south, and v the zonal velocity, measured positive from west to east, then the convective derivative of the velocities becomes

$$\frac{Du}{Dt} = \omega \frac{\partial u}{\partial \phi} + \frac{1}{r}\left(u \frac{\partial u}{\partial \theta} + \frac{v}{\sin\theta}\frac{\partial u}{\partial \phi} - v^2 \operatorname{ctn}\theta\right)$$

$$\frac{Dv}{Dt} = \omega \frac{\partial v}{\partial \phi} + \frac{1}{r}\left(u \frac{\partial v}{\partial \theta} + \frac{v}{\sin\theta}\frac{\partial v}{\partial \phi} + uv \operatorname{ctn}\theta\right)$$

with θ the co-latitude, ϕ the longitude and r the distance from the center of the earth.

We have made use in these equations of the steady-state assumptions, i.e. we have assumed all variables to depend only upon local time and not explicitly upon longitude and universal time. For this reason we had to exclude all non-steady-state phenomena from our computations, like the explicit longitudinal dependence of ion density (as separate from the local time dependence) and the effects caused by the tilt of the magnetic field of the earth with respect to its axis.

3. The Driving Force

We have calculated the driving force from the Jacchia [6] model. In accordance with this model the driving force at a height of 120 km is zero; therefore, we have chosen this height as the lower boundary of our computations. In the Jacchia model the pressure is a continuous function of the coordinates, but the horizontal pressure gradient is not: it has a discontinuity at the poles. For this reason we have slightly modified the Jacchia model without changing the value of the variables themselves by more than 1.8% [8]. A consequence of the Jacchia model is the almost linear increase of the driving force with height in the isothermal region. We have chosen a height of 500 km as the upper boundary of our computations.

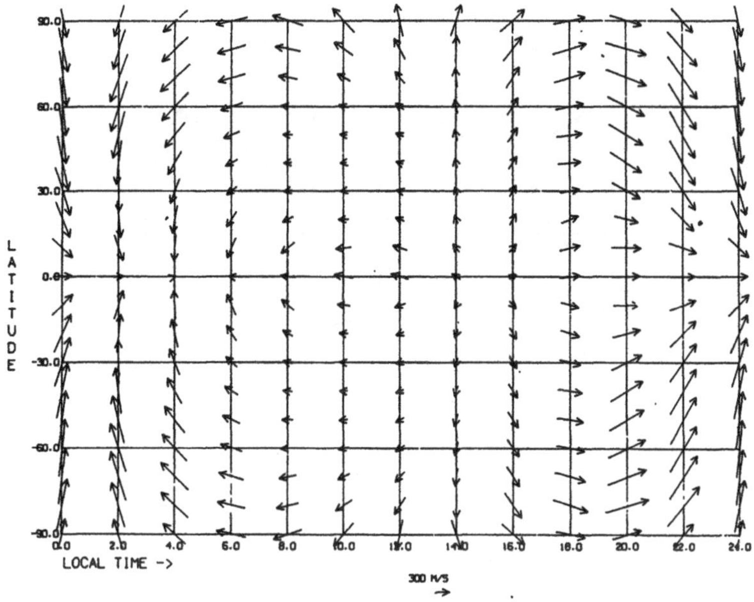

Fig. 1. Global wind field at a height of 300 km, equinox conditions. Penn State ionospheric model.

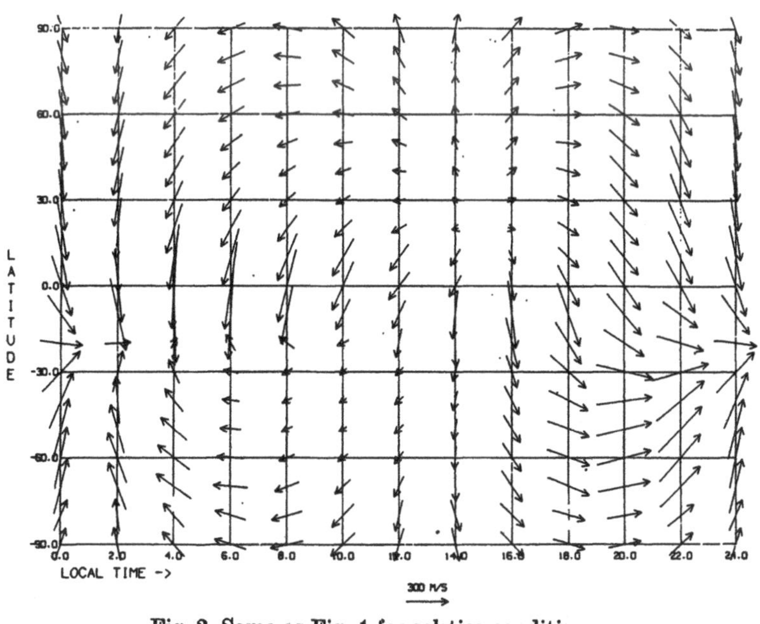

Fig. 2. Same as Fig. 1 for solstice conditions.

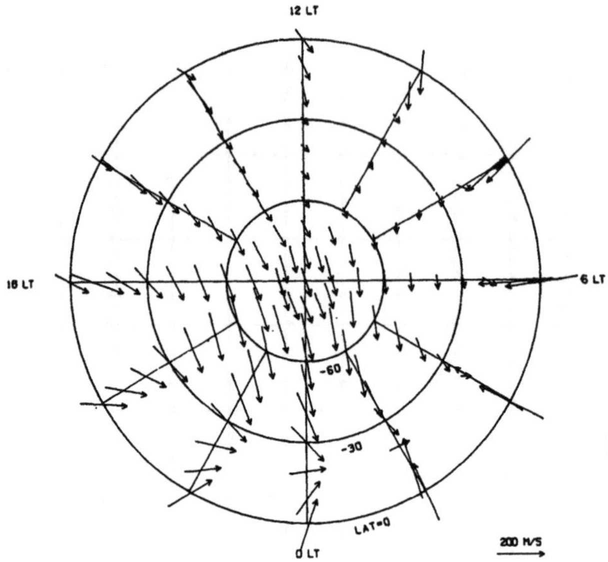

Fig. 3

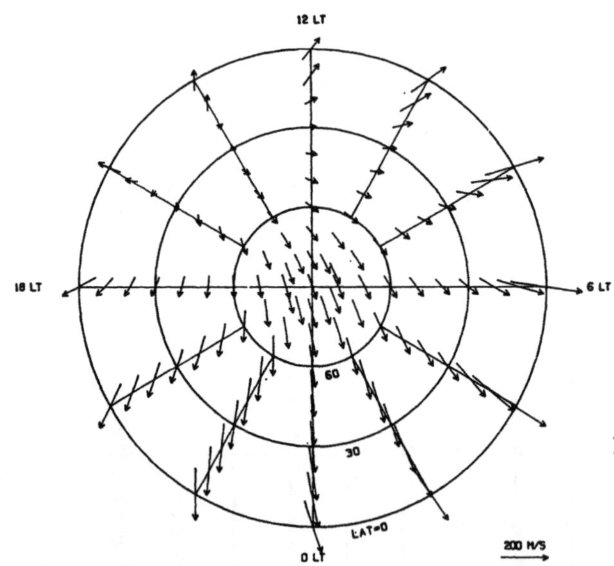

Fig. 4

Fig. 3—4. Global wind field at a height of 300 km for each hemisphere for solstice conditions. Penn State ionospheric model. Fig. 3, southern, Fig. 4, northern.

4. Ion Drag

We have used the Penn State Ionospheric Model to determine the electron density. This model includes an explicit dependence upon longitude and local time and was therefore in its original form not directly applicable to our computations. In order to obtain the steady-state components of the electron densities from this model we have averaged for a given local time the electron densities over various longitudes. A result of this averaging process is a considerable reduction of the semidiurnal and higher components of electron density as compared with electron densities derived for a specific longitude.

In line with previous investigations we have assumed that the ions are fixed with respect to the field lines in the zonal direction and move with the neutral air velocity component along the field lines. We have not included ion movement due to electrostatic fields. This yields ion-drag expressions given by Kohl and King [9]. As we have assumed that the geomagnetic dipole is coincident with the axis of rotation of the earth, there results zero ion drag in the meridional direction at the equator. However, because of the tilt of the geomagnetic field there exists a non-zero value for the mean ion drag of the meridional component at the equator. By averaging over all longitudes one obtains a minimum value for the meridional ion drag in the equatorial region, which can be shown to correspond to its value at a latitude of 5.8°. This minimum value was used in our calculations.

5. Viscosity

We have calculated the viscosities of the various atmospheric constituents as a function of temperature as derived from calculations [10], and then used the Jacchia model to compute the contribution of each constituent to the total viscosity. By this method we could include local time variations of viscosity. For the viscous force we simply used the second derivative with respect to altitude for the vertical shears and also included viscous terms due to horizontal velocity shears. The horizontal shears became important only in the equatorial region ($\pm 20°$); there they tend to reduce the considerable horizontal velocity gradients.

6. Method of Solution

The method of solution adopted is to decompose the equations with respect to Fourier modes and construct a finite difference scheme with respect to latitude and altitude. We have included the zero or average Fourier mode, the diurnal mode and the semidiurnal mode. The local time dependence of all terms and coefficients, like ion drag, viscosity, and driving force was taken into account by this decomposition. In our mesh are used differences in height of 20 km and in latitude of 5°. Taking into account the ten Fourier components included, there results a system of simultaneous non-linear equations of 7030 unknowns. We used for the solution a double iterative procedure. A Newton—Raphson sweep was performed in latitude and a Gauss—Seidel iterative scheme in altitude. In the Newton—Raphson sweep the matrix was brought to tri-diagonal form with blocks of size 10×10 corresponding to the Fourier modes. More than a hundred complete sweeps were required to obtain the non-linear solution.

7. Results

In order to determine conclusively the effect of the non-linear terms we have also executed a computation of the linear system. This involved a slightly different scheme of computations. As in the linear case there is no coupling between different latitudes, but strong coupling between altitudes due to viscosity, we have computed separately for each latitude the set of simultaneous equations that involves all altitudes. For this reason no iterative procedure was necessary. The result of this comparison was that the non-linear terms cause little changes in the velocity pattern at mid and high latitudes greater than about 30 to 40 degrees. On the other hand their influence is very marked in the equatorial region.

Their effects may be summarized:

(i) The very steep velocity gradients in the meridional direction are somewhat reduced. Part of this reduction may also be due to the inclusion of the horizontal viscous terms and not only be caused by the non-linear terms of the convective

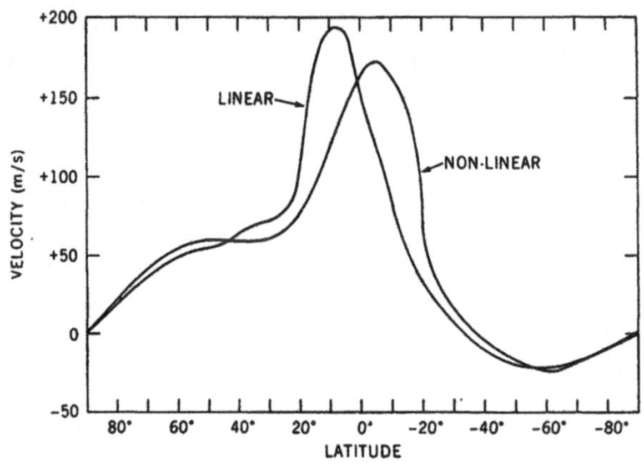

Fig. 5. Comparison of linear and non-linear solution of the average meridional velocity for solstice conditions at the height of 300 km.

derivative. (In line with the absence of latitudinal coupling, horizontal viscous terms were not introduced into the linear solution.)

(ii) Generally the velocities of the non-linear solution are lower than in the linear solution; this is especially true for the average meridional velocity. The magnitude of the average meridional flow is strongly dependent upon the latitudinal coupling in the equatorial region.

(iii) The maximum of the average meridional flow is shifted appreciably in latitude in the solstice case with respect to the linear solution (Fig. 5).

The general properties of the global wind pattern (Figs. 1—4) which we have found are similar to the winds computed by Geisler [3] and Cho and Yeh [5]. We found appreciable deviations form the wind pattern given by Challinor [4]. The following details can be mentioned for the wind field at a height of 300 km, but these results apply actually to the whole isothermal region: (i) The wind over the poles blows generally from the night side about 01 h local time to 13 h local time with a velocity of about 100 m s^{-1}. (ii) At equinox the average meridional component is directed generally towards the equator. (iii) The amplitude of the winds is lowest at noon and highest at midnight. The meridional component has its maximum slightly later than midnight at about 01 h local time, whereas the zonal component has its maximum at 20 h with a tendency towards later hours at low latitudes. (iv) The average zonal velocity, which can be thought to represent super-rotation [11] is at equinoxes positive with a maximum in the equatorial region. A super-rotation effect may be deduced from this; its magnitude is about 5% of the rotational velocity of the earth and therefore much smaller than the super-rotation found by King-Hele from satellite data. Similar results to ours were also found by Bailey et al. [12]. It has been suggested by Chiu [13] that super-rotation can be induced by the non-linear terms. Thus our results do not bear out this contention. (v) There are only small vertical velocity shears in the isothermal region, an effect of the strong coupling between altitudes due to the very high viscosity. The viscosity also influences the amplitude of the winds, but has little effect on the phases.

Acknowledgment

We express our gratitude to Professor John Nisbet of the State University of Pennsylvania Ionospheric Research Laboratory who put the computer program of the Penn State Mark I ionospheric model at our disposal.

References

[1] R. S. Lindzen, Geophys. Fluid Dynam. **1**, 303 (1970).
[2] H. Volland and H. G. Mayr, Annls Géophys. **27**, 907 (1970).
[3] J. E. Geisler, J. Atmos. Terr. Phys. **29**, 1469 (1967).
[4] R. A. Challinor, Planet. Space Sci. **18**, 1485 (1970).
[5] H. R. Cho and K. C. Yeh, Radio Sci. **5**, 881 (1970).
[6] L. G. Jacchia, Smithson. Contr. Astrophys. **8**, 215 (1965).
L. G. Jacchia, Smithson. Astrophys. Obs. Spec. Rep. No. 332 (1971).
[7] J. S. Nisbet, Pennsylvania State Univ. Ionospheric Research Scient. Rep. No. 335 (1970).
[8] P. W. Blum and I. Harris, NASA/GSFC Rep. No. X-641-71-367 (1971).
[9] H. Kohl and J. W. King, J. Atmos. Terr. Phys. **29**, 1045 (1967).
[10] K. S. Yun, S. Weissman and E. A. Mason, Phys. Fluids **5**, 672 (1962).
[11] D. G. King-Hele and R. R. Allan, Space Sci. Rev. **6**, 248 (1966).
[12] G. J. Bailey, R. J. Moffett and H. Rishbeth, J. Atmos. Terr. Phys. **31**, 253 (1969).
[13] Y. T. Chiu, Planet. Space Sci. **19**, 1201 (1971).

Dynamics of the Thermosphere and Ionosphere

Space Research XIII — Akademie-Verlag, Berlin 1973

VARIATIONS IN DENSITY AND CHEMICAL COMPOSITION AT 120 km FROM CHEMICAL AND DYNAMICAL PROCESSES*

F. S. JOHNSON[a] and P. BAUER[b]

[a]University of Texas at Dallas, Dallas, Texas, USA
[b]Centre National d'Etudes de Communications, R.S.R., Issy-les-Moulineaux, France

Atmospheric parameters show both systematic and random patterns of behavior at 120 km altitude. Variations in density, temperature, pressure, and especially atomic oxygen concentration are important. The variations are particularly significant because constant boundary conditions at this altitude have commonly been assumed in atmospheric model making. It is surprising how well the assumption of constant boundary conditions has served over the past decade. However, their use has probably introduced erroneous concepts into the field of atmospheric structure, or at least held off the recognition and introduction of important new concepts. The variations at 120 km are caused by changes in energy input into the upper atmosphere, including internal gravity waves and tidal energy from below, and changes in the transport processes within the atmosphere; the relative importance of the various effects in producing changes is not very well known, but knowledge is improving.

1. Introduction

It has become increasingly clear over the past few years that atmospheric parameters at 120 km altitude vary significantly in a systematic way. This is very important in connection with atmospheric models because constant boundary conditions have usually been assumed to prevail at 120 km. Variations in atmospheric parameters at 120 km are to be expected on the basis of physical processes that are known to occur in the atmosphere. It is in fact surprising that the simplifying assumption of constant boundary conditions at 120 km has served so well and so long in atmospheric studies. However, it is also probable that the use of this simplifying assumption has introduced errors into our concepts of atmospheric structure and behavior, or at least held off the recognition and introduction of some important new concepts.

There seems to be nothing new to report at this time concerning the processes that affect the density and chemical composition of the atmosphere at 120 km. Recent advances in our understanding of variability at 120 km have come mainly from new measurements and whatever inferences can be drawn from these concerning the relative importance of the various physical processes that affect atmospheric structure.

In examining the causes of atmospheric variations, one must regard dynamical processes as primary. Chemical processes may somewhat modify the variations that are produced, but chemical processes do not appear to initiate the variations. Generally, the changes in atmospheric chemistry that occur in conjunction with

* This paper summarizes the discussion of this topic at the COSPAR Panel 4B Joint Workshop Meeting on Thermosphere and Ionosphere Dynamics, Madrid 1972.

dynamical processes appear to be stabilizing, reducing the magnitude of the physical changes that would otherwise occur. This might be contrasted with the destabilizing effect of water vapor condensation in the lower atmosphere, which enhances convection beyond that which would take place without the release of latent heat of condensation.

2. Variable Factors

There are at least three variable factors that influence the structure of the atmosphere. These are the variable geometry of solar heat input (diurnally and seasonally), variable energy deposition in the upper atmosphere by dynamical processes and energetic particle influxes, and variable transport in the upper atmosphere.

2.1. Geometry

The variability in solar heat input due to geometric factors is straightforward and does not require further amplification here. However, it must be recognized that the altitude dependence of solar heating on solar zenith angle is an important factor in addition to the variable heat input per unit area.

2.2. Dynamical and Energetic Particle Sources

There are many different sources of energy deposition in the upper atmosphere including the following: (i) energetic particle influx, (ii) Joule heating by ionospheric currents, (iii) hydromagnetic waves from the magnetosphere, (iv) internal gravity waves and tidal waves from below. These various energy sources do not all deposit their energy in the upper atmosphere at the same altitude; most of them individually do not deposit their energy in a constant profile of heat input, but change depending upon physical conditions.

The dissipation of gravity waves has been studied by Hines [1] and by Pitteway and Hines [2]. Taking into account the probable wavelength ($\lambda \approx 20$ to 30 km), Spizzichino [3] finds the region of dissipation to be above 110 km and argues that it probably does not extend above 120 km. For the semidiurnal tide generated at altitudes below 100 km, Hines [4] and Richmond [5] found an altitude of dissipation 115—130 km. The dissipation of the diurnal tide takes place mainly between 70 and 90 km, but Elford and Roper [6] and Spizzichino [3] suggest that it is first transformed into gravity waves, which in turn generate turbulence and otherwise undergo viscous dissipation.

The altitude of energy deposition is important because it causes the temperature profile through the atmosphere to vary depending upon the importance and characteristics of the various heat sources. Swartz and Nisbet [7] have noted the importance of the shape of the temperature profile and the fact that the exospheric temperature is not a reliable guide to thermospheric temperatures in the altitude region of rapid temperature rise. If the altitude of predominant heat input into the thermosphere is raised, the exospheric temperature is raised without increasing the maximum temperature gradient; this is brought about by prolonging the altitude region over which the high temperature gradient prevails. Many combinations of circumstances of heat input can be imagined including cases where the average temperature up to a given altitude may be depressed while the exospheric temperature is elevated compared with conditions found in models, or vice versa. Also, by varying the shape of the temperature rise in the thermosphere over the polar region relative to low latitudes, it is possible to produce reversals with altitude in the meridional pressure gradients, and therefore presumably also in the meridional winds. Complications of this sort have normally not been included in atmospheric models, but they undoubtably sometimes occur.

Alcayde et al. [8] using incoherent scatter observations have found temperature profiles in the altitude range 100—200 km that are characterized by steeper gradients around 130 km in summer than in winter; this could be due to the fact that solar heating is greater in summer than in winter. They have also observed temperature variations indicative of tides.

There is an asymmetry between the northern and southern hemispheres, and this is of course associated with the annual variation in the thermosphere, at least partly due to changing earth—sun distance. Volland et al. [9] indicate that the mechanism involves variable tidal energy associated with variable heat input into the ozone layer. Keating et al. [10] show a definite asymmetry in the seasonal variation of thermospheric helium between the north and south characterized by a larger amplitude in the south. Barlier et al. [11] have observed that the oxygen bulge tends to stay over the southern hemisphere. A closer look at the data of Barlier et al. leads to the conclusion that the atmosphere is more variable over southern mid-latitudes than over northern. Recognizing that helium is a tracer of the circulation while oxygen is more nearly a tracer of the temperature, it can tentatively be stated that both the effects of global circulation and the amplitude of the temperature variation are larger in the southern hemisphere.

The semi-annual variations in the thermosphere probably arise from dynamical effects within the atmosphere [12]. Volland et al. [9] attribute them to semi-annual variations in the stratospheric winds and in Joule heating associated with geomagnetic activity. In this view, the semi-annual variation in the stratosphere, associated with the breakdown of the winter polar vortex, serves as a variable source of internal gravity waves that in turn contribute to the thermospheric variation.

2.3. Variable Transport

There are a great variety of transport phenomena that occur in the upper atmosphere, and for purposes of discussion here the phenomena are divided into several categories that are not entirely distinct from one another. These are turbulence, internal gravity waves and tides, diurnal circulation, seasonal circulation, and random geophysical disturbance variations.

2.3.1. Turbulence

Perhaps the most surprising property of turbulence in the upper atmosphere is its omnipresence. Knowledge of the properties of turbulent motions in the upper atmosphere is far from adequate, but a few points may be put forth tentatively. The largest sized turbulent eddies appear normally to be smaller than 200 m with lifetimes less than 10^3 s. Larger eddies are probably occasionally present; Zimmerman and Trowbridge [13] have deduced scale sizes as large as a kilometer, but examination of vapor trails suggests that the presence of such large eddies is not the general rule. Turbulence in the upper atmosphere apparently is usually isotropic in character [6], at least for eddies smaller than 200 m, suggesting that the buoyancy subrange usually does not occur in the lower thermosphere. However, Blamont and Barat [14] state a contrary view — that the buoyancy subrange is often the only range present and that turbulence is mainly anisotropic.

The largest sized eddies provide most of the eddy transport, but eddies of lesser size, down to the smallest that can survive the dissipative effects of molecular diffusion, must normally also be present. The smallest size that can survive gets larger with increasing altitude, so eddies of a single scale size are to be anticipated at the highest altitude before eddy activity ceases altogether. The dissipation of eddies represents a significant energy input into the upper atmosphere, yet it is not sufficient to attribute this energy source to turbulence itself; one must look instead for the energy source that produces the turbulence, and this is almost surely tides and other internal gravity waves.

It is natural to look to strong shear zones as the means by which turbulence is produced, using the Richardson criterion to establish the value of shear required. However, examination of vapor trails does not in general lead to the conclusion that the strong shear zones are the only or even the main regions of turbulence, a point noted long ago by Greenhow and Neufeld [15] on the basis of radio echo

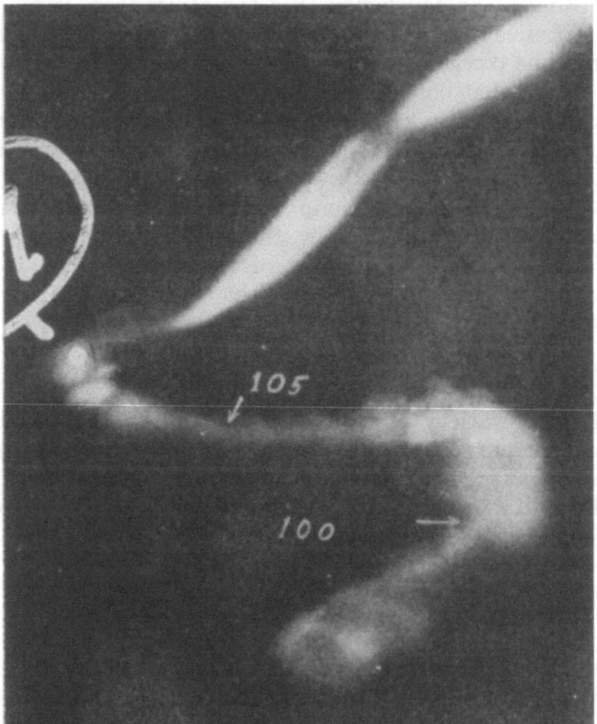

Fig. 1. A tri-methyl aluminum vapor trail with altitudes marked at 100 and 150 km. The dot at the left is a reference inserted in the analysis. This photograph was selected to show the necking down of the trail above the turbulent region (photo supplied by S. P. Zimmerman).

observations from meteor trails. Thus, if one accepts that the strong shear zones generate turbulence, it seems necessary to assume that turbulence propagates in some manner away from the strong shear zones into regions of lesser (or zero) shear. Although the mechanism by which this propagation occurs is not known, evidence for it seems strong, and the distance over which the turbulence can propagate must be comparable with the distance between strong shear zones, a distance that is of the same order of magnitude as the scale height. Occasionally turbulence is reported as being enhanced in regions of strong vertical or horizontal shear [16].

There are important conceptual differences about the way in which turbulence finally ceases with increasing altitude. One view is that eddy activity ceases suddenly at a turbopause, with the rate of eddy diffusion just below the turbopause generally considerably larger than the rate of molecular diffusion just above it so that a sudden reduction occurs at the turbopause in the total diffusivity, eddy plus molecular. The other view is that eddy activity does not cease so suddenly, that molecular diffusivity, which increases exponentially with height, finally exceeds eddy diffusivity above some altitude, and that eddy activity persists about a scale height or so into the region in which the molecular diffusivity exceeds the eddy diffusivity. In this latter view, there is no discontinuity in the eddy diffusivity and the turbopause is defined as the altitude above which the molecular diffusivity exceeds the eddy diffusivity (and below which the eddy diffusivity exceeds the molecular diffusivity). Some evidence in favor of this latter view is found in the measurements of diffusion above the turbopause that are too large to be explained in terms of molecular diffusion alone [17—19]; the excess is presumably due to a contribution from eddies that cannot be identified in the vapor trail photograph because of the larger effects of molecular diffusion. Further, Rees (private communication) has observed sharp variations in vapor trails up to 130 km very early in the lifetimes of trails before molecular diffusion obliterates the effect; these variations may be associated with turbulent motions well above the turbopause. Evidence in support of the opposing view is provided by occasional vapor trail photographs that show a sudden constriction at and immediately above the turbopause; although a few striking photographs of this sort do exist, it seems that this behavior is not representative of conditions most of the time. Fig. 1, provided through the courtesy of S. P. Zimmerman, was selected as a good example favoring the sudden cessation hypothesis for the turbopause.

Turbulence and the associated eddy diffusion produce strong transport effects in the atmosphere — most notably a downward transport of heat [20] and of atomic oxygen [21, 22]. Philbrick et al. [23] have measured atomic oxygen concentrations on different days and obtained rather different results that might be attributed to a change in the intensity of eddy diffusivity between the two measurements. Their results are shown in Fig. 2. The lesser atomic oxygen concentrations on 20 November 1970 might be explained in terms of greater eddy diffusivity on that day as compared with 5 October 1971.

2.3.2. Internal Gravity Waves and Tides

Internal gravity waves, including tides, provide an upward energy flux into the upper atmosphere. They grow in amplitude as they propagate upward and they produce strong shear zones in the lower thermosphere which generate turbu-

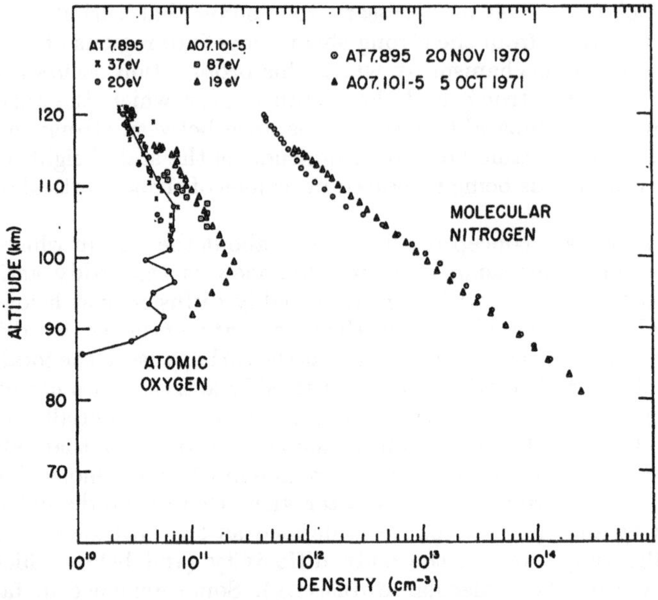

Fig. 2. Measurements of atomic oxygen and molecular nitrogen concentrations on two days showing a marked difference in the atomic oxygen concentrations [23] probably due to different eddy transport conditions.

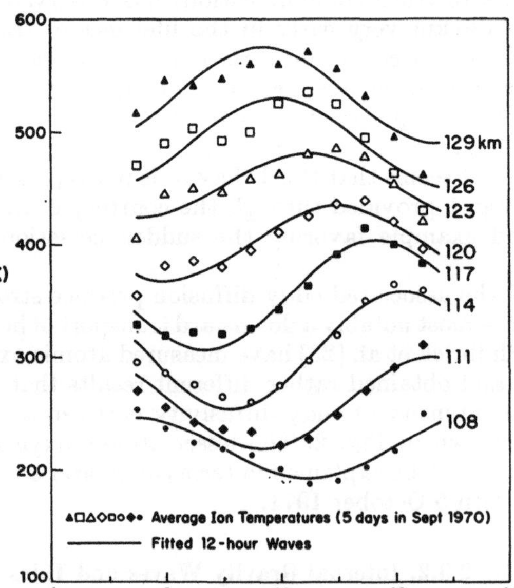

Fig. 3. Pattern of semidiurnal temperature changes as observed at Arecibo by Wand [29].

lence. The waves undergo viscous dissipation and deposit their energy in the form of heat in the upper atmosphere.

It is frequently supposed that internal gravity wave activity is stronger over the winter polar region than elsewhere and that this leads to greater eddy diffusivity there [24]. The supposition is eminently reasonable, though experimental evidence in its support is rather weak. The matter is of great importance in establishing the physical and composition characteristics of the atmosphere at 120 km. An equatorial source has also been suggested for the production of gravity waves [3].

Incoherent scatter has proved an effective means of observation of internal gravity waves and tides. Carru et al. [25], Wand [26], and Salah and Evans [27] have found strong evidence of temperature fluctuations in the E region due to tidal oscillations. Bernard and Spizzichino [28] have shown by means of a statistical analysis of wind and temperature measurements in the altitude range 100—140 km that a semidiurnal tide prevails in this region. Fig. 3 shows the pattern of semidiurnal temperature changes at 108—129 km at Arecibo as observed by Wand [29] in September 1970. Below 100 km, the diurnal tide, the semidiurnal tide and gravity waves all have comparable energies [3].

2.3.3. Diurnal Circulation

The diurnal circulation of the upper atmosphere might be included with the tides. It is discussed separately here only because it does not depend upon an upward flux of wave energy, at least not in the same sense as with internal gravity waves from the lower atmosphere. The energy input for the diurnal circulation is into the atmosphere in the same altitude range where the resulting circulation is of interest. However, it is recognized that the diurnal circulation and the tides cannot be completely separated when examining the wind fields. Amayenc and Reddy [30] give some support to this approach; by making a harmonic analysis of wind measurements in the 100—450 km region, they showed that the diurnal component can be viewed as the sum of trapped and propagating modes. The relative importance of these still has to be determined.

The diurnal circulation provides a transfer of heat from the hotter to the cooler regions of the upper atmosphere, and as a consequence must shift the phase of the temperature wave from what it would have been in the absence of the circulation. It must also cause some changes in composition, and these may in turn cause some phase difference between the thermal wave and the density wave. However, motion at and just above the turbopause is most important in producing composition changes: the diurnal wind patterns near that altitude are very poorly known and generally weak compared with those at higher altitudes. Tidal motions probably dominate the wind patterns near the turbopause [31] and hence are more important in determining composition changes than are the diurnal winds caused by solar heating.

Newton et al. [32] have presented composition measurements from the San Marco 3 satellite that indicate diurnal composition changes. Their results are shown in Fig. 4. It should be noted that it takes 40 days for the line of nodes on San Marco 3 to complete an orbital regression and so to sample a diurnal cycle; during this time other systematic changes may occur. The measurements show a clear tendency for atomic oxygen and helium to decline relative to nitrogen and argon during the hours of daylight — a result that might reasonably be expected due to transport associated with a diurnal circulation.

A similar behavior was observed by Alcayde et al. [33, 8] for atomic oxygen concentrations at 400 km as determined by incoherent scatter measurements; the observed variation did not simply follow the temperature variation but was characterized by a rapid decrease in the afternoon. Diurnal and semidiurnal

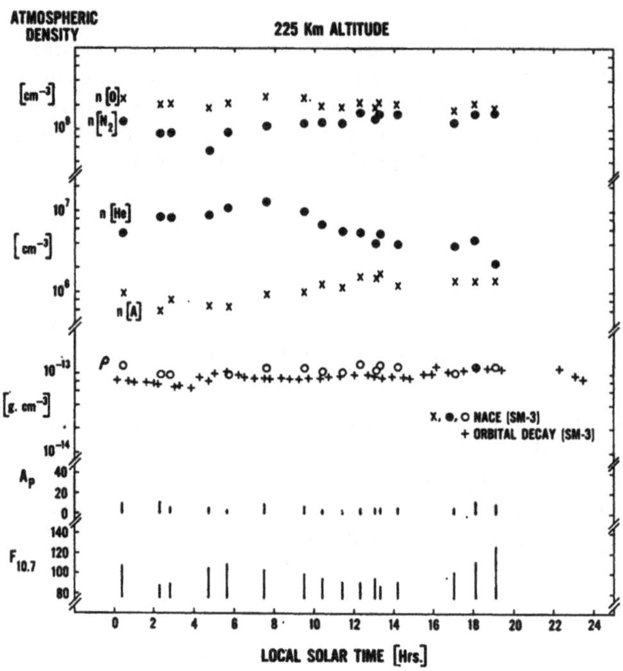

Fig. 4. Measurements of atmospheric composition and density made in San Marco 3, reduced to an altitude of 225 km [32].

changes in the lower boundary values for oxygen could account for the observations. The apparent discrepancy between the phases of satellite-drag and incoherent-scatter temperatures are mainly due to the variation in the oxygen concentration yielding an earlier maximum for density than for temperature.

In this connection it should be mentioned that winter wind observations in the 100—200 km region by Reddy [34] indicate a general tendency for poleward winds at altitudes between 120 and 200 km, associated with a large-amplitude wind oscillation around zero at 100 km. While the mean behavior at the higher altitudes might be associated with the seasonal circulation to be discussed in the next section, the 100 km behavior could be linked with the diurnal movement of the oxygen superimposed on the seasonal movement.

It is also interesting to note (Amayenc, private communication) that when the summer winds at 100 km are quite small, the corresponding summer diurnal variation of atomic oxygen at 225 km appears also to be considerably reduced [35].

2.3.4. Seasonal Circulation

There is strong evidence for an average seasonal circulation in the thermosphere from the summer hemisphere to the winter polar region; this is the relative warmth of the winter polar region and the presence there of atomic oxygen and excess helium. This circulation is not as strong as the diurnal circulation in which it is imbedded, which makes it more difficult to detect observationally. There is, however, some indication of this seasonal circulation in the data of Reddy and Vasseur [36]. Meridional wind measurements in the 150—225 km region at middle latitudes indicate a general tendency for poleward winds, this tendency being more accentuated in winter than in summer.

Hedin et al. [37] have presented data from OGO 6 indicating a small reduction in atomic oxygen and a larger reduction in molecular nitrogen at 450 km over the winter polar region, as shown in Fig. 5. Also shown is the winter enhancement in

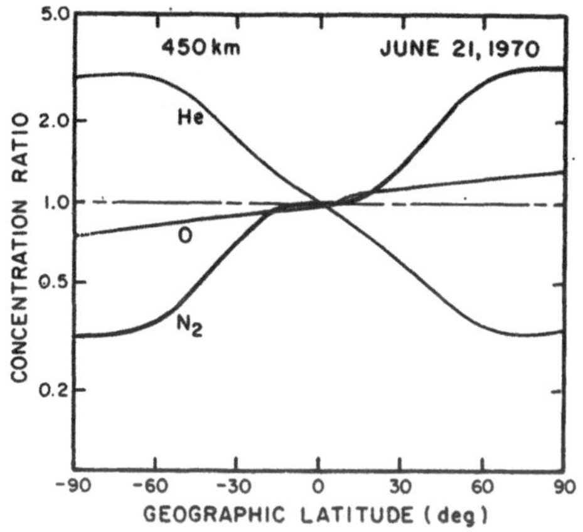

Fig. 5. Concentrations at 450 km of O, N_2 and He relative to equatorial concentration, inferred from measurements on OGO 6 [37].

the helium concentration. The enhancement of atomic oxygen relative to nitrogen can be explained, up to a factor somewhat less than two, by the average meridional (i.e. seasonal) circulation. The means by which this enrichment occurs is the same as for helium except that the effect is much less pronounced because the atomic weight of oxygen is not so much below the average molecular weight of air as is the case for helium. The enhancement of the lighter constituents is due to the fact the transport associated with poleward winds just above the turbopause is proportional to the scale height of the constituents in question [38].

Seasonal variations in atomic oxygen and molecular nitrogen have been observed by incoherent scatter [39, 40]. Instead of an absolute increase in atomic oxygen concentration at 200 km and above in winter, it is rather a decrease in the molecular nitrogen that occurs, thus producing a relative enrichment of atomic oxygen. The combined effects of a changing temperature profile in the 120—200 km region and of the global circulation control the concentration of atomic oxygen. While the circulation tends to enhance the atomic oxygen in winter, the temperature profile variation tends to reduce the concentration of atmospheric constituents in winter compared with summer. The opposing effects tend to compensate the atomic oxygen concentration at 200 km while the molecular nitrogen is only affected by the temperature-profile effect. Alcayde et al. [40] have found a seasonal variation in atomic oxygen at 120 km and 45° latitude amounting to a factor of 1.5.

2.3.5. Geophysical Disturbance Variations

Although it has long been clear that dramatic events associated with geomagnetic disturbances occur in the upper atmosphere over the polar regions, some of the most impressive evidence of the magnitude of the atmospheric perturbations has come from OGO 6 observations. The molecular nitrogen concentration at 500 km may increase by over an order of magnitude, as shown in Fig. 6 for a mildly disturbed day (7 March 1970) and a highly disturbed day (8 March 1970) [41]. Strangely, the atomic oxygen concentration does not change much, and this

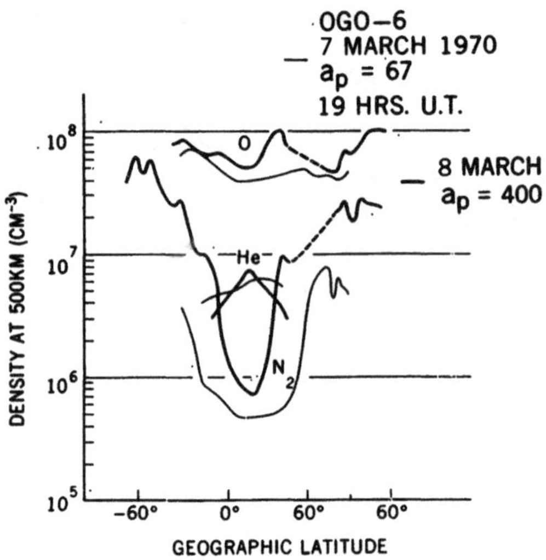

Fig. 6. Concentrations at 500 km of O, N_2 and He, inferred from measurements on OGO 6 [41] for a day of moderate magnetic disturbance (7 March 1970) and a highly disturbed day (8 March 1970).

is the reason that the magnitude of the perturbation was not recognized from satellite drag data. Apparently there is a very substantial increase in thermospheric temperature — enough to cause the large increase in molecular nitrogen at 500 km. The average meridional flow into the winter polar region should then be cut off, since the heat input deficit has become a surplus as a result of polar thermospheric heating associated with the geomagnetic disturbance. The source of atomic oxygen over the winter polar region is thereby cut off and the total atomic oxygen content should start to decrease. The near constancy of the atomic oxygen content at 500 km during the event apparently results from competition between opposing effects — the cutting off of the source on one hand and a reduction in the loss rate and an expansion upward of the atomic oxygen atmosphere as a result of temperature increase on the other; the reduction in the loss rate is due to the upward expansion of the atmosphere near the turbopause, the upward velocity of the atmosphere countering the downward diffusion of atomic oxygen through the remainder of the atmosphere.

Reddy [34], using incoherent scatter data from France, found evidence for equatorward winds from the winter polar region at 200 km during a geomagnetic disturbance. This constitutes direct evidence for the reversal of the seasonal average meridional circulation at that altitude during a geomagnetic disturbance. He also found evidence for a much weaker poleward wind at 100 km. This might be regarded as a return circulation associated with the equatorward wind at higher altitudes or as a continuation of the normal seasonal poleward circulation. In either case, if the observation is typical of disturbed conditions, it may help explain why the atomic oxygen concentration does not fall when the normal poleward circulation at higher altitudes is reversed; poleward winds near the turbopause are most effective in transporting atomic oxygen to the winter polar region and their continuance would provide a continuing source of atomic oxygen. However, unless the poleward winds near the turbopause are reduced from the normal values during the disturbance, the difficulty becomes one of explaining why the atomic oxygen concentrations do not increase along with the nitrogen concentrations at higher altitudes.

In addition to the substantial temperature increases observed at times of geomagnetic storms at middle latitudes, aurorally induced gravity waves have been observed on several occasions [42—44]. These waves, while of different characteristics from those generated at meteorological levels, could also initiate chemical changes. More observations are needed to make this point. It can be stated, however, that at these latitudes, in a similar fashion to that involving OGO 6 at high latitudes, a relative decrease in atomic oxygen with increasing geomagnetic activity has been observed [44].

3. Discussion

Direct measurements of atmospheric composition from eccentric orbit satellites do not provide information on global distributions of atmospheric parameters very satisfactorily because of the difficulty in extrapolating the measurements to any constant reference altitude, especially so for one as low as 120 km. The Atmospheric Explorers that are planned for launch in 1974 will have the capability of adjusting their orbits to circular and hence will be able to determine the distribution of atmospheric composition very effectively, at least at those circular-orbit altitudes within the orbital adjustment capability of the spacecraft and for the orbital plane of the spacecraft. However, it is unlikely that circular orbits below about 250 km will be practical because of the magnitude of the drag at lower altitudes, and hence the difficulty in extrapolating to 120 km or lower will remain. The Explorers will be able to make measurements at lower altitudes, perhaps down to 130 km, while in eccentric orbits, but the usual difficulties will be encountered in extrapolating to reference altitudes. This is not to downplay the value of the measurements at lower altitudes where many important processes of great importance to ionospheric physics take place; it is merely to state that the most valuable data obtained in eccentric orbit will be the investigation of processes of ionospheric physics rather than the observation of patterns of atmospheric composition at fixed reference levels.

The need to measure vertical profiles of atmospheric parameters remains very great, especially in view of the difficulties in extrapolating satellite data downwards from the lowest altitudes at which data on geographic distributions can be gathered effectively. The traditional way of obtaining such data has been by means of rockets, but incoherent scatter is rapidly emerging as a competitive source of such data with the added advantage of making long time sequences of measurements at relatively low incremental cost. High priority ought to be assigned to both rocket programs and incoherent scatter programs for obtaining vertical profiles of atmospheric parameters, as both are needed to improve the value of the measurements made by the other, and both are greatly needed to improve the value of the measurements made in satellites. Measurements at a good selection of latitudes are required. The combination of Atmospheric Explorer, rocket, and incoherent scatter measurements should provide good information on the patterns of atmospheric parameters in the thermosphere over the coming years and will probably disclose some important processes whose existence is not now even suspected.

While vapor trail measurements have provided most of our knowledge of turbulent processes, it should be stressed that ground-based techniques, like optical observations of the 5577 Å green line [45] and meteoric radars [15], offer a complementary and powerful approach to the study of turbulence because of the extended periods of observation provided by these means.

The need for coordinated programs including vapor-trail experiments and ground-based observations in order to study gravity waves and tidal waves should also be emphasized. The possibilities of night-time coverage with good height resolution provided by vapor trails would nicely complement the daytime coverage of incoherent scatter and the continuous coverage of meteoric radar over a limited altitude range.

The current problem on degree of oxygen dissociation as measured by mass spectrometers is an instrumental one that must be solved experimentally. Most mass spectrometer measurements have indicated atomic-to-molecular-oxygen ratios near unity at 120 km, but Offermann and von Zahn [46], using a liquid helium cooled ion source on their mass spectrometer, have obtained a value of 3.5. The 1971 atmospheric models of Jacchia [47] use a ratio near 2.5. The difference between the Jacchia models and the Offermann and von Zahn data at 120 km is not too serious since seasonal changes seem to be responsible for a factor of 1.5 variation in the atomic oxygen concentration at 120 km. A point of interest in this connection is that the meridional wind system required to compensate the asymmetry in heat input into the upper atmosphere cannot transport enough atomic oxygen to the winter polar region to produce the observed concentrations if the atomic-to-molecular-oxygen ratio at 120 km is unity. If it is 2.5, as indicated by the Jacchia 1971 models, the atomic oxygen deficiency is avoided. However, a ratio higher than 2.5 is favored in order to produce an atomic oxygen enhancement over the winter polar region.

Acknowledgment

The authors gratefully acknowledge the financial support of the National Aeronautics and Space Administration under Grant NGL 44-004-026 and of the Centre National d'Etudes des Communications and the Centre National de la Recherche Scientifique.

References

[1] C. O. Hines, Canad. J. Phys. 38, 1441 (1960).
[2] M. L. V. Pitteway and C. O. Hines, Canad. J. Phys. 41, 1935 (1963).
[3] A. Spizzichino, Annls Géophys. 25, 693 (1969).
[4] C. O. Hines, J. Geophys. Res. 70, 177 (1965).
[5] A. D. Richmond, Radio Sci. 6, 175 (1971).
[6] W. G. Elford and R. G. Roper, Space Research VII, 42 (1967).
[7] W. E. Swartz and J. S. Nisbet, J. Geophys. Res. 76, 185 (1971).
[8] D. Alcayde, P. Bauer, C. Jaeck and J. L. Falin, J. Geophys. Res. 77, 2378 (1972).
[9] H. Volland, C. Wulf-Mathies and W. Priester, J. Atmos. Terr. Phys. 34, 1953 (1972).
[10] G. M. Keating, D. S. McDougal, E. J. Prior and J. S. Levine, paper presented at XV COSPAR Plenary Meeting, Madrid 1972.
[11] F. Barlier, J. L. Falin, M. Ill and C. Jaeck, Space Research XIII, 349 (1973).
[12] F. S. Johnson, Space Research VII, 262 (1967).
[13] S. P. Zimmerman and C. A. Trowbridge, Space Research XIII, 203 (1973).
[14] J. E. Blamont and J. Barat, Annls Géophys. 24, 375 (1968).
[15] J. S. Greenhow and E. L. Neufeld, Proc. Phys. Soc. 74, 1 (1959).
[16] N. W. Rosenberg, D. Golomb, S. P. Zimmerman, W. K. Vickery and J. S. Theon, Space Research XIII, 435 (1973).
[17] D. Rees, R. G. Roper, K. H. Lloyd and C. H. Low, Phil. Trans. Roy Soc. A 271, 631 (1972).
[18] R. E. Good and D. Golomb, Space Research XIII, 249 (1973).
[19] K. H. Lloyd and L. M. Sheppard, Aust. J. Phys. 19, 323 (1966).
[20] F. S. Johnson and B. Gottlieb, Planet. Space Sci. 18, 1707 (1970)
[21] F. D. Colegrove, F. S. Johnson and W. B. Hanson, J. Geophys. Res. 71, 2227 (1966).
[22] T. Shimizaki, J. Atmos. Terr. Phys. 29, 723 (1967).
[23] C. R. Philbrick, G. A. Faucher and E. Trzcinski, Space Research XIII, 255 (1973).
[24] T. Shimizaki, Space Research XII, 1039 (1972).
[25] H. Carru, M. Petit, G. Vasseur and P. Waldteufel, Annls Géophys. 23, 455 (1967).
[26] R. H. Wand, J. Geophys. Res. 74, 5688 (1969).
[27] J. E. Salah and J. V. Evans, Space Research XIII, 267 (1973).
[28] D. R. Bernard and A. Spizzichino, J. Atmos. Terr. Phys. 33, 1345 (1971).
[29] R. H. Wand, paper presented at USNC/URSI spring meeting, 1972.
[30] P. Amayenc and C. A. Reddy, Planet. Space Res. 20, 1269 (1972).
[31] H. Volland, Planet. Space Sci. 17, 1581 (1969).
[32] G. P. Newton, D. T. Pelz and W. T. Kasprzak, Space Research XIII, 287 (1973).
[33] D. Alcayde, P. Bauer, C. Jaeck and J. Falin, J. Geophys. Res. 76, 7814 (1971).
[34] C. A. Reddy, Evidence of a meridional circulation cell in the lower thermosphere during a magnetic storm, paper presented at 15th COSPAR Symposium 1972.
[35] P. Bauer, P. Waldteufel and D. Alcayde, J. Geophys. Res. 75, 4825 (1970).
[36] C. A. Reddy and G. Vasseur, Space Research XII, 951 (1972).
[37] A. E. Hedin, H. G. Mayr, C. A. Reber, C. R. Carignan and N. W. Spencer, Space Research XIII, 315 (1973).
[38] F. S. Johnson and B. Gottlieb, Space Research IX, 442 (1969).
[39] L. P. Cox and J. V. Evans, J. Geophys. Res. 75, 6271 (1970).
[40] D. Alcayde, P. Bauer and J. Fontanari, paper presented at the 15th COSPAR Meeting, Madrid 1972.
[41] G. R. Carignan and C. A. Reber, Trans. Am. Geophys. Un. 52, 871 (1971) (abstract).
[42] J. Testud, J. Atmos. Terr. Phys. 32, 1763 (1970).
[43] R. M. Harper, J. Geophys. Res. 77, 1311 (1972).
[44] D. Alcayde, J. Testud, G. Vasseur and P. Waldteufel, J. Atmos. Terr. Phys. 34, 1937 (1972).
[45] J. Barat, J. E. Blamont, M. Petitdidier, C. Sidi and H. Teitelbaum, Annls Géophys. 28, 145 (1972).
[46] D. Offermann and U. von Zahn, J. Geophys. Res. 76, 2520 (1971).
[47] L. G. Jacchia, Smithson. Astrophys. Obs. Spec. Rep. No. 332 (1971).

Space Research XIII — Akademie-Verlag, Berlin 1973

WAVES AND TIDES AND THEIR OBSERVATION FROM GROUND AND SPACE*

J. Testud

Centre National d'Etudes des Télécommunications, Issy-les-Moulineaux, France

Planetary waves are probably generated by meteorological phenomena and seem to be correlated with the occurrence of stratospheric warmings. Both planetary waves and tidal waves may be investigated by meteor radar or incoherent scatter radar. Large scale gravity waves show correlation with magnetic activity and may be caused by auroral events. Medium scale gravity waves show no correlation with magnetic activity; they may originate in the lower atmosphere.

1. Introduction

This report covers the open session of Working Group 4 devoted to waves and tides; three main subjects were discussed: planetary waves, tidal waves, gravity waves. Discussion was based on a paper presented by Dr. C. O. Hines.

Since extensive reviews of these subject can be found in recent publications [1—5], this report is limited to a discussion of the main problems concerned with waves in the upper atmosphere.

2. Planetary Waves

Planetary waves consist of global oscillations of the neutral atmosphere with periods larger than one day.

The detection of these waves is in fact difficult: it requires observations lasting several days; the analysis of the data must eliminate carefully shorter period waves, especially tidal waves. However, the existence of such waves in the lower thermosphere (80—100 km) seems clearly established by meteor radar results similar to those of Greenhow and Neufeld [6], who observed long-period variations of the prevailing wind during an experiment lasting several days. Incoherent scatter results obtained at Saint Santin—Nançay (France) [7] suggest that planetary waves are absent or of small amplitude in the upper thermosphere, above 200 km.

Planetary waves are presumed to be generated at tropospheric heights by meteorological phenomena. As they propagate, they carry a large amount of energy upwards, quite huge indeed when compared with the thermobaric energy of the whole thermosphere. However, only a very small part of this energy reaches the 100 km level, as shown by the experimental results mentioned above. Most

* The paper reports a discussion based on a paper of above title presented by Dr C. O. Hines at the COSPAR Panel 4 B Joint Workshop Meeting on Thermosphere and Ionosphere Dynamics, Madrid 1972.

of it is dissipated below this height through damping processes such as thermal conduction, viscosity and turbulence. Stratifications of the horizontal wind reflect energy back to the ground. Planetary waves seem to be correlated with certain large scale temperature increases observed at low level and called "stratospheric warming".

It seems that the most fruitful tool for observing planetary waves in the upper atmosphere is meteor radar, thanks to its ability to make continuous measurements of neutral wind during several consecutive days in the altitude range 80—100 km. Long campaigns of observation with incoherent scatter radars would also provide a contribution to the observation of planetary waves above 250 km (only above this height are continuous measurements possible).

3. Tidal Waves

Tides are global oscillations of the atmosphere, the period of which is an integral fraction of one day. There are four main techniques for the observation of tides:

Fading: this early technique brought evidence of the presence of tides in the upper atmosphere.

Rocket launched artificial clouds: these experiments provide neutral wind measurements in the altitude range 90—160 km. However, this method is restricted to limited portions of the day: sunrise and sunset for sodium cloud experiments, and night for trimethylaluminium. Another disadvantage of these rockets experiments lies in their high cost.

Meteor radar: one (or two) component of the neutral wind can be measured at any time the day and in the altitude range 80—110 km. The analysis of the data in relation to time and space allows us to separate the different tidal modes which are present and in principle to identify them, knowing the vertical wavelengths associated with each mode [8].

Incoherent scatter sounder: this technique provides simultaneous measurements of several parameters: ion temperature and velocity, electron temperature and density, ion composition, etc. Through a sophisticated analysis of these data, the neutral gas temperature and velocity can be obtained with a good accuracy in the altitude range 100—500 km in day time, and 250—500 km during the night. The papers by Reddy et al. [9] and by Bernard [10] clearly illustrate the power of the incoherent scatter technique for the study of tidal waves in the thermosphere: the phase and amplitude of the 24-hour and 12-hour tides were determined independently for the neutral temperature at altitudes between 100 and 450 km.

Identification of the tidal modes without ambiguity requires observations at several latitudes; cooperation between many observatories using incoherent scatter radar, meteor radar, or rocket launches has become indispensable. The analysis of the data collected in coordinated experiments will certainly bring significant new knowledge of thermospheric tides.

Another interesting problem is the propagation of the tides above 120 km altitude. Computational models of their damping up to F-region heights already exist [11]. Recent experimental results obtained by incoherent scatter [9] should give a new impetus to such theoretical studies.

4. Gravity Waves

Two categories of experimental results are interpreted in terms of gravity waves:

(i) Direct measurements of neutral wind: the height profile of neutral wind determined by rocket launch sometimes exhibits a wave-like structure which is interpreted as gravity waves; analysis versus time and space of meteor radar data has demonstrated in many cases the existence of gravity waves with periods in the range 1—3 hours [12].

(ii) Observation of travelling ionospheric disturbances: irregularities of the electron density in the F2 region have been extensively observed using various techniques: ionosondes, virtual height records, hf Doppler shift, backscatter, total electron content and, more recently, incoherent scatter. In many cases, the wave-like behaviour of these irregularities has suggested an interpretation in terms of neutral gravity waves; in particular incoherent scatter observations, allowing simultaneous determination of many parameters, are able to yield a complete description of the phenomena; the results lead us to state definitely that large scale ionospheric disturbances are a manifestation of neutral atmospheric waves [13, 14].

The origin of one type of gravity wave called "large scale gravity waves", which correspond to "large scale ionospheric disturbances" seems clear. The characteristics of this type of wave are: period 1—3 hours; vertical wavelength 200—500 km; horizontal wavelength 1000—3000 km; horizontal propagation speed $\sim$500 m s^{-1}; propagation from auroral regions towards the equator.

Correlation of the occurrence of large scale gravity waves with magnetic activity, and their equatorwards propagation, have suggested for several years that they were generated by the intense energy input occurring during magnetic substorms in the auroral regions. Recent theoretical computations have shown that such an energy input could actually generate neutral waves with the characteristics of the observed ones. However, up to now, no definitely convincing experimental verification has been published. For that purpose it would be necessary to observe simultaneously auroral events occurring at high latitudes and their effects at mid-latitudes.

An experiment involving an incoherent scatter radar and a large scale network of ionosondes constitutes a very good tool for the study of large scale gravity waves; however, the 1-hour spacing of routine ionosonde data is quite insufficient; a 5-minute spacing is highly desirable.

At the same time, a good description of auroral events can be found in the routine magnetogram data proceeding from observatories located along the auroral oval. An experiment such as the above has been set up using the French incoherent scatter equipment and the European ionosonde network. Results will be published shortly. Preliminary results presented during the workshop session showed that the delay between an auroral event and the presence of large scale gravity waves at mid-latitude is about 1 hour.

The origin of the second type of gravity waves present in the thermosphere, called "medium scale gravity waves", is not clear. The characteristics of this type of wave are: period less than 1 hour, 20 minutes being very common; horizontal wavelength less than 500 km; horizontal propagation speed less than 200 m s^{-1};

lack of correlation with magnetic activity; propagation equatorwards during winter, polewards during summer, which is critically influenced by stratifications of neutral horizontal wind which are not themselves very well known.

It seems accepted that medium scale gravity waves originate in the lower atmosphere: a stratospheric origin, such as the jet stream, or a tropospheric origin, such as orographic waves, are often mentioned. In particular, studies of certain well defined sources of orographic waves like a mountain chain or an isolated island, would be of considerable interest for a better understanding of medium scale waves.

As a concluding remark, all types of waves present at thermospheric heights are subject to more or less intense dissipation there. Subsequent effects of this dissipation — dynamical heating, momentum deposition — certainly play an important role in the thermal balance of the thermosphere and in the general circulation of neutral air in the upper atmosphere.

References

[1] C. O. HINES, Phil. Trans. Roy. Soc. A **271**, 457 (1972).
[2] C. O. HINES, Report on workshop session of IUCSTP Working group 10, International Symposium on Waves in the Upper Atmosphere, Toronto 1970.
[3] C. O. HINES, Nature, Lond. **239**, 73 (1972).
[4] R. S. LINDZEN, J. Atmos., Terr. Phys. **31**, 449 (1969).
[5] G. VASSEUR, C. A. REDDY and J. TESTUD, Space Research XII, 1109 (1972).
[6] J. S. GREENHOW and E. L. NEUFELD, Quart. J. Roy. Met. Soc. **87**, 472 (1961).
[7] P. AMAYENC, private communication, 1972.
[8] A. SPIZZICHINO, Annls Géophys. **25**, 5 (1969).
[9] C. A. REDDY and P. AMAYENC, Planet. Space Sci. **20**, 1269 (1972).
[10] R. BERNARD, J. Atmos. Terr. Phys. **33**, 1345 (1971).
[11] R. S. LINDZEN, Geophys. Fluid Dynamics **2**, 89 (1971).
[12] I. REVAH, Thèse de doctorat d'état, Paris 1969.
[13] J. TESTUD and G. VASSEUR, Annls Géophys. **25**, 525 (1969).
[14] R. M. HARPER, J. Geophys. Res. **77**, 1311 (1972).

Space Research XIII — Akademie-Verlag, Berlin 1973

ORIGIN OF CURRENTS AND ELECTRIC FIELDS IN THE DYNAMO REGION*

S. MATSUSHITA[a] and F. S. MOZER[b]

[a]High Altitude Observatory, National Center for Atmospheric Research, Boulder, Col., USA
[b]Space Sciences Laboratory, University of California, Berkeley, Calif., USA

Recent theoretical and observational results of dynamo-region electric currents and fields are briefly reviewed. The variations with latitude of the diurnal and semidiurnal components of the field are considered and their possible origins in ionospheric dynamo or magnetospheric processes are discussed.

1. Quiet days

Electric currents responsible for the external part of geomagnetic solar quiet (*Sq*) daily variations (e.g. Fig. 1) seem to flow mainly in the dynamo region (90—150 km altitude), although some may flow along field lines from the dynamo region in one hemisphere to the other. Magnetospheric electric currents (such as front magnetopause, tail, and quiet-time eccentric ring currents) are not sufficient to explain *Sq* variations [1]. Rocket measurements of the geomagnetic field, solar flare and eclipse effects on geomagnetic and ionospheric variations, and height distribution of electric conductivity are good evidence of the dynamo-region electric current.

To explain the *Sq* current system, the dynamo mechanism caused by the lowest order solar diurnal tidal mode (1, —1) in the E region has been discussed [e.g. 1, 2]. An alternative suggestion [1] is that the necessary electric field in the dynamo region is produced originally by the plasmaspheric motion and propagated down along magnetic field lines. The electrostatic fields produced by the E-region dynamo and by the plasmaspheric dynamo mechanism are compared in Fig. 2a. Horizontal and vertical components of $\boldsymbol{E} \times \boldsymbol{B}$ drifts in the F region [3] caused by those two different electrostatic fields in the dynamo region are compared in Fig. 2b.

Another current system corresponding to lunar effects on the geomagnetic field (L) seems to flow in the dynamo region. To explain this L current system, a lunar semidiurnal tidal mode (2, 2) has been applied to the dynamo theory, and electrostatic fields for two different lunar phases are compared in Fig. 3a. Two components of $\boldsymbol{E} \times \boldsymbol{B}$ drifts in the F region in each case are presented in Fig. 3b. In order to apply these L results to *Sq*, an order of magnitude intensity increase for both $\boldsymbol{E}$ fields and drifts is required if the semidiurnal component at mid-latitudes is the main variation in *Sq*, as discussed later.

Horizontal $\boldsymbol{E} \times \boldsymbol{B}$ drifts in the E and F regions observed by the incoherent scatter radar system at Millstone Hill (42.6° N, 71.5° W geographic; dip latitude

---

* This paper summarizes the discussion of this topic at the COSPAR Panel 4B Joint Workshop Meeting on Thermosphere and Ionosphere Dynamics, Madrid 1972.

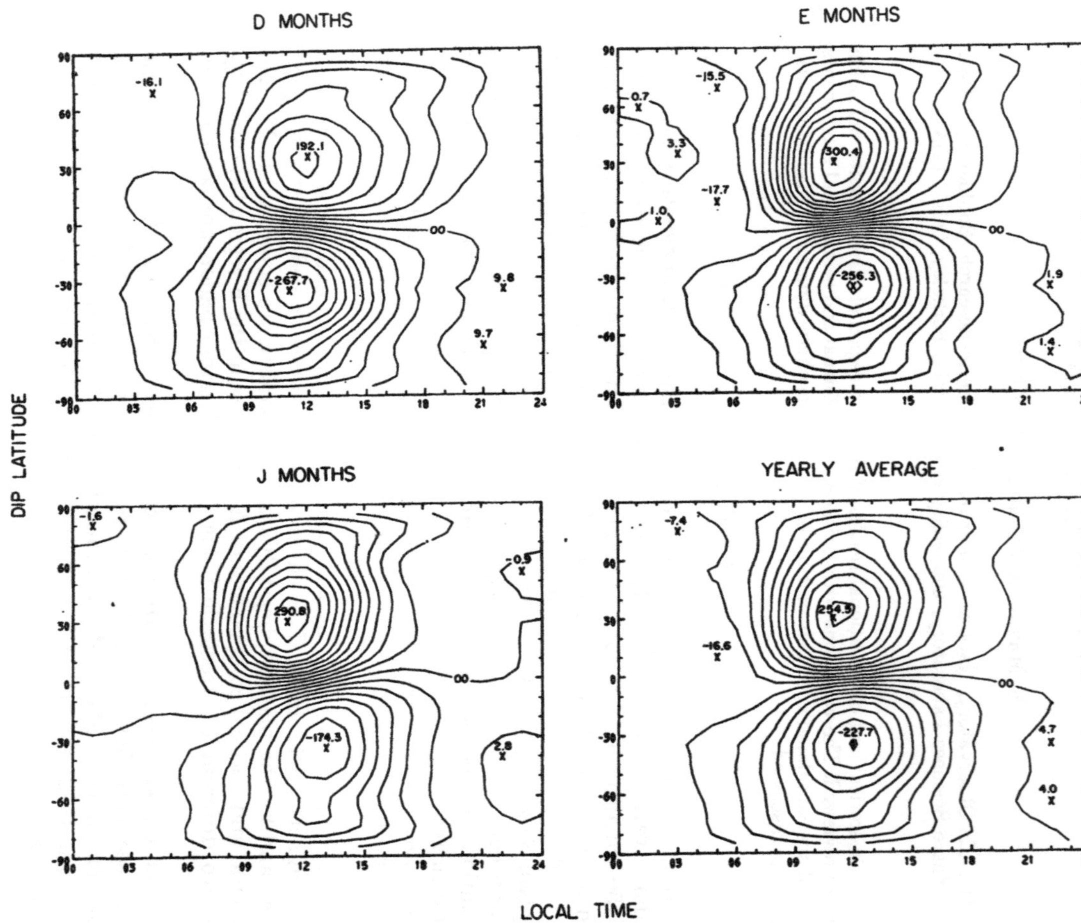

Fig. 1. External *Sq* current systems for December solstitial, equinoctial and June solstitial seasons and the yearly average during the IGY. The current intensity between two consecutive lines is 25×10^3 amp, and the numbers near the cross marks indicate the total current intensities of vortices (positive is counter-clockwise) in units of 10^3 amp.

57°) are shown in Fig. 4. These are averaged results of nine measured days in different seasons of 1968—1969: two of them are geomagnetically disturbed days and three are quiet [4]. Thus they may be assumed to represent a yearly average behavior on a moderately quiet day. From the vector rotation and the time lag between the two levels it appears that the electric field responsible for the drift in the F region is generated in the E region by the dynamo action of the solar semidiurnal tidal oscillation, such as the (2, 2) mode [4, 5]. However, F-region vertical drifts observed at Jicamarca (12.0° S, 76.9° W geographic;

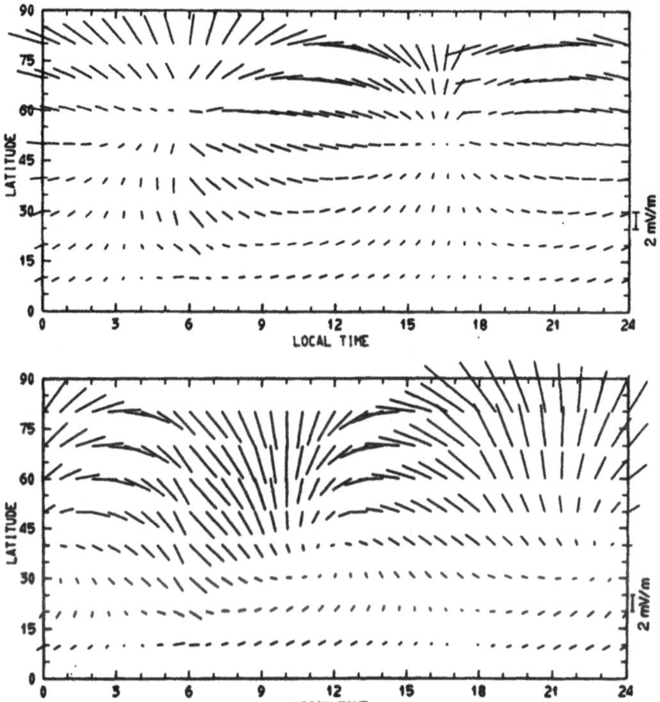

Fig. 2(a). Dynamo-region electrostatic field for the *Sq* produced by E-region dynamo due to the wind of (1, —1) tidal mode (top) and that produced by plasmaspheric dynamo (bottom).

dip latitude 1°) shown in Fig. 5 suggest that the daily variation of the $\boldsymbol{E}$ field is mainly diurnal, although an additional increase often appears near dusk and a small increase occasionally appears near dawn [6].

The incoherent scatter radar observations of wind at St. Santin (44.7° N, 2.2° E geographic; dip latitude 42°) in summer and winter indicate predominance of a diurnal component at 100—450 km altitude [7], and a semidiurnal component with a mode of (2, 4) at 100—135 km altitude and (2,2) at 135—225 km [7, 8]. The (2, 4) mode is also assumed to explain the night-time intermediate ionosphere at 110—140 km [9]. However, F-region diurnal neutral winds deduced from St. Santin data present a better theoretical agreement when a *diurnal* electric field from the dynamo region is taken into consideration during summer and equinoxes [10]. Also, the F-region $\boldsymbol{E} \times \boldsymbol{B}$ drifts observed at Arecibo (18.3° N, 66.8° W geographic; dip latitude 31°) suggest an approximate agreement with the diurnal

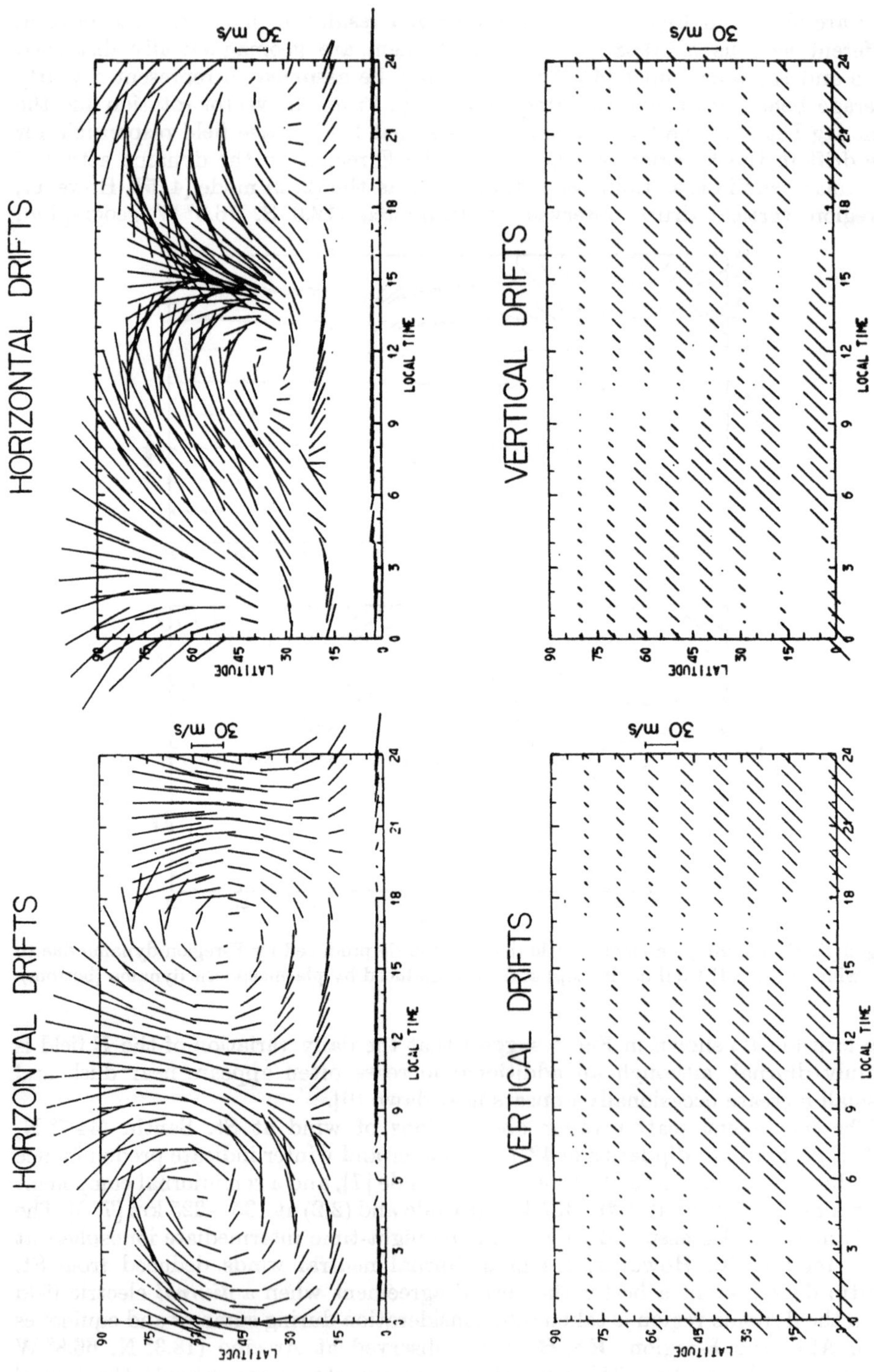

Fig. 2(b). Horizontal and vertical components of $\boldsymbol{E}\times\boldsymbol{B}$ drifts at 300 km altutude due to the $\boldsymbol{E}$ shown at the top of Fig. 2a (left) and due to the $\boldsymbol{E}$ shown at the bottom of Fig. 2a (right).

E shown in the bottom of Fig. 2a (Harper, private communication). Measurements of chemiluminous clouds released by rockets give the E-region wind shears [11] and even the vertical wind speed of about 10 m s^{-1} (e.g., Project Aladdin, Rosenberg, private communication), but provide almost no information in the daytime. Haerendel (private communication) and Rosenberg (private communication) have each reported ten to twenty barium releases at mid-latitudes in which they found that the barium cloud ionospheric drift at sunrise was generally westward, in agreement with the data of Fig. 4. However, the barium cloud data generally

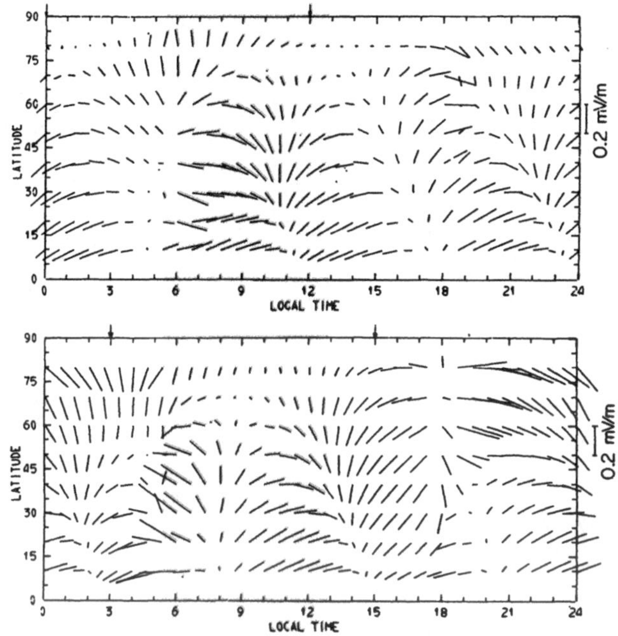

Fig. 3 (a). Dynamo-region electrostatic field for the L produced by the E-region dynamo due to the wind of (2, 2) tidal mode when the moon is over the 0- or 12-hour meridian (top) and over the 3- or 15-hour meridian (bottom).

indicated an eastward drift near sunset, which disagrees with the results of Fig. 4 but is consistent with the diurnal variation of the lower latitude data. Equatorial electric field measurements from observations of whistler duct motion have shown that mid-latitude variations like those of Fig. 4 occur fairly regularly (Carpenter, private communication). However, diurnal patterns also appear and the typical variation depends on magnetic conditions. For example Morgan (private communication) has observed a drift pattern exhibiting a complex dependence on local time during one particular day. Various theories are presented on thermospheric wave dynamics [12], gravity waves [13], and thermospheric wind effects on the F region [14, 15], but no definite answer is given to the question of the dynamo-region currents.

Based on the incoherent scatter observations at four places (Millstone Hill, St. Santin, Arecibo, and Jicamarca), a clue is the predominant semidiurnal component at dip latitudes higher than 35° and the predominant diurnal component

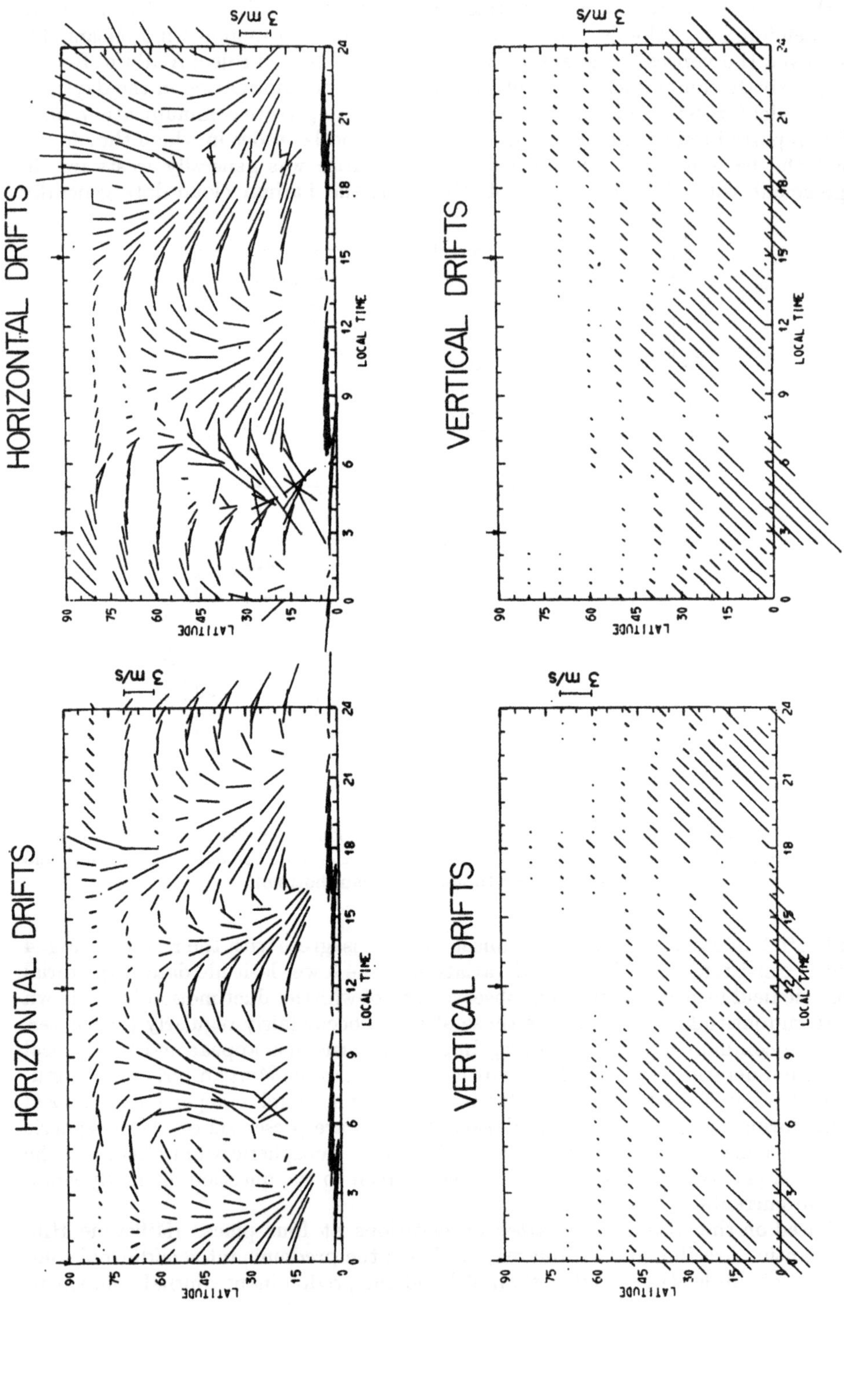

Fig. 3 (b). Horizontal and vertical components of $\boldsymbol{E} \times \boldsymbol{B}$ drifts at 300 km altitude due to the $\boldsymbol{E}$ shown at the top of Fig. 3a (left) and due to the $\boldsymbol{E}$ shown at the bottom of Fig. 3a (right).

at low and equatorial latitudes. In Fig. 6 the middle diagram is the *Sq* yearly average current in the northern hemisphere and the bottom one is the current system estimated from the (1, —1) tidal mode. When the schematic current pattern shown by broken lines in the top diagram of Fig. 6 is added to the bottom one, the results become much more similar to the *Sq* current. These additional currents are in the region higher than dip latitude 35°, and combined currents may show a stronger semidiurnal component, as observed at Millstone Hill and St. Santin. To explain an increase in ***E*** over Jicamarca around dawn and dusk, two small currents may add to the equatorial region (solid line in the top diagram of Fig. 6). However,

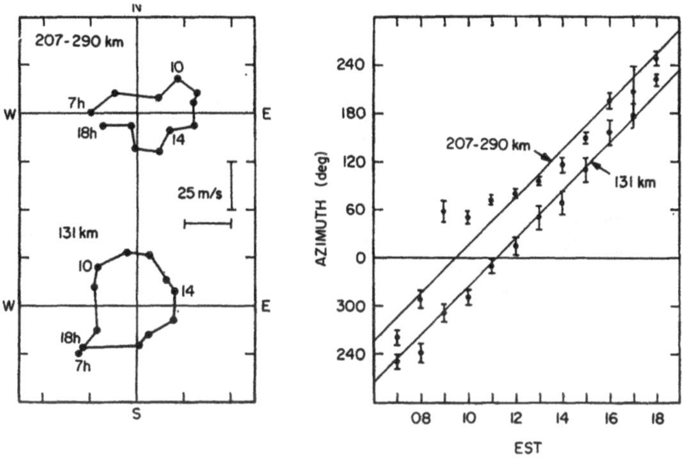

Fig. 4. Vector diagrams (left) and azimuth variations with time (right) in 75° WMT of average hourly horizontal drifts at 207—290 km and 131 km heights observed by Millstone Hill incoherent scatter radar system (arranged from Evans [4]).

this is not particularly essential, since the increase can be explained by other mechanisms. A sudden drop of conductivity, different location of the dip equator and the geographic equator, and the F-region dynamo need to be considered. At higher latitudes than the equatorial region, electrodynamic F-region drift caused by conjugate point sunrise has also been suggested [16]. Schieldge, Richmond and Venkateswaran (private communication) simulated an *Sq* pattern on 21 September 1958 using (2, 2), (2, 3), (2, 4) tidal modes in addition to (1, —1). Since the diurnal wind speed is greater than that of semidiurnal winds, the eastward ***E*** field obtained presents a diurnal variation with an increase in the evening (19 h at the equator). This type of calculation with various combinations of different modes may produce an ***E*** field distribution to satisfy both incoherent scatter observations and an *Sq* current system deformed greatly near 60°. The problem is to make arbitrary selections of the phase and amplitude for different tidal modes. At present, the (1, —1) mode still seems to be the main tide responsible for producing low-latitude *Sq* current by the E-region dynamo action. However, if Arecibo or low-latitude stations regularly observe a semidiurnal component stronger than the diurnal component, the E-region dynamo action will not be very hopeful. The question of the typical pattern of the local time variation of the higher mid-latitude electric field must be considered as open until further

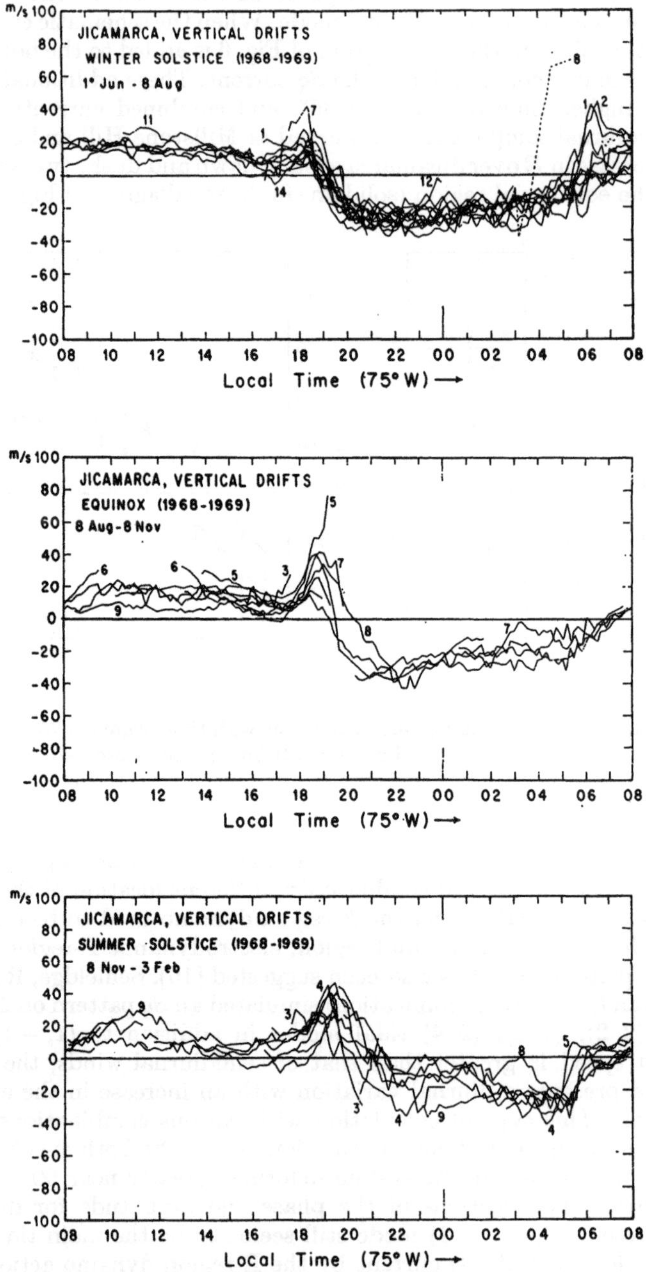

Fig. 5. Daily variation of vertical ionospheric drifts observed at Jicamarca in three seasons (after Woodman [6]).

electric field data obtained by the many different techniques can be merged into a single consistent, typical pattern.

Concerning the L-current system, the previously suggested lunar semidiurnal (2, 2) mode is satisfactory. The estimated $\boldsymbol{E}$ field (Fig. 3a) causes a reasonable foF2 variation [17] as shown in Fig. 7. The eastward $\boldsymbol{E}$ field for Sq is based on Jicamarca observations and the $\boldsymbol{E}$ for L is superposed on that. It may safely be concluded that the L current system and the $\boldsymbol{E}$ field in the dynamo region are produced by the E-region dynamo action.

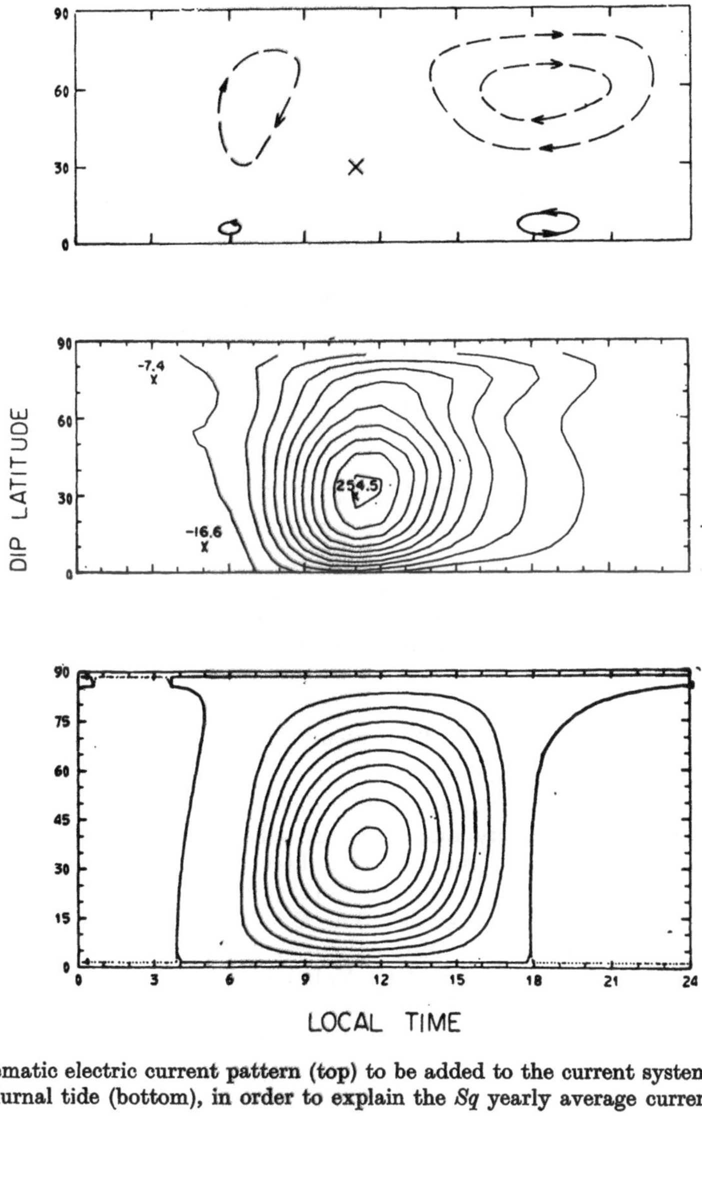

Fig. 6. Schematic electric current pattern (top) to be added to the current system estimated from the diurnal tide (bottom), in order to explain the Sq yearly average current (middle).

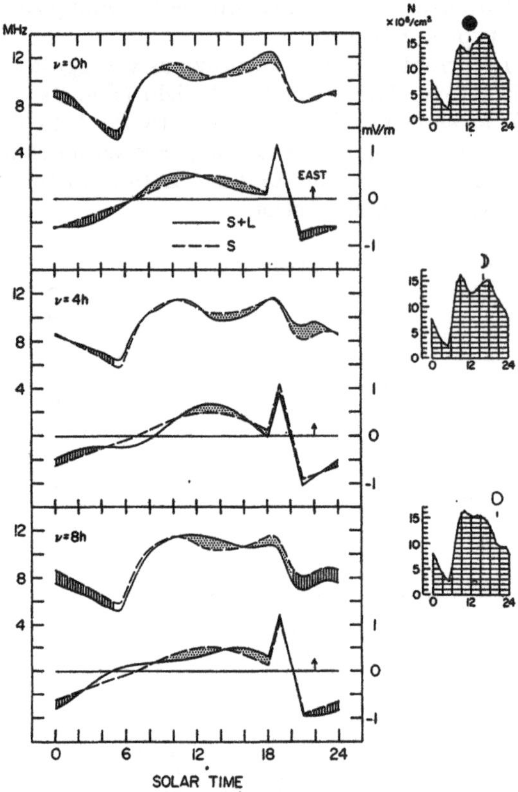

Fig. 7. Daily variations of fo F2 estimated theoretically from a model of electrostatic field distribution for *Sq* (broken curves) and *Sq* with L (solid curves) over the dip equator for three lunar phases. Daily variations of *N* obtained at Huancayo corresponding to the three lunar phases are shown at the right.

2. Disturbed periods

As reviewed recently [18, 19], electric fields associated with substorms are complicated, although the dynamo-region $\boldsymbol{E}$ field at high latitudes during disturbances is mainly caused by the magnetosphere. Effects of magnetospheric convective electric fields on the plasmasphere [20] and on polar thermospheric winds [21] have also recently been discussed. The $\boldsymbol{E}$ field observations by balloon probes were made in high latitudes by Mozer [22—24], and mapping of the $\boldsymbol{E}$ field into the lower atmosphere is discussed by Volland [25]. Also the $\boldsymbol{E}$ field has been measured by rocket probes [26] and satellites [27] and by barium clouds [27, 28]. Combined results of previous observations and recent OGO 6 probe measurements [29] are presented in Fig. 8. It must be noted that the field in the polar cap is from dawn to dusk, and the field in the auroral latitudes is, as expected, poleward before midnight and equatorward after that time. A very interesting measurement [4] has been made at Millstone Hill at the beginning of a sudden commencement storm (see Fig. 9): a strong westward $\boldsymbol{E} \times \boldsymbol{B}$ drift (hence, poleward $\boldsymbol{E}$) occurred

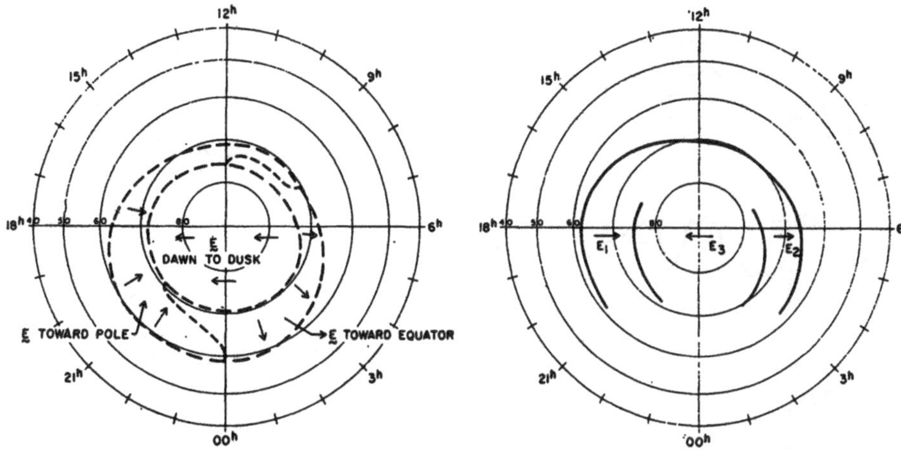

Fig. 8. Electric field distribution in high latitudes during substorms obtained by various techniques (magnetic disturbance analysis, OV 1-10 satellite, Ba+ releases) (left) and by OGO 6 observations (right) (after Heppner [29]). If W is width of region, $\bar{E}_3 W_3 \approx \bar{E}_1 W_1 + \bar{E}_2 W_2$; typically $\bar{E}_3 W_3$ in center of range 20—100 kV·

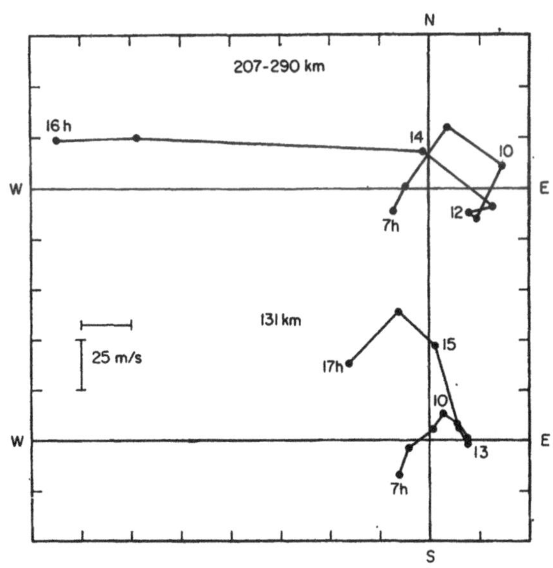

Fig. 9. Vector diagrams in 75° WMT of hourly horizontal drifts (207—290 km and 131 km heights) observed at Millstone Hill on a disturbed day, 14 May 1969 (after Evans [4]). A large sudden commencement occurred at 1429 in 75° WMT. $\sum Kp = 50$.

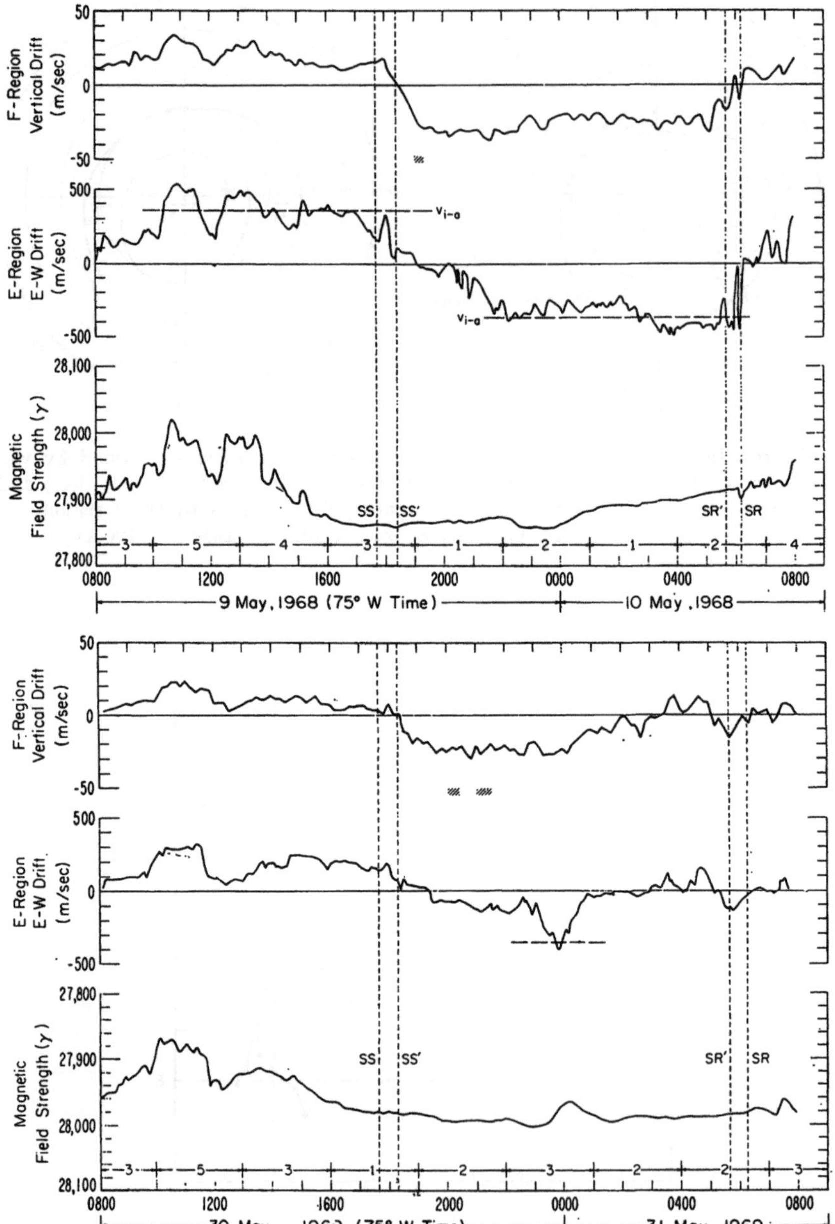

Fig. 10. F-region vertical upward drifts and E-region electron westward speeds observed at Jicamarca are compared with the geomagnetic H variations and three-hour local K indices at Huancayo (after Balsley and Woodman [30]). The horizontal broken lines indicate the nominal ion—acoustic velocity, V_{i-a}.

at the F level in the afternoon. At the E altitude, the vector rotation changed from clockwise to counterclockwise. This kind of observation at Millstone Hill and low latitudes is extremely helpful for discussing dynamo-region currents and $\boldsymbol{E}$ field during disturbances. At mid-latitudes in the immediate vicinity of an intense sporadic $\boldsymbol{E}$ layer, long-boom probe measurements detect large excursions of the electrostatic field amounting to 6.5 mV m^{-1} (Burrows and Rees, private communication). To show equatorial behavior, two examples of Jicamarca F-region vertical drifts and E-region electron speeds (roughly speaking, electric current) compared with Huancayo geomagnetic variations are presented in Fig. 10. In the daytime, a 25γ decrease of H disturbances seems to be caused by an approximately 100 m s^{-1} drop of electron speed, which is in turn a westward $\boldsymbol{E}$ field causing a drop of F-region vertical upward drift of a few meters per second. Causes of low latitude $\boldsymbol{E}$-field changes are an interesting subject [31]. In addition to the plasmaspheric and magnetospheric dynamo, E- and F-region dynamo due to disturbed wind and wave systems must be studied. Electric currents leaked from the auroral region toward middle and low latitudes on the daylight side should be carefully examined.

3. Discussion

The interaction of the solar wind with the terrestrial magnetic field produces important electric fields that map along magnetic field lines into the ionosphere. It is the purpose of the present section to compare such field with those from the ionospheric dynamo in order to assess their effects on the Sq current system.

Magnetospheric electric fields are known to vary by factors of about two to four between quiet and active periods [22, 24]. Because changes of ionospheric electrical conductivities at high latitudes also occur during active periods as a result of particle precipitation, very large differences in ionospheric equivalent current systems are produced by geomagnetic activity [18, 19]. During reasonably quiet periods the magnetospheric electric field is thought to be roughly independent of altitude in the equatorial plane and to have a magnitude of about 0.4 mV m^{-1} [22]. Such a field maps along magnetic field lines to produce an ionospheric field whose latitude dependence is given in Fig. 11. Since the average magnitude and latitude dependence of this field are known to only a factor of two to four, and since its time variations are similar in magnitude, Fig. 11 should not be interpreted quantitatively. Nevertheless, qualitative comparisons of the magnetospheric electric field with dynamo and co-rotation fields are important. For this purpose the range of dynamo-field strengths given in Fig. 2a is reproduced in Fig. 11, as is the co-rotation electric field that would be observed in a non-rotating frame of reference. Several comments can be made from comparison of these rough estimates of field strengths:

(1) *The plasmasphere nearly co-rotates with the earth.* At latitudes below 45° or so, the dynamo and magnetospheric electric field strengths are less than $\sim 20\%$ of the co-rotation field strength. Thus, the main motion of the plasmasphere is clearly rotational. At a latitude of about 60° the magnetospheric electric field is comparable with the co-rotation field so the ionospheric plasma motion differs greatly from co-rotational. This fact has been used to explain the plasmapause in many theories.

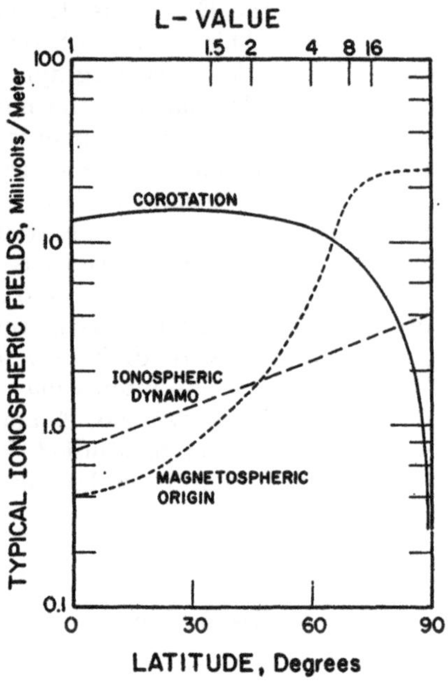

Fig. 11. Typical values of electric-field strengths resulting from co-rotation, magnetospheric processes, and the ionospheric dynamo presented in a non-rotating frame of reference as a function of latitude (after Mozer [32]).

(2) *At high latitudes, the dominant electric field is due to magnetospheric processes.* At latitudes above about 65° the ionospheric electric field resulting from magnetospheric processes is about an order of magnitude larger than that due either to co-rotation or dynamo processes. Thus, ionospheric currents and motions are controlled by external processes which might interact back on the neutral atmosphere through ion—neutral collisions to create large neutral winds (LOGACS result by DeVries and Wu, private communication). Such effects may account for some of the latitude dependent phenomena discussed in the previous section.

(3) *At low latitudes, electric fields of magnetospheric origin can be comparable with those arising from dynamo winds.* The major uncertainty in this conclusion involves the extent to which magnetospheric electric fields penetrate the plasmasphere. Many recent theories have shown that the magnetospheric electric field sets up a ring current at the plasmapause and the charges in this current produce an opposing field that efficiently shields the magnetospheric field from the plasmasphere [33]. When inertial effects are included in such theories, the conclusion is reached that low-frequency field variations are effectively shielded while the finite time response of the magnetosphere prevents the above process from effectively shielding high-frequency variations. The critical time response of the magnetosphere is estimated in various models as 1 to 10 hours.

Experimentally it is clear that $\lesssim 1$ hour fluctuations in the magnetospheric electric field efficiently penetrate into the low latitude ionosphere. An example of such an observation in conjunction with a sudden commencement storm is given in Fig. 9, where the drift deviated from the normal semidiurnal pattern in the afternoon and a large $\boldsymbol{E} \times \boldsymbol{B}$ motion was observed. At the present level of understanding it is fair to conclude that, even during magnetically quiet times, magnetospheric electric fields comparable with dynamo fields can be present in the low-latitude ionosphere for periods as long as one to several hours.

4. Conclusion

To answer the question whether the *Sq* dynamo currents are mainly caused by the E-region dynamo action or by the plasmaspheric dynamo, further detailed observations of the E-region winds and ***E*** field, particularly in low latitudes, are essential. If these results in low latitudes clearly show a predominant diurnal variation the E-region dynamo is satisfactory. If these present a semidiurnal component stronger than the diurnal, the plasmaspheric dynamo may be more satisfactory. An observation, such as shown by Figs. 4 and 9, will greatly assist in separating the ***E*** effect caused by the field from the E region or by that from the plasmasphere. More sensitive probe measurements of the ***E*** fields together with barium experiments are also very helpful. Daytime wind observation by lithium clouds [34, 35] may be another future technique. Theoretical investigations of thermospheric tidal and wave motions should certainly be encouraged. Simulation of *Sq* current by combining various tidal modes is worthwhile to try, even if it does not give any conclusive result. Day-to-day fluctuations of *Sq*-type current and correlated thermospheric variations are another worthwhile project.

There remain questions on the electric current and field in the dynamo region at middle and low latitudes during disturbances. Extensive studies of substorms have been made for the high latitudes, but little has been done for the low latitudes. Although the disturbance mechanism in high latitudes is not yet very clear, and hence needs further investigations, the disturbance mechanism in middle and low latitudes should be relatively simpler than that in high latitudes, and hence would be more fruitful. Detailed observational and theoretical investigations of thermospheric, ionospheric and magnetospheric dynamics during disturbances should be strongly encouraged.

Acknowledgments

One of the authors (S. M.) would like to thank Drs. J. D. Tarpley and D. N. Anderson for their helpful discussion and Mrs. Elaine L. Miller for her great assistance in preparing the present paper. The other author provided the third section in addition to revising several places in other sections. The National Center for Atmospheric Research is sponsored by the National Science Foundation.

References

[1] S. Matsushita, Radio Sci. **6**, 279 (1971).
[2] H. Volland, Cosmic Electrodynamics **1**, 428 (1971).
[3] S. Matsushita, Space Research XII, 1087 (1972).
[4] J. V. Evans, J. Geophys. Res. **77**, 2341 (1972).
[5] J. V. Evans, J. Atmos. Terr. Phys. **34**, 175 (1972).
[6] R. F. Woodman, J. Geophys. Res. **75**, 6249 (1970).

[7] P. Amayenc and C. A. Reddy, Planet. Space Sci. **20**, 1269 (1972).
[8] R. Bernard and A. Spizzichino, J. Atmos. Terr. Phys. **33**, 1345 (1971).
[9] K. Fujitaka and T. Tohmatsu, J. Atmos. Terr. Phys. **35**, 425 (1973).
[10] P. Amayenc and G. Vasseur, J. Atmos. Terr. Phys. **34**, 351 (1972).
[11] N. W. Rosenberg and S. P. Zimmerman, Radio Sci. **7**, 377 (1972).
[12] S. Kato, Space Sci. Rev. **12**, 421 (1971).
[13] J. D. Whitehead, Radio Sci. **7**, 403 (1972).
[14] D. Eccles, J. W. King and H. Kohl, J. Atmos. Terr. Phys. **33**, 1371 (1971).
[15] H. Rishbeth, J. Atmos. Terr. Phys. **34**, 1 (1972).
[16] H. C. Carlson, Jr. and J. C. G. Walker, Planet. Space Sci. **20**, 141 (1972).
[17] D. N. Anderson, S. Matsushita and J. D. Tarpley, Trans. Am. Geophys. Un. **53**, 475 (1972).
[18] E. N. Bramley and M. I. Pudovkin, in: Solar-Terrestrial Physics, D. Reidel Publ. Co., Dordrecht, Holland 1972.
[19] G. Rostoker, Rev. Geophys. Space Phys. **10**, 157 (1972).
[20] A. J. Chen and R. A. Wolf, Planet. Space Sci. **20**, 483 (1972).
[21] J. A. Fedder and P. M. Banks, J. Geophys. Res. **77**, 2328 (1972).
[22] F. S. Mozer and R. H. Manka, J. Geophys. Res. **76**, 1697 (1971).
[23] F. S. Mozer, J. Geophys. Res. **76**, 3651 (1971).
[24] F. S. Mozer, J. Geophys. Res. **76**, 7595 (1971).
[25] H. Volland, J. Geophys. Res. **77**, 1961 (1972).
[26] M. C. Kelley, F. S. Mozer and U. V. Fahleson, J. Geophys. Res. **76**, 6054 (1971).
[27] N. C. Maynard, NASA-GSFC Rep. No. X-645-71-231 (1971).
[28] J. P. Heppner, J. D. Stolarik and E. M. Wescott, J. Geophys. Res. **76**, 6028 (1971).
[29] J. P. Heppner, NASA-GSFC Rep. No. X-645-72-10 (1972).
[30] B. B. Balsley and R. F. Woodman, WDC-A Rep. No. UAG-17 (1971).
[31] R. F. Woodman, D. L. Sterling and W. B. Hanson, Radio Sci. **7**, 739 (1972).
[32] F. S. Mozer, Rev. Geophys. Space Phys., **11**, in press (1973).
[33] V. M. Vasyliunas, in: Particles and Fields in the Magnetosphere, D. Reidel Publ. Co., Dordrecht, Holland 1970.
[34] J. F. Bedinger, GCA Rep. No. 71-4 (1971).
[35] J. F. Bedinger, GCA Rep. No. 72-3 (1972).

Space Research XIII — Akademie-Verlag, Berlin 1973

PLANETARY SCALE MOTIONS AT F LAYER HEIGHTS*

R. E. Dickinson[a] and H. Rishbeth[b]

[a]National Center for Atmospheric Research, Boulder, Col., USA
[b]Radio and Space Research Station, Slough, UK

This paper describes some current ideas on three topics that were described at the Workshop Meeting on Thermosphere and Ionosphere Dynamics: Super-rotation of the thermosphere (Section 2 of the paper); the phase of the diurnal density, temperature and pressure variations (Section 3); and the mean meridional circulation in the thermosphere (Section 4). Various points that were brought out in discussion are incorporated in the text.

1. Introduction

At the 1971 COSPAR meeting in Seattle, thermospheric dynamics was viewed as the problem of deducing the composition, temperature and motion fields of the neutral constituents, the only specified parameters (besides fundamental physical constants) being the incident solar radiation and the coupling with the lower atmosphere [1]. The neutral composition of the thermosphere is determined by balances between photodissociation of molecules, transport of constituents by vertical molecular and eddy diffusion and transport by large-scale wind systems. The distribution of temperature depends on balances between solar heating, vertical molecular and eddy conduction and adiabatic heating by vertical motion. In turn, the three-dimensional distribution of constituents and temperature creates the large-scale horizontal pressure gradients which are responsible for driving the global atmospheric wind systems.

While the overall problem treating all the mutual interactions is not entirely intractable with current high-speed computers, it is far from solved. Further understanding can be gained by studying individually the various linkages involved, such as that between motions and ionization, which is the central theme of this COSPAR workshop. Motions affect the distribution of ionization directly by transport and indirectly through their influence on temperature and composition. Conversely, the transfer of momentum between the neutral gas and the ions is of major importance in the equations for the horizontal motion of the neutral gas.

One approach, developed within the past decade to study this particular linkage, starts by assuming a global distribution of pressure (based on empirical density models) as a driving force in the equation of motion. Wind velocities are then calculated, and the ion drifts they produce are incorporated in models of the F layer. In some of the past studies, the winds have been determined using an

* Based on introductory remarks prepared for Topic A, 'Coupling of ion-drag with planetary-scale motions at F layer heights' of COSPAR Panel 4B Workshop Meeting on Thermosphere and Ionosphere Dynamics, Madrid, Spain, 18 May 1972.

assumed distribution of ion-drag while in others the winds and ionization have been calculated jointly in a self-consistent fashion: see [2] and [3] for recent surveys. These studies of the coupling between ionization and winds using, for input, empirical density models, have not only provided solid interpretation of many features in the distribution of F-region ionization, but have also greatly increased our overall understanding of the problem of thermospheric dynamics. An alternative approach makes use of experimentally determined ion drifts to deduce information about the neutral air motion, e.g. [4, 5].

The remainder of this paper deals with three related problems of particular current interest.

2. Super-rotation of the Thermosphere

By observing the changes in orbital inclination of a satellite, one can deduce the mean rotational rate Λ (rotations/day) of the atmosphere, roughly in the vicinity of the satellite's perigee [6]. Detailed analysis shows that the observed inclination changes are not attributable to other factors, such as meridional winds, except in very special circumstances [7]. On average Λ increases from about 1.0 at 150 km altitude to a peak of 1.4 at 350 km, and then decreases to 0.8 at 550 km [6, 8]. However, the rotation is not constant; e.g. at 200—250 km there is a westward motion by day, and a faster eastward motion at night. Other evidence comes from incoherent scatter observations of zonal ion drifts at Jicamarca [9] although the inferred value of Λ is only about 1.1 at 300 km. Hence there seems to be a discrepancy between the satellite and incoherent scatter data, which may possibly be due to uncertainty in the relation between the ion drift and neutral air wind. We note that $\Lambda = 1.4$ implies a mean west-to-east wind of 200 m s^{-1}.

Theoretically, super-rotation must satisfy the constraints that (i) the sources and sinks of angular momentum about the earth's axis must be in balance, and (ii) the north—south Coriolis force must balance other meridional forces. If we regard super-rotation as a mean zonal wind largely confined to latitudes between $\pm 30°$ (where it is believed to be strongest) we can express (i) in terms of the zonal momentum equation

$$\varrho \left(\frac{\partial u}{\partial t} + ud\right) = \varrho F - \frac{\partial p}{\partial x} \tag{1}$$

where ϱ is density, u is west-to-east wind velocity, d is the ion-drag parameter, $\partial p/\partial x$ is the zonal pressure gradient and F represents all other forces including the non-linear advection terms and a Coriolis term due to north—south motion. For simplicity assume u, d to be periodic in t, x or in longitude λ, measured eastward from the noon meridian, such that ε being a phase angle

$$d = d_0 + d_1 \cos \lambda; \quad u = u_0 + u_1 \cos (\lambda + \varepsilon). \tag{2}$$

Dividing (1) throughout by ϱ and averaging over all longitudes we get

$$u_0 d_0 + {}^1/_2\, u_1 d_1 \cos \varepsilon = \overline{F} \tag{3}$$

where we neglect second-order effects due to the diurnal variation of ϱ, which is relatively much less at 300 km than that of d and u. Hence

$$u_0 = (\bar{F} - {}^1/_2 u_1 d_1 \cos \varepsilon)/d_0. \tag{4}$$

The ion-drag term can therefore give super-rotation if there is a suitable phase relationship between the ion drag $d(\lambda)$ and wind $u(\lambda)$, such that $\cos \varepsilon < 0$. (The wind variation lags the pressure variation by 90° so, if one assumes the pressure bulge to be centred in the afternoon at $\lambda = 45°$, say, it turns out that $\varepsilon = 225°$ so that $\cos \varepsilon \simeq -0.7$). If the ion-drag is merely assumed proportional to the ion concentration N_i, then in the low-latitude F2 layer the day/night variation is quite small ($d_1/d_0 \ll 1$); but the "polarization field" mechanism [10], which largely depends on E-layer conductivity variations, can give $d_1/d_0 \sim 0.8$ so that if $u_1 = 100\ \mathrm{m\ s^{-1}}$ then the mean zonal wind $u_0 \simeq 30\ \mathrm{m\ s^{-1}}$. A more detailed calculation might give a larger u_0, though $30\ \mathrm{m\ s^{-1}}$ seems compatible with the Jicamarca ion drift data.

Other possible causes of super-rotation may also be discussed in terms of Eq. (4). Suppose the thermospheric pressure distribution leads to a prevailing meridional wind v_0, say $40\ \mathrm{m\ s^{-1}}$; this will make a contribution to $\bar{F}$ of $f v_0$ (+ if poleward, − if equatorward), where the Coriolis parameter $f = 7 \times 10^{-5}\ \mathrm{s^{-1}}$ at latitude 30°. Thus if $d_0 = 3 \times 10^{-4}\ \mathrm{s^{-1}}$, a typical F2 layer value, we get $u_0 = \pm 10\ \mathrm{m\ s^{-1}}$, which does not suggest that reasonable meridional winds will cause much super-rotation. Whether the required poleward decrease of pressure exists in the thermosphere has still to be established.

Now consider a process — which could be a gravity wave, or the convection set up by the "moving flame effect" [11] — that transports zonal momentum ($\varrho u'$) upwards with vertical velocity w'. If this momentum is deposited within a vertical scale distance H, say 30 km, then the resulting mean zonal wind $u_0 = \overline{u'w'}/H d_0$. With horizontal and vertical eddy velocities $u' = 100\ \mathrm{m\ s^{-1}}$, $w' = 1\ \mathrm{m\ s^{-1}}$ and d_0 as before, we only get $u_0 = 10\ \mathrm{m\ s^{-1}}$.

Magnetospheric electric fields, if producing a net southward component in the ionosphere, would cause eastward ion drift which would entrain the air [12, 13]. Even if this motion were transmitted to low latitudes, considerations of angular momentum would render it inefficient as a cause of super-rotation. In any case, chemical release experiments do not indicate that there would be a *mean* eastward ion drift, even though eastward drift occurs in particular substorms.

We seem to be faced with an impasse. Some possible mechanisms for producing super-rotation (Coriolis force, gravity waves, "moving flame" convection) are severely hindered by ion-drag. The numerical value of the ion-drag parameter used above is not excessive: it corresponds to $N_i \simeq 5 \times 10^{11}\ \mathrm{m^{-3}}$, typical of the F2 layer, though N_i is smaller at other heights. Of course N_i varies with the solar cycle, but it is noteworthy that no particular solar cycle variation of super-rotation seems to exist. Magnetospheric processes, operating at high latitudes, seem inefficient in causing super-rotation at low latitudes. The remaining mechanism discussed above is the "polarization field" effect, which depends on ion-drag; but according to currently available incoherent scatter data, this mechanism does not give super-rotation as strong as is implied by the satellite observations. It would be useful to determine, e.g. by means of incoherent scatter, the height-integrated and diurnally integrated quantity

$$\iint N_i V_i \, dh \, dt$$

which gives a measure of the total momentum of the F-layer ions. Better spatial and temporal resolution of the satellite data should help to define the roles of the possibly significant mechanisms, which are described in more detail elsewhere [14].

Another problem is the "sub-rotation" at around 550 km [6, 8], which requires a source of east-to-west momentum in contrast to the west-to-east source needed at lower levels. Since the levels of "sub-rotation" are near the base of the exosphere, some additional non-hydrodynamic effect, such as preferential escape of the fastest molecules to the exosphere, may also be of importance.

3. Phase of Diurnal Variations in the Thermosphere

There have been several incoherent scatter studies of the diurnal variation of thermospheric temperature that indicate that the maximum daytime temperature at 300 km lags behind the maximum density of Jacchia's model by several hours [15]. Furthermore, studies of the diurnal variation of the F layer, and now more directly of field-aligned ion velocities by incoherent scatter, suggest that the phase of the diurnal pressure wave may lead the phase of the density models by an hour or more [16, 17]. Temperature and pressure variations on the other hand derived from Jacchia's density distribution are in phase with the model densities.

These discrepancies undoubtedly arise from some combination of two circumstances. First, the incoherent scatter data give much finer resolution in time than the drag data but the available time series are much shorter. There may result spurious phase differences between drag and incoherent scatter data, due to incompatible averaging and smoothing of these two different data sources. Ideally, the same averaging and smoothing implicit in the drag data should also be applied to the scatter data before comparison is made. Second, there may be differences in phase between density, pressure, and temperature due to sensitivity of these parameters to variations in the lower thermosphere, where data on the phases are not readily available.

The features of Jacchia's empirical density models [18], as derived from satellite drag, most relevant to the present considerations are as follows:

(a) Since the local time of perigee changes only slowly with the precession of the satellite orbit, up to several hundred days of drag data are required to obtain a diurnal cycle.

(b) As a consequence of the necessary averaging, only the grossest global-scale horizontal variation of densities can be obtained by analysis of satellite drag.

(c) The most reliable drag information on diurnal variation of density derives from low-inclination satellites with perigees between 350 and 700 km. We can obtain with confidence from these data the amplitude and phase of the diurnal component, that is, a single 24-hr sine wave. Phases of minimum and maximum densities can be separately determined within ± 0.5 hr and indicate a definite departure from pure sinusoidal behaviour.

(d) There is sufficient unexplained variance in the drag data to allow a semi-diurnal component or random amplitude fluctuation in the diurnal component of as much as 30% of the amplitude of the diurnal component. Furthermore, there may well be day-to-day fluctuations in the phase of the diurnal component by as much as several hours from the mean.

(e) Owing to less definitive drag information on the density phase at higher latitudes and lower levels, the actual mean phase could vary with latitude by an hour or two. Indeed, a latitudinal variation appears in the satellite drag model of Marov and Alpherov [19] and in the incoherent scatter data [15]. Below 200 km, the phase of the diurnal oscillation is not at all determined by drag information.

(f) It is now established that from 350 km up to say 600 km there is no detectable shift with height of the 1400 LT phase of the density maximum.

Clearly, not much attention should be paid to "phase discrepancies" between the satellite models and small samples of other data. For a single day, the time difference between the late afternoon bump in temperature and the maximum according to Jacchia's model is not significant, for there is no way the drag data could resolve such a feature. What is significant are phase differences, even of only $^1/_2$—1 hour, that stand out at several locations for many days of data and with smoothing compatible with that implicit in the model densities. A suitable measure of a real phase difference which would be useful to a theoretician would be the difference in phase between the first Fourier harmonics of these two time series.

If a real phase difference exists, its explanation will require fuller data on the variations and better understanding of how temperatures, motions and composition interact. It is noteworthy that satellites have detected differences between the local-time variations of He, O and N_2 [20, 21]. The conditions that might allow phase differences according to hydrostatic arguments include:

(i) oscillations of pressure or density at a lower boundary [22]; (ii) departures from diffusive equilibrium of the atomic oxygen distribution [23]; (iii) diurnal variation of the temperature gradient in the lower thermosphere [24]. With these mechanisms (all of which require redistribution of mass by transport in the lower thermosphere [25]) it is feasible to produce theoretical models that at 300 km have a density maximum around 1400 LT and a temperature maximum around 1600 LT. But, because of the high thermal conductivity, the temperature phase can vary little above 300 km: with increasing height diffusion becomes more and more dominant, the density and pressure phases must approach the temperature phase, and by 600 km all phases must be about the same. Hence models with a 1600 LT temperature maximum at 300 km are likely to have a 1600 LT density maximum at 600 km, contrary to the satellite data, though this is not necessarily true of the helium distribution, because of its large scale height.

4. Mean Meridional Circulation

In considering the winds deduced from empirical density models, one striking feature stands out. The night-time equatorward winds are much stronger than the poleward daytime winds. This feature is due in part to the night-time decrease of ion drag and in part to the pattern of pressure gradients inferred from the Jacchia model. The mean equatorward flow implied by the strong night-time southward winds — if real — has very serious consequences for the momentum and thermal budgets of the thermosphere.

As for momentum, the equatorward flow of air has a "spindown" action which tends to produce westward rotation (see Section 2), and so aggravates the already difficult problem of explaining super-rotation. As for the thermal balance, the requirement of continuity implies that the equatorward flow of air must be accompanied by upward motion at high latitudes and downward motion at low latitudes. Although the lower atmosphere can amply fulfil the role of "source" and "sink" of air, the energetics poses a problem: these vertical motions would tend to refrigerate the poles and heat the tropics [26], which is impossible unless there is a very strong heat source in high latitudes.

Under quiet conditions, net upward flow in high latitudes seems unacceptable. Even without such a flow, it is difficult to balance the heat budget in high latitudes: in particular, the energy required to maintain the observed high temperatures in the winter polar mesosphere is usually attributed to downward air motion [27]. The probable explanation is that the model pressure distribution is inaccurate, because either the shape of the vertical temperature profile [28] or the chemical composition varies with latitude. It should be possible to alter the pressure distribution without destroying the agreement with satellite density data: we need more theoretical and experimental information on latitude variations.

The situation is different for storm conditions, in that a strong energy source does exist at high latitudes, i.e. in magnetic substorms and particle precipitation from the magnetospheric cusp regions. For limited periods the total energy input at high latitudes may even exceed the global EUV input, though OGO 6 data suggest that the heating is generally confined to latitudes above 50°. Auroral substorms act as powerful source of atmospheric gravity waves which carry energy to lower latitudes within a few hours. With the heating at high latitudes, the global pressure distribution is disturbed and strong nocturnal equatorward winds (of up to 500 m s^{-1}) have been measured by airglow techniques in the USA [29] and incoherent scatter at Arecibo [30]. Recent French data show that the normal daytime poleward wind in the F region is suppressed on a disturbed day.

We need more information on thermospheric structure of less than global scale, i.e. of smaller scale than is contained in current models. Satellites such as OGO 6 have detected localized variations of neutral gas concentration both in the auroral zone, which is relevant to the storm effects mentioned above, and in low latitudes: it seems that N_2 and O concentrations show latitudinal variations similar to the F2 layer "equatorial anomaly" [31]. This appears to be an instance of neutral thermospheric structure being influenced by the ions, presumably through ion-drag effects: localized ion variations can set up localized pressure gradients in such a way as to ensure that the neutral air flow satisfies continuity requirements [32]. Incidentally, this implies that in detailed dynamical calculations one must use air distributions and ion distributions that have similar spatial resolution horizontally; otherwise spurious results may ensue. All the problems discussed in Sections 2, 3 and 4 of this paper might be brought nearer solution if we had more information on localized variations in the thermosphere.

Acknowledgments

Contributions to the discussion were made by P. Amayenc, P. Bauer, J. E. Blamont, P. W. Blum, S. A. Bowhill, K. S. W. Champion, G. Haerendel, A. E. Hedin, C. O. Hines, L. G. Jacchia, F. S. Johnson, D. G. King-Hele, K. Moe, G. P. Newton, C. R. Philbrick, D. Rees, P. Vila, S. P. Zimmerman. Chairman, J. W. King.

The National Center for Atmospheric Research is sponsored by the National Science Foundation.

This paper is published by permission of the Director of the Science Research Council, Radio and Space Research Station at Slough.

References

[1] R. E. Dickinson, Space Research XII, 1015 (1972).
[2] H. Kohl, Space Research X, 550 (1970).
[3] H. Rishbeth, J. Atmos. Terr. Phys. **34**, 1 (1972).
[4] J. V. Evans, J. Atmos. Terr. Phys. **34**, 175 (1972).
[5] P. Amayenc and G. Vasseur, J. Atmos. Terr. Phys. **34**, 351 (1972).
[6] D. G. King-Hele, Nature, Lond. **233**, 325 (1971).
[7] D. G. King-Hele, Space Research X, 537 (1970).
[8] E. J. Prior, Trans. Amer. Geophys. Un. (EOS) **53**, 486 (1972).
[9] R. F. Woodman, Space Research XII, 969 (1972).
[10] H. Rishbeth, Planet. Space Sci. **19**, 357 (1971).
[11] G. Schubert and R. E. Young, J. Atmos. Sci. **27**, 523 (1970).
[12] K. D. Cole, Planet. Space Sci. **19**, 59 (1971).
[13] H. Volland, Nature. Lond. **232**, 248 (1971).
[14] H. Rishbeth, Rev. Geophys. **10**, 799 (1972).
[15] P. Waldteufel, P. Bauer and J. P. McClure, Space Research XII, 899 (1972).
[16] G. Vasseur, J. Atmos. Terr. Phys. **31**, 397 (1969).
[17] D. Eccles, J. W. King and H. Kohl, J. Atmos. Terr. Phys. **33**, 1371 (1971).
[18] L. G. Jacchia, Smithson. Contr. Astrophys. **8**, 215 (1965).
[19] M. Ya. Marov and A. M. Alpherov, Space Research XIII, 337 (1973).
[20] G. P. Newton, D. T. Pelz and W. T. Kasprzak, Space Research XIII, 287 (1973).
[21] C. R. Philbrick, R. S. Narcisi, D. W. Baker, E. Trzcinski and M. E. Gardner, Space Research XIII, 321 (1973).
[22] S. Chandra and P. Stubbe, Planet. Space Sci. **18**, 1021 (1970).
[23] H. G. Mayr and H. Volland, J. Geophys. Res. **77**, 2359 (1972).
[24] W. E. Swartz and J. S. Nisbet, J. Geophys. Res. **76**, 185 (1971).
[25] H. Rishbeth, Annls Géophys. **25**, 495 (1969).
[26] R. E. Newell, Nature, Lond. **211**, 700 (1966).
[27] F. S. Johnson and B. Gottlieb, Planet. Space Sci. **18**, 1707 (1970).
[28] R. E. Dickinson and J. E. Geisler, Mon. Weather Rev. **96**, 606 (1968).
[29] A. F. Nagy, P. B. Hays and K. D. McWatters, Space Research XI, 919 (1971).
[30] R. M. Harper, J. Geophys. Res. **77**, 1311 (1972).
[31] A. E. Hedin and H. G. Mayr, Trans. Amer. Geophys. Un. (EOS) **52**, 872 (1971).
[32] R. E. Dickinson, R. G. Roble and E. C. Ridley, J. Atmos. Sci. **28**, 1280 (1971).

The Ionosphere

Space Research XIII — Akademie-Verlag, Berlin 1973

REVIEW OF RECENT PROGRESS IN IONOSPHERIC RESEARCH

S. A. Bowhill

Aeronomy Laboratory, University of Illinois, Urbana, Ill., USA

Six topics in ionospheric research are reviewed. These are ion chemistry, neutral chemistry, stratosphere—ionosphere coupling, ionospheric thermal structure, steady F-layer motions, and travelling ionosphere disturbances (TIDs).

1. Ion Chemistry

Most of the interest in ion chemistry has centered recently on problems of the D region, below 90 km altitude. These have been recently reviewed by Thomas [1], by Ferguson [2], and more recently by Donahue [3], in his paper presented at the 1971 COSPAR Symposium on Ion Chemistry of the D and E Regions. Many excellent contributions from that meeting appear in the same issue of *Radio Science*, and will be of considerable interest to members of COSPAR.

The most important discovery on the D region in recent years has been the identification of hydrated ions, both positive and negative, as the dominant form of ionization below 82 km altitude. The positive ions are hydrates of hydronium, H_3O^+, and have masses of 37, 55, 73, etc., corresponding to the addition of successive water molecules. The formation of these hydrates is by no means simple. Fig. 1, due to Donahue [3] illustrates the two basic mechanisms by which they can be generated from O_2^+ and from NO^+. It happens that the back reaction by which O_4^+ is disintegrated by atomic oxygen is sufficiently rapid to make the upper part of the reaction chain more or less ineffective, so it appears that the hydrated ions are formed via the chain involving NO^+. It is to be noted, however, that this latter set of reactions gives rise to doubly hydrated hydronium, mass 55, as the lightest hydrate present. In fact, most experimental mass spectrometric determinations of D-region ion composition have shown the presence of masses 37 and 19 as well; however, it is possible that these may be produced by breakdown of heavier ions, either inside or outside the mass spectrometer, by the electric field applied. Even so, there is considerable difficulty in explaining the observed mass distribution of hydrates. New reactions are continually being discovered; for example, it is now known that clustering to N_2, followed by switching with water vapor, is several times faster than direct clustering of hydrates with water vapor.

An even more formidable problem relates to negative ions in the D region. Fig. 2 shows the scheme of Ferguson [2] for generating so-called "terminating ions", starting with O_2^-. The latter ion, formed by three-body attachment, can follow two principal routes: charge exchange with ozone to give O_3^-, and reaction with atomic oxygen by associative detachment to reverse the attachment process. Atomic oxygen can also participate in several other back reactions as shown. Ultimately, this scheme arrives at terminating ions such as NO_3^-, which have a high electron affinity and therefore find it difficult to undergo further charge exchange. They disappear by ion-ion recombination with positive ions.

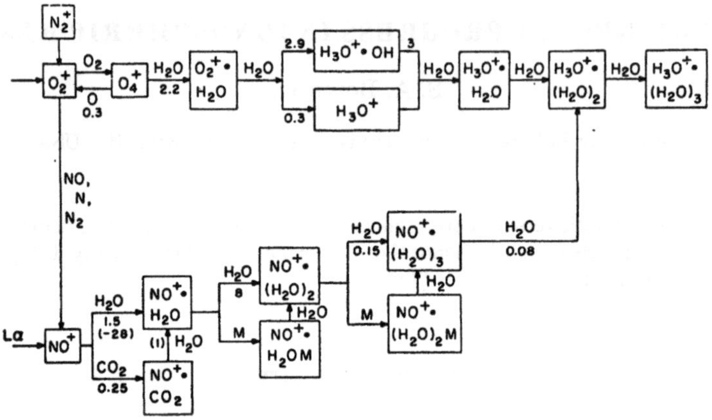

Fig. 1. Positive-ion chemical scheme in the D region (after Donahue [3]).

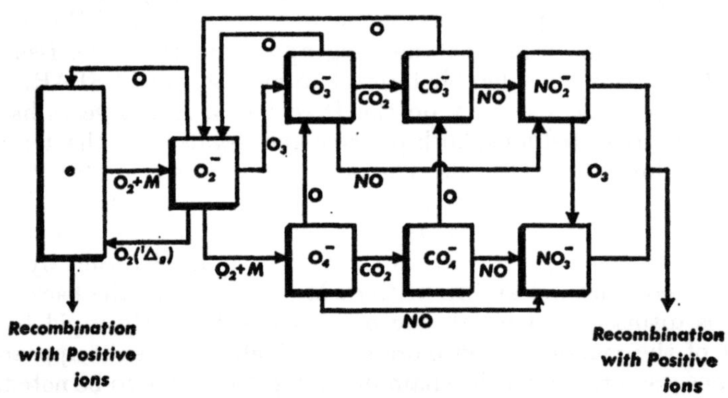

Fig. 2. Negative-ion chemical scheme in the D region (after Ferguson [2]).

Obviously, atomic oxygen plays a critical role in determining the rate at which these ions are formed. This fact may well account for the rapid disappearance of electrons below 70 km at night and during eclipse conditions, when much of the atomic oxygen has recombined, and the back reactions have less effect. Arnold and Krankowsky [4] have measured the ion composition of the night-time D region and obtained good agreement with the calculated composition. However, Narcisi et al. [5] find a predominance of larger ion masses, 78 amu and greater, probably multiple hydrates of CO_3^- and NO_3^-. The reason for the discrepancy between these results is not known, but may be associated with breakup of the delicate hydrated negative ions by the vehicle shock wave or the draw-in field of the mass spectrometer.

One important consequence of hydrated positive ions in the D region is for the dissociative recombination coefficient of the ionization, which is more than an order of magnitude faster for hydrated hydronium ions than for NO^+ or O_2^+. Striking evidence for this has come from measurements made by two groups during the 7 March 1970 solar eclipse. Belrose et al. [6] have deduced recombination coefficients by comparing the electron densities at each altitude at maximum obscuration with those obtained prior to the eclipse; the recombination coefficients deduced are about 5×10^{-7} cm^3 s^{-1} at 90 km, but dissociative recombination coefficients of 10^{-5} or greater are needed to explain the rapid disappearance of electrons below 87 km. In the same eclipse at Wallops Island, Mechtly et al. [7] found that in less than one minute during totality, the D-region electron concentration decreased by about 30%, though that in the E layer stayed nearly constant. Values of recombination coefficient as high as 5×10^{-5} cm^3 s^{-1} were deduced by Mechtly et al. from these data, in agreement with similar data for a previous eclipse in Brazil on 12 November 1966.

During the 1966 eclipse, Narcisi et al. [5] have measured the positive-ion composition of the D and lower E region. Mass 30, NO^+, was found down to about 78 km, at which altitude the concentrations of the first two hydrates, 37 and 55, were about 800 and 200 cm^{-3}. At totality, the same ions had concentrations of about 400 and 180 cm^{-3}, while the NO^+ ions had disappeared altogether. Evidently, masses 37 and 55 must be maintained by a more effective formation mechanism at totality, perhaps due to a decreased atomic oxygen concentration, making the back reaction for O_4^+ formation less effective. At totality, a predominance of the NO^+ over O_2^+ was found above 94 km, whereas O_2^+ predominated for full-sun conditions. Dissociative recombination would not bring this about, since O_2^+ recombines less rapidly than NO^+. The answer lies in charge exchange between the O_2^+ and NO converting the latter to NO^+, indicating that the appropriate recombination rate for E-region ionization should be that of NO^+, rather than a weighted average including O_2^+. During eclipse totality, Narcisi et al. [8] found abundant multiple hydrates of NO_3^-, with masses of 62, 80, 98, 116, 134 and 152. The mechanism of formation of these hydrates is not understood.

Measurements recently published by Brace et al. [9] from the 7 March 1970 solar eclipse at Wallops Island illustrate the way in which rocket measurements can confirm many features of E- and F-region ion chemistry. Ion concentration profiles were measured for about 50% and about 90% obscuration. In the E region, at about 140 km, the ions were primarily molecular, and the fraction of molecular oxygen ions decreased as the eclipse progressed, in spite of the fact that its dissociative recombination coefficient is less than that of nitric oxide ions. This discrepancy is, of course, accounted for by charge exchange reactions which convert O_2^+ into NO^+. Between 160 and 180 km, the transition between the molecular ions and atomic oxygen ions rises in altitude by nearly 20 km as the eclipse progresses, confirming the common observation that the F1 layer (produced by molecular ions) becomes more pronounced on sweep-frequency ionograms during an eclipse.

Finally, there is an increasing recognition of the usefulness of the incoherent scatter technique for studies of E- and F-region chemistry. Not only can very detailed electron-density profiles be determined, but the different thermal velocities of the various ion species give different spectral widths to the returned scattered signal, so the ion composition (or at least the relative amounts of atomic and molecular ions) may be determined. Evans and Oliver [10] have used this technique successfully to determine the nature of the E-F1 layer transition and its variability with time of day.

2. Neutral Chemistry

Progress in upper-atmosphere neutral-chemistry theory has been rapid, though in experiment it has been disappointingly slow. Strobel [11] has recently summarized the minor-constituent status. He has also shown that the concentration of one particular constituent of great aeronomic interest, nitric oxide, may be much less abundant than suggested by the measurements of Meira [12]. The technique is to measure the emission rate of fluorescence in the gamma bands of nitric oxide, the slope of the emission-rate curve with altitude giving the concentration. Particularly after correction for ozone absorption, the results seem consistent with theory.

Complete studies of the photochemistry of minor neutral species, the latest of which is by Hunt [13] involve many interlocking reactions, and some parameters, such as eddy diffusion coefficient, which are very poorly known. New techniques for the measurement of minor neutral species in the mesosphere and lower thermosphere are sorely needed. Two examples of successful new approaches may be cited. The first of these is a set of measurements by Carver et al. [14] of the estimation of near-ultraviolet light reflected from the moon, and absorbed by atmospheric ozone. The ozone profiles obtained agree well with most previous measurements, but extend over a range of height. A second type of approach has been used by Giraud et al. [15]. They have used a combination of mass spectrometry and ionospheric incoherent scatter to devise precise density models for comparison with existing model-atmosphere syntheses.

3. Stratosphere-Ionosphere Coupling

The subject of stratosphere-ionosphere coupling is built on some experimental observations showing a correlation between the stratospheric temperature or pressure and the electron density in the lower ionosphere, particularly in winter. Schwentek [16] has shown that three types of winter anomaly occur in the D region. Fig. 3 shows daily values of the ionospheric absorption index L for two winters, at a fixed solar zenith angle χ. Firstly, the base level of L (open circles) is seen to rise in winter to about double its summer value; secondly, the average value of L is seen to rise substantially above the base level; and thirdly, there is a large short-term variability with a period from 2 to 7 days. It is the latter variability which has been found [17, 18] to correlate with stratospheric parameters.

Speculations concerning the origin of this coupling effect are available; Geller and Sechrist [19], for example, recently suggested it might be associated with changes in the relative abundance of hydrated ions at 80 km, brought about indirectly by changes in neutral transport. Rocket campaigns were mounted both in Europe and in the USA this winter for the intensive study of this phenomenon, but the results are not yet available.

Some rather striking additional experimental evidence has recently been found. Deland and Friedman [20] have found strong correlation between zonal Fourier components of the 50 and 10 millibar thickness and the value of f_{min} — a measure of ionospheric absorption. The fluctuations of wave number 1 are approximately in phase for the two quantities, which they attribute to planetary waves. Also striking are some correlations by Brown and Williams [21], shown in Fig. 4, between the pressure at the 10-millibar level (about 30 km altitude) and the height of reflection of the E layer.

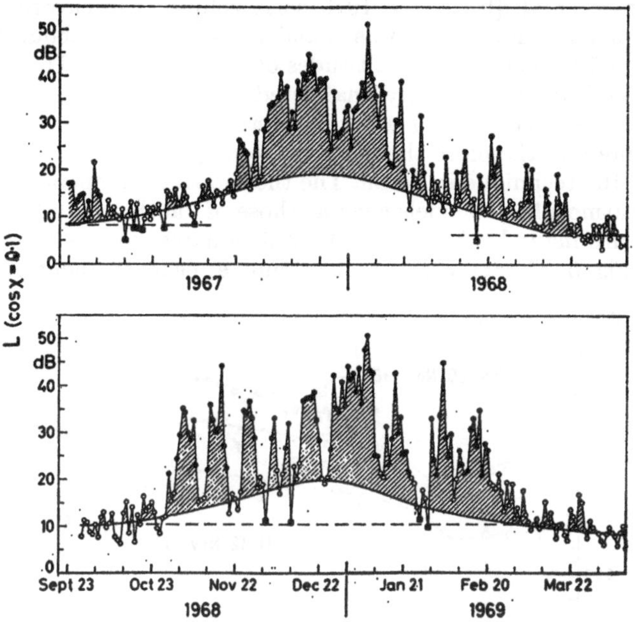

Fig. 3. Ionospheric absorption in winter (after Schwentek [16]).

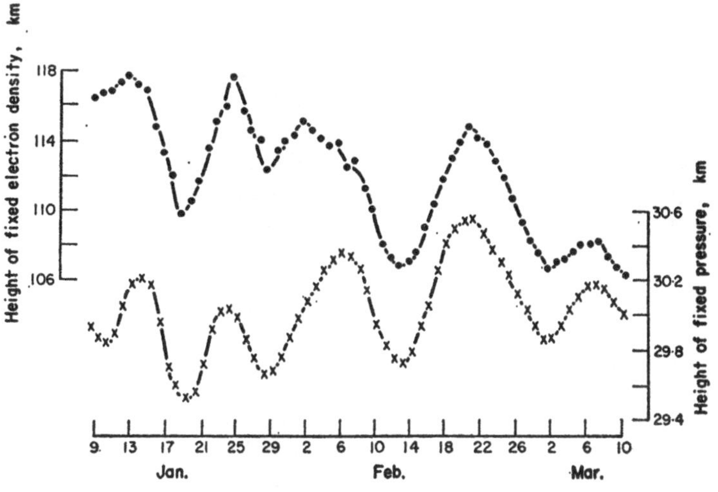

Fig. 4. Correlation between heights of the E layer and of a stratospheric isobar (after Brown and Williams [21]).

4. Ionospheric Thermal Structure

Nothing very new has come recently from satellite Langmuir probes; the discrepancy between them and the incoherent scatter measurements of electron temperature has still not been resolved. However, an interesting result was obtained by Brace et al. [9] in the 1970 solar eclipse; the electron temperature measured by a Langmuir probe was found to decrease by about 12% between obscurations of 50% and 90%, at altitudes of 130 to 280 km. The neutral temperature was found to be completely unaffected.

Important work is continuing using incoherent radar scatter to determine the ion temperature and therefore the neutral thermopause temperature. Evans [22] has described the technique in detail. The thermopause temperature is a primary parameter of atmospheric models such as those of Jacchia [23]. Measurements of this temperature made by McClure [24] at Jicamarca, over a three-year period, are shown in Fig. 5. The temperature maximum appears at about 1600 local time.

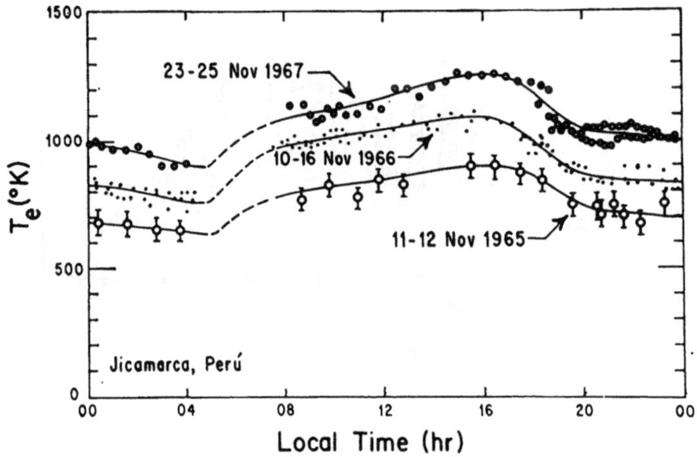

Fig. 5. Thermopause temperature over Jicamarca, deduced from incoherent scatter (after McClure [24]).

Similar data have been obtained by Waldteufel and Cogger [25] at Arecibo over a five-year period. Comparison of these data with atmospheric models will lead to great improvement in the accuracy and consistency of these models.

5. Steady F-Layer Motions

Motions of the ionization and neutral gas of the F layer were treated in detail in the three-day symposium on Dynamics of the Thermosphere and Upper Ionosphere at the 1971 COSPAR meeting in Seattle (published in *Space Research XII*), and a further review paper [26] on the subject is found elsewhere in this volume, so it is not appropriate to give a detailed summary here. Generally, the key questions are the sources of energy for the vertical and horizontal movements of the thermosphere and ionosphere, and how they interact. Heating of the thermosphere by ultraviolet radiation produces thermal expansion of the upper atmosphere, resulting in a horizontal flow over the poles from the afternoon side to the early-morning side of earth. This wind has about 40 m s^{-1} velocity at 300 km. It tends to be impeded by collisions between the neutral air and F2-layer ions, which are held in place by polarization forces; in turn, the ion-drag force exerted on the ionization distorts the F layer and produces overall motion of the F-layer plasma.

Consideration of this interaction is further complicated by the effect on the ionization of electric fields conducted along magnetic field lines both from other altitudes as the E region, where the winds may be quite different, and from the magnetosphere, where magnetic-field distortions may arise from other causes.

The principal purpose in reviewing this topic is to draw attention to some recently developed capabilities of the incoherent scatter technique for the measurement of F-region ionization velocities. Fig. 6, from a recent review [27] by

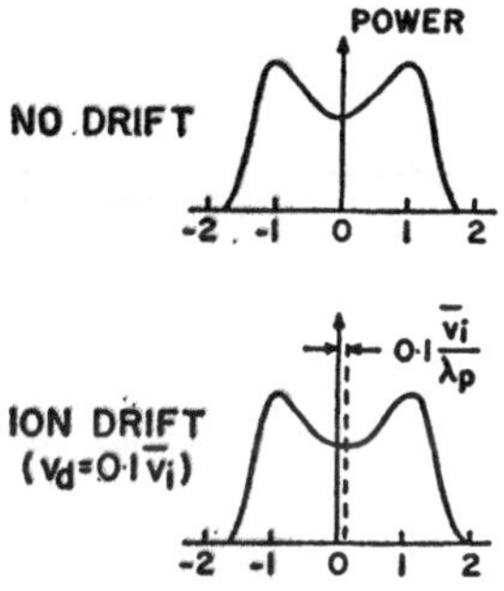

Fig. 6. Incoherent-scatter radio-frequency spectra with and without ion drift (after Evans [27]).

Evans, shows the power spectrum of the radio signal scattered back to a radar by F-layer ionization. In the absence of drift, the spectrum is determined by thermally excited plasma waves whose wavelengths match the Bragg reflection condition for the radar frequency. If the plasma is in motion, its line-of-sight velocity produces a small Doppler shift in the spectrum as shown. From this shift the velocity may be calculated.

Fig. 7 shows measurements [27] of the vertical velocity of F-region ionization at four heights throughout a 24-hour period. From these data and from electron-density profiles obtained simultaneously the flux of ionization into and out of the F region can be calculated. Fig. 8 illustrates the typical daytime situation in the F layer [27]. Ionization below 600 km altitude is diffusing downwards towards a sink below the F-layer peak, while above 600 km it is moving upwards into the plasmasphere, replacing ionization which leaked into the F region from the plasmasphere during the previous night.

The form of Fig. 8 is strongly influenced by neutral winds, and Amayenc and Vasseur [28] have demonstrated that neutral winds may be deduced from these incoherent-scatter data. Evans [29] has measured horizontal motions by means of an obliquely pointing radar, and has found velocities of about 100 m s^{-1}, presumably associated with electric fields in the magnetosphere, such as those measured by whistler-duct attenuation.

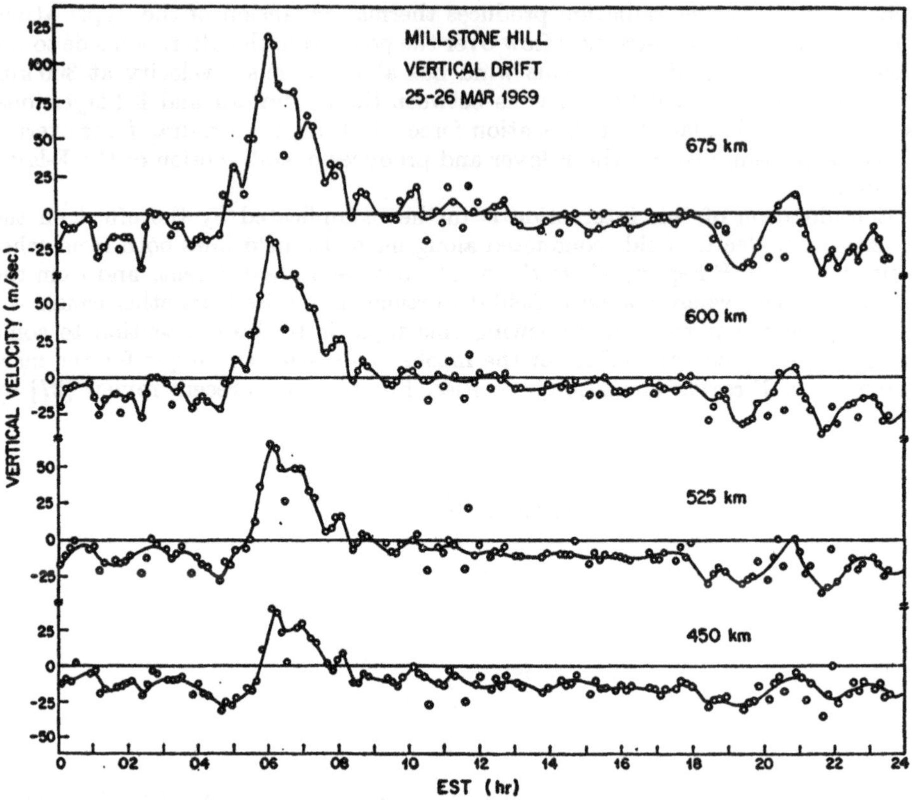

Fig. 7. Vertical drifts of the F layer over Millstone Hill (after Evans [27]).

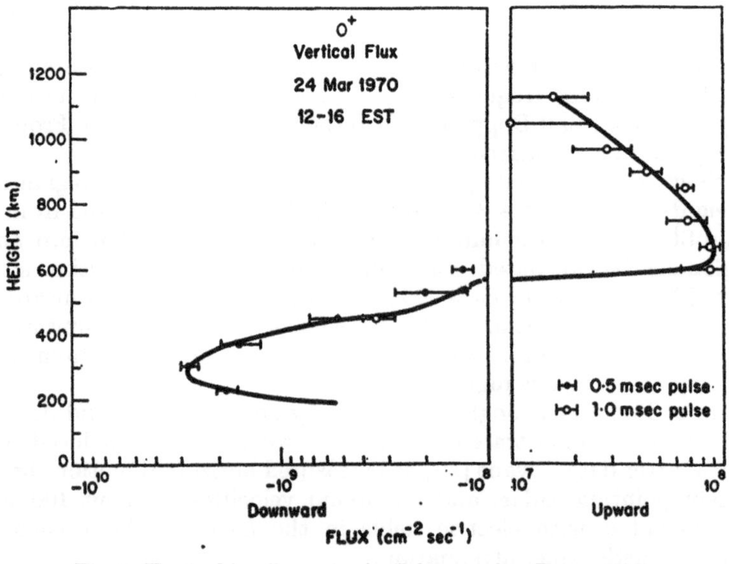

Fig. 8. Vertical ion fluxes in the F layer (after Evans [27]).

6. Travelling Ionosphere Disturbances

Travelling ionosphere disturbances (TIDs) are large-scale, long-period undulations of the F2 layer that travel meridionally in an equatorward direction. They are believed to result from internal gravity waves in the neutral atmosphere, probably excited by heat deposited in the auroral E region. Their detection and explanation is reviewed in a paper elsewhere in this volume, so the present review will discuss only three recent examples of the experimental evidence. Fig. 9,

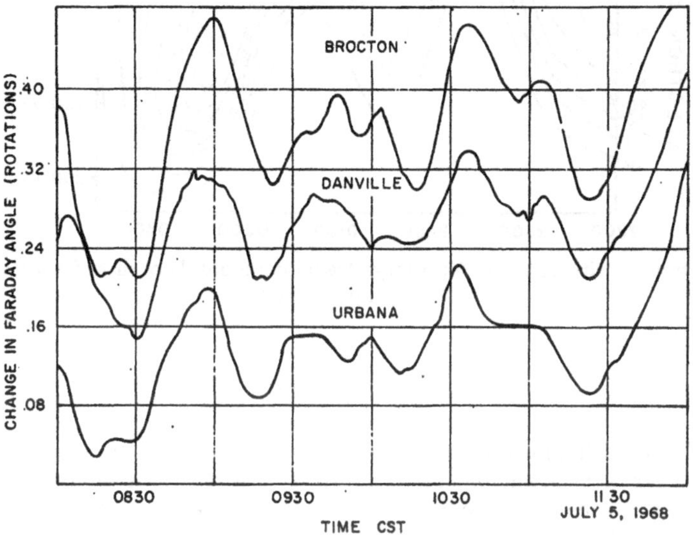

Fig. 9. Travelling ionosphere disturbances showing on satellite Faraday rotation at three spaced receivers (after Yeh [30]).

from a recent paper by Yeh [30], shows the effect of a TID on the Faraday rotation from a geostationary satellite signal. The stations shown are about 50 km apart. Spectral analysis of records such as this give the dispersion relation for the waves, since the velocity and scale size can be deduced.

Use of the incoherent-scatter technique can provide more detailed information; Fig. 10 shows the appearance of the electron density distribution found by Harper [31] for a TID passing over Arecibo on 6—7 November 1970. The downward phase progression of the wave is characteristic of internal gravity waves. The vertical component of the ionization velocity associated with the same disturbance [31] gives times of maximum enhancement of the ionization which are found to correspond to the times of upward reversals in direction of the velocity, showing that the enhancements come from ionization blown down into the F layer by neutral winds from the topside.

Sterling et al. [32] have made similar measurements of a TID at the geomagnetic equator. The general appearance of the contour plots is similar to that at higher latitudes, but the interpretation is more difficult.

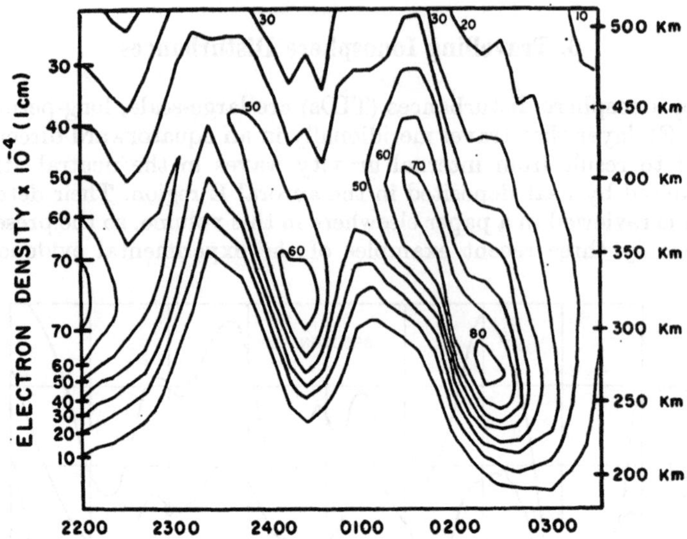

Fig. 10. Electron-density contours for a large travelling ionosphere disturbance over Arecibo (after Harper [31]).

7. Conclusion

Many other exciting new developments have taken place in ionosphere research in the past year or two, such as artificial heating of the ionosphere and the interesting physical processes it reveals. However, it is evident that the increasing sophistication of techniques of ionospheric investigation, and the coordinated use of rocket, satellite and ground-based methods, is opening significant new horizons for future research.

References

[1] L. Thomas, J. Atmos. Terr. Phys. **33**, 157 (1971).
[2] E. E. Ferguson, Rev. Geophys. Space Phys. **9**, 997 (1971).
[3] T. M. Donahue, Radio Sci. **7**, 73 (1972).
[4] F. Arnold and D. Krankowsky, J. Atmos. Terr. Phys. **33**, 1693 (1971).
[5] R. S. Narcisi, A. D. Bailey, L. Della Lucca, C. Sherman and D. M. Thomas, J. Atmos. Terr. Phys. **33**, 1147 (1971).
[6] J. S. Belrose, D. B. Ross and A. G. McNamara, J. Atmos. Terr. Phys. **34**, 627 (1972).
[7] E. A. Mechtly, C. F. Sechrist, Jr. and L. G. Smith, J. Atmos. Terr. Phys. **34**, 641 (1972).
[8] R. S. Narcisi, A. D. Bailey, L. E. Wlodyka and C. R. Philbrick, J. Atmos. Terr. Phys. **34**, 647 (1972).
[9] L. H. Brace, H. B. Mayr, M. W. Pharo III, L. R. Scott and N. W. Spencer, J. Atmos. Terr. Phys. **34**, 673 (1972).
[10] J. V. Evans and W. L. Oliver, Radio Sci. **7**, 103 (1972).
[11] D. F. Strobel, Radio Sci. **7**, 1 (1972).
[12] L. G. Meira, Jr., J. Geophys. Res. **76**, 202 (1971).
[13] B. G. Hunt, J. Atmos. Terr. Phys. **33**, 1869 (1971).

[14] J. H. Carver, B. H. Horton, R. S. O'Brien and B. Rofe, Planet. Space Sci. **20**, 217 (1971).
[15] A. Giraud, G. Scialom and A. A. Pokhunkov, J. Geophys. Res. **77**, 1251 (1972).
[16] H. Schwentek, J. Atmos. Terr. Phys. **33**, 1647 (1971).
[17] M. Bossolasco and A. Elena, C. R. Acad. Sci. Paris **256**, 4491 (1963).
[18] A. H. Shapley and W. J. G. Beynon, Nature, Lond. **206**, 1242 (1965).
[19] M. A. Geller and C. F. Sechrist, Jr., J. Atmos. Terr. Phys. **33**, 1027 (1971).
[20] R. J. Deland and R. M. Friedman, J. Atmos. Terr. Phys. **34**, 295 (1972).
[21] G. M. Brown and D. C. Williams, J. Atmos. Terr. Phys. **33**, 1321 (1971).
[22] J. V. Evans, Proc. Inst. Elect. Electron. Engrs **69**, 496 (1969).
[23] L. G. Jacchia, Smithson. Astrophys. Obs. Spec. Rep. No. 332 (1971) (p. 113).
[24] J. P. McClure, J. Geophys. Res. **76**, 3106 (1971).
[25] P. Waldteufel and L. Cogger, J. Geophys. Res. **76**, 5322 (1971).
[26] R. E. Dickinson and H. Rishbeth, Space Research XIII, 413 (1973).
[27] J. V. Evans, J. Atmos. Terr. Phys. **34**, 175 (1972).
[28] P. Amayenc and G. Vasseur, J. Atmos. Terr. Phys. **34**, 351 (1972).
[29] J. V. Evans, J. Geophys. Res. **77**, 2341 (1972).
[30] K. C. Yeh, J. Geophys. Res. **77**, 709 (1972).
[31] R. M. Harper, J. Geophys. Res. **77**, 1311 (1972).
[32] D. L. Sterling, W. H. Hooke and R. Cohen, J. Geophys. Res. **76**, 3777 (1971).

Space Research XIII — Akademie-Verlag, Berlin 1973

THE ALADDIN EXPERIMENT — PART I, DYNAMICS

N. W. Rosenberg[a], D. Golomb[a], S. P. Zimmerman[a], W. K. Vickery[a] and J. S. Theon[b]

[a]Air Force Cambridge Research Laboratories, Bedford, Mass., USA
[b]NASA Goddard Space Flight Center, Greenbelt, Md, USA

A coordinated rocket experiment was carried out on 20 November 1970 at Eglin AFB, Fla. (30°N Lat) to determine Atmospheric Layering and Density Distributions of Ions and Neutrals. The program measured vertical profiles between 70 and 150 km altitude of total neutrals, ions and electrons, individual species, temperature, density, diffusion coefficients, horizontal winds, and turbulent structure. A dynamic computer code model using these parameters was compared with the observed profiles.

Wind fields were measured from seven chemical trails separated up to 60 km horizontally and 600 seconds in time; rms shears of the horizontal winds were found of the order of 10^{-4} s^{-1} horizontal, 10^{-2} s^{-1} vertical and 2×10^{-2} m s^{-2} in time. Kinetic temperatures were measured from fluorescence of chemically released AlO, ranging from 440°K to 680°K between 126 and 161 km. Densities derived from diffusion of the AlO ranged from 10 to 5×10^{-12} g cm^{-3} between 126 and 139 km. A rocket-borne pitot probe showed wavelike density deviations from current standard atmosphere models, amounting to −25% at 80 and 110 km and to +25% at 100 and 125 km. The derived temperature profile from the pitot probe showed a double minimum at 74 and 102 km. Horizontal shear and Coriolis forces have been included in a modified Richardson number stability criterion to obtain correlation between predicted and observed turbulent and laminar layers.

1. Introduction

The simultaneous observation of different parameters in the upper atmosphere makes it possible to define their interrelations and to test computer code models of these interrelations. More specifically, the dynamics of tidal and gravity waves perturb any static model, directly affecting temperature, density and winds [1]. These indirectly result in ionized layer formation and movement [2], and in formation of regions of turbulence with increased vertical transport of neutral species [3].

The ALADDIN experiment was carried out on 20 Nov. 1970 at Eglin AFB, Fla (30° N) to determine Atmospheric Layering and Density Distributions of Ions and Neutrals. The program measured vertical profiles (between 70 and 150 km altitude) of total neutrals, ions and electrons, individual species, temperature, density, diffusion coefficients, horizontal winds, and turbulent structure. All measurements were made in a 13 minute period (1722—1735 CST) with a local solar depression of 8° to 10.5°.

This paper reports on measurement and interpretation of winds, density, temperature and turbulence, i.e. the dynamics of the region. The following paper [4] reports the composition measurements and their comparison with computer codes incorporating the observed dynamical parameters as inputs.

2. Ionospheric Winds

Seven vertical profiles of horizontal winds were measured from chemical releases along up- and down-legs of two rockets launched along different azimuths, followed by two more rockets launched 450 s later. The maximum and rms separation of the seven legs were 660 and 250 s in time, 61 km and 23 km in km E, 45 and 19 km in km N. The mean time of observation was 90 s and the rms triangulation position error was 120 m. The wind at each altitude for each trail was measured from a linear fit of position against time. The rms discrepancy from a linear fit was 160 m, and the rms velocity uncertainty 2 m s^{-1}. (Vertical motions were measured but were dominated by thermal convection of released material, reaching a maximum upward motion of 30 m s^{-1} at 100 km, obscuring observations of local vertical winds.)

Velocity differences in the wind field from the mean velocity of all trails at a given altitude were then analyzed as a linear combination of constant shears (S_T, S_E, S_N, S_Z) in separations (δT, δE, δN, δZ) of the individual trails from the mean time and space at each altitude. Combining each three adjacent altitudes provided $7 \times 3 = 21$ equations in five unknowns:

$$V(\delta T, \delta E, \delta N, \delta Z) = V(0, 0, 0, 0) + S_T\,\delta T + S_E\,\delta E + S_N\,\delta N + S_Z\,\delta Z.$$

The standard deviation in the mean velocity was reduced from 14 m s^{-1} to 6 m s^{-1} by this fit. Total velocity and individual rms shears plotted against altitude are shown in Fig. 1. Shears in Fig. 1 are reported as m s^{-1} per rms separation ($\sim$1 km altitude, 25 km N or E, 240 sec time) since we have no knowledge that the shears would be constant over greater separations. The rms velocity was about 30 m s^{-1} and the "normalized" shears were on the order of 10^{-2} s^{-1} in the vertical, 10^{-4} s^{-1} in horizontal separation, and 2×10^{-2} m s^{-2} in time. The dominant scale size

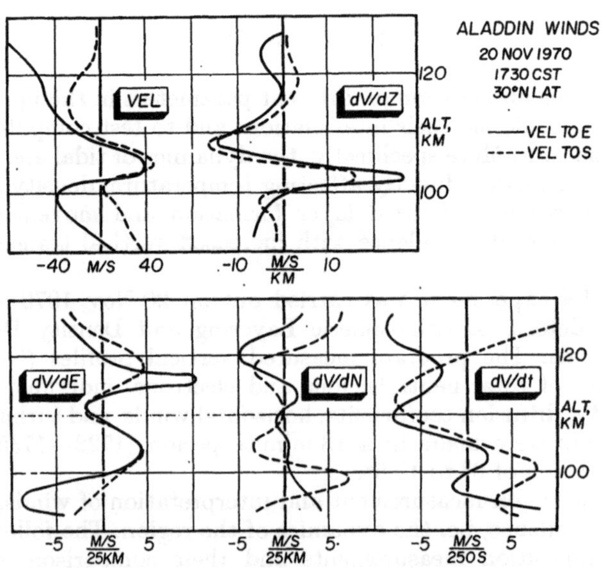

Fig. 1. Horizontal wind and shear profiles.

being measured can be estimated as $2\pi \times$ rms velocity/rms shear, leading to values of 20 km in the vertical, 2000 km in the horizontal and 3 hours time. This time scale is much shorter than the semidiurnal tidal period. The wind field serves as an input into the transport equations used for prediction of local concentrations and comparison with observed layer formation [4]. The shears serve to define local turbulence stability criteria and ion convergence velocities.

3. Temperatures and Derived Densities from Chemical Releases

On the downleg of one of the chemical releases (1719 LT) a string of spherical TMA (trimethyl aluminum) puffs was released. The TMA produces AlO and the resonance fluorescence of AlO was observed from the ground by time-lapse photography and by spectrophotometry. Diffusion coefficients were obtained from the rate of growth of the puffs [5]. The temperature of the radiating molecules

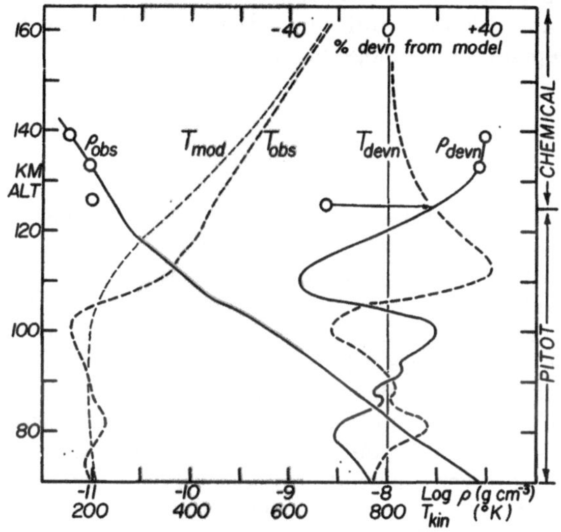

Fig. 2. Temperature and density profiles.

was obtained from the fluorescent spectrum [6]. Assuming that the ground state of the molecules is in thermal equilibrium with the ambient, the temperature of the ambient can be inferred. Atmospheric densities can be computed from diffusion coefficients and temperatures, using reasonable values for the collision cross section and mean molecular weights at each altitude [5]. Unfortunately only three puffs were photographed between 126 and 139 km for a sufficiently long period to obtain diffusion coefficients; however, good temperature data were obtained from six puffs between 126 and 161 km. The temperatures and densities are plotted in Fig. 2. Estimated precision of the temperature measurements near 500 °K is to about 5%; the derived density measurement also requires a measurement of the diffusion coefficient, with an estimated error of 20% in regions where radial growth is not distorted by shears. However the 126 km density estimate is apparently 40% low, probably due to shear distortion and therefore a poor diffusion coefficient measurement.

4. Densities and Derived Temperatures from Pitot Probe

A rocket carrying a pitot probe payload launched at 1730 LT provided a vertical profile of the neutral density in the 34—125 km region, permitting the molecular scale temperature and the pressure profiles to be derived from the hydrostatic equation and the equation of state. The pitot probe technique utilizes a radioactive ionization gage and a hot filament gage mounted in the forward portion of the payload to measure impact pressure. Impact pressure is related to density by the Rayleigh equation in continuum flow below 85 km and by a modified thermal transpiration equation in free molecular flow at higher altitudes [7]. A sealed tip which prevented contamination of the gage before launch was ejected at about 33 km and continuous measurements were made to an altitude in excess of 140 km. The uppermost 15 km of data were discarded so that at 125 km the profiles become independent of the initial temperature which must be assumed at 140 km. The errors in the pitot probe measurement of density are: 2% below 85 km; 4% between 85 and 100 km; and 10% above 100 km. The temperature errors are similar but depend on the slope of the density profile.

5. Discussion of Profiles

Fig. 2 shows the observed density and temperature profiles from 70 to 161 km. It also shows an atmospheric model temperature, and on the right side of the figure, deviations of the observed temperature and density from the model. (The atmospheric model used in Fig. 2 was taken from Jacchia [8] for altitudes above 100 km, with an exospheric temperature of 900 °K. The atmospheric model below 100 km was taken from Groves [9] for 1 December 1970 at 30° N.)

The wavelike nature of the density and temperature fluctuations is seen from the deviation curves, with amplitude 20—40% and a vertical wavelength of about 30 km. Nulls occur at 70, 90, 105 and 125 km with density and temperature 180° out of phase. However the wind profile of Fig. 1 is unusual in that the winds, which are 90° out of phase from 90 to 100 km as expected, are shifted to in phase in the 100—120 km region. This phase shift is not predicted by tidal theory and probably represents a contribution from short period gravity waves. It is significant that the phase shift does not appear in the density and temperature profiles, since a constant phase difference of 90° between wind and temperature is predicted both by linear internal gravity wave theory and by tidal theory.

6. Turbulence

Turbulence was observed on inspection of the photographic records of the chemical trails. One trail showed turbulence at all altitudes below 113 km, except for laminar layers from 99—101 km, 103—106 km, and 109—111 km.

The Richardson number was computed based on horizontal shears and local temperature gradient, and is shown as Ri_{obs} in Fig. 3. It may be noted that only near 103 km does this stability parameter approach the 0.25 criterion usually accepted for onset of turbulence.

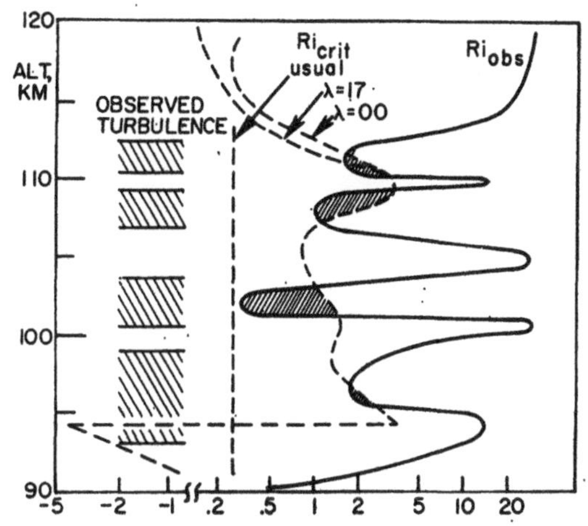

Fig. 3. Turbulence structure.

However an analysis incorporating horizontal shears of the horizontal zonal wind, Coriolis forces and kinematic viscosity results in a criterion for instability

$$Ri_{\text{cr}} = f^2[\nu^2 m^4 - f(\partial U/\partial y - f)]^{-1},$$

where f is the Coriolis parameter, $\partial U/\partial y$ is the transverse gradient of the east—west wind, ν is the kinematic viscosity and m is the vertical wavenumber $= 2\pi/\lambda_v$. Inclusion of these measured parameters results in a better correspondence between Richardson number and observed scalar turbulence as shown in Fig. 3.

References

[1] C. O. Hines, in: Physics of the Earth's Upper Atmosphere, Prentiss Hall, Englewood Cliffs, New Jersey 1965 (p. 145).
[2] T. J. Keneshea and M. A. MacLeod, J. Atmos. Sci. 27, 981 (1970).
[3] T. J. Keneshea and S. P. Zimmerman, J. Atmos. Sci. 27, 5 (1970).
[4] C. R. Philbrick et al., Space Research XIII, 441 1973.
[5] D. Golomb et al., J. Geophys. Res. 72, 2365 (1967).
[6] D. Golomb et al., Space Research VIII, 705 (1968).
[7] J. J. Horvath, R. W. Simmons and L. H. Brace, Univ. Michigan Scientific Rep. NS-1, Ann Arbor, Michigan (1962).
[8] L. G. Jacchia, Smithson. Astrophys. Obs. Spec. Rep. No. 332 (1971).
[9] G. V. Groves, AFCRL 71-0410 Envir. Res. Paper 368 (1971).

Space Research XIII — Akademie-Verlag, Berlin 1973

THE ALADDIN EXPERIMENT — PART II, COMPOSITION

C. R. Philbrick[a], R. S. Narcisi[a], R. E. Good[a], H. S. Hoffman[a],
T. J. Keneshea[a], M. A. MacLeod[a], S. P. Zimmerman[a] and B. W. Reinisch[b]

[a]Air Force Cambridge Research Laboratories, Bedford, Mass., USA
[b]Lowell Technological Institute Research Foundation, Lowell, Mass., USA

Vertical profiles of N_2, O_2, Ar, O, CO_2, O_3, $O_2(^1\Delta_g)$, total neutral density and temperature were measured in the ALADDIN experiment. The measured turbulent and molecular diffusion coefficients, total mass density and temperature were used as inputs to model the profiles of O, O_2, and other neutral constituents which are compared with the corresponding measurements. Emission of $O_2(^1\Delta_g)$ showed a peak density at 82 km of 1.3×10^9 cm^{-3}, corresponding to an ozone density peak of 6×10^7 cm^{-3}. Oxidation of lithium released as a tracer decreased rapidly from 85 to 100 km, requiring a Li—Li—M reaction to account for the observations. A blanketing Es layer was observed to increase from 3.5 MHz at launch to 8 MHz two hours later, possibly associated with Al^+ ions formed during the chemical releases used for wind and temperature profiles. Positive ion composition measurements showed a meteoric ion layer at 108 km composed mainly of Si^+ with Mg^+, Na^+, Fe^+, Ca^+, Ni^+, and SiO^+ which corresponds to a 3.5 MHz blanketing sporadic-E layer seen with a ground-based ionosonde. Al^+ associated with the chemical release of $Al(CH_3)_3$ was measured in two regions of the flight; calculations of the effect of wind fields on the ion density distributions show ion density profiles which generally agree with the measurements.

1. Introduction

The atmospheric dynamics measurements of the ALADDIN 1 program have been presented by Rosenberg et al. [1]. Here a brief summary of the electron and ion density and ion composition measurements are presented and compared with a theory for the E region [2—4]. An explanation of the inventory of lithium after its release as a vapor in the 80—100 km region is given [5]. Photometer measurements of $O_2(^1\Delta_g)$ have been used to infer O_3 and O densities [6]. The neutral atmospheric composition measured with a mass spectrometer has been compared with an atmospheric model which determines the vertical time dependent profiles of the constituents using the measured parameters relating to the dynamical processes [7, 8].

2. Ion Composition and Electron Density

Close to the launch site at Eglin AFB a Digisonde 128 was monitoring the ionosphere. E-region ionograms were recorded with 1.5 km height range sampling and 25 kHz frequency increments. These ionograms indicated a moderate activity of sporadic E during the month of November, sometimes with up to three layers occurring simultaneously. In the ten days surrounding the launch day, i.e. from 13 to 22 November 1970, the *f*oEs values always remained below 4 MHz. On 20 November, foEs reached a peak of 5.8 MHz and decreased to 3.5 MHz before the launch. Two hours after the chemical release foEs reached 8 MHz and this high electron density is probably associated with the released chemicals. The high Es activity carries on into the next day and then levels off.

The strong Es occurrence during the launch time made the detection of the normal E-region ionization very difficult. The resulting E-region electron density profiles, therefore, contain large error bars, see Fig. 1. Sporadic E layers were located at 100 km and 108 km before the launch. These layers moved down by one kilometer during the launch program and another layer appeared at 102 km.

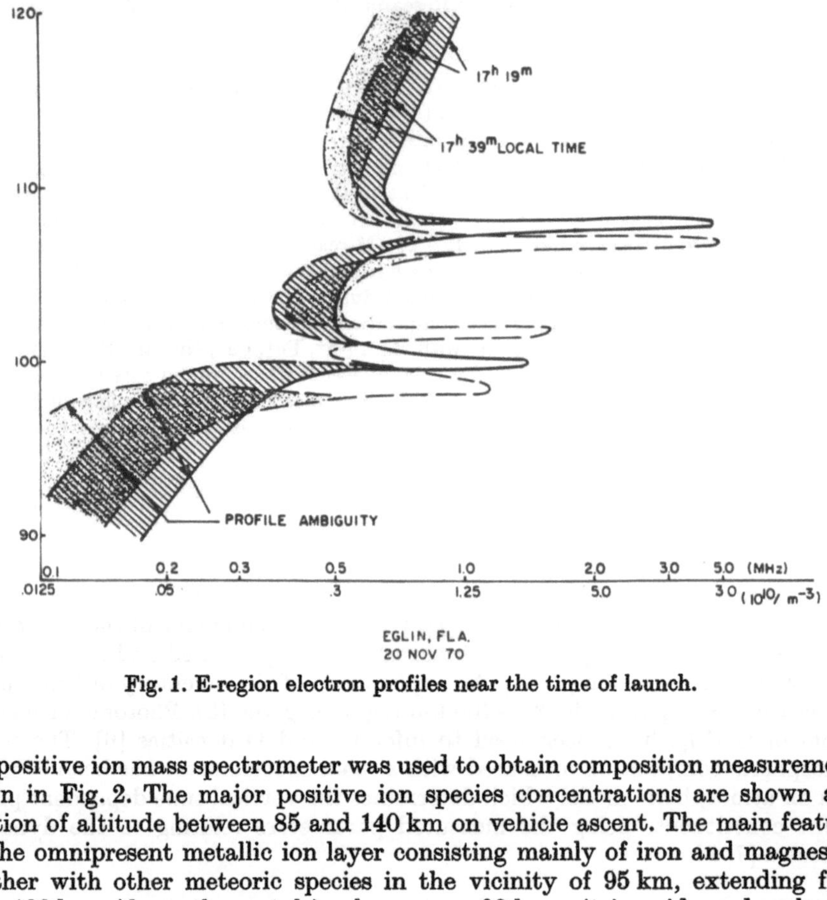

Fig. 1. E-region electron profiles near the time of launch.

A positive ion mass spectrometer was used to obtain composition measurements shown in Fig. 2. The major positive ion species concentrations are shown as a function of altitude between 85 and 140 km on vehicle ascent. The main features are the omnipresent metallic ion layer consisting mainly of iron and magnesium together with other meteoric species in the vicinity of 95 km, extending from 88 to 102 km. Above the metal ion layer near 98 km, nitric oxide and molecular oxygen ions become dominant. At 108 km, a thin meteoric ion layer was observed both on upleg and downleg. This layer was composed of mainly Si^+ with Mg^+, Fe^+, Na^+, Ca^+, Ni^+, and SiO^+. Where the meteoric ions were dominant, NO^+ was greatly reduced. The altitude resolution of the mass spectrometer is limited by the 1.5 seconds required for a mass spectrum scan, and thus the peak density of the meteoric ions in this layer may not be well defined by the spectrometer measurement. The digisonde showed peak density for a thin layer at this altitude, indicated by the "X" in Fig. 2. At higher altitudes, two layers of Al^+ were observed. The lower one, centered around 122 km and extending from 114 km to 128 km, can be attributed to Al^+ ions formed after the release of TMA, $Al(CH_3)_3$, from a chemical wind trail rocket launched four minutes before the mass spectrometer payload. The motion of the Al^+ ions formed in the vicinity of the chemical release has been calculated using the sporadic E theory of MacLeod [9]. Although the detailed chemistry resulting in the formation of the Al^+ is not understood, the Al^+ layers have been observed previously when measurements were made with chemical releases [10]. The distribution of the low altitude Al^+ ion profile shows reasonable agreement with the Al^+ ion trail position calculated by utilizing the TMA trail

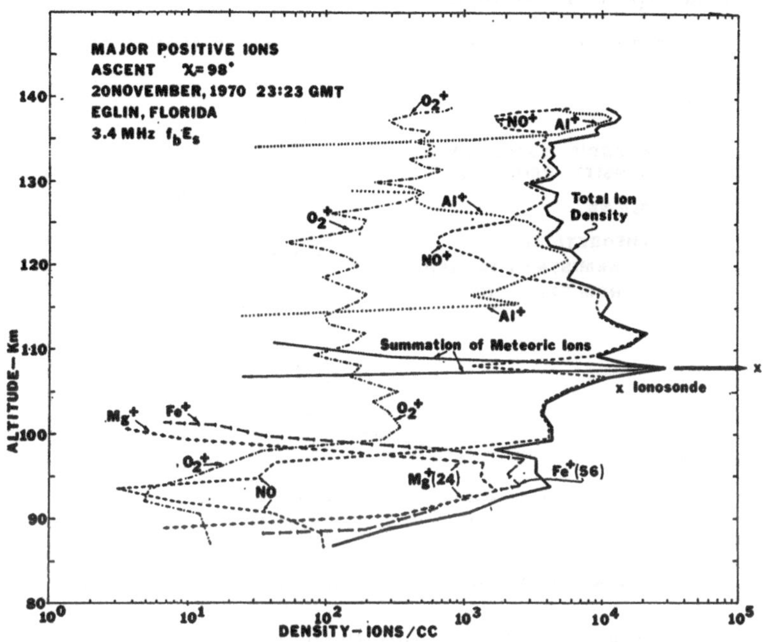

Fig. 2. Mass-spectrometer measurements of positive ion composition.

position and the ion motion deduced from ion—neutral collision forces and geomagnetic Lorentz forces. The higher altitude Al^+ layer near 138 km probably resulted from the TMA release also; however, the above forces do not provide sufficient motion of the Al^+ to move the ions into the flight path of the mass spectrometer. If some additional force were acting on the ion distribution, it would have to be significant. For instance, if an electric field was assumed as a driving force a field strength of about 10 mV m^{-1} would be required. In the region where these two Al^+ ion layers occur, a significant depletion of the NO^+ is also observed which occurs because of recombinative loss of NO^+ with electrons and perhaps charge transfer to metal atoms as well. At altitudes where the atomic—ion species are either absent or present in relatively small concentrations, NO^+ is the dominant species with O_2^+ varying from 1% to 10% of the NO^+ concentration.

3. E-Region Model

Theoretical ion density profiles were computed by integrating the time-dependent coupled continuity equations for an ionospheric model containing the species NO^+, O_2^+, N^+, O^+, M^+ and e. The calculation builds up the background ionization (except the metal M^+) from solar photoproduction beginning at noon on the experiment day. As the model time advances into the evening, the photoproduction is reduced, and the chemistry is dominated by the charge exchange and dissociative recombination reactions. At a preset time (here taken to be 2000 sec before the positive ion mass spectrometer flight time) the effect of the ion winds is introduced

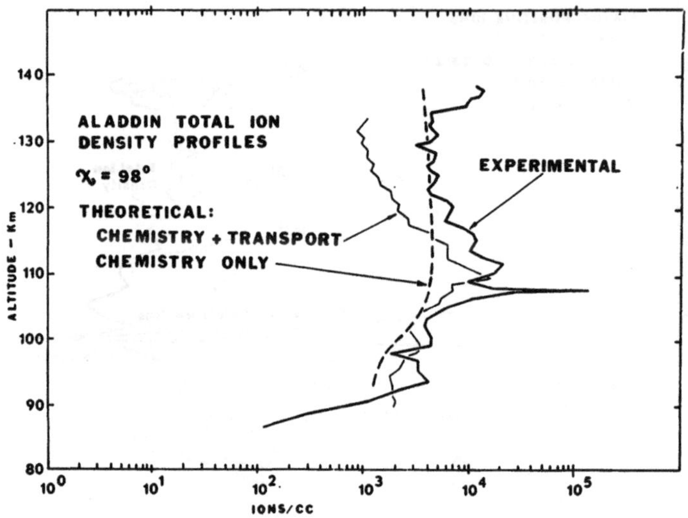

Fig. 3. Comparison of theoretical and measured positive ion density.

by adding the divergence of the ion flux to the continuity equations (using ion winds calculated from the measured neutral winds). Simultaneously metal ions with a density of 1000 cm^{-3} are added throughout the range 90—137 km. The integration is then carried on until the experiment time (2323 GMT, $\chi = 98°$). The total ion density profile obtained in this manner is shown in Fig. 3, together with the experimental profile and a profile calculated leaving out the ion transport contribution. Although the effect of ion transport is clearly seen and results in a better prediction than the chemistry only profile, the agreement between theory and experiment is poorer than was obtained previously [11]. The peak at 110 km is due to the layering of the metal ions at the zero of the vertical ion velocity profile, and is 2 km higher than the experimental peak. The (chemistry + transport) curve does not reproduce well the small-scale variations of the experimental profile. Both discrepancies may reflect the influence of short period motions which cannot be taken into account by the model at the present time. Although the sources of the differences between the measured and calculated curves may be sought among a variety of effects which are not presently included in the calculations, some improvement will undoubtedly result when refinements of the present model are introduced.

4. Lithium Chemistry in the Upper Atmosphere

Lithium vapor was released at altitudes of 80—100 km. The measurements of the 6708 Å Li atom emission spectrum were analyzed to determine the lithium inventory as a function of altitude and time after release. It was found that lithium is consumed extremely rapidly at the lower altitudes. At 85 km, only 1% of the lithium released was observed after 3 seconds, while at 100 km, 40 % of the released lithium was observed after 3 seconds. At these altitudes simple oxidation reactions of monatomic lithium by atomic and molecular oxygen or by ozone cannot account for such rapid disappearance of the vaporized lithium. It is necessary to include reactions involving dimerized lithium or termolecular reactions in which two lithium atoms participate. High density conditions exist during the initial expansion at lower altitudes which enhance such reactions significantly. The simple oxidation reactions become the dominant mechanism of lithium consumption a minute or so after release.

5. Twilight $O_2(^1\Delta_g)$ Measurements

The volume emission rate profile of the measured $O_2(^1\Delta_g)$ is shown in Fig. 4. The instrumentation consisted of a 120 Å width IR filter centered at 1.279 μ. The signal was detected with a PbS detector with a 0.1 second time constant using synchronous detection. A peak in the emission rate profile was observed near 83 km.

The $O_2(^1\Delta_g)$ observations in the mesosphere can be used to calculate the densities of atomic oxygen and ozone under the assumption of steady-state photo-

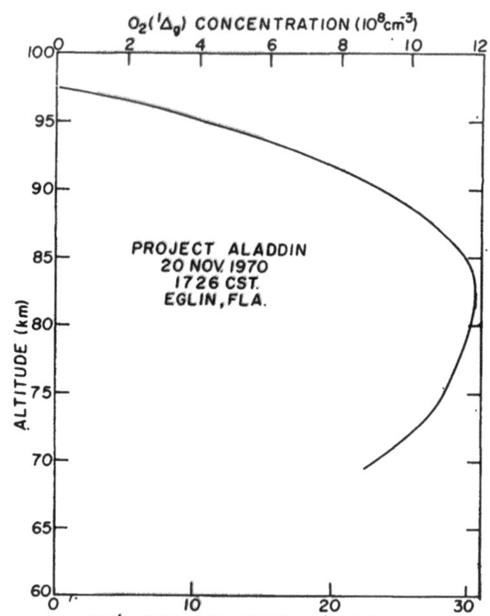

Fig. 4. Measured volume emission rate profile of $O_2(^1\Delta_g)$.

chemical equilibrium. It can be shown that the rate of loss of ozone by hydrogen is only 4% of the daytime dissociation rate and can be neglected. Another factor which is important to the computation of atomic oxygen density during twilight is that the relationship is independent of the solar dissociation rate. Using the mass-spectrometer measurements of N_2 and O_2 and the photometer measurements of $O_2(^1\Delta_g)$, together with the measured temperature and density, the following species concentrations were calculated.

Altitude (km)	$O_2(^1\Delta_g)/cm^3$	O/cm^3	O_3/cm^3
85	1.2×10^9	7.2×10^{11}	4.5×10^7
90	8×10^8	1.1×10^{11}	2.8×10^7

6. Neutral Composition

Measurements of the neutral composition, shown in Fig. 5, were obtained with a cryosorption pumped RF quadrupole mass spectrometer [12, 13]. Two ionization energies were used to extend the range of the atomic oxygen measurement to lower altitudes. The lower altitude points on the O profile may be in error because of the large signal from dissociative ionization of O_2 which was removed from them. At the 90 km level, the ground level mixing concentrations are shown in Fig. 5 and the measurements are in good agreement. Comparison of the Ar and N_2 densities shows a mixed atmosphere to an altitude about 105 km. The variations shown in the O profile generally reproduce on ascent and descent. The atomic oxygen density should be viewed as a lower limit for the present time. Tests are planned over the next several months which should help establish the accuracy. The total density profile was obtained from the measured mass density of Theon

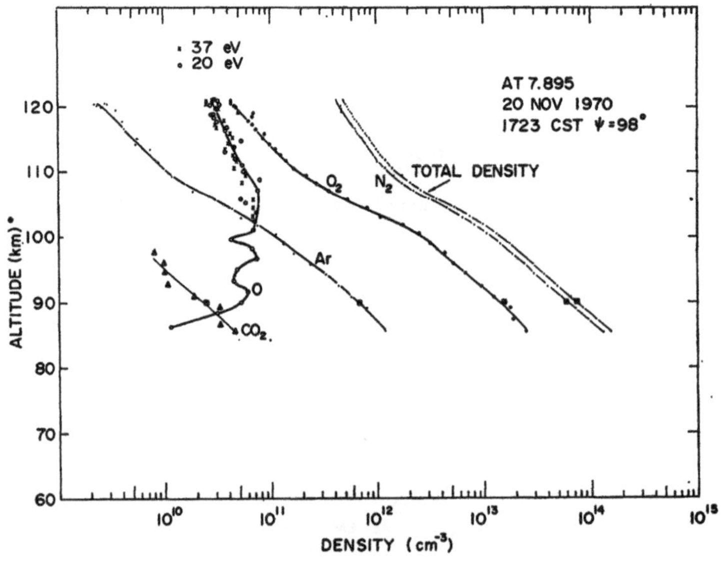

Fig. 5. Mass spectrometer measurements of the neutral atmospheric composition.

and Horvath [14] and the mean molecular mass measured by the mass spectrometer.

A theoretical computer calculation of the distribution of the species O, O_2, and Ar follows directly from the procedure given by George et al. [15]. The major difference is that here the mass density, temperature, and both the turbulent and molecular diffusion coefficients were determined directly from measurements during the ALADDIN 1 program. The calculation was performed for 30° N latitude with a solar declination angle of —15° and the O and O_2 densities at the upper boundary of 150 km are those determined by flux continuity.

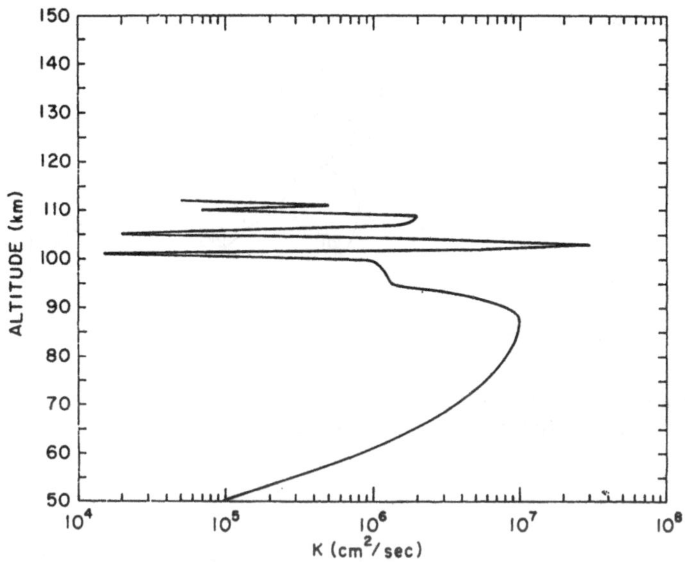

Fig. 6. Measured turbulent diffusion coefficients from the chemical releases.

The determination of the vertical turbulent diffusion coefficients is based upon measurements of the rate of dissipation of turbulent kinetic energy and the measured temperature of Theon and Horvath [14], and yields only a lower limit. The results of these diffusion calculations are shown in Fig. 6, where the turbulent layers are separated by laminar regions. The diffusivity in the altitude region below 93 km, where turbulence could not be measured, was estimated to fit Johnson and Wilkins [16] calculations of turbulent heat transfer.

The calculated profiles of the minor species are shown in Fig. 7. Two values of the vertical turbulent diffusivity were used, the lower limiting value K and an arbitrary factor of three times this value, $3K$. The agreement between the mass-spectrometer measurements and the calculated profile is fairly good, with the exception of the argon distribution above 105 km and the atomic oxygen distribution below 90 km. The discrepancy in the argon profile can be rectified if the turbulent diffusivity in the layer at 110 km was amplified over the other layers. This change may also improve, still further, the calculated O distribution. Also, the calculated O profile below 90 km may be in error due to our lack of measurements of turbulence in this region and the assumptions about the turbulent diffusivity there. It is interesting to note that the dips in the measured profile near 93 and 100 km are also indicated in the calculated profile.

Although the analysis and comparison of these measurements is not yet complete, the importance of comparing the dyamical processes with the distribution of ion and neutral species is apparent. This experiment has been followed by the recent ALADDIN 2 program (April 1972) and will be continued and expanded in the ALADDIN 3 program in late 1973.

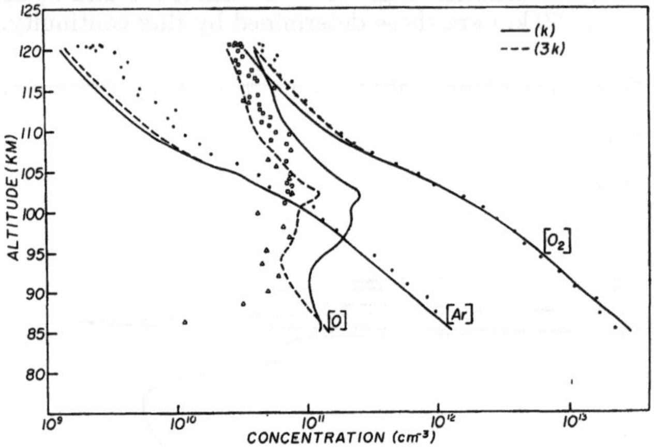

Fig. 7. Comparison of the theoretically computed neutral composition for the conditioning of the 20 November experiment with the measured points.

References

[1] N. W. ROSENBERG, D. GOLOMB, S. P. ZIMMERMAN, W. K. VICKERY and J. S. THEON, Space Research XIII, 435 (1973).
[2] B. W. REINISCH and K. BIBL, Trans. Am. Geophys. Un. **53**, 462 (1972).
[3] R. S. NARCISI, C. R. PHILBRICK, M. A. MACLEOD and N. W. ROSENBERG, Trans. Am. Geophys. Un. **53**, 462 (1972).
[4] M. A. MACLEOD, T. J. KENESHEA and R. S. NARCISI, Trans. Am. Geophys. Un. **53**, 464 (1972).
[5] H. S. HOFFMAN, Trans Am. Geophys. Un. **53**, 463 (1972).
[6] R. E. GOOD, Trans Am. Geophys. Un. **53**, 463 (1972).
[7] C. R. PHILBRICK and G. A. FAUCHER, Trans. Am. Geophys. Un. **53**, 463 (1972).
[8] T. J. KENESHEA, J. D. GEORGE and S. P. ZIMMERMAN, Trans. Am. Geophys. Un. **53**, 464 (1972).
[9] M. A. MACLEOD, J. Atmos. Sci. **23**, 96 (1966).
[10] R. S. NARCISI, D. M. THOMAS, A. D. BAILEY and L. DELLA LUCCA, Trans. Am. Geophys. Un. **50**, 654 (1969).
[11] T. J. KENESHEA and M. A. MACLEOD, J. Atmos. Sci. **27**, 981 (1970).
[12] C. R. PHILBRICK, G. A. FAUCHER and R. A. WLODYKA, AFCRL Rep. No. 71-0602 (1971).
[13] C. R. PHILBRICK, G. A. FAUCHER and E. TRZCINSKI, Space Research XIII, 255 (1973).
[14] J. S. THEON and J. J. HORVATH, Trans. Am. Geophys. Un. **53**, 463 (1972).
[15] J. D. GEORGE, S. P. ZIMMERMAN and T. J. KENESHEA, Space Research XII, 695 (1972).
[16] F. S. JOHNSON and E. M. WILKINS, J. Geophys. Res. **70**, 1281 (1965).

Space Research XIII — Akademie-Verlag, Berlin 1973

INVESTIGATION OF MID-LATITUDE IONOSPHERIC CURRENTS BY COMBINED ROCKET TECHNIQUES

D. Rees[a], T. L. Aggson[b], K. Burrows[c], G. Haerendel[d], E. Rieger[d], K. H. Lloyd[e], J. W. G. Wilson[f], E. B. Dorling[g] and G. L. Wrenn[g]

[a]Department of Physics, University College London, London, UK
[b]Goddard Space Flight Center, Greenbelt, Md, USA
[c]Radio and Space Research Station, Slough, Bucks, UK
[d]Institute for Extraterrestrial Physics, Max-Planck Institute, Munich, FRG
[e]Aerospace Division, Weapons Research Establishment Salisbury, S. Australia
[f]Department of Space Research, University of Birmingham, Birmingham, UK
[g]Mullard Space Science Laboratory, University College London, London, UK

A comprehensive range of neutral atmospheric and simultaneous ionospheric structure measurements were obtained from a Skylark rocket launched at Woomera during evening twilight on a quiet magnetic day. Above 150 km good agreement was found between the results obtained from three different methods of electric field measurement despite the low field strength ($\sim$ 1.5 mV m^{-1}). In the immediate vicinity of an intense sporadic E layer the probe measurements indicated large excursions of the electrostatic field amounting to 6.5 mV m^{-1} at 105.3 km and 2.7 mV m^{-1} at 104.8 km. The calculated ionospheric current system in the vicinity of the rocket trajectory was of similar magnitude to that indicated by local ground-based magnetometers but was diametrically opposed in direction.

1. Introduction

A Skylark rocket was launched at Woomera (31° S 137° E) on 2 March 1971 at 1008 UT (1938 CST, evening twilight). The payload included experiments to make a comprehensive range of neutral atmospheric structure measurements and to obtain simultaneous ionospheric structure including electric and magnetic fields.

2. Electric Field Measurements

Three independent sets of electric field measurements were made during the flight: (i) from the drift of two barium ion clouds released near 210 km altitude [1]; (ii) from a double spherical probe system operated in a stepped current mode [2]; (iii) from a long boom antenna system operated in a floating potential mode [3]

2.1. Electric Field Measurements above 130 km

Electric field probe data both before and after apogee indicate relatively good agreement with the data obtained from the barium releases, values varying around 1.5—2.0 mV m^{-1} directed approximately west to north-west. The barium clouds showed a mean electric field of 1.5 mV m^{-1} directed 30° west of north. Both probe experiments were severely perturbed following the two barium cloud releases, although these occurred several hundred metres from the rocket following delayed initiation. However, both instruments recovered rapidly, although with permanent changes of contact potential.

2.2. Electric Field Measurements in the Vicinity of Sporadic E

Fig. 1 shows the electric field data from the long boom experiment on descent between 115 and 92 km. (The $\boldsymbol{V} \wedge \boldsymbol{B}$ of the rocket has not been removed from these data.) The most striking feature is the remarkable excursion near 402 seconds (105 km). Also shown is the output of the positive ion experiment on the flight, which shows clearly the coincidence of the electric field excursion with the peak of the sporadic E layer.

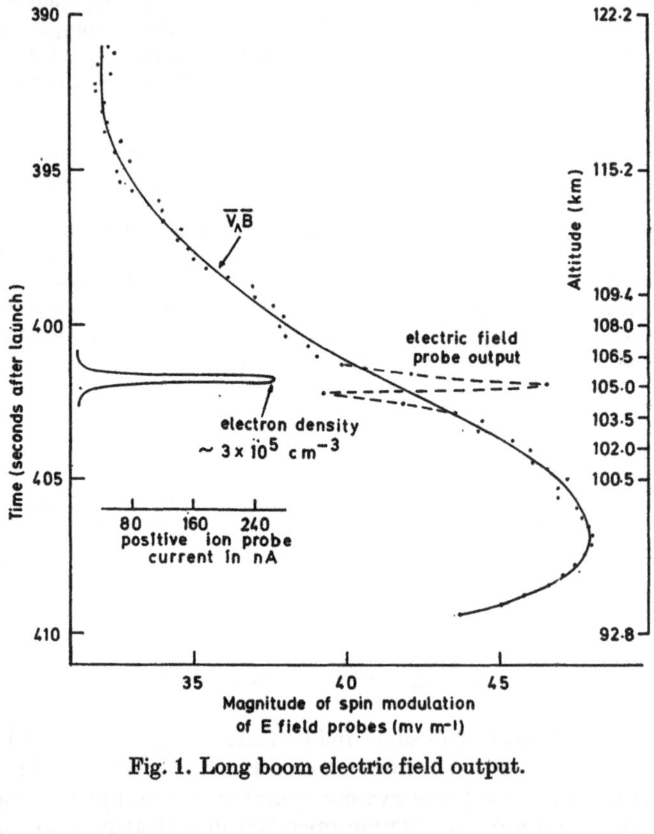

Fig. 1. Long boom electric field output.

Due to the near coincidence of the spin rate (3 Hz) and the vertical wavelength of the electric field structure (1 km, $\sim$0.6 second), our normal analysis, of vector subtraction of $\boldsymbol{V} \wedge \boldsymbol{B}$ of the rocket, is inappropriate. The positive excursion was accompanied by a 6° phase rotation of the total field as seen by the probes, and, from the magnitude and phase variations, two electric field values with $\boldsymbol{V} \wedge \boldsymbol{B}$ removed of 6.5 ± 1 mV m^{-1} at 045° and 2.9 ± 1 mV m^{-1} at 265° have been deduced at altitudes of 105.3 and 104.8 km respectively.

The neutral wind profile (Fig. 2) shows that the electric field variations coincided with a region of high wind velocity but relatively low wind shear. It seems remarkable that a long lasting and intense sporadic E layer should be located within a wind velocity maximum. A second rocket launched two hours later confirmed the coincidence, and the ground ionosonde showed that an intense sporadic E layer was stable near this height (105 km) for at least three hours either side of the dusk firing. At the time of the firing the local electric field values, quoted above, were quite strong enough to perturb severely the convergence or divergence of ions due to the neutral wind in the vicinity of the sporadic E layer. Additionally, the two measured values refer only to horizontal components of the electric field. The vertical gradient of this field ($> 2 \times 10^{-5}$ V m^{-2}) indicates that significant local parallel electric fields existed in association with this particular sporadic E layer.

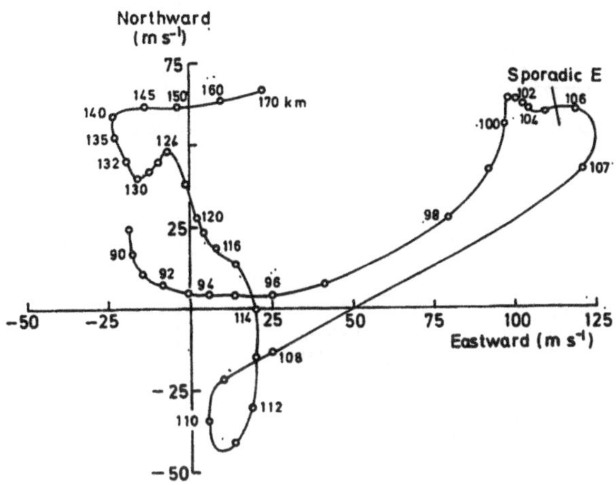

Fig. 2. Neutral wind profile.

We have investigated whether the large apparent ionospheric electric fields seen near 105 km could have been due to any interaction of the rocket or probe system with the steep electron density gradient in the sporadic E layer. Possible errors due to a local electron temperature gradient amount to the order of microvolts per metre only. It would also appear from a comparative analysis of the spherical probe data and the Langmuir probe and positive ion probe data that wake effects, caused by the descending vehicle, perturbing the conductivity in the vicinity of the rocket should not have significantly perturbed the long boom experiment.

3. Ionospheric Current Structure

Fig. 3 shows ground magnetometer records for the second half of the day of the experiment from Mirikata, a site some 50 km NW of the downleg rocket trajectory and from Mount Torrens, some 450 km SW. At the time of launch these give a ground perturbation of the order of $+5\gamma$ eastward component (Y) and -5γ northward component (X) taken from conventional base lines (quiet night for X, equal average southward and northward currents for Y). Records for the purpose of obtaining base lines were available for about two weeks either side of the experiment. The day of the launch was magnetically quiet (C9 = 0).

Fig. 4 shows the computed local ionospheric current system using the measured neutral wind profile, the F-region electric field (from the barium clouds) and the conductivity, which, for this purpose, was computed from ionosonde data only and did not use the Langmuir probe and positive ion probes on the rocket. The presence of the sporadic E layer was ignored in this computation. A vertically integrated current of about 1.2×10^{-2} amp m^{-1} was directed 40° east of north. At ground level such an ionospheric current in a homogeneous sheet would cause a magnetic field perturbation of the order of $10-15\gamma$ directed towards 50° west

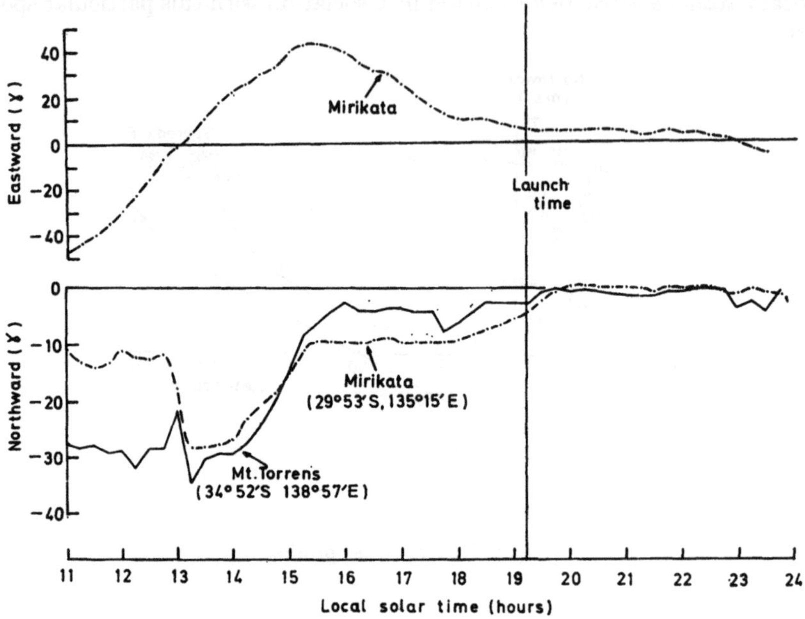

Fig. 3. Ground magnetometer records.

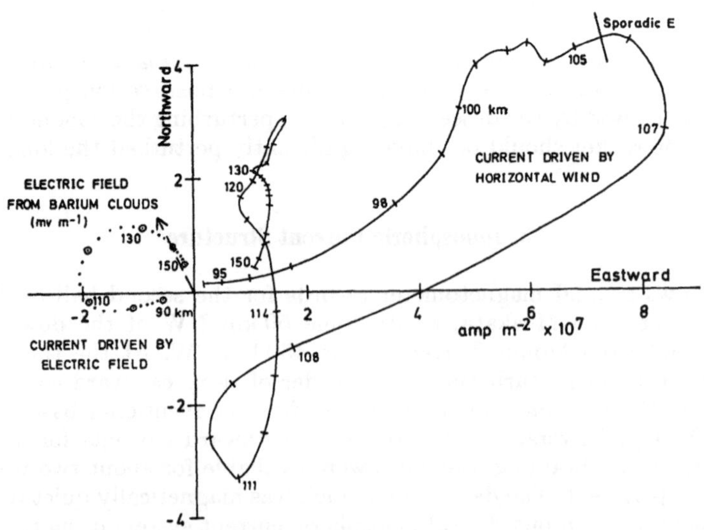

Fig. 4. Computed ionospheric currents.

of north, directly opposed to that indicated by the local ground magnetometer system.

In Fig. 5 the ionospheric current structure in the immediate vicinity of the sporadic E layer has been considered. The $\boldsymbol{V} \wedge \boldsymbol{B}$ dynamo field of the neutral wind was substantially modified between 106.0 and 104.0 km, the peak of the sporadic E layer being at 105.3 km. The direction of the total $(\boldsymbol{E} + \boldsymbol{V} \wedge \boldsymbol{B})$ field was modulated by 90°, from south-east above 105.5 km, to east at 105.3 km, south at 104.8 km and back to south-east below 104.5 km.

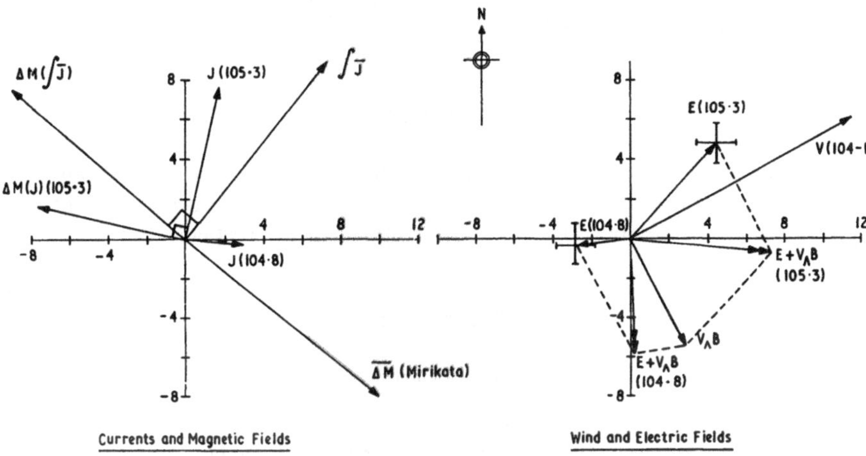

Fig. 5. Vectors associated with ionospheric currents. Numbers in parentheses refer to corresponding heights in kilometres.

J = computed current of Es layer at 105.3 km (A m$^{-1} \times 10^2$)
$\int \bar{J}$ = vertically integrated current = $\int \sigma \cdot (\boldsymbol{E} + \boldsymbol{V} \wedge \boldsymbol{B})$ (A m$^{-1} \times 10^2$)
$\Delta M(\int \bar{J})$ = computed magnetic fields at ground level due to vertically integrated current (gamma)
$\overline{\Delta M}$ (Mirikata) = measured magnetic field perturbation at ground level at time of flight (gamma)
$\boldsymbol{E}$ = electric fields measured by probe (mV m^{-1})
$\boldsymbol{V}$ = neutral wind velocity (m s$^{-1} \times 10^{-1}$)

The predominant current system, which may have a limited horizontal scale, should be close to the sporadic E layer maximum electron density. The local electrostatic field (i.e. as seen by the probes) did not cancel the $\boldsymbol{V} \wedge \boldsymbol{B}$ field due to the neutral wind in this region (105.3 km) so that we infer a fairly large local current due to the high electron density flowing in a direction approximately 20° east of north. It is interesting that such a current, which, in a uniform horizontal sheet, could cause a ground-level magnetic perturbation in excess of 10γ directed approximately NW, was in the same direction as the vertically integrated current and opposed to that inferred from the local ground magnetometers.

We conclude therefore that the ground magnetic perturbation at the launch time was dominated by current systems external to the ionosphere. Recent results of Sugiura et al. [4] indicate that a sufficient ground-level perturbation (20γ) would be quite characteristic of magnetospheric currents even under quiet geomagnetic conditions.

4. Ionospheric Current Measurements

Unfortunately, the rubidium magnetometer flown in this experiment did not operate correctly below 110 km on the upleg or on most of the downleg. The data, with spin heading error residuals removed, showed only a coning heading error modulation of about 3γ and no indication of any field discontinuities above 110 km, in agreement with our computed current system. (The southward current maximum near 111 km would not contribute significantly to the perturbation as seen by a rocket-borne scalar instrument.)

Acknowledgments

The Skylark rocket was launched in the British National Space Programme. Our sincere thanks are due to those who organized the collaborative programme and provided technical and data analysis support at our respective institutes.

References

[1] G. HAERENDEL and R. LÜST, in: Earth's Particles and Fields, Reinhold Book Corp., New York 1968 (p. 271).
[2] J. W. G. WILSON to be published (1973).
[3] T. L. AGGSON, in: Atmospheric Emissions, Van Nostrand Reinhold Co., New York 1969 (p. 305).
[4] M. SUGIURA, B. G. LEDLEY, T. L. SKILLMAN and J. P. HEPPNER, J. Geophys. Res. **76**, 7552 (1971).

Space Research XIII — Akademie-Verlag, Berlin 1973

N AND NO DISTRIBUTION BASED ON E-REGION ION COMPOSITION

A. D. Danilov and M. N. Vlasov

Institute of Applied Geophysics, Hydrometeorological
Service of the USSR, Moscow, USSR

A summary of the rocket mass-spectrometer measurements of the E-region ion composition was used to obtain the nitric oxide and atomic nitrogen concentrations at altitudes 100 to 120 km. The value of p (branching ratio for the dissociative recombination of NO^+ ions) was calculated on the basis of the same ion data. Using this value of $p = 0.9$ and rocket data on the F1-region ion composition the values [NO], [N] and [N*] were calculated at altitudes 150—200 km.

Recently views on the photochemistry of nitrogen oxide in the upper atmosphere have changed radically [1—5]. At present it is generally accepted that the process

$$O_2 + N(^2D) \rightarrow NO + O \tag{1}$$

suggested by Norton [1] is the principal source of NO. Thus, to construct a photochemical model of NO it is necessary to consider the photochemistry of excited atoms of $N(^2D)$. The production of the atoms may result from the following processes:

1) dissociative recombination of NO^+ [1—5]:

$$\begin{aligned} NO^+ + e &\rightarrow N(^2D) + O(^3P) \\ &\rightarrow N(^4S) + O(^1D) \end{aligned} \qquad p_1 \tag{2}$$

2) ion—molecular reaction [2—5]:

$$\begin{aligned} N_2^+ + O &\rightarrow NO^+ + N(^2D) \\ &\rightarrow NO^+ + N(^4S) \end{aligned} \qquad p_2 \tag{3}$$

and 3) dissociation of molecular nitrogen by photoelectrons [5]:

$$N_2 + e_{ph} \rightarrow N(^2D) + N(^4S). \tag{4}$$

The main process of $N(^2D)$ atom disappearance is reaction (1). The probability of spontaneous radiation for ^2D state of atomic nitrogen is negligible ($A = 2.3 \times 10^{-5}\ s^{-1} \ll \gamma_1[O_2]$).

For the E region the scheme of mutual transformations of N, NO, N* and NO^+ based on reactions (1)—(4) and:

$$O_2^+ + NO \rightarrow NO^+ + O_2 \qquad \gamma_5 \tag{5}$$

$$O_2^+ + N \rightarrow NO^+ + O \qquad \gamma_6 \tag{6}$$

$$NO + h\nu \rightarrow N + O \qquad j_{NO} \tag{7}$$

$$N + NO \rightarrow N_2 + O \qquad \alpha_8 \tag{8}$$

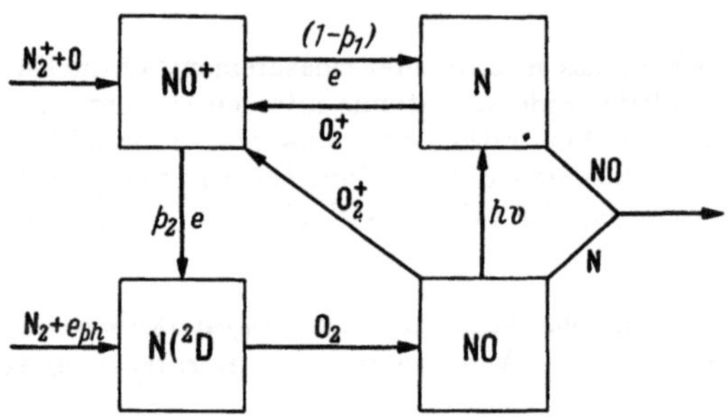

Fig. 1. The scheme of mutual transformations of N, NO, N* and NO^+ in the E region.

is given in Fig. 1. Corresponding equilibrium concentrations are given by the expressions

$$[N^*] = \frac{p_1[NO^+][e]\alpha^*_{NO^+} + p_2[N_2^+][O]\gamma_3 + j_{ph}[N_2]}{\alpha_1[O_2]} \tag{9}$$

where $j_{ph} = \int \frac{d\phi_e}{dE}\sigma_e\, dE$ is the dissociation probability by a photoelectron flux per molecule of N_2:

$$[NO] = \frac{\alpha_1[O_2][N^*]}{j_{NO} + [O_2^+]\gamma_5 + [N]\alpha_8}, \tag{10}$$

$$[N] = \frac{(1-p_1)\alpha^*_{NO^+}[NO^+][e] + j_{NO}[NO] + (1-p_2)[N_2^+][O]\gamma_3}{[O_2^+]\gamma_5 + [NO]\alpha_8}, \tag{11}$$

$$[NO^+] = \frac{[O_2^+][NO]\gamma_5 + [N_2^+][O]\gamma_3 + [O_2^+][N]\gamma_6}{\alpha^*_{NO^+}[e]}. \tag{12}$$

Using data on ion composition from Eqs. (9)—(12) not only concentrations of N, NO and N* can be obtained but also the value of p_1. As far as the value of p_2 is concerned the contribution of reaction (4) in the E region is small; therefore (as will be shown below) the uncertainty of p_2 is of no importance.

If $p_2 = 0$, i.e. reaction (3) contributes only to N production, then the solution to Eqs. (9)—(12) is

$$[NO] = \frac{(1-K_1)[NO^+][e]\alpha^*_{NO^+}}{\gamma_5[O_2^+]}, \tag{13}$$

$$[N] = \frac{(K_{ph} + 2K_1)\gamma_5[O_2^+]}{2\alpha_8(1-K_1)}, \tag{14}$$

$$[N^*] = \frac{(1 + K_{ph}/2)[NO^+][e]\alpha^*_{NO^+}}{\alpha_1[O_2]}, \tag{15}$$

$$p_1 = 1 - K_{ph}/2 \tag{16}$$

where $K_1 = \dfrac{[N_2^+][O]\gamma_3}{[NO^+][e]\alpha^*_{NO^+}}$ and $K_{ph} = \dfrac{j_{ph}[N_2]}{[NO^+][e]\alpha^*_{NO^+}}$.

Eqs. (13)—(16) were used for calculations of [NO], [N], [N*] and p_1 at heights 100, 110 and 120 km. The following constants of the reactions were assumed:

$$\gamma_1 = 6 \times 10^{-12}\ \mathrm{cm^3\,s^{-1}}\ [6, 7];\quad \alpha^*_{NO^+} = 4.5 \times 10^{-7}\,\frac{300}{T}\ \mathrm{cm^3\,s^{-1}}\ [8]$$

$$\gamma_3 = 2.5 \times 10^{-10}\ \mathrm{cm^3\,s^{-1}}\ [9];\qquad \gamma_5 = 8 \times 10^{-10}\ \mathrm{cm^3\,s^{-1}}\ [9];$$

$$\gamma_6 = 1.8 \times 10^{-10}\ \mathrm{cm^3\,s^{-1}}\ [9];\qquad \alpha_8 = 2 \times 10^{-11}\ \mathrm{cm^3\,s^{-1}}\ [10].$$

According to Bates [11] the contribution of reaction (7) at $j_{NO} = 10^{-7}\ \mathrm{s^{-1}}$ to destruction of NO and N production may be neglected. However Strobel et al. [4] found the much higher value $4 \times 10^{-6}\ \mathrm{s^{-1}}$ for j_{NO} taking into account a predissociation process by radiation in γ and β bands. At E-region heights this value should be much less due to radiation absorption by molecular oxygen, which is difficult to calculate accurately. Therefore, following [4], we neglected reaction (4) in our calculations.

Data on the composition were taken for $Z_0 = 60°$ from the review of all mass-spectrometer experiments in the E region [12]. Dissociation rates for N_2 by photoelectrons for the same value of Z_0 were taken from [5]. So far N_2^+ concentrations in the E region are small and below the mass-spectrometer limit; values for $[N_2^+] \cdot [O] \cdot \gamma_3$ were found from the balance equation for N_2^+ ions [13]:

$$[N_2^+] = \frac{q_{N_2^+}}{[O]\gamma_3 + 10^{-10}[O_2]}. \tag{17}$$

Values $q_{N_2^+}$ were taken for $Z = 60°$ according to G.S. Ivanov-Kholodny [14, 15] (110—120 km) and Swider [16] (100 km).

The calculated results for [NO], [N], [N*] and p_1 are given in Table 1 together with the initial data. As seen from the table for heights 100 and 110 km very close values of p_1 0.93 and 0.91 were obtained. A slightly smaller value $p_1 = 0.82$ was obtained for 120 km. Results for 100 and 110 km are more reliable since at 120 km dissociation of N_2 by photoelectrons increases (4), the accurate accounting of which is difficult. We therefore consider the value $p_1 = 0.9$ which is

Table 1

h (km)	$[NO^+]$ (cm^{-3})	$[O_2^+]$ (cm^{-3})	[e] (cm^{-3})	$[O_2]$ (cm^{-3})	$j_{ph}[N_2]$ ($cm^{-3}s^{-1}$)	[NO] (cm^{-3})	[N*] (cm^{-3})	[N] (cm^{-3})	p_1
100	3.2×10^4	8.0×10^3	4×10^4	1.8×10^{12}	1.3×10^2	1.2×10^8	8.3×10^1	4.0×10^4	0.92
110	7.5×10^4	2.5×10^4	1.0×10^5	2×10^{11}	6.8×10^2	1.8×10^8	3.6×10^3	1.6×10^5	0.91
120	6.6×10^4	3.3×10^4	1.0×10^5	3×10^{10}	9.1×10^2	7.7×10^7	1.6×10^4	6.8×10^5	0.82

close to Meira's result [17] ($p_1 = 0.87$) obtained from simultaneous measurements of [NO], $[NO^+]$ and $[O_2^+]$ to be more correct.

Concentrations of NO obtained are consistent with the most reliable experimental data by Meira [18]. According to these data the value of [NO] at 100—120 km is about 10^8 cm^{-3}. As the analysis of Eqs. (9)—(12) shows that calculated values of [NO] do not depend on the value of p_2, i.e. it does not depend on whether reaction (3) produced $N(^2D)$ or $N(^4S)$ atoms. At the same time the value of p_1 obtained from Eqs. (9)—(12) will depend on the value of p_2 assumed. In the extreme case $p_2 = 1$ the value of p_1 is 0.88 at 100 km and 0.75 at 110 km. Thus an inaccurate value of p_2 leads to a decrease of p_1 by not more than 0.1 compared with the assumed value 0.9.

This value of p_1 was used to determine [N], [NO] and [N*] at heights 150—200 km where it is impossible to use Eq. (12) since reactions (5) and (6) are not the main source of NO^+ ions. For $h = 150$—200 km the contribution of the reaction

$$O^+ + N_2 \rightarrow NO^+ + N \tag{18}$$

to N and NO^+ formation was taken into account. An empirical model of neutral composition [19] and charged particle concentrations from the review of mass-spectrometer experiments [13] were used in the calculations.

The results of the calculation are shown both in Table 2 and Fig. 2. As seen from the figure the theoretical profile of [N*] is in good agreement with experimental results [20] in the total height range 100—200 km.

Table 2

h (km)	150	170	200
[N*]	1.8×10^5	6.8×10^5	2.3×10^6
[NO]	4.0×10^7	1.1×10^7	1.3×10^7
[N]	1.6×10^6	1.3×10^7	9.4×10^6

[N] values obtained as a result of the calculations described smaller are than those obtained in the experiments of Ghosh et al. [21] (Fig. 2) and in usual mass-spectrometer experiments [22—26]. Strobel [5] obtained a similar strong difference between calculated values of [N] and experimental values [21]. In view of present photochemistry high values of [N] are impossible. Indeed if it is assumed that, say, at 120 km (where there are reliable experimental data of [NO]) [N]/$[N_2]$ in the atmosphere is 1% (see [22—24]), i. e. $[N] = 2\times10^9$ cm^{-3}, then the rate of the reaction (8) will be (at $[NO]\approx10^8$ cm^{-3}) 4×10^6 $cm^{-3}s^{-1}$. This will be the rate of N and NO destruction per $cm^3 s^{-1}$. In view of the modern physics of the upper atmosphere it is impossible to imagine a source which could maintain such a high rate of N or NO production. Therefore the simultaneous existence of both high [NO] and high [N] is impossible. So far recent measurements of [NO] are reliable (and are consistent with theoretical estimates); one must therefore suppose that the [N] measurements available are erroneous. The values of [N]/

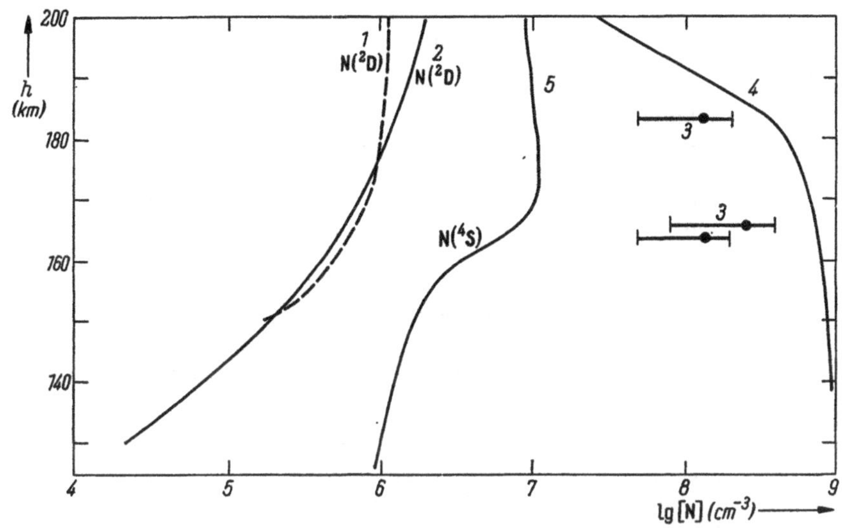

Fig. 2. A comparison of calculated atomic nitrogen concentrations with experimental data:
1, [N (^{2}D)] according to experiment [20].
2, calculated [N (^{2}D)].
3, [N (^{4}S)] measured in experiment [21].
4, [N (^{4}S)] recalculated according to the values of [N]/[N_2] obtained in experiment [21].
5, calculated [N (^{4}S)].

[N_2] in the rocket mass-spectrometer experiments are sometimes several times higher than calibration values in the laboratory and may approach tens per cent [25, 26]. But it is quite possible that these high values of [N]/[N_2] do not reflect the true [N] concentrations and are the result of the dissociation of excited N_2 molecules in the ion source of the device. If so, the mass-spectrometer data on [N]/[N_2] should be used, not for [N] determination, bur for the determination or at least estimation or at least estimation of [N_2*].

Nitrogen oxide concentrations obtained in the calculations (Table 2) are on the whole in good agreement with the calculated results of Strobel [4, 5].

An attempt was made to study [NO] diurnal variations in the E region using data on the ion composition. In this case there was no need in the photoelectron flux variation with Z_0 to determine the value of [NO] since as seen from Eq. (13) [NO] (contrary to [N] and [N*]) does not depend on K_{ph}. Data on ion composition at 100, 110 and 120 km for $Z_0 = 30, 60, 90$ and $120°$ taken from the above mentioned review of mass-spectrometer experiments [12] were used for the analysis

The results of the calculations are given in Table 3. As seen from the table [NO] values have practically no diurnal variation at heights of 100 and 110 km. At a height of 120 km a sharp decrease of [NO] from the approximately constant day-time value of $(5-10) \times 10^7$ cm^{-3} to 1.5×10^7 cm^3 at 120° is observed. However one should bear in mind that there are few measurements of ion composition in the night-time E region (see [12]), therefore for 120° the value of $[NO^+][e]/[O_2]$ in expression (13) is taken conventionally and so night-time [NO] values obtained characterize only a tendency for [NO] variations from day to night.

Table 3

h (km) $Z_\odot$	30	60	90	120
100	1.2×10^8	1.2×10^8	1.2×10^8	1.2×10^8
110	1.8×10^8	1.8×10^8	1.9×10^8	1.5×10^8
120	5.6×10^7	7.7×10^7	1.0×10^8	1.5×10^7

References

[1] R. B. NORTON, Ph.D. Thesis, Univ. Colorado 1967; ESSA Technical Memorandum IERTM-ITSA 60, 1967.
[2] R. B. NORTON and C. A. BARTH, J. Geophys. Res. 75, 3203 (1970).
[3] M. NICOLET, Aeronom. Acta, Brussels, A-N 71-1970.
[4] D. F. STROBEL et al., J. Geophys. Res. 75, 4307 (1970).
[5] D. F. STROBEL, J. Geophys. Res. 76, 2441 (1971).
[6] C. L. LIN and F. KAUFMAN, paper presented at DASA Symp. on Physics and Chemistry of the Upper Atmosphere, Stanford Research Institute, 1969.
[7] G. BLACK, J. Chem. Phys. 51, 116 (1969).
[8] M. A. BIONDI, Canad. J. Chem. 47, 1711 (1969).
[9] W. L. FITE, Canad. J. Chem. 47, 1797 (1969).
[10] L. F. PHILLIPS and H. I. SCHIFF, J. Chem. Phys. 36, 1509 (1962).
[11] D. R. BATES, The Earth as a Planet, Univ. of Chicago Press, Chicago 1954.
[12] A. A. DANILOV, paper presented at Symp. on Chemistry of D and E regions, Illinois 1971.
[13] A. D. DANILOV, Space Research X, 725 (1970).
[14] G. S. IVANOV-KHOLODNY, Geomagn. i Aeronom. 6, 832 (1966).
[15] G. S. IVANOV-KHOLODNY, Dokl. Akad. Nauk USSR, 170, 831 (1966).
[16] W. SWIDER, Rev. Geophys. 7, 573 (1969).
[17] L. G. MEIRA, Ph.D. Thesis, Univ. Colorado 1970.
[18] L. G. MEIRA, J. Geophys. Res. 76, 202 (1971).
[19] A. D. DANILOV, Space Research XII, 1299 (1972).
[20] L. WALLACE and M. B. MCELROY, Planet. Space Sci. 14, 677 (1968).
[21] S. M. GHOSH et al., J. Geophys. Res. 73, 4425 (1968).
[22] A. O. NIER et al., J. Geophys. Res. 69, 979 (1964).
[23] A. A. POKHUNKOV, Space Research I, 101 (1960).
[24] E. J. SCHAEFER, J. Geophys. Res. 74, 3488 (1969).
[25] A. A. POKHUNKOV, Space Research XII, 657 (1972).
[26] G. M. MARTYNKEVICH and E. D. BYURO, Space Research XII, 691 (1972).

Space Research XIII — Akademie-Verlag, Berlin 1973

SEASONAL VARIATION OF ATMOSPHERIC COMPOSITION IN THE F REGION AS A FUNCTION OF SOLAR ACTIVITY

T. R. Tyagi[a], K. K. Mittal[a], A. P. Mitra[a] and S. Prakash[b]

[a]Radio Science Division, National Physical Laboratory, New Delhi, India
[b]Physics Department, Kurukshetra University, Kurukshetra, India

The sunrise increase of the electron content measured from the Faraday fading records of satellite transmissions during the period 1964—1968 have been used to calculate the integrated electron production rates Q for the various values of solar activity. It has been found that for $F_{10.7} > 100$, Q is considerably higher in winter than in summer.

The integrated production rates due to 1—100 Å X-rays and 100—1300 Å EUV rays for O^+ ions have been calculated for different values of $F_{10.7}$ using average particle densities from CIRA 1965, Jacchia 70 and Jacchia 71 models for 06 h. It is noted that from solar minimum to solar maximum the EUV flux should increase by a factor of 2.8, 2.1 and 1.5 respectively depending on whether CIRA 65, J 70 or J 71 is used.

The seasonal variation of Q has been utilized to derive the seasonal variation of atmospheric composition for different models. It is noted that at 200 km the ratio $n(O)/n(O_2)$ in summer is higher than that in winter by a factor 1.3, 1.35 and 1.8 for CIRA 65, J 70 and J 71 and $n(O)$ increases from summer to winter by a factor of 1.09, 1.1 and 1.25 respectively.

1. Introduction

Measurements of the electron content of the ionosphere give information on the gross behaviour of the ionosphere as a function of time of day, season and solar activity. During the sunrise period, for $95° > \chi > 87°$ where χ is the solar zenith angle, the rate of production of ion—electron pairs in the ionosphere is larger than the rate of loss of electrons. An accurate measurement of variation of electron content versus time in this interval permits the determination of electron production rate [1]. Several other workers [2—5] have also used the observed variation of the electron content during the sunrise period for calculating the electron production rate.

The electron content measurements have been made at Delhi (28.6° N, 77.2° E) for the period Oct. 1964 to March 1969 from satellite Explorer 22 and for the period June 1965 to Sept. 1967 from satellite Explorer 27. Tyagi and Mitra [6], using sunrise increase of electron content obtained from satellite Explorer 27, calculated the integrated electron production rate for summer season, and derived the solar cycle variation of EUV flux from these calculations. Subsequently when the integrated electron production rates were derived for winter (and these are presented here) it was clear that there is a considerable difference in the total production rate in the two seasons. It is the purpose of this paper to show that these observations can be used to derive seasonal changes in the atmospheric composition, a range of atomic oxygen profile that will fit with the observations and to give the solar cycle variation of EUV flux.

2. Method of Analysis

The continuity equation for electrons can be written as

$$\frac{dN}{dt} = q - L(N) - \operatorname{div}(N\boldsymbol{v}) \tag{1}$$

in which q is the production rate of electron—ion pairs, L is the loss rate, N is the electron density and $\boldsymbol{v}$ is the drift velocity produced by diffusion, electrostatic or other forces. We assume that horizontal gradients are sufficiently small and that only vertical derivatives are important in the divergence expression. When Eq. (1) is integrated over all heights, the divergence terms drop out, since $N \approx 0$ at both low and very high altitudes. Physically this is equivalent to the statement that although electrons may move in altitude, their integrated number remains unchanged. Thus we get

$$\frac{d}{dt}\int N\,dh = \int q\,dh - \int L\,dh. \tag{2}$$

It has been observed that the electron content is nearly constant between 00 and 06 h LT. This fact may be explained in several ways: (a) The integrated loss rate is very small, perhaps produced by an upward drift of the ionization. (b) There is downward drift of ionization from the exosphere (where it contributes little to the observed Faraday rotation), which sustains the night-time ionosphere, in spite of relatively normal loss rates. (c) There is a steady night-time production of electrons, perhaps by a particle flux, which maintains the ionosphere in equilibrium. Although these alternatives are not equally likely, the important thing for the purpose of this paper is that whichever choice is made, the integrated loss rate in Eq. (2) may be omitted at night and probably for at least 15 minutes after ground sunrise. In alternative (a) the loss rate is assumed small, while in (b) and (c) there is an extra production term not shown in Eq. (2), which just balances the extra loss. Therefore, at night and for a short while after sunrise, Eq. (2) will reduce to [1]

$$\left(\frac{dI}{dt}\right)_{\chi=90^\circ} = Q \tag{3}$$

in which I is the integrated electron density or electron content. Thus the integrated electron production rate can be calculated by measuring dI/dt at $\chi = 90°$ from sunrise observations of electron content.

Integrated production rates for different ions can be theoretically calculated using appropriate ionizing flux. It may be noted that the change in total electron content around sunrise, i.e. the integrated production rate, will be represented by the total production of O^+ ions only, as the production of O_2^+ ions is negligible in comparison with O^+ ions at $\chi = 90°$ and N_2^+ ions are supposed to be in quasi photo equilibrium [7]. Hinteregger [8] has given flux values for the wavelength range of 100—1300 Å for the low solar activity period. X-ray flux measurements in the range 1—100 Å are given by various workers for almost all conditions of solar activity [9, 10]. It is possible to estimate how the 100—1300 Å flux varies with solar activity by using the observed Q-values in combination with the above cited flux measurements, of course assuming that the X-rays (1—100 Å) and ultraviolet rays (100—1300 Å) are the chief sources of production. Since the neutral particle density is a third variant with the solar cycle, the following procedure is adopted. The integrated production rate for O^+ ions has been calculated for different values of solar activity using appropriate X-ray flux values, but using minimum solar activity flux values for wavelengths 100—1300 Å and using appropriate average particle density from CIRA 1965, J 70 and J 71 models for 06 h at $\chi = 90°$. Contribution from X-rays in the 1—100 Å range has also been calculated separately. The X-ray contribution is then removed from the measured and calculated values of production rates. The comparison of resultant production rates will give the variation of 100—1300 Å flux with solar activity and the atmospheric composition changes as explained in the following sections.

3. Results and Discussion

3.1. Integrated Electron Production Rate

Integrated electron production rates Q, calculated from sunrise increase of electron content measured from Faraday fading records of satellite radio transmissions during the period 1964—1968, are plotted as a function of solar activity in Fig. 1. It is seen that for $F_{10.7} > 100$, Q is considerably higher in winter than in summer. During low solar activity, there is some indication of a reversal of this seasonal variation, the summer values being higher than winter. However,

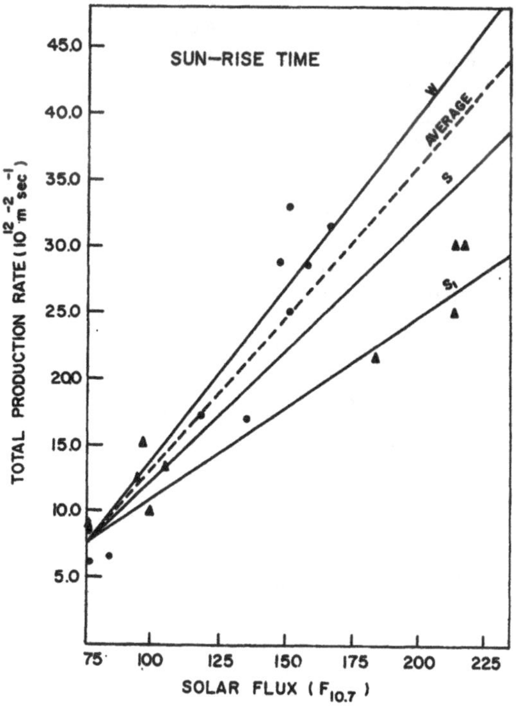

Fig. 1. Variation of observed Q with $F_{10.7}$.

the accuracy with which these estimates are made does not allow a definitive conclusion at this stage, and so the values during solar minimum have been assumed to show no seasonal dependence. It was mentioned in the previous section that $\int N\,dh$ values are found to remain almost constant between 00 and 06 h. However, during summer this is not true and hence L cannot be taken to be negligible at sunrise in this season. Therefore L values were estimated from $\int N\,dh$ variation between 00 and 06 h LT. These were found to increase from 1×10^{12} m^{-2} s^{-1} to 8.5×10^{12} m^{-2} s^{-1} from low to high solar activity. The modified summer Q variation with $F_{10.7}$ is also shown in Fig. 1 (curves S). The dotted line shows the average Q variation with $F_{10.7}$.

Such measurements of Q have been reported by several workers [1—5]. Their results are summarized in Table 1. It may be noted that all these Q values are quite comparable with our estimates.

Table 1

Q (10^{12} m^{-2}s^{-1})	$F_{10.7}$	Season	Ref.
8.5	89	Summer	[2]
14.9	155	Summer	
24.0	146	Winter	
32.0	175	Winter	
7.5	75	Average	[1]
3.0	75	Winter	[3]
4.0	75	Summer	
5.8	75	Average	[4]
17.0	low (75)	Average	[5]
7.6	75	Average	Present results
36.7	225	Summer	
46.2	225	Winter	

3.2. Variation of 100—1300 Å Ionizing Flux with Solar Activity

The integrated production rate due to 1—100 Å X-rays and 100—1300 Å EUV rays for O^+ ions has been calculated for different values of 10.7 cm solar flux using average particle densities as explained in the previous section. Calculated Q for 100—1300 Å EUV rays using CIRA 1965, J 70, and J 71 models are plotted as a function of exospheric temperature (T_∞) in Fig. 2. To plot observed Q_{EUV} against T_∞ we have taken the J 71 model to obtain exospheric temperature for different values of $F_{10.7}$ solar flux, then calculated $Q_{X\text{-rays}}$ for average densities for the corresponding T_∞ were subtracted from observed Q to get Q_{EUV} which was thus plotted against T_∞ in the same diagram. Comparison of calculated and measured values of Q during solar minimum shows that CIRA 1965 fits the results quite well, provided T_∞ for solar minimum ($F_{10.7} = 75$) is taken to be 630° as given by J 71, while J 70 and CIRA 65 models give higher values. The observed variation in Q_{EUV} from solar minimum to solar maximum is a factor of ≈ 5; of this the density change alone accounts for about 1.9 for CIRA 65, 1.6 for J 70 and 1.3 for J 71. Therefore, the flux would change by a factor of 2.5, 3 or 3.7 respectively for different models. Now the ratio $Q_{EUV(obs)}/Q_{EUV(calc)}$ at given solar activity would give the factor by which the solar EUV flux should increase for

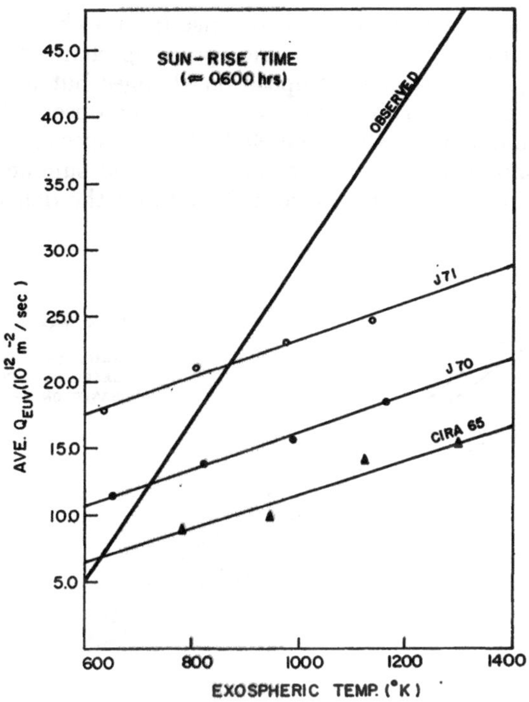

Fig. 2. Calculated Q for 100—1300 Å EUV rays using CIRA 1965, J 70 and J 71 models as a function of exospheric temperature T_∞.

that particular value of solar activity (if no matching is done in Q values at solar minimum). It is noted that from solar minimum to solar maximum the EUV flux increases by factors of 2.8, 2.1 and 1.5 for CIRA 65, J 70 and J 71 respectively. As at present this factor is considered to be around 2.5, it is inferred that J 71 gives very high density values and probably one has to use something between CIRA 65 and J 70 for high solar activity atmospheric densities.

3.3. Seasonal Variation of Atmospheric Composition

We have been hitherto discussing the solar cycle (or T_∞) variation of average Q_{EUV}. If we compare measured Q_{EUV} values for winter and summer during high solar activity, we find that winter values are substantially higher than summer values. If the CIRA 65, J 70 and J 71 models have to explain this seasonal change of Q_{EUV} there have to be substantial seasonal changes in the atmospheric densities given by these models. Though the J 70 and J 71 models have tried to incorporate seasonal changes in atmospheric composition, they give a much smaller seasonal change than is observed and also a wrong trend of the seasonal change. Now the required density changes have to be incorporated either through the seasonal changes in exospheric temperature or the changes in atomic to molecular oxygen density ratio $n(O)/n(O_2)$ arising from a change in the turbopause level. However, as no experimental evidence (from incoherent scatter measurements) is available of seasonal variation of exospheric temperature by more than 50 °K, variation of $n(O)/n(O_2)$ would then be expected to arise from a change in the turbopause level. In the present work we are not incorporating changes in turbopause level to derive the required seasonal composition changes but are doing so through an artificial change in T_∞. To do this, we divide the observed Q_{EUV} for summer and winter by the factor by which the EUV flux changes separately for different models. Corresponding to these values for winter and summer now we can read two values of T_∞ on the model curve in Fig. 2 and the densities corresponding

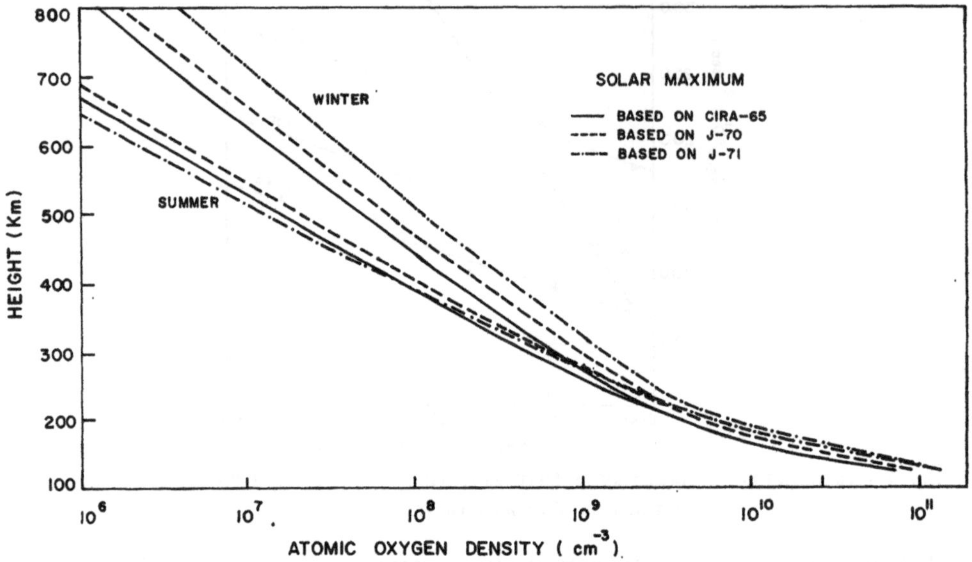

Fig. 3. Seasonal variation of $n(O)$ values derived by the present method based on CIRA 65, J 70 and J 71 models.

to these artificial temperatures are the required ones which can explain the observed results. Fig. 3 shows the winter and summer $n(O)$ values for high solar activity derived by the above method based on these three models.

Values of $n(N_2)$, $n(O_2)$ and $n(O)$ are given in Tables 2, 3 and 4 for selected heights. The important conclusion from this figure is that during high solar activity $n(O)$ increases from summer to winter by a factor of 1.1, 1.1 and 1.25 for CIRA 1965, J 70 and J 71 at 200 km and by a factor of 4.3, 4.8 and 14.0 respectively at 600 km, and the ratio $n(O)/n(O_2)$ in summer is higher than that in winter by a factor 1.3, 1.35 and 1.8 for CIRA 65, J 70 and J 71 at a height of 200 km. It may be mentioned that Danilov [11] on the ground of the data obtained from the experiments of Nier at Whitesands and Fort Churchill concluded that this ratio in summer is higher than winter by a factor of 1.5—2.

Acknowledgments

One of the authors (S. P.) is a senior research fellow of the National Oceanic and Atmospheric Administration, USA, and another (K. K. M.) is a junior research fellow of the Council of Scientific and Industrial Research.

Table 2

Representative average atmospheric densities during solar minimum derived from present study

Height	n(O)	n(O_2)	n(N_2)
120 km	7.6 + 10	7.5 + 10	4.0 + 11
200 km	2.49 + 09	1.18 + 08	1.35 + 09
600 km	1.2 + 05	—	4.0 + 01

Table 3

Representative atmospheric densities for winter during solar maximum following different models

Height	n(O)			n(O_2)			n(N_2)		
	CIRA 65	J 70	J 71	CIRA 65	J 70	J 71	CIRA 65	J 70	J 71
120 km	7.6 + 10	9.2 + 10	1.41 + 11	7.5 + 10	6.2 + 10	5.7 + 10	4.0 + 11	3.91 + 11	3.92 + 11
200 km	3.6 + 09	4.8 + 09	6.4 + 09	5.0 + 08	5.02 + 08	3.65 + 08	4.37 + 09	4.85 + 09	4.11 + 09
600 km	1.47 + 07	2.0 + 07	3.6 + 07	8.74 + 03	1.07 + 04	1.43 + 04	2.89 + 05	3.86 + 05	5.53 + 05

Table 4

Representative atmospheric densities for summer during solar maximum following different models

Height	n(O)			n(O_2)			n(N_2)		
	CIRA 65	J 70	J 71	CIRA 65	J 70	J 71	CIRA 65	J 70	J 71
120 km	7.6 + 10	9.2 + 10	1.4 + 11	7.5 + 10	6.3 + 10	5.3 + 10	4.0 + 10	3.83 + 04	3.75 + 11
200 km	3.3 + 09	4.3 + 09	3.1 + 09	3.45 + 08	3.4 + 08	1.6 + 08	3.2 + 09	3.56 + 09	2.13 + 09
600 km	3.3 + 06	4.2 + 06	2.3 + 06	4.02 + 02	3.68 + 02	3.68 + 01	2.01 + 08	2.09 + 04	3.16 + 03

References

[1] O. K. GARRIOTT and F. L. SMITH III, Planet. Space Sci. **13** 839 (1965).
[2] G. N. TAYLOR, Planet. Space Sci. **13**, 507 (1965).
[3] J. E. TITHERIDGE, J. Atmos. Terr. Phys. **28**, 1137 (1966).
[4] N. N. RAO, J. Geophys. Res. **72**, 2929 (1967).
[5] F. L. SMITH III, J. Geophys. Res. **73**, 7385 (1968).
[6] T. R. TYAGI and A. P. MITRA, J. Atmos. Terr. Phys. **32**, 1807 (1970).
[7] A. A. GRAN, J. Atmos. Terr. Phys. **27**, 1807 (1965).
[8] H. E. HINTEREGGER, in: Annals of IQSY, Vol. **5**, Solar Terrestrial Physics, Terrestrial Aspects, MIT Press, Cambridge, Mass. 1969 (p. 305).
[9] S. L. MANDELSTAM, Space Sci. Rev. **4**, 587 (1965).
[10] J. H. UNDERWOOD, Science **159**, 383 (1968).
[11] A. D. DANILOV, Space Research XII, 891 (1972).

Space Research XIII — Akademie-Verlag, Berlin 1973

REDUCTION OF SOLAR X-RAY FLUX MEASUREMENTS FOR USE IN IONOSPHERIC RESEARCH

R. W. Kreplin, D. M. Horan and K. P. Dere

Naval Research Laboratory, Washington, D.C., USA

Previously derived values of solar X-ray flux based on an assumed 0.5×10^6 °K gray body spectrum are corrected for use in ionospheric electron density calculations by removing the contribution of radiation below 20 Å and by using a spectrum based on an actual rocket measurement. The results show that the actual 44—60 Å flux is three times lower than that originally calculated.

The efficiency $\varepsilon(\lambda)$ of the 44—60 Å ionization chamber [1] shows a strong response to radiation below 20 Å. Since, during solar flares, there is considerable radiation in this region it is necessary to correct the published values of "44—60 Å" flux before using the data to calculate ion production rates. And since it has been shown [2] that in the 44—60 Å region line emission predominates this too must be taken into account.

In order to make a correction for radiation below 20 Å a separate photometer is flown whose response matches that of the 44—60 Å unit very closely below 20 Å. An evaporated aluminum coating on the Mylar window effectively eliminates sensitivity in the 44—60 Å range. It can be shown that the current produced in the 44—60 Å chamber (I_5') is related to the current produced in the 1—20 Å chamber (I_6) by a simple proportionality constant, i.e. $I_5' = FI_6$ so that the correction can be made by subtraction of one current value from the other. Assuming identical efficiency curves below 20 Å the value of F is found to be 0.845. Horan and Kreplin [1], taking into account the small differences in efficiency and using gray body approximations to describe the solar spectrum below 20 Å, found values of F ranging from 0.922 at 0.5×10^6 °K to 1.00 at 2.0×10^6 °K. Thus the assumption of a value $F = 1.0$ seems justified for practical calculations.

Application of this correction procedure to measurements made during a small flare (19 November 1971, ≈ 1400 UT) indicates that the true 44—60 Å flux increased only by about 60% over the pre-flare background level. The uncorrected calculation showed an increase of a factor of three. In another case (30 August 1971, 0410 UT) the corrected increase was reduced to a factor of three from the previously calculated factor of five.

There is little information available concerning the flare spectrum in the 44—60 Å region with the exception of a spectrum made by Freeman and Jones [3] during a small event. Manson [4] has measured the spectrum during quiet periods. Comparison of their data indicates that it is emission mainly from the high ionization states of silicon and from Fe XIV, XV, and XVI that is responsible for the major part of the flare X-ray flux. Using the spectrum of Freeman and Jones and assuming that the line ratios remain constant during the event and that there is no continuum emission, we calculate for the 30 August 1971 flare a maximum flux of 0.47 erg cm^{-2} s^{-1}. The older calculation using the 0.5×10^6 °K gray body approximation in which no correction was made for radiation below 20 Å gave 1.4 erg cm^{-2} s^{-1}. This difference, a factor of about three, is due to the inclusion of radiation below 20 Å in the older calculation. There is little difference in the integrated flux value depending on the shape of the solar spectrum.

References

[1] D. M. Horan and R. W. Kreplin, NRL Rep. 7408 (1972).
[2] R. Tousey, W. E. Austin, J. D. Purcell and K. G. Widing, Annls Astrophys. 28, 755 (1965).
[3] F. F. Freeman and B. B. Jones, Solar Phys. 15, 288 (1970).
[4] J. E. Manson, Astrophys. J. 147, 703 (1967).

Space Research XIII — Akademie-Verlag, Berlin 1973

OBSERVATIONS OF THE LOWEST IONOSPHERE

J. D. MITCHELL and L. C. HALE

Ionosphere Research Laboratory, The Pennsylvania State University, University Park, Penn., USA

Electrical conductivity data obtained using the subsonic blunt-probe technique in the mesosphere and upper stratosphere during the winter and summer of 1971 at White Sands Missile Range, New Mexico, and the winter of 1972 at Wallops Island, Virginia, are presented and discussed.

In the cosmic ray ionization-dominated region, variations in positive ion conductivity are correlated with variations in temperature. The observation of relatively altitude-independent electron density and positive conductivity between 50 and 60 km, implying a positive ion density proportional to neutral air density, leads to the conclusion that dissociative recombination is the principal positive ion loss mechanism in this altitude range. Decreased electron densities in this region distinguish days of low radio-wave absorption from days of moderate to high absorption. Below 40 km, the electron density is negligible and the positive ion density is essentially constant with altitude, which is consistent with cosmic ray ionization and a three-body mutual neutralization process.

1. Introduction

During the past two years, ten parachute-borne blunt probes for measuring positive and negative ionospheric conductivities have been flown in coordinated rocket experiments at White Sands Missile Range (WSMR), New Mexico, and NASA Wallops Station, Wallops Island (WI), Virginia. Discussion in this paper is limited to data collected on six of these flights (see Table 1). The first three rocket shots in 1971 were part of a series conducted with the Atmospheric Sciences Laboratory of WSMR to study the wintertime variability in D-region ionization [1]. The data obtained on 6 January 1972 are from a similar experiment at WI in conjunction with NASA. The two 1971 summer launches were conducted at WSMR in a series with the Atmospheric Sciences Laboratory and Utah State University.

In conjunction with our probe measurements in the lower D region, radio-wave absorption experiments were used to monitor the relative amount of ionization at higher altitudes. At WSMR, A3 absorption measurements were conducted using station WWV at 5 MHz (approximately 930 km away). Recent A3 absorption calculations using ray-tracing techniques indicate that this path should serve as a good daytime monitor of relative changes in D-region ionization [2]. The A1 absorption experiment (1.8 MHz) conducted by S. Gnanalingam at NASA Goddard Space Flight Center, Greenbelt, Maryland, and A3 absorption measurements (3.36 MHz) by J. N. Rowe, A. J. Ferraro, and H. S. Lee of the Ionosphere Research Laboratory, University Park, Pennsylvania, provided information concurrent with the WI launch. Fig. 1 shows hourly relative A3 absorption values on days of the rocket launches at WSMR. The arrows identify the curves and indicate the launch times.

Table 1

Parameters for WSMR and WI rocket launches

Date	Time (hours)	Solar zenith angle (°)	Location	Absorption
22 Jan. 1971	1300 MST	54	WSMR	High
26 Jan. 1971	1430 MST	62	WSMR	High
1 Feb. 1971	1215 MST	50	WSMR	Low
9 June 1971	0809 MST	53	WSMR	Moderate
28 July 1971	0705 MST	68	WSMR	Moderate
6 Jan. 1972	1337 EST	64	WI	Low

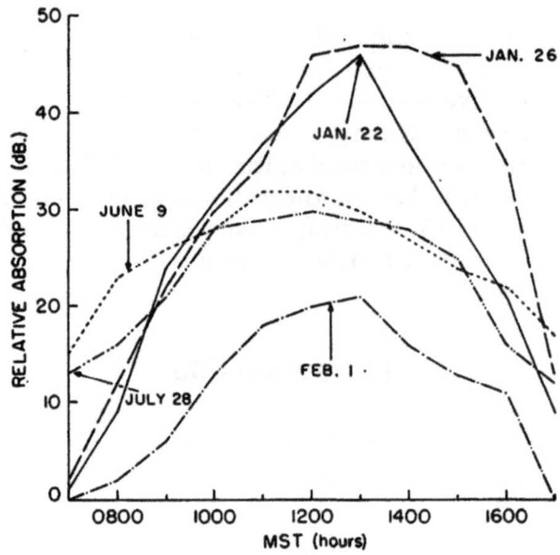

Fig. 1. A3 absorption relative to the night-time level for 1971 WSMR rocket launches.

2. Data

The parachute-borne subsonic blunt probe measures polar electrical conductivity [3]. Smoothed curves representing conductivity data measured on six different days are given in Fig. 2.

2.1. Conductivity Measurements

Excluding the data measured on 22 January and 28 July 1971, the profiles for positive conductivity are nearly the same below 60 km. On the day of enhanced positive conductivity (22 Jan. 1971), mesospheric temperatures measured at WSMR were significantly warmer than the mean monthly values for January [1] and similarly on the day of reduced positive conductivity (28 July 1971), the mesosphere was cooler than the monthly mean for July [4]. In the region where $\sigma_+ \simeq \sigma_-$ all of the curves display an altitude dependence approximately pro-

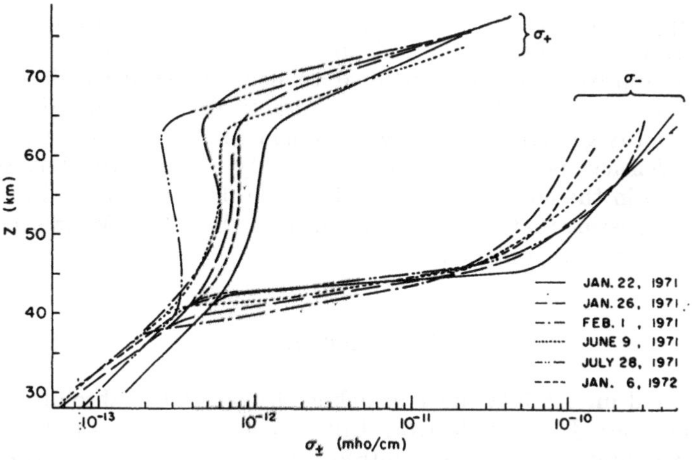

Fig. 2. Blunt probe conductivity curves from rocket launches at WSMR and WI.

portional to the inverse of neutral air density. Above this region positive ion conductivity is fairly constant with height up to 60 km.

All of the negative conductivity curves have a very steep gradient in the altitude region of 40—45 km. Above 50 km, the curves for the two days of low absorption begin to diverge from those for the other four days on which the absorption was moderate to high.

2.2. Positive Ion Densities

Assuming a profile for positive ion mobility [3], it is possible to deduce positive ion density profiles (Fig. 3) from the blunt probe conductivity data.

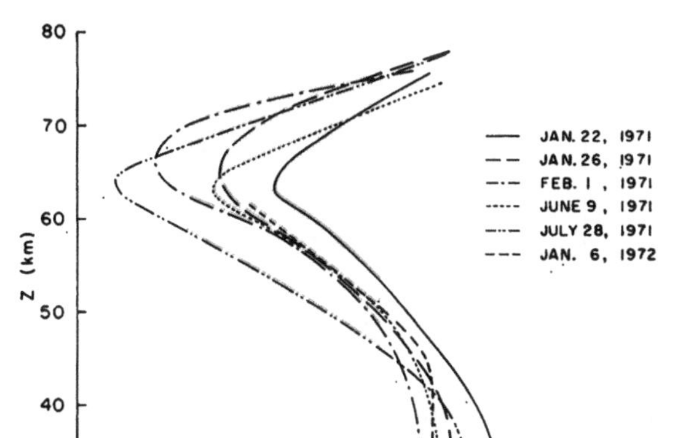

Fig. 3. Positive ion density profiles from rocket launches at WSMR and WI.

Below approximately 40 km, the positive ion density profiles do not vary significantly with height. Between 50 and 60 km, all of the curves have a height dependence approximately proportional to neutral density and, excluding the data for 22 January and 28 July 1971, are practically the same. Since production in this region (by galactic cosmic rays) also has a height dependence proportional to neutral density, positive ion number density is approximately proportional to production in the 50—60 km range. Between 63 and 66 km all of the profiles turn over, suggesting that in this region ionization by solar UV radiation supplants production by galactic cosmic rays.

2.3. Electron Densities

An admitted uncertainty must be acknowledged in the absolute values of the electron densities. Limited electron drift velocity data for the low E/P values at which the blunt probe operates introduce an unresolved error. Also, an uncertainty in probe theory is introduced by the non-linear dependence of electron drift velocity on E/P. However, all of the data were reduced using the same techniques and thus a higher level of confidence is placed in comparing the profiles from one day to another.

The electron density curves (see Fig. 4), corresponding to the negative ion conductivities, display a steep gradient between 40 and 45 km. Above 50 km, the curves are relatively constant with altitude and the electron density values on days of moderate to high absorption are two to three times larger than those for the low absorption days.

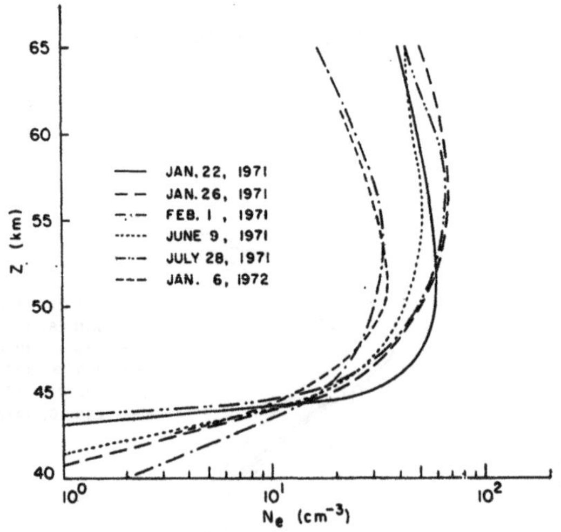

Fig. 4. Electron density profiles from rocket launches at WSMR and WI.

No real distinction is apparent below 65 km between the electron density profiles on days of moderate absorption and those for high absorption. Thus it may be concluded that the variation in electron density below 65 km distinguishes days of low radio-wave absorption from those of moderate to high absorption.

Presumably, the difference between moderate and high absorption days must be reflected in electron density variations at higher altitudes.

3. Discussion

Consider the lumped effective positive ion continuity equation:

$$\frac{dN_+}{dt} = Q - \alpha_d N_+ N_e - \alpha_i N_+ N_- \tag{1}$$

where $Q =$ ion production rate, $\alpha_d =$ dissociative recombination coefficient, $\alpha_i =$ mutual neutralization coefficient.

Below 65 km production is primarily attributable to galactic cosmic rays and thus $Q = I\varrho$ where ϱ is the atmospheric density and I is the ionizing coefficient, which is essentially independent of altitude to below 30 km.

Where electrons can be neglected, below 40 km, the steady state solution of (1) becomes

$$N_+ = N_- = \left(\frac{I\varrho}{\alpha_i}\right)^{1/2} \tag{2}$$

If N_+ is to be independent of height, as indicated by the data, then α_i must be proportional to ϱ below 40 km. Such is the case when three-body mutual neutralization is the dominant ion loss process.

Between 50 and 60 km the steady state solution to Eq. (1) agrees with the observation of constant N_e and $N_+ \propto \varrho$ if the mutual neutralization term is small compared with dissociative recombination.

The positive ion conductivity curves are generally in good agreement below 60 km, excluding those for 22 January and 28 July 1971. The variations observed on these two days may be attributed to differences in ion density and/or ion mobility. The difficulty of postulating processes to explain the former leads us to suspect that the variations in positive ion conductivity, which correlate with temperature variations, are more likely due to changes in the mobility of the ions. Such variations are much larger than can be explained by a direct influence of temperature on ion mobility. (They could, for instance, be due to a rapid variation with temperature of the size of water vapor—ion clusters.) In any case, the decrease in electron density on 1 February 1971 and 6 January 1972 can be interpreted as implying an increase in the electron loss rate.

Maintenance of the observed electron density values requires a detachment mechanism which is approximately proportional to neutral atmospheric density in the 65—50 km range, but which cuts off sharply between 50 and 40 km. (It is interesting to observe that this sort of behavior has been reported for the distribution of $O_2(^1\Delta_g)$ by a number of workers, as summarized by Evans and Llewellyn [5]).

4. Conclusions

The study of ionization processes below 65 km is of interest because the primary ionizing radiation, galactic cosmic rays, remains relatively constant, with some modulation due to solar activity. Thus, large variations in ionization at these altitudes must be due to variations in the atmosphere, either meteorological or chemical. The observations of ionization in the 50—65 km region are of particular interest because the altitude distributions of positive conductivity σ_+ and electron density N_e enable us to draw the following conclusions:

(i) The relative constancy with altitude of σ_+ and N_e indicate that these quantities are not strongly related to atmospheric density ϱ in this altitude range.

(ii) The altitude distribution of positive ion density, implied from the conductivity distribution $N_+ \propto \sigma_+ \varrho$, and the relatively constant N_e suggest that dissociative recombination is the principal positive ion loss mechanism.

(iii) A relatively large variation in positive ion conductivity with mesospheric temperature was observed. It is not known whether this conductivity variation is due to changes in ion density or ion mobility, but a somewhat stronger case can be made for the latter at this time.

(iv) The greatest variation in electron density in this altitude range was between days of very low radio-wave absorption and other days of moderate to high absorption. Two different days of low absorption, at different geographical locations, showed nearly identical electron density profiles.

(v) The observed values of electron density require an effective detachment coefficient which increases with decreasing altitude between 65 and 50 km, approximately proportional to neutral air density, but cuts off sharply below 45 km.

Acknowledgments

This research was supported by the US Army Research Office, Durham, N.C., under Grant DA-ARO-D-31-124-71-G110, and by NASA Grant NGR 39-009-218. Assistance is gratefully acknowledged from the Atmospheric Sciences Laboratory at White Sands Missile Range, N.M., and the National Aeronautics and Space Administration at Wallops Island, Va.

References

[1] J. D. Mitchell, L. C. Hale, R. O. Olsen, J. Randhawa and R. Rubio, Radio Sci. 7, 175 (1972).
[2] J. S. Nisbet and M. K. Lee, private communication (1972).
[3] L. C. Hale, D. P. Hoult and D. C. Baker, Space Research VIII, 320 (1968).
[4] R. O. Olsen, private communication (1972).
[5] W. J. F. Evans and E. J. Llewellyn, Annls Géophys. 76, 167 (1970).

Space Research XIII -- Akademie-Verlag, Berlin 1973

A STUDY OF THE EQUATORIAL D REGION

Y. V. SOMAYAJULU, M. B. AVADHANULU and K. S. ZALPURI

Radio Science Division, National Physical Laboratory, New Delhi, India

Electron density profiles in the D region obtained by rocket sounding at the geomagnetic equator are briefly reviewed. The study shows that the transition from cosmic ray to Lyman alpha as the domination ionization source occurs at an altitude between 65 and 70 km. The observed behaviour is found to be consistent with the current concepts of ionization production by Lyman alpha radiation and cosmic rays in the D region. These experimental electron density profiles are utilized to develop a diurnal model of the equatorial D region.

1. Introduction

Although the mid-latitude D region has been systematically explored by rocket sounding technique [1], the exploration of the equatorial D region has started only recently. The earliest rocket measurements of the electron densities in the equatorial D region have been conducted by the University of Illinois group [2] under relatively low solar activity conditions. The electron densities were measured by a combination of radio absorption and dc probe techniques. These electron density profiles in the 70–90 km range exhibit an anomalous latitudinal behaviour similar to the equatorial anomaly in *f*oF2, i.e. low values at the equator flanked by two peaks located at about $\pm 15°$ geomagnetic latitude. The existence of such an equatorial anomaly in radio absorption has been reported using ground-based A1 techniques [3, 4].

Table 1

Details of the D region rocket flights from Thumba, India

Flight No.	Date	Time (LT)	(χ)	Ref.
Brahma I	8 Mar. 1968	Noon	14	[5]
Brahma II	8 Mar. 1968	Sunset	86	[5]
ISRO 30.01	2 Jan. 1970	1212	30	[6]
NASA 14.425/ ISRO 17.04	19 Mar. 1970	0804	53.2	[7]
NASA 14.424/ ISRO 17.05	19 Mar. 1970	0954	27.8	[7]
NASA 14.372/ ISRO 17.06	19 Mar. 1970	1446	48.3	[7]
ISRO 20.07	29 Aug. 1968	1353	28	[8]

An extensive series of rocket soundings has been made at the Equatorial Rocket Range at Thumba, India, situated on the magnetic equator [5–8]. The details of the various flights are given in Table 1. In all these flights, excepting the last one, the electron densities were measured using the ordinary and extraordinary wave absorptions obtained from a radio propagation experiment devised by Kane [5]. In this paper the results from all the flights are summarized and an attempt is made to generate theoretical electron density profiles for daytime conditions at the equator.

2. Electron Density Profiles at the Equator

The electron density profiles obtained from the various flights from Thumba are shown in Fig. 1. The general features of these profiles are similar to those obtained at other latitudes. The electron density rapidly increases with altitude up to about 70 km. Above 70 km the electron densities vary relatively slowly with altitude showing a peak at 75—78 km and reach a sharp minimum in the range 80—85 km. Above this minimum the electron density increases very rapidly with altitude. Satyaprakash et al. [8], from the electron density profile derived from the dc-probe current, had noted that the profile does not show any layering and that no significant difference between the equatorial and mid-latitude profiles could be observed.

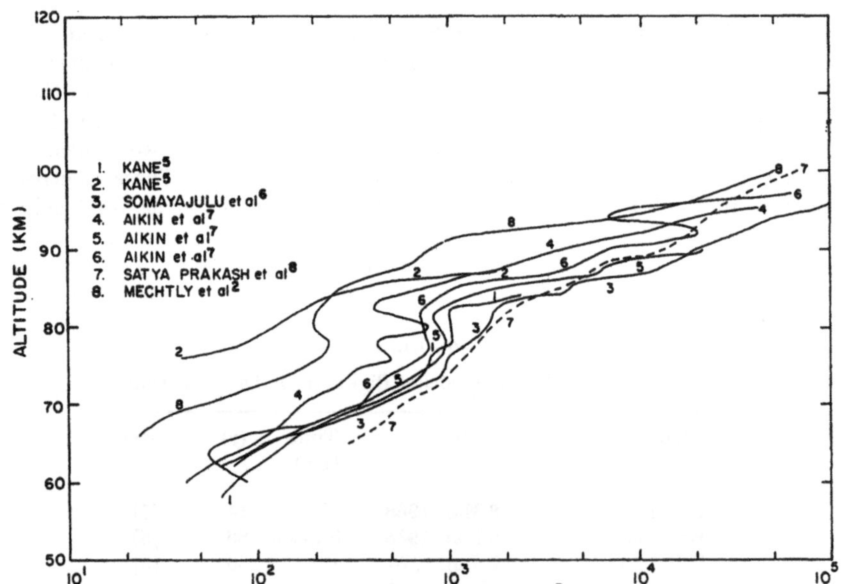

Fig. 1. Electron density profiles near the geomagnetic equator obtained from radio propagation experiments. (After [19].)

The present concept of production of electrons in the D region is based on three principal ionizing sources, namely, the galactic cosmic rays, the solar Lyman alpha radiation (1216 Å) and solar X-rays (< 10 Å) [9]. The galactic cosmic rays and solar X-rays ionize all the atmospheric constituents while solar Lyman alpha radiation produces ionization of the important trace constituent nitric oxide. It has been shown that the excited molecular oxygen in the $^1\Delta_g$ state can be ionized by the EUV radiation (< 1118 Å) [10] but recent measurements [11] show that this would be an important production mechanism only above 80 km. The solar X-rays are important at altitudes above 80 km and the galactic cosmic ray production is believed to dominate below 70 km.

The transition from Lyman alpha to X-rays as dominant ionization source occurs in the range 75—85 km depending on the solar activity because, while the solar Lyman alpha flux is relatively constant over a solar cycle (with a variation by about 50%), the solar X-ray flux is highly variable, increasing by some three orders of magnitude from solar minimum to solar maximum. Thus while the solar Lyman alpha radiation may be the principal production source in the altitude range 70—80 km, considerable contamination due to solar X-rays may occur below 80 km depending on the solar activity and the zenith angle. Any latitudinal dependence of the production due to the sources should then come only through the variation of the neutral atmospheric structure with latitude. The neutral atmosphere in the equatorial latitudes is considered to be more stable compared with the mid-latitude atmosphere, in that the seasonal variability is least in low latitudes. So one may expect that the seasonal variability in the electron density profiles above 70 km altitude, which is characteristic of the mid-latitudes [12], would be minimum in the equatorial latitudes.

In Fig. 2, all the equatorial electron density measurements are used to depict the variation of electron densities as a function of solar zenith angle in an attempt to identify the relative importance of the solar Lyman alpha and galactic cosmic

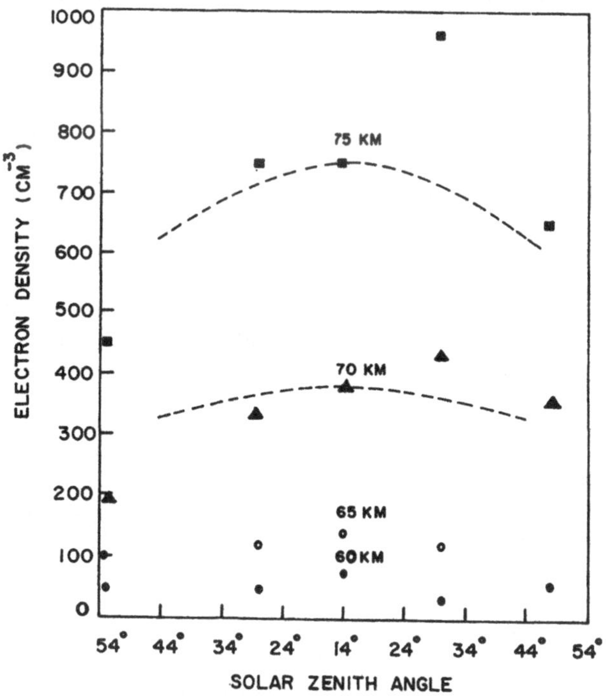

Fig. 2. Variation of electron density with solar zenith angle. (After [19].)

rays as production sources. It may be seen from the figure that at altitudes below 65 km the electron densities are practically independent of solar zenith angle thus showing that solar Lyman alpha makes little contribution to the ionization and that galactic cosmic rays are the dominant production source. The electron densities at heights of 70 and 75 km show a strong solar zenith angle dependence. The dashed curve in the same figure represents the computed electron density variation with zenith angle based entirely on Lyman alpha radiation as production source, calculated assuming a $(\cos \chi)^{1/2}$ dependence. The experimental points seem to follow fairly well the theoretical variation with zenith angle. The values for $\chi = 30°$ are somewhat high, possibly due to the contamination from high X-ray flux [6]. Thus we may infer that in the altitude range 70—75 km, the solar Lyman alpha is the principal ionization source. Since below 65 km cosmic rays are the main ionizing source, the transition from cosmic rays to Lyman alpha seems to occur in the altitude region 65—70 km at equatorial latitudes.

It was shown by Aikin et al. [7] that the profiles for $\chi = 53°$, 28° and 48° are consistent with the latest nitric oxide density profiles measured by Meira [13], the presence of hydrated ions and the recombination coefficients measured in the laboratory. Above this altitude, the profiles are shown to be consistent with nitric oxide profiles of the type given by Meira and the relevant solar X-ray flux measured by proportional counters aboard the same rockets and by satellites.

The principal difference between mid-latitudes and the equatorial regions comes in the ionization below 70 km. While the solar Lyman alpha production is fairly constant, the cosmic ray flux has both latitudinal and solar activity dependence. For example, the cosmic ray production at the equator compared with that at mid-latitudes (60° geomag. lat.), is decreased by a factor of ~ 10. The cosmic ray flux decreases with increasing solar activity, decreasing by a factor of ~ 2 from solar minimum to solar maximum so that the electron density at the equator should be lower by a factor of about 3—4, compared with values at mid-latitudes. Thus, in principle, a comparison of the equatorial electron density profiles with those at mid-latitudes should also be useful in identifying the contribution due to, and the importance of, the cosmic rays as a source of ionization.

Thus it would be of interest to compare the equatorial electron densities with those at mid-latitudes for similar solar zenith angles at an altitude of 75 km since the Lyman alpha flux would be the same for the same zenith angle. However, most of the available measurements for mid-latitudes are for different solar activity conditions compared with those corresponding to the equatorial measurements, so that a direct comparison for altitudes below 70 km is difficult because of the solar cycle dependence of cosmic rays. In Table 2 are given the values of electron densities at 65, 70 and 75 km altitudes from equatorial and mid-latitude measurements.

It is seen that for $\chi = 53°$ the electron densities at 75 km at the equator and mid-latitudes are almost the same. However, for $\chi = 14°$ the mid-latitude value is considerably higher. At 70 km the electron densities for $\chi = 14°$ are comparable while for $\chi = 53°$ the value at the equator is considerably higher than the mid-latitude value. At 65 km, most of the measurements are comparatively less reliable as regards the absolute magnitudes. However, even after making allowance for the experimental uncertainties, it would seem that at the equator the electron densities are too great compared with the mid-latitude values to be explained on the basis of latitudinal dependence of cosmic ray production; Kane [5] also noted a similar discrepancy.

Table 2
Comparison of electron density values at equator and mid-latitudes

Alt. (km)	8 Mar. 1968 $F_{10.7} = 124.1$	24 Jul. 1968 $F_{10.7} = 152.7$	19 Mar. 1970 $F_{10.7} = 130.2$	9 Apr. 1963 $F_{10.7} = 82$
	Equator $\chi = 14°$	Mid-lat $\chi = 18°$	Equator $\chi = 53°$	Mid-lat. $\chi = 55°$
75	710	1200	470	500
70	375	450	185	110
65	145	40	100	100

Aikin et al. [7] have noted an unexplained asymmetry between the forenoon and the afternoon profiles above 70 km. Above 75 km, contamination due to solar X-rays begins to be appreciable. In this region, above 80 km, the atomic oxygen density rises rapidly; this may have important consequences on the ion composition and the effective loss rate [14].

3. Theoretical D Region Electron Density Profiles for Equatorial Latitudes

To develop theoretical electron density profiles for the equatorial region the following procedure is adopted. We assume that: (i) quasi-equilibrium conditions are valid so that dN_e/dt may be neglected; (ii) the electron density profiles are symmetrical about noon in relation to the solar zenith angle. Thus the electron density values at any altitude are then derived from the continuity equation and are given

$$N_e = (q_{total}/\alpha_{eff})^{1/2} \qquad \ldots \quad (1)$$

by where q_{total} is the total production rate due to all ionization sources and α_{eff} is a lumped (or effective) recombination coefficient; α_{eff} is taken to be independent of the solar zenith angle. The height distribution of α_{eff} adopted here is that derived by Aikin et al. [7]. This is consistent with positive ion composition measurements of Goldberg and Aikin [15]. Now it is, therefore, necessary to specify the total production rate for one solar zenith angle say, overhead sun ($\chi = 0$). This production rate can then be scaled for any other zenith angle using the $\cos\chi$ dependence. The electron density N_e can then be computed using Eq. (1).

4. Specification of the Production Rates

For computing the production rate of solar Lyman alpha radiation, an average flux of 3×10^{11} photons cm^{-2} s^{-1} is assumed. The density distribution of nitric oxide is taken from the experimentally derived profile by Meira [13] extrapolated between 70 km and 65 km. The production rate due to galactic cosmic rays is calculated essentially following Velinov [16]. Since the X-ray flux is highly variable from solar minimum to solar maximum conditions we have adopted an integrated 0—8 Å flux of 10^{-4} erg cm^{-2} s^{-1} as a unit representative of solar minimum conditions and a unit integrated flux of 10^{-2} erg cm^{-2} s^{-1} corresponding to solar maximum conditions with an effective temperature of 2×10^6 °K.

For computing the cosmic ray and solar X-ray production rates a neutral atmosphere model due to Jacchia [17] and UV and X-ray absorption ionization and absorption cross sections reviewed by Swider [18] are adopted.

The various production rate profiles thus computed are shown in Fig. 3. The electron density values and their computed diurnal variation are described in Fig. 4. It should be pointed out, however, that these represent a preliminary attempt and more refinements to account for departures from the simplified picture are currently in progress and will be reported later.

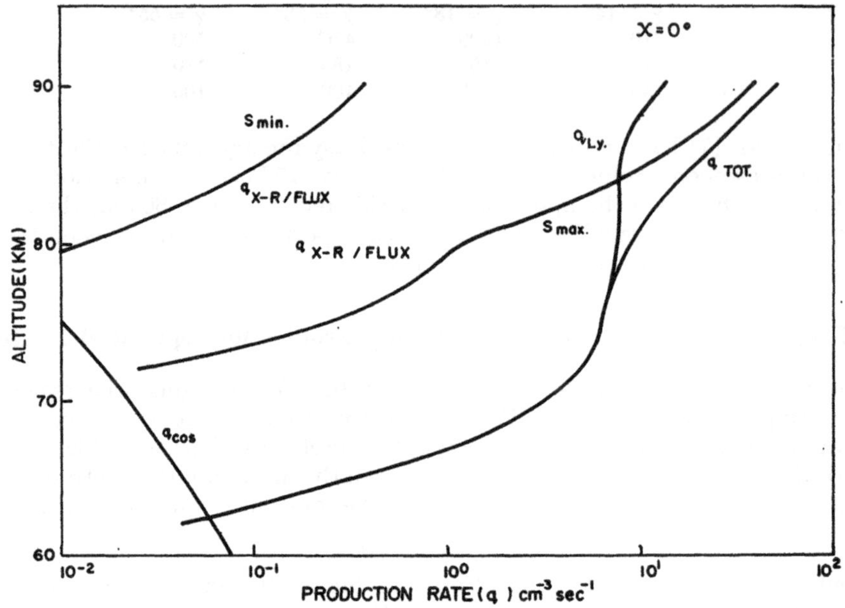

Fig. 3. Computed production rates profiles for $\chi = 0°$,

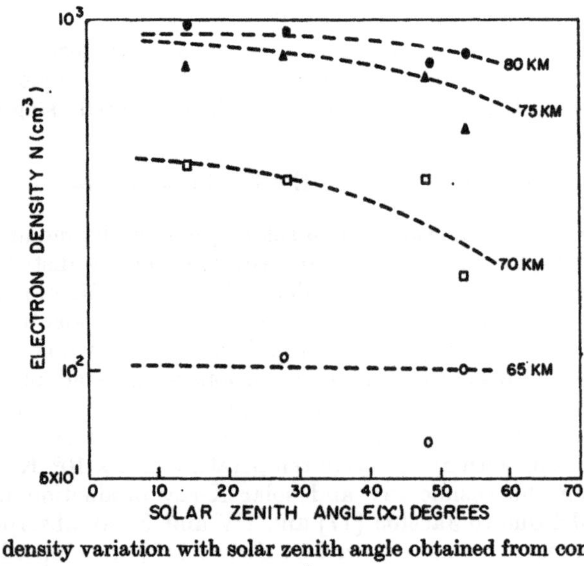

Fig. 4. Electron density variation with solar zenith angle obtained from computed production rates.

References

[1] E. A. Mechtly, S. A. Bowhill, L. G. Smith and H. W. Knoebel, J. Geophys. Res. **72**, 5239 (1967).

[2] E. A. Mechtly, M. Mukunda Rao, D. O. Skaperdas and L. G. Smith, Radio Sci. **4**, 517 (1969).

[3] J. S. Shirke and G. W. Henry, Annls Géophys. **23**, 517 (1967).

[4] S. Gnanlingam, paper presented at 3rd Int. Symp. on Equatorial Aeronomy, Ahmedabad 1969.

[5] J. A. Kane, NASA-GSFC Rep. No. X-615-69-499 (1969).

[6] Y. V. Somayajulu, M. B. Avadhanulu, K. S. Zalpuri and S. C. Garg, Space Research XI, 1131 (1971).

[7] A. C. Aikin, R. A. Goldberg, Y. V. Somayajulu and M. B. Avadhanulu, J. Atmos. Terr. Phys. **34**, 1483 (1972).

[8] Satyaprakash, B. M. Subbaraya and S. P. Gupta, J. Atmos. Terr. Phys. **33**, 129 (1971).

[9] M. Nicolet and A. C. Aikin, J. Geophys. Res. **65**, 1469 (1960).

[10] D. M. Hunten and M. B. McElroy, J. Geophys. Res. **73**, 2421 (1968).

[11] R. E. Huffman, D. E. Paulsen, J. C. Larrabee and R. B. Cairns, J. Geophys. Res. **76**, 1028 (1971).

[12] G. C. Reid, J. Geophys. Res. **75**, 2551 (1970).

[13] L. G. Meira, J. Geophys. Res. **76**, 202 (1971).

[14] A. P. Mitra and J. N. Rowe, paper presented at COSPAR Symp. on D and E Region Ion Chemistry, Univ. Illinois 1971.

[15] R. A. Goldberg and A. C. Aikin, NASA-GSFC Rep. No. X-625-71-73 (1971).

[16] P. Velinov, J. Atmos. Terr. Phys. **30**, 1891 (1968).

[17] L. G. Jacchia, Smithson. Astrophys. Obs. Spec. Rep. No. 332 (1971).

[18] W. Swider, Jr., Rev. Geophys. **7**, 573 (1969).

[19] Y. V. Somayajulu and M. B. Avadhanulu, Ind. J. Radio Space Phys. **1**, 81 (1972).

Space Research XIII — Akademie-Verlag, Berlin 1973

ROCKET OBSERVATIONS IN THE EQUATORIAL IONOSPHERE

E. Neske and R. Kist

Arbeitsgruppe für physikalische Weltraumforschung, Freiburg, FRG

In French Guyana a phase measuring type of swept RF probe with dc biased sensor has been flown twice at the same local time with in a few days. Electron density profiles up to 210 km have been determined and are found to be largely different in shape between 130 and 180 km. While a profile shows an almost monotonic increase of electron density with height as mostly assumed for low latitudes, the other profile shows a deep valley near 165 km. Data are compared with ionograms and found to be compatible. The dc bias current can also be used to deduce a profile which agrees with that determined by the probe. The data permit an estimation of the relaxation time of the ion sheath when changing the bias potential.

In June 1971 two Veronique rockets were launched from Kourou, French Guyana. Together with six scientific and two technical experiments an impedance probe (IP) was flown in each payload. This swept frequency impedance probe was conceived as preparation for an impedance probe experiment in the German satellite project Aeros. Previously it had always been tested in a series of plasma tank measurements. Two points of view had been of particular interest: the influence of the dc bias of the sensor on the phase curve and the behaviour at frequencies above the upper hybrid frequency (up to 10 MHz). With these aims only good relative accuracy of the measurement was needed. Using ionograms taken simultaneously, however, an absolute accuracy of the electron density determination of about 20% could be obtained.

The principle of measurement is that the frequency dependent behaviour of the unknown impedance of a plasma condenser is investigated by comparing the RF voltage from a generator with that across a coupling impedance. In the Veronique project the plasma condenser was formed by a 1 m conical aluminium sensor supported on an insulating nosecone and the rocket skin, with a free space capacity of 11 ± 0.5 pF. In a plasma this pure capacitance changes to a complex impedance, the imaginary part of which can become zero, meaning that resonances occur.

The technical concept of the experiment is indicated in Fig. 1. By means of a mixing technique two frequency bands are produced. One of them is fed to the sensor. By mixing again two signals of constant frequency are obtained, in which the information about the frequency dependence of the plasma condenser is contained (superhet technique). In this version of the IP only the phase response is determined as a function of frequency. The dc biasing of the sensor (SPC) was simply made with a constant voltage of ± 6 V applied across a resistor to the sensor. The actual sensor potential was measured thus allowing the current collected from the sensor in the plasma and flowing to the rocket to be determined. In order to see the influence of the biasing the experiment was operated consecutively in two modes, namely without and with biasing applied. Mode switching was triggered by the scan pulse of the experiment Retarding Potential Analyser (RPA).

The measured data are shown in Figs. 2 and 3. In both cases only the ascents are considered since effects of rocket wake and outgassing during descent make it very difficult to obtain quantitative results on the density of the undisturbed plasma. Considering first the second launch from 12 June 1971 1544:00 LT,

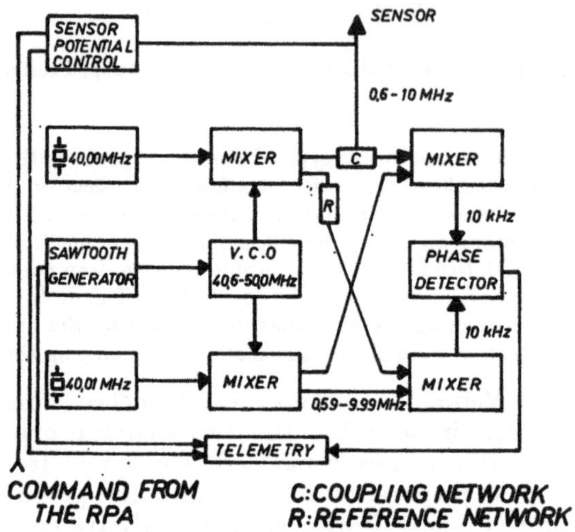

Fig. 1. Block diagram of the impedance probe Veronique.

an overall monotonic increase of the electron density in the height range 95—210 km is seen, which is typical for a day time profile in the equatorial ionosphere. The sensor current shows, after a rapid increase at 90 km, a nearly constant shape. The small decrease of current with increasing density indicates the influence of other parameters such as vehicle potential and electron temperature. This problem will be treated after the RPA results become available.

The first launch of 8 June 1971 at nearly the same local time shows a completely different structure of ionosphere. First of all there are two well pronounced sporadic E layers at 93 km and 105 km. At about 120 km the profile shows the beginning of a deep and wide density "valley" with its minimum at 163 km. This valley is also reflected in the collected sensor current profile. From this the thickness of the lower Es layer was determined with high accuracy to be 1.2 ± 0.1 km. At the second Es the sensor potential control was switched off. However, the thickness of this layer is found by the variation of resonance frequency to be 2.5 ± 0.5 km. Comparison of the two density profiles shows (after calibration with the ionogram data) nearly identical density values for the normal E layer at about 120 km and for the lower side of the F layer, here at 200 km. The electron density profile in the deep valley, which corresponds to a change in density by about one order of magnitude can not be deduced from the ionogram data.

It is known that such a density valley between the E and F layers can occur from time to time, but its density variation and its generating mechanism are not known up to now.

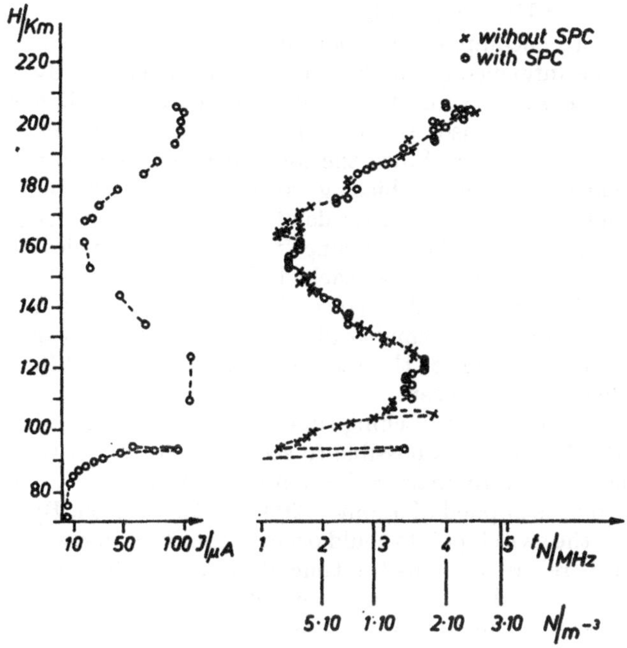

Fig. 2. Electron density and sensor current (ascent), Veronique-FU 194. Kourou 8 June 1971 1545 : 48 LT.

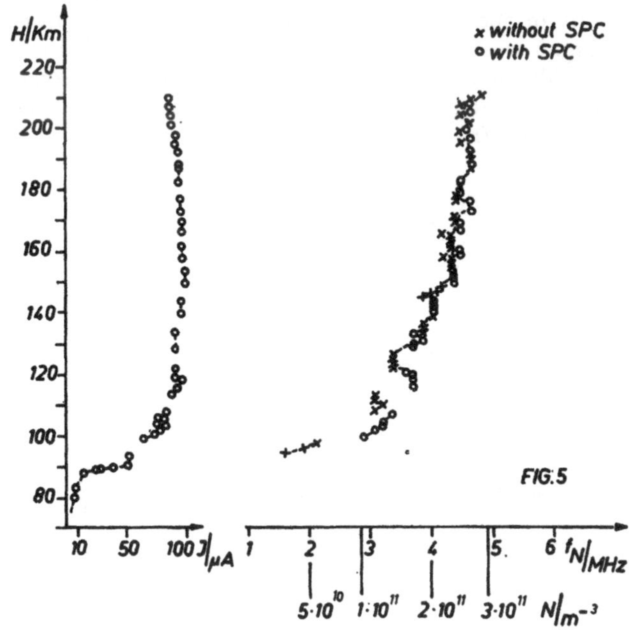

Fig. 3. Electron density and sensor current (ascent), Veronique-FU 194. Kourou 12 June 1971, 1544 : 00 LT.

This version of the IP was investigated over a wide frequency range from 0.6 to 10 MHz in order to encompass the upper hybrid frequency. A second point was to investigate the influence of the SPC on the phase curve. By shifting the dc potential of the sensor towards the plasma potential, the diameter of the ion sheath surrounding the sensor is more and more decreased. This means an increasing ion sheath capacity. Since the ion sheath capacity influences strongly the series resonance frequency a biasing causes a shift of the series resonance to lower frequencies. In Fig. 4 some records of the phase versus frequency are presented confirming this behaviour: coming from high frequencies first the parallel and then the series resonances are reached. From the slope of the phase curve a damping related to the partial collapse of the ion sheath can be observed. The damping leads additionally to a smaller phase shift, here from nearly 180° to about 110°. For the determination of the resonance frequencies a different evaluation process must therefore be chosen.

As already mentioned the switching times for the SPC were related to the scan time of the RPA. Thus it was possible to see the influence of the sensor bias on rocket potential. In a preliminary reduction of RPA data a change of the rocket potential when bias is applied of about −200 mV can be seen. By this synchronization, however, the switch points could occur just at the time when the resonance curve of the IP is recorded. Thus the time constant for the relaxation of the ion sheath could be estimated. From ten such events (corresponding to the given time resolution) a relaxation time for the increase and decrease of the ion sheath shorter than 10 ms was found. This is compatible with the value of about 1 ms given in the literature.

A further effect could be observed, namely the influence of gyrofrequency harmonics on the phase curve. This influence, causing a small dip and indicating a small phase shift in the opposite direction is pronounced near the resonance frequencies. In Fig. 5 we see the dip at a fixed frequency being crossed by the parallel resonance. Therefore at heights where the parallel resonance and a harmonic of the gyrofrequency are close together care must be taken when determining the upper hybrid frequency.

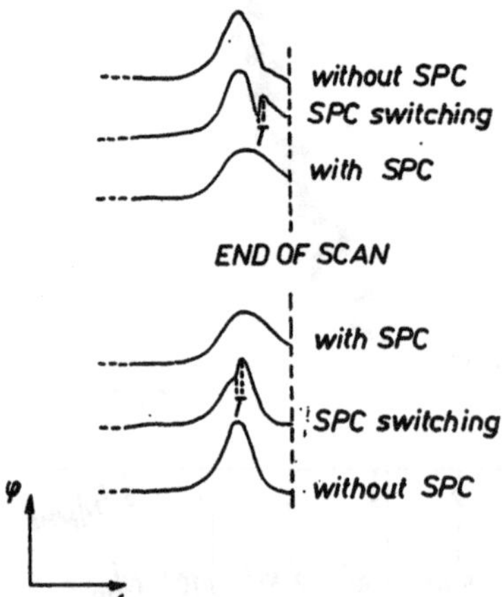

Fig. 4. Examples of time constant determination and curve shape.

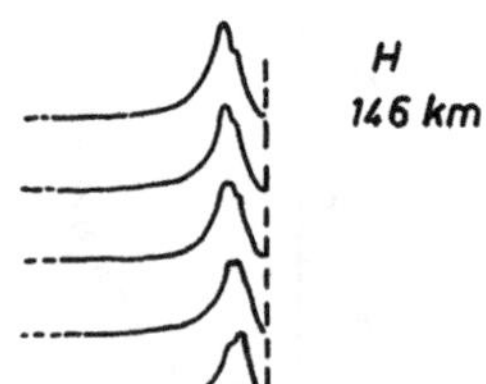

Fig. 5. Effect of gyrofrequency harmonic.

Space Research XIII — Akademie-Verlag, Berlin 1973

COMPARISON OF CLEANED AND UNCLEANED PROBES ON BOARD A ROCKET

K. Hirao and K. Oyama

Institute of Space and Aeronautical Science, University of Tokyo, Komaba, Tokyo, Japan

In situ measurements of electron temperature with a Langmuir probe on a rocket are seriously affected by contamination of the probe surface. To investigate this effect, two Langmuir probes, which had been sealed in evacuated glass tubes after baking for a long time, and one exposed probe were installed on board the S-210-7 rocket. All these probes were made of stainless steel cylindrical rods. The rocket went into the ionospheric E region, where the effect of contamination was serious. An electron temperature probe was also installed in the same rocket; there was a noticeable discrepancy between the electron temperatures measured by cleaned and uncleaned probes. It is experimentally verified that the uncleaned Langmuir probe is generally unreliable for measurements in the lower ionosphere. However, it is expected that a high sweep rate Langmuir probe may give an approximate value in some cases.

1. Introduction

There is a well-known argument about the reliability of the Langmuir probe for the measurement of electron temperatures in the ionosphere, especially in the lower regions, arising from the comparison of values obtained with rocket-borne Langmuir probes and incoherent backscatter. The electron temperature obtained by the Langmuir probe is almost always higher than that obtained by the backscatter radar.

In laboratory experiments the reliability of the probe technique for the measurement of electron temperatures has been discussed by many investigators. Wehner and Medicus [1] summarized this problem and concluded that the Langmuir probe measurement may give some erroneous results due to the change of the work function of a probe, the insulating layers on a probe surface or the random distribution of work function on the surface of a probe.

Brace, Carignan and Findlay [2] reported their laborious investigation and concluded that there was no significant source of error for electron temperature measurements in the ionosphere with the Langmuir probe except in the E region. They also discussed the possibility of a non-Maxwellian nature of the E region. Recently, Hoegy [3] has also considered temperature differences caused by the non-equilibrium plasma in the E layer.

We have also carried out some laboratory and space experiments on the reliability of the Langmuir probe [4] and concluded that the current—voltage characteristic of a Langmuir probe with a bias sweep rate of several sweeps per second or less exhibits a remarkable hysteresis effect and the electron temperature deduced is generally too high. Consequently, the electron temperatures determined from increasing bias and decreasing bias conditions differ significantly.

After the probe surface has been cleaned by ion bombardment, the hysteresis phenomenon diminishes and the electron temperature decreases to a constant value.

In order to investigate the effect in space measurements, we made a comparison experiment between cleaned and uncleaned probes with the S-210-7 rocket which was launched at 1700 JST on 17 February 1972 from the Kagoshima Space Center, attaining a peak altitude of 118 km in the ionospheric E region.

2. Instrument and Experimental Results

2.1. Probe

In our laboratory experiment, the probe electrode of the Langmuir probe system is always ion-bombarded by supplying the probe with about −1000 volts negative bias in test plasma. The cleaning of the probe, which is identified by the diminishing hysteresis in the current—voltage characteristic of the probe, is almost completed by bombardment of one to two hours duration. The same effect can be achieved by baking the probe at a temperature of several hundred degrees for a longer time, such as twenty to thirty hours. The material of the probe is not a serious matter as reported by Brace et al. [2].

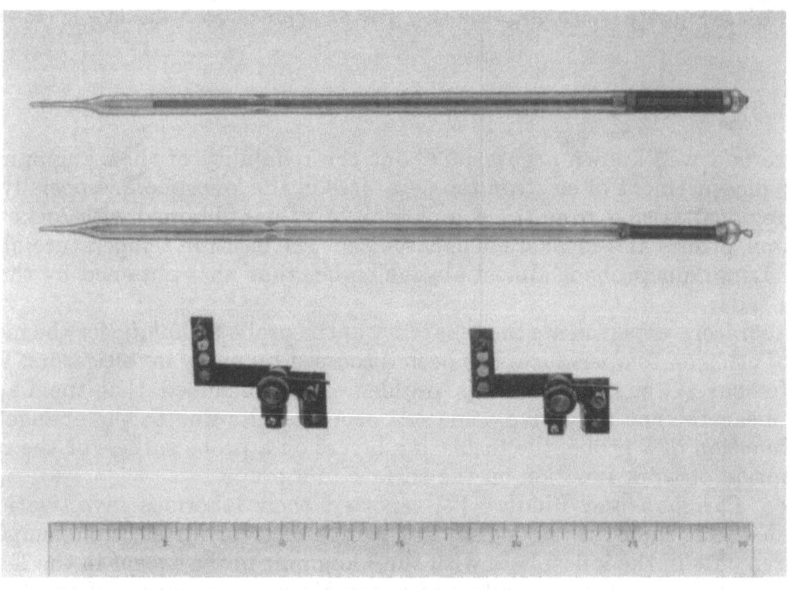

Fig. 1. Glass sealed probes and cutter.

The probes used in our rocket experiment, 1 mm and 3 mm in diameter and 18 cm in length, are sealed by evacuated glass tubes after baking in a vacuum. They are shown in Fig. 1; the glass tube is cut at the end of the supporting metallic tube (Covar) by a cutter which is also shown. The cutter is actuated by a sequence timer after the nose cone is ejected. The uncleaned probe is of the same size, and of the same material, stainless steel, as one of the cleaned probes, but it is not covered by a glass tube.

2.2. Experimental Results

The S-210-7 rocket was launched at 1700 JST on 17 February 1972 from the Kagoshima Space Center, attaining a maximum altitude of 118 km. Though the sequence of probe erection was satisfactory, the attitude of the rocket was disturbed considerably by abnormal ejection of the nose cone. The data obtained were, therefore, much disturbed due to the probe being in the rocket wake. However, some good data were obtained.

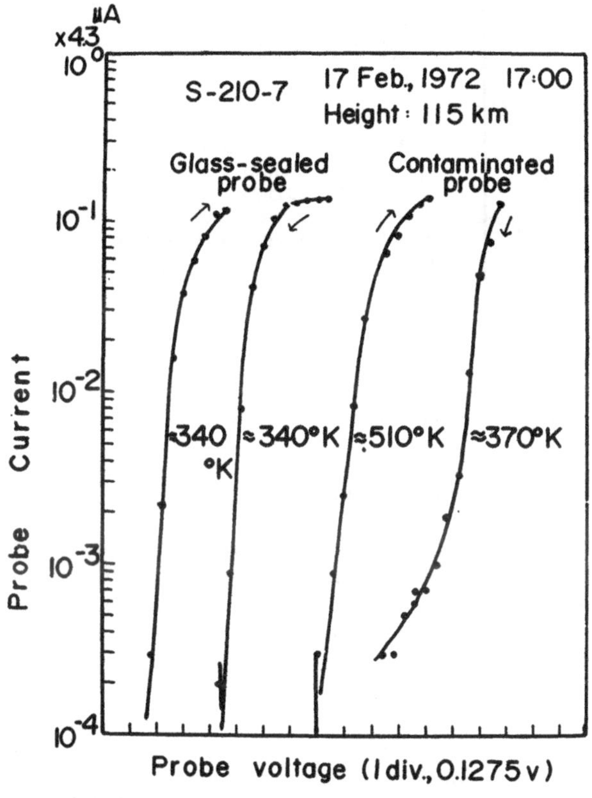

Fig. 2. Comparison of characteristics of the probes.

Fig. 2 shows current—voltage characteristics of both cleaned and uncleaned probes corresponding to increasing and decreasing sweeps of the biasing probe potential. In spite of slight modulation due to the effect of the wake, it is clearly seen that the hysteresis appears only on the uncleaned probe, i.e. the gradient of the semi-log plot of the characteristic changes differs for increasing and decreasing bias: in the case of the cleaned probe no difference is recognized between the curves of increasing and decreasing bias.

In Fig. 3, electron temperatures obtained with both cleaned and uncleaned probes are plotted with a model profile of neutral temperature of Jacchia's 1971 model atmosphere. In this figure, one can see that the cleaned probe gives very similar values in both increasing and decreasing bias conditions, but not the uncleaned probe. Horizontal bars in this figure show the difference between these two bias conditions. Also, one can see that the uncleaned probe gives generally higher electron temperatures than the cleaned probe. Superposing these plots, with one data point of the rectification electron temperature probe shown as a cross, good agreement with the data of the cleaned probe is achieved.

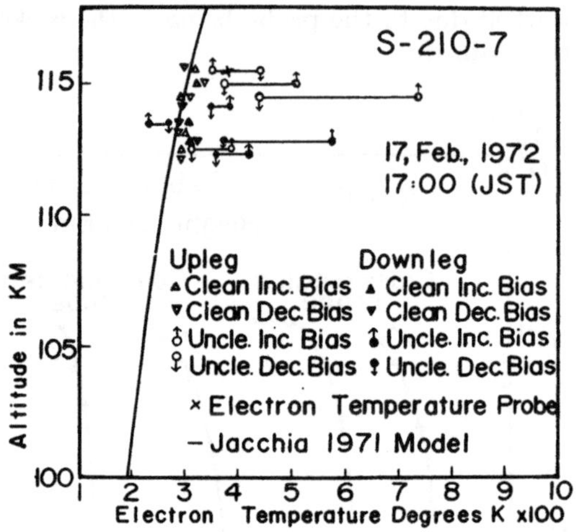

Fig. 3. Comparison of electron temperature profiles.

3. Discussion and Conclusion

As we have already mentioned [4], the surface of the probe is generally contaminated, and this contamination cannot be removed by ordinary chemical or mechanical cleaning processes. The contaminated layer acts like a parallel resistance/capacitance circuit which is connected to the probe. Therefore, it shows various effects on the characteristics corresponding to various sweep rates, sweep direction and density of the ambient plasma (the hysteresis effect being greater at lower densities). In the case of high sweep rates, such as more than ten per second, however, the hysteresis phenomenon again diminishes and the electron temperature obtained becomes very close to the correct temperature when the ambient plasma density is adequate. Experimental and theoretical studies on these effects are almost complete and will be published in the near future.

In conclusion, it can be stated that a Langmuir probe on board a rocket sometimes gives incorrect values of ionospheric plasma parameters, if the probe has an uncleaned surface.

References

[1] G. WEHNER and G. MEDICUS, J. Appl. Phys. **23**, 1035 (1952).
[2] L. H. BRACE, G. R. CARIGNAN and J. A. FINDLAY, Space Research XI, 1079 (1971).
[3] W. R. HUEGY, J. Geophys. Res. **76**, 8333 (1971).
[4] K. HIRAO and K. OYAMA, J. Geomag. Geoelect. **23**, 161 (1971).

Space Research XIII — Akademie-Verlag, Berlin 1973

ATMOSPHERIC ION COMPOSITION MEASUREMENTS UP TO A HEIGHT OF 170 km

A. D. DANILOV and V. K. SEMENOV

USSR

The results of measurement of the upper atmospheric ion composition by an RF mass spectrometer, installed at geophysical rocket MR-12, launched at Volgograd on 2 March 1971 at 1333 LT, on 1 October 1971 at 1245 LT, on 7 October 1971 at 1230 LT and on 26 December 1971 at 1054 LT are presented. The ion peaks with masses 16^+, 28^+, 30^+ and 32^+ were detected at heights of 100—170 km. The narrow layer of ions with mass 28^+ was found in the E region, during the flight of 2 March 1971, and the metallic ions were detected in the E region during the flight of 6 October 1971.

In a number of geophysical rocket MP-12 flights in 1971 while studying photochemical processes directly in the ionosphere (cf. [1, 2]) ion composition measurements by radio-frequency mass spectrometer MX 6407П at heights of 100—170 km were carried out. The sweep time of the device was 0.7 sec, and the mass range measured was 12—48 amu.

Data on the flights are given in Table 1; flights Nos. 1—3 were carried out at Volgograd ($\phi = 49°$ N $\lambda = 44°$ E), flight No. 4 from the scientific research ship *Professor Zubov*. Flights Nos. 2 and 3 were part of complex experiment "Sun—Atmosphere" [3]; a more complete analysis of the results combined with other data obtained in this complex experiment will be published later.

In flights Nos. 1—4 ions with mass numbers given in Table 2 were recorded. In the present paper data on three main positive ions: $16^+(O^+)$, $30^+(NO^+)$ and

Table 1

No	Date	LT	Place	$Z_\odot$
1.	2 March 1971	13.33	Volgograd	60°
2.	1 October 1971	12.45	Volgograd	52°
3.	7 October 1971	12.30	Volgograd	54°
4.	26 December 1971	10.54	0°, 24°3 W	30°

Table 2

Flight No.	Ion mass number
1	16, 18, 19, 28, 30, 32, 44
2	14, 15, 16, 18, 19, 24, 26, 28, 30, 32, 34
3	11, 13, 15, 16, 18, 19, 22, 24, 26, 28, 30, 32
4	16, 18, 19, 28, 30, 32

$32^+(O_2^+)$ are presented. Since the authors have no data on simultaneous measurements of electron concentration profiles the results are presented as diagrams (Figs. 1 and 2) giving relative concentrations of the main ions.

A comparison of two flights (Nos. 1 and 3) carried out from one place at nearly the same time and at nearly the same solar zenith angle is made in Fig. 1. However, flight No. 3 was made during disturbed conditions (magnetic storm SC); therefore a comparison of the two parts of Fig. 1 gives an opportunity to judge the influence of disturbances on ion composition. Thus at 120 km the main ion ratio NO^+/O_2^+ was ≈ 1 in flight No. 1 and 1.8 in flight No. 3. This fact supports the conclusion

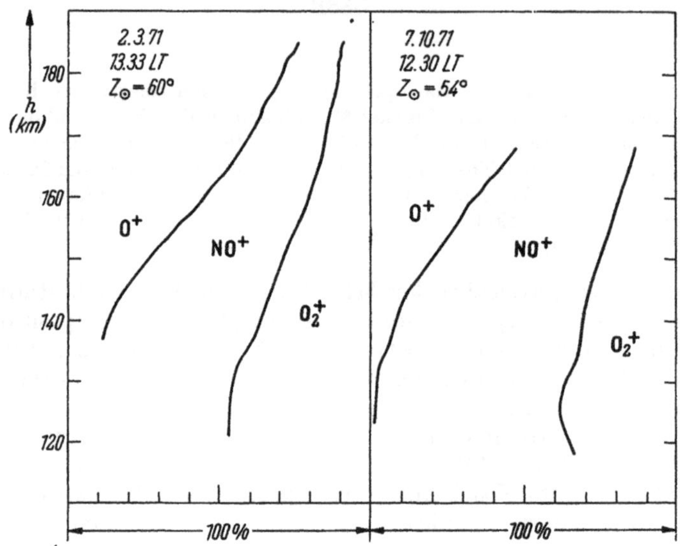

Fig. 1. Ion composition measured in flights No. 1 and No. 3.

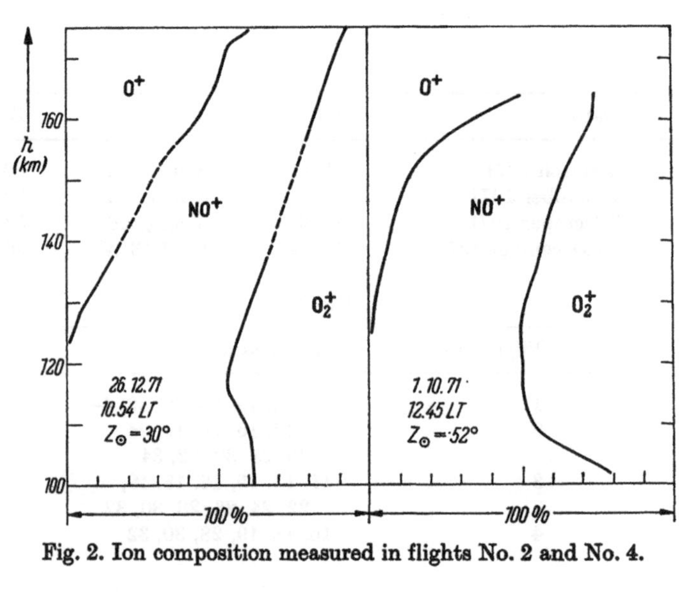

Fig. 2. Ion composition measured in flights No. 2 and No. 4.

[4] drawn on the basis of the analysis of total ion data that $[NO^+]/[O_2^+]$ increases in the E region during geomagnetic disturbances. As seen from Fig. 1 an ion composition variation leading to an increase of relative fraction of NO^+ occurs during geomagnetic disturbances not only in the E region, but at considerably higher altitudes. Thus at heights 150 and 168 km the relative concentration of NO^+ is 0.45 and 0.27 for flight No. 1 and 0.55 and 0.40 for flight No. 3 respectively.

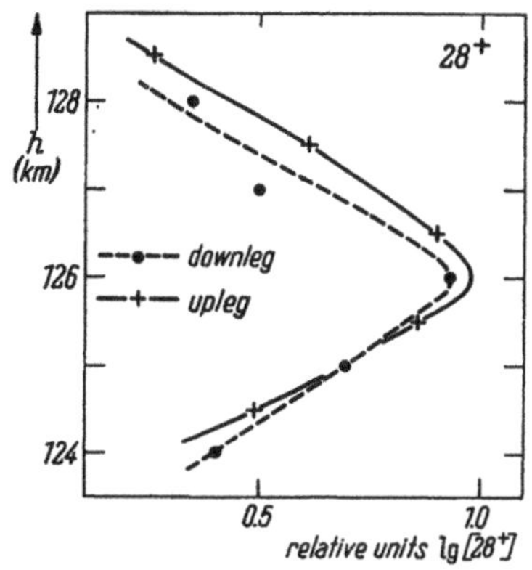

Fig. 3. Ion 28^+ profile recorded in flight No. 1.

In flight No. 1 a narrow layer of ions of mass 28 amu was recorded (Fig. 3). The concentration of 28^+ in the layer maximum is 4×10^{-2} of total ion concentration. Although the layer maximum during descent was situated slightly higher (≈ 131 km) than on ascent (≈ 126 km) (the downleg profile has been displaced by 5 km in Fig. 3), the coincidence of the shape of 28^+ in both measurements is evidence that one and the same layer was recorded. However, the accuracy of altitude determination does not permit us to judge of the fact whether the height difference mentioned was due to errors in trajectory determination or to real height difference of the layer at the points of crossing by the rocket on ascent and descent (the difference in the horizontal plane is about 70 km). It should be noted that the narrow layer of ions 28^+ with maximum concentration of 10^{-2} of the electron concentration was recorded by Istomin [5] at $h = 120$ km. However due to relatively long (2.6 sec) sweep time of the instrument [5] it was impossible to obtain the layer structure. The analogous layer of 28^+ ions was also detected at $h = 120$ km in the recent experiment [6]. The problem of identification of recorded mass 28^+ is left open. In paper [5] this mass was identified with N_2^+. However it seems more probable that the layer of ions 28^+ is caused by silicon ions which may be connected with layers of meteor ionization often recorded at heights 100—120 km.

References

[1] A. D. Danilov and V. K. Semenov, Geomagn. i Aeronom. **11**, 248 (1971).
[2] A. D. Danilov and V. K. Semenov, Space Research X, 736 (1970).
[3] L. A. Aleksandrov et al., paper presented at COSPAR Meeting, Madrid 1972.
[4] A. D. Danilov, Space Research XII, 1299 (1972).
[5] V. G. Istomin, Dokl. Akad. Nauk SSSR **137**, 1102 (1961).
[6] A. A. Pokhunkov, private communication.

Space Research XIII — Akademie-Verlag, Berlin 1973

THE DETERMINATION OF IONOSPHERIC CHARGED PARTICLE TEMPERATURES FROM *IN SITU* MEASUREMENTS

J. R. Sanmartin

Instituto Nacional de Tecnica Aeroespacial "Esteban Terradas", Madrid, Spain

The recently noticed disagreement between ionospheric charged-particle temperature values obtained from ground-based (incoherent backscatter) and *in situ* (Langmuir probe type) measurements is considered; it is suggested that a main cause of disagreement lies in the poor theoretical basis of present *in situ* measurements. It is pointed out that the usually neglected geomagnetic field influence may result in too high an electron temperature. It is also shown that the theory used at present to interpret data from ion retarding potential analyzers has serious pitfalls, and that these devices greatly disturb the surrounding plasma when measuring ion temperature. Finally, it is shown how the ion temperature can be accurately obtained from the characteristic of a cylindrical Langmuir probe in a rarefied plasma flow.

1. Introduction

The two prime methods of determination of ionospheric temperatures are *in situ* measurements using Langmuir-type probes on board satellites or rockets [1], and ground-based, radar measurements of incoherent backscatter [2]. Radar results are limited both geographically and in altitude resolution and range; moreover, when several ion species, and differing electron (T_e) and ion (T_i) temperatures are considered, interpretation of experimental data becomes difficult.

Significant comparisons between radar and probe T_e results have been achieved recently. Hanson et al. [3] and McClure and Troy [4] found probe-to-radar T_e ratios of 1.7 and 1.5 respectively. Similar results have been reported by Brace and McClure [5] and Carlson and Sayers [6]. A review of such comparisons has been given by Booker and Smith [7]. On the other hand, much smaller, although often systematic, disagreements have also been found [8—10]. In a check of self-consistency, Donley et al. [11] found that differences between three types of probes did not exceed 20%.

T_i data, and *a fortiori* radar and probe T_i comparisons, are much less abundant. Donley et al. [11] found a systematic, large (up to 100%) discrepancy between T_i values from planar and from spherical ion trap probes. Taylor and Wrenn [8], in a very limited comparison, found moderate, unsystematic differences between radar and spherical trap results. McClure and Troy [4] found a planar trap to radar T_i ratio of 1.4.

Carlson and Sayers [6] suggested that T_e differences could be due to a drift in the work function of the probes. Brace et al. [12] discussed and rejected this and several other sources of error in T_e probe results. Hoegy [13] observed that if the electron distribution function is not Maxwellian, probe and radar measurements should lead to different T_e values.

These discrepancies must be understood before real temperature variations of the ionospheric plasma can be studied.

2. Sources of Error in Present Temperature Measurement Analysis

2.1. Electron Temperature

T_e measurements are based on standard Langmuir probe theory. The I_p (plasma-to-probe current) versus V_p (probe-to-plasma potential) diagram is recorded for V_p negative. If the ion component of the current can be neglected, I_p is, under quite general conditions, independent of probe shape; for a Maxwellian electron distribution function [14]

$$I_p \propto \exp(e V_p / k T_e) \tag{1}$$

where $-e$ is the electron charge and k is Boltzmann's constant. A variety of probe shapes, both of the plain Langmuir and multiple grid (electron trap) types, have been used in the past.

We wish to show now that the geomagnetic field can lead to a systematic error in this method of finding T_e. Magnetic field effects have been largely ignored in probe theory because of the difficulty of their analysis. It is generally agreed nonetheless that an ambient magnetic field decreases the current in the $V_p > 0$ range, unless the electron Larmor radius l_e is large compared with the probe size; most probes do not satisfy this condition in the ionosphere, where $l_e \sim 1$ cm. On the other hand, it is frequently assumed that a magnetic field does not affect Eq. (1) ($V_p < 0$), even if l_e is not larger than the probe size. Recently, it has been shown that this is not generally true [15]. It was found there that the potential field did not increase monotonically from V_p on the probe to zero far from it; instead, a potential hill appeared along the magnetic field lines intersecting the probe. This is equivalent to a decrease of V_p by an amount equal to the potential overshoot (the positive value of potential at the top of the hill), the important point being that the overshoot *depends* on V_p. As in the case of the V_p-dependent drift of the probe work function [6], that dependence leads to an error in the value of T_e obtained in the usual manner. Since the potential hill is greater as V_p increases, it is clear that the value of T_e obtained will be larger than its real value.

2.2. Ion Temperature

To measure T_i in the ionosphere, multiple grid probes (ion traps) are used. The entrance grid G1 and the collector C are either biased negative relative to the vehicle, or grounded to it; the vehicle itself has a potential $V_s < 0$ relative to the ionosphere. The suppressor grid G3 is biased highly negative to turn back all electrons able to get past G1, and to inhibit photo and secondary emission from the collector. G2 is the retarding grid; its potential V_R is swept from slightly negative to highly positive to analyze the energy spectrum of incoming ions. The trap is oriented facing the vehicle velocity $-\boldsymbol{U}$ (planar trap).

To interpret trap measurements, a formula derived by Whipple [16] is always used. Whipple assumed that since the Debye length λ_D is smaller than the trap frontal dimension, and U is larger than the ion thermal velocity, a one-dimensional approach could be used to calculate the current I_c to the collector, as a function of V_R. If v_z is the velocity toward the trap of an ion, in the vehicle frame of reference, the undisturbed ion distribution function is proportional to $\exp[-m_i(v_z - U)^2/2kT_i]$, where m_i is the ion mass. Whipple then integrated over v_z between $(2eV_R/m_i)^{1/2}$ and $+\infty$ to get

$$I_c \propto 1 + \operatorname{erf} x + (2kT_i/\pi m_i U^2)^{1/2} \exp(-x^2), \tag{2}$$

$$x = [U - (2eV_R/m_i)^{1/2}](m_i/2kT_i)^{1/2}, \qquad V_R > 0.$$

As V_R increases, the right-hand side of Eq. (2) goes from about two to zero; as $(2kT_i/m_iU^2) \to 0$, I_c becomes a step function.

Although possible errors of Eq. (2) have been discussed in the literature [17—19], the most basic objection to the above formulation seems to have been overlooked: Consider V_R so large that practically no ions reach the collector. Obviously the ions repelled by G2 are ejected out of G1 with velocities close to $-U$. Inside this ion jet, and a few Debye lengths ahead of the trap, the ion density is $N_i \simeq 2N_e \simeq 2N_0$, where N_0 is the undisturbed plasma density. Taking into account that G1 has a transparency factor $\alpha < 1$, we would have $N_i \simeq (1 + \alpha^2)N_e \simeq (1 + \alpha^2)N_0$. As V_R is decreased, some ions begin to get through G2; for $V_R = m_iU^2/2e$, say, only half the ions are turned back and we would have $N_i \simeq (1 + \alpha^2/2)N_0$. The excess charge density inside the jet goes from $e\alpha^2N_0$ to zero, as V_R decreases to zero. It should be noted that Eq. (2), for $(2kT_i/m_iU^2)^{1/2}$ small, is practically independent of T_i, unless a non-negligible fraction of the ions are turned back by G2; in fact, even at $V_R = m_iU^2/2e$, I_c only changes by 1% as T_i changes by 20%, for $(2kT_i/m_iU^2)^{1/2} = {}^1/_5$. (This point is rarely made clear in papers that show curve-fitting of Eq. (2) to experimental data.) Thus, over most of the time that the V_R sweep lasts, and certainly during the T_i-sensitive part of this sweep, enough ions are repelled by the trap to produce an excess ion density $N_i - N_0$ that is a non-negligible fraction of N_0. For typical V_R sweep times of the order of seconds, the length of the cylindrical tube (with the same cross section of grid G1) filled by the ion jet, is of the order of several kilometres! Obviously such a situation is impossible to maintain. Probably a large potential field will build up, spreading in *all* directions far away from the trap. Analysis of this problem is difficult, but certainly the real situation cannot resemble that envisaged in Eq. (2).

We have thus seen that an ion trap, measuring T_i, greatly disturbs the plasma far ahead of it. This is a general characteristic of the motion of a body moving through a plasma at a speed larger than $(2kT_i/m_i)^{1/2}$, but smaller than $(2kT_e/m_e)^{1/2}$, and *reflecting* ions. The ions in excess ahead of the body are those missing from the wake behind. When the body absorbs the ions, the perturbation ahead of the body disappears, and only the wake behind remains.

3. A Method to measure Ion Temperature

The current to a thin cylindrical probe immersed in a rarefied plasma flow is correctly given by orbital motion theory [14] when the probe radius r_p is small compared with the Debye length λ_D. For a flow velocity U such that

$kT_i/m_i \ll U^2 \ll kT_e/m_e$, and V_p large and negative $(-eV_p \gg kT_e, kT_i)$, the current is given by

$$I_{p\infty} \simeq 2 r_p l_p N_0 e (U^2 \sin^2\theta - 2eV_p/m_i)^{1/2} \tag{3}$$

where l_p is the probe length and θ is the angle between $\boldsymbol{U}$ and the probe axis; the ∞ subscript means that l_p is assumed so large that any end (finite l_p) effect will be negligible. Eq. (3) shows that $I_{p\infty}$ *decreases* smoothly as θ goes from $\pi/2$ to zero.

Recently, both satellite [20] and laboratory [21] probe measurements have detected a significant experimental disagreement with Eq. (3); the current measured was found to follow closely that equation, except at small values of θ, where the current showed a large *peak*. Although long probes were involved (l_p up to 820 r_p and 20 λ_D), it has been found that the disagreement is due to an end effect [20, 22]. An analysis of this effect has been developed recently [23].

The important point about the end effect is that both the height and the angular half-width of the peak are sensitive to T_i, thus allowing a measurement of this quantity. The single ion species analysis of [23] can be trivially extended to the case of several ion species, and it is also possible to choose values of V_p, l_p and r_p such that all constraint conditions derived in that paper are approximately satisfied; these constraints arise from the existence of the end effect itself, and from the simplifications required by the analysis. We find optimum values $l_p \sim 7$ cm, and V_p about $-20kT_e/e$; for r_p the value 0.028 cm, used frequently for ionospheric cylindrical-probe measurements [12], can be adopted.

For these values, and under the conditions prevailing in the ionosphere, the formulae of [23] can be simplified. For the height of the I_p peak we get

$$I_{p\max} = I_p(\theta = 0) \simeq 4\bar{\delta} r_p l_p N_0 e \left(\frac{-eV_p}{kT_i}\right)^{1/2}$$
$$\times \sum_\alpha \frac{p_\alpha (-2V_p/m_\alpha)^{1/2} (\ln \sigma_\alpha^2 - 1)(l_n \sigma_\alpha^2)^{-1/2}}{[1 + 0.16 (l_p/U)^2 4\pi N_0 e^2/m_\alpha]^{1/2}} \tag{4}$$

where the summation is over all ion species, and m_α and p_α are the mass and the relative concentration of species α, respectively (obviously $\Sigma p_\alpha = 1$); for the region of interest (350—1000 km), only H^+, H_e^+ and O^+ ions need be considered (N^+ being lumped together with O^+). We also have

$$\sigma_\alpha^2 = \frac{(2/\pi)(l_p/r_p)^2 \bar{\delta} (-eV_p/m_\alpha U^2)}{1 + 0.16 (l_p/U)^2 4\pi N_0 e^2/m_\alpha}$$

and $\bar{\delta} = [\ln(\lambda_D/r_p) + Y(r_p/\lambda_D, -eV_p/kT_e)]^{-1}$ where Y is given in Fig. 1 of [23]; $4\bar{\delta}$ is always close to unity, and $\ln \sigma_\alpha^2$ lies between 5 and 10. For the angular half-width $\theta_{1/2}$ (the width of the peak at half-value of its maximum) the T_i dependence can be eliminated and we get

$$\frac{I_{p\max}}{2} \simeq \frac{(8/\pi)^{1/2}}{\theta_{1/2}} 4\bar{\delta} r_p l_p N_0 e \sum_\alpha \left(\frac{-eV_p}{m_\alpha U^2}\right)^{1/2}$$
$$\times \frac{p_\alpha (-eV_p/m_\alpha)^{1/2} (\ln \sigma_\alpha^2 - 1)(\ln \sigma_\alpha^2)^{-1/2}}{[1 + 0.16 (l_p/U)^2 4\pi N_0 e^2/m_\alpha]^{1/2}}. \tag{5}$$

Before obtaining T_i from Eq. (4) it is necessary to know N_0, the satellite potential (and thus V_p), and p_α ($\alpha = H^+, He^+, O^+$). The first two quantities can be obtained by means of a second cylindrical probe, using a well-known method (see [12] for references on this method). That probe should be oriented approximately perpendicular to $\boldsymbol{U}$ and should be biased highly negative for part of the time, so that Eq. (3) can be applied to give a linear relation between the concentrations. Two other linear relations are given by $\Sigma p_\alpha = 1$ and Eq. (5). T_i follows then from Eq. (4). It should be noted that to obtain this quantity it is only necessary to make an order of magnitude estimate of T_e.

References

[1] K. I. Gringauz, in: Solar Terrestrial Physics, Academic Press, New York 1967 (Chap X).
[2] J. V. Evans, in: Solar Terrestrial Physics, Academic Press, New York 1967 (Chap IX).
[3] W. B. Hanson et al., J. Geophys. Res. **74**, 400 (1969).
[4] J. P. McClure and B. E. Troy Jr., J. Geophys. Res. **76**, 4534 (1971).
[5] L. H. Brace and J. P. McClure, paper presented at 52nd Annual Meeting of the AGU, Washington D.C., 1971.
[6] H. C. Carlson and J. Sayers, J. Geophys. Res. **75**, 4883 (1970).
[7] H. G. Booker and E. K. Smith, J. Atmos. Terr. Phys. **32**, 467 (1970).
[8] G. N. Taylor and G. L. Wrenn, Planet. Space Sci. **18**, 1663 (1970).
[9] L. H. Brace et al., J. Geophys. Res. **74**, 1883 (1969).
[10] R. C. Sagalyn and R. H. Wand, J. Geophys. Res. **76**, 3783 (1971).
[11] J. L. Donley et al., Proc. Inst. Elect. Electron. Engrs **57**, 1078 (1969).
[12] L. H. Brace et al., Space Research **XI**, 1079 (1971).
[13] W. R. Hoegy, J. Geophys. Res. **76**, 8333 (1971).
[14] J. D. Swift and M. J. R. Schwar, Electrical Probes for Plasma Diagnostics, Iliffe Books, London 1970.
[15] J. R. Sanmartin, Phys. Fluids **13**, 103 (1970).
[16] E. C. Whipple, Proc. Inst. Radio Engrs **47**, 2023 (1959).
[17] W. C. Knudsen, J. Geophys. Res. **71**, 4669 (1966).
[18] S. J. Moss and E. Hyman, J. Geophys. Res. **73**, 4315 (1968).
[19] L. W. Parker and E. C. Whipple, J. Geophys. Res. **75**, 4720 (1970).
[20] R. T. Bettinger and A. A. Chen, J. Geophys. Res. **73**, 2513 (1968).
[21] S. D. Hester and A. A. Sonin, in: Rarefied Gas Dynamics, Vol. II, Academic Press, New York 1969.
[22] S. D. Hester and A. A. Sonin, Phys. Fluids **13**, 1265 (1970).
[23] J. R. Sanmartin, Phys. Fluids **15**, 1134 (1972).

Space Research XIII — Akademie-Verlag, Berlin 1973

BURNT-OUT ROCKET PUNCHES HOLE INTO IONOSPHERE

B. W. Reinisch

Lowell Technological Institute Research Foundation, Lowell, Mass., USA

Continuous monitoring of the ionosphere in Eglin, Fla, USA, with a ground-based Digisonde showed disturbances in the F region seconds after a burnt-out Black Brant rocket (0.5 m × 7 m) penetrated this region. Evaluation of the ionograms reveals at 250 km altitude an electron depletion of about 25% in the wake of the rocket. The fact that electromagnetic waves, propagated in this depleted cylinder, return with amplitudes which are 10–15 dB below the normal echo returns suggests an average depletion cross section of 2 km². These ducted waves propagate to slightly higher altitudes than the ambient electron density could permit. The longer propagation time required results in a separate echo trace on the ionogram, 30 km above the regular trace. The effect is very short-lived since after ten minutes the extra trace was no longer visible. For all rocket-borne electron density measurements using *in situ* probes or radio propagation experiments this depletion effect must be taken into account.

1. Ionogram Observations

During a rocket launch at Eglin AB, Fla, on 4 April 1970 we were monitoring the ionosphere with a Digisonde 128. Since the rocket payload was geared towards measurements in the E region, the ionosonde scanned three E-region and one F-region ionogram every ten minutes. Digital phase coherent integration over 2 seconds for each transmitted frequency greatly increased the signal to noise ratio. The amplitudes of 128 adjacent height ranges, spaced by 1.5 km for the E and 4.5 km for the F ionograms, were recorded with a resolution of 1 dB on digital magnetic tape and simultaneously printed out in 2 dB steps. The digionogram scanning from 3 to 14 MHz ten minutes before launch time is shown in Fig. 1. Each tenth frequency, i.e. 4.0, 5.0 ... 13.0 MHz, is replaced by a preface line which contains, besides a list of all operational parameters, the exact time with one second accuracy of each frequency radiated.

At 1128:01 CST a small rocket of type Black Brant 5B was launched. It had a diameter of 0.5 m and a length of 7 m after ejection of the nosecone. The rocket burned out after 32 seconds at an altitude of 35 km, and the burnt-out vehicle reached its apogee at 385 km altitude 325 seconds after launch. The E region is reached 30 seconds and the F region 100 seconds after burn-out.

It is surprising to see in Fig. 2 the disturbance which the rocket created in the F region, while the E-region ionograms are undisturbed. A third echo-trace besides the ordinary and extraordinary component is clearly visible in Fig. 2. The next F ionogram recorded ten minutes later is again undisturbed. The peculiarity shown in Fig. 2 was only seen this one time in many days of observation so that we can safely assume that the third trace was caused by the rocket. The iono-

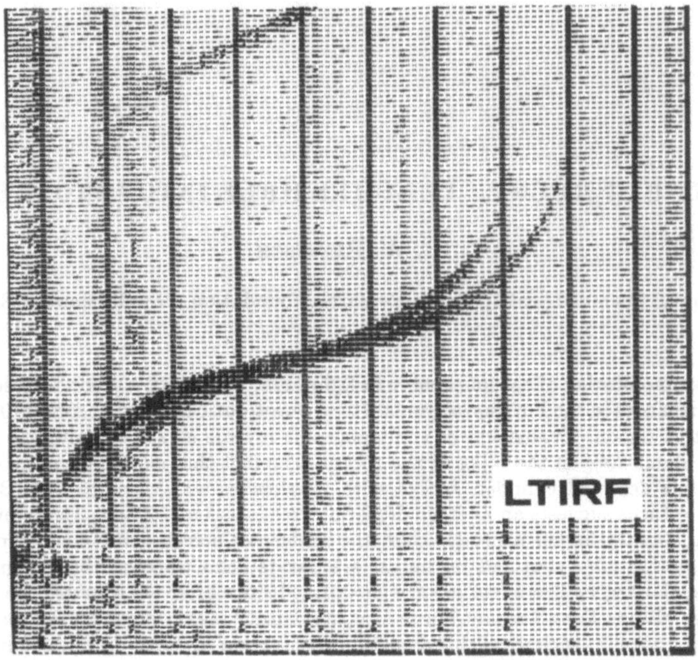

Fig. 1. Digionogram 10 minutes before launch.

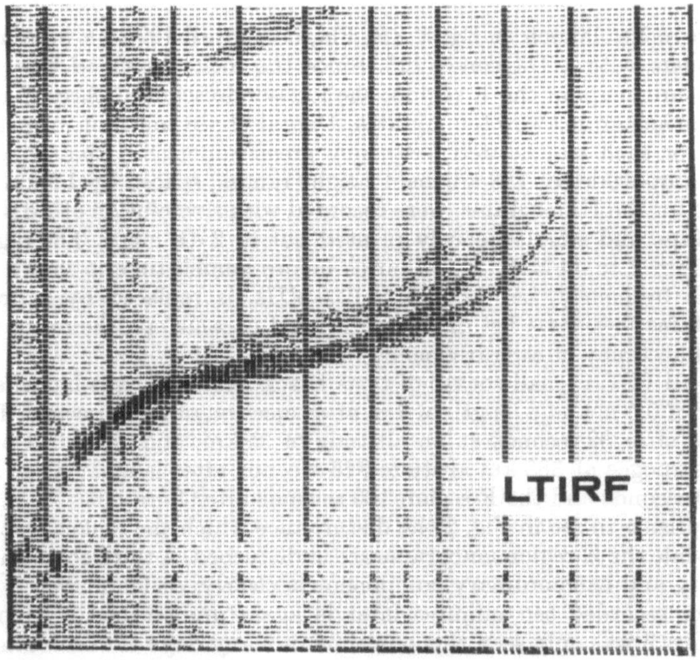

Fig. 2. Digionogram during rocket flight.

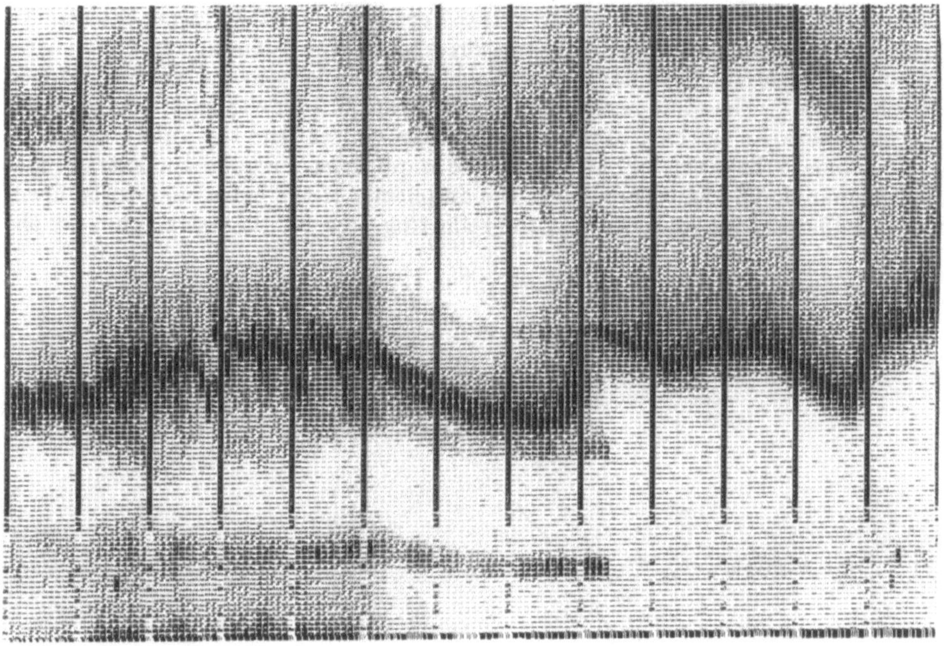

Fig. 3. $h'(t)$ for E and F region, 4/5 April 1970, 07h—04h CST.

spheric characteristics in Fig. 3, $h'(t)$ for E and F layer, show that the ionospheric conditions were relatively quiet during the launch time. Four hours after the launch a rapid depression of the F layer occurs lasting until 2040 CST when h' increases by 63 km within 20 minutes. These characteristics were obtained in a computer by compressing the tape-recorded ionograms.

The difference in virtual heights between regular and third echo varies between 22 and 44 km in the frequency range from 7 to 10 MHz. The amplitudes of the extra trace are 10—15 dB smaller than those of the regular trace.

2. Rocket Wake

Bourdeau [1] and Henderson and Samir [2] report about wakes of reduced electron density for satellites Explorer 7 and Ariel 1. The observed third trace cannot simply be interpreted by oblique reflections from the rocket disturbance, since the shape of the extra trace is very similar to the regular traces. But if we postulate a wake of depleted electron density a ducted mode of propagation will exist which is similar to the vertical sounding path. The radio waves propagate upwards in this duct until they reflect at an altitude slightly higher than the reflection height in the ambient plasma. The vertical gradient in electron density determines the difference in the reflection heights. The ionogram in Fig. 2 shows a mean difference of 30 km of which 10 km are accounted for by the obliqueness of the duct.

In Fig. 4 the electron density profile calculated by using the regular echo trace is plotted together with the rocket trajectory. In the altitude range from 200 to

250 km the profile shows a mean vertical gradient of 0.6 MHz per 10 km. An increase in reflection height of 20 km corresponds to a reduction in plasma frequency of 1.2 MHz. This means that the electron density in the wake is reduced by about 25%. The same result can also be deduced from the difference of about 1.2 MHz between the critical frequencies of the vertical and the wake echoes.

To estimate the diameter of the wake the amplitudes of regular and extra echoes are compared. The ionogram shows the amplitude of the extra echo to be lower by 10—15 dB. This suggests that the reflecting area for the extra echo, i.e. the

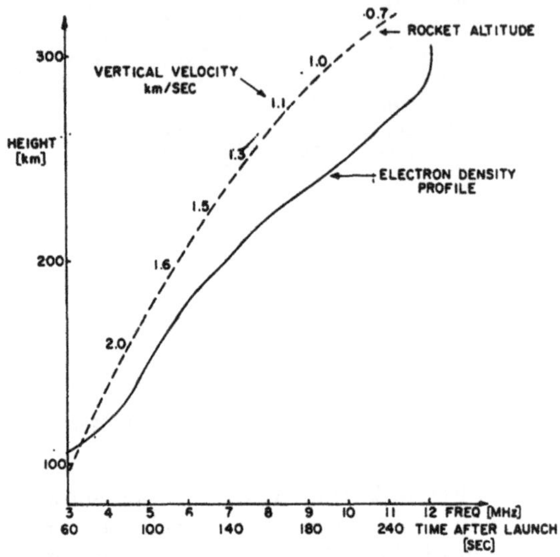

Fig. 4. Electron profile and rocket trajectory.

cross section of the wake, is approximately a quarter of half a Fresnel zone which is reflecting most of the energy of the vertical echo. Since the radius of the first Fresnel zone for 10 MHz in 200 km altitude is 2400 m the wake has a diameter of about 1600 m.

3. Conclusion

From the fact that no other ionosonde observations about the wake of a small burnt-out rocket are reported it must be concluded that the depletion zone has a short lifetime. For all wave propagation experiments between ground and rocket the wake must be taken into account. *In situ* experiments must carefully consider the placement of the sensors. Direct interaction between vehicle and ambient atmosphere as well as some remaining outgassing [3] must be considered as possible causes of the wake.

References

[1] R. E. Bourdeau, Space Sci. Rev. **4**, 373 (1963).
[2] C. L. Henderson and U. Samir, Planet. Space Sci. **15**, 1499 (1967).
[3] W. Pfister and J. Ulwick, Radio Sci. **69** D, 1219 (1965).

Space Research XIII — Akademie-Verlag, Berlin 1973

TROPOSPHERIC AND STRATOSPHERIC RESPONSE TO SOLAR INFLUENCE DURING GEOMAGNETIC DISTURBANCES

V. V. Mikhnevich and R. F. Usmanov

USSR

In September—October 1971 a complex rocket experiment was carried out to study variations of atmospheric parameters and wind characteristics during periods of geomagnetic disturbances. Variation of meteorological parameters at altitudes up to 40 km during the development of recurrent and sporadic geomagnetic disturbances is considered here.

1. Introduction

A complex experiment was carried out in the USSR in September—October 1971. One of the principal scientific aims of the experiment was the study of the response of the atmosphere and ionosphere to the sun during periods of geomagnetic disturbances. Ground, upper-air and rocket measurements using various methods allowed the pattern of variations of thermobaric parameters and winds, up to high altitudes ($\sim$ 100 km), to be studied. The results of the wind, temperature and pressure measurements in the troposphere and stratosphere in the region of the rocket station "Volgograd" ara analysed in the paper.

2. Conditions of the Experiment

The year 1971 fell in a period of decreasing solar activity when both strong solar disturbances and geomagnetic disturbances seldom occurred. During the period of the experiment, from 20 September to 10 October, three groups of sunspots of comparable intensity passed over the central solar meridian. On 3 October at 1700 Moscow time one of the spots resulted in a 2+ flare which existed approximately for 2 hours.

At the same period there were recorded three magnetic storms: one with a sudden commencement and two with a gradual start. Maximum geomagnetic disturbances were observed in the period of 30 September—3 October in the first case and 25—27 September and 8—9 October in the other two cases. The last storm after the solar flare of 3 October developed slowly. Minimum geomagnetic indexes were recorded on 23 and 29 September and on 4 and 10 October.

The characteristics of solar and geomagnetic activity are given in Fig. 1.

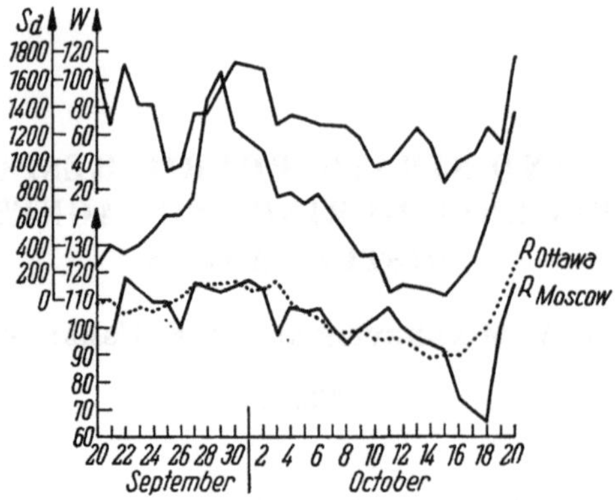

Fig. 1. Characteristics of solar and geomagnetic activity.

3. Wind

As is seen from Fig. 2, where a temporal vertical section of the meridional wind component is given, on 22—23 and 28—29 September and on 8—9 October a change of wind direction took place from northern to southern, and on 25—26 September and 4—5 October from southern to northern. It is concluded that meridional wind reconstruction correlates with the geomagnetic field variation. The periods of maximum wind velocity and of variation in wind direction correspond to periods of minimum values of the geomagnetic index, see Fig. 3.

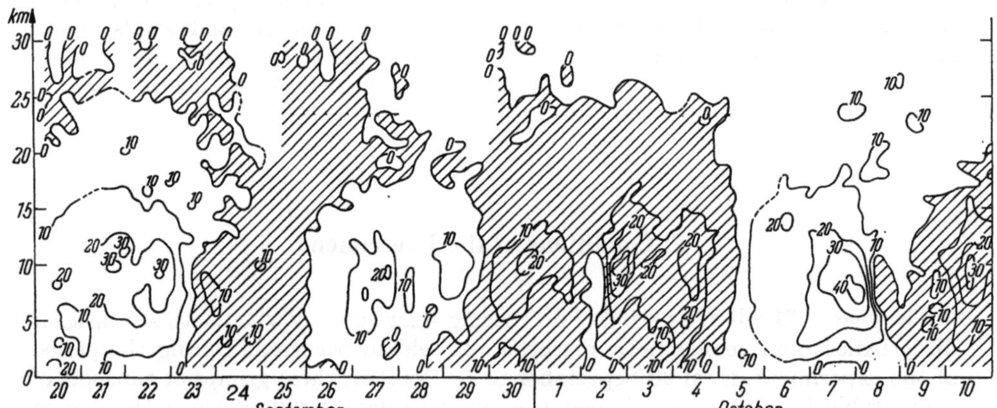

Fig. 2. Temporal vertical section of the meridional component of the wind velocity.

4. Temperature

Let us consider the temperature variation in the troposphere and lower stratosphere. As is seen from the charts of isothermal levels (Fig. 4) in the periods of 21—23 and 27—30 September and 7—10 October there occurred a remarkable reconstruction of the temperature stratification of the atmosphere. Temperature

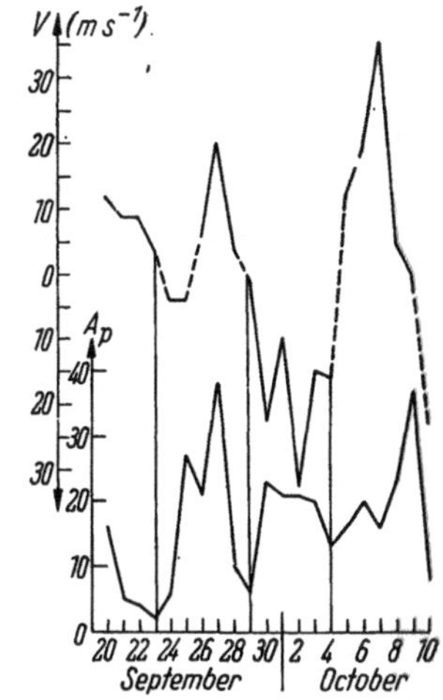

Fig. 3. Variation of the meridional component of the wind velocity at the height of the tropopause (at 15h) and the geomagnetic index.

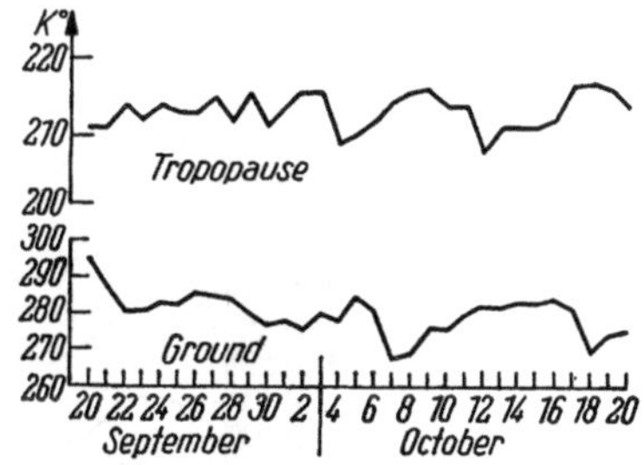

Fig. 5. Temperature variation at the tropopause level and near the ground (at 15h).

variations in the region of the tropopause and above had a reversed character in comparison with those in the troposphere, Fig. 5 [1]. The most powerful reconstruction was observed on 7—10 October after the solar flare. The maximum temperature in the region of the tropopause was recorded during the same period.

In the period 7—9 October a general temperature decrease took place below the level of 8 km. Temperature variations of a periodic nature exceeded diurnal variations in amplitude. The amplitude variation in general decreased with decreasing height (with the atmosphere density increasing); maximum (or minimum) temperature values were observed slightly later. Thus on 8 October the maximum temperature value was approximately at 08—09 h at an altitude of 8 km, at 10 hr at 7 km, at 15 hr at 5 km. These temperature variations indicate that the disturbances propagate from high altitudes.

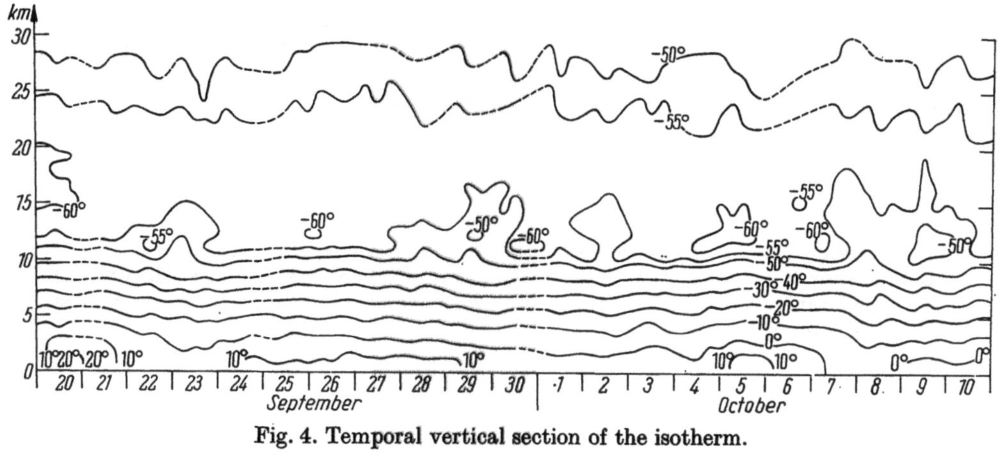

Fig. 4. Temporal vertical section of the isotherm.

5. Pressure

Small pressure variations at altitudes up to 30 km were observed after the solar flare during the period 7—12 October (Fig. 6). In contrast to temperature, the pressure varied synchronously at all altitudes. After 5 October the pressure

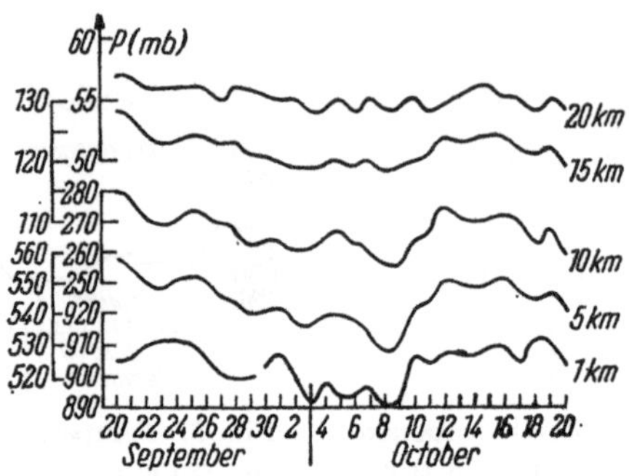

Fig. 6. Pressure variations at altitudes of 1, 5, 10, 15, 20, 25 km at 15h.

decreased, approaching a minimum value on 9 October and then increased and approached maximum value on 12 October. The maximum of temperature at the altitude of the tropopause was observed on 9 October and the minimum on 4 October.

6. Conclusions

The development of the phenomena studied in the atmosphere during the period of the experiment is now considered. Solar-disturbance-induced variations of the magnetic field of the earth preceded the variation of atmospheric circulation by several hours. At the same time a reconstruction of the temperature stratification of the atmosphere took place, a feature studied earlier [2]. A considerable pressure variation is observed some days later. Maximum temperature and pressure variations in the troposphere and stratosphere were observed after solar flare. Temperature decrease was observed in the troposphere; at the tropopause and above, a temperature increase was observed. The disturbance which caused the reconstruction of the temperature stratification propagated downwards.

References

[1] R. F. Usmanov and A. B. Bondarenko, Solnechnye Dannye, No. 8 (1968).
[2] G. S. Ivanov-Kholodny, T. V. Kazatschevskaya, G. A. Kokin and V. V. Mikhnevich, in: Issledovaniya atmosphery i ionosphery v period povyshennoy solnechoi aktivnosti, Gidrometeoizdat, Leningrad 1970, (p. 143).

Space Research XIII — Akademie-Verlag, Berlin 1973

DIURNAL AND SEASONAL VARIATIONS OF NEUTRAL WINDS AND ELECTRIC FIELDS ABOVE 90 km IN THE VICINITY OF THE AURORAL ELECTROJET

D. Rees[a], T. L. Aggson[b], K. Burrows[c], G. Haerendel[d] and J. W. G. Wilson[e]

[a]Department of Physics, University College London, London, UK
[b]Goddard Space Flight Center, Greenbell, Md, USA
[c]Radio and Space Research Station, Slough, Bucks, UK
[d]Institute for Extraterrestrial Physics, Max-Planck Institute, Munich, FRG
[e]Department of Space Research, University of Birmingham, Birmingham, UK

The response of the atmosphere above 90 km altitude to auroral electric fields was examined during several rocket flights of combined payloads making electric field and neutral wind measurements from Kiruna and Andenes in early 1971.

In the winter period very high neutral wind speeds are observed above 100 km following significant substorm activity. These winds, up to an altitude of about 150 km, can be explained as being predominantly due to collisional acceleration by ions driven by a strong meridional electric field associated with the substorm. Above 150 km the gas acceleration, by ion—neutral collisions, is augmented in a rather complex manner by acceleration due to anomalous pressure gradients caused by auroral heating.

During the summer period when the local auroral ionosphere is continuously sunlit, the response of the neutral wind to auroral activity is apparently reduced. The seasonal change appears to occur quite abruptly, and happened within a week about 20 April at Kiruna in 1971. Possible mechanisms for the change are discussed.

1. Introduction

This paper presents the results from rockets launched at Andenes (69° 18′ N 16° 01′ E) and ESrange, Kiruna (67° 53′ N 21° 06′ E) during January and April 1971. Table 1 gives details of the launch contitions and payloads..

Table 1
Data on the firings

Round	Launch site	Date	Time (CET)	Sol. dep. (°)[a]	Payload	Magnetometer perturb. (γ)[b] 2 hrs before	at launch
ANT 38	Andenes	28 Jan. 71	1655	−12	TMA and Ba	+150	+150
P66K	ESrange Kiruna	15 Apr. 71	0130	−10	TMA and long boom electric field probe	−900	−600
P64K	ESrange	21 Apr. 71	1959	−2	twin spherical electric field probe		
P67K	ESrange	21 Apr. 71	2243	−9	TMA and long boom electric field probe	−125	−150
P63K	ESrange	22 Apr. 71	0004	−8	barium	−275	−500

[a] Approximate solar depression angle at trajectory.
[b] Average horizontal magnetometer perturbation (γ) 2 hrs before launch and at launch.

2. Electric Field Data

Ionospheric electric fields were measured by barium cloud releases on ANT 38, by a long boom antenna system operated in a floating potential mode on P66K and P67K and from a double spherical probe system operated in a stepped current mode on P64K. The techniques are described in [1—3].

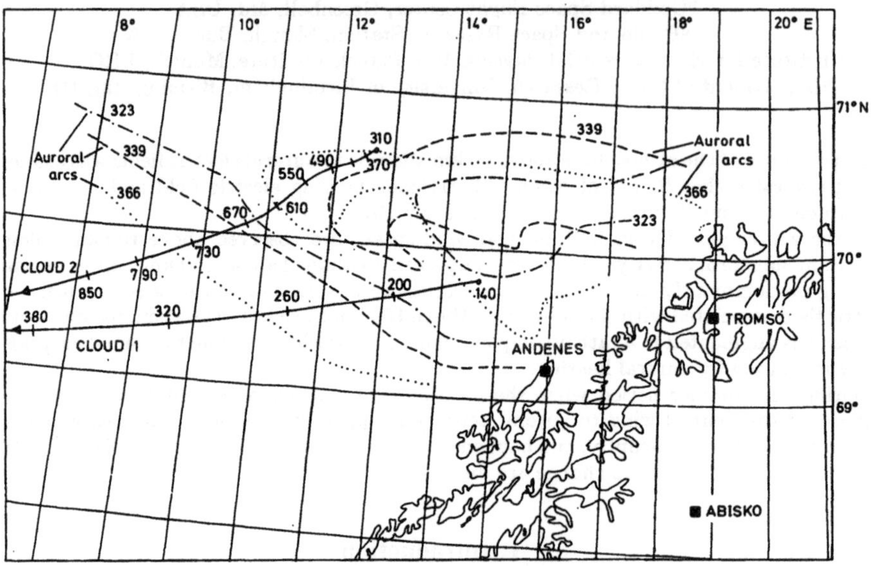

Fig. 1. Three successive positions of auroral arcs belonging to a westward travelling surge and the trajectories of two barium clouds for ANT 38 projected to the 100 km level. Times in seconds after launch.

Fig. 1 [4] shows the trajectories of the two barium releases from ANT 38 mapped in invariant magnetic coordinates. Northward electric fields of the order of 50 mV m^{-1} were deduced from the $\boldsymbol{E} \wedge \boldsymbol{B}$ drift measurements of the two ion clouds above 200 km, well above the maximum of the horizontal component of the conductivity tensor. These field measurements were to the south of a westward travelling auroral surge, indicated at three successive times in Fig. 1 and described in detail in [4].

Fig. 2 shows the electric field data (with $\boldsymbol{V} \wedge \boldsymbol{B}$ of the rocket removed) from P67K, one of the two probe experiments flown on the night of 21 April 1971. During both flights there were considerable variations of both the magnitude and direction of the observed fields. At the time of the first flight (P64K) auroral forms were not visible because of the sky brightness. However, P67K travelled through a highly structured, bright and rapidly changing aurora.

Three general points can be made about the probe measurements:

(i) Both experiments seem to produce consistent results above 75 km on upleg and downleg. The P67K electric field magnitude decreased as the rocket fell below 80 km on descent as has happened on previous flights using the long boom system [2]. Below 85 km the probes show a systematic change in the direction of the electric field probably due to aerodynamic bending of the probes.

(ii) Where, during the P67K flight, it is possible to determine a positive concurrence of aurora on the same field line as the rocket (in particular near 260 s), the magnitude of the electric field was variable and generally decreased. However, the field strength did not decrease to zero. The change in magnitude was associated with a deviation of the field direction from southward to westward.

(iii) The 2 kg of trimethyl aluminium (TMA) released in a continuous downleg trail from the P66K and P67K payloads caused only a slight perturbation of the electric field measurements for a few seconds after the start of the trail.

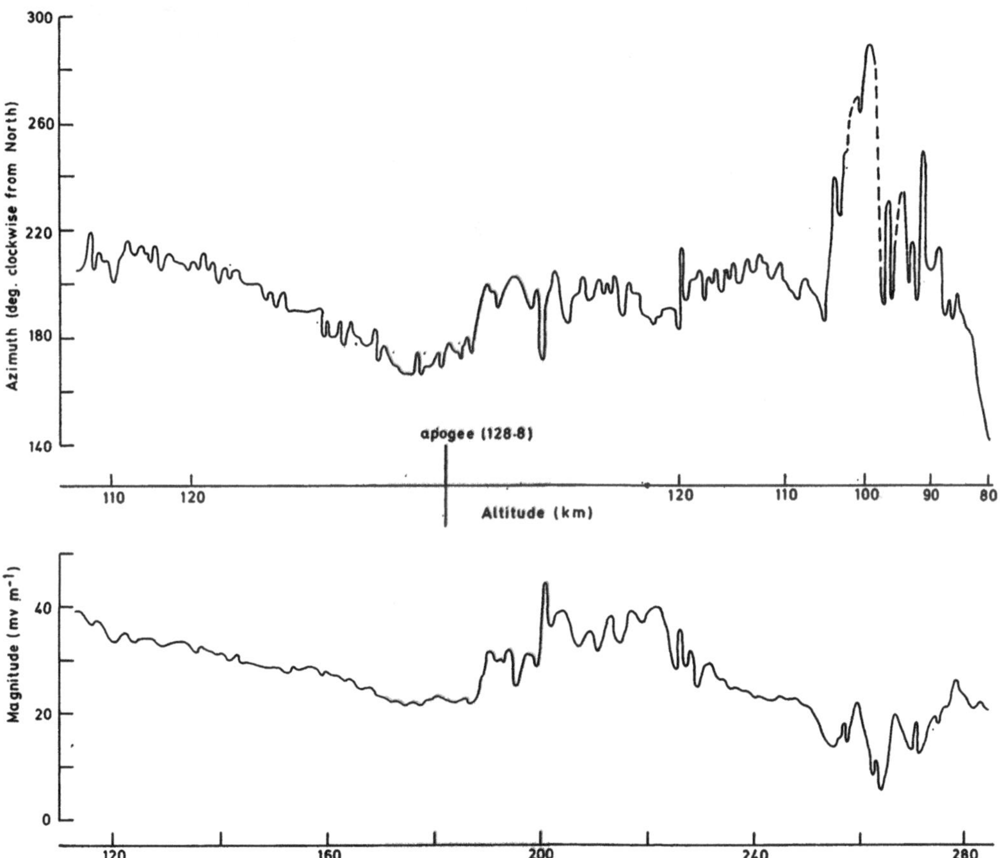

Fig. 2. Electric field data from P67K.

3. Neutral Wind Observations

Neutral wind profiles were obtained from the TMA trail releases on three of the experiments discussed here. The W—E neutral wind component, taken as an average between 130 and 150 km altitude for ANT 38 and at 130 km (apogee) for P66K and P67K, is plotted against the mean S—N ground magnetometer perturbation for two hours prior to the launch in Fig. 3 and compared with previous results from Kiruna [5].

The ANT 38 data are consistent with the previous results, although the measurements were from a downleg release from a rocket launched at Andenes, a point some 500 km northwest of Kiruna. That the same correlation exists between the W—E wind component and the Kiruna magnetic field perturbation gives some indication that 500 km is within the latitudinal scale of the E-region neutral wind disturbance set up by the positive phase of an auroral substorm.

P66K was launched following an intense negative bay disturbance. The data may confirm a prediction [6] that E-region winds should not be as great following intense negative bay disturbances as might otherwise be indicated by the relationship in Fig. 3. The wind velocity, 300 m s^{-1} near 130 km, was increasing linearly

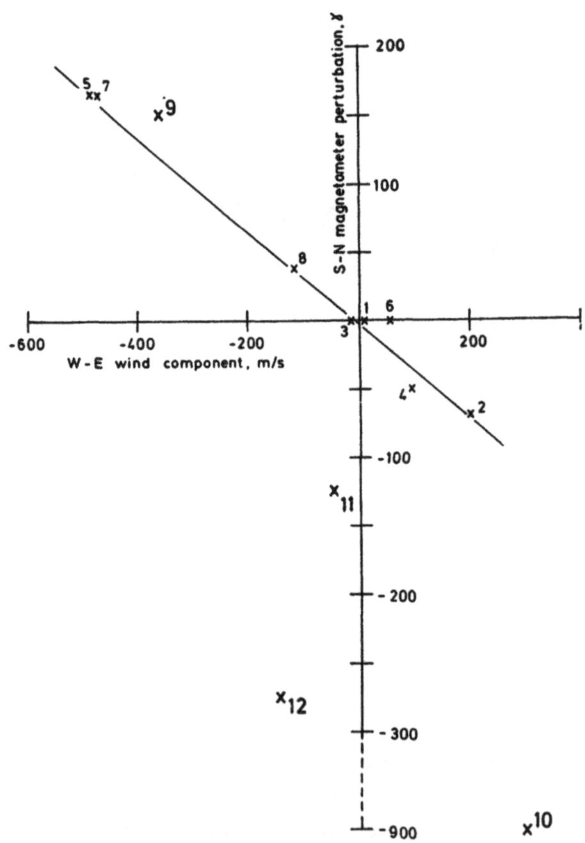

Fig. 3. W—E neutral wind component (near 150 km generally) plotted against S—N ground magnetometer perturbation (mean of 2 hours prior to launch).

1, 1 Feb. 1968 0642 CET, 130—150 km; 2, 4 Feb. 1968 0633 CET, 130—150 km; 3, 15 Oct. 1968 1732 CET, 130—150 km; 4, 1 Nov. 1968 0540 CET, 130—150 km; 5, 15 Mar. 1969 1855 CET, 150 km; 6, 16 Mar. 1969 1905 CET, 160 km; 7, 17 Mar. 1969 1910 CET, 150 km; 8, 18 Mar. 1969 1916 CET, 165 km; 9, 28 Jan. 1971 1655 CET, 130—150 km; 10, 15 Apr. 1971 0130 CET, 130 km; 11, 21 Apr. 1971 2243 CET, 130 km; 12, 22 Apr. 1971 0004 CET, 165 km.

with height. We have computed a model of neutral wind speed above 100 km from the acceleration of the neutral gas due to ion—neutral collisions, given the 30—40 mV m^{-1} southwards field measured on this experiment, and a mean E-region electron density profile deduced from ionograms for the whole substorm period. This gives a maximum E-region wind of not more than 500—600 m s^{-1} near 150 km. If we extrapolate the earlier data of Fig. 3, then for a two-hour mean disturbance of about 1000γ (peak at -1700γ in the northward component at Kiruna), a neutral wind speed of about 2000 m s^{-1} would be indicated.

It is concluded that, above 110 km, the mean ion density, and therefore electron density and conductivity, which governs, with the electric field, the acceleration of the neutral gas, was not significantly greater during the extreme -1700γ bay than during an average positive bay of the order of 200—300γ. The additional conductivity required to carry most of the ionospheric currents causing the -1700γ ground perturbation must have been below 110 km, where, since the ion—neutral collision frequency is higher than the ion gyro frequency, there would be little momentum transfer to the neutral gas. It seems likely that, on average, the neutral wind disturbances caused by positive and negative bays in the region above about 110 km are of similar magnitude and duration and do not impart significant net eastward or westward momentum to the upper atmosphere.

During the P67K flight, which was launched after several hours of moderate negative bay activity, the neutral wind near 130 km was about 80 m s^{-1} southwestward. Ninety minutes after P67K, when the bay had intensified to -500γ, a barium release experiment (P63K) at 165 km altitude showed variable southward electric fields of about 50 mV m^{-1}, but again a neutral wind directed southwestward of about 200 m s^{-1}. For both P67K and P63K an eastward wind component of at least 100—200 m s^{-1} would have been expected from Fig. 3. These two results seem indicative of the variability of the neutral wind response to auroral activity found by other experimenters at Heiss Is. (80° 37′ N 58° 03′ E) [7].

As potential causes of the anomaly, mechanisms are required which will either decrease the rate of momentum transfer from the ions to the neutral gas or mix the momentum into a larger region of the thermosphere. In this latter way the local wind acceleration may be proportionately reduced, perhaps to the extent where, for a single measurement during a moderate disturbance, the "auroral" contribution is inextricably merged with winds arising from a combination of tidal, gravity wave and thermospheric pressure gradient sources.

The total transferred momentum did not appear to be appreciably lower during the night of the second launch compared with that of the first launch since: (i) The electric field was predominantly southward through a five hour period with an average magnitude of 20—50 mV m^{-1}. (ii) The auroral electron density was consistently enhanced to values of about 10^5 cm^{-3} at 110—130 km as shown by ionograms and the magnitude of the negative ground magnetic perturbation (200—500γ). (iii) There did not appear to be an anomalous auroral E-region neutral temperature, which could lead to winds driven by the pressure gradient modifying winds driven by ion drag. The neutral temperatures at 120—130 km were quite normal during both experiments (see § 4).

Possible sources of enhanced momentum coupling of the neutral gas within the region of auroral acceleration to that outside are: (i) A summertime thermospheric wind pattern [8], which should have become established by the time of the second experiment, leading to an equatorward wind component of the order of 150 m s^{-1} near 70° N at midnight, which would decrease the time a particular air parcel spent in the accelerating region to less than about 1 hour. (ii) The presence of sunlight throughout 24 hours, at the time of the second experiment, over a substantially greater region of the auroral oval [9, Figs. 11a and b]. The resulting higher electron density throughout the ionosphere surrounding the auroral activity could substantially increase the coupling of the atmosphere in the non-auroral E region to that inside the auroral oval.

Fedder and Banks [10] have recently demonstrated that in the summer polar cap a neutral wind may be induced in an anti-solar direction by convection electric fields. The process could enhance both mechanisms above and be very difficult to distinguish from them.

4. Neutral Temperature Results

Neutral temperature in the lower thermosphere has been measured on several high-latitude experiments using observations of aluminium oxide (AlO) spectra obtained from TMA trails and aluminium grenade releases [6, 11]. The observation period has often exceeded thirty minutes and occasionally the AlO clouds have been on the same magnetic field lines as bright auroral forms including the three experiments reported here. It has not been possible, however, to distinguish any local neutral gas heating correlated with auroral precipitation.

Taking account of the neutral wind speeds prevalent at times of substorms, the geometry of the auroral forms and experimental limitations regarding accuracy and rate of individual temperature measurements [11], we can put a limit on the short term (less than one minute) and long term (greater than ten minutes) heating rates. Associated with bright, discrete auroral forms with a N—S dimension of perhaps 200—500 metres, and given a neutral wind with a component of about 200 m s^{-1} perpendicular to the form, our observations indicate that neutral gas

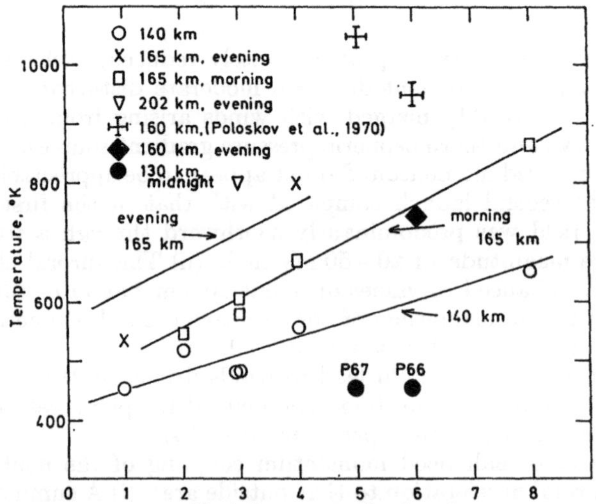

Fig. 4. Temperatures at 140 and 165 km with corresponding C 9 values.

heating rates in the immediate vicinity of auroral forms are less than 5—10 °K s^{-1}. Associated with the mean precipitation over a large area (periods of ten minutes corresponding to horizontal movements of the neutral atmosphere of up to 300 km), the maximum mean neutral gas heating rate between 120 and 180 km altitude seems to be less than 0.2 °K s^{-1}.

The temperature data of the three experiments are plotted in Fig. 4 along with previous data [11]. The measurements confirm a correlation with the C9 index of the 24 hours prior to the measurements [6]. The P66K temperature measurement, in particular, does not reflect in any way the extreme local auroral electrojet activity of the previous six hours. (Note that the P66K and P67K measurements are limited to below 130 km, the rocket apogee.)

Acknowledgments

We wish to express our sincere gratitude to the groups at each of our institutes whose technical and computing assistance made these experiments possible. The ANT 38 Nike Tomahawk was launched in the Federal Republic of Germany's National programme, while the other rockets were launched in the UK National Space Programme. The experiments required a great deal of international co-operation and we acknowledge our debt to those who expedited the necessary agreements.

References

[1] G. Haerendel and R. Lüst in: Earth's Particles and Fields, Reinhold Book Corp., New York 1968 (p. 271).
[2] T. L. Aggson in: Atmospheric Emissions, Van Nostrand Reinhold Co., New York 1969 (p. 305).
[3] J. W. G. Wilson to be published (1973).
[4] G. Haerendel, paper presented at ESRO Summer School, Cortina (1971).
[5] D. Rees, J. Brit. Interplan. Soc. **24**, 233 (1971).
[6] D. Rees, Phil. Trans. Roy. Soc. London A **271**, 563 (1972).
[7] L. A. Andreeva, L. A. Katasyev and D. B. Uvarov, Phil. Trans. Roy. Soc. London A **271**, 559 (1972).
[8] K. H. Lloyd, C. H. Low, B. J. McAvaney, D. Rees and R. G. Roper, Planet. Space Sci. **20**, 761 (1972).
[9] S.-I. Akasofu, Space Research VIII, 213 (1968).
[10] J. A. Fedder and P. M. Banks, J. Geophys. Res. **77**, 2328 (1972).
[11] D. Rees, Planet. Space Sci. **19**, 233 (1971).

Space Research XIII — Akademie-Verlag, Berlin 1973

SUBAURORAL X-RAYS AND CONJUGATE ELECTROJETS

J. P. TREILHOU[a], G. GASSET[a] and I. ZHULIN[b]

[a]Centre d'Etude Spatiale des Rayonnements, Université Paul Sabatier, Toulouse, France
[b]IZMIRAN, Moscow, USSR

Two balloon X-ray experiments have been launched from a subauroral region ($L \sim 3.8$) in the northern hemisphere, on 11 March 1969 and 8 March 1970. During the 11 March 1969 flight a decrease in the e-folding energy was found to be correlated with the local electrojet, but anti-correlated with the conjugate electrojet. During both substorms microbursts were recorded; they disappeared when the electrojet moved over the detector.

1. Introduction

The purpose of the present paper is to present bremsstrahlung X-ray measurements over subauroral latitudes ($L \sim 3.8$). X-ray events analysed by balloon techniques in the energy range 15—150 keV are investigated during two disturbed days: 1969 March 11 ($Kp = 6$) and 8 March 1970 ($Kp = 9$). Balloon flights were carried out from the Arkhangelsk region (USSR), in relation with many auroral studies performed by Russian ground stations.

2. Instruments and Flight Conditions

The X-ray detectors consisted of NaI (Tl) scintillation counters, 7.6 cm in diameter and 0.5 cm thick, collimated to provide a geometrical factor of 15.12 cm^2 sr in order to view a circle of 43 km in diameter at the 100 km stopping height for electrons. Energy spectrum information was obtained by use of a pulse height analyser and fast time response analog rate meters were also employed, giving the possibility of investigating the structure of microbursts.

In Fig. 1 are plotted the trajectories of the balloon flights. The stations of Karpogory and Sogra are also shown: it is important to note that Sogra is the conjugate point of Port aux Français (Kerguelen) during disturbed conditions [1]. The distance between Sogra and Karpogory is about 160 km.

For the 11 March 1969 flight the balloon reached a ceiling altitude of 3.5 millibars at 2257 UT and was carried north with a speed of 200 km h^{-1}. During this time, magnetic bays of 230 γ at 2300 UT and 165 γ at 2335 were recorded at Karpogory. The balloon was located in the early morning sector in local time, travelling under multiple auroral structures and bands which are known to occur usually in this sector.

For the 8 March 1970 flight the balloon reached a ceiling altitude of 3 millibars at 1910 UT and was following approximately a geomagnetic parallel. The sector of investigation here is the mignight sector, where magnetic bays were as strong as 2000 γ.

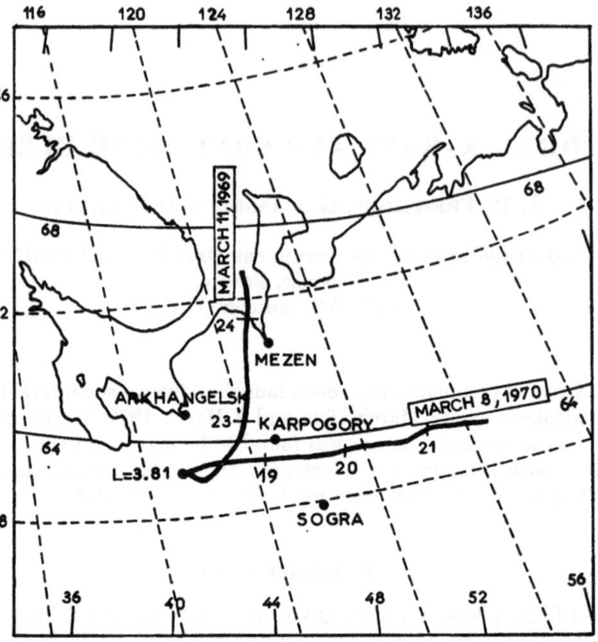

Fig. 1. Positions of balloon trajectories during the 11 March 1969 and 8 March 1970 flights. Dotted lines correspond to corrected geomagnetic coordinates.

3. Event of 11 March 1969

In Fig. 2, the thirty-second averaged X-ray temporal profile is plotted and the mean tendency of the precipitation is outlined by a dotted line. Many regular bursts can be seen, with duration of about 5 minutes. They are believed to be due to the multiple bands encountered successively by the detector during its motion towards north. If so, the spacing between neighbouring auroral structures would be approximately equal to 17 km. A clear decrease of the e-folding enery E_0, of about 8 keV, appears between 2315 and 2330 UT.

We have also computed the ratios $\Delta Z/\Delta H_t$ and I/h for both near-conjugate stations of Karpogory (full lines) and Port-aux-Français (Kerguelen Islands) (dotted lines), to be able to discuss either motions of electrojets or intensity variations. ΔZ and ΔH_t are respectively the magnetic components, in the radial direction and in the horizontal direction, created by the electrojet assumed to be a line of current at the altitude h and with intensity I. After 2315 the intensity variation of the electrojet at Karpogory agrees well with the mean profile of the precipitation; however, these are anti-correlated with the temporal behaviour of the conjugate electrojet during the time interval where the precipitation shows a decrease in the e-folding energy.

Both electrojets being situated polewards of the conjugate stations (as revealed by the sense of the ΔZ variations in Fig. 4), a decrease of $\Delta Z/\Delta H_t$ at Karpogory would correspond to an equatorwards motion of the northern electrojet whereas an increase of $\Delta Z/\Delta H_t$ at Port-aux-Français would correspond to a polewards motion of the southern electrojet. Such motions would be produced by an east—

west ionospheric electric field in the northern hemisphere and by a west—east ionospheric electric field in the southern hemisphere. The oppositely directed electric fields so obtained are in contradiction with a mapping of an east—west electric field from the northern hemisphere to the southern one along a field line supposed to be equipotential.

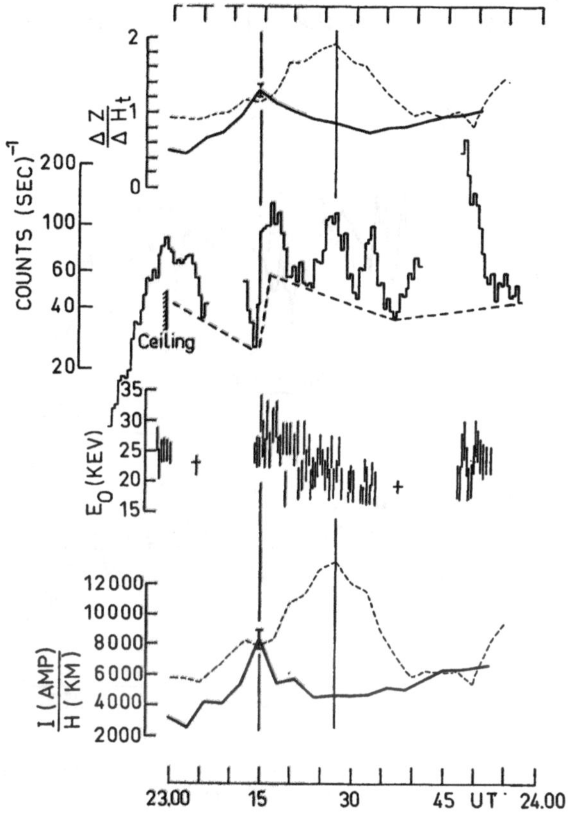

Fig. 2. Temporal characteristics of the X-ray precipitation for the 11 March 1969 event. Also shown are the ratios $\Delta Z/\Delta H_t$ and I/h for Karpogory (full lines) and for Port-aux-Français (Kerguelen) (dotted lines).

So the fact that I/h diminishes near Karpogory whereas it increases near Kerguelen suggests the arrival of different particles in the two hemispheres after 2315 UT. This out-of-phase behaviour of conjugate electrojets is in good agreement with Hargreaves' results using conjugate riometer records [2]. In Fig. 3 the ratios I/h and $\Delta Z/\Delta H_t$ have been plotted during the entire substorm of 11 March 1969. It can be seen that in-phase relations exist at some times, for example at 1905 UT and some minutes after, whereas at other times completely antiphase structures occur. Certainly, we must not ignore a possible coastal effect, principally on the Z component of Kerguelen (Fig. 4). Nevertheless, a new intensification of the electrojet nearer or further from the coast of Kerguelen, in a brief time (let us say 15 or 40 minutes), can also explain the changing from in-phase to out-of-phase in the behaviour of the conjugate electrojets detected by magnetometers. And indeed such intensifications have been shown to occur usually in the development of the polar electrojet [3].

As regards the variation in the e-folding energy between 2315 and 2330, we believe that the detector sees an intensification of the electrojet at 2315 and then moves away from this intensified part of the electrojet. New computations performed by Saint-Marc [4] using Monte-Carlo methods in order to take into account Compton scattering show that when the X-ray detector moves away

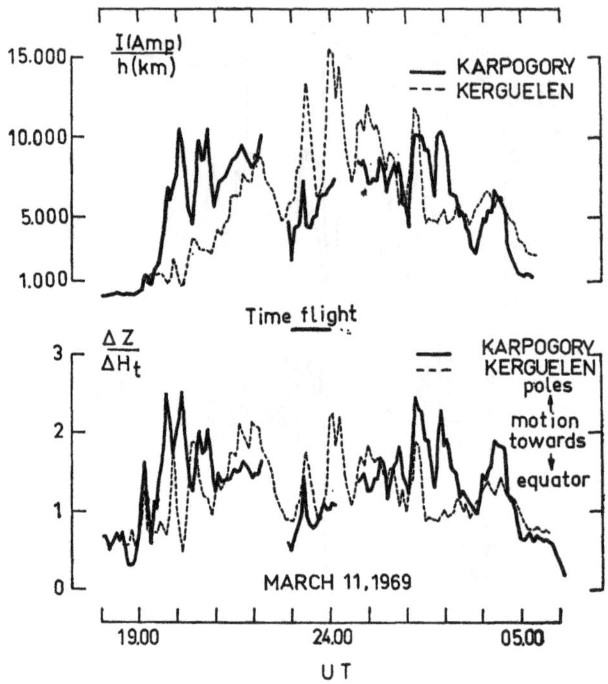

Fig. 3. Temporal behaviour of conjugate electrojets during the whole substorm of 11 March 1969.

from the source, Compton scattering can become preponderant and affect the low energy part of the spectrum, producing an apparent decrease in the e-folding energy.

4. Event of 8 March 1970

The time—intensity profile of the X-ray precipitation is shown in the middle part of Fig. 5 for the event of 8 March 1970. Many papers have been published about this strong event [5—8] as regards its interrelation with other geophysical phenomena. In spite of the fact that the trajectory of the balloon was situated along a geomagnetic parallel, the magnetic storm was so large that the auroral oval reached very low latitudes (as low as Moscow). We can see, on the $\Delta Z/\Delta H_t$ curve of Fig. 5, the changing position of the electrojet from north to south. At 2050 UT, the electrojet is completely south of Arkhangelsk. This point is of crucial importance for the microburst aspects to be discussed now.

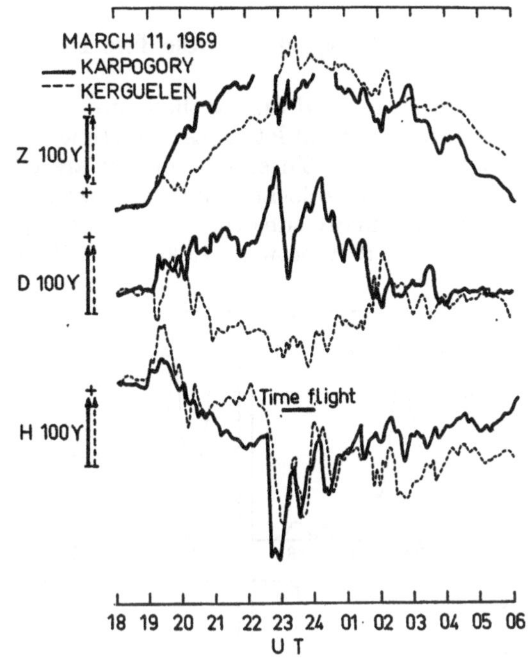

Fig. 4. Temporal variations of the magnetic components of the conjugate stations during the whole substorm of 11 March 1969.

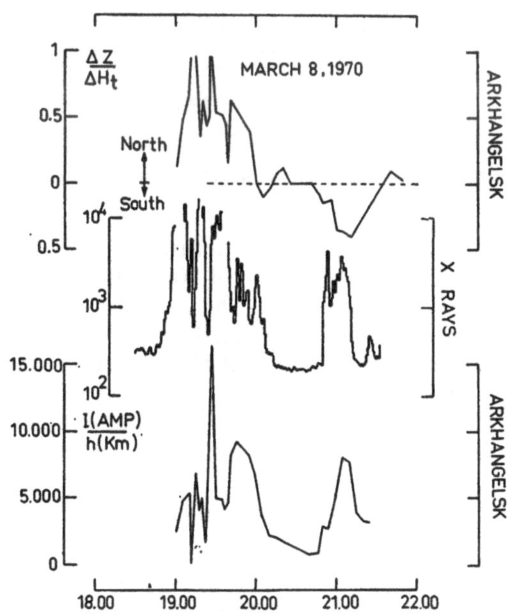

Fig. 5. Temporal characteristics of the X-ray precipitation (in counts s^{-1}) for the 8 March 1970 event and behaviour of the associated electrojet.

5. Microburst Aspects of these Two Events

Another interesting feature noticed during these two disturbed days was the presence of microbursts at subauroral latitudes, sometimes of very great amplitude in the case of 8 March 1970. These microbursts are observed during the course of magnetic bays and around local midnight. They are seen during all the flight of 11 March 1969, with many substructures, and are present only during a part of the flight of 8 March 1970 between 1900 and 2010 UT. During this last

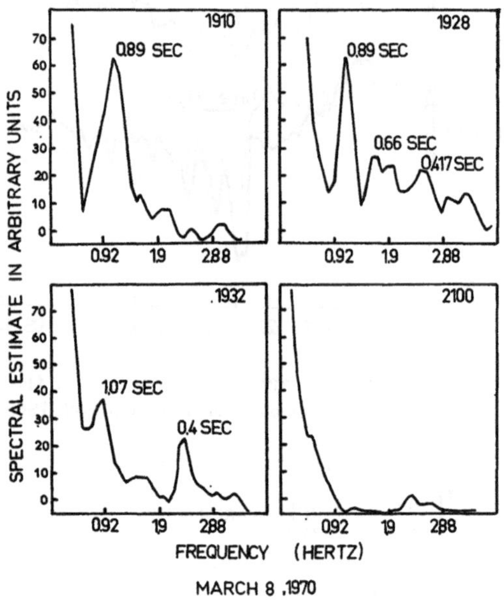

Fig. 6. Spectral analysis of the microbursts at different times during the 8 March 1970 flight.

flight, 335 examples of isolated and well structured microbursts have been selected and gave a half-amplitude mean duration of 200 milliseconds. But after 1950 UT no microbursts at all are detected until the end of the flight; this can be seen from the spectral analysis presented in Fig. 6, which shows predominant periods during the flight of 8 March 1970. The fact that microbursts disappear when the auroral electrojet travels over the detector towards the south strongly supports the suggestion of the key role played by the most active part of aurorae in controlling microburst production.

6. Conclusion

As a conclusion, the new results presented in this paper show how interesting are X-ray measurements in subauroral regions. Simultaneous conjugate measurements of stratospheric electric fields and X-rays by balloon, and of the magnetic effects of electrojets by ground magnetometers, would be of invaluable help in understanding magnetospheric processes.

Acknowledgments

We are grateful to Professor F. Cambou for his first suggestion of subauroral studies and constant encouragement; we thank Dr. Schlich and Dr. Kanonidi for kindly providing the magnetograms for Kerguelen and Karpogory.

References

[1] Kh. D. Kanonidi, CNFRA Report, No. 25, p. 31 (1970).
[2] J. K. Hargreaves, Planet. Space Sci. 19, 513 (1971).
[3] J. L. Kisabeth and G. Rostoker, J. Geophys. Res. 76, 6815 (1971).
[4] A. Saint-Marc, private communication, 1972.
[5] G. Gasset, I. Zhulin (Jouline), F. Cambou, Kh. D. Kanonidi, I. M. Kopylov, A. Saint-Marc and J. P. Treilhou, Geomagn. i Aeronom. 11, 541 (1971).
[6] F. Cambou, G. Gasset, I. Zhulin (Jouline), Kh. Kanonidi, V. Roldugin, A. Saint-Marc, G. Starkov, J. P. Treilhou, E. Vasilkova and L. Yevlashin, Annls Géophys. 28, n° 2 (1972).
[7] F. Cambou, G. Gasset, I. Zhulin (Jouline), Kh. Kanonidi, N. Kleimenova, O. Raspopov, A. Saint-Marc and J. P. Treilhou, Geomagn. i Aeronom., to be published (1972).
[8] G. Gasset, Thèse 3ème cycle, Toulouse, 1971.

Space Research XIII — Akademie-Verlag, Berlin 1973

THE DISTRIBUTION FUNCTION OF ENERGETIC ELECTRONS MEASURED DURING A SLOWLY VARYING IONOSPHERIC ABSORPTION EVENT

K. Wilhelm

Max-Planck-Institut für Aeronomie, Institut für Stratosphärenphysik, Lindau, FRG

Electron fluxes in the energy range from 1 to more than 90 keV have been measured by experiments on board two sounding rockets launched from ESRange, Kiruna, during a slowly varying absorption event in the morning of 27 November 1969. The time resolution of the measurement was 50 ms. Flux variations often occurred on a time scale of 0.5 s with a burst-like appearance. It was found to be a characteristic feature of these bursts that the electron energy spectrum hardened with increasing intensity and that this process was accompanied by an energy dependent decrease of the anisotropy of the flux. Detailed studies of the time history of the bursts revealed that the energy spectra were considerably harder during their growth phase compared with those during their recovery phase. The observations strongly support the assumption that a precipitation mechanism was operating intermittently on the field line of observation.

1. Introduction

Electron precipitation occurring in the morning sector of the auroral zone during times of magnetospheric substorms gives rise to characteristic ionospheric absorption events of the cosmic radio noise, commonly called slowly varying absorption (SVA) events [1—5]. Although this name rightly suggests a smooth temporal amplitude variation of the amount of absorption, high time-resolution observations by means of balloon-borne X-ray experiments and electron detectors on board sounding rockets and satellites have conclusively demonstrated that fast variations of the electron flux on a time scale of less than a second are a predominant feature of SVA events [6—11]. The purpose of this study is the analysis of these microburst phenomena with a view to describing the flux of energetic electrons as a function of time, energy and pitch angle.

Measurements were made during two sounding rocket flights (ESRO Payload No. C-62/1 and C-62/2) that were launched into a single SVA event from ESRange, Kiruna (geomagnetic latitude 65° N), on 27 November 1969 at 0529 and 0631 UT respectively. Both rockets reached approximately 160 km altitude. In an earlier paper [12], the electron flux measurements obtained were used to elaborate on the differences in the electron energy parameters during the two flights. Time averages over 1 second were adequate to achieve that goal, whereas the full time resolution of the measurements will be utilized for this investigation.

In the light of present theoretical models on the generation of discrete very low frequency (VLF) emissions through a coherent cyclotron resonance interaction with energetic electrons [13—16], it is tempting to assume that the microburst activity represents the precipitated particle aspect of such a process.

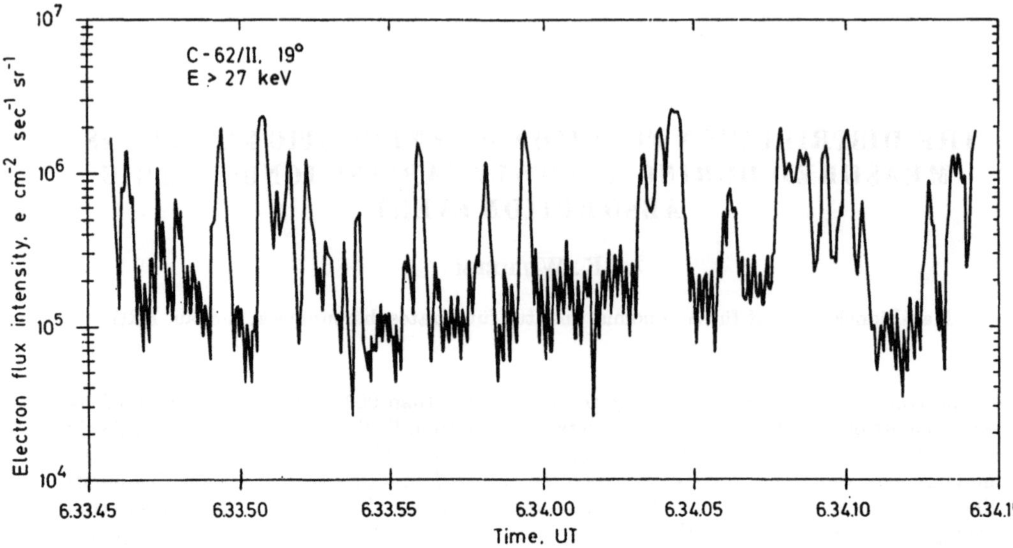

Fig. 1. Flux intensity variations of electrons with energies $E > 27$ keV during the second flight near apogee.

2. Description of the Experiments

Both rockets were equipped with the same set of experiments. Only data from electron detectors in the energy range from 1 to more than 90 keV will be used in this paper. Measurements in the range above 25 keV were performed by plastic scintillation counters mounted under 19° and 60° with respect to the rocket spin axis. A description of this experiment is given by Petrou [17]. The information on counting rates in four integral energy channels per viewing direction was sampled every 50 ms and transmitted to the ground by a PCM system. Reference will also be made to a low energy electron experiment in the range from 1 to 13 keV that employed magnetic energy analysers and continuous channel electron multipliers as detection devices [18].

3. Observations

3.1. Temporal Variations of the Electron Flux

As can be seen from Fig. 1, rapid temporal variations of the electron flux were superimposed on a more slowly varying background flux. Increases normally reached up to 2×10^6 cm^{-2} s^{-1} sr^{-1} for electrons with energies $E > 27$ keV thereby surpassing the background flux by more than an order of magnitude. The time interval depicted in Fig. 1 exhibits the typical behaviour of flux increases at low pitch angles for the second flight and the second part of the first flight. High pitch angle fluxes and those with energies of 1—13 keV were much less structured in time. For this analysis 15 isolated bursts were selected and studied in detail with a view to establishing the distribution function that could be used to describe the energetic electron flux in excess of the background flux.

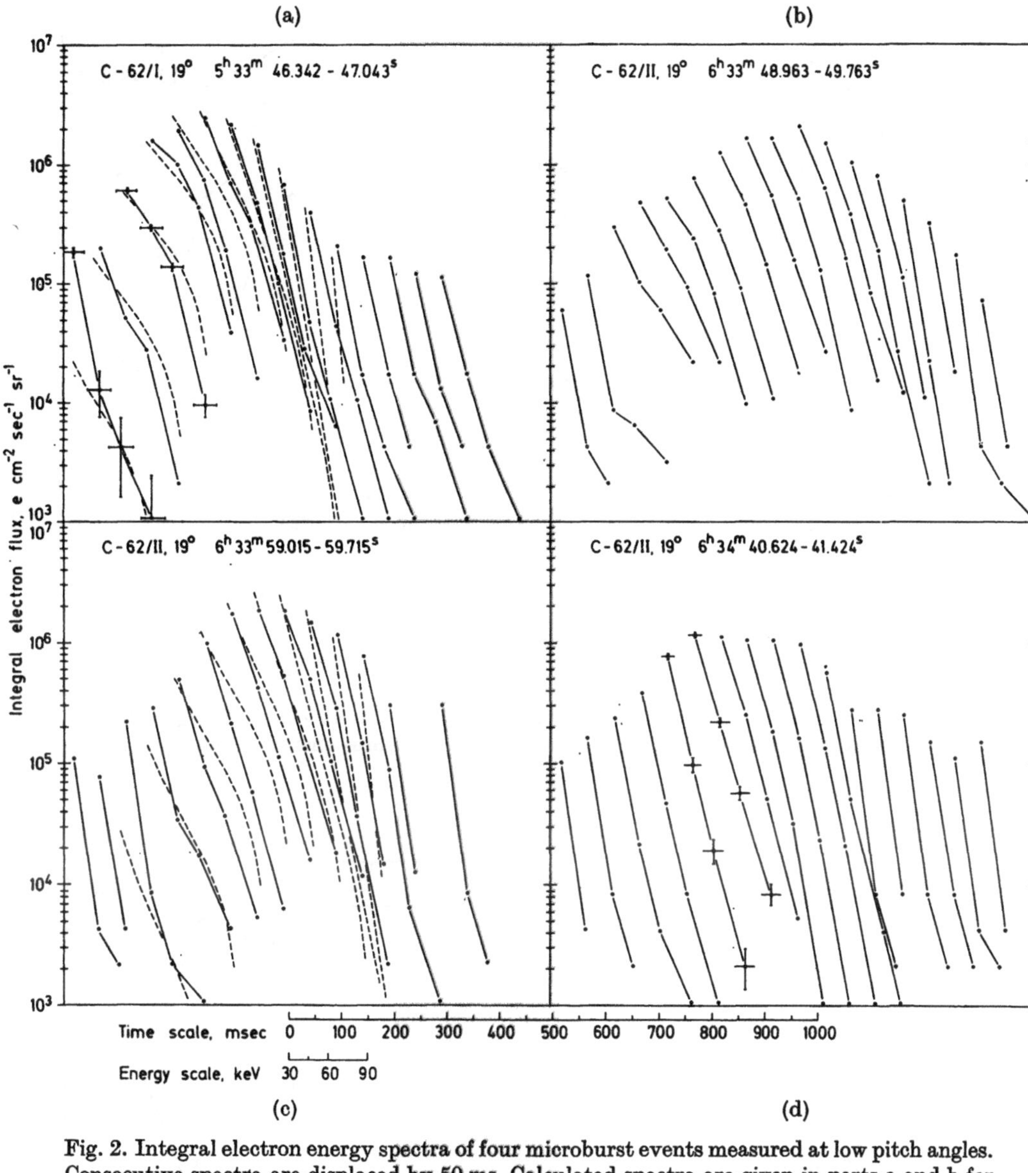

Fig. 2. Integral electron energy spectra of four microburst events measured at low pitch angles. Consecutive spectra are displaced by 50 ms. Calculated spectra are given in parts a and b for comparison.

3.2. Energy Spectra

The temporal evolution of the electron energy spectra was studied by plotting the integral spectra with a time resolution of 50 ms. Four representative examples out of the 15 events studied are shown in Fig. 2. The pitch angle range in which these measurements were performed extended roughly from 0° to 40°. The corre-

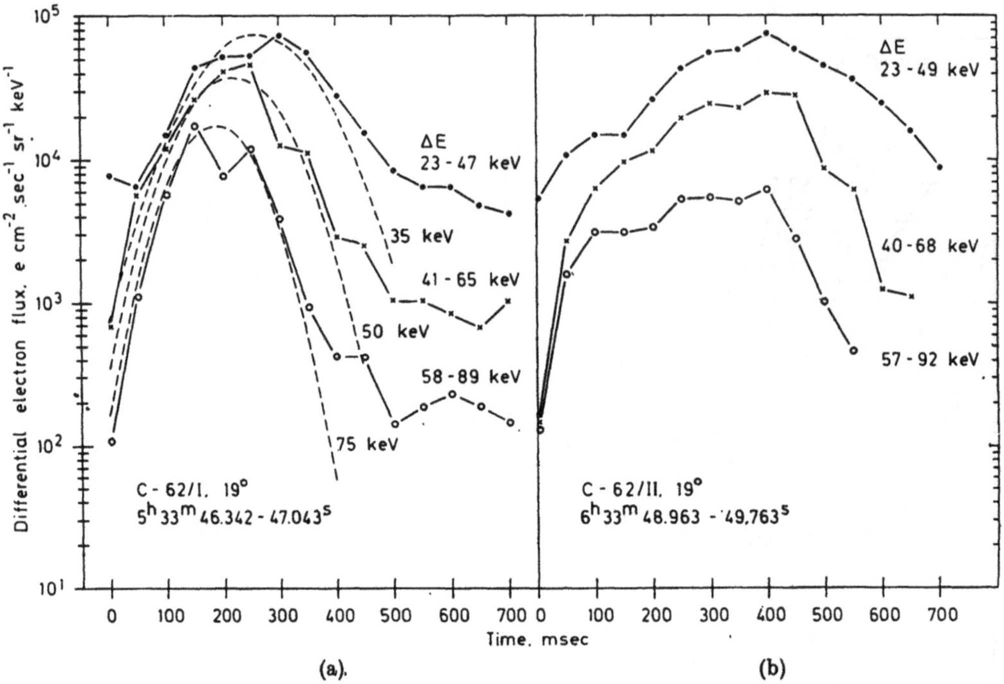

Fig. 3. Two microburst events as seen in different energy bands at low pitch angles. Broken lines in part a show $f_1(E, \alpha, t)$ for $\alpha = 20°$ and three energy values.

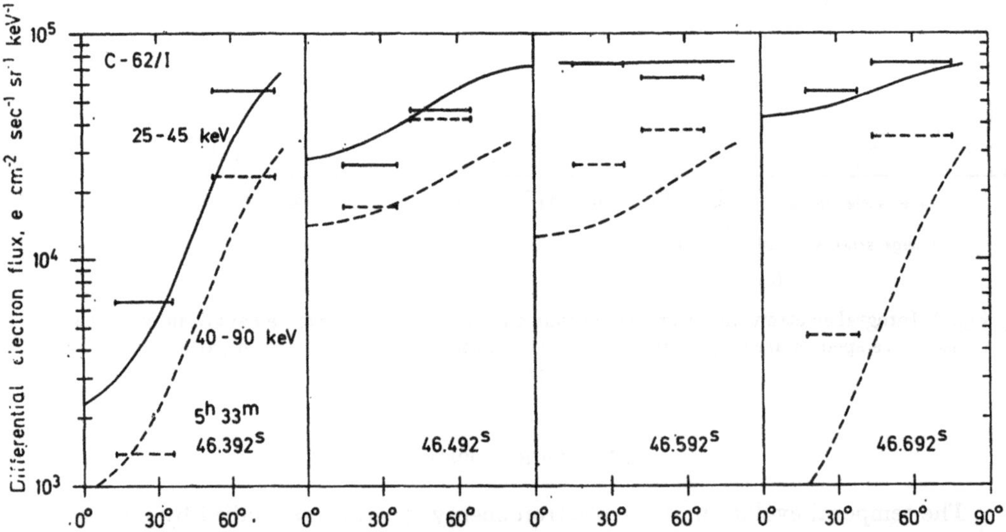

Fig. 4. Evolution of the pitch angle distribution during a microburst event as measured and calculated from $f_1(E, \alpha, t)$ for two different energy ranges at four values of t.

sponding spectra in the high pitch angle range were observed to be considerably flatter.

The following points should be noticed: (i) energy spectra before and after microburst events are comparable with each other; (ii) all events commence with flux increases in the high energy channels; (iii) the spectra flatten considerably with increasing flux intensity; (iv) at the maxima of the bursts or shortly after, the spectra steepen again, thus leading to an asymmetric distribution of spectral steepness versus flux intensity. Two microbursts are presented in Fig. 3 in different energy ranges in order to demonstrate more clearly that high energy channels reach their maxima prior to low energy ones.

3.3. Angular Distribution

The orientation of the spin axis of the rocket with respect to the magnetic field direction during the event of 05 h 33 m 46—47 s (Fig. 2a) was such as to permit flux measurements with high time resolution in two pitch angle ranges. It follows from Fig. 4 that before and after the microburst event a strong anisotropy prevails with peak fluxes at high pitch angles. During the rise time of the event, the flux of all energies became more isotropic in the pitch angle range under consideration. Shortly after the maximum, however, high and low energy fluxes behaved quite differently insofar as the low energy flux stayed isotropic much longer than the high energy part. It should be added that no anisotropy was observed for very low energy particles.

4. Distribution Function

With the particular event of Fig. 2a as point of departure, an attempt was made to find a description of the distribution of the electron flux as a function of energy E, pitch angle α and time t. As no information on the spatial extent can be derived from these measurements, it was assumed that the dimensions of typical microburst precipitation were larger than the distance the rocket travelled perpendicular to the magnetic field direction during a single event. It was found that the event in question could be approximated by

$$f_1(E, \alpha, t) = 7.5 \times 10^4 \left(\frac{35\ \text{keV}}{E}\right)^{1+\cos\alpha} \exp\left\{-\left(\frac{t - t_M}{\tau}\right)^2\right\} (\text{e cm}^{-2}\ \text{s}^{-1}\ \text{sr}^{-1}\ \text{keV}^{-1})$$

with

$$\tau = \frac{t_M}{2.3 \cos\alpha} \quad \text{and} \quad t_M = \frac{0.62\, T_m}{4}$$

where T_m is the period of the mirror oscillation of electrons as given by Hamlin et al. [19] and calculated for Kiruna ($L = 5.5$).

In order to demonstrate the degree of correspondence, $f_1(E, \alpha, t)$ was also plotted in Figs. 3 and 4 for suitable values of the variables. For a comparison with integral energy spectra such as those of Fig. 2 the integral over energy

$$F_1(E, \alpha, t) = \int_E^{90} f_1(\eta, \alpha, t)\, d\eta$$

was evaluated numerically and plotted in Fig. 2a for $\alpha = 20°$. It should be pointed out that the correspondence critically depends on the integration limit of 90 keV. At angles near 60° the approximation could be significantly improved by integrating up to 120 keV. By choosing the energy exponent as $2 + \cos\alpha$ and without any further modification, the function $f_2(E, \alpha, t)$ describes reasonably well the microburst events that occurred during the second flight as can be seen from Fig. 2b, c and d.

5. Discussion

Before entering the discussion on possible physical interpretations of some of the characteristics of $f(E, \alpha, t)$, it is worth mentioning the limitations of the approximation; these are: (i) the background flux was not taken into account; (ii) the time history of the events at different energies was approximated by symmetric Gaussian functions, although there is a tendency for those events to have slightly longer fall than rise times; (iii) indications were neglected that the duration of microbursts was on the whole longer during the second flight; (iv) any variation in the flux level for different events was not incorporated. For constant E and α $f(E, \alpha, t)$ attains a maximum at $t = t_M$. The time t_M depends only on E in such a way that the arrival times of electrons with different energies can be explained by energy dispersion. If electrons of all energies were to start at the same time and location, this point would have to lie approximately 25° north of the geomagnetic equator on the field line of observation. Both time and location of the process that precipitates a given particle may, however, depend on its energy.

The width of the microbursts depends both on the range of energy and pitch angle in which the event is observed. Regarding the energy dependence, one has to remember that the measurements were made in rather wide energy bands. Calculations show that only one half of the width in time could be attributed to energy dispersion. It must, therefore, be concluded that at least a part of the energy dependent width was a consequence of the generation process. Similarly, the weak dependence of travel times on the mirror points of electrons [19] excludes the possibility of explaining the influence of the pitch angle as a propagation rather than a generation effect.

In addition to the Gaussian term, there is a term representing the general shape of the energy spectrum. Although details of this term might still be subject to modifications when more data become available, it can be stated that the energy spectra were flatter during the first flight and at higher pitch angles. Wilhelm et al. [12] and Stüdemann [20] obtained similar results for this and another SVA event by using time averaged data.

The integral energy spectra could only be reasonably well approximated by assuming an upper integration limit near 100 keV. It thus appears as if electrons of more than 100 keV energy do not significantly contribute to the flux of precipitated particles during these events. On the other hand, very low energy particle fluxes observed during the flights were by a factor of 5—10 smaller than an extrapolation of $f(E, \alpha, t)$ would yield.

Acknowledgments

The author thanks Messrs. J. P. Legrand, N. Petrou, G. Pfotzer, G. Kremser, J. Münch and W. Stüdemann both for their cooperation during the experiment development phase and for discussions on the observations.

The experiments were supported by the German Bundesministerium für Bildung und Wissenschaft under Grant WRK 163 and the French Centre National d'Etudes Spatiales.

The payloads were integrated and launched under the responsibility of the European Space Research Organisation.

References

[1] R. R. Brown, J. Geophys. Res. **69**, 2315 (1964).
[2] Z. A. Ansari, J. Geophys. Res. **70**, 3117 (1965).
[3] A. Bewersdorff, G. Kremser, W. Riedler and J. P. Legrand, Arkiv Geofys. **5**, 115 (1966).
[4] D. Jelly and N. Brice, J. Geophys. Res. **72**, 5919 (1967).
[5] A. Bewersdorff, G. Kremser, J. Stadsnes, H. Trefall and S. L. Ullaland, J. Atmos. Terr. Phys. **30**, 591 (1968).
[6] K. A. Anderson and D. W. Milton, J. Geophys. Res. **69**, 4457 (1964).
[7] K. A. Anderson, L. M. Chase, H. S. Hudson, M. Lampton, D. W. Milton and G. K. Parks, J. Geophys. Res. **71**, 4617 (1966).
[8] M. Lampton, J. Geophys. Res. **72**, 5817 (1967).
[9] L. M. Chase, J. Geophys. Res. **73**, 3469 (1968).
[10] D. Venkatesan, M. N. Oliven, P. J. Edwards, K. G. McCracken and M. Steinbock, J. Geophys. Res. **73**, 2333 (1968).
[11] M. N. Oliven and D. A. Gurnett, J. Geophys. Res. **73**, 2355 (1968).
[12] K. Wilhelm, G. Kremser, J. Münch, M. Schnell, J. P. Legrand, N. Petrou and W. Riedler, Space Research XII, 1437 (1972).
[13] M. N. Brice, J. Geophys. Res. **68**, 4626 (1963).
[14] R. A. Helliwell, J. Geophys. Res. **72**, 4773 (1967).
[15] R. A. Helliwell, Rev. Geophys. **7**, 281 (1969).
]16] T. J. Rosenberg, R. A. Helliwell and J. P. Katsufrakis, J. Geophys. Res. **76**, 8445 (1971).
[17] N. Petrou, Thèse, Conservatoire National des Arts et Métiers, Paris 1972.
[18] H. Schütz, M. Schnell and K. Wilhelm, Mitt. a. d. Max-Planck-Institut für Aeronomie **40**, 71 (1970).
[19] D. A. Hamlin, R. Karplus, R. C. Vilz and K. M. Watson, J. Geophys. Res. **66**, 1 (1961).
[20] W. Stüdemann, Mitt. a. d. Max-Planck-Institut für Aeronomie **111**, 49, 1 (1972).

Space Research XIII — Akademie-Verlag, Berlin 1973

ROCKET MEASUREMENTS OF ELECTRON CONCENTRATION AND ELECTRON TEMPERATURE IN THE POLAR IONOSPHERE

N. M. Klyueva, V. G. Khryukin, Yu. K. Chasovitin, G. P. Komrakov and I. V. Popkov

USSR

In order to investigate the electron density and electron temperature in the polar ionosphere up to 185 km four MR-12 rocket experiments were carried out from Heiss Island in March—April 1971. The electron density N_e was determined with a high-frequency impedance probe and the electron temperature was measured using a cylindrical Langmuir probe. In the experiment carried out on 29 March the electron density changed gradually with height. In three other experiments (on 14, 28 March and 2 April) the N_e profiles are characterized by a number of maxima and minima. On 28 March a distinct electron density maximum was registered at 106 km, with $N_{e(max)} = 1.0 \times 10^5$ cm^{-3}, which coincided with an Es layer of type f observed on the ionogram.

Four RM-12 [1] rocket experiments to measure electron density N_e and electron temperature T_e in the polar ionosphere were carried out in March—April 1971 from Heiss Island (geographical coordinates 80°37′ N 58°03′ E; geomagnetic coordinates 71° N, 155° E). Heiss Island is within the auroral zone but close to midnight hours it is situated in the polar cap [2]. The basic information on the rocket experiments is given in Table 1.

Table 1

Rocket experiments on N_e and T_e measurements

Launch No.	Date 1971	Time		Solar zenith angle (χ)	Magnetic activity at moment of launch				Max. height of rocket ascent (km)
		UT	LMT		Kp	ΣKp	Heiss Island K	Heiss Island ΣK	
1	14 March	1626	2018	97°	3+	32	7	41	158
2	28 March	1942	2334	96°	1	4	3	19	183
3	29 March	0845	1237	77°	1+	4—	2	14	183
4	2 April	2240 1 April	0232	93°	0+	13	3	29	186

Systematic magnetic, ionospheric and riometric measurements were carried out by members of the observatory under the guidance of Geophysical Department of the Arctic and Antarctic Science Research Institute (AASRI).*

The electron temperature was measured using a cylindrical Langmuir probe and electron density was determined with a high-frequency impedance probe. A

* The authors wish to thank the Geophysical Department of AASRI for guidance and scientific workers of the observatory for ground geophysical data kindly given to them.

detailed description of the measurement procedure and equipment used is given in [3, 4]. The accuracy of measurements was about 15% for electron density and ±100 °K for electron temperature. Electron density profiles in the 70–185 km region obtained during these experiments are given in Fig. 1. The first flight, carried out during significant geomagnetic disturbance showed a complex electron density profile (Fig. 1, curve 1). For the second flight, carried out in quieter conditions, the electron density values (Fig. 1, curve 2) at heights greater than 85 km are essentially lower than in the first experiment although both flights were carried out at similar solar zenith angles (97° and 96° respectively). Such an N_e

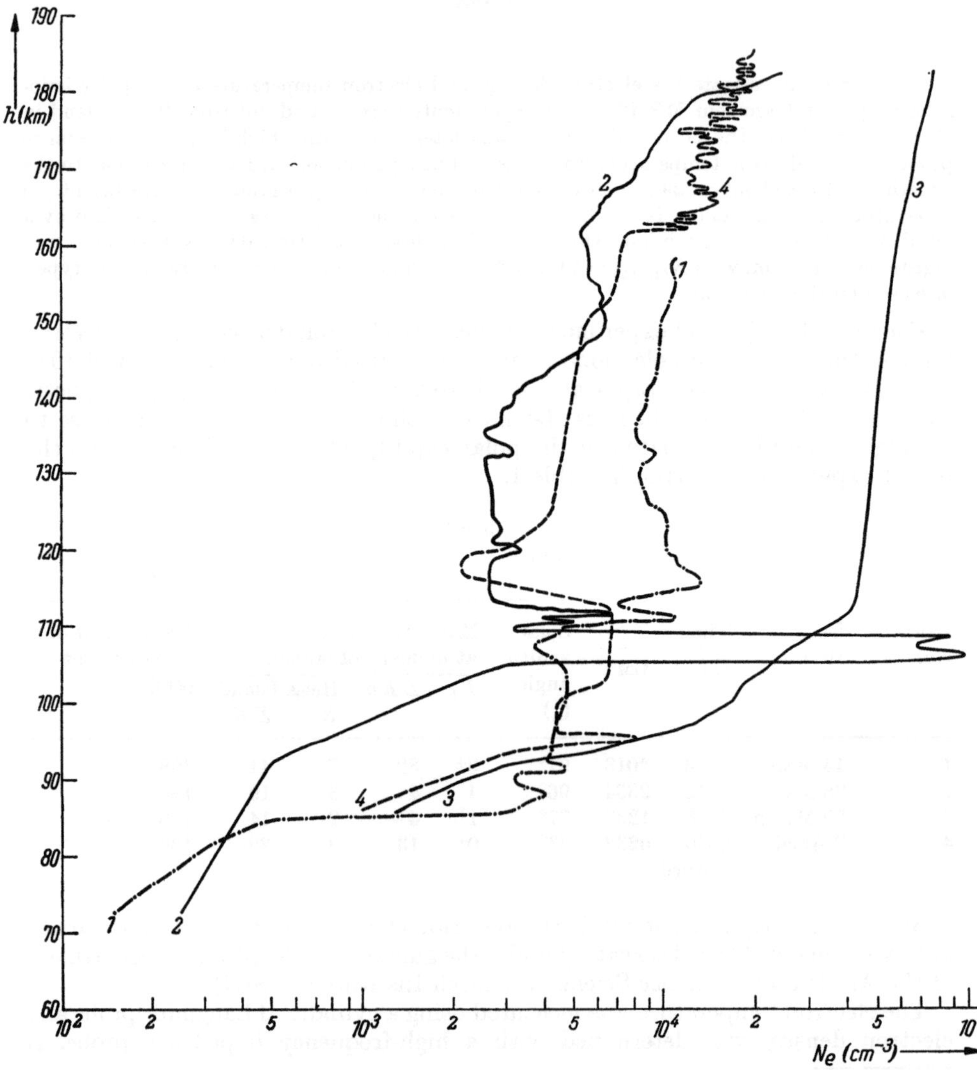

Fig. 1. Electron density profiles, Heiss Island, 1971: 1, 14 March, 2018 LMT; 2, 28 March, 2334 LMT; 3, 29 March, 1237 LMT; 4, 2 April, 0232 LMT.

increase during the magnetic disturbance is related to the corpuscular ionizing agent influencing the ionosphere [5]. At the same time an increased absorption of cosmic noise was registered by the 30-MHz riometer. This absorption was equal to 0.7—0.9 dB with a median value of 0.2—0.3 dB. In the second flight a well developed electron density maximum was recorded at 106 km with $N_{e(max)} = 1.0 \times 10^5$ cm^{-3}, which corresponds to an Es layer of type f registered on the ionogram taken during the flight. This sporadic E layer thickness at half the electron density is equal to about 2.5 km. N_e values for this layer calculated according to maximum frequency and blanketing frequency taken from the appropriate ionogram are 2.1×10^5 cm^{-3} and 1.1×10^5 cm^{-3} respectively; from all this it is seen that the blanketing frequency better characterizes the maximum electron density in the Es layer [6]. It should be noted that the structure of this layer is similar to night sporadic layers registered at mid-latitudes [7, 8].

The third flight was carried out in the daytime under quiet geomagnetic conditions. The electron density during this flight increased monotonically with height (Fig. 1, curve 3). From the comparison of results of the second and third

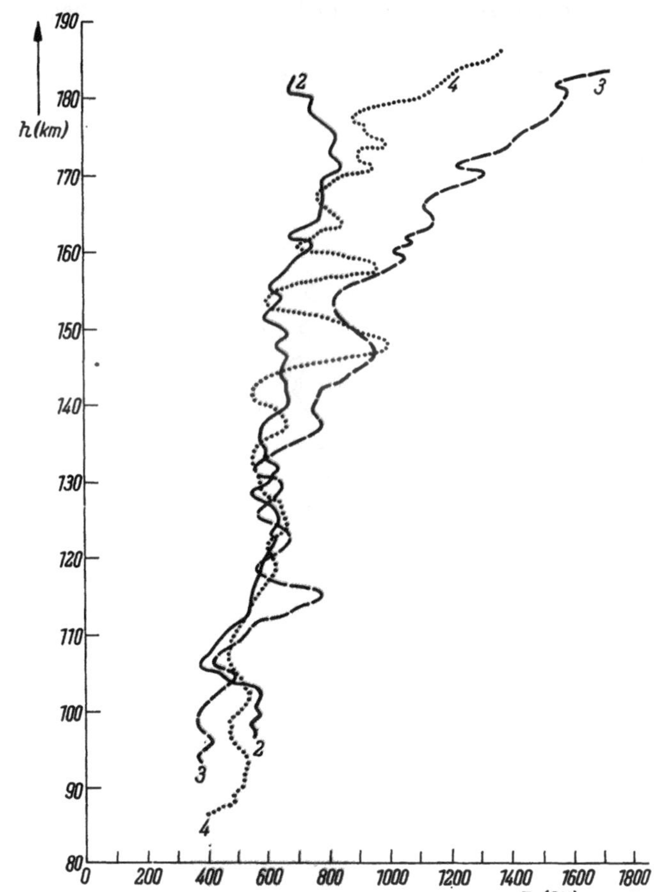

Fig. 2. Electron temperature profiles, Heiss Island, 1971. Curves numbered as in Fig. 1.

flights carried out on one day at different solar zenith angles the daily electron density variations are clearly seen: at heights greater that 100 km the N_e values at noon are almost an order of magnitude greater than those at midnight.

During the fourth flight the electron density variation with height was rather complicated (Fig. 1, curve 4). A thin sporadic layer with an electron density in maximum 8.7×10^3 cm^{-3} was observed at 96 km; the N_e minimum of 2.2×10^3 cm^{-3} was situated at 118 km. Above 160 km weak small-scale irregularities with $\Delta N_e/N_e \sim 10\%$ were registered; the mean dimensions along the vertical were equal to about 1 km. At the same time scattering-type reflections are seen from the ionogram.

From comparison of the profiles it is clear that at high latitudes, along with short-wave ionizing solar radiation, the irregular ionization sources are of great importance, the role of these sources being increased during disturbed periods.

Experimental profiles of electron temperature in the 85—185 km region are given in Figs. 2 and 3. To remove irregular changes caused by measurement errors, the individual T_e values were averaged by the method of running averages over 5—15 separate measurements, which corresponds to height intervals of

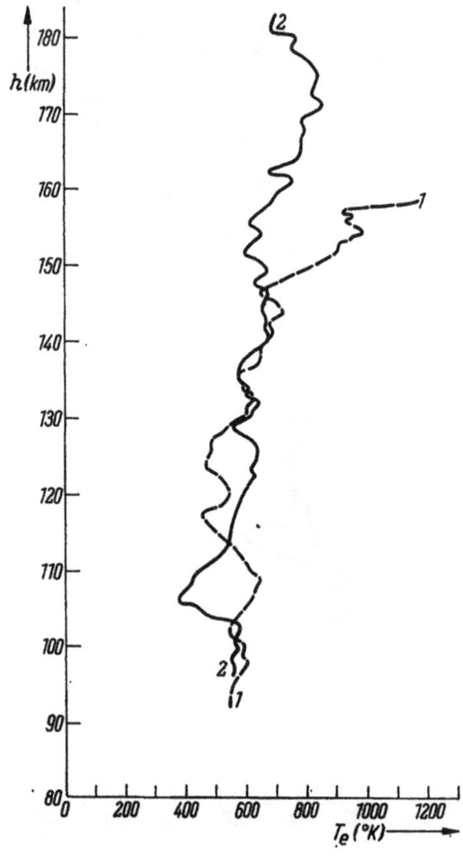

Fig. 3. Electron temperature profiles in magnetically disturbed (1) and magnetically quiet (2) periods, Heiss Island, 1971. Curves numbered as in Fig. 1.

about 4 km in the lower part of the trajectory and about 2 km at its top. In all experiments the electron temperature increased with height, the most rapid increase being observed in the third flight (curve 3), i.e. at the smaller solar zenith angle. In this daytime experiment the electron temperature was much higher than in those carried out at $\chi > 90°$. For the second and third flights carried out during one day the T_e values at 180 km differ by 800 °K.

Fig. 3 illustrates the electron temperature profiles obtained during the magnetically disturbed period (curve 1) and in much quieter conditions (curve 2). It is seen that up to 145 km the electron temperature values in both flights were practically the same. Only above 145 km is a rather distinct T_e increase registered for the disturbed period. It should be noted that electron temperatures recorded during the second and fourth experiments are close to T_e values obtained at mid-latitudes at appropriate solar zenith angles [9, 10].

Irregular structure of the electron temperature profiles, mentioned earlier [11, 12] and evident in these data, needs further experimental checks and theoretical substantiation. It should be noted that in the second flight a noticeable electron temperature minimum was registered at 106 km, i.e. in the region of intense Es which may be connected with the increased heat loss of the electron gas at the N_e increase. It possible to conclude that above 140 km the electron gas heating during the second and fourth experiments was mainly affected by short-wave solar ionizing radiation whereas during the first launch corpuscular fluxes may have given an additional contribution. To explain the electron temperatures registered below 140 km it is necessary here, as well as in mid-latitudes, to consider some additional energy sources; however the question of their origin is not yet settled [11, 13, 14].

References

[1] A. D. Zhlood'ko, S. M. Poloskov, A. A. Pokhunkov, V. P. Teslenko, A. A. Shidlovsky and Yu. F. Ivanov, Meteorol. i Gidrol., No. 3, 72 (1970).

[2] S. Akasofu, Polar and Magnetospheric Substorms, D. Reidel Publ. Co., Dordrecht, Holland 1968.

[3] Yu. B. Burakov, V. I. Styazhkin and Yu. K. Chasovitin, Proc. Inst. Exp. Met., Hydrometeoizdat, Moscow, No. 16, 46 (1970).

[4] G. P. Komrakov, V. P. Ivanov, I. V. Popkov and V. N. Tyukin, Kosm. Issled. 8, 278 (1970).

[5] G. S. Ivanov-Kholodny and G. M. Nikol'sky, The Sun and the Ionosphere, Nauka, Moscow 1969.

[6] C. A. Reddy and M. J. Rao, J. Geophys. Res. **73**, 215 (1968).

[7] L. G. Smith, Radio Sci. **1**, 178 (1966).

[8] L. A. Andreeva, L. A. Katasev, V. P. Nesterov, D. B. Uvarov, Yu. K. Chasovitin and G. P. Komrakov, Space Research XII, 1321 (1972).

[9] N. W. Spencer, L. H. Brace, G. R. Carignan, D. R. Taeusch and H. Niemann, J. Geophys. Res. **70**, 2665 (1965).

[10] L. H. Brace, H. G. Mayer and G. R. Carignan, J. Geophys. Res. **74**, 257 (1969).

[11] K. I. Gringauz, Advanc. Phys. Sci., USSR Acad. Sci. **92**, 207 (1967).

[12] K. Hirao, Space Research VII, 477 (1967).

[13] P. M. Banks, Proc. Inst. Elect. Electron. Engrs **57**, 258 (1969).

[14] J. C. G. Walker, Planet. Space Sci. **16**, 321 (1968).

Space Research XIII — Akademie-Verlag, Berlin 1973

THERMAL POSITIVE IONS IN THE DAYSIDE POLAR CUSP MEASURED ON THE ISIS 1 SATELLITE

M. Ahmed[a] and R. C. Sagalyn[b]

[a]Regis College, Weston, Massachusetts, USA
[b]Air Force Cambridge Research Laboratories, Bedford, Mass., USA

A spherical electrostatic analyzer aboard the polar orbiting ISIS 1 satellite measured thermal positive ions over the altitude range 580 to 3525 km. Simultaneous ISIS 1 measurements of low energy electrons and protons have established that soft particle fluxes observed at high latitudes on the dayside originate in the magnetosheath and enter the ionosphere via the cusp regions. The ionospheric effects resulting from this precipitation are clearly evident in the positive ion measurements which exhibit irregular enhancements in the cusp region. The peak densities, their latitudinal extent and the location of the equatorward and poleward boundaries show a strong dependence on magnetic activity variations, and follow a diurnal pattern. In the noon sector, the location of the equatorward boundary Λ_E correlates better with the solar wind velocities V (correlation coefficient $\sim$0.97) than with the magnetic indices Kp (correlation coefficient $\sim$0.65). The following empirical relation is found by averaging results obtained in the time period 1000—1230 MLT: $\Lambda_E = 85.93 - 0.026\,V$. In the morning sector, centered around $\sim$0630 MLT, Λ_E correlates better with the Kp indices (correlation coefficient 0.95) than with solar wind velocities (correlation coefficient 0.65); averaging these results gives $\Lambda_E = 80.16 - 2.52\,Kp$.

1. Introduction

Magnetospheric models deduced by Fairfield [1] show that the magnetic field lines crossing the earth above invariant latitudes 78° $\pm$ 3° and 68° $\pm$ 2°, in the noon and midnight meridians, respectively, are open and extend into the geomagnetic tail. The observations [2, 3] have demonstrated that there is direct access of magnetosheath protons and electrons to the lower ionospheric regions via the dayside cusp region. The present work shows that the interaction of low energy magnetosheath particles with the ambient gases in the topside ionosphere often produces a significant increase of thermal ions and electrons during magnetically disturbed periods. The magnitude and location of the ionization enhancements is found to vary during magnetic storms and with changes in local time. Under quiet magnetic conditions the ionization produced by the magnetosheath particles is small compared with that produced by solar ultraviolet radiation.

2. Observations

High resolution measurements of thermal positive ions were obtained on the polar orbiting ISIS 1 satellite. The sensor consisted of a Spherical Electrostatic Analyzer (SEA) designed to measure the flux, density, temperature and energy distribution of positive ions in the energy range 0—15 eV by operating the experiment in various electrical configurations. In this report we discuss only the ion density measurements. The ion density was sampled 60 times per second giving an effective spatial resolution of 100—150 meters along the orbit. The ISIS 1 satellite was launched on 31 January 1969 into a near polar orbit with an inclination of 88.4° and an orbital period of 128.3 minutes. The perigee and apogee at launch were at 578 and 3526 km, respectively. In addition to the SEA experiment there were charged particle measurements on board the ISIS 1 satellite [4].

The density measurements used in this study cover the period February—March and 23—26 October 1969. During these periods, at invariant latitudes greater than 65° the satellite orbit traversed the range 0600—1800 MLT. The observational period covered eight magnetic storms with the maximum Kp varying between 5 and 8. The locations and magnitudes of density irregularities at high latitudes on the dayside have been compared with Kp and, when available, with solar wind velocities measured simultaneously aboard the Explorer 33 satellite by the MIT plasma probe.

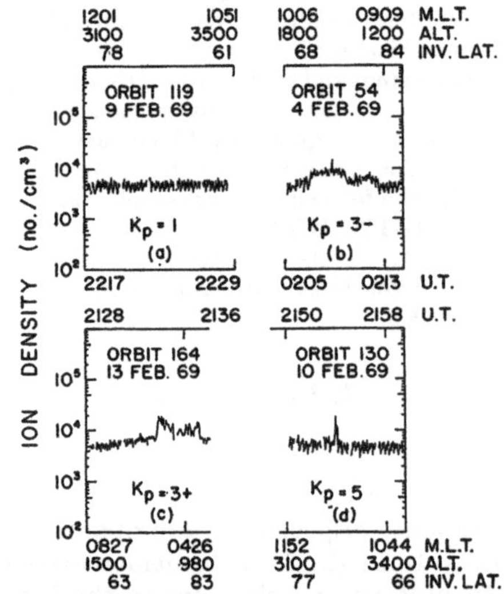

Fig. 1. Some daytime thermal positive ion density profiles at high latitudes from ISIS 1; 4—13 Feb. 1969.

Dayside density profiles for latitudes greater than 65° for various levels of Kp are illustrated in Fig. 1. Fig. 1a shows a density profile obtained for $Kp = 1$. No influence of magnetosheath particles on the ionization profile is observed. Fig. 1b shows the density profile for $Kp = 3^-$. A 60% increase in density above the background is observed over the invariant latitudes 71—80°. Fig. 1c shows the profile over the same latitude range for $Kp = 3^+$. The ionization distribution shows a well defined equatorward and poleward boundary. The profile obtained on orbit 130 (Fig. 1d) shows a sharp increase in density at 71° latitude over a width of less than 1°, during a period when Kp was equal to 5.

Variations in the magnitude of these ionization irregularities with Kp and solar wind velocity V have been studied. Results obtained from the magnetic storm of 2—5 February 1969 in the noon sector are shown in Fig. 2. The magnitude

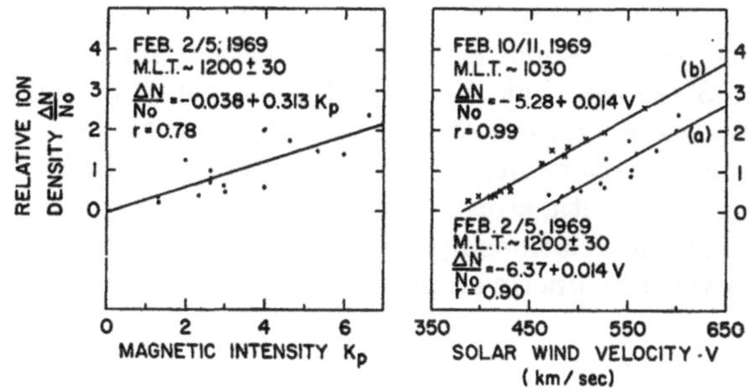

(a) (b)

Fig. 2. Variation in the magnitude of density in the ionization features with Kp and V in the noontime sector from ISIS 1.

of the ionization increase above background is found to increase with increasing magnetic activity, Kp (Fig. 2a) and with increasing V (curve a, Fig. 2b). The correlation coefficient r obtained from these results is found to be higher for solar wind velocity data ($r = 0.90$) than for the magnetic intensity ($r = 0.78$). The correlation analysis with the solar wind data for the storm of 10—11 February, MLT 1030 hrs (curve b, Fig. 2b) gives a correlation coefficient of 0.99. Data from subsequent storms give similar results. Thus relative changes in ion density in the high latitude precipitation zones on the dayside are found to correlate better with solar wind velocity than with Kp.

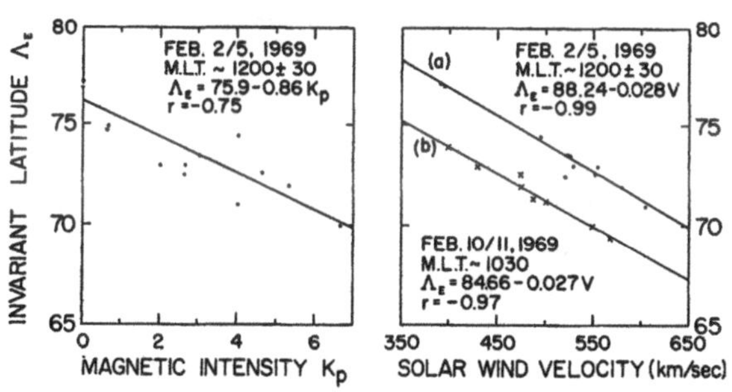

(a) (b)

Fig. 3. Variation in the location of the equatorward boundary Λ_E of the ionization region with Kp and V in the noontime sector from ISIS 1.

The variation of the location of the equatorward boundary of the ionization irregularity region Λ_E with changing magnetic activity and solar wind velocity during two disturbances is shown in Figs. 3 and 4. Λ_E is approximately at 78° invariant latitude for $Kp = 0—1$ as seen in Figs. 3a and 4a. Fairfield [5] gives a similar value for the equatorward boundary for low magnetic activity. With increasing Kp and solar wind velocity, Λ_E moves toward the equator.

In the noontime sector, for the storm of 2—5 February as illustrated by curve a in Fig. 3b the correlation between Λ_E and solar wind velocity ($r = \sim 0.99$) is also found to be higher than that between Λ_E and Kp ($r = \sim 0.75$) in Fig. 3a.

The storm of 10—11 February (curve b, Fig. 3b) shows very close agreement with curve a (with almost equal slopes); the vertical shift of the two curves in Fig. 3b by about 3° shows that Λ_E has a higher value at noon than at earlier times. In the dawn sector, on the other hand, the variations in Λ_E show a higher correlation with the variations in Kp than with variations in V. This is illustrated in the results of the magnetic storm of 21—24 March shown in Figs. 4a and 4b. Table 1 summarizes the relationships between Λ_E and Kp and V, in three local time sectors for several magnetic storms.

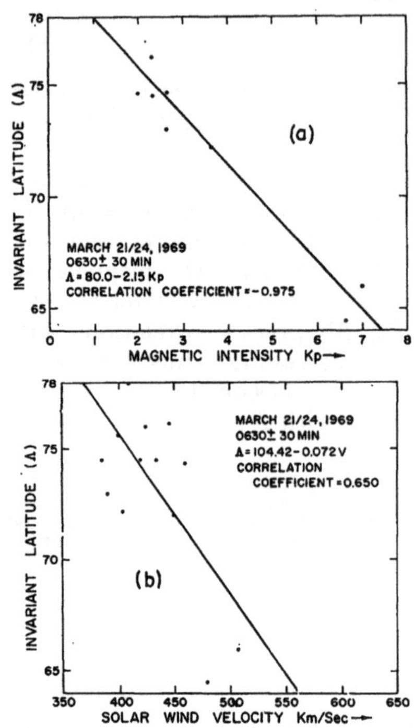

Fig. 4. Variation in the location of the equatorward boundary of the ionization region with Kp and V in the early morning sector from ISIS 1.

Recent observations [6] of the effects of polar cusp precipitation on F layer electron densities made aboard the AFCRL flying ionospheric laboratory are in excellent agreement with the relationship between Λ_E and Kp in the noon and early morning sectors reported earlier by Ahmed and Sagalyn [7] and as illustrated in Figs. 3a and 4a.

The latitudinal extent of ionization irregularities during a disturbed period may be summarized as follows: for $Kp \sim 2$—3 the latitudinal extent varies between 8° and 10° as illustrated in Figs. 1b and 1c. Under conditions of high magnetic activity, $Kp \geqq 5$, the width is much reduced and varies between 0.2° and 1.5° as illustrated in Figs. 1d, 5a and 5b. This corresponds to a distance of 15—40 km along the orbital path. These results for high Kp are similar to results reported for the width of the dayside cusp [8] from the analysis of Injun 5 data.

Table 1

Summary of position of Λ_E in different time sectors

Date (1969)	MLT (~hrs)	Λ_E vs	DAWN r	DAWN Λ_E	NOON r	NOON Λ_E	DUSK r	DUSK Λ_E
2–5 Feb.	1200	Kp			0.75	75.9 − 0.86 Kp		
		V			0.98	88.2 − 0.028 V		
10–11 Feb.	1030	Kp			0.67	74.8 − 0.64 Kp		
		V			0.97	84.6 − 0.027 V		
27–28 Feb	0630	Kp	0.94	82.4 − 3.25 Kp				
		V	0.52	104.4 − 0.072 V				
6–7 Mar.	0630	Kp	0.94	79.1 − 2.22 Kp				
		V	~0.05	74.3 − 0.004 V				
	1930	Kp					0.95	75.3 − 1.97 Kp
		V					0.62	63.9 + 0.012 V
15–18 Mar.	0630	Kp	0.98	81.6 − 3.25 Kp				
		V	0.44	47.4 + 0.057 V				
	1100	Kp			0.52	126.8 − 1.19 Kp		
		V			0.97	85.9 − 0.025 V		
19–20 Mar.	0630	Kp	0.97	80.2 − 2.72 Kp				
		V	0.39	93.3 − 0.042 V				
23–24 Mar.	0630	Kp	0.98	80.0 − 2.15 Kp				
		V	0.65	104.4 − 0.072 V				
23–26 Oct.	1330	Kp			0.55	76.9 − 0.61 Kp		
	1630	Kp			0.77	77.2 − 1.35 Kp		
	1900	Kp					0.98	77.3 − 1.74 Kp

3. Discussion

In a recent paper McDiarmid et al. [9] discuss changes in the geomagnetic field configuration in the vicinity of the dayside polar cusp region using high energy electron and proton flux data recorded aboard the Alouette 2 and ISIS 1 satellites. They identify the 'knee latitude' λ_{kn} as the latitude below which the field lines are closed. This is based on pitch angle distributions of both electrons and protons in the dayside sector ~0800–1400 MLT. λ_{kn} may also be defined as the latitude at which a sharp decrease with increasing latitude occurs in the flux of electrons with energies greater than 20 keV. Experimentally it is found to be nearly identical with Λ_E, the equatorward boundary of the ionization irregularity discussed in this paper, and with Λ_{cl}, the latitude [1] at which sharp increases are found in the soft particle spectrometer flux. A general decrease in λ_{kn} with increasing Kp, similar to that shown in Fig. 3a has been found [9].

Observations [9–11] have established that solar protons and electrons can scatter or diffuse across closed field lines into the lower latitude regions during magnetically disturbed periods. The solar wind velocity V and the dynamic pressure of the solar wind P ($P = {}^1/_2\,\varrho V^2$, where ϱ = particle density), are significantly enhanced during magnetic storms, and, acting directly across the closed field lines, could lead to higher diffusion. The explanation [9] that this enhanced diffusion or scatter of particles across the field lines could result in the opening of hitherto closed field lines at lower invariant latitudes seems to be a good qualitative explanation for the higher degree of correlation between Λ_E and V than between Λ_E and Kp in the noontime sector as shown in Table 1. This suggests that solar wind velocity (or the pressure) is the more meaningful parameter to use in the noontime sector as it is a direct measure of the physical conditions prevailing across the field lines and also implies a magnetosheath origin for the ionizing flux. Results given in Table 1 show Λ_E increasing with time between 1000 and 1230 MLT. Averaging values of Λ_E and V in the noon sector gives the relationship: $\Lambda_E = 85.93 - 0.026\,V$.

In the early morning sector, ~0600 MLT, the higher correlation of Λ_E with Kp rather than with V suggests that the relation to solar wind particles is less direct. It implies a different origin for the source of the observed thermal ionization. This source could be ionizing particles from the plasma sheet. Since all the observations in the dawn sector are centered around 0630 MLT, it is possible to deduce a single relationship for the average position of Λ_E as a function of Kp in the dawn sector: $\Lambda_E = 80.16 - 2.52\,Kp$.

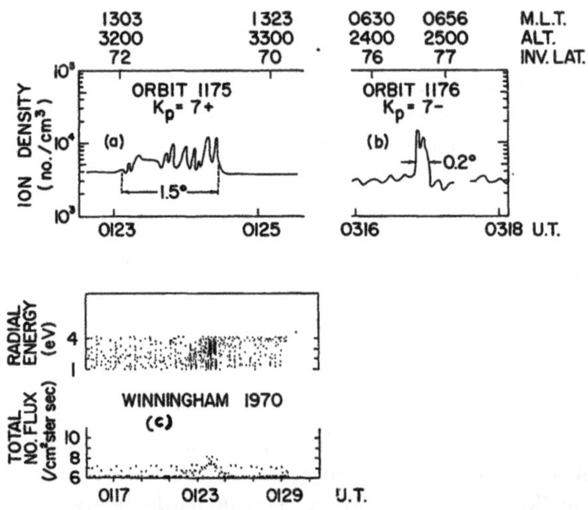

Fig. 5a and b. Two examples of narrow ionization features from ISIS 1 SEA experiment. Fig. 5c. Corresponding measurements of low energy protons by the soft particle spectrometer of Heikkila [12] on ISIS 1, 15 May 1969.

Simultaneous ISIS 1 observations of thermal positive ions (SEA) and low energy proton fluxes from Heikkila's soft particle spectrometer (SPS) [12] are shown in Figs. 5a and 5c, respectively. Fig. 5a shows an expanded time plot of an SEA ionization enhancement over the invariant latitude range 70—72°. The simultaneously measured SPS fluxes shown in Fig. 5c show an increase in the total number flux of about two orders of magnitude over the energy range 300—5000 eV. Excellent agreement is found between the locations of Λ_E and the widths of the corresponding features in Figs. 5a and 5c. The observed ionization increase in the thermal energy range is thus shown to be a direct consequence of the influx of magnetosheath electrons and protons into the topside ionosphere.

The relative contributions to the thermal ion densities of locally produced ionization and of diffusion along the field lines from lower altitudes are being studied.

References

[1] D. H. Fairfield, J. Geophys. Res. 73, 7329 (1968).
[2] W. J. Heikkila and J. D. Winningham, J. Geophys. Res. 76, 883 (1971).
[3] L. A. Frank, J. Geophys. Res. 76, 5202 (1971).
[4] J. H. Whittaker, T. R. Hartz, J. R. Burrows, L. H. Brace, W. J. Heikkila, R. C. Sagalyn and D. M. Thomas, J. Geophys. Res. 77, 6121 (1972).
[5] D. H. Fairfield, The Polar Ionosphere and Magnetospheric Processes, Gordon and Breach Science Publishers Inc., New York 1970 (p. 1).
[6] C. P. Pike, EOS, Trans. Am. Geophys. Un., 53, 360 (1972).
[7] M. Ahmed and R. C. Sagalyn, EOS, Trans. Anm. Geophys. Un., 52, 877 (1971).
[8] L. A. Frank and K. L. Ackerson, J. Geophys. Res. 76, 3612 (1971).
[9] I. B. McDiarmid and M. D. Wilson, J. Geophys. Res. 77, 1102 (1972).
[10] L. J. Lanzerotti, Phys. Rev. Lett., 21, 929 (1968).
[11] I. B. McDiarmid and J. R. Burrows, J. Geophys. Res. 74, 6239 (1969).
[12] J. D. Winningham, Ph. D. Dissertation, A & M Univ., Texas 1970.

Space Research XIII — Akademie-Verlag, Berlin 1973

OBSERVATIONS OF ELECTRON FLUXES AND RELATED VARIATIONS OF IONOSPHERIC PLASMA PARAMETERS IN THE SOUTH POLAR CUSP

K. I. Gringauz, G. L. Gdalevich, M. Z. Khokhlov, A. P. Remizov, Yu. I. Logachev, V. G. Stolpovsky, B. N. Gorožhankin, V. V. Afonin and S. M. Sheronova

Institute for Space Research, Academy of Sciences for the USSR, Moscow, USSR

Measurements of charged particles in a broad energy range (in particular of fluxes of electrons with $E > 40$ keV and $E = 0.8-10$ keV) and of the density and temperature of ionospheric electrons and ions were made at the satellite Cosmos 378, launched 17 November 1970 into an orbit with perigee 240 km, apogee 1700 km and inclination 74°. These measurements could be performed continuously along the whole orbit with the help of a memory device installed on the satellite. In the present paper only a part of the data obtained during one pass of the satellite over the southern hemisphere near noon is given.

The steady influx into the magnetosphere of solar wind thermalized plasma from the magnetosheath via the sheath was first observed at ionospheric altitudes from the satellite ISIS by Heikkila and Winningham [1]. This influx occurs through the regions near the neutral lines of the dayside magnetosphere. Frank [2] with the help of the satellite IMP-5 first discovered and examined these influxes of solar wind plasma both at the magnetopause and at different altitudes in the magnetosphere. Indications of the polar cusp were observed by Frank and Ackerson [3, 4] in the range of ionospheric altitudes from the satellite Injun 5 at local time ~ 10 h. The information about polar cusps [1—5] is mainly concerned with observations over the north hemisphere of the earth; data on the southern cusp are significantly more scanty (only in the dissertation by Winningham [5] have we found the results of observations of the south day-side cusp from the satellite ISIS-1 at ~ 10 h LT). There are no data in this information on simultaneous measurements of local parameters of ionospheric plasma.

Therefore the Cosmos 378 results of observations of low energy electrons precipitating into the high-latitude ionosphere over the earth's southern hemisphere at a local time near noon and simultaneous measurements of the ionospheric plasma seem to be interesting. An example of measurement results of electron fluxes with energies > 40 keV, electron fluxes with energies < 10 keV, the ionospheric ion density and the ionospheric electron temperature obtained during the flight of the satellite through the high-latitude ionosphere on 27 November 1970 near the apogee (1500—1750 km) are shown in Fig. 1. During 13 minutes of the flight (from 0812 to 0825 UT) the orbit of the satellite lies in the range of altitudes from 1700 to 1750 km; at this time the invariant geomagnetic latitude Λ of the satellite at first increased from 70° to 79.5° and then decreased to 75°. This corresponds to variations of the local time from 09 h to 16 h.

Thus the part of the satellite orbit under consideration is related to the same local time and the same values of Λ at which the sharp enhancement of ionospheric electron fluxes with energies < 1 keV is observed over the northern hemisphere from ISIS 1 [1] and Injun 5 [3, 4] and identified by the authors of these observations with the polar cusp.

The detailed description of the apparatus installed on Cosmos 378 will be published elsewhere. In the present paper we notice only that electrons with energy > 40 keV are detected with the help of a properly screened gas-discharge counter [6]. Spectra of electrons with energies < 10 keV were recorded once each 15 seconds by means of two cylindrical layered electrostatic analysers, which were a modification of the device described in [7], in energy ranges with $\Delta E/E \cong 0.4$ at the following values of E: 1, 2, 4 and 8 keV. Measurements of the density of ionospheric positive ions were made with the help of the spherical ion trap, a photograph of which is shown in [8]. The measurements of the ionospheric plasma temperature were performed in two different ways: by means of a spherical Langmuir probe and a spherical high-frequency probe. The external grid of the trap, which is analogous to the grid described in [8], was used as a spherical Langmuir probe.

The decrease of the counting rate of electrons with energies > 40 keV is an indication of the boundary of the trapped radiation zone, i.e. the boundary of closed geomagnetic field lines and the low-latitude polar cusp boundary at the dayside magnetosphere.

Let us consider the results of measurements, shown in Fig. 1. From the top part of this figure it is seen that the trapped radiation boundary ($\Lambda \sim 75°$) was crossed by the satellite at ~ 11 h MLT (magnetic local time); fluxes of electrons with energies 0.8—10 keV did not exceed 4×10^6 cm^{-2} s^{-1} keV^{-1} sr^{-1} inside the radiation belt (values of fluxes indicated in this figure are maximum values, recorded in each energy range). In this region the fluxes are regular enough. Immediately after crossing this boundary the flux of electrons with $E = 800$—1200 eV sharply increases and sometimes exceeds 3×10^7 cm^{-2} s^{-1} $\times$ keV^{-1} sr^{-1}. A certain increase of the electron fluxes are also observed in the energy range 1600—2400 eV. Fluxes of electrons with higher energy increase essentially less and their values never exceed 6×10^6 cm^{-2} s^{-1} keV^{-1} sr^{-1}. It should be noted that, in this region outside the trapped radiation boundary, fluxes of low-energy electrons are greatly variable with typical times apparently less than the measurement time constant (0.2 s). Values of fluxes in this region are also probably underestimated because of a certain influence of ultraviolet solar radiation and these values are distorted to some extent owing to interference which seems to be connected with rapid variations of the satellite potential. However this does not change the general character of the behaviour of electron fluxes. Curves α shown in the figure give the angle between the magnetic field direction and the axes of the analysers.

$\Lambda = 75°$ at ~ 15 h LT, the counting rate of electrons of > 40 keV sharply increases, the satellite is again in the trapped radiation zone and the fluxes of electrons with $E = 800$—1200 eV decrease sharply, reaching the value $< 10^5$ cm^{-2} s^{-1} keV^{-1} sr^{-1} at $\Lambda \sim 70°$. The enhancement of the density of ionospheric positive ions from 3×10^4 cm^{-3} to 5×10^4 cm^{-3} at the altitude 1700 to 1754 km shown in Fig. 1 is observed in the zone of existence of intense fluxes of electrons with energies 0.8—10 keV. Finally, the results of the measurements of ionospheric electron temperature T_e are also shown in Fig. 1; the solid curve corresponds to averaged values obtained by means of the high-frequency probe and triangles to those obtained by means of the spherical Langmuir probe in the period of time from 1 to 5 minutes. Both probes showed the significant growth of T_e (from $\sim 3000\,°K$ to $6000-7000\,°K$) during the transition from the trapped radiation zone to the precipitation zone of electrons with energies 800 to 1200 eV.

It is necessary to note that the data of the Langmuir probe gave us the opportunity to draw some conclusions about the distribution function of ionospheric electrons. Besides the recording of the probe curve it was analysed by the modulation method similar to that used by Bowen et al. [9]. The analysis of the results showed that after the exit of the satellite from the zone of closed geomagnetic field lines (after the fall of the counting rate of electrons with $E > 40$ keV),

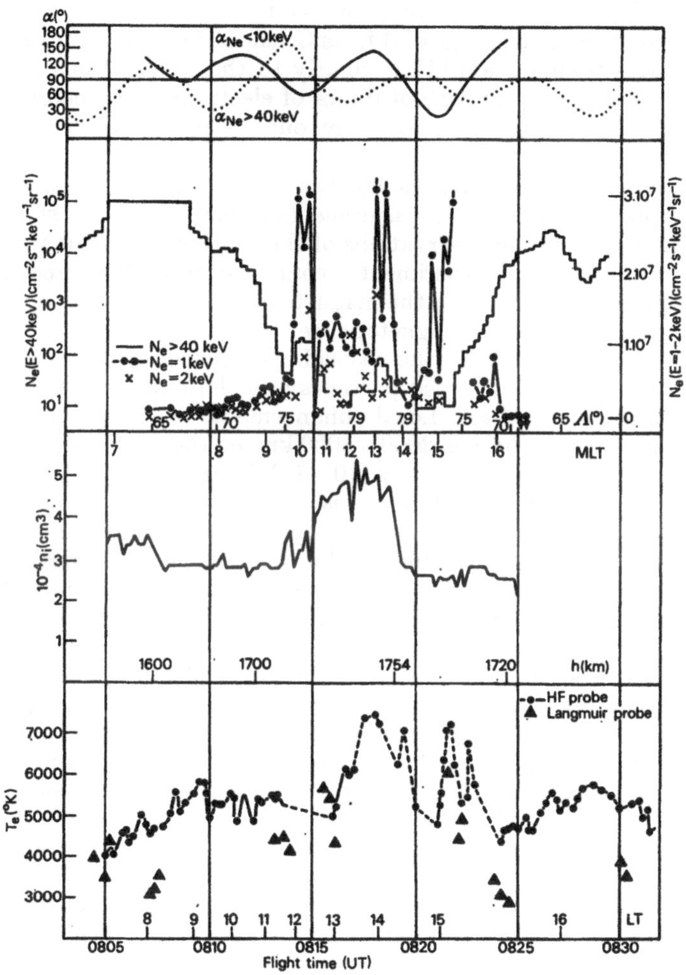

Fig. 1. Fluxes of electrons with $E < 10$ keV and $E > 40$ keV. Positive ion density n_i, electron temperature T_e; $\alpha_{Ne} < 10$ keV, angle between the magnetic vector and the normal to the aperture of the electrostatic analyser; $\alpha_{Ne} > 40$ keV, the angle between the magnetic vector and the normal to the aperture of the gas-discharge counter. Λ, invariant latitude; h, altitude of satellite over the earth's surface. Cosmos 378, November 1970. —·— high frequency probe; ▲ Langmuir probe.

the velocity distribution of ionospheric electrons was often not Maxwellian. This appeared to be caused by thermal plasma disturbances accounting for fluxes of more energetic particles. So values of T_e in the region of $\Lambda > 75°$ are measures of a certain "effective temperature", which represents the intensity of the disturbance of thermal ionospheric plasma in the region under consideration.

Measurements from ISIS 1 showed that, in the range of Λ from 74° to 79° the sharp growth of fluxes of electrons was observed with energies of the order of hundreds of eV (approximately from 10^8 to 10^9 cm^{-2} s^{-1} sr^{-1}) and the highest energies (up to 1 keV) were observed near $\Lambda = 77°$. Since the electrostatic analyser on Cosmos 378 could record only the most energetic part of the electrons, recorded at $\Lambda \sim 74-79°$ by means of ISIS 1, one can consider that values of electron fluxes recorded in the same range of Λ from Cosmos 378 do not contradict the data by Heikkila and Winningham [1]. Similarly to the observations by Frank and Ackerson [3, 4], the sharp growth of fluxes of electrons with energy of the order of hundreds eV takes place immediately out of the boundary of the trapped radiation zone.

Taking into account these points and the range of local time including noon, one can interpret the results of measurements of electron fluxes shown in Fig. 1 as the results related to the observations of the southern day-side polar cusp and therefore as the additional experimental confirmation of the direct penetration of thermalized solar plasma from the magnetosheath into the ionosphere through the cleft of the south dayside magnetopause.

It is seen from Fig. 1 that in the day-side polar cusp at altitudes $\sim$ 1700 km the density of ionospheric charged particles and the temperature of ionospheric electrons significantly increase. Local enhancements of charged particle density at different altitudes including 2000—3000 km in the high-latitude ionosphere have been observed by many authors [10—12]. Morse et al. [12], on the grounds of the analysis of data on simultaneous observations of the precipitation of electrons with $E = 2.5$ keV and data of the ion trap from the satellite OV1-15 concluded that local enhancements of the charged particle density caused by electrons with the energy of the order of 1 keV and less penetrating into the ionosphere, can arise both beneath the maximum of the F region and above it. The experimental data cited in [12] are, however, obtained only at night-time. As for day-side polar cusps, the systematical increase of F-region critical frequencies in the region of the geomagnetic pole near the magnetic noon [13] can serve as indirect evidence of its influence on the ionosphere.

Present results of simultaneous local measurements of electron fluxes penetrating into the ionosphere from the day-side magnetosphere cusp and the enhancements of the ion density and the electron temperature are indicative of the ionizing effect of solar plasma electrons on the upper atmosphere. Apparently the ionization and the heating of the ionosphere takes place at lower altitudes, beneath the F-region maximum (according to [14] the altitude of maximum ionizing effect of electrons with $E \sim 1$ keV is $\sim$ 160 km and with $E \sim 0.5$ keV is $\sim$ 240 km) and positive ions detected by Cosmos 378 reached the altitude of $\sim$ 1700 km as the result of vertical diffusion.

In conclusion it should be noted once more the significant peculiarity of the orbit of Cosmos 378: this satellite covered the range of local time $\sim$ 5 hours during one 10-minute real time flight across the day-side cusp. Owing to this point the results of measurements from Cosmos 378 give the most convincing confirmation of the magnetospheric model by Winningham [5] and Heikkila [15], constructed on the ground of the data of many flights of the satellite ISIS 1 across the day-side cusp; according to this model the day-side cusp extends from the noon meridian on both sides for several hours in local time.

References

[1] W. J. Heikkila and L. D. Winningham, J. Geophys. Res. **76**, 883 (1971).
[2] L. A. Frank, J. Geophys. Res. **76**, 5202 (1971).
[3] L. A. Frank and K. L. Ackerson, J. Geophys. Res. **76**, 3612 (1971).
[4] L. A. Frank and K. L. Ackerson, "Local time survey of plasma over auroral zones", Univ. of Aiova 71:40, preprint, August 1971.
[5] J. D. Winningham, Ph. D. Thesis, Texas, A & M University, 1970.
[6] Yu. V. Gozeluk, S. N. Kuznetsov, Yu. I. Logachev and V. J. Stolpovsky, private communication.
[7] E. L. Lein, L. T. Oldekop, B. V. Polenov, A. P. Remizov, B. I. Hazanov, M. Z. Khokhlov and A. V. Shifrin, PTE, No. 6, p. 44 (1971).
[8] K. I. Gringauz, G. L. Gdalevich and V. A. Rudakov, Trudy Radiotehn. Inst. Akad. Nauk SSSR, No. 1. p. 112 (1970).
[9] P. J. Bowen, R. L. F. Boyd, C. L. Henderson and A. P. Willmore, Proc. Roy. Soc. A **281**, 526 (1969).
[10] T. Sato and L. Colin, J. Geophys. Res. **74**, 2193 (1969).
[11] J. L. Donley, Space Research VIII, 381 (1967).
[12] F. A. Morse, H. H. Hilton and P. F. Mizera, J. Geophys. Res. **76**, 6099 (1971).
[13] A. S. Besprozvannaya and L. A. Udovich, Geomagn. i Aeronom. **9**, 691 (1969).
[14] Rees, Planet. Space Sci. **11**, 1209 (1963).
[15] W. J. Heikkila, preprint, the University of Texas at Dallas, August 1971.

Space Research XIII — Akademie-Verlag, Berlin 1973

FIELD ALIGNED ELECTRON ANISOTROPIES OBSERVED BY THE ESRO 1A (AURORAE) SATELLITE

L.-Å. Holmgren and B. Aparicio*

Kiruna Geophysical Observatory, Kiruna, Sweden

Field-aligned electron anisotropies have been observed at high latitudes for energies 1 keV and 6 keV with a typical frequency of occurrence of 1—3% and in some rare cases also at 13 keV. It is found that angular distributions with maxima along the geomagnetic field lines occur most frequently at lower energies. The occurrence frequency increases with increasing *Kp*. Two major regions of maximum occurrence frequency have been found. One is located in the evening—midnight local magnetic time sector closely poleward of the experimentally defined limit of stably trapped electrons $E > 40$ keV. The second region, identified as the dayside low altitude polar cusp, shows a maximum in occurrence frequency in the noon—evening sector. For $Kp = 1^+ - 3^0$ another region is found at auroral zone latitudes in the morning—noon sector.

1. Introduction

In the light of present knowledge of auroral phenomena it is quite clear that the auroral particle acceleration is not caused by one well localized mechanism, but that different processes distributed over the magnetosphere may contribute. A detailed study of particle pitch angle distribution and energy dependence can provide important information as to the nature and location of the mechanisms involved. Of particular interest are the type of angular distributions peaking along the geomagnetic lines of force, since they clearly indicate the existence of accelerating forces directed along the geomagnetic field lines or, possibly, decelerating forces perpendicular to the field lines. The fact that such field aligned pitch angle distributions have been observed not only in the plasma sheet [1] and at the geostationary orbit [2], but also in and near the high latitude ionosphere, see e.g. [3—12], indicates that acceleration takes place, also, in or near the polar ionosphere. This study is an attempt to establish, from a large number of data, the morphology of this interesting phenomenon as seen at the energies 1 keV, 6 keV and 13 keV by the low altitude polar orbiting satellite ESRO 1A (Aurorae).

2. Satellite and Experiment

ESRO 1A (Aurorae) was put into a polar orbit with an initial apogee of 1533 km and a perigee of 258 km on 3 October 1968. The satellite was magnetically stabili-

* On leave from Laboratorio de Fisica Cosmica UMSA, La Paz, Bolivia.

zed, with the symmetry axis nominally directed along the geomagnetic field line through the satellite. Experiment S-71B (channel multiplier detectors with cylindrical electrostatic energy analysers) had the characteristics given by Table 1. Fields of view were rectangular, 16° × 16°, for all sensors. The minimum counting time was 41.8 ms, equivalent to a maximum spatial resolution of around 300 m. For details of the instrument and its calibration see [13, 14].

Table 1

Characteristics of experiment S 71B

Type of particles	Electrons	Positive ions	Electrons	Positive ions
Nominal pitch angle (deg)	10	10	80	80
Centre energy (keV)	1.3 2.9 5.8 13.3	5.8	1.4 6.3 13.1	1.4 6.3
Energy resolution (%)	20 10 20 10	20	20 20 10	20 20

3. Data Analysis

The ratios of differential number fluxes at 10° to the flux at 80° have been computed for corresponding energies, i.e. 1.3—1.4 keV, 5.8—6.3 keV and 13.1—13.3 keV for the electrons. The ratios, which in the following will be denoted 1 keV, 6 keV and 13 keV, have been computed on the basis of 4 seconds average flux values. Because of the logical operation of the instrument this really means, in most situations, that electron fluxes are averaged over a time no longer than 90—140 ms for each energy (except for 10° electrons of 5.8 keV, which are measured most of the time). In general average electron fluxes are established for all energies (with the exception mentioned above) within 0.4 s followed by proton (positive ion) and 10° 5.8 keV electron measurements for 3.2 s. A lower limit was set for the flux level, allowing for a maximum statistical error of 32%. Every ratio greater or equal to 3 was recorded as a significant field-aligned anisotropy. A factor of 3 was considered to be a fair margin for statistical errors and the uncertainty introduced by measuring with different detectors in the two directions. The events of field-aligned anisotropies were accumulated in "boxes" of "size" one hour local magnetic time (based on the eccentric dipole model) times one degree of invariant latitude, and further classified into one of four classes of *Kp*. By normalizing against the sampling density, the occurrence frequency has been found and expressed as a percentage for "boxes" with at least 50 samples.

4. Observations

$Kp = 0^0—1^0$ (Fig. 1).

The frequencies of occurrence are very low. A strong energy dependence is evident with field-aligned electron anisotropies occurring more frequently at 1 keV than at higher energies. Only rarely is the frequency above 3% for 1 keV, and for 6 keV it is mostly below 1%.

The distribution in eccentric dipole time and invariant latitude for 1 keV shows a semi-circular structure from early dawn over the entire dayside up to 19—20 h

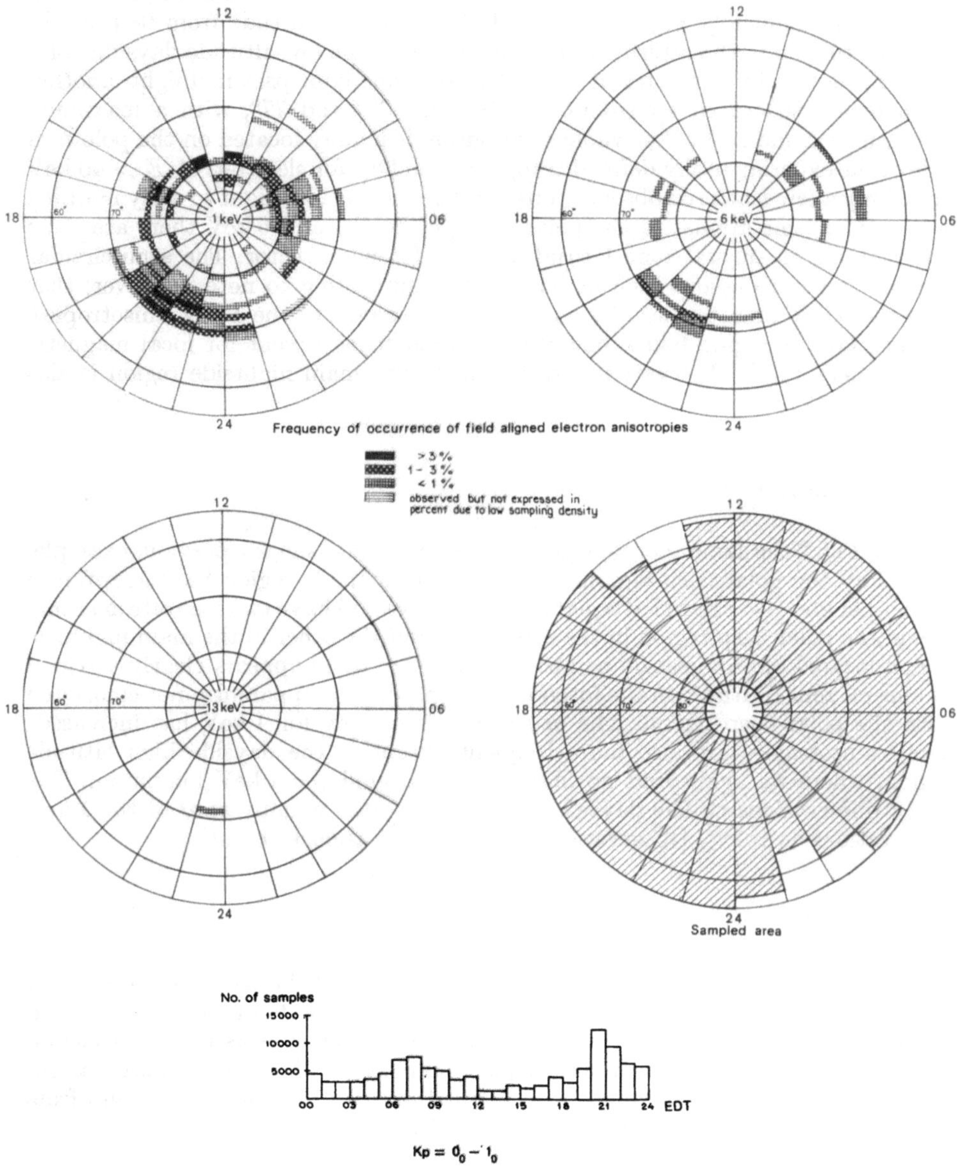

Fig. 1. Frequency of occurrence of field-aligned electron anisotropies for 1, 6 and 13 keV, $Kp = 0^0 - 1^0$, together with sampled area and sampling density per hourly interval.

in the evening, where a discontinuity is observed. This dayside region, which lies almost completely between 78° and 80° invariant latitude from 08 h in the morning to 16 h in the afternoon, is identified with the low altitude dayside polar cusp. From the discontinuity at 19—20 h up to one hour past midnight, another main region is located approximately between 69° and 77°, with a maximum between 21 h and 23 h. The whole distribution is mainly located on the poleward side of the experimentally defined trapping boundary for electrons of $E > 40$ keV [15]. Only a few scattered observations are found in the late night to early morning sector. The same applies to the polar cap. For both 1 and 6 keV there are some observations at auroral zone latitudes in the forenoon sector. The semi-circular dayside structure is also seen at 6 keV. It is interesting to note, however, that here the 6 keV anisotropies obviously lie equatorward of the 1 keV anisotropies. In this dayside region 6 keV observations tend to be absent for local magnetic times where 1 keV observations are frequent. The main nightside region is also seen at 6 keV.

$Kp = 1^{+}—3^{0}$ (Fig. 2).

From Fig. 2 the same energy dependence is apparent as in the previous example. For 1 keV the polar cusp region and the main nightside region have grown into each other and also expanded equatorward, but we still have two separate maxima, one in the polar cusp region in the afternoon, and the other in the main nightside region located approximately as above. Now, too, there appears clearly a region at auroral zone latitudes in the forenoon sector, which is apparently not connected with the polar cusp region. The occurrence frequency for 1 keV has increased; 6 keV anisotropies are also more frequent except in the dayside high latitude region, where we again find no observations around the 1 keV maximum. The 6 keV observations on the dayside tend to be located equatorward of the 1 keV observations as above. Observations at 13 keV are almost absent.

$Kp = 3^{+}—5^{0}$.

From the data in this Kp class we may conclude that the occurrence frequency has increased. Even at 13 keV some observations are found at auroral oval latitudes in the evening. However, the energy dependence is as before. No significant further expansion towards the equator is observed. Most of the observations are now in the dusk to midnight sector, but there is still a clear maximum from 14 h to 17 h in the polar cusp region for 1 keV.

$Kp = 5^{+}—9^{0}$.

For this Kp class the number of data were not sufficient to apply this type of analysis with any success. However, we can state that field-aligned electron anisotropies do occur under these very disturbed conditions. There are indications of the existence of field-aligned electron anisotropies in the polar cusp region as well as in the main nightside region, and that occurrence frequencies are as high as for the previous Kp class.

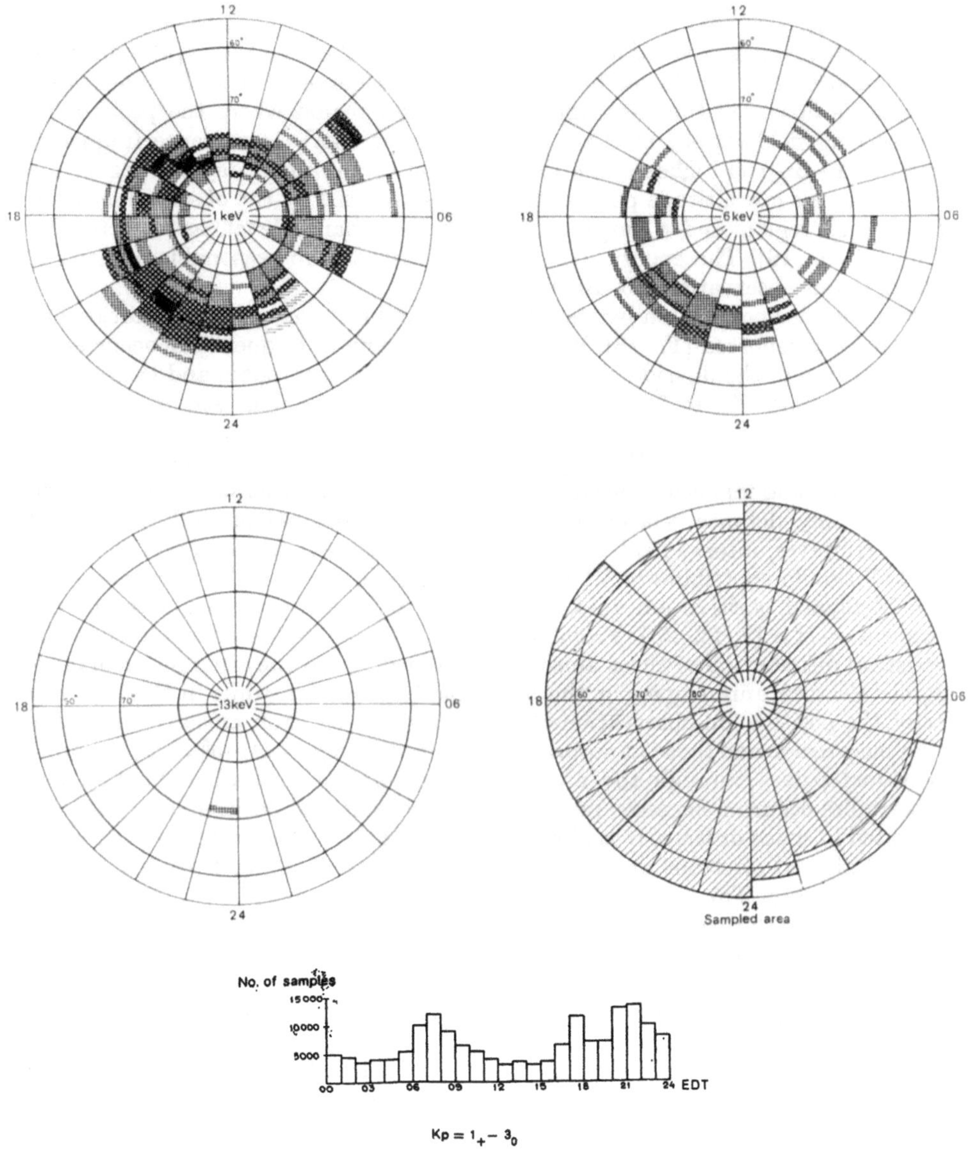

Fig. 2. Frequency of occurrence of field aligned electrons, $Kp = 1^+ - 3^0$. For interpretation of scale, see Fig. 1.

5. Discussion

The energy bands covered by experiment S-71 B lie above the typical energies (some hundred eV) for plasma observed in the dayside polar cusp. It is thus quite remarkable how well the cusp appears here, as a region where the probability of observing field-aligned electron anisotropies for 1 keV (occurrence frequency typically 1—3%) is highest. The identification of the main dayside region for 1 keV with the low altitude polar cusp is well supported by the observed location of the cusp region under varying magnetic disturbance level, as reported by Winningham [16]. The energy dependence observed is in reasonable agreement with previous observations [6, 10, 11].

Quite detailed correlation between VLF hiss and soft electron precipitation has recently been reported [17, 18]. We find two maxima in frequency of occurrence. One is located in the dayside polar cusp region, in the afternoon, and the other is observed around 70° to 72° in the late evening to midnight sector. Similarly located maxima have been found for occurrence frequency of VLF hiss in the frequency range 5.5—8.8 kHz [19], for average whistler mode noise intensity at 0.2 MHz [20], and for occurrence frequency of VLF emissions at 3.2 kHz [21]. We do not infer any definite relationship between our observations and those cited above. However, intense field-aligned electron fluxes with an energy of the order of 1 keV should contribute to VLF emission through the Cerenkov process.

Field-aligned positive ion anisotropies have also been observed with the experiment S-71B [7, 22, 23]. From Riedler and Borg [22] we conclude that the statistical distribution of these anisotropies coincide with the distribution of field-aligned electron anisotropies in the main nightside region, while in the forenoon sector the positive ion observations fall between the electron anisotropies in the polar cusp region and those at auroral zone latitudes ($Kp = 1^{+}—3^{0}$). Inspection of individual passes through the nightside region shows that when field-aligned anisotropies occur in the same pass for both electrons and positive ions, they are separated in space or time. It is also important to note that occurrence frequencies for field-aligned electrons are much lower than for field-aligned positive ions of the same energy. This is in good agreement with the concept of a local electric field, parallel to the geomagnetic field lines, as the one discussed by Hultqvist [23]. One must of course remember that occurrence frequencies are very dependent on the factor chosen for a significant anisotropy. Our results are relevant to fairly strong anisotropies (ratios greater than 3). Extreme ratios up to 200 have been observed.

There have been a number of reports on field-aligned angular distributions of charged particles [1—12, 22]. Various mechanisms, e.g. electric fields parallel to the geomagnetic field lines, field-line merging, Fermi acceleration, wave—particle interactions, and plasma instabilities have been proposed to explain the observations. A more detailed measurement of the energy dependence of field-aligned angular distributions together with electric field and VLF measurements is, however, needed before one can judge the relative importance of the various mechanisms.

The source mechanism(s) for field aligned angular distributions observed in the high latitude ionosphere is likely to be located fairly close to, or in, the ionosphere. This is so since the high mirror ratio of the geomagnetic field would require an unrealistically narrow peak in the angular distribution in the region of acceleration, if the source were located far from the ionosphere. A forthcoming study of the height dependence of these events will give further insight into this problem. Field aligned electron anisotropies observed in the dayside polar cusp region may be a result of strong wave—particle interactions or of field-line merging in the magnetosheath—polar cusp interface, which according to Scarf and Fredricks [24] is a region of intense electric field wave activity. Our observation of a semicircular high latitude region, clearly seen for $Kp = 0^0 - 1^0$ and also for $Kp = 1^+$ to 3^0, should have a bearing on the topology of the magnetosphere. It supports a magnetospheric model of the type depicted by, e.g., Mishin et al. [25].

6. Summary

We may summarize the more important of our observations as follows. At the energies 1 keV, 6 keV and 13 keV, fairly strong field-aligned electron anisotropies (i) occur at high latitudes, mainly poleward of the experimentally defined limit of stably trapped electrons of $E > 40$ keV [15], (ii) are minority phenomena with typical frequencies of occurrence of 1—3% for 1 keV, less than 1% for 6 keV and only a few observations at 13 keV if a factor of 3 is chosen for a significant anisotropy, (iii) are strongly energy dependent, with the highest frequency of occurrence for lower energies, (iv) are more frequent for higher Kp, (v) occur most frequently in the late evening to midnight sector and in the dayside polar cusp region. Both regions expand equatorward with increasing Kp.

Acknowledgments

This work has been sponsored by the Swedish Natural Science Research Council and the Swedish Board for Technical Development. Lic. Phys. Bruno Aparicio acknowledges the financial support of the International Atomic Energy Agency. We are indebted to the whole staff of the Kiruna Geophysical Observatory for valuable suggestions and technical assistance in the preparation of this report.

References

[1] E. W. Hones et al., J. Geophys. Res. **76**, 63 (1971).
[2] S. E. DeForest and C. E. McIlwain, J. Geophys. Res. **76**, 3587 (1971).
[3] W. Riedler, Kiruna Geophys. Obs. Rep. No. 663 (1966).
[4] M. Lampton et al., paper presented at Birkeland Symposium, Sandefjord, Norway 1967.
[5] R. A. Hoffman et al., J. Geophys. Res. **73**, 6201 (1968).
[6] G. Paschmann et al., Trans. Am. Geophys. Un. **50**, 654 (1969).
[7] B. Hultqvist et al., Planet. Space Sci. **19**, 279 (1971).
[8] L. A. Frank and D. A. Gurnett, J. Geophys. Res., **76**, 6829 (1971).
[9] H. Reme and J. M. Bosqued, J. Geophys. Res. **76**, 7683 (1971).
[10] B. J. O'Brien and D. L. Reasoner, J. Geophys. Res. **76**, 8258 (1971).
[11] B. A. Whalen and I. B. McDiarmid, J. Geophys. Res. **77**, 191 (1972).
[12] J. R. Sharber and W. J. Heikkila, J. Geophys. Res. **77**, 3397 (1972).
[13] W. Riedler et al., Arkiv f. Geofys., **5**, 619 (1971).
[14] P. Christophersen, Arkiv f. Geofys., **5**, 659 (1971).
[15] D. E. Page and M. L. Shaw, Space Research XII, 1471 (1972).

[16] J. D. Winningham, in: Earth's Magnetospheric Processes, D. Reidel Publ. Co., Dordrecht, Holland 1972 (p. 68).
[17] D. A. Gurnett and L. A. Frank, J. Geophys. Res. 77, 172 (1972).
[18] R. A. Hoffman and T. Laaspere, J. Geophys. Res. 77, 640 (1972).
[19] D. A. Gurnett, J. Geophys. Res. 71, 5599 (1966).
[20] R. E. Barrington et al., J. Geophys. Res. 76, 5278 (1971).
[21] A. R. W. Hughes and T. R. Kaiser, in: The Radiating Atmosphere, D. Reidel Publ. Co., Dordrecht, Holland 1971 (p. 336).
[22] W. Riedler and H. Borg, Kleinheubacher Ber. 15, 167 (1971).
[23] B. Hultqvist, Planet. Space Sci. 19, 749 (1971).
[24] F. L. Scarf and R. W. Fredricks, in: Earth's Magnetospheric Processes, D. Reidel Publ. Co., Dordrecht, Holland 1972 (p. 329).
[25] V. M. Mishin et al., J. Geophys. Res. 75, 797 (1970).

Space Research XIII — Akademie-Verlag, Berlin 1973

A3 ABSORPTION ASSOCIATED WITH ARTIFICIAL RADIATION BELTS

J. Oksman

Department of Electrical Engineering, University of Oulu, Oulu, Finland

It is suggested that, in addition to the D-region effects associated with nuclear explosions, discussed by other authors, long-lasting A3 absorption was caused by the Russian explosions on 22 and 28 October 1962. This absorption is suggested to have been caused by electrons in the energy range of a few hundred keV, injected into trapped orbits by the explosions and precipitated relatively rapidly thereafter. Longitudinal asymmetries and radial diffusion of trapped particles may have played a role in determining the temporal variation of absorption. Quantitative estimates show that the average precipitation on the days following the explosion on 28 October seems to have been intensive enough to present a dominant source of ionization in the D region. The possible role of precipitated electrons as a source of D-region ionization in general is discussed.

1. Introduction

The role of precipitated charged particles as a cause of absorption of radio waves at high latitudes is reliably established and generally accepted (see e.g. review papers [1, 2]). The situation at middle latitudes is not so clear, although many authors have considered the possibility of explaining at least a part of middle-latitude absorption as the effect of charged particles [3].

Particles in the energy range required to explain the electron production in the D region (electrons of about 40—300 keV energy have usually been considered) cannot penetrate the geomagnetic field directly at the latitudes in question but must have been injected into the radiation belts and stored there for some time before being precipitated. A similar situation occurred many times in the era of high-altitude nuclear explosions (1958—1962): the radiation belts were intensified by electrons released by the explosions, with a subsequent decay through precipitation. The effects of these artificial belts on the ionosphere have been the subject of a great number of studies earlier, but evidence is now presented on long-lasting increases of short-wave absorption in Middle Europe following some of the most powerful high-altitude nuclear detonations in 1962. This evidence has, for some reason, escaped attention earlier.

The following discussion on the storage and precipitation of high-energy particles following nuclear explosions might shed some additional light also on the problem concerning the role of precipitated particles as a source of D-region ionization in general. The merit of the artificial radiation belts in this respect is that their intensities, extents and decay times have been measured extensively, allowing some quantitative estimates to be made of the expected effects on the ionosphere. In addition, the times of injection are known precisely, allowing the time history of the effects to be studied reliably.

2. Observations on Artificial Radiation Belts

In this study, only the Soviet megaton or near-megaton nuclear explosions above the main part of the ionosphere in the year 1962 will be considered. There were three of them, viz. on 22 October, 28 October and 1 November 1962. These explosions probably took place near Semipalatinsk, Central Asia, at L values between 1.8 and 2.0.

Following the first explosion, Telstar 1 measured an increase by a factor of 60 in the omnidirectional counting rate (OCR) of > 0.5 MeV electrons at $L = 2.0$ to 2.2, followed by a decay by almost an order of magnitude in 5 days [4]. Following the second test, Telstar 1 detected again an increase in OCR by a factor of about 60, with a subsequent decay by an order of magnitude in 5 days. The third explosion did not cause any discernible effect at $2 \leqq L \leqq 2.2$. West [5] gives the results of a thorough analysis of the measurements made by the STARAD satellite after the second explosion. The mirror point distribution for electrons with $E > 955$ keV about 7.3 hours after the detonation reveals that the artificial belt was located between the L shells 1.8 and 2.8. The corresponding distribution for the epoch 41 hours after the explosion shows a decrease in the peak density by a factor of about 2.5. West estimates the total number of artificially injected electrons as $(2.0 \pm 1.0) \times 10^{25}$ for $E > 200$ keV at 7.3 hours after the detonation.

From the foregoing results it can be expected that the explosion on 28 October offers the best chances for a study of possible effects in middle latitudes.

3. Absorption observed at Lindau, Germany

As an uninterrupted series of A3 measurements using the 296 km path Norddeich—Lindau and a frequency of 2.614 MHz exists at Lindau since 1956, the recordings were examined for possible effects of the nuclear explosions discussed above. A thorough analysis of the results is to be found in [3]; the most important findings will be discussed here. The field strengths in the period in question are shown in Fig. 1 and 2, together with some other relevant data.

The effect on 22 October is interesting because of the long-lasting quasiperiodic variation in absorption. The mean period (about 90 minutes) corresponds to the drift period of ~ 300 keV electrons around the earth at $L = 2.4$ [6]. It can be assumed that a longitudinal asymmetry persisted the whole day in the artificial radiation belt causing quasiperiodic variations in the precipitation and absorption. The possibility that the variations in absorption were caused by a wave motion in the ionosphere seems to be ruled out by the fact that the ionograms do not reveal any corresponding variations in the parameters of higher layers, which should also have been affected by the waves. The effect on 22 October lasted the whole day. After sunset the effect obviously decreased due to the lack of photodetachment which reduced the electron density appreciably. Also the precipitation may have decreased towards the evening: McDiarmid and Burrows [7] have noticed that the precipitation at middle latitudes has a diurnal variation with a maximum by day. On the following day, 23. October, only a slight effect was seen in absorption but the more sensitive VLF method [8] indicated increased ionization at the underside of the ionosphere until the next explosion on 28 October. It seems probable that precipitation from the artificial radiation belt produced by the explosion was for many days intense enough to cause a discernible increase in the D-region ionization.

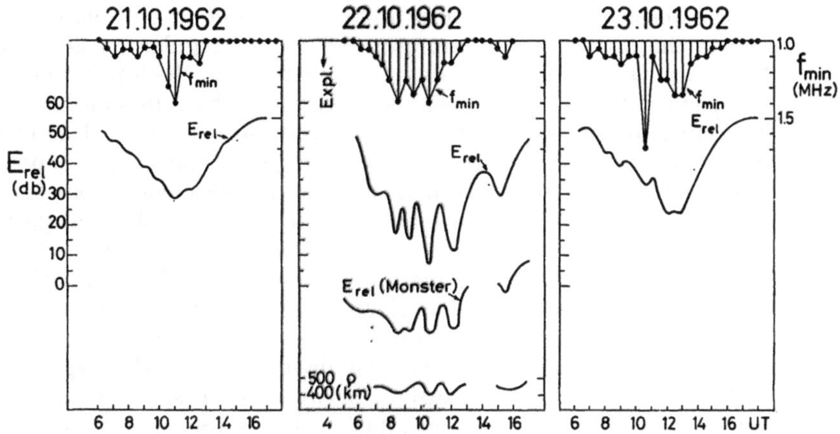

Fig. 1. Variations of the relative field strength on the Norddeich—Lindau path and f_{min}, as measured at Lindau, on 21—23 October 1962. The variations of the relative field strength of Rugby, as measured at Monster, and the estimated distance of the VLF field strength minimum from Rugby (ϱ) are also shown for 22 October [8]. The time of the Soviet nuclear explosion on 22 October is shown with an arrow.

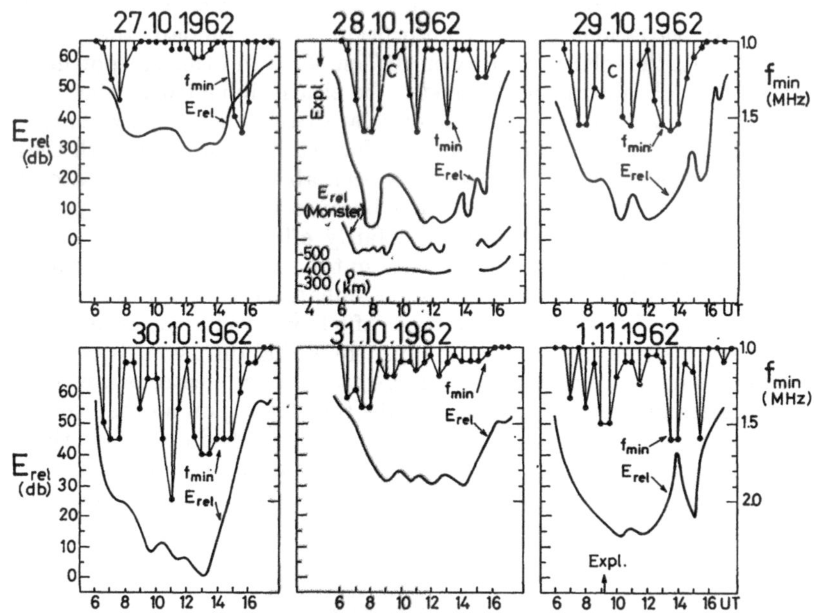

Fig. 2. Variations of the relative field strength on the Norddeich—Lindau path and f_{min}, as measured at Lindau, in the period 27 October—1 November 1962. E_{rel} (Monster) and ϱ have the same meanings as on 22 October. The times of the Soviet nuclear explosions on 28 October and 1 November are shown with arrows.

The effect following the explosion on 28 October differs from the previous one. The delay in the onset of increased absorption is the same in both cases (about 3 hr 20 min) but the absorption is stronger and more irregular on 28 October than on 22 October. Obviously this is due to a difference in the injection of electrons into trapped orbits in the two cases, possibly caused by different detonation altitudes. The effect of the explosion on 28 October seems also to last longer: strong absorption is observed on three days following the explosion. It is possible that radial diffusion of electrons outwards delivers new electrons from lower L shells where the precipitation losses are lower [9] to the outer shell in question.

Using the information given by West [5] on the total number of electrons (2×10^{25} electrons with $E > 200$ keV) and the extent of the artificial radiation belt ($1.8 < L < 2.8$), as well as the decay time given by Braun (one order of magnitude in 5 days, i.e. decay to 1/e in 2 days $\simeq 1.6 \times 10^5$ s) one can estimate the average flux of precipitated electrons during the first days following the explosion on 28 October. The area on the earth's surface between the L shells in question is about 7×10^7 km^2, and the average precipitation flux is 180 cm^{-2} s^{-1}. When this number is compared with that given by Zmuda [10] for the required flux of 300 keV electrons to present a dominant source of ionization at 74 km throughout the day (10 cm^{-2} s^{-1}), it is noticed that the flux obtained is of the right magnitude to produce a drastic effect.

Summing up, it is believed that enhanced absorption on the Norddeich—Lindau path was observed following the high-altitude nuclear explosions considered in this investigation. The possibility that the absorption effects discussed above would be linked with geomagnetic or solar disturbances was checked by studying the day-to-day variations of noon absorption A_f at Lindau, the geomagnetic activity index A_p, the solar radio flux on 10.7 cm $\Phi_{10.7}$ and the sunspot number R_Z. The days with high absorption seemed to be clearly linked with the nuclear explosions but no link with geomagnetic or solar disturbance was evident.

The fact that the artificial radiation belts formed in connection with the nuclear explosions of megaton class seem to have been able to affect the state of the D region for many days is very interesting because it suggests that precipitation from the radiation belt may really be a fairly effective source of ionization in the D region. This might be true especially in the slot region of the radiation belt ($L = 2.5 - 3.5$) where the precipitation is enhanced through the inherent instability of the belt [9] leading to lifetimes of a few days for the excessive electron population transported into this region, e.g. by radial diffusion. It is known that after major geomagnetic disturbances large amounts of electrons are transported from outer parts of the magnetosphere into the slot region so that the counting rates observed there are of the same order of magnitude as after the nuclear explosions discussed here [11]. It can be expected that precipitation at these times is intense enough to exert a pronounced effect on the D region, especially in the winter time when other ionization sources are weak.

Acknowledgments

The main part of this investigation was carried out when the author was at the Max-Planck-Institut für Aeronomie, Lindau, Germany, with the aid of a grant from Alexander von Humboldt foundation. Fruitful discussions with many colleagues there, especially with Dr. Schwentek, are gratefully acknowledged.

[1] D. K. Bailey, Planet. Space Sci. **12**, 495 (1964).
[2] B. Hultqvist, Space Sci. Rev. **5**, 771 (1966).
[3] J. Oksman, Department of Electrical Engineering, University of Oulu, Report No. S 21 (1971).
[4] W. L. Brown, in: Radiation trapped in the Earth's Magnetic Field, Reidel Publ. Co., Dordrecht 1966 (p. 610).
[5] H. I. West, Jr., in: Radiation trapped in the Earth's Magnetic Field, Reidel Publ. Co., Dordrecht 1966 (p. 634).
[6] J. S. Lew, J. Geophys. Res. **66**, 2681 (1961).
[7] I. B. McDiarmid and J. R. Burrows, Canad. J. Phys. **42**, 6161 (1964).
[8] J. Frisius et al., J. Atmos. Terr. Phys. **26**, 251 (1964).
[9] B. A. Tverskoy, Rev. Geophys. **7**, 219 (1969).
[10] A. J. Zmuda, J. Geophys. Res. **71**, 1911 (1966).
[11] J. R. Winckler, in: Particles and Fields in the Magnetosphere, Reidel Publ. Co., Dordrecht 1970 (p. 332).

Space Research XIII — Akademie-Verlag, Berlin 1973

SIMULTANEOUS SATELLITE AND RIOMETER MEASUREMENTS OF PARTICLES DURING SOLAR COSMIC RAY EVENTS

M. B. Baker, P. R. Satterblom, A. J. Masley and A. D. Goedeke

Space Sciences Department, McDonnell Douglas Astronautics Company-West, Huntington Beach, Calif., USA

The expected 30 and 50 MHz riometer absorptions have been calculated for three events in 1969 using data from the MDAC-W charged particle experiment on OGO 6. Several times during each event the satellite passed over the MDAC-W Arctic and Antarctic Geophysical Observatories. The calculated total absorption (using 2 minute averages of the data) agrees well with the measured absorption for the overpasses. The alpha particle and electron contributions usually amount to less than a few per cent of the proton absorption. During the large 2 November 1969 event, hower, the electrons produced the major part of the absorption up to the peak and a significant contribution during virtually its entire duration. With the two frequencies for which simultaneous riometer data were taken it is possible to detect the softening of the particle spectra during the events and the relative hardness differences between events.

1. Introduction

The McDonnell Douglas Astronautics Company-West (MDAC-W) charged particle experiment on OGO 6 measures the energy spectra of protons between 5.3 and 78 MeV and alpha particles between 17.5 and 84 MeV, and the intensity of electrons having energies greater than about 350 keV. The instrument has been

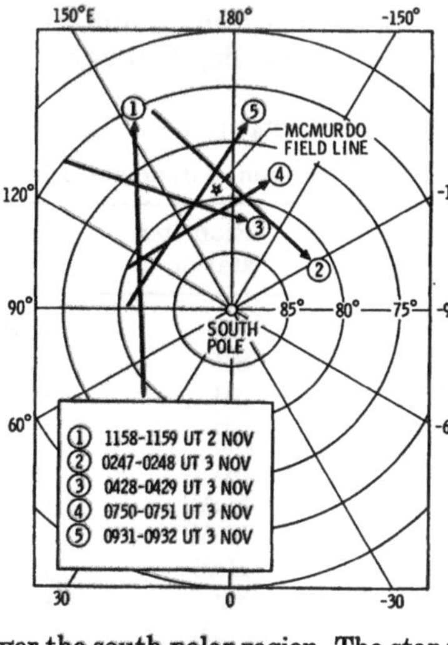

Fig. 1. OGO 6 trajectories over the south polar region. The star marks the position at 1000 km altitude of the geomagnetic field line passing through the McMurdo sound riometer station.

described by Masley and Satterblom [1]. Between June 1969 and September 1970 measurements were made during solar cosmic ray (SCR) events simultaneously with riometer absorption measurements at 30 and 50 MHz at the MDAC-W conjugate geophysical observatories located at Shepherd Bay, N. W. T., Canada, and McMurdo Sound, Antarctica. The geomagnetic field line at Shepherd Bay is almost vertical at 68.8° N 93.9° W. Throughout this period the satellite altitude was near 1000 km over the south polar cap; the field line originating at McMurdo is situated at 78.8° S 170° E at 1000 km altitude. Since the maximum latitude reached by OGO 6 was 82°, it passed within a few hundred kilometers of the McMurdo field line on four or five consecutive orbits each day; an example of some passes is shown in Fig. 1. On north polar passes, however, the satellite could come close to the Shepherd Bay field line at most twice per day. These near overpasses were examined during the three largest events (i.e. those having the highest peak measured riometer absorption) between June 1969 and December 1969: 7 June, 25 September and 2 November.

2. Discussion

A computer program was written, based on the work of Adams and Masley [2], which calculates the riometer absorption from satellite-measured proton and alpha particle spectra. The absorption due to electrons was calculated using the methods of Bailey [3]. The electron energy spectra measured on Explorer 41 [4] were used. The antenna correction factor (to convert from a pencil beam to the actual riometer antenna pattern) used was 1.4.

Two-minute averages of the OGO 6 particle data were used to calculate the 30 and 50 MHz absorption for all possible near overpasses. Since the analysis of Adams and Masley was carried out for the sunlit ionosphere, only daytime overpasses were used for this study. Other overpasses were deleted because of lack of either particle or riometer data. The results for 30 MHz for the remaining near overpasses during the events of 7 June 1969 and 25 September 1969 are shown in Table 1. "Distance" is the approximate minimum distance between the OGO 6 orbit and the geomagnetic field line passing through the riometer station.

Table 1

Simultaneous 30 MHz absorption, 8 June and 25 Sept. 1969 events

Date	UT	Distance (km)	Riometer station	Measured abs. (dB)	Calculated abs. (dB)
8 June	0218	70	SB	0.56	0.48
9	1143	190	SB	0.90	0.98
10	0231	190	SB	0.67	0.62
11	0145	170	SB	0.39	0.35
25 Sept.	1208	110	SB	0.50	0.98
26	0714	300	McM	0.30	0.28
26	0855	160	McM	0.30	0.25
28	0722	170	McM	0.80	0.76
28	2140	360	SB	0.65	0.92
29	0635	360	McM	1.5	0.94
29	0817	240	McM	0.70	0.61
29	2054	40	SB	0.20	0.15

The electron intensities during the events were quite small, and the absorption produced by them insignificant. Similarly in all cases the alpha-particle contribution to the calculated 30 MHz absorption never amounted to as much as 1% of the total. This confirms the conclusion given by Adams and Masley [5] that the alpha particle constituent in SCR events usually does not contribute significantly to measured riometer absorption. The agreement for the June and September 1969 events is very good.

The 2 November 1969 event was by far the largest during the period studied, and considerable intensities of solar electrons were detected [6]. During this event the sun was continually above the horizon at McMurdo Sound. There were seven

Table 2

Simultaneous 30 MHz absorption, A, 2 Nov. 1969 event

Date	UT	Distance (km)	A_p (dB) OGO 6	$A_{p+\alpha}$ (dB) OGO 6	A_e (dB) OGO 6	A_t (dB) calc.	A_t (dB) meas.
2 Nov.	1158	800	4.3	4.3	7.3	8.2	6.1
3	0248	160	6.3	6.4	2.3	6.8	7.2
3	0428	240	5.4	5.5	2.0	5.8	6.2
4	0201	400	2.3	2.3	1.1	2.6	2.6
4	0341	100	2.2	2.2	1.0	2.4	2.5
4	0523	340	2.1	2.1	0.8	2.3	2.2
4	0703	360	2.0	2.0	1.1	2.2	2.2
4	0844	120	1.9	1.9	0.6	2.0	2.1

p = protons; α = alpha particles; e = electrons; t = total

near overpasses of the McMurdo riometer for which good data were obtained, all of which were late in the event, past the time of peak absorption. One earlier overpass occurred, during the initial rise in absorption, which was somewhat further away from the McMurdo field line than the others. There were no usable overpasses of Shepherd Bay. The results for this event are shown in Table 2. "A_t (meas)" is the absorption measured at McMurdo when OGO 6 was at its closest approach.

The proton and alpha perticle absorption, electron absorption, and total 30 MHz absorption calculated from OGO 6 particle data, and the simultaneous measured 30 MHz riometer absorption are shown in Fig. 2. The tabulated values (the near overpasses) are open circles. The intervening points are calculated from averages of the particle data during southern polar passes. The early absorption before the peak is produced primarily by electrons, while the broad shoulder in the absorption curve during the second seven hours of the event is due primarily to protons. The agreement is excellent. The apparent large differences between 1200 UT and 1800 UT on 2 November are no doubt due to large variations in particle intensities (especially protons) over the polar cap as observed by OGO 6. Changes larger than those shown in Fig. 2 were seen in much shorter times in both the riometer and OGO 6 data.

Both 30 and 50 MHz absorptions were calculated for a large number of other times during polar passes during the events investigated. The ratio of 50 to 30 MHz absorption was examined to see if information on the particle energy spectra could be determined from absorption at two frequencies. The theoretical basis for this is discussed by Adams and Masley [2]. The large 2 November 1969 event is the only one investigated during which the proton and alpha particle spectra softened appreciably. The 50 to 30 MHz absorption ratio (for the protons and alpha particles) decreased by more than 10%, which could have been detected by the riometers if the electron component of absorption had not been present. The minimum unambiguously detectable difference in energy spectra would probably be that which causes a change of 10% in the 50 to 30 MHz absorption ratio. For comparison a change in γ, the spectral index of a differential energy spectrum, from $\gamma = 5$ to $\gamma = 3$, causes an increase in this ratio of about 8%, and from $\gamma = 3$ to $\gamma = 2$ a further increase of 8%. A change from $\gamma = 7$ to $\gamma = 5$, however, changes the ratio by less than 1%.

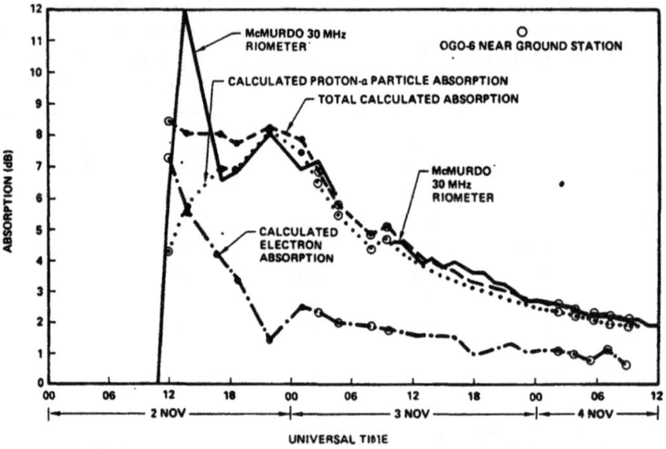

Fig. 2. Calculated (OGO 6) and measured 30 MHz absorption, 2 November 1969 event. Open circles are two minute averages at the times of near overpasses of the McMurdo Sound riometer station; other points are averages over the intervening south polar passes.

3. Summary

Excellent agreement is found between measured polar cap riometer absorption during three solar cosmic ray events and absorption calculated from detailed particle data measured on a satellite passing over the riometer. Electrons are found to contribute the greater part of the 30 MHz absorption before the peak of the 2 November 1969 event; at all other times protons produce most of the absorption. For proton events it would be possible to estimate the hardness of the energy spectrum by examining the ratio of 50 to 30 MHz absorption.

Acknowledgments

The OGO 6 experiment was supported by the National Aeronautics and Space Administration (Contract NAS 5-9324). The operation of the Arctic and Antarctic Geophysical Observatories was supported by the National Science Foundation (Contract NSF-C393). This study has been partially supported by the McDonnell Douglas Independent Research and Development Program.

References

[1] A. J. Masley and P. R. Satterblom, in: Proc. 11th Int. Conf. on Cosmic Rays, Budapest, 1969; Acta Physica Acad. Sci. Hungaricae **29**, Suppl. 2, 513 (1970).
[2] G. W. Adams and A. J. Masley, Planet. Space Sci. **14**, 277 (1966).
[3] D. K. Bailey, Rev. Geophys. **6**, 289 (1968).
[4] L. J. Lanzerotti and C. G. Maclennan, Presented at COSPAR Symp. on November 1969 Solar Polar Particle Event, June 1971.
[5] G. W. Adams and A. J. Masley, in: Recent Advances in Cosmic Ray Research **11**, Western Periodicals Co., N. Hollywood, Ca., 1966 (p. 201).
[6] A. J. Masley and P. R. Satterblom, in: Conf. Papers, 12th Intl. Conf. on Cosmic Rays, Hobart, Australia **5**, 1971 (p. 1849).

Space Research XIII — Akademie-Verlag, Berlin 1973

ROCKET MEASUREMENTS OF PRODUCTION AND IONIZATION DURING A PCA EVENT

W. F. GRIEDER[a] and D. A. BURT[b]

[a]Air Force Cambridge Research Laboratories, L. G. Hanscom Field, Bedford, Mass., USA
[b]Utah State University, Logan, Utah, USA

Three rockets instrumented with RF impedance probes and scintillators were launched on 6 and 7 April 1971 during a polar cap absorption (PCA) event at Thule, Greenland. Electron density and proton flux were measured on two rockets launched at local noon when the incident proton flux was maximum and the 30 MHz riometer indicated 2 dB of absorption. Electron density was also measured twelve hours later near local midnight ($\chi = 97°$). Electron density profiles during the PCA are compared with profiles measured by identical probes during quiet daytime conditions on 22 March 1971, and show an order of magnitude enhancement below 85 km. Riometer absorption computed from measured electron density profiles agrees within 20% of measured absorption. The twilight electron density profile compared with daytime results shows a marked gradient around 80 km even though the solar zenith angle did not exceed 97°. Ion production rates as a function of altitude are computed from measured proton spectra. The electron density computed using these production rates and electron loss rates obtained from previous PCA results are compared with the rocket measurements.

1. Introduction

On 6 April 1971 at 0935 UT an importance-1 brilliance flare occurred near solar coordinates S19W87 and caused a polar cap absorption (PCA) event observable at Thule, Greenland. Riometer (30 MHz) measurements at Thule indicated a maximum absorption of 2.2 dB at 1800 UT on 6 April 1971 with typical recovery characteristics during the next day. Ionosonde data taken at Qanaq, Greenland, 90 miles north of Thule, indicated "blackout" conditions during the daytime on 6 April. Disturbed traces appeared during the night and "blackout" resumed at sunrise on 7 April (private communication, V. Neble Jensen, Det Danske Meteorologiske Institut). Three Air Force Cambridge Laboratories Super Arcas rockets were launched during the period for measurement of proton and α-particle flux (scintillator probe) and electron density (RF impedance probe). Near the peak of the event at 1719 UT on 6 April a scintillator payload was launched followed by an impedance probe payload at 1753 UT. Another impedance probe payload was launched near local midnight (0430 UT) on 7 April. Results of an identical impedance probe flown at Thule on 22 March 1971 during quiet conditions are used for comparison.

2. Proton Measurements

Proton flux measured on the satellite Explorer 41 [1] are shown in Fig. 1 and compared with the Thule 30 MHz riometer data (riometer data courtesy of R. J. Cormier, AFCRL). The satellite flux data show onset and recovery characteristics typical of a minor PCA event (note the low levels of high energy particles). Riometer absorption also shows typical minor PCA characteristics with a maximum absorption of only 2.2 dB. A slight but positive ionospheric night-time recovery is indicated by the dip in the riometer absorption curve at 0600 UT on

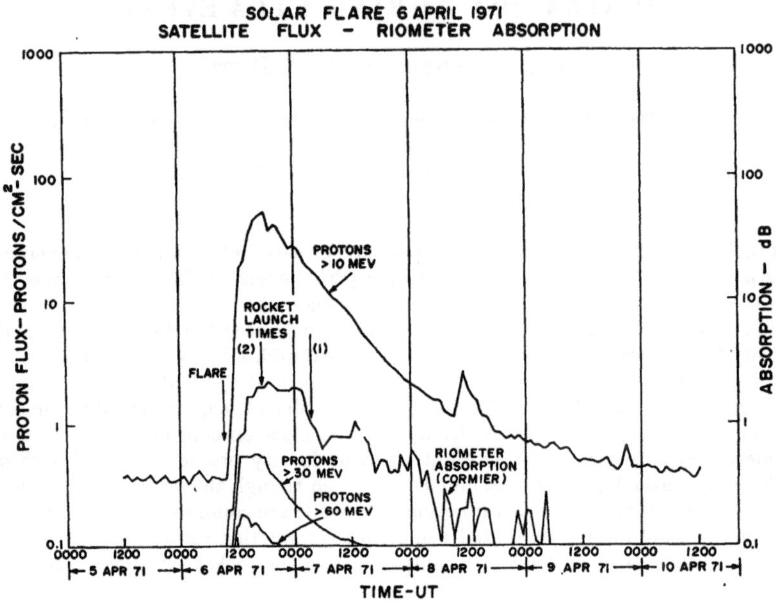

Fig. 1. Proton flux measured during the 6 April 1971 PCA by satellite Explorer 41 compared with 30 MHz riometer measurements at Thule, Greenland, and showing rocket launch times.

7 April. Ionosonde records from Qanaq, Greenland, confirm this as a night-time recovery phenomenon. During the period of this PCA the solar depression angle χ did not exceed 97°; consequently the maximum earth shadow height did not rise above approximately 50 km. This accounts for the presence of only a weak ionospheric night-time recovery. A summary of statistics relative to the rocket flights

Table 1

Summary of rocket launch data and ionospheric conditions during the 6 April 1971 PCA at Thule, Greenland

Rocket No.	Launch Date Time (UT)	Apogee (km)	Measurement	Conditions	χ (°)	R_z[a]	A[b]
70-1E	22 Mar. 71 1640	111	Electron density	Quiet	76	55	0
70-4P	6 Apr. 71 1719	89	Proton flux	Daytime PCA	71	58	2
70-2E	6 Apr. 71 1753	111	Electron density	Daytime PCA	71	58	2
70-3E	7Apr. 71 0430	122	Electron density	Twilight PCA	97	59	0.7

[a] R_z = Sunspot number; [b] A = riometer absorption (30 MHz).

and existing ionospheric conditions is presented in Table 1. In particular it should be noted that the daily sunspot number was essentially the same for all the rocket flights.

A scintillator payload for the measurement of proton flux was launched during the peak of the event at 1719 UT, see Fig. 1. The probe was capable of measuring proton energies from 2 to 100 MeV. The incident differential energy spectrum was determined by adjusting an assumed spectrum until the count rate variation of the measured flux with altitude agreed with the assumed distribution. An equation of the form $J(>E) = J_0E^{-n}$, where J is the integral proton flux greater than energy E, was derived from a fit to the differential spectrum. The rocket measured data were obtained between 50 km (altitude of tip ejection) and 89 km (apogee). This corresponds to an energy range of 2—30 MeV protons at Thule assuming vertical incidence. A similar technique was used to fit integral flux measurements taken on satellites Explorer 41 and 43. The daytime rocket measured flux and the daytime satellite data agree to within a factor of two.

With the technique described in [2] ionization rates were derived for daytime from the rocket flux data and for twilight from the satellite data (rocket measurements of proton flux were not accomplished during twilight). These data are shown in Fig. 2 and indicate a flux decrease of lower energy protons as the event progressed.

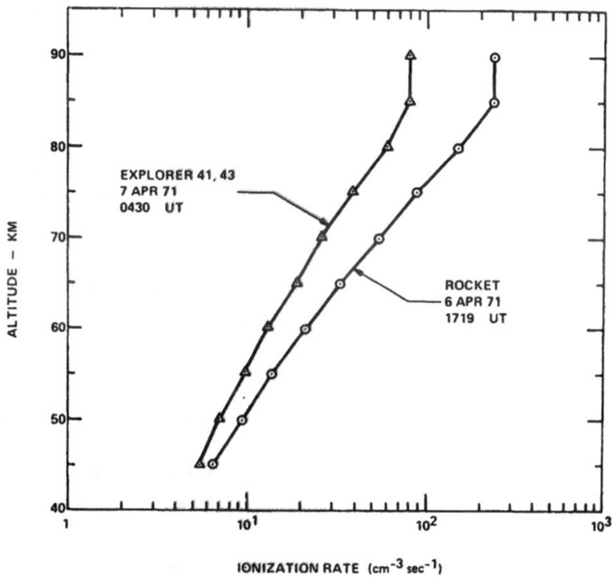

Fig. 2. Ionization rates derived from proton flux measurements of satellites Explorer 41 and 43 and a rocket probe launched from Thule, Greenland, during the 6 April 1971 PCA event.

3. Electron Density Measurements

Electron density was measured with an RF impedance probe technique in which the skin of the payload and the metal case of the rocket motor (electrically isolated) were employed as the arms of a dipole antenna. This antenna configuration was driven alternately by a 3.0 and a 7.2 MHz signal throughout the rocket flight. Since the payload was not separated from the motor, measurements were possible on rocket ascent and descent. During flight the total impedance and phase angle of the antenna were measured and telemetered to a ground recording station. These data were used to compute antenna reactance as a function of altitude and with the equations described in [3] were reduced to electron density profiles. Fig. 3 shows the electron density measured during the peak of the 6 April 1971 PCA compared with a quiet condition profile measured on 22 March 1971 using an identical probe at Thule, Greenland. Data for both ascent and descent are shown. Quiet day density profiles from a midlatitude—spring—fall model reported in [4] and a model profile based on chemistry described in [5] show excellent correlation with the Thule quiet day results. As shown in Fig. 3 there

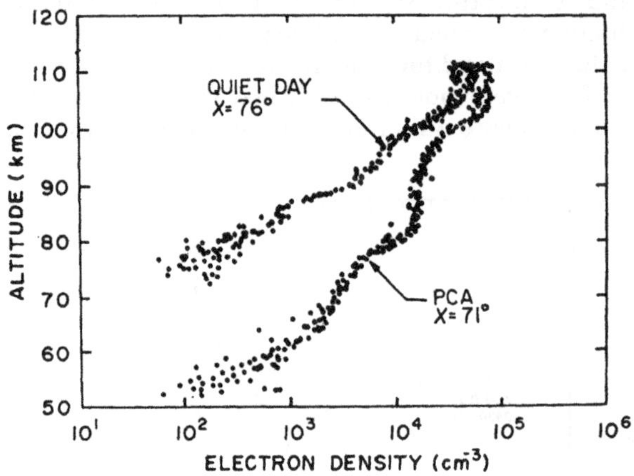

Fig. 3. Rocket measured electron density profile during daytime PCA conditions compared with a quiet day profile.

is a marked enhancement of electron density below 85 km during the PCA (at 80 km the enhancement exceeds an order of magnitude). The lower value of the quiet day electron density at 100 km compared with the PCA results is attributed to the difference in solar angles during the measurements rather than to proton production.

Data for twilight PCA conditions ($\chi = 97°$) and the quiet day profile merge at 85 km; previous PCA results [6] indicate this condition should exist at approximately 80 km. The discrepancy may be attributed to uncertainties in the trajectory data for the twilight rocket flight. As shown in Table 1 this rocket reached an apogee 11 km higher than the other electron density payloads even though effective launch angles and payload configurations were the same. Above 100 km the profiles are considered to be solar dominated. It should be pointed out that Thule twilight conditions are markedly different from those for midlatitude. At Thule the solar angle during the twilight case ($\chi = 97°$) did not change more than one degree for two hours prior to rocket launch.

4. Electron Production

The twilight and day rocket measured electron density profiles corrected for collisional frequency [3] at the lower altitudes are shown in Fig. 4. The sharp gradient in the twilight profile near 80 km (less pronounced in the daytime profile) has been observed previously during night-time PCA conditions [6].

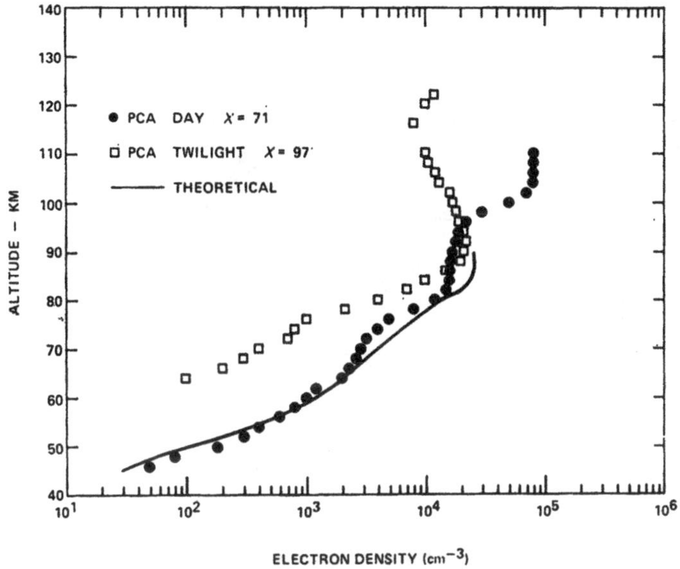

Fig. 4. Comparison of PCA twilight and daytime electron density profiles and a theoretically computed profile for daytime.

Also shown in Fig. 4 is a theoretical daytime electron density profile derived from the daytime production rate q shown in Fig. 2. The theoretical profile was computed from the relation $(q/\alpha_{eff})^{1/2}$ where α_{eff} was obtained as a function of altitude from previous PCA results [7]. Agreement of the theoretical profile with the measured data is well within the accuracy of the measurements. Total dB absorption was also calculated using a method described in [8] and the measured electron density for both day and twilight. The computed values agree within 20% of the measured 30 MHz riometer data (1.7 dB for the day measurement and 0.8 dB for the twilight measurement).

An attempt was also made to derive a theoretical density profile for the twilight case using appropriate loss rates given in [7] and the twilight production shown in Fig. 2. Results of these calculations did not agree with the measured twilight electron density. The basis for the difficulty is evident in Fig. 5. Here the loss rate q/N_e^2 computed from the twilight production and the measured electron density are shown with loss rates from previous events for sunset, sunrise and night. Comparison of these results shows that the Thule twilight data more nearly match the night-time case than sunset or sunrise. Whether this discrepancy is due to a dependence of the loss coefficient on the production rate during twilight or to the uncertainties in the rocket trajectory cannot be firmly stated.

However, additional measurements are planned at Thule during twilight PCA conditions to resolve this problem.

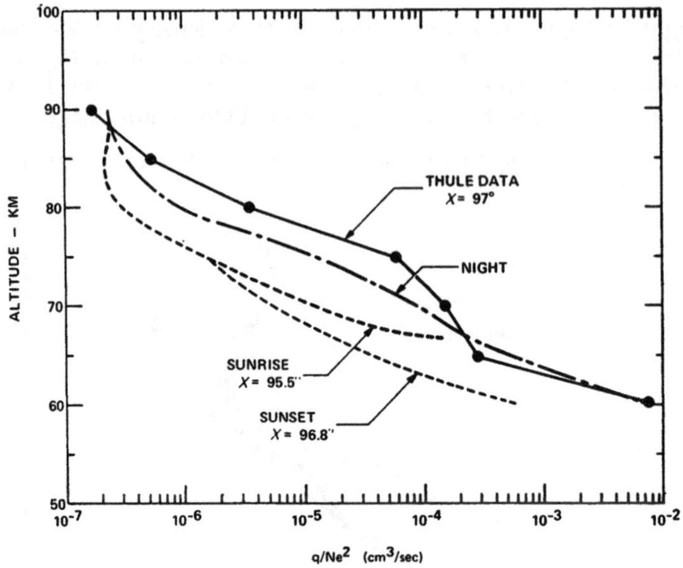

Fig. 5. Twilight electron loss coefficient derived from Thule data for the 6 April 1971 PCA compared with other PCA results.

Acknowledgments

We wish to thank the Air Force Air Weather Service Detachment 48 at Thule, Greenland, for conducting the rocket launches and the AFCRL Computation Services Division, especially R. McInerney, for data reduction.

References

[1] Solar-Geophysical Data, No. 328, Part II, 1971 (p. 88).
[2] W. SWIDER and M. E. GARDNER, in: Proc. COSPAR Symp. on November 1969 Solar Particle Event, AFCRL (unpublished).
[3] J. C. ULWICK, W. PFISTER, O. C. HAYCOCK and K. D. BAKER, in: Electron Density and Temperature Measurements in the Ionosphere, COSPAR Technique Manual Series, March 1967 (p. 66).
[4] K. MAEDA, J. Geomag. Geoelect. 23, 133 (1971).
[5] T. J. KENESHEA, R. S. NARCISI and W. SWIDER, J. Geophys. Res. 75, 845 (1970).
[6] J. C. ULWICK, Proc. Meeting on Operation PCA 69, AFCRL-70-0625, 1970 (p. 135).
[7] J. C. ULWICK, Space Research XIII, 581 (1973).
[8] W. PFISTER, Handbook of Geophysics and Space Environments, AFCRL 12.17—12.18 (1965).

Space Research XIII — Akademie-Verlag, Berlin 1973

STEADY STATE COEFFICIENTS IN THE D REGION DURING SOLAR PARTICLE EVENTS

J. C. Ulwick

Air Force Cambridge Research Laboratories, L. G. Hanscom Field,
Bedford, Mass., USA

Rocket and satellite measurements were made during the 2 November 1969 solar particle event of the proton spectra and of the altitude profiles of electron N_e and ion N_+ concentrations. Ion production rates q between about 40 and 80 km are presented for daytime, night-time and twilight conditions. The ratios λ of the negative ion density to the electron density are shown for corresponding times. The steady state electron coefficients $\Psi_e = q/N_e^2$ determined from the measurements for night-time and daytime conditions are compared and discussed. The daytime Ψ_e's, obtained over a three-day period where the ion production rate changed by almost two orders of magnitude, differ by less than a factor of two at all altitudes. The steady state ion coefficients $\Psi_i = q/N_+^2$ in two daytime measurements, where q changed by an order of magnitude, are within a factor of two at all altitudes. Variations of Ψ_e with solar zenith angle for both sunrise and sunset are discussed.

1. Introduction

Solar particle events (sometimes called polar cap absorption (PCA) events) offer a good opportunity to calculate the relation between the rate of production of electron—ion pairs and the steady state electron and ion density as a function of altitude in the D region. The intense ionization of the region, in contrast to the undisturbed situation, allows more accurate measurements of the ion and electron densities and to a much lower altitude. Also, the energy influxes (protons and α-particles) are slowly decaying over a period of days and the ionization effects of these particles can be calculated more accurately and easily. From the continuity equations, the relationship between the rate of ionization q and the electron N_e and ion density N_+ can be expressed under quasi-equilibrium conditions (as in a solar particle event) by

$$q = (1 + \lambda)(\alpha_D + \lambda \alpha_i) N_e^2$$

and

$$q = \frac{1}{(1 + \lambda)} (\alpha_D + \lambda \alpha_i) N_+^2$$

where λ is the ratio of the negative ion density to the electron density, α_D is the ion—electron recombination coefficient and α_i is the ion—ion recombination coefficient. The coefficient $\alpha_D + \lambda \alpha_i$ is often called the effective recombination coefficient α_{eff} and $(1 + \lambda)\alpha_{eff}$ the steady state coefficient Ψ [1]. In this paper, since both ion and electron densities will be used, the designation will be

$$\Psi_e = (1 + \lambda)\alpha_{eff}; \quad \Psi_i = \frac{1}{1 + \lambda}\alpha_{eff}$$

and therefore

$$q/N_e^2 = \Psi_e; \quad q/N_+^2 = \Psi_i.$$

Rocket and satellite measurements [2, 3] have been made at various times during the 2 November 1969 PCA event of the proton spectra for the determination of the rate of ion production q. Simultaneous rocket measurements were made of the altitude profiles of electron N_e [4, 5] and ion N_+ [6, 7] concentrations, and the steady state electron Ψ_e and ion Ψ_i coefficients calculated.

2. Steady State Electron Coefficients

Since α_D, α_i and λ may all be functions of q, Ψ_e is not necessarily characteristic of the lower ionosphere under all conditions. In fact Ulwick and Sellers [8] pointed out the large differences between three different night-time Ψ_e profiles derived from experimental observations with three distinct ionization sources of auroral electrons [9], β-particles [10] and protons [8]. For daytime conditions Adams and Masley [1] show experimental results during PCA events that suggest Ψ_e is only a function of altitude. During the 2—5 November 1969 PCA event the ion production rate as a function of altitude (Swider, private communication) was determined from four rocket measurements of the proton flux [2] and electron flux [11] and is shown in Fig. 1. From these measurements and nearly simultaneous electron density measurements [4, 5], the Ψ_e profiles shown in Fig. 2 were derived for daytime conditions on 2, 3 and 4 November. The results are in agreement within a factor of about two at all altitudes. The factor of two difference is within the experimental error (40%) of our electron density measurements at altitudes below 65 km. This means that even though the production rate changed by over two orders of magnitude in the three-day period, the Ψ_e profile is essentially constant and Ψ_e is only a function of altitude. Thus the electron density profile can be computed ($N_e = (q/\Psi_e)^{1/2}$) for daytime PCA events if the production rate (viz. the proton spectrum) is known. It should be pointed out that the value of Ψ_e at the lowest altitude (the highest value of Ψ_e) in each case is the value with the greatest error since the electron density is lowest and therefore derived from measurements near the limit of instrumentation sensitivity. Therefore, at 55 km the Ψ_e derived from the 2 November measurements ($N_e = 5.5 \times 10^3$ cm^3 s^{-1}) is less in error than the Ψ_e for 3 November ($N_e = 1.5 \times 10^3$ cm^3 s^{-1}) and 4 November ($N_e = 4 \times 10^2$ cm^3 s^{-1}). Comparing these results with other experimentally derived Ψ_e values under PCA conditions, there is excellent agreement with Potemra et al. [12], but Megill et al. [13] report values which differ by almost an order of magnitude at some altitudes.

From the ion production rates (Fig. 1) and positive ion density measurements from rocket-borne Gerdien condensers [6], the Ψ_i for 2 and 3 November were calculated and are also shown in Fig. 2. These two curves show similar shape and magnitude, irrespective of the ion production rate decrease by over an order of magnitude in the two days and the altitude of maximum production changing

from 55 km on 2 November to 65—70 km on 3 November (as shown in Fig. 1).
The relationship between the two coefficients is

$$\Psi_i(1+\lambda)^2 = \Psi_e.$$

Thus when $\lambda \to 0$ the two coefficients are equal, the positive ion density is equal to the electron density, and Ψ_e is the ion—electron recombination coefficient α_D. This is at approximately 75 km altitude as shown in Fig. 2.

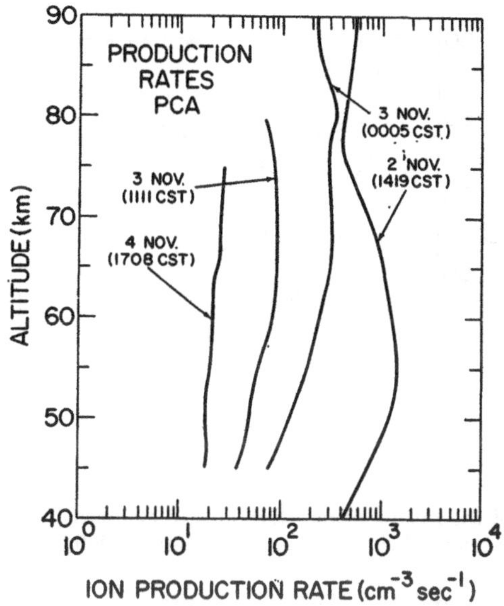

Fig. 1. Ion production rate as a function of altitude during 2 November 1969 PCA event from rocket proton and electron flux measurements.

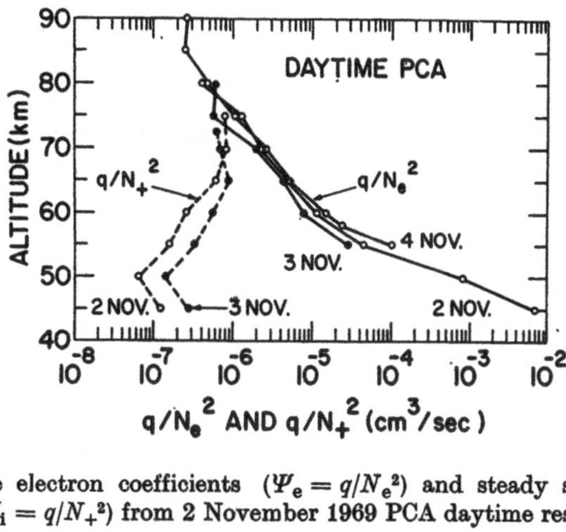

Fig. 2. Steady state electron coefficients ($\Psi_e = q/N_e^2$) and steady state ion coefficients ($\Psi_i = q/N_+^2$) from 2 November 1969 PCA daytime results.

Fig. 3 shows the variation of λ as a function of altitude calculated from the positive ion and electron density measurements for the daytime, sunset and night-time rocket flights. The night-time values were from two different PCA events: rocket 17.757 was launched during the 18 November 1968 PCA while 17.758 and the other results are from the 2 November event. The ion production rates at the time of the two night-time launches were quite similar. Using the 17.757 results, Swider et al. [14], gave a simple expression for λ for night showing the dependence on the production rate. In daytime, to a first approximation, λ is regarded as a characteristic parameter of the atmosphere and independent

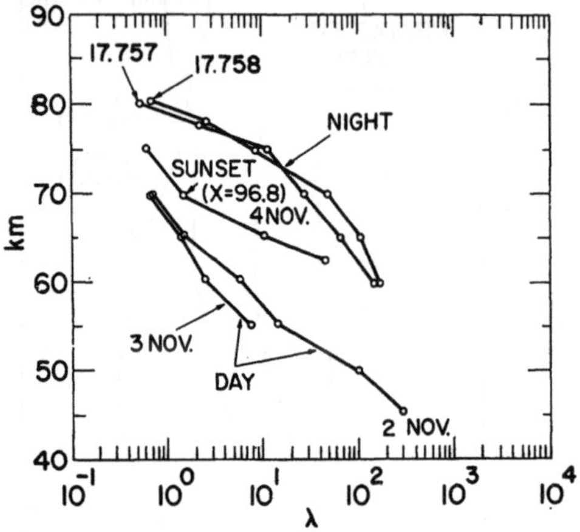

Fig. 3. The variation of the ratio λ of negative ions to electrons for daytime, night-time and sunset conditions during 2 November 1969 PCA event.

of the production rate and the electron density. The daytime λ values in Fig. 3 show agreement at least within a factor of two even though the production rate changes by over an order of magnitude. The change from night-time, sunset transition, and daytime is evident, with $\lambda = 1$ at approximately 80 km, 72 km and 68 km respectively.

Fig. 4 shows the 2 November 1969 daytime (17.602) Ψ_e values compared with the 3 November 1969 (17.758) and 19 November 1968 night-time values. Both night-time results are quite similar, reflecting the similarity of the ion production rates for the two events which were within about 30%. Differences, such as the sharper gradient between 75 and 80 km on the 19 November 1968 results, are essentially due then to different electron density concentrations. For night-time the parameter Ψ_e can only be used below about 75 km associated with a specific production rate as mentioned. Swider et al. [14] using the 19 November 1968 rocket results were able to develop, instead, a simple expression to determine the electron concentration. The electron loss rate above 75 km was taken to be due to dissociative recombination and below 75 km, by means of three-body attachments, to produce O_2^- with no significant detachment processes. They point out, however, that detachment by the process $O_2^- + O \rightarrow O_3 + e$ may be operative in the vicinity of 80 km since there is a sharp decrease in the electron density at this altitude (reflected in Fig. 4 as the sharp decrease in Ψ_e) that appears to be comparable with the steep gradient in atomic oxygen calculated by Shimazaki and Laird [15] between 80 and 75 km. In the figure, above 80 km, the three Ψ_e

values are about the same, indicating that photodetachment processes are not significant, Ψ_e is not a function of the production rate and Ψ_e is the ion—electron recombination coefficient.

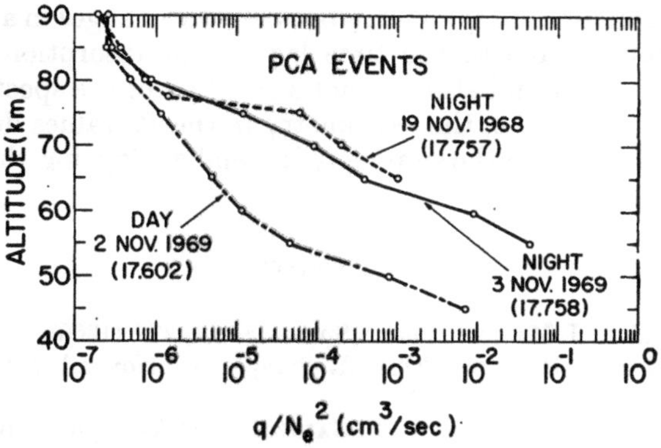

Fig. 4. Comparison of night-time and daytime steady state electron coefficients ($\Psi_e = q/N_e^2$). The night-time results are for two different PCA events.

Figs. 5 and 6 show the q/N_e^2 variation for sunrise and sunset on 3 and 4 November respectively. Electron densities were from rockets launched at the designated solar zenith angles noted on the figures [4, 5]. The ion production rates shown in Fig. 1 were used for the corresponding days and were considered unchanged during the time interval of rocket launching. In both figures the variation with solar zenith angle is clear. At each altitude where increase in Ψ_e is significant, the Ψ_e is always largest for the largest solar zenith angle and all approach the daytime values as a limit at different altitudes dependent upon the solar zenith angle.

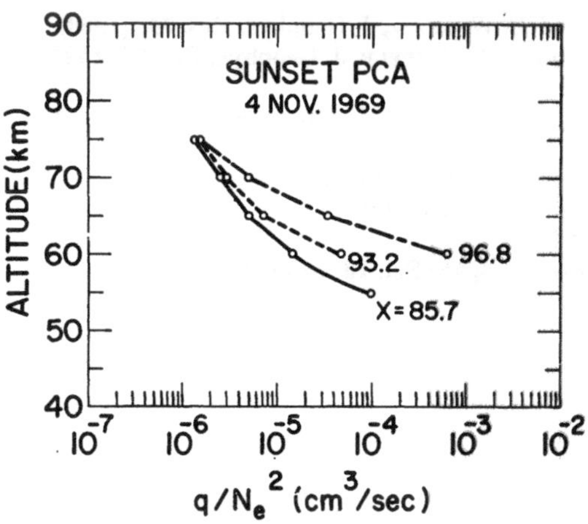

Fig. 5. The variation of steady state electron coefficients ($\Psi_e = q/N_e^2$) with solar zenith angle χ from rocket results during sunset.

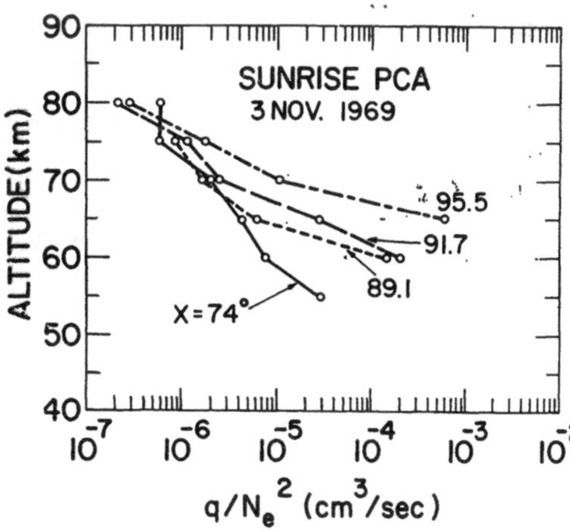

Fig. 6. The variation of steady state electron coefficients ($\Psi_e = q/N_e^2$) with solar zenith angle χ from rocket results during sunrise.

Both figures exhibit the so-called "twilight anomaly" effect [16, 17], i.e. the day/night and night/day transitions take place when the shadow of the solid earth is traversing the altitude range 0—50 km and not when the visible light is removed from the ionized altitudes. Chivers and Hargreaves [18] using riometer data showed that the pronounced asymmetry between the changes in absorption during the sunset and sunrise due to the time lag in the absorption increase during sunrise was a time effect and also was not a function of the spectrum of the solar flux. Comparing Figs. 5 and 6, the asymmetry in the Ψ_e values during the sunrise and sunset is apparent with the sunrise (3 November) lagging the sunset (4 November) values.

References

[1] G. W. Adams and A. J. Masley, Planet Space Sci. **14**, 277 (1966).
[2] B. Sellers and F. A. Hanser, COSPAR Symp. on 2 November 1969 Solar Particle Event, AFCRL-72-0474, 12-1 (1972).
[3] J. B. Reagan, W. L. Imhof and E. E. Gaines, COSPAR Symp. on 2 November 1969 Solar Particle Event, AFCRL-72-0474, 10-1 (1972).
[4] J. C. Ulwick, COSPAR Symp. on 2 November 1969 Solar Particle Event, AFCRL-72-0474, 27-1 (1972).
[5] W. Dean, COSPAR Symp. on 2 November 1969 Solar Particle Event, AFCRL-72-0474, 20-1 (1972).
[6] T. Conley, COSPAR Symp. on 2 November 1969 Solar Particle Event, AFCRL-72-0474, 24-1 (1972).
[7] E. Richards, COSPAR Symp. on 2 November 1969 Solar Particle Event, AFCRL-72-0474, 23-1 (1972).
[8] J. C. Ulwick and B. Sellers, Space Research XI, 1173 (1971).
[9] I. B. McDiarmid and E. E. Budzinski, Can. J. Phys. **42**, 2402 (1964).
[10] R. Whitten, I. G. Poppoff, R. S. Edmonds and W. W. Berning, J. Geophys. Res. **70**, 1737 (1965).
[11] R. Hegblom, COSPAR Symp. on 2 November 1969 Solar Particle Event, AFCRL-72-0474, 26-1 (1972).
[12] T. A. Potemra, A. J. Zmuda, C. R. Haave and B. W. Shaw, J. Geophys. Res. **74**, (1969).
[13] L. R. Megill, G. W. Adams, J. C. Haslett and E. C. Whipple, J. Geophys. Res. **76**, 4587 (1971).
[14] W. Swider, R. Narcisi, T. Keneshea and J. C. Ulwick, J. Geophys. Res. **76**, 4691 (1971).
[15] T. Shimazaki and A. R. Laird, J. Geophys. Res. **75**, 3221 (1970).
[16] G. C. Reid and C. Collins, J. Atmos. Terr. Phys. **14**, 63 (1959).
[17] D. K. Bailey, Planet. Space Sci. **12**, 495 (1964).
[18] H. J. A. Chivers and J. K. Hargreaves, Planet. Space Sci. **13**, 583 (1965).

Space Research XIII — Akademie-Verlag, Berlin 1973

PHASE METHOD OF INVESTIGATION OF SHORT TIME SCALE DISTURBANCES AND MOTIONS IN SPACE PLASMA

G. S. Ivanov-Kholodny, A. A. Korchak, I. P. Stakhanov
and B. M. Chikhachev †

USSR

A phase method of investigation of short-time-scale processes and motions in a space plasma is considered. The effects of the ionosphere and its slow variations are cancelled by a special frequency conversion method proposed earlier to cancel the Doppler effect on satellite measurements. Taking into account the time-lag of the signal propagation while using stable quantum frequency oscillators one can record small-amplitude rapid phenomena in the ionosphere caused by solar flares and internal gravity waves; effects of tropospheric and gravitational frequency displacement and of the earth's rotation are also estimated. The most valuable supplement of the method is the measurement of the motion of the ionized medium. As a result the large-scale electric fields in both ionosphere and magnetosphere causing particle acceleration can be measured.

1.

By improving the usual phase method it is possible to investigate short-time-scale disturbances and movements of space plasma. First let us consider the phase method which has been successfully used for studying the stationary and slowly varying ionosphere.

The variation with time t of phase ψ of a radio wave of frequency ω_0 along the path r with the electron content of the column $N = \int_0^z n_e \, dr$ where n_e is the electron concentration, is defined by the equation

$$\psi = -\omega_0 t + \frac{\omega_0 r}{c} - \frac{\gamma N}{\omega_0}, \tag{1}$$

where constant $\gamma = 2\pi e^2/mc = 0.053$ cm s^{-1}. The frequency difference between the emitted frequency ω_0 and that received $\omega = -d\psi/dt$ is

$$\Delta\omega = \omega - \omega_0 = -\frac{\omega_0 v}{c} + \frac{\gamma}{\omega_0}\left[v n_e + \nabla_\perp N + \frac{\partial N}{\partial t}\right], \tag{2}$$

where v is the vehicle velocity, $\nabla_\perp N$ is the gradient of N perpendicular to the line of sight. Using (1) and (2) it is assumed that $\omega^2 \gg \omega_e^2 \gg \omega_H^2$ where $\omega_e^2 = 4\pi e^2 n_e/m$ is the electron plasma frequency and $\omega_H = eH/mc$ the electron gyrofrequency.

There are four special cases when the phase method can be applied to the ionosphere.

(i) For a vertical sounding rocket $\nabla_\perp N = 0$; it is also assumed that $\partial N/\partial t = 0$. If the rocket velocity is known the Doppler effect is estimated and the local electron concentration n_e is estimated using Eq. (2). Difficulties of the method are connected with difficulties of accounting for ionospheric changes due to gas release from the rocket as well as accounting for rapid variations of N.

(ii) For a geostationary satellite $\nabla_\perp N = 0$ and $v = 0$. This case permits the total electron content N and its relative variation to be found. Difficulties arise in connection with the elimination of tropospheric effects and the determination of the absolute value of N, if there are no continuous measurements during the day. This method fails to measure n_e.

(iii) In detecting SFD phenomena [1] radio-wave reflection from the ionosphere is used. In this case $\nabla_\perp N = 0$ and the frequency variation is caused by both N variations and the change of height of radio-wave reflection in the period of additional ionospheric ionization due to the solar flare. These two effects may be distinguished if two frequencies are used. The main deficiency of the method is connected with difficulties of correctly accounting for effects arising near the reflection level in the ionosphere. This deficiency may be overcome if a satellite is used.

(iv) With satellites two frequencies are usually used to eliminate Doppler effect:

$$\psi_2 - \frac{\omega_2}{\omega_1}\psi_1 = \frac{\gamma}{\omega_1} N \frac{(k^2-1)}{k}, \tag{3}$$

where $k = \omega_2/\omega_1$. In the measured value $\Delta\omega$ it is difficult to distinguish the separate effects of the three terms in the square brackets in Eq. (2). New possibilities arise if a special transformation of the received frequency ω' on board the vehicle is used

$$\omega'' = 2\omega_0 - \omega'. \tag{4}$$

This transformation was suggested by Badessa et al. [2, 3] to cancel the Doppler effect in an attempt to measure the effect of the theory of relativity [4, 5].

In the case of a stationary or slowly varying ionosphere transformation (4) allows cancelling of ionospheric effects. The phase difference $\Delta\psi = \psi'' - \psi_0$ recorded by a ground station in the presence of rapid disturbances in the ionosphere along the line of sight is determined by the following expression

$$\varkappa\Delta\psi = -\frac{\gamma}{\omega_0}[I_1 - \varkappa^2 I_2] + A, \tag{5}$$

where

$$I_1 = \int_0^r n_e\left(r', t - \frac{r'}{c}\right) dr', \qquad I_2 = \int_0^r n_e\left(r', t - \frac{2r - r'}{c}\right) dr'$$

are integrals which take into account variations of n_e during radio wave propagation, and $\varkappa = \omega_0/\omega''$ takes into account the frequency difference of radio waves transmitted and received by a satellite. A in Eq. (5) defines any anisotropic effects which do not depend on the direction of propagation of the ray, such as gravitational frequency shift, movements and drifts of plasma along the ray path, and uncompensated electric fields [6]). To measure small displacements described by expression (5) at $\varkappa \approx 1$ it is necessary to use highly stable molecular frequency generators. Possibilities of measuring both short-time-scale ionospheric disturbances and space plasma are considered in detail below.

If the ionosphere is stationary then the difference between the integrals in Eq. (5) is equal to zero. If, at the height range h to $h + \Delta h$ between the satellite and the ground station , a local disturbance rapidly changing with time occurs, then neglecting A in (5) we obtain

$$\varkappa \Delta \psi = -\frac{\gamma \Delta h}{\omega_0}\left[n_e(t) - n_e\left(t - \frac{2h}{c}\right)\right] = -\frac{\gamma \Delta h}{\omega_0}\frac{\partial n}{\partial t}\,\delta t \tag{6}$$

where δt is twice the time of radio propagation from the disturbance to the satellite. The higher the satellite orbit the slower the variations of n_e that may be recorded by a device of a certain measurement accuracy. The considerations made above may be easily summarized and applied to a disturbance extended in space.

Let us consider two satellites which are in cosmic space. If charged particle concentrations do not change in space then the frequency difference $\Delta\omega$ between radio waves transmitted by one and received at the other of the satellites when anisotropic effects are neglected is

$$\Delta\omega = \frac{\gamma}{\varkappa\omega_0}\left\{\int_0^r dr' \left.\frac{\partial \bar{n}}{\partial t}\right|_{t-\frac{r'}{c}} - \varkappa^2 \int_0^r dr' \left.\frac{\partial \bar{n}}{\partial t}\right|_{t+\frac{r'}{c}-\frac{2r}{c}} + (1-\varkappa^2)\frac{\partial r}{\partial t}\,\bar{n}\left(t - \frac{r}{c}\right)\right\}, \tag{7}$$

where r is the distance between the satellites.

Local terms being neglected and after integration we obtain

$$\Delta\omega = \frac{\gamma c}{\varkappa\omega_0}\left\{\left[\bar{n}(t) - \bar{n}\left(t-\frac{r}{c}\right)\right] - \varkappa^2\left[\bar{n}\left(t-\frac{r}{c}\right) - \bar{n}\left(t-\frac{2r}{c}\right)\right]\right\}. \tag{8}$$

If the variation of $\bar{n}$ in the time r/c is small, then, expanding the right side of Eq. (8) in a series, we obtain

$$\Delta\omega \approx \frac{\gamma(1-\varkappa^2)}{\omega_0\varkappa}\, r\, \frac{\partial \bar{n}}{\partial t}. \tag{9}$$

Since in the case considered $\bar{n}r = N$, this result coincides with (3).
At $\varkappa = 1$ we find that

$$\begin{aligned}\Delta\omega &= \frac{\gamma c}{\omega_0}\left\{\left[\bar{n}(t) - \bar{n}\left(t-\frac{r}{c}\right)\right] - \left[\bar{n}\left(t-\frac{r}{c}\right) - \bar{n}\left(t-\frac{2r}{c}\right)\right]\right\}\\ &\approx \frac{\gamma r^2}{\omega_0 c}\frac{\partial^2 \bar{n}}{\partial t^2}.\end{aligned} \tag{10}$$

Now let us consider the situation when the local Doppler effect may be neglected. If T is the characteristic time of electron density variation in the plasma, such that $T\,\partial\bar{n}/\partial t \sim \bar{n}$, then the ratio of the local Doppler effect given for example in Eqs. (7)—(9) is $\sim vT/r$. Thus the local Doppler effect may be neglected if the satellite separation is significantly greater than the distance the satellite covers during the characteristic time of the disturbance. On the other hand expansion in r/c used in obtaining Eqs. (9) and (10) may be made only if $cT/r \gg 1$. Since $v \ll c$ this condition is consistent with the requirement $vT/r \ll 1$.

2. General Characteristics of the Method

A peculiarity of the proposed method is in the combination of three characteristic features, namely, usage of stable quantum frequency standards, a unique technique of eliminating Doppler shift [2—5] and utilization of the radio signal delay. This combination improves the usual phase method. Indeed at $\varkappa \neq 1$ the frequency shift, as in the dispersion method, is proportional to the first derivative with time of the electron content along the ray path of the signal. On the contrary at $\varkappa = 1$ the influence of the first derivative is completely eliminated and the frequency shifting of the signal is defined by the second time derivative enabling short-time-scale non-stationary processes to be studied.

In addition there is a possibility of studying wave processes in the medium. Since the signal delay $2r/c$ depends on the distance r to the satellite any oscillations with a period $4rn/c$ and its harmonics will be amplified (here n is an integer). Oscillations with periods $(2n + 1)2r/c$ will not cause a frequency shift (see, for example, Eq. (8)). Since the time delay depending on distance to the satellite varies from fractions of a second to several minutes, then as the satellite moves off the earth one may investigate a wide spectrum of oscillations. Thus a space vehicle moving to the moon, Mars and other planets may be used as a peculiar analyser of oscillations in the interplanetary medium.

To investigate the ionosphere the simpliest variant of the method is to use one satellite and one ground station. Estimations of the expected shift of 100 MHz radiation for three main types of non-stationary ionosphere are calculated using Eq. (10) and given in Table 1, where ΔN is the electron content variation in a column during the time interval Δt. At $\varkappa = 1$ (taking into account the second

Table 1

	ΔN (cm^{-2})	Δt (s)	$\partial N/\partial t$ ($cm^{-2}\,s^{-1}$)	$\Delta\omega/\omega_0$ at $\varkappa = 5/6$	$\partial^2 N/\partial t^2$ ($cm^{-2}\,s^{-2}$)	$\Delta\omega/\omega_0$ at $\varkappa = 1$
Diurnal variation	3×10^{13}	10^4	3×10^9	10^{-10}	3×10^5	3×10^{-14}
Internal gravity waves	10^{13}	6×10^2	2×10^{10}	10^{-9}	3×10^7	3×10^{-12}
Solar flare	3×10^{12}	30	10^{11}	5×10^{-9}	3×10^9	3×10^{-10}

derivative), the distance r to the satellite is assumed to be 1.5×10^5 km. For an assumed accuracy $\Delta\omega/\omega \sim 10^{-12}$ it is possible to record both the first and the second derivative of N for all cases but the diurnal variation at $\varkappa = 1$. In this case, to determine the second derivative satellites at considerable distances from the earth are necessary. Use of not one, but several, satellites and a ground station network considerably widens the possibilities of investigation of non-stationary phenomena in the ionosphere or in interplanetary space. To study interplanetary space it is possible to instal an observation station on the lunar surface.

References

[1] R. F. Donnelly, ESSA Tech. Rept IER 19, IISA 19, Boulder 1967.
[2] R. S. Badessa, R. L. Kent and I. C. Nowell, Phys. Rev. Lett. **3**, 79 (1959).
[3] R. S. Badessa, R. L. Kent, I. C. Nowell and C. L. Searle, Proc. Inst. Radio Engrs **48**, 758 (1960).
[4] N. T. Basov, B. M. Chikhachev et al., Uspekhi Fizich. Nauk **75**, 1 (1961).
[5] N. T. Basov, B. M. Chikhachev et al. Kosmicheshie Issledovaniya, **5**, 608 (1967).
[6] I. P. Stakhanov, A. A. Korchak and B. M. Chikhachev, Geomagn. i Aeronom. **10**, 1001 (1970).

Space Research XIII — Akademie-Verlag, Berlin 1973

PLASMA PARAMETER DETERMINATION BY MEASUREMENT OF ELECTRIC NOISE FIELDS DUE TO THERMODYNAMIC PLASMA FLUCTUATIONS

R. Grabowski

Arbeitsgruppe für physikalische Weltraumforschung, Freiburg, FRG

Thermodynamic density fluctuations in plasmas are associated with fluctuating electric fields. The root mean square potential difference induced by these fluctuations in a dipole-like double probe system has been calculated. In the upper ionosphere and in the magnetosphere values of about 10^{-7} to 10^{-6} volt are expected within a 1 Hz bandwidth. Such levels should be measurable with a suitable probe system on a spacecraft. Since the noise spectrum depends on electron density, electron temperature and ion temperature its measurement should be useful for plasma analysis if complementary measurements of plasma parameters are available.

1. Introduction

Stochastic fluctuations of a physical quantity, usually described by the covariance function or its Fourier transform, the quadratic (or power) spectrum, will be considered in most cases as a hindrance when measuring the quantity. On the other hand, one technique in space research is a direct effect of fluctuations, namely, the incoherent backscatter of radio signals by density fluctuations in the ionosphere. As thermodynamic density fluctuations and the associated local fluctuations of currents, electric and magnetic fields are present at any probe measurement on board a spacecraft, one may ask whether a complementary measurement of the spectrum of the fluctuating quantities by the probe system is possible, and whether this method will deliver worthwhile information.

In the following we will look upon the effect of fluctuations in a dipole-like double probe system, which is often used to observe dc and ac electric fields. We expect the following fluctuation phenomena to cause a statistically varying potential difference at the input of the amplifier of such a probe system: (i) fluctuations of the electric field inducing a fluctuating potential difference between the two probes separated by an effective length l, (ii) fluctuations of the particle flux hitting the probe surface, (iii) the surface structure statistically varying in time and thus associated with a fluctuating distribution of areas with different work function, (iv) fluctuating electric fields in the environment of the spacecraft due to the technical installations. It is clear that only the fluctuation phenomena (i) and eventually (ii) will provide information about the ambient plasma.

2. External Noise in a Double Probe System

A basic theory on plasma fluctuations has been already established [1]. Using the basic relationship given in [1] the author has recently derived an expression for the quadratic spectrum of the fluctuating potential difference

$$\Delta U = \int_0^l \boldsymbol{E} \cdot d\boldsymbol{s}$$

between two points separated by a distance l [2]. The distance l may be identified with the effective length of a double-probe system. The quadratic spectrum, i.e. the Fourier transform of the covariance function

$$C(\tau) \equiv \langle \Delta U(t) \, \Delta U(t+\tau) \rangle ,$$

is given by

$$S(\nu) = 2\bar{\theta} \left\{ \frac{1}{\omega_{\mathrm{Pe}}} \left(\frac{\bar{\theta}}{\theta_{\mathrm{e}}} \right)^{1/2} + \frac{1}{\omega_{\mathrm{Pi}}} \left(\frac{\bar{\theta}}{\theta_{\mathrm{i}}} \right)^{1/2} \right\} \frac{F(l)}{8\pi^{3/2} \varepsilon_0 \lambda_{\mathrm{D}}} . \qquad (1)$$

The expression contains an average temperature

$$\bar{\theta} = 2\theta_{\mathrm{e}}\theta_{\mathrm{i}}/(\theta_{\mathrm{e}} + \theta_{\mathrm{i}}),$$

measured in units of energy, the plasma frequencies ω_{Pe} and ω_{Pi}, the Debye length λ_{D}, and a function of the effective length l:

$$F(l) = 1 - \frac{2\lambda_{\mathrm{D}}}{l} \int_0^\infty d\varkappa \, \frac{\sin(\varkappa l/\lambda_{\mathrm{D}})}{(1+\varkappa^2)^2} ,$$

which is approximately unity, if the effective length is large compared with the Debye length.

This result may be modified if the influence of an external magnetic field is included. The appropriate expression for this case is not available as yet, mainly because only numerical solutions are possible. We assume that at least for electric field components parallel to the magnetic field Eq. (1) is a good approximation.

The expression (1) represents "white" noise. The theory is restricted to fluctuating Coulomb fields. Therefore at very high frequencies the character of the noise spectrum may change. But it is usually assumed that at least up to the plasma frequency the Coulomb fields will prevail and determine the character of the noise.

2.1. Now we have to examine whether a measurement of the noise spectrum is feasible at all, i.e. whether the noise induced by the fluctuations in the ambient plasma is strong enough to cover the internal noise of the actual measuring device. With the help of Eq. (1) we have estimated the root mean square potential difference

$$\langle \Delta U^2 \rangle^{1/2} = (2\,|S|\,\Delta\nu)^{1/2}, \qquad (l \gg \lambda_{\mathrm{D}}),$$

for a bandwidth of $\Delta\nu = 1\,\mathrm{Hz}$. $(\langle \Delta U^2 \rangle / \Delta\nu)^{1/2}$ should have values of about 1.0×10^{-7} to $1.0 \times 10^{-6}\,\mathrm{V\,Hz^{-1/2}}$ throughout the ionosphere and magnetosphere. There is only little variation with height, as the influence of decreasing density and of increasing temperature will roughly compensate each other. Measurement of an external noise with a magnitude indicated above should be technically feasible.

2.2. Problems may arise due to the external fluctuation phenomena (ii), (iii) and (iv) of § 1. Probably the best way to estimate the effect of statistically varying particle fluxes upon the surface of the probe would be to start again with the basic relationship for current fluctuations given in [1]. At present an investigation of this kind seems not to have been undertaken. However, a first estimate may be obtained by considering them in the same way as the electron current fluctuations in the case of thermionic emission ("shot noise") in electron valves [3]. Then the covariance function for the electron current fluctuations is proportional to the mean value of the electron current. A corresponding statement can be made for the ion current. We consider unbiased probes, so that a variation in the ion current is accompanied by an equal variation in the electron current. Assuming that the currents are stochastically independent, the covariance function for the total electron current variation is simply the sum of both contributions:

$$\langle \Delta I_e(t)\, \Delta I_e(t+\tau)\rangle = 2e(|I_e| + |I_i|)\delta(\tau).$$

This expression is valid for each of the two probes.

According to

$$I_e = eN\, 4\pi r^2 \left(\frac{\theta_e}{2\pi m_e}\right)^{1/2} \exp\left(\frac{eV}{\theta_e}\right), \qquad (V < 0),$$

the probe potential V (with respect to the potential of the undisturbed plasma) must change, if there is a variation in the electron current. Here N is the plasma density, m the particle mass and $4\pi r^2$ the probe surface area. For small variations we find

$$\Delta V = R\Delta I_e; \quad R = \theta_e/(eI_e).$$

Now it is easy to deduce the covariance function for the variation of the potential difference between the two probes:

$$\begin{aligned}\langle \Delta U(t)\, \Delta U(t+\tau)\rangle = {} & R_1^2 \langle \Delta I_{e1}(t)\, \Delta I_{e1}(t+\tau)\rangle \\ & + R_2^2 \langle \Delta I_{e2}(t)\, \Delta I_{e2}(t+\tau)\rangle \\ & - 2R_1R_2\langle \Delta I_{e1}(t)\, \Delta I_{e2}(t+\tau)\rangle.\end{aligned}$$

A correlation between the electron currents to probes 1 and 2, i.e. a non-vanishing term $\langle \Delta I_{e1}(t)\, \Delta I_{e2}(t+\tau)\rangle$, would reduce this kind of noise. Though such a correlation will be present due to large scale components of the particle flux variations in the ambient plasma, in a first approximation we will treat the currents as uncorrelated. Then the quadratic spectrum of the potential difference is given by

$$S(\nu) = 4e(R_1^2 |I_{e1}| + R_2^2 |I_{e2}|).$$

We point out again that each probe is always at its floating potential, a situation which allows us to set $I_e = I_i$. If the probe is in the electron retarding region, the currents are given by the expression for the ion saturation current. For spherical probes of equal size finally we find the following expression

$$S(\nu) = 8\theta_e^2/(e\,|I_i|)$$
$$= \frac{8\theta_e^2}{e}\,\frac{1}{4\pi r_s^2\,eN}\left(\frac{2\pi m_i}{\theta_i}\right)^{1/2}. \qquad (2)$$

The variable r_s is the radius of the ion sheath surrounding the probe. Numerical values of the root mean square potential difference $\langle \Delta U^2\rangle^{1/2} = (2\,|S|\,\Delta\nu)^{1/2}$ are represented in Table 1. It is seen that the stochastic variations in the particle flux are the source of noise with order of magnitude equal to that induced by the field fluctuations in the ambient plasma.

Table 1

Root mean square potential difference between two probes due to stochastic variations of the particle fluxes

	r (m)	r_s (m)	N (m^{-3})	θ_e (eV)	θ_e/θ_i	$(\langle \Delta U^2\rangle/\Delta\nu)^{1/2}$ (V Hz$^{-1/2}$)
Ionosphere	0.1	1×10^{-1}	1×10^{11}	0.2	2	2×10^{-7}
Magnetosphere	0.1	1×10^{1}	1×10^{8}	1.0	1	2×10^{-7}

We have to remark, however, that the relations used for the particle flux will not be fully adequate in the outer regions of the magnetosphere. The photoeffect will cause a photoelectron current, which probably will dominate, and which will produce a photoelectron cloud [4]. On the other hand the extension of the perturbed plasma around a spacecraft may be comparable with the length of any technically feasible probe system. The theory developed for the homogeneous case may fail. Thus generally a more refined theory is needed for the outer regions of the magnetosphere.

2.3. Now we have to examine whether there will occur a disturbing contribution by a fluctuating work function distribution. Experimental results have not yet been reported. In electron valves this kind of fluctuation is responsible for electron current fluctuations with a quadratic spectrum decreasing with increasing frequency (Funkel effect). For most electrode structures the spectrum will fall below the level of the "shot noise" at a frequency of about 1 kHz. We expect the same behaviour for probes in the ionospheric or magnetospheric plasma. Coating the probe surface with special metals or with an amorphous (liquid-like) material seems to be a successful step to get a more uniform work function distribution. A higher degree of uniformity is accompanied by lower spectral amplitudes, and we feel that the noise level caused by a statistically varying work function distribution can be kept smaller than the "shot noise" already at a frequency of about 100 Hz.

2.4. Man-made noise in the spacecraft environment depends on the actual spacecraft installation and has to be examined separately for each spacecraft disign. Presumably solar cells will be the main source. Although spacecraft builders have investigated this aspect, final results concerning the electric fields have not been published. The following data have to be seen only as preliminary ones. Electric fields generated by solar cells do not exceed about 1.0×10^{-7} V m^{-1}

at a point 1 m distant from the cell. The spectral distribution seems to depend on frequency as a $1/\nu$ law. These data do not allow a determination of the noise spectrum, but they indicate that at higher frequencies probably a sufficiently low level can be obtained. Coating the solar cells by a double layer of insulating and conducting material may be an effective way of reducing this kind of noise.

3. Conclusion

After a review of the different sources of externally generated noise in a double-probe system we must state that many aspects need further investigation. Especially for the outer region of the magnetosphere a more refined theory has to be developed. For regions, however, where the ion sheath around the spacecraft will be small compared with the effective length of the probe system, and where the photoelectron current is not important, two main sources have been found. The quadratic spectra for the fluctuations of the potential difference induced by these sources are given by Eqs. (1) and (2).

We assume that the two contributions are not correlated. Then the quadratic spectrum for the total noise is given simply by the sum of Eqs. (1) and (2). This spectrum does not depend upon frequency; it represents "white noise". It is directly related to three essential plasma parameters, and to the effective area of the ion sheath surrounding the probes. The knowledge of the spectrum will support plasma analysis if two of these plasma parameters and the effective sheath area are determined by different methods. Thus the measurement of the stochastically varying potential difference between two probes of equal size should be a useful tool in plasma diagnostics, complementing other techniques.

References

[1] N. Rostoker, Nuclear Fusion **1**, 107 (1961).
[2] R. Grabowski, Planet. Space Sci. **20**, 571 (1972).
[3] J. D. Swift and M. J. R. Schwar, Electrical Probes for Plasma Diagnostics, Iliffe Books Ltd, London 1970.
[4] R. J. L. Grard and J. K. E. Tunaley, ESRO Sci. Note No. SN-108 (1970).

Bei Fragen zur Produktsicherheit wenden Sie sich bitte an:
If you have any questions regarding product safety,
please contact:

Walter de Gruyter GmbH
Genthiner Straße 13
10785 Berlin
productsafety@degruyterbrill.com